THE HISTORY OF TECHNIQUES

THE HISTORY OF TECHNIQUES

Volume 1
Techniques and Civilizations

Written and Edited by
Bertrand Gille

Translated from the French by
P. Southgate and T. Williamson

Technical terminology revised by
A. Keller, University of Leicester

Additional bibliography compiled by
E.F. Kranakis, Yale University

GORDON AND BREACH SCIENCE PUBLISHERS
New York London Paris Montreux Tokyo

Gordon and Breach Science Publishers

P. O. Box 786
Cooper Station
New York, NY 10276
United States of America

P. O. Box 197
London WC2E 9PX
England

58, rue Lhomond
75005 Paris
France

14-9 Okubo 3-chome
Shinjuku-ku
Tokyo 160
Japan

Originally published in French in 1978 as *Histoire des Techniques l'Encyclopédie de la Pléiade* by Editions Gallimard. French edition © Editions Gallimard 1978.

Library of Congress Cataloging-in-Publication Data

Histoire des techniques. English.
History of techniques.

Translation of: Histoire des techniques.
Bibliography: p.
Includes index.
Contents: v. 1. Techniques and civilizations—
v. 2. Techniques and sciences.
1. Technology—History. 1. Gille, Bertrand,
1920–   . II. Title.
T15.H5713 1986 609 86–14295
ISBN 2–88124–144–1 (France: v. 1)
ISBN 2–88124–146–8 (France: v. 2)
ISBN 2–88124–155–7 (France: set)

ISBN Volume 1 2-88124-144-1 Two volume set 2-88124-155-7 Printed in Great Britain by W. & G. Baird Ltd., Antrim.

CONTENTS OF VOLUME 1

CONTENTS OF VOLUME 2

PREFACE

It is perfectly logical that an encyclopaedia† of the knowledge and activity of man should devote a volume to techniques, which is concerned both with activities and knowledge. However, neither would it have been surprising if one had not been included, since the material world is still not considered to be part of the more noble preoccupations of mankind. Which writer today would prepare a discourse on the ball-point pen? Workers do not refer to the machines they work, they prefer to talk about alienation, fatigue and the class struggle, and this in spite of the long chapter which Marx devoted to machines. In everyday conversation this topic is limited to the strictly domestic: to the performance of a car or the qualities of a self-cleaning oven, or to the comparative effectiveness of washing powders. Diderot's *Encyclopédie* may be admired; it is rarely emulated.

Any attempts to reconcile man and everday objects, those which are used, and those of which one uses oneself would probably be futile. So too would an attempt to decipher the true relationships between man and everyday objects, and their long history of quarrels and reconciliations.

The main difficulty was to write a work which did not duplicate others. There are three major histories of techniques already in existence, not including innumerable popularised works, and works limited in historical time, space and subject. The first of these works was in English, published in five large volumes. It was followed by a Russian, and then a French work, which is being completed at the very moment these lines are being written. Three histories were devoted almost exclusively to pure techniques, and are in a sense technical histories of techniques. Was it a good idea to add a fourth one, even if there are still unexplored fields and obscurities?

It seemed to us that it would be more fruitful to tackle this problem from quite a different angle, using a new method. This implied a new work hypothesis, which is explained at length at the beginning of the book. The idea was based on two major principles.

†The French edition is part of the *Encyclopédie de la Pléiade*. (Translator's Note.)

The first of these led to a plan almost diametrically opposed to that of preceding works. These in effect presented the evolution of techniques in a fragmentary manner, each technique isolated one from the other, and furthermore, divided into somewhat dubious chronological divisions. This fragmentation seemed regrettable to us on more than one account: it inevitably lead to a sort of mosaic where colour and design were confused, creating juxtapositions without formal clarity. The authors only knew their own period or speciality, ignoring the works of others. As a result no general overview, no global explanation could emerge. The documentation was worthwhile and interesting, but one was left feeling somewhat lost at the end of it. One found the Parthenon and Descartes in this history mosaic, but one would certainly never find Trafalgar and the *Pont des Arts* together in the same paragraph.

The notion of a technical system: a coherent ensemble of compatible structures, seemed to us to be necessary to render in a more interesting way the material world of mankind since its origins. This was a more sensitive approach to the problem, even if its formulation seemed to some readers a little excessive. By exaggerating contrasts, the painter or the photographer permits one to perceive more clearly the reality of certain things. Daily life is a close mixture of constraint and material acts, even if one does not intuitively perceive the multiple relationships which undoubtedly exist between the elements which surround us in this technical environment, a world sometimes strange, always ineluctable. This environment only has meaning or value in so far as it has a specific order, and our manuals are more than reticent on the order of technical knowledge, even in the narrowest sense of the concept. Those who do not care about this problem are led directly to complete misapprehensions: to say that Heron of Alexandria, with his eolipyle, came close to inventing the steam engine is striking proof of this.

This way of tackling the technical phenomena has another advantage: that of starting a dialogue with specialists of other disciplines, with economists, linguists, sociologists, politicians, lawyers, scientists, and philosphers, for example. This is a dialogue which is difficult to organise these days, but in this way it would be possible to highlight the indispensable coordination of human activities, and the evident concordance of results which are its products.

Such a proposition implies at its very least a dividing into chronological periods. Technical systems succeed one another. The dynamics of such systems thus conceived, gave a new value to what is sometimes called in a vague and ambiguous way 'industrial revolutions'. From system to system, it was now possible to scan the centuries, indeed millenia, according to rhythms quite other than those to which we have been introduced by a more conventional history. We can already hear critical voices: but in writing a history, is it not inevitable that some aspects will be emphasised at the expense of others? We are aware of a certain viewpoint somewhat scornfully known as battle-history, this being the history of a long period, where events and humanity were submerged beneath descriptions of secular movements. What we propose here is a history which is in a sense bound in by the material world.

Once we took this position, the work plan fell into place of its own accord. The first part of the book was conceived according to a plan which was almost completely chronological. The stages of technical progress needed to be examined, and put into the context of the other great movements of history. Of course, several compromises were made, but this did not unduly affect the chosen plan. It was the vestiges of a classical education which made us separate the Greeks and the Romans (who apart from a few exceptions possessed almost the same technical system) and made us link the Byzantines with Rome. Vitruvius was only the last of the architects and engineers of the School of Alexandria, and the Latin agronomists that have so often been highly praised are in fact only the descendants of the agronomists from the whole of the Mediterranean Basin, less well known because their works have disappeared. In the same way we have grouped together in a somewhat arbitrary fashion three independent civilizations, both in terms of time and space. Situated outside the Western domain, their only characteristic in common was that they all found themselves limited at various levels. The chapter on the first great technical civilizations (as far as the great empires are concerned) presents a juxtaposition of these civilizations, which were similar to one another but had major differences.

In our descriptions there are gaps, which the extent of the subject made inevitable. India is not mentioned, neither are some belated civilizations of Africa and Asia. Is it really necessary to explain that we are perfectly conscious of these faults?

The most important objective from our point of view was to understand, if not explain completely, the transition from one technical system to another, because it is there that the problem of technical progress truly lies. What might seem complex but clear and strictly ordered in the field of scientific progress, will appear infinitely less clear in the field of technical progress. The gap which exists between what has been conventionally called theory and practice is such that invention assumes a singular importance, and this has often led to a collection of inventors' biographies. However, there is something else to consider: the development of the stages of scientific progress is logical and autonomous. And above all a certain coherence is necessary (which we have mentioned above) because a technique does not exist in isolation, and is always linked to other techniques. 'In 1814, Collier, the great engineer from Paris, began his attempts to make shears with helicoidal blades for linen, the basic principle of which came from the United States, but where no attempts at actually making such an implement had been made. Everything remained to be done, and it took him five years. Collier had to create alloys, smelting and machining processes, anti-friction bearings, a transmission chain, a control system using a mitrometric screw, in fact all the elements required to make a machine of unprecedented precision.' In the field of invention this coherence of techniques is an inescapable necessity. And last but not least, remains the economic viewpoint. Recently a writer quite correctly made clear the difference between the progress of techniques, and technical progress, the first being of a purely technical order, in the sense that there are some inventions which are not immediately useful, but which can lead to 'innovations'. The second represents exactly the point at which the invention enters everyday or industrial life. It is thus possible to put invention on two planes: that of technical thought representing technique in its pure form, and that of economic need in the wider sense of the term, in order to respond to technical problems or to follow purely economic needs. The latter may be in the form of increased demand, in the need for reduction in production costs, etc, all factors more or less linked with each other. Lastly we note (and we shall return to this point throughout the book) the compatibilities which exist between a technical system which is being developed and the contemporaneous social system.

Internal technical problems, the necessary relationships between various techniques, and compatibilities with other systems, appear to be the principal dynamic elements existing in the field of technical systems. The concept of a technique developed to its limits, and the resulting impasse followed quite naturally from this approach. There has been very little research into the mechanisms of technical progress, and it seemed to us that research would eventually have to develop in this direction so we have consistently promoted this method of study. Certainly we have dealt with what is known as growth: the causes of technical progress and the acceleration of the latter according to universally known lines of development. The works of Rostow, although sometimes vague and ambiguous, provided us with a model to follow, develop and refine. Despite some surprising gaps, it fills a large vacuum in historical thought. It is far from our intention to fill these gaps; what was more important was the line of research to be taken.

The second volume of this work was to our mind just as fundamental, and complementary to the first part. Our aim here was to introduce technical progress in a certain number of disciplines representing other human activities: science and law, politics and geography, sociology and economics. The language of technology was substituted for the "language of the arts", examined by Diderot in the *Encyclopédie,* and an essay on technical knowledge was the final chapter in this part of the book.

Of course there was no question of giving techniques a privileged position, and of giving it pre-eminence in the formation of all types of structures' in the dynamics of global evolution. It was only necessary to show that techniques as well as influencing the external world were also affected by external influences. We thus hoped to integrate techniques with disciplines which have to date not given techniques much notice.

Generally this work was written for people with an enquiring mind, the peculiar characteristic of all encyclopaedias, and in particular for those who come into contact daily with techniques without ever realising it, such as the office worker, or the teacher. We also believe that technicians of all kinds should be aware of the history of their techniques. They are rarely taught this, and that is why in most cases, they have false ideas about them. It would no doubt be interesting for such people to become aware of all the various links between techniques, and to discover in the final analysis their implications for a world which is becoming more and more complex. We must take care to avoid however the self-satisfaction of the author of the *Prospectus* of Diderot's *Encyclopédie* 'too much has been written on the sciences, not enough has been written on the liberal arts: how little of worth we see in the authors' contributions compared with the extent and diversity of the subject'.

Having established the broad lines of the general idea, the next step was to find some collaborators, and collaborators who would accept that way in which the problem was to be posed. This proved to be the most difficult task, since the association between sciences and techniques is a difficult one to make, and requires a particular kind of effort. Of course there are many people today who are interested in technical progress, and in its implications and consequences. Since the last war an abundant literature proves this fact, even though it is of very uneven quality. But the principal aim of the work was historical.

Meetings between historians and technicians at the end of the 19th century were frequent and fruitful. A large number of highly skilled technicians had been provided with a good classical education which enabled them to consider the origins of their 'art' with more objectivity. Such relationships, apart from a few very notable exceptions, have become all too rare today. The technician of today is totally cut off from ancient techniques, so great are the differences between the two. Only a few decades ago, retiring technicians could easily throw themselves into historical research: Quenedey, an engineering officer, studied the construction of wooden houses during the Middle Ages; Lefebvre des Noëttes, a cavalry officer, researched the problem of harnessing horses; Masson, an engineer of the *Ponts et Chaussées,* studied the gothic vault. Today's generation has burnt its bridges and treats history with the fleeting attention of the dilettante – this can be a dangerous exercise where techniques are concerned.

For their part, historians tend to feel ill at ease in a field where they hardly ever dare venture knowingly for fear of ridicule. History manuals reveal this only too clearly. The unfortunate result of this attitude is that the historian often neglects this aspect of history, and consequently this affects their interpretation of events, especially in the field of economic history. The writing of history throughout the world is riddled with misunderstandings and simple mistakes of this kind.

Too often the technician is ignorant of historical methods, the tracing of source material, and the critical commentary accompanying them, which can result in the wildest anachronisms. Historians, on the other hand, shy away from such a task, or at best, they have already written and published what they know, and have no desire to start all over again.

For these reasons, the editor of this work has had to take the unusual step of writing the whole of the first part himself. Since the main outline was perfectly systematic, this made the work more homogeneous, and a mosaic of more or less divergent opinions was replaced by a coherent, unified discourse. But this advantage was accompanied by numerous difficulties and some dangerous stumbling blocks. One person could not possibly be a specialist on all periods and all techniques discussed. As with Diderot (if one can make such an immodest comparison) when he wrote most of the technical articles in his *Encyclopédie,* much had to be learnt in order to be able to describe the techniques, and complete success would have been too much to expect. In the final analysis, the reader will have to judge whether the trouble that was taken was worth it.

The second volume, which consisted essentially of the work of specialists, posed the same problems. The economist and the geographer, the scientist and the lexicographer responded to the call, and understood the project perfectly. All the others made themselves scarce, some of them because they were too busy, others because they were somewhat nervous and felt ill-prepared to venture into foreign territory, something to which they admitted in all honesty. We were left with the option of leaving the second half incomplete with only some of the aspects covered. After a great deal of hesitation, and not without misgivings which still persist, the same solution was adopted, and the editor of the work had to force himself into another exercise of mental contortion and assume the role of sociologist, lawyer and politician. Again the reader will have to judge for himself whether this solutions works. All this explains the slow pace at which this work was written. One can imagine how many new books have been published in the last ten years dealing with the periods and questions that have already been written about in this book. Short of constantly revising the text, it was practically impossible to take all these new works into account. The same difficulty occurred with the covering of the newest techniques and their implications. The rapidity of progress in certain techniques is often greater than the speed of writing and publishing a book. On the other hand, there are other areas of development, such as maritime law for example, which develop very slowly. We hope that readers will take this into account when they pass their final judgement on this book.

Nothing evokes the spirit in which this work was written better than this text by Raymond Queneau: "There should be no question of considering the volumes of this encyclopaedia as treatises from which one could learn a science. It goes without saying, however, that the reader stands to learn much, since it consists after all of a complete course of study. This work aims to educate, to assess, and to provide an opening into the future. Nowhere does this undertaking conceal the extent of our uncertainty, or the immensity of our ignorance. The reader will learn to ignore, to doubt, and that, too, is a critical enterprise."

Raymond Queneau, who entrusted us with this work, is no longer with us. When things became difficult his sometimes sardonic but always gentle smiles proved to be the most effective form of encouragement. We should like this work, or at least the very best parts of it, to be considered as a homage to his memory.

BERTRAND GILLE

PUBLISHERS' NOTES

NOTE TO THE ORIGINAL EDITION

To help further research, the reader will find at the end of the book the following:

1. Bibliography
2. A table of the major contemporaneous technical events from the earliest times to the present.
3. A detailed summary of each chapter.
4. A list of illustrations.
5. A comprehensive index of names of individuals mentioned in the book's chapters.
6. A title index of books quoted.
7. A subject index.

NOTE TO THE ENGLISH EDITION

The reader already familiar with the original French edition will notice that the contribution by Bernard Quemada, "Technique et langage" has not been included here. This decision was approved by the original publisher: the content was devoted to the French language only, which made it unapplicable to a wider context, and indeed untranslatable.

The Bibliography has also been modified. Its sections have been regrouped toward the end of the book. Each has been extended and updated by Dr Eda F. Kranakis of Yale University, whose contributions will appear under the headings "Additional Bibliography".

PROLEGOMENA TO A HISTORY OF TECHNIQUES

In 1935, Lucien Febvre was still able to write: "The History of techniques, one of those numerous disciplines which has still yet to be created." A few years before a work was published which proved to be a milestone in history – it raised a particular issue, that of harnessing and the saddle horse, and its relationship with one of the great watersheds in history, the disappearance of slavery. However much Captain Lefèbvre des Noëttes' ideas might have been disputed later, there was no arguing the fact that his work had opened up new channels, and original explanations and perspectives.

In the *Annales* of 1935, Lucien Febvre and Marc Bloch do not appear to have just discovered the history of techniques, but rather they were emphasising the interest of the subject, and the little attention paid to it by historians. This does not mean that it had been completely neglected until this point, but by its very nature it was placed outside the main body of history. It was as difficult to integrate technical history into history as a whole as it was to integrate the techniques themselves into a general economic theory, to take but one example. Lucien Febvre clearly discerned that there was a conflict of spheres: "A technical history of techniques, necessarily the work of technicians, risks serious errors, confusion, and a complete misunderstanding of the general conditions of how such a task should be executed." But he qualified this: "a work by technicians who look beyond their period or sphere of interest and are thus capable not only of understanding and describing, but also of reconstituting the tools in the manner of exacting and ingenious archaeologists, and also of interpreting the texts as wise historians." This was the problem: to link these different types of knowledge, and to use different methodologies.

It is not surprising, therefore, that historians were hesitant to enter such unfamiliar territory. The technicians, for their part, held little interest for techniques which were no longer in existence, and when they did broach the subject, they did so with an attitude of mind which had only tenuous links with history. Some wrote a history which did not include techniques at all, while others concentrated on research concerned solely with this aspect, such that history became mere chronology. "Technical activity cannot be isolated from other human activities" noted Lucien Febvre, and synthesis was less necessary than concordance. In any global historical explanation, it was essential to include techniques. It is interesting to note that when economics finally found its way into global explanations, techniques still remained apart, by the very fact of its slow and difficult integration into general economic theory which we have just mentioned.

There were numerous traps, not least within the history of techniques itself. We had to avoid too many divisions, which were essential however for the initial exposition of the facts, but which could rapidly lead to every history of each particular technique closing in on itself. Beyond this, it was vital to reintegrate each history of techniques into a larger historical grouping, which could itself be easily linked to economic or demographic history and the history of science or ideas, and finally also factual history the importance of which is often underestimated. This was our aim, then. But before we approach the very heart of this subject, and because this attempt is a relatively original one, certain precautions should be taken. It is the discussion of these precautions which constitutes the essence of this long introduction.

It would be useful to sketch a brief picture of the history of techniques, thereby enabling us to see the development, with all its wealth and indeed with all its problems, of a discipline which can already lay claim to being established.

The oldest history of techniques is that by the German Beckmann, *Beiträge zur Geschichte der Erfindungen*, published in Leipzig between 1780 and 1805. As the title indicates, this is a history of inventions, that is the very type of fragmented history which I spoke of above. It is very similar to a work which is almost contemporary, the *Geschichte der Technologie seit der Wiederherstellung der Wissenschaften bis an das Ende des 18 Jahrhunderts*, the three volumes of which were published in Göttingen, between 1807 and 1811. However this work did at least take into account some major historical events and the notion, albeit still a vague one, of technical systems.

It was not before the middle of the 19th century that the history of techniques underwent a marked development as attempts were finally made to integrate it with other disciplines. It was at this time, during the period corresponding to the Second Empire, that techniques and technology were demanding more attention. Several viewpoints emerged from this.

The first viewpoint aimed at providing a response to the growing public interest in techniques. This meant writing popular works, which entailed not only popularising existing techniques, but also demonstrating the amount of progress which had been made. Two works are particularly worth mentioning in this category: L. Figuier's *Les Merveilles de l'industrie*, and *Merveilles de la science*, which are still useful even today. Each industrial sector is dealt with separately but they are nonetheless placed in some kind of historical context.

The second viewpoint responded to the need felt by certain technicians to know the history of their own technique. Some authors of technical manuals were quite happy to devote several pages to the history of a single technique. Percy's long work, *A Manual of Metallurgy*, not only provides information about the history of metallurgical techniques, but also about the techniques used in some exotic countries.

The third viewpoint reveals concerns which are already more truly historical, and was propounded by archaeologists and technicians who combined their efforts to reconstruct ancient techniques. Two types of work are distinguishable within this sizeable amount of research. The first deals with the reconstruction of ancient military techniques. It is well known that Napoleon III himself was possibly its instigator: he began research on this subject before 1848, when he was imprisoned in Ham castle. At the emperor's request, Colonel Favé began work on the history of artillery. He also reconstructed certain weapons which were used in trials. This method was devised and developed and after a while, proved to be quite efficient. The second type of work was born of necessity. The restoration of ancient monuments had been undertaken on a large scale, and it was felt that the techniques actually used to build the original monuments were the only ones which could make the restored monuments look authentic. The efforts made to understand these techniques by Viollet-le-Duc are well known and his dictionaries of architecture and furniture are still an interesting source for historians of techniques today.

Those adhering to the third viewpoint went even further to say that techniques should be integrated into global explanations. Most people are familiar with the importance that Marx attached to techniques as part of his general theory and it is not surprising that for the historical part of his works he used histories of techniques, such as he was able to find at that time. It was also around this time that some economists began to include technical progress in their general theory.

This proved to be the launch-pad for the history of techniques, and during the last decades of the 19th century, some works were written which are still useful today. Usually they deal with particular techniques: Ludwig Beck's work on techniques used in the iron and steel industry, Thurston's on the history of the steam engine, and Theodor Beck's work on the construction of machines are all examples of this. Similarly, many technical treatises, especially those from Antiquity, were studied, published and translated, and Berthelot's research on alchemists and other technicians, and the numerous works on the Greek engineers from the Alexandria school bear witness to this. T. Henri Martin's work on the life and works of Hero of Alexandria were in fact written in 1854. Sometimes, the expositions were even more general, thus, in 1897, A. Espinas published his work on *les Origines de la technologie*.

The movement was consolidated during the first years of the 20th century, when the inestimable importance of original sources and reconstructions finally began to be appreciated, and when the first museums devoted to the history of techniques were opened. This was not without a certain amount of chauvinism – the Science Museum opened in London in 1857 was created in honour of British science and techniques. The Deutsches Museum in Munich was established in 1906. The traditional history of techniques by sector and histories of inventions continued to be written. The first historical dictionaries of techniques appeared at this time with works such as Blümner's dictionary of the techniques of classical Antiquity, and Feldhaus' dictionary of the techniques of Antiquity, the Middle Ages and the modern period.

The most important development was certainly that of the first attempts to integrate the history of techniques into global historical explanations. The publication in 1906 of the thesis by Mantoux on the

English industrial revolution of the 18th century was an important stage in the historiography of techniques. This was followed by Ballot's research on the introduction of mechanisation in French industry, which sadly was interrupted by the Great War, with the result that it was not published until 1922. The French and the English were to become the specialists on the techniques of the Renaissance period, while the Germans concentrated on the techniques of the Ancient and Medieval eras.

During the period after the First World War, the progress of the history of techniques was temporarily halted, its vigour only renewed with the arrival of the 1930's. Usher's work on mechanical inventions published in 1929, and Captain Lefèbvre des Noëttes' work on harnessing and the saddle horse breathed new life into the discipline and strengthened it. It was now clear that writing history without including techniques was impossible. The *Annales* of Bloch and Febvre in 1935, which devoted an entire volume to the history of techniques, demonstrated just how important this discipline had become. By expounding the aims and assessing the scale of the enterprise, *Annales* encouraged historians to continue on the path they had already traced. Before the Second World War, the history of techniques had already assumed the character which it still has today. Dickinson's history of the steam engine, published in 1939, and Captain Quenedey's history of wood construction in Rouen are evidence of two different types of methodology, each adapted to its own particular field. At the same time, new museums were being founded and centres of research were being established, in Vienna in 1931 and at the university of Paris in 1932.

However, we should note that this research was badly coordinated, partial and sometimes narrow in outlook. Some technicians did show a true historical sense, but historians were reluctant to begin studying techniques, fearing problems in an area which was relatively new and uncertain. The links made between the various fields were unsatisfactory, and sometimes they were not attempted at all.

After the Second World War, the history of techniques became an established discipline. It was not yet perfectly coherent: there were still conceptual differences. But one factor was immediately obvious – the museums and centres specialising in the history of techniques were springing up everywhere.

The history of techniques remains an isolated discipline, however, still considered as the domain of the "scientists." It has still not managed to find its way into the international history congresses, and it was not included in the recent international congress of economic history [Transl. Note: although B. Gille himself delivered the General Report: "Progrès scientifique, progrès technique et progrès économique" for the Section on "Technology and Economic Growth" at the 4th International Congress on Economic History, Bloomington, Indiana, 1968; there was also a Theme devoted to "Science and Technology" at the 6th International Congress on Economic History, Copenhagen, 1974]. It is quite significant, however, that the *Comité Français des travaux historiques et scientifiques* has established a joint commission for the history of science and techniques. The same indispensable collaboration exists in some other countries, although usually it is limited to specific areas of research. A work on medieval metallurgy published recently in Hungary was written with the collaboration of an engineer, a metallurgist, an archaeologist and an historian. It would not be impossible to create such cooperation on an institutional level.

The latest general histories of techniques reveal other gaps which in fact have already been mentioned. But this is a difficult problem to resolve. As Lucien Febvre noted, we do need a technical history of techniques, and it is difficult to deny the interest afforded by the small technical monographs which deal with a single process, or the birth of an invention. It is from an accumulation of such works that we can gain a true knowledge of the history of techniques. However, this knowledge needs to be extended, and this first of all within the technical domain itself. Both the personality of the inventor and the genesis of the idea are integral parts of the birth of an invention. A successful invention is often preceded by a set of objectives which presuppose that they are technically possible, and also by various needs, which we shall consider later. One should consider next the moment when the invention is made, and the person who makes it. And finally, the success of the invention – which thus becomes innovation – is an important factor, because of course the true significance of techniques is in their concrete application in the social, economic, institutional and political structures, without which it is almost impossible to understand them. In our opinion, all these recent works were too insular: they are useful, but still incomplete.

This was our aim, then. We did not simply want to write another technical history of techniques, because of the attraction afforded by any research which is slightly esoteric. Essentially, we wished to include the technical world in general history. Our main obstacles were the gaps in our own knowledge, and our reliance on histories which have tended to be descriptive rather than interpretative. We found it useful to construct what economists call a model, but which we would prefer to be known as an explanatory schema. To do this, we had to clarify and agree on our concepts, allow for all the variables – which could be either extrinsic or intrinsic – and be conscious of all the links within this structure. The discussion that follows is a major examination of this possible explanatory schema. Of course, this is only a provisional schema and it is certainly not immutable.

Since we were approaching this problem from a different standpoint, we felt it would be useful to devote some of the space to a discussion and critique of the nature of the various sources available to us. In order that we should not have to quote the same titles again and again, there is a general directive bibliography for this chapter, to be found at the beginning of the Bibliography.

CONCEPTS AND METHODOLOGY

No science or discipline would merit these names if they did not possess the conceptual and methodological means necessary to any analysis. The reader will not be surprised, therefore, that a large part of this long introduction will be concerned with these aspects of the problem.

Techniques are best analysed as a scientific entity. This is only possible on a global level if one can work with an appropriate language, and also with models based on precise concepts. These models should reflect two levels of reality, the static structures and systems, and the dynamic factor which we have termed "technical progress." Working on the contributions of our precursors, we shall try to construct such a method of analysis.

There is so much literature about the notions of systems and structures that there seems little point in discussing them again. It is true that there is still some uncertainty about the substance of these two notions which are often applied within very different areas. It seemed necessary to highlight them here, however, since a comprehensive treatment of this angle of research has not yet been done in this field. As for the study of political economy, the use of these notions "seems to be the only method which science has found to date capable of bridging the two spheres of research which are separated all too often, historical research and theoretical analysis." This seems all the more desirable since the history of techniques is still a young discipline, which should be provided right from the beginning with clearly defined concepts. Some of these, incidentally, are already controversial and have been the object of rigourous methods of research. However, the lack of specialised research means that we must confine ourselves to general themes and only occasionally provide examples.

The task seems beset with difficulties right from the start. The term "techniques" itself is usually used in the plural – there are textile techniques and iron and steel making techniques, for example. Even at the simplest level, take for example, the technique of the clog-maker, it is soon obvious that this "technique" is in fact a group of operations which often require different tools. So what about the "technique" of the locksmith, as it was described by Mathurin Jousse at the beginning of the 17th century or by Duhamel du Monceau in the middle of the 18th century? It is thus almost impossible to clearly define the object of research. It is very rare that a technique involves only a single operation, and even if it does, this simple action still entails a necessary coupling of matter and energy. These two elements are combined precisely by the technical act, which very often requires some sort of support. Even at the most elementary level, and even for the most primitive techniques, there is a technical combination which, as far as more complex techniques are concerned, can be called a technical ensemble. The support, at the simplest level, is a tool or a process. Cutting down a tree presupposes the following: the existence of the raw material; the materials to be used; for a specific purpose, the finality of the technical act, energy, and what is known as an instrument or instruments, an axe, saw, rope, wedge and sledgehammer, etc.

We can distinguish several important notions from these observations, namely that there are several types of technical combinations which can be studied from several different points of view.

At the bottom of the scale, we can effectively talk of structures, even though this word is still somewhat ambiguous. A structure is a unitary combination. There are elementary structures, such as the tool, for instance, and assembly structures, such as the machine. Some examples will help to explain this point.

A. Leroi-Gourhan has clearly demonstrated that it is possible to define structures even within elementary operations. He used the example of cutting by percussion, of which there are three different methods:

A) Static percussion – the knife applied to the wood – which gives a precise but low evergy cut.

B) Dynamic percussion – the bill hook, the woodcutter's axe, the carpenter's adze – which gives an imprecise but strong energy cut.

C) Static percussion with the aid of a striker –scissors, hammer or sledgehammer – which combines the advantages of both of these processes, and which Bachelard called "administered force."

The saw should be added to this, because it is slightly different to the three other processes. It is a more complex instrument, consisting of a series of knives arranged in such a way to give a precise cut, and to enable greater force to be applied than with one knife alone.

The term "structure" is also used to denote other combinations, which in spite of their complexity, in fact represent a single act. Take J.-L. Maunoury's example: "Heat engines have characteristics which can be defined on two levels. As motors, their function is to create work, as heat engines, they have in common the functioning principle of using heat derived from the fuel of certain bodies." Using this definition, the author attempts to demonstrate the corresponding elementary structures "the combinations of which explains the different types of heat engines." Firstly, he distinguishes two different series of structures: work structures and heat structures. The table below will explain this better than a long discussion.

The table is a perfect example of an assembly structure. Maunoury was led to distinguish between structural groups and sub-groups, but one could almost certainly improve this model: the type of fuel and its conditions of use, which require annexed elements (carburettor, electric spark). Certain formulas could also be clarified, and the actual structure of the energy converter could be constructed: the cylinder and piston which with the crank-shaft is able to provide rotary motion, paddle-wheels, etc.

Table 1

Structural Group	Structural Sub-Group	Type
Work	Functioning of fluid motor	Action
		Reaction
	Movement created	Rotary
		Alternating
Heat	Method used to obtain heat	Fuel
		Fission
	Location where heat is obtained	Internal
		External

(Table taken from Maunoury, *La Genèse des innovations*, Paris, 1968)

This still leaves a great deal to be said about the structure of tools, depending on their function, the material which they are used to work, the material of which they are made, and also on the traditions of size and shape. We will examine two very basic examples.

The number of different types of saw is clearly evident in Charles Frémont's study of this tool. There is no need to discuss the differences between saws made of bronze, iron or steel, they are self-explanatory. But Frémont distinguished several other types: (*A*) the hand saw; (*B*) the cross-cut saw for pit sawyers; (*C*) the bow saw; (*D*) the frame saw; (*E*) the circular saw; (*F*) the band saw.

Fig. 1 shows some pictures of bill hooks which appear in a tool maker's catalogue. There are one hundred and six different types of this single tool, the names of which are toponymic, related, naturally only to France. These varieties correspond to local traditions and usage, and also according to the work and the main type of vegetation.

Systematic research is needed in order to achieve as fine an analysis as possible. The same catalogue has a similar collection of pictures of axes, scythes and even anvils. Listings of tools would certainly be very

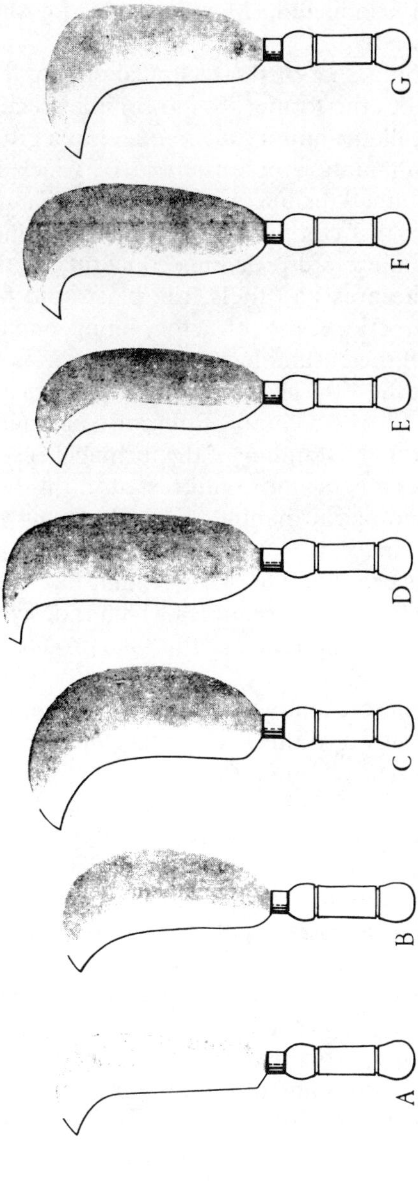

Figure 1. — Some types of billhook. A, billhook from Baden; B, Moselle model; C, Mirecourt model; D, Chaumont model; E, Parisian model; F, Saulieu model; G, Dijon model.

useful as would a collection of ancient catalogues and inventories of tools kept in museums. Some such studies at the most primitive level, i.e. into the types of tools or instruments used for a given operation, have already been conducted. Most notable are those by Leroi-Gourhan.

The second notion is that of the technical ensemble, which is at a quite different level to the former. Some complex techniques require not what could be called a unitary technique, but a group of tributary techniques, the combination or ensemble of which is designed to perform a clearly defined technical act. Take the manufacture of cast iron, for example, the complexity of which is illustrated in the adjoining diagram: there are problems concerned with energy, with the components, the ores and fuels, the blast, and finally with the instrument itself–the frame, the refractory lining, and the shape. This is a good example of the ensemble whose every part is essential to the required result (see fig. 3). There are other examples of this in the chemical industry, all of which are different combinations.

There is a greater understanding of the technical ensemble, because technological research is far more interested in this aspect, with the result that all technological manuals have diagrams which are very useful to the historian.

The final notion is possibly that of the technical concatenation. This is a combination of technical ensembles which is designed to produce the final product. The manufacture of this final product is usually done in several successive stages.

One of the simplest examples of this is clog-making. Figure 2 illustrates the breakdown of this process as conceived by researchers from the *Musée des Arts et Traditions populaires*. This diagram in fact represents only a part of the process, but here six different operations are apparent, which require three different types of tool. The process continues beyond this, however. From the initial rough casting to the scythe, only one instrument is used, the water-powered tilt hammer, which is adapted using a series of panels to tools as varied as hammers and anvils.

This schema is more complicated where a more elaborate product is concerned (fig. 3). Cast-iron can be used to produce iron or steel, and the latter can be used to provide the power needed to give the product its final form. There is thus an entire hierarchy of diverse techniques, which contribute to the working of the technical complex. The same is

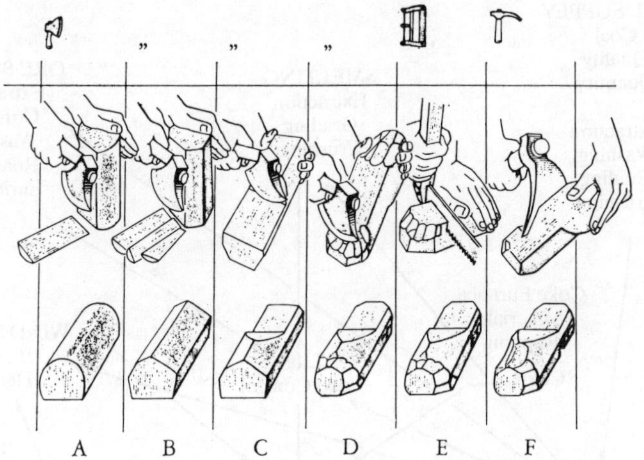

Figure 2. — The different stages in the clog-making process at Cuse (Doubs, France).

true of the textile industry: the first ensemble is the production of the raw material (animal, vegetable or synthetic); the second is to prepare this material for use (washing, retting, degreasing). After this comes the spinning, the weaving, and finally the sucessive finishings, all of which are integrated at different stages of the manufacturing process (fulling, shearing, bleaching and dyeing).

These structures, except for the technical ensembles, are best considered as vertical chains. However it is possible to conceive of them on a horizontal level as well i.e. a technical structure can be used in several different series. This is true of course of tools, which although having a given structure, can take on different forms or aspects. Two simple examples are the hammer or hammers and the pincer or pincers. The same is true at a more complicated level. The cylinder-piston structure is used in heat engines, but also in suction pumps and even in cigarette lighters.

A technical ensemble or series could not function normally without certain conditions being fulfilled. Usually these conditions are those concerning quality and quantity (if we confine ourselves to the technical domain, a restriction that we shall discuss further on). The interac-

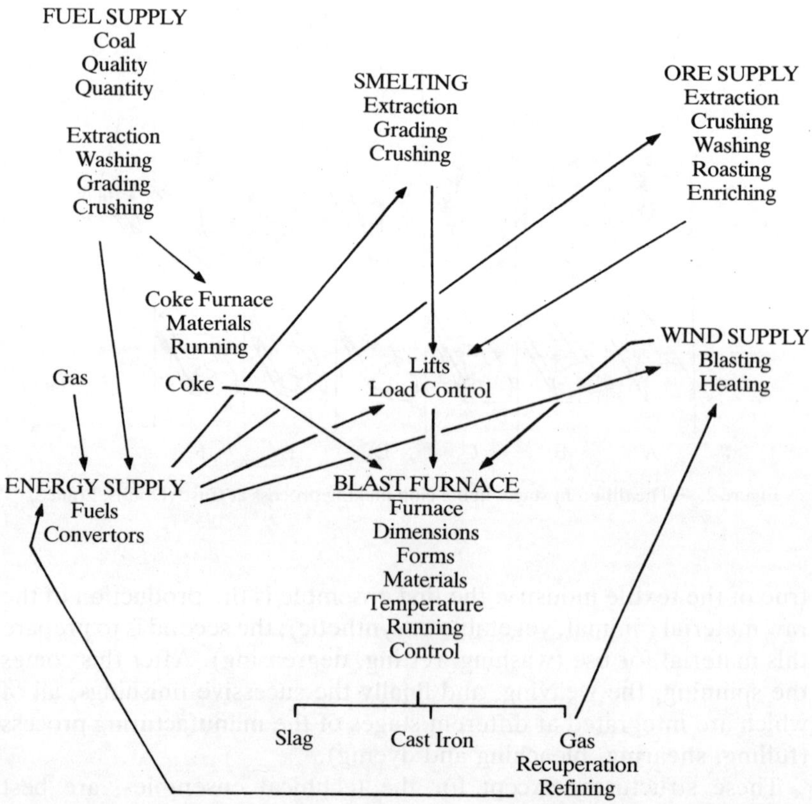

Figure 3. — A technical complex: the blast furnace.

tions between quality and quantity are many. For example, in terms of quality, the working of a given type of material requires tools of a given type. But the product of given quantities can also require precise qualities in the means of production. The influence of quantity on quality is less marked, but it is nevertheless apparent in a great number of cases.

Beyond this, the links are not only within the linear process, but also with more indirect actions. Each part of a technical ensemble in order to function properly may need one or more of the products of the same ensemble. This relationship is obvious where materials are concerned: iron production requires a steam engine, which in turn needs to be made of a metal strong enough to withstand high pressure and superheating. This relationship exists, although in more diluted forms, in many other areas. In the diagram illustrating iron production, it is immediately apparent that both cast and wrought iron are necessary parts of the sub-ensemble. Many other examples could be brought into play and the schema made more complex by including the techniques of transport and the associated problem of energy for example. In some cases, the links between each of the sub-ensembles and the latter and the global ensemble are very close; such is the case for the chemical industry, for example. In other cases, these links are far less apparent, less precise, and in fact relatively weak –the textile industry is an example of this.

This means therefore, that in general, all techniques are dependent upon the others, and this necessarily requires a certain coherence: the coherence within the structures, ensembles and series constitute what could be called a technical system (fig.4). The internal links so vital to these technical systems become greater in number as the systems evolve and become more complex. These links are only effective if the technical ensembles have reached a common level, save perhaps for a few techniques which remain above or below the general level. The second state is naturally more favourable than the first.

Once such an equilibrium has been obtained, the technical system is viable. Those enamoured with chronological divisions are thus able to assess history in terms of a series of different successive technical systems. These can be analysed beyond the extent of the monographs which deal only with one technique: the links between the techniques can be examined, and the nature and conditions of such links.

Research still has many gaps to fill and uncertainties to resolve: most of the studies available to us were hardly adequate. In *Technics and Civilisation*, Lewis Mumford, like many others before him, did not deal with technical systems but with ill-defined and very long periods, such as the eotechnic and paleotechnic phases. However, he does seem to have grasped the problem: "Each phase represents a period of human history, but they are better characterised by the fact that they each constitute a separate technological complex… Each phase has its origins in certain well-defined regions, and tends to use certain particular resources and raw materials; each one has its own specific means of creating and utilizing energy, and its own particular forms of production. Almost every part of a technical complex (this term is used in a different way to our use of it above) tends to accentuate and symbolize, even within the complex itself, entire series of interconnections." In

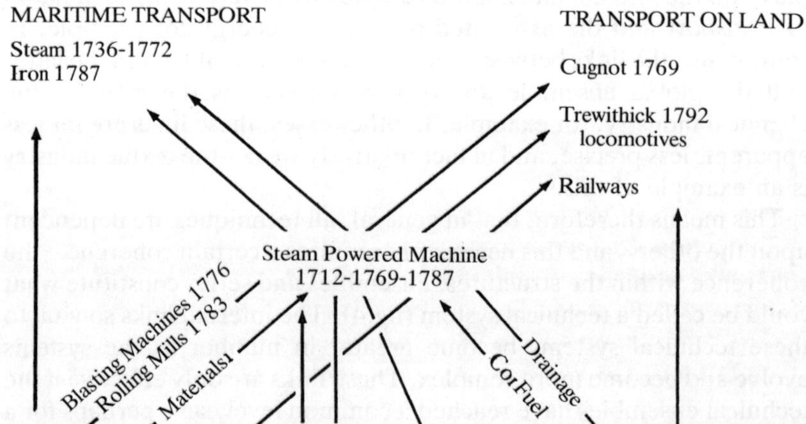

Figure 4. — Simplified scheme of the technical system of the first half of the 19th century.

order to distinguish between these various phases, Mumford isolated what could be called the dominant techniques which exercised – precisely because of their general importance – a commanding influence on all the others. "The eotechnic phases is a complex of water and wood, the paleotechnic is one of coal and iron, and the neotechnic phase is a complex of electricity and alloys." The idea itself was an interesting one, but the treatment of what we call technical systems was unsatisfactory. Even though Mumford did talk about the overlapping of one system into another, the overall treatment of these systems was vague and arbitrary.

A more refined analysis taking into account a greater number of elements would create a tighter chronology, so that the history of techniques would not simply represent three very long periods. This is all the more necessary if technical history is to be related to other areas of human activity, without which it cannot be understood. It is vital, therefore, that technical and non-technical systems should be compared, and the aspects which unite them outlined.

It is only natural that economists, insofar as they also think in terms of structures and systems, should have found this problem interesting: we shall return frequently to this point. However, this does not seem to have done much to advance the present state of research. Following in part the line taken by A. Lewis in *The Theory of Economic Growth*, André Marchal divided up the phases of technical evolution, not in technical terms, but in terms of the ideas held about technics at various times. This greatly facilitated the introduction of technics into economic thought, but it also created a mixture of varied notions rather than general concepts, which would have enabled him to define a technical system. In pre-literate societies, technical achievements remained roughly on the same level: "the same instruments, the same techniques for cutting and polishing, the same metallurgical processes, the same methods of cultivation and irrigation, and the same hunting tricks were used by populations which were separated by different continents and even by different milleniums." It was the knowledge of writing, the development of mathematics and the accumulation of knowledge which was to "distinguish the technically archaic societies from the technically primitive ones." Marchal places the Middle Ages in the technically archaic division: "This period is distinguished from the modern age by the fact that a small group of privileged men were at leisure to indulge in abstract thought and some experimentation, but with no practical aim in mind. Any technical achievements came from skilful but nearly illiterate artisans who improved their methods by a process of empirical trial and error." The Renaissance is merely said to have brought with it curious mind and the experimental spirit, and to have filtered these through to other social layers. From the 17th to the beginning of the 19th century, most revolutionary inventions were due to professional artisans, or expert amateurs. "The 20th century, however, marked the change to the concept of full-time experts and professional technicians working in teams, either for large enterprises, but most often for governments." The emergence of new ideas is not enough to compensate for the insufficient analysis, laced with those clichés so dear to economists, and some unfortunate areas of confusion. Consequently, the divisions are still unsatisfactory.

In fact, economic analysis, even when it is structural, completely neutralises the technical phenomenon, or else deals with it only from certain angles – take Adam Smith's concept of the division of labour for example. Perhaps the only economist to rank it highly was Marx, who quite rightly felt it was extremely important and dealt with it at length. Most economists have included a "technical variable" only when dealing with the dynamics of systems and structures. The founders of the English classical school and their successors, including Marx, perhaps intuitively sensed the necessary relationships of compatibility and coherence between the various systems, but it has long been, and still is, difficult to integrate techniques into global economic theory – or at least to close the gap between the two areas of activity. If the ultimate aim of every discipline is to be of central importance, it is up to the disciples of political economy to incorporate technics into their theory. Similarly, historians of techniques must not neglect neighbouring "forces."

All economists refer to technical structures, but they do not ususally discuss them explicitly. Marchal writes, in a rather vague fashion, that "the economic system is characterised by the particular arrangement of various types of structures," some of which naturally are technical. François Perroux is more precise: "the economic system is the combination of a technical apparatus, an apparatus of juridical and social relations, together with fundamental economic forces." Technical structures were thus considered to be a fundamental element of the "economic system".

Detailed and comprehensive research is required into the system of relations between technics and the economy; current research still concentrates more on the dynamics of these two types of system. Johan Akerman was well aware of the importance of techniques in "autonomous, or motor forces," and even though his successors reduced the number of these, technics was still considered to be an important part of the most dynamic ones.

The effect of technical systems on economic ones is clearly evident today, but it needs to be properly understood. Akerman's concept of "autonomous forces" helps us to define the problem more clearly, but it is obvious that there is interaction between the two systems, and that no force is ever completely autonomous or completely induced. The size of an enterprise, the costs of production and the rate of investment are closely dependent on the current level of technology. In other words, the main objective is to determine how economic and technical systems interact, not which one is more important than the other, even without determining whether or not there are obligatory routes. It would also be very useful to compare the economic and technical systems of various different periods. Like many others before him, Payen symbolised modern capitalism by the steam engine (this is clear from the title of his work, *Machine à vapeur et capitalisme*): this does not provide us with an entirely accurate picture – there are different forms of capitalism – but at least it paved the way for more specialised and fruitful research which had not yet been undertaken. The issue of the reciprocal requirements of technical and economic systems would be an interesting topic of research; many think that technical systems have always been more constricting and demanding than economic ones.

On the other hand, techniques must also be incorporated into a price system, and into the organisation of production, without which they would have no economic interest, which is after all their final purpose. We know that ultimately, artisan techniques only survived because of a particular demand, and that techniques in general were accepted or rejected according to how they affected the market and profit margins. On a worldwide or at least an international scale, legal measures can maintain the use of techniques which would not survive in a free market.

A system can evolve in several different ways; consequently the perspective of dynamics is paramount. If a technical system appears to be imposing on the economy a course which it is unable to support, even temporarily, then the latter can be changed. It was possible to substitute for the liberal capitalism of Western Europe, a form of state capitalism, even collectivism. Colbert, for example, conceived of several formulas to introduce as yet unknown techniques into France. Then, in the 19th century, the governments in France and elsewhere found it in their interests to fund the building of several railway networks. Perhaps even more significant here are the protectionist policies pursued by European powers from around 1820–1821, designed not only to avoid competition from the technically superior English industries, but also to give their own industries a chance to undergo this technological mutation.

Even if his claims are somewhat arguable, by linking the harnessing of horses with the disappearance of slavery, Captain Lefèbvre des Noëttes showed he understood of the close relationship between technical and social systems.

One question in particular has been the subject of many discussions: the relationship between the general level of technology and the optimum population. Léon Buquet has made a detailed analysis of the various interpretations of this problem since the time of John Stuart Mill, dealing in particular with the issue of dynamics which we shall consider later. It seems that there is a correlation between the two factors: any progress in the level of technology necessarily alters the curve of the population optimum. With regard to income per capita, the form of the curve itself remains unchanged, but its position within the graph is raised.

The close links between social and technical systems are obvious. The adoption of a particular technical system requires the adoption of a corresponding social system to preserve coherence. But does this mean that technical systems have a dominant influence over social ones? As with the economic system, the answer is two-fold. Yes, in the sense that the technical system imposes itself from the outside. But there are numerous cases where new techniques have been refused. The imposition of a technical system requires some adaptations in the social system. Since the Middle Ages, for example, corporate bodies have refused to include in their regulations which usually consist of lists of prohibitions, rather than collections of positive instructions, many new techniques which would have destroyed their organisations – they refused to accept new preparation and dyeing products, the spinning wheel, mechanical fulling and carding. This continued until the 17th century. In the late Middle Ages, the coastal communities resisted the new technique of preserving herring in barrels because it threatened to disrupt their established social order. Another case which has been closely studied is the mechanisation of shoe-making in Yankee City during the second half of the 19th century. The nature of technical change – substituting a traditional activity with a completely different one, for example – greatly affects the degree of social change necessary.

The factors which must be considered here are not only social habits, community traditions and ways of thinking, but also the distribution of labour, qualifications – and hence standards of education. And if one considers the rigidity of some social structures, it is easier to understand the difficulty of reconciling the two systems. The fear of technological unemployment in more industrialised and advanced countries and the incidences of luddism of which there have been several examples since the Middle Ages – has existed for centuries. It is no wonder therefore, that technology has progressed so little in underdeveloped countries, where social structures are sometimes completely inadaptable.

The structures of consumption are certainly more flexible, but even here there have been some surprising cases of resistance. In 1829, French "forge masters" replying to a government inquiry stressed that their customers refused to accept the new iron made with coal, and remained steadfast to charcoal iron. But as we move towards the 20th century, so we find less such consumer resistance. At any rate, there is still a great deal of research to be done on the relations between technical and other non-technical systems.

Technical systems succeed one another both in time and in space, and studying the history of these systems' dynamics would greatly help to solve the problem of static factors, and static analysis. The dynamic analysis raises several important questions, one of which concerns the limits of a technical system, and the effect this has on technical progress.

The notion of structural limits is now a familiar one, and is naturally concerned with movement. These structural limits usually become apparent at the end of a period of expansion and are evident in the difficulties of increasing the diversity and quantity of production, or lowering production costs. These various elements are all concerned primarily with economics, with questions of quantity, quality and cost, and there are many interactions between them.

It is most unfortunate that so little research has been done on this issue, because it would have been very instructive. It has been suggested that the major crises of the 14th century were precipitated by the failure of existing techniques to meet the demands of population growth as well as the increased demands of the individual consumer. But it would help to study in more detail the limits of what we have termed as "classical techniques," that is those of the Renaissance. The rise in prices as a result of the influx of precious metals from the New World, must have been followed by a far greater demand for luxury goods. It is possible that the problems experienced from the mid 16th to the mid 17th centuries were as a result of this, as was the population decline at the end of the 16th century. Such phenomena could reflect the inability of classical technology to respond to such a demand and we shall see that these techniques were prevalent until the first third of the 18th century. The population decline towards the mid 17th century would also have made technical innovation less necessary. It is also true that this period was still marked by famine and plague, and these seem to accompany periods of technical stagnation: the plague of 1629–1631, the famine of 1709, and the plague of 1720 are examples of this. It is noteworthy that it was not until the middle of the 18th century that both technical progress and demographic growth were resumed.

It appears that a new technical system emerged during the middle of the 19th century. The technical limits of the periods preceding the 18th century caused a halting of growth and several crises, more or less serious, but when growth was resumed, this led to the formation of a new technical system. It is probable that without this new system, the economically most advanced countries could not have continued to develop. Furthermore, those countries where industrialisation was just beginning were able to benefit from the new system without having to face the difficulties of abandoning the old one. It is not impossible that one of the reasons for the great depression of 1929 was a technical blockage.

In the report *Limits to Growth*, D.H. Meadows and other researchers at M.I.T. gave us a gloomy picture of the future, using a model based on five different factors: the state of non-renewable resources (e.g. iron and other ores, oil, etc.); the world population; the amount of food available per head; industrial production per head; the level of pollution. They then examined the relationships between these factors (using the loop process), and the resultant model was analysed using the method of systems dynamics (fig. 5). The result was a standard model comprised of a network of curves representing the period from 1900 to 2100. In the first phase, a large part of the technical progress is neutralised. In the second phase, the authors of the report do take into account some cumulative elements of technological progress, but they are not seen to constitute a system, or to have any relationship with the other systems. We shall return to this point later. At any rate, it seems that one could just as easily have produced a model between 1850 and 1855 for the next two centuries, and come to exactly the same pessimistic conclusions. Let us consider why.

This problem needs to be considered not only at the general level, but also at the level of each technique. For example, land can only become more fertile and productive if plant selection is practised, and more effective fertilisers are used – apart from the increase of land cultivated, which was a major factor after the discovery of America, and particularly so in the 19th century. It is conceivable, therefore, that limits do exist as long as botany does not lend its support to seed selection or chemistry to the creation of artificial fertilisers. All these elements have a part to play: extensive and intensive culture, methods of cultivation and methods of fertilisation. It should be possible to create schemas which would assess technological limits, those which historical facts relate. The increase in meat consumption was blocked for a long time by the difficulty of keeping animals alive over the winter period: the adoption of artificial fodder went a long way to removing this block, and was more significant than the breed selection. Similarly, when it finally became possible to conserve grain, the effects of bad harvests were not nearly so devastating.

The industrial sector also affords examples of such limits, particularly in the field of energy. As machines became more and more powerful, they also became more and more cumbersome, which posed many problems for steam transport, both on land and sea (for a ship, the power necessary as a function of the speed increases rapidly, approximately as the third power of the speed). Another difficulty was the relative inflexibility of the steam engine in relation to the variable demand for power. Then, finally there is the delicate problem of transporting that energy over distances. Apart from all these problems, the design, the construction and the productive capacity of the

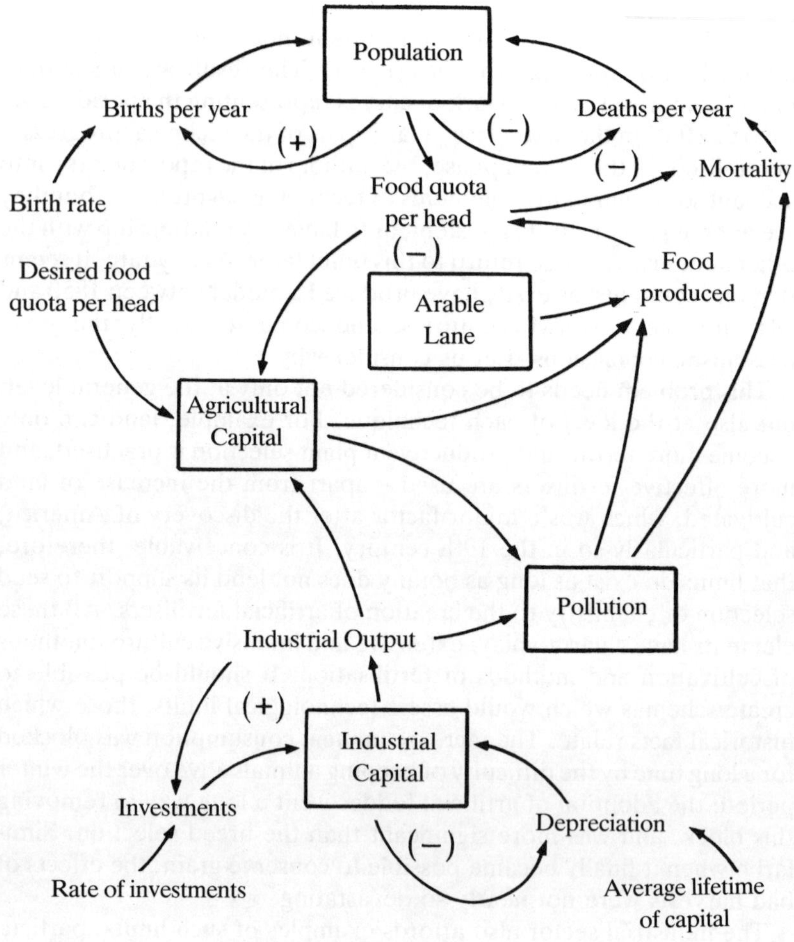

A

classical steam engine – i.e., the old alternating engine – imposed several limits. Its power capacity did not go above 5000 horse power. Below this, it was not profitable, but it was technically impossible to go beyond it. Consider the curve in figure 6. It shows that at a certain level, the increase in power, P1 to P2, can be achieved at a reasonable cost, C1 to C2. But beyond this, a proportionately lesser increase, P2

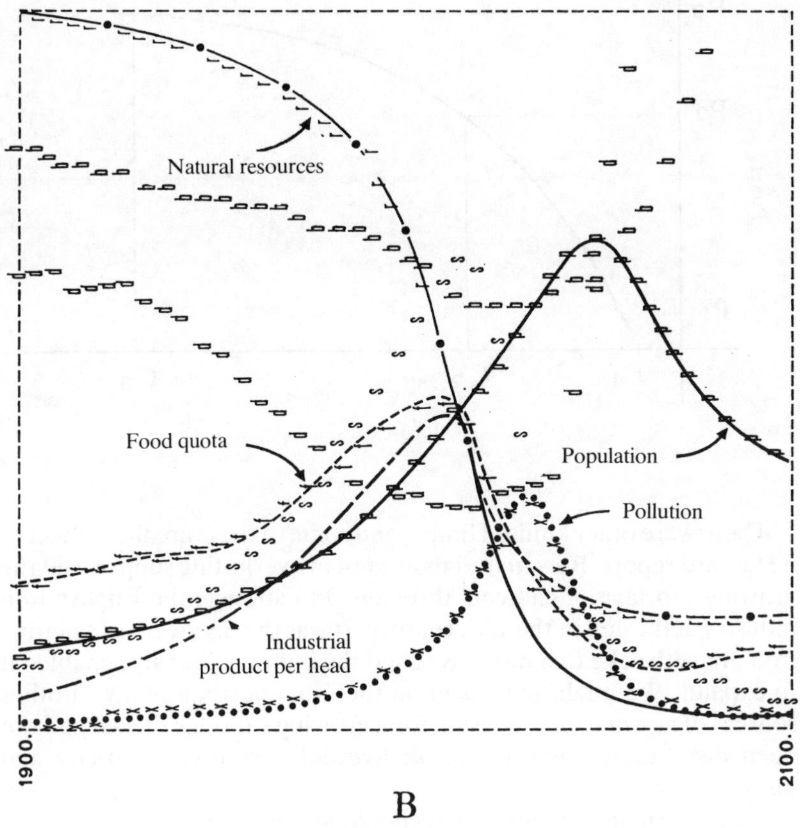

Figure 5. — Feedback loops on population, capital, agricultural production and pollution [A] and behaviour of the world model [B]. (From D.H. Meadows and Others, *Limits to Growth*, 1972)

to P3, can be achieved only with a proportionately much greater increase in cost, C2 to C3. In other words, beyond a certain threshold of power, no increase is economically feasible: the dimensions, the efficiency and the costs all imposed limits which were impossible to overcome.

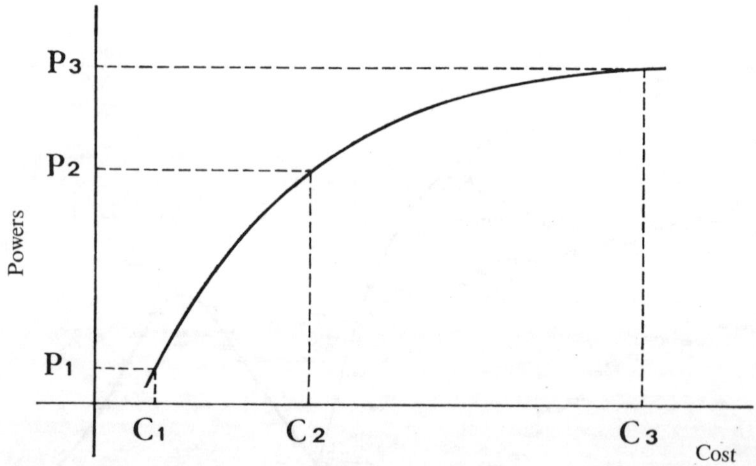

Figure 6.

There were other kinds of limits, and some of these are the subject of a Harvard report. Raw materials are not in everlasting supply, and the growing shortage of charcoal threatened to strangle the English iron industry at the end of the 15th century. It was the discovery of smelting iron ore with coke that not only saved the industry, but also enabled it to expand. We might have been in the same position today, if other sources of power to compensate for a growing shortage of coal had not been discovered, such as petrol, hydraulic energy, electricity and nuclear power.

There are many examples of technological limits which can block an entire system, and create the kind of imbalances which generate major crises. There are the integral limitations of a given technological lineage and the limitations of raw materials. Then there are the limitations within a technical sequence imposed by the distortions within a manufacturing process – this is evident in the English textile industry at different stages of the 18th century – and finally there are economic limitations, such as those affecting the building of the railways during the 19th century. Here again, we need precise and detailed analyses and inventories if we are to be able to study technical progress effectively and universally.

Since all techniques are interdependent in any one system, it only needs one sector to reach its limit for the whole system to be blocked. The railway is a good example of this. Around 1850–1855, rails were having to be replaced so frequently, that they threatened to become a major financial liability if trains continued to increase in weight and speed. Only the steel rail made it possible to solve this problem. Steel had been too costly to use in quantity until after 1860 when cheap steel began to be produced by the Bessemer process – which itself was to be superseded by the Thomas-Gilchrist process twenty years later.

The very concept of technical synthesis shows the importance of structural limits in this field, but their importance varies according to whether or not the problems they create can be solved using substitutes. One example of the latter is the intensive culture of land, as long as this was politically feasible. Some countries are able to do without some technical innovations because their particular social and economic circumstances are such that the limits of certain techniques have yet to be reached. This is a case of external factors being used to correct, on an external level, obvious imbalances. The imposition of tariffs as a national defence against a more advanced technology is an example of this. Thus, in 1821–22, the French placed a tariff on British iron – this was to enable France to maintain her traditional structures in that sector. It is true that there were other reasons for adopting such a policy, such as the inaccessibility of natural deposits, the narrow outlook of the industry's leaders, and the influence of pressure groups. It was for similar reasons that France took so long to adopt the steam engine: coal deposits were difficult to mine, and transport was also beset with problems. In fact, it was not until 1881 that more energy was produced by steam than by water. The nature of France natural deposits, coal and iron ore, meant that they were obliged to continue with traditional methods of smelting – it was only in 1864 that smelting with coke became more widespread than smelting with charcoal. Thus, as long as France and other countries in the same position kept to a certain level of production, and as long as they did not form an important part of international trade, they were able to keep within the limits of their technical system. This would explain the small role played by France for so long in the international market.

A study of this subject would be of invaluable importance. There are several courses open to a country whose technical system has reached its limits in the face of more advanced techniques in other countries. Some countries rich in raw materials can survive on the export of these materials, and still maintain the structures of an older system. Elsewhere, if these materials are vital to the economy, the problem is solved by technical means, and from this stems the complex history of replacement products or "ersatz," of which coke was perhaps the first example, followed by beet sugar and artificial rubber and a whole host of other products.

Dynamic analysis is perhaps the best method of studying this relatively new area of research, because it throws into sharp relief technical structures and systems, and clearly shows the structural limits which encourage invention and precipitate the mutation of systems. These are the two essential poles of "technical progress": the technological lineage on the one hand, and technical blockages, on the other. The study of the dynamics of technical structures is still in its infancy, even though several major works have been written on the subject. It will only be possible for us to deal with it on a fairly superficial level therefore, especially since we will have to consider exterior systems and structures as well if we are to understand the complexity of the problem. There are four distinct concepts which enable us to explain the dynamics of the technical system in relation to other systems. We shall discuss these briefly.

Scientific progress is perhaps the best known and most widely studied of these, but its relationship with technical progress still needs to be considered in more detail: an entire chapter has been devoted to this problem. Nevertheless, a few points warrant a mention here. Scientific progress is not always autonomous: it is often precipitated by attempts to solve a technical problem. And it is not impossible that some forms of mathematics and modern physics were also the fruit of technical preoccupations. Geometry, for example, was a technique used in land-surveying before it became a "pure" science. It is the same with the instruments used in science itself which depend largely on technical progress.

Two points of convergence between science and techniques should be emphasised here, each of which appears to be the inverse of the other. Science can solve some technical problems, and technics can help the development of science.

It is clear that up to a certain level, there is hardly any link between techniques and science. However, the more complicated techniques become, and the more they need to be rationalised, the more science becomes necessary, and even indispensable. This concerns not only the so-called manufacturing processes (tools, machines and the processes used), but also the exact definition of the required product. For a long while, the textile industry did not need the support of science at all. But there are some cases more doubtful than this. It is not true, for example, that the steam engine was devised completely outside the scientific context. It is often said that it was developed and in widespread use before Carnot elaborated his theory of the heat engine in 1824, but in fact, it could not have been devised without the scientific discoveries made in the 17th and 18th centuries. The discovery of atmospheric pressure, the Magdeburg experiments with hemispheres, coupled with the increased understanding of condensation and latent heat finally enabled engineers to translate what had only been an idea into reality. In the same vein, siderurgy became a precise method only in 1788, when Berthollet, Monge and Vandermonde determined "the different states of iron."

The two curves in the following diagram represent scientific and technical progress. Curve "a" represents the scientific knowledge required at the different chronological stages of a given technique. Curve "b" represents the level of scientific knowledge which is at the disposal of a given population. The technique can be adopted at any moment before the point T1, but point T2 must necessarily precede level C2. Obviously, all types of curves can be used.

This leads us directly to an important point. In the past, inventors are considered not to have had any great knowedge of science, but to have been gifted with a particular type of genius, and they themselves have often perpetuated this idea. Leonardo called himself an "uneducated man," and Lenoir, who invented the internal combustion engine, and Gramme, the inventor of the dynamo, were no doubt self-taught. But at this time, much of the scientific knowledge they required was easily popularised and readily available. From public lectures, their own reading, conversations and consultations, these self-taught inventors acquired all the science they needed. In fact what needs to be studied further is precisely the level of convergence between science and technology on the one hand, and on the other, the general level of scientific knowledge in a given population. The curves

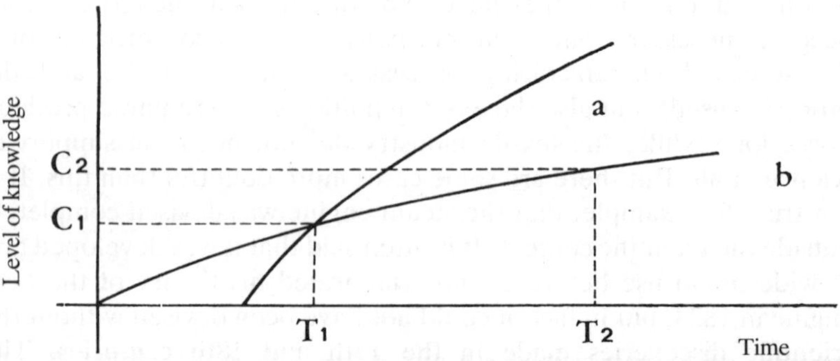

Figure 7.

in figure 7 reveal that there is no chronological determinism, but they also reveal that certain inventions would have been impossible before a clearly determined period. The analysis could be further refined by allowing for the body of specialists whose level of knowledge would be above that of the general level and also by defining the scientific bases of a given population according to its elementary education. It would be very interesting to compare the latter in 18th centuty England, with what it was in other countries during the same period.

The opposite problem to this can also arise, whereby scientific development has reached such a point that a certain invention seems possible, and yet it occurs only after a considerable lapse of time. Historians still wonder at the fact that the Greeks were unable to develop their technology, despite having the required amount of scientific knowledge. They have even criticised the Greeks for failing to invent the steam engine. But this only shows that we are still unsure as to the extent of scientific knowledge required for such an invention, and for many other inventions besides. Furthermore, technical construction is a complex process, and technical contexts can be very restrictive. It may be that there is no suitable material, or else that a suitable mechanism is not available. It was an entire century after the physical phenomenon used in photography was discovered that the first actual photograph was produced (1727–1829): this is evidence of the technical problems which can delay the realisation of a scientific idea. Here again, it is not enough simply to consider this one convergence, there is a whole ensemble of convergences which should be dealt with. The discrepancies between scientific discovery and technical invention became less significant as science became better equipped to explain more phenomena, and as technical progress placed more extensive means at the disposal of the inventor: this discrepancy was 56 years for the telephone (1820–1876); 35 years for radio (1867–1902); 12 years for television (1922-1934); 14 years for the radar (1926–1940); 6 years for the uranium bomb (1939–1945); and 5 years for the transistor (1948–1953).

The phenomenon of obsolescence is relevant here, but we shall discuss this later. It concerns many areas, from manufactured products to industrial installations.

The inventive process has its own rules, its own logic, and consequently, it has its own history. In *Théorie générale de l'invention*, René Boirel has attempted to systematise the concepts and language used to analyse this subject.

Boirel speaks of a "diffuse rationality" to be found in technical progress, but he does not thereby believe that the latter is determined. He feels that is is necessary to "retrace the steps of the inventor in order to discover the dialogue between his intellectual intentions, and the technical structures which he is manipulating." Rather than "expounding the necessary thought processes involved in passing from one structure to another," it is necessary to "demonstrate the possible courses of action available to anyone who clearly understands the potential of the structures they are dealing with." Boirel points out that it would be interesting to draw up a table of all the possible transitions from one technical form to another: for example, the advance from wooden planks to metal rails. The "creation of machines through the combination of pre-existing mechanical structures" should also be studied in this way. One example of such a machine is the steam engine which used the principle of the suction pump. There were three ways in which the inventor could use steam: its direct action, i.e. the force of a jet of steam; its pressure; or the difference in pressure created by the vacuum produced by condensation. Boirel calls this the "operative project." It is implemented by means of a structure, either a steam turbine or a piston and cylinder. Finally, they could also use the force of atmospheric pressure, either directly, as in Savery's machine, or indirectly as in the engine invented by Newcomen. Thus it is possible to establish a kind of genealogy of techniques, which determines the nature of inventions made in successive generations. Of course, a technical structure used in a given type of machine can be used in another type of machine: when Lenoir created the first internal combustion engine, he used the basic parts of the classic steam engine. Generally, a technological transfer represents the use of a given structure for a technique other than the one for which it was originally conceived. Coke was originally used in malthouses, which means that Darby's use of it in blast furnaces really represents a technical transfer.

We are certainly better placed now to discuss "diffuse rationality." Invention and progress are conditional upon pre-existing elements. A technological lineage of course has a certain logic, but it is diffuse in that choices can be made and different combinations of these elements can be constructed. What historians need to establish is how wide this area of choice actually is, for it seems to be much more restricted than has been supposed.

Boirel and Usher's work on the history of techniques, the latter of which is incomplete since it deals only with mechanical techniques, have both been assessed by J.-L. Maunoury, and he concludes that the theory is not yet complete.

According to this theory, technical progress consists of a succession of strategic inventions which synthesise disparate and relatively new elements. There is a clearly pronounced rhythm of technical evolution, but this periodic discontinuity does not supress the fact that this process is a social one, and that need plays a particularly dynamic role. Perhaps W. James' term "soft determinism" is a more useful one, since it preserves the idea of individual liberty and initiative, whilst leaving room for an important margin of uncertainty and chance. But this chance is not the contingency of a shapeless and arbitrary universe, but rather the kind of chance which manifests itself through a system of relations and interdependencies. The succession of events is an ordered and logical one, therefore, such that it is possible to trace quite closely, a posteriori, the genesis of innovations through their technical filiation. However, it is still impossible to predict a priori what the nature of these inventions will be, and when they will appear.

To illustrate this, Maunoury traced the genesis and evolution of heat engines from the 18th century to the present day, the schema of which is produced here.

This simplified schema incorporates only two types of structure as variables, and seems to support the concept of "soft determinism," in that the inventor does seem to have had a certain number of choices. Of course, if more real and constraining variables were incorporated, then there would be fewer choices. These variables range from the types of fuel and energy sources – for the engine is after all only an energy convertor – to the quality of the materials used (high speed turbines can only be made with certain specified materials). Furthermore, the heat engine could be combined with other types of convertors. The electric motor is either connected directly to the energy source, as in hydraulic turbines, for instance, or it is connected indirectly, as in the case of diesel, which is the intermediary between the energy source, and the energy produced.

Certain points can be made on the basis of these texts and our above discussion. Both of these concepts, diffuse rationality and determinism, suggest a reluctance on the part of their authors to compromise the liberty of the inventor. There is rationality because any invention is necessarily situated in one technical line, even though lines may be broken and new ones may appear. Such was the case with the steam engine and hydraulic energy, for the classic steam engine was in fact only a more refined suction pump. The development of the idea is not difficult to understand: the action of steam had been known since Ancient Times, but only as a force, and the action of a vacuum and atmospheric pressure had been demonstrated by scholars, but whose use at this time was not necessarily a logical one. In fact the real invention was the use of atmospheric pressure simultaneously with the vacuum created by the condensation of steam. The object of this was not the raising of water directly, as was the case with Savery's machine, but rather the indirect action of the pressure on the other side of the piston. In other words, the actual pump was finally separate from the engine producing the energy, the latter using the structure of the former. So from this stage onwards, from Watt's machines to the most recent ones, the technical chain followed a quite logical process. "Any analysis of a filiation of inventions, must therefore be conducted in terms of structural development."

But the concept of "soft determinism" invoked by Maunoury to avoid choosing between free-will or determinism is not so easily justifiable. We shall take issue once more with Maunoury, the originator of the formula:

> It is easy to understand how the inventor changed from one structural combination to another; what is not so easy to understand is why this precise combination should have been chosen and realised, because an invention is not simply a logical progression from antecedent structures. The inventor must have a problem to solve and certain objectives

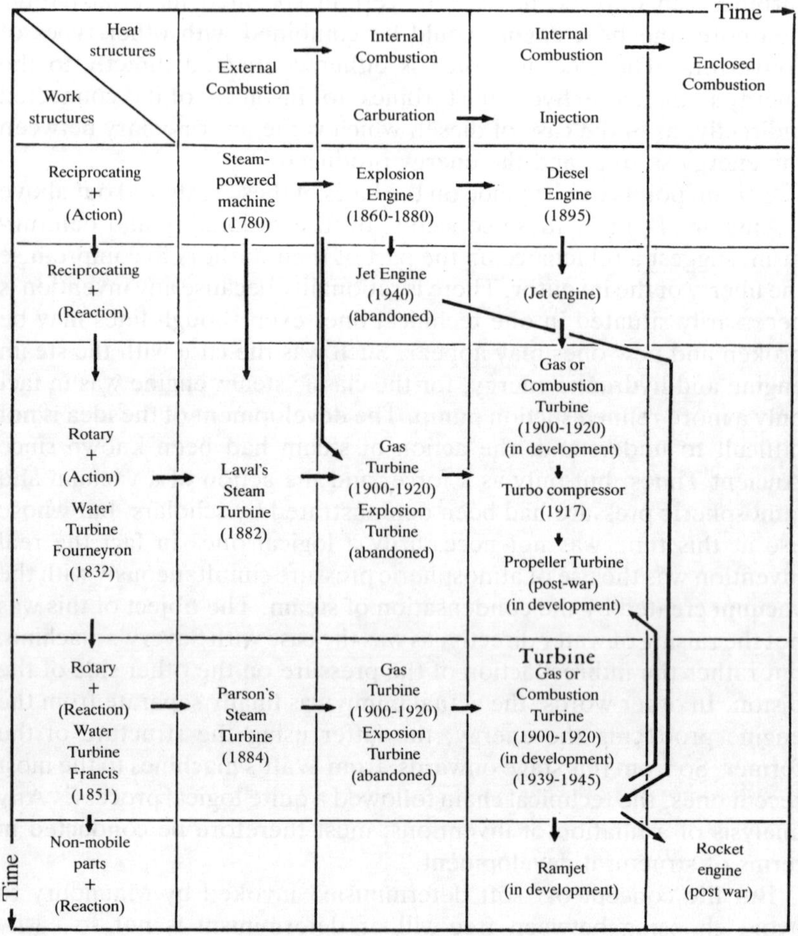

Figure 8. — Genesis and structural evolution of heat engines from the 18th century to the present day. (From Maunoury, *la Genèse des Innovations*, Paris, 1968)

to fulfil. It is from the interaction of these objectives with the technical structures available to the inventor that invention is borne.

There are various factors, technical, economic, social and others, too (armament techniques for example), which should be considered in relation to the concepts of rationality and determinism. Neither Boirel or Maunoury are perhaps sufficiently aware of this, because they are both approaching this issue from a relatively narrow standpoint. It is only natural for the philosopher to support the notion of rationality and reject determinism, and similarly, the economist speaking in terms of means and ends is bound to come down in favour of determinism at the expense of rationality. Furthermore, they draw their examples only from the domain of mechanical techniques, and especially from the heat engine, which has become the almost mythical symbol of technical progress and invention. If this problem is to be dealt with properly, the analysis needs to be refined still further.

Firstly, the invention needs to be placed in its technical context. Some inventions could be termed "simple" when the progress is autonomous – take John Kay's flying shuttle for example. But in fact the flying shuttle is related laterally to another much older technique, the weaving loom. The invention of the hot blast for the blast furnace was an identical case. The term "development" should be used, moreover, to denote the successive improvements of a technique of which the fundamental principles have remained unchanged. The introduction of automation in many industries, for example, left intact the basic principles of existing techniques.

Invention can also be an assembly operation, as in the case of the internal combustion engine, for example. The essential idea of this invention was to place the heat source actually inside the apparatus, and this depended on a combination of innovations: the carburettor for the fuel mixture; the dynamo or the battery; the coil, and the spark plug to produce sufficient electricity for the spark; and finally all the components of the classic steam engine. We are deliberately simplifying this description. This does not represent a single technical line, but a whole series of lines combined together. The structure of the piston and cylinder did not change – what did change was the type of fuel, the source of heat (coal, gas, and oil). The next step was the choice of heat conductor which was either steam, or an explosive mixture of air and petrol, or air and gas. This in turn determined the fuel-type used.

This enables us to move onto other levels. The first of these undoubtedly concerns scientific knowledge: we have dealt with this above. Scientific rationality may often be superimposed on technical rationality, as in the case of chemical techniques for example, and some branches of physics.

At the second level, beyond the structures, lie the systems. In other words, insofar as technical invention is not only the product of intellectual speculation but also its practical realisation, it is necessarily part of a system since the interdependence of techniques is one of the greatest constraints on technical progress. There is no better proof of this than the problem of materials. The complete mechanisation of the textile industry was inconceivable as long as machines were made of wood. Similarly, high pressures, superheating, and high-speed turbines are only possible with suitable materials. And there are many more examples of this.

The last two cases are examples of a kind of "negative determinism" in the sense that invention is conditional upon certain diverse factors. Positive determinism is quite different, and involves the consciousness of a problem to be solved. There is more than one variable contained in this concept.

Most needs are purely technical, although they do have economic consequences, particularly in terms of cost. This occurs in technical filiations, where an invention at a given level causes imbalances within the whole series. The English textile industry – especially weaving and spinning – affords many examples of this throughout the 18th century. The discoveries of Bessemer, Martin, and Thomas and Gilchrist also belong in this group. And the same happens when an element of a given technique is lacking: this was the case just before coke was used in smelting, only coke coal (which was already being used in forging), could make up for the insufficient production of charcoal. The numerous patents taken out between 1570 and 1710 are evidence of the fact that this was the only possible solution at the time. Darby's discovery of using coke in blast furnaces was made possible by borrowing from other techniques which used coal prepared in exactly the same way (probably malting and the refining of copper). This is an example of a technical transfer.

Economic needs – expressed in problems of quantity and cost – are just as important, since the problems of quality are of a technical order. Most of these needs are usually divided roughly into three categories: to produce the same quantity at a lower cost, or a greater quantity at the same cost, or to produce a greater quantity at a lower cost. These objectives are usually achieved only through technical invention. We shall discuss the necessary links between techniques and the economy further on. The idea of invention can only be analysed correctly if both technical and economic factors are taken into account.

No matter what else has been said of it, the history of inventions is still largely built on myth and idealisation, which in turn are often prompted by nationalist tendencies. The myth arises from the incorporation of vague autonomous and impersonal forces, and the idealisation quite simply from the endowment of the inventor with supernatural powers. The two are closely linked because any person who figures in such a mythology must necessarily have the qualities of a saint, or at least be in direct contact with the divine. Recent writers on the subject have tried to demystify the subject by concentrating only on the technical aspects of invention, since the economic analysis of technical progress, or at least the incorporation of the technical variable in economic theory was only coming into practice, and not without some reluctance.

Our aim, then, is to tighten the definitions of the expressions used in this type of analysis. We cannot avoid the concept of rationality, since the number of technical combinations is definitive and since any inventor must rely on existing structures and therefore follow largely pre-determined paths. The extent to which these paths are pre-determined needs to be analysed. This rationality is actually evident on many occasions when it seems to be unnoticed by the inventor. The explanations for example, put forward by Bessemer to rationalise his invention do not warrant serious consideration. The logic exists, certainly, but it is quite different from Bessemer's notion of it. The successive patents taken out by Bessemer enable us to follow his steps quite closely, and it appears that he was trying to produce large quantities of steel at low cost, using existing techniques. The injection of air into a bath of molten iron was already practised during the first stage of refining cast iron (*mazéage*). The essential factor was to produce a convertor and a suitable apparatus, and to create all the conditions necessary for its constant function (there were problems with the blast of air, the refractory materials, the addition of manganese, etc). Bessemer relied on pre-existing structures, so the main problem is to establish how he solved the problem of the apparatus, for the final development of the process was in fact the work of several people. It is possible to conceive of another chain of logic, which was not the result of sometimes misleading research into works on ancient technology, or patents, but which arose – because of the demands of the time – from the beginnings of a true spirit of scientific, or quasi scientific speculation.

The notion of determinism is no less evident, in all kinds of forms, technical, scientific, economic, and even social and political. The latter two have not been discussed, but we have already mentioned the necessary links which exist between their systems. Social determinism is evidenced in the English textile inventions of the 18th century, which, rightly or wrongly, have been attributed to inventors who wanted to reduce the misery of the working classes. The "Jenny" symbolises this by its very name. Political determinism is far more obvious, especially in the field of military techniques: there is no need to insist on this, but Nef's *Technical progress and War* is worth reading again. As far as Bessemer's discovery is concerned, one could study the patents to establish the range of "needs" felt most strongly during this period. These needs were mainly technical and were imposed by the development of certain activities mainly concerned with defence and communication: shells and steel cannons, rails and tyre wheels for railways, and finally cheap, plentiful steel.

In fact, whatever the period, and at whatever level one chooses to place the liberty of the inventor, it is always severely circumscribed by the needs that the invention is designed to meet. The inventor is constrained by choice, then and this range of choice is limited (the choice of possible channels for nuclear power, or of processes which can be used in colour television, for example), and furthermore, the timing of inventions depends largely upon scientific and technical progress, and on economic necessity, etc.

It would also be useful to study the institutional context within which the invention is created and developed. The main function of these institutions is to encourage invention and its application.

In earlier times, this institutional context was vital – and this is proof of more than one type of determinism – since it was felt necessary to publicise the need for inventions, and point out the most promising approaches, which represent incidentally, various different types of determinism and rationalism. An example of the latter is the transformation of the nature of technical knowledge. Symbolised by the change from the exhibitions of machines in the 17th century, to the collections of models and experiments in the 18th, technical instruction no longer consisted of the diffusion of facts, instructions on the better use of existing techniques and the techniques of invention, but of instilling a more rational scientific approach. The same institutions sought to increase an awareness of technical and economic need, but without always distinguishing between the two. The foundation of the great science academies, whose field was technical as much as scientific, and the creation of the industrial or agricultural societies as well as the offering of prizes, such as those given in the early 19th century for the discovery of sugar beet or for the development of mechanisation in spinning, were all forms of encouragement which have remained important ever since.

Another aspect to consider was the fact that the inventor would naturally want to reap the financial benefits of his invention without giving away any secrets. It was on account of this that whole system of patenting arose. The government in Venice is said to have promulgated a law in 1474 which granted certain privileges to the inventors of "new techniques or machines." The first real patent rights were introduced in England after the statute of monopolies was promulgated by James II in 1623. In France these privileges, which in fact rarely resulted in monopolies, were granted only to those who were importing foreign techniques: this gave rise to a new method of diffusion, the patent for the importation of new techniques, which clearly had nothing to do with invention. Then in 1699, the *Académie royale des sciences* was authorised to examine and approve new inventions, introducing firstly control of the invention, then gradually, the idea of priority. Legislation in France developed gradually, from the declaration of 24 December 1762, to the law passed on 7 January 1791, which protected both technical innovations and the importation of foreign techniques. From this point, the law on invention was extremely rigourous. (The first patent laws in the United States were passed in 1792.)

In fact, the only element of freedom left is that of the inventors themselves. But this is only interesting insofar as it enables us to study more precisely the conditions of invention. i.e. in Maunoury's words, "the encounter between the objectives and the possibilities offered by the technical structures available." Individual motive, be it selfish or altruistic, is of little importance.

We have examined the framework of invention, we should now examine the idea of invention itself in a little more detail. We have defined simple and complex inventions, but the phenomenon itself still needs to be clearly defined. Many writers have tried this, but few have succeeded. This may not be very important to the engineer, but to the historian it is vital. We shall leave the question of indispensable scientific discoveries aside for the time being, since this appears only at a certain level.

A concrete definition of invention, writes Maunoury, should help us resolve a number of delicate problems. An invention is one event among many which leads to the launching of a new product on the market, it is not simply an activity or a type of intellectual process. It is distinguished, therefore, by three other events, which generally precede, accompany and follow it, and which tend to "smother" the invention itself:

a) technical discoveries which affect invention directly, by summation, accumulation, etc., or by improving tests and insufficient outlines;

b) simple improvements;

c) the operations of commercial development.

According to Maunoury, the harmony is based on three sets of criteria: "an invention is a technical device or schema which either resolves a technical problem in a new way, or resolves a new technical problem; next, it may possibly be useful in production; and finally, it is the result of creative processes and above-average intellectual effort." Similarly, S. Kuznets defines technical inventions as "new combinations of existing knowledge in the shape of devices potentially useful for economic production and which are the result of a mental act above the average."

But these formulas are too simple and too general to accurately reflect what is doubtless a complex reality. One can hardly talk of the invention of the car or television, except in terms of a combination of a number of older inventions that had appeared separately for use in other combinations. So in fact, the only "invention" is in the originality of the combination, which is mainly due to a single new element completing an assembly of older structures. Furthermore, the expression "it may possibly be useful in production" is not sufficiently precise: the terms "possibly" and "useful" are both difficult to apply.

However, there are some terms which warrant more precise definition, such as the term "development" for example. Development is generally accepted as meaning: "the creation, testing and perfecting of prototypes; the creation of models on different scales; trials; the construction of pilot installations; research based on these pilot schemes with a view to mass production; the attempts to find solutions to all the various problems raised during all these stages; new trials, etc., until finally a temporarily definitive solution is found which enables the product to be launched into the market. The problem is certainly an interesting one, if only because it raises the question of when the term "invention" actually becomes meaningful. In fact, it seems to be at the stage which was formerly known as "improvement." It is true that Bessemer's invention dates from 1855, but it is equally important to note the date when the first true Bessemer convertor was actually used in industry, which was 1862. It is noteworthy that very few historians have concerned themselves with this aspect of invention; we read everywhere that Bessemer's invention dates from 1855, but nothing is ever said about its development. Similarly, the steps taken to perfect the invention, taking it to its full potential, are also worthy of study. The discovery of Thomas and Gilchrist, for example, which was followed by a "development" process making its practical application possible, could also be considered a development. We have decided to use the term "improvement" to denote the phase which leads to the practical application of the invention, and the term "development" for the phase which takes the invention to its limits. Some historians have talked of "associated inventions" with regard to the latter stage.

Innovation is quite different to invention, but the two are necessarily linked. "The dynamic entrepreneur," writes F. Perroux, "is an economic innovator, in that he takes a technical invention, or more broadly, a new combination of techniques, and places it in the real world of the market place."

The principles of innovation are better understood thanks to the interest of economists in this subject. It would still be useful to clarify certain concepts at this point, however. Innovation is chiefly an economic factor, so we are dealing here with a different type of link from those discussed above: the relation between technical progress and economic need, of which some examples have already been given. In general, and as we have already stated several times, this relationship between techniques and economics is based on quantity and cost. Thus, for example, economic calculation can assess the usefulness of a new technique as opposed to an old one. There are several new problems involved with the notion of cost, as opposed to that of quantities, however, all of which obviously influence the solutions adopted.

Innovation, like invention and for the same reasons, responds to certain technical needs. This may be the need to reestablish the balance of a technical concatenation, or the coherence of a technical system, etc. But unlike invention which is in a sense a gratuitous act, there are limits to innovation. In effect, there is one major factor which cuts short the period of profitability of a technical structure. This is obsolescence, which according to A. Hayek, "can be applied to any element of capital whose physical durability outlives its usefulness." The appearance of new techniques cuts short the economic life of techniques in an older system.

The threat of innovation can be avoided in certain cases, as it was in Spain by monopolies, for example. Twenty years ago, Spanish railways still used some of their first locomotives, and some of these were therefore almost a hundred years old. Again, protectionist policies are useful for protecting outdated techniques rendered obsolete by international competition. Beyond this, there is the problem of newly-industrialised countries equipped with the latest techniques who find themselves in advance of countries whose industrialisation is long-established. This may have been the case with England at the end of the 19th century – this point is made in André Siegfried's work on the subject. In countries ravaged by war, the differences between those regions affected, and those left untouched are obvious.

All this inevitably weakens the spirit of the entrepreneur, understandably, because there might seem little point in changing something which can still be used. Furthermore this attitude is sustained by tariff protection and the rising costs of manufacturing. Mimerel, the renowned spinner of Roubaix (France), noticed this in the textile industry in northern France after 1834. This attitude can quickly turn into a fear of technical progress to the extent that if an invention threatens to overthrow a manufacturing process, some entrepreneurs may refuse that innovation in case a new invention might appear which would render their changes obsolete. There were some striking examples of this just after the invention of the Bessemer process. Apart from the iron-masters' usual hesitation and scepticism, there were others who really feared that this invention might lead to others which would soon overthrow current practises. Studies of reports from the boards of directors to general meetings of companies in existence at this time reveal that this fear was widespread. The 1834 survey of the French textile industry also revealed clear proof of this. A Parisian spinner, Griolet, declared: "when no advance is made in industry, it declines." Vayson, a carpet-manufacturer in Abbeville changed his industrial equipment twice in under ten years, while Dablaing, a spinner from Douai changed his three times within the same period. Yet Mimerel, whom we have just quoted, expressed that fear of seeing his latest equipment rendered obsolete by yet another round of inventions: "we are reluctant to replace our old machines with new ones, preferring instead to patch up the old ones." Quite recently, the large French firm Alsthom expressed exactly the same fear in a letter to the press: "if we get orders for two nuclear plants per year, we hope to make sufficient profit on our investment (in equipment), based on the hope that the present level of technology will remain stable for the next few years."

Similar reactions are produced when having gone beyond the level of a given technique, one tries to integrate it into the entire technical system. Some entrepreneurs have been quick to see the birth of a new technical system and adopt its structures within their own particular industry. For example in 1822–23, François de Wendel sought to introduce new English processes into his iron works in anticipation of the growing market for iron created by innovations in other areas such as canals, steam engines, railways, etc. Such innovation on the part of de Wendel was justified, therefore, in the light of innovations made in other areas. By contrast, Péchiney from Salindres, was doubtful about the market for aluminium, which he was beginning to produce: "The market for aluminium is limited: it is used to make tubes for opera glasses, and whether you sell it for ten francs or a hundred, you will not sell a kilogram more or less."

The economic requirements of innovation are no less important, and they have been well exploited. Economists have constructed abstract models which do not necessarily coincide with reality, but they do at least provide us with some precious elements of analysis. A given technique is expressed by a production function of a general type:

$$P = f(a,b, \dots n)$$

P expressing the quantity produced and $a,b, \dots n$ the quantities of production factors $A,B, \dots N$ which have been necessary to make this amount of merchandise. As G. Hosmalin noted, whose presentation of the model we have drawn from here, the technical level of a branch defines itself by its "technical horizon," its "average production function" and a series of particular production functions, each one of which is characterised by a given enterprise or a "sector" in the branch. Each particular function describes a type of relations between factors and production. It allows for the evolution of proportions according to which different factors are combined in correlation with the variations of the volume of production. For each function, these proportions are represented by independent coefficients or parameters, all of which modify the terms *a,b, ...n*. There are two categories of technical coefficients, the flux coefficients, which relate to variable factors of production, and the capital coefficients which relate to fixed factors of production. Their values constitute the specific aspect of each function.

The production function of an industrial branch is a "series of technical relations between its factors and its products." It resembles the totality of particular functions which can be used at any one given moment, in any one given society. Two graphical representations can be drawn from this. In Figure 9, the two axes of coordinates serve to measure the quantities of production factors necessary for the fabrication of product *M*. To simplify this, the number of combined factors has been reduced to two, *l* and *k*. The lines *A, B, C* and *D* represent the different techniques used by particular enterprises or groups of enterprises which constitute the branch manufacturing the product *M*. Each line expresses, therefore, a particular function of production, i.e., the "law" of a combination of factors according to the variations in the amount being produced. Any point on one of these lines corresponds to a volume of the production of *M*. *Va* represents 1000 objects, *V'A* 1200. The points *l* and *k* have been associated for the first value, *l'* and *k'* for the second. One can see that $l/k = l'/k'$. No matter what the volume of production, for each technique the factors are always combined according to the same proportion.

Figure 10 introduces a new method of production into the branch. This enables us to define the conditions on which the newly introduced technique can qualify as technical progress. Points *VA, VB, VC* and *VD* have been joined by line segments. These points each correspond to the techniques used in producing 1000 objects. In this way, an "indifferent production curve" or "isoquant curve" is obtained. Line

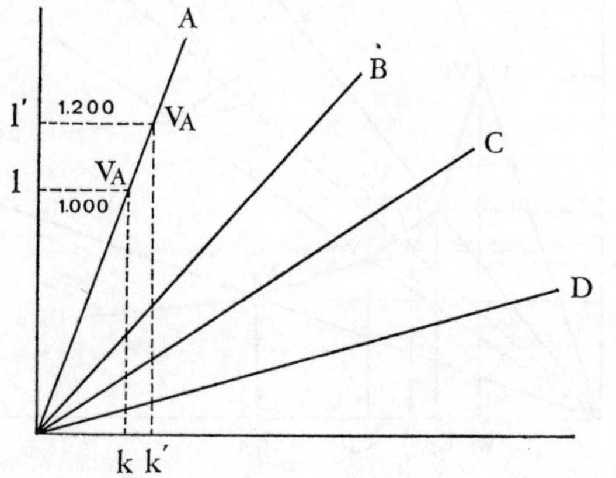

Figure 9. — Production function in an economic branch. (From G. Hosmalin, *Investissements, rentabilité et progrès technique*, Paris, 1956)

C′ (in dots) represents a new technique which associates *l* and *k* according to new proportions. It therefore expresses a new particular production function. It is presumed that it lies between lines *B* and *C*. Point *VC′* on line *C′* represents the combination of factors required for producing 1000 objects. In this way, the isoquant curve can be completed.

The estimated volume of production (1000 objects in this case) is obtained for each technique by bringing a number of factors into play which are expressed by the following indices:

method *A*	$l1 + k1$
method *B*	$l2 + k2$
method *C*	$l3 + k3$
method *D*	$l4 + k4$
method *C′*	$lx + kx$

Consequently, method *C* requires the quantity *lx* in factor *l* to be smaller than *l1* and *l2*, and greater than *l3* and *l4*.

In order that the newly introduced method can be unequivocally considered as technical progress, it must enable the same volume of

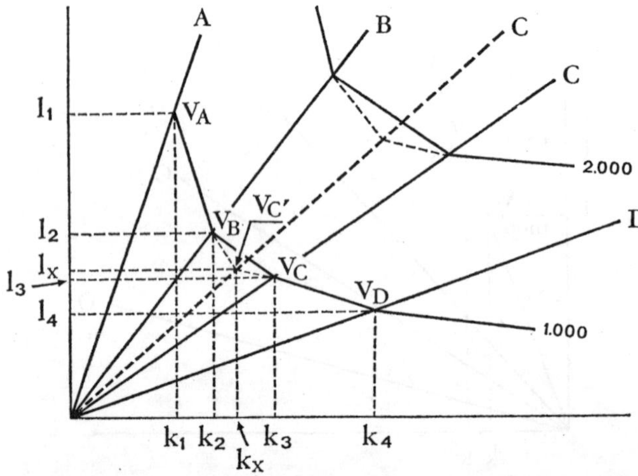

Figure 10. — Technical progress and production function in an economic branch. (From G. Hosmalin)

merchandise to be produced, incurring on average in each of the factors *l* and *k* a smaller expense than the other known techniques. This is not the case with this example. We need to specify, therefore, if the use made by method *C′* of one of the factors is greater than the subsequent increase in expense regarding the other factor. Method *C′* only represents real progress if:

l1 – *lx* greater than *kx* – *k1* in relationship to *A*
l2 – *lx* greater than *kx* – *k2* in relationship to *B*
lx – *l3* smaller than *k3* – *kx* in relationship to *C*
lx – *l4* smaller than *k4* – *kx* in relationship to *D*

If the four conditions thus expressed are not met, then method *C′* does not represent true technological progress, but rather technological change. It is possible that a discovery of this type could be effectively adopted, but it would not be in order to increase actual productivity, because it does not carry this advantage. The exploitation of a new method is sometimes used for other reasons: to make the best use of a country's resources in one of the production factors, in order to actually use national resources, or there are sometimes commercial considerations linked to price changes in factors according to the fluctuations in the market. Every model has its limits, and this is clearly evident in this case. First of all, the actual number of production factors is much greater than two, and furthermore, these are of a varied nature; Bessemer's convertor needed pure cast-iron for example. There is no perfect identity between production factors for all the available methods, and the possibilities of substitution entail changes in the proportions according to which they are combined. Besides this, clearly defined units of measurement are needed to make comparisons. In the preceding model, progress was described in terms of the increase in the productivity of labour or global productivity, or in the framework of a production function, by measuring the quantities of factors within the same group of coordinates: one would think, therefore, that the problem of the unity of measure would be resolved. There are some well-founded objections to this however, which extend even to price measures. Some have attempted, with little success, to solve this problem, with productivity per man and per hour, or the National Bureau of Economic Research index "man-hour."

It is up to the historian to make the transition from theory to reality and to examine all the factors of innovation, even if only to produce a catalogue. This means studying the formal analysis of the economists, but also outlining the other factors involved. These may be economic or technical, but they are also sometimes part of exterior systems, which may be important. It is in this sense that the history of technical progress contributes as much to general theory as to the means of analysis, in that it offers a more refined method. It is hardly necessary by now to state that research has only just begun on this subject, so that anyone reading this work is bound to be left with some unanswered questions.

Shortly after the Imperial period, a French iron-master named Rambourg revealed his awareness of all these variables when he replied to the criticism directed against his industry for not adopting the more advanced English processes:

Those people who propose so lightly that we should substitute coal for charcoal, seem to be unaware of the number of changes that this would require, in the furnaces, the refineries, the machines, and the workshops. They seem to forget as well that it is necessary to be close to the right type of coal mine, to have the ore close by the fuel, and to train all the workers for this new type of work.

Thus, for a given date, 1815, we know all the variables that were considered important in this particular sector: well-placed raw materials, the other elements in the technical complex, and a trained workforce. Rambourg also considered the situation in England:

> We would be deluding ourselves if we thought that most of our forges could compete with the English in the price of iron. Mother Nature has favoured them with a close proximity of fuel and ore. They convert the former into coke, and with steam powered machines they are able to produce enough power to stretch the iron into bars. They are then able to transport the iron easily down the many canals to the sea. All these advantages, which we so sorely lack, are for them an immense source of economy.

An iron-master from Bourbonnais, also a contemporary of Rambourg's, included the lack of invested capital among these problems.

Of course, it would be useless to apply models which are too rigid. The variables which represent production factors are not necessarily identical, be it from one sector to another, or, where the same technique is concerned, from one concatenation to another. In the latter case, they cannot be reduced to a single common denominator (of price, for example) and nor are they interchangeable. In the case of the changeover from iron smelting with charcoal to iron smelting with coke, besides other problems of price (and transport may be included in this), there was also the question of the different properties of the two fuels. Coke was not as easily crushed and could furnish higher temperatures than charcoal.

We must also consider the notion of time in this discussion. The graphs which have been reproduced suppose that the problem exists at one particular date or during a specific period of time. If the passage of time is taken into account the production function particular to each technique can very well change and no longer remain the same. According to the descriptions made, especially by the National Bureau of Economic Research, one may notice that the rate of technical progress affects: (a) a rapid advance in the middle of a period of expansion; (b) a slowing down at the end of this phase; (c) an acceleration at the beginning of the cyclical contraction; (d) a new slowing down in the middle of the period of depression. Yet once again, a more detailed analysis demonstrates the flaws besetting these measures when they try to explain reality.

Consequently, a list should be drawn up illustrating the whole typology of innovation, both a typology with relation to purely technical elements, which would correspond fairly well with a typology of invention, and a typology according to production factors which are not of a technical nature, and finally a typology according to a chronology yet to be determined.

Disagreement arises at the decision-making level. While the individual entrepreneur was still in command, he was not subjected to the pressure of a technical framework. Indeed, the appearance of these technical frameworks within a business literally creates the problem. The problem is further complicated on the management level of the firm, that is to say, when not only technicians but also financiers intervene. History often cites financiers' fears in the face of innovations proposed by technicians. Take the example of Fourneyron developing the hydraulic turbine despite the misgivings of the bankers in charge of the business. Yet the position of the financiers is understandable. Had Péchiney been refused by the bankers before whom he presented his projects, the liaison between the enterprise and the financiers would have ceased, for the banker must consider not only the profitability of a proposed innovation, but he must choose among the various profitable uses to which the capital he manages might be put. One need hardly multiply examples of this. A curve (fig.11) has been drawn from the observations made by Shelby Cullom Davis regarding the case of the U. S. Steel Corporation. It concerns the evolution of the relationship between the investments recommended by the engineers, on the basis of scientific and technical considerations, and those adopted which were advocated by the "financial commission" who were enlisted to assess the eventual profitability of the projects. This curve is very similar in nature to the one illustrating the applications for patents put before the Department of Commerce in the United States between 1924 and 1939 (fig.12).

As innovations exist on the individual level, there is little need for explanation or numerous examples. Obviously innovations also exist on the collective and national level. There is also the problem of what might be called the channels of innovation, which may affect the individual, collective and national levels. We will shortly return to the purely technical problems of these channels. The most important problems are those of a financial nature, as they affect both capital structures and financial mechanisms. The notions of structures and convergence must also be included. A coherence of structures must necessarily exist between technical innovation on a certain level of

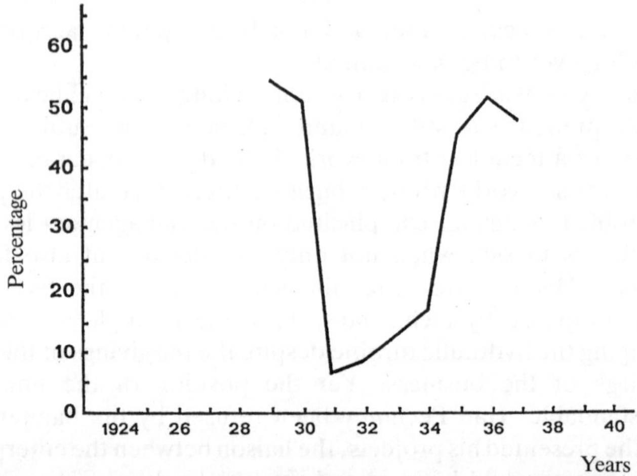

Figure 11. — Percentage of investments recommended by the Financial Commission of the U.S. Steel trust in relation to the Bureaux of Techniques' propositions from 1929 to 1937. (From G. Hosmalin)

investment and available capital structures, that is to say the mobile capital available for investment in new types of business. It is then necessary to create other forms of societies. This is indeed the innovation which led to the creation of individual companies in France in the middle of the 18th century, which had nothing to do with the legal structures imposed by the *Ordonnance sur le Commerce* (ordinance on trade) taken in 1673 but still in force. This is why the evolution towards the modern limited liability corporation began to take place, which was regularized in France by the *Code de commerce* of 1807. This action made possible the mobility of investment, while maintaining the permanence of capital. Yet the capital for industrial innovation still had to be found, but for most of the first half of the 19th century, the structure of French banking made it incapable of meeting this growing demand for industrial investment capital. It was in 1825 that the Parisian banker Jacques Laffitte realized the necessity for the convergence between available capital and the innovative industrialist who needed it. "This tendency for available capital to be productively invested comes together naturally with another tendency, that being to

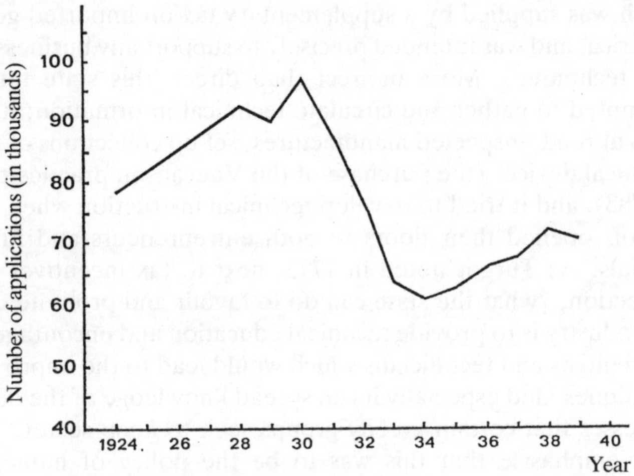

Figure 12. — Annual number of applications for patents filed by the United States Department of Trade from 1924 to 1939. (From G. Hosmalin)

innovate and improve, which is no less favourable to the progress of industry than investment." Laffitte proposes the same propensities defined by Rostow for the take-off of growth. He also suggested a "Société commanditaire de l'industrie" (an industrial investment bank) whose object would be "to contribute to and participate in the success of any enterprise, any invention, and any improvement concerning agriculture, industry, or commerce." This investment company was also to be advised by a commission of applied scientists, on matters of mathematics, physics, chemistry, and mechanics, but for political reasons Laffitte's project for an investment bank never saw the light of day.

Group innovation is relatively recent. It only took place with the aid of state intervention. Colbert's many efforts to introduce and establish various foreign techniques in France have already been mentioned. The invervention of the state was on the level of promoting the search and importation of the best foreign techniques and craftsmen, of encouraging qualified workers and entrepreneurs with subsidies and tax exemptions, and to put pressure on capitalists to invest. This policy was no doubt less in effect in the 18th century. One must also cite the *Caisse du demi pour cent* (the half-percent fund) of the 18th century, which was supplied by a supplementary tax on imported goods from America, and was intended precisely to support any business adopting new techniques. More indirect than direct, this state intervention attempted to gather and circulate technical information; it sent missions abroad, inspected manufactures, set up collections of models of technical devices (the purchase of the Vaucanson practice took place in 1783), and it tried to develop technical instruction when the larger schools opened their doors to both entrepreneurs and future state officials. As Turgot noted in 1772, next to tax incentives and tariff protection, "what the state can do to favour and promote commerce and industry is to provide technical education and encourage research by scientists and technicians which would lead to the improvement of techniques, and especially it can spread knowledge of the devices and processes that certain greedy groups seek to keep secret." And one must emphasise that this was to be the policy of numerous governments in France up until the end of the 19th century.

Innovation thus necessarily takes place only at a point where many factors converge. The foremost of these convergences is purely technical. Indeed it is essential that, on the level of production itself, innovation must fit in a technical system that is in equilibrium. It is useless to innovate at a given stage of a technical procedure, if the network of techniques upstream as well as downstream are not soon developed and coordinated at the same level. Although having said this, it is true that some old techniques can maintain a satisfactory equilibrium with innovations for a certain time. Such was the case on the European continent for ironmaking for a while after the English technical revolution: the most advanced techniques were imported and applied for puddling and rolling iron, while pig iron continued to be produced to a large extent by traditional techniques. But sooner or later the coexistence of old techniques and advanced techniques, for which there are various reasons, will eventually have the effect of checking innovation. This phenomenon is particularly true of some exploitation techniques, agriculture being the main one.

Beyond that, such convergences must exist within the principal systems themselves. We have already mentioned above the problem of equilibrium between the technical system and the economic one, but we should also point out that harmony must similarly be achieved between the technical and the social system, and between the political and the institutional system. There are consequently many sorts of constraints which inevitably weigh upon innovation, and the power of judgment which the successful entrepreneur must exercise before embarking upon innovation should also be considered. And conversely, the evolution of the other systems may call for technical innovation. History, as we will see, is teeming with examples of both: obstacles and resistance to technical innovation as well as stimulants, indeed even pressure, for the adoption of a new technique, or a whole new technical system. From the prohibitions, in the name of quality, imposed by medieval guilds on the spinning wheel, or certain dyeing and other textile processes in the Middle Ages, to the first printers' strikes against the introduction of new presses, to the first machine smashing, as in the case of Lee's machine for knitting stockings in the 16th century, one encounters obstacles of all sorts to innovation. The more a society is rigidly structured, the more difficult it is to initiate innovation. It has recently been noted that African tribal societies without an urban base, as is frequently the case in Africa, offer less resistance to Western industrial influence than the traditional Asian societies founded upon an urban network. In extreme cases of social rigidity, only an authoritarian state can bring about innovation.

It is in this way that the historian can explain certain delays in innovation, certain blockages of technical progress which may appear astonishing, and rightly so, such as those in Ancient Greece or in China in the 15th and 16th centuries. Once again, we require a catalogue of well-defined innovations, the study of which would have to be carried out on all levels. Unfortunately all we can do is to bewail the absence of such a catalogue.

Today economic growth is a problem much studied by economic theorists, and historians have begun to analyse it with some success. But both economic theory and historical analysis have as yet produced an inadequate account of technical progress. They come up against the same problem of the difficulty of measuring it. According to some authorities, "technical progress is the relative variation of overall productivity in a given sphere, between two given epochs." The overall productivity is defined by the following ratio:

$$R = \frac{\text{production}}{\text{factors of production}} = \frac{Q}{F}$$

If one calls *R1* the productivity at the period *T1*
R2 the productivity at the period *T2*
technical progress is measured by the expression:

$$\frac{R2 - R1}{R1}$$

But in actual production technical progress manifests itself according to much more varied modalities than would be supposed by the concepts of real productivity growth, or the concept of the functional relation between the factors and the products:

I. – From the micro-economic point of view.

A) Increase in the dimensions of the firm in order to produce the same product by means of the same factors of production. The productivity of certain factors increases and then decreases. This is the well-known phenomenon of "economies or diseconomies of scale."

B) The substitution among factors within each of the two large categories of capital and labour, in order to produce the same product.

C) The introduction of new factors to produce the same objects (new machines, modifications in the composition of the work force, introduction of certain specialists, etc.).

D) A change in the quality or even in the nature of the products made by the firm (for example, the substitution of steel for iron at the end of the 19th century).

II. – From the macro-economic point of view.

A) Variation of the volume of investment necessary both within the branch and in relations between branches of an economy. This is where calculations of profitability are involved.

B) In a more general manner, "the technical evolution is principally an indogenous variable oriented in directions determined by economic forces."

But in fact even the most recent analyses show a certain confusion, which arises precisely from the way they define a technical variable: technical variables defined only in terms of productivity cannot account for the overall reality. This is where historical analysis sheds much light on the subject although for the most part this analysis remains to be done.

The first point to establish concerns growth or growths. Attention has mainly been attracted to the periods of growth in the 19th century, beginning in England about 1780 – 1800, then gradually spreading through the 19th century to most other Western countries. The many studies which have been done on this subject have emphasized the role of technical progress in the take-off stage, and it is in this perspective that the notion of an "industrial revolution" has taken on its definitive image. Using the terminology of the systems' approach, we may say that the formation of the new system was one of the causes of the beginning of growth, and that this beginning could not have taken place without a new and viable technical system whose diverse techniques had attained a certain coherence. It is well known that this equilibrium was only established during the years 1780 – 1800.

What must be explained is the role of technology in the continuation of economic growth, whatever the shape of the curve the latter may take. To be sure, the balanced development of the various techniques in an existing system constitutes technical progress that contributes to the maintenance of economic growth. But at some point the moment must arrive when the technical system reaches its limits. It may even be enough for one technique to attain its limits for disequilibrium to be brought about within the whole system. This disequilibrium may then lead to one of two results: either economic growth will cease, which can of course happen for other reasons, and which will simultaneously stop technical progress, or a new technical system will substitute the old, which will permit economic growth to continue.

Some authors attribute the great crisis of the beginning of the 14th century to tensions provoked in this way when the medieval technical system reached its limits. At the end of the 16th century another period of growth also came to an end for various reasons accompanied by a stagnation of technical progress. So what seemed to ensure, both the growth of the Middle Ages and the Renaissance, and the growth of the English technical system at the end of the 18th century, was the organization, each time, of a new technical system which, to a certain degree, enabled new economic growth to take place.

We have already noted the fact that economic growth in the 19th century, particularly in England and France, would have faced a similar stagnation had not a new technical system completely different from the one prevalent in the first half of the century taken its place in the second half of the century. It is possible to answer some of the questions asked in the recent works by François Crouzet and Maurice Lévy-Leboyer. These two historians, following many other studies and observations, in particular the works of the *Institut des Sciences Economiques Appliquées*, emphasize the slow rate of French economic growth between 1815 and 1914, and the absence of any real take-off. It is precisely owing to reasons ranging from the location of natural resources to social and political preoccupations, that the technical progress developing in England since the end of the 18th century was introduced extremely slowly into France under very strong protection from customs. We have already mentioned traditional iron smelting which continued until after 1864. Similarly, in order to avoid social conflicts and to maintain wages at the lowest possible level, a large part of weaving was not mechanized, and according to the declarations of an Alsatian industrialist of the first half of the 19th century, weaving continued to be practiced for a long time in the countryside on more or less primitive looms. Such a policy meant that the conquest of foreign markets was impossible. It was completely different in England where innovators had been required to respond rapidly to certain production demands such as the shortage of wood for iron smelting. There was also a large and relatively cheap labour force and the general political evolution made available immense markets for English industry. This explains why the English take-off into growth was so much faster.

But one should be cautious of the growth curves described by these authors, for while they do represent the overall movement very well, they do not provide all the explanations. For example, in one of his curves Crouzet puts together iron and steel, saying that "steel was one of the substitutes for iron." This is precisely the kind of thing the historian of technology cannot accept. There is a very clear division between the properties of the two metals, as steel has very different qualities from iron, and as a result permitted technical developments and hence economic growth which would not have been possible otherwise.

In fact, in order to get back to it in terms of theoretical analyses of growth, technical progress must result both in productivity and in capital stock. In the case of capital stock, analysis is difficult. "We escape from this difficulty," an economist wrote recently, "by supposing that several types of capital goods always exist: each type belongs to a different generation, a generation being defined by the date at which the corresponding equipment was built. So a unit of capital of a given generation represents a certain production capacity, which requires the use of a given quantity of workers in order that it may be exploited." This is in effect much like the notion of the technical system which we have tried to clarify a few pages above.

Linked to technical progress, growth causes distortions to appear at very diverse levels of the economy. Firstly, it affects the individual level, that is the level of the individual enterprise. Not only does technical progress cause the disappearance of individual enterprises it has rendered marginal, but it also tends to reduce the total number of enterprises. In France in 1815, there were approximately 600 iron making enterprises; today there are only two, and in order to create a new unit of production, these two have to come to an agreement in order to make it a reality. In most cases, it is indeed technical progress that makes necessary the concentration of enterprises. Between 1837 and 1845, the problem of pumping water out of the mines forced all the coal mines in the Loire basin to merge into one company.

The higher level involves the nation. Here again, technical progress causes distortions between regions. For as an industry advances technically, it moves closer to its natural resources and the best transport facilities. Natural resources are imperative, even if transport is less important for some sectors (such as the transport of energy), so the tendency to concentrate geographically is none the less very restricting. The same problem appears on the global level. This is why economic growth and technical progress have always left depressed zones, in both agriculture and industry. The phenomenon of depressed zones was evident during the 19th century in England and France, and can be seen all over the world today. Just as technical progress leaves depressed geographical zones in its wake, so it also leaves depressed groups in the structures of societies and sectors of the economies.

We have thus emphasized the importance of these four principal factors: scientific progress, invention, innovation, and economic growth, to put it as generally as possible. In each section we have discussed two by two the connections that can exist between these various factors. We have stressed that the influences among them are reciprocal, and that these influences vary by sector and by epoch, and that in the end they must arrive at points of convergence with each other. It is from this ensemble of convergences that the final convergence results: technical progress-economic progress.

We should also study the links which are established in three of the factors thus defined. These may be categorized into two groups:

scientific progress-invention-innovation.
invention-innovation-economic growth.

But of course here again multiple combinations occur. In order to explain how this phenomenon is produced, one may, depending on the case, start at one side or at the other, or in the middle, and follow the effects in both directions. Let us take the most famous English example from the 18th century, the steam engine. It was partly under the influence of scientific progress that Watt made his first discoveries. But subsequently it was the industrialist, Boulton, who offered him a partnership that enabled him to develop his second series of inventions. Similarly, in the 19th century, the great firm of Saint-Gobain secured as its director a talented chemist, Clément-Desormes. On the other hand, it was a professor of chemistry, Kuhlmann, who founded in the north of France the important enterprise that was to bear his name.

In the second group of factors, invention-innovation-economic growth, there is no clear relationship between scientific progress and invention. Here it is rather the entrepreneur himself who is encouraged to innovate by his own perception of economic need, whereupon he seeks the appropriate invention with which to innovate. Clearly, in the first group the median term is invention, while in the second it is innovation. In other words, in the first type of connection technology is the driving force, while in the second it is the economy.

Let us recapitulate our aim: invention is the median term. That is to say that invention cannot be realized without pressure from one side and a point of convergence on the other. More precisely, whenever scientific progress has advanced far enough to render a new invention possible, science may be said to exert pressure on technology. But an invention is only feasible if it accords with an economic, social or other need, which is to say, when innovation is desirable. Conversely, the pressure of a particular requirement, in other words the need for an innovation, can stimulate invention only if there is a convergence between technology and science.

This is the most simple case. In reality, these diverse conditions rarely occur except in individual cases, such as inventions and innovations which bear upon one specific aspect of technology. One might even suggest that it is the pattern of a type of developing technical system which is progressing towards a definitive result by the accumulation of small improvements. The history of English technology during the 18th century is teeming with examples where the mechanisms of invention were of this individual type: the search for additional net profit, fresh capital taken out of turnover, and costs transferred to third parties. The absence of risks, or to put it a better way, the minimization of certain risks were compensated by a temporary monopoly. Here the advance of technical progress appears to be more rapid vertically than horizontally.

The connections between the last three elements, invention-innovation-growth, are more difficult to imagine. In this group, innovation is the median term. The direction of pressures and the level of convergences are more difficult to situate. It would seem, although our analyses are still insufficient, that the pressure comes above all from the necessities of economic growth, whether in order to accelerate the development and completion of a new technical system or to correct distorsions in technical progress, and in the economic or social system. In all these domains collective efforts are involved. The inventor now counts less than the entrepreneur, who decides and initiates the junction between families of innovations. The time of the struggling inventor such as Watt is past, for henceforth it is Boulton who creates not only an improved steam engine, but also the whole industry of machine building associated with it. Watt who was at the beginning the central element, at least the Watt of the patents of 1769, is no longer anything more than one cog among many others, and now it is only as a part that he can put his patents of 1787 into use. With Watt and Boulton, clearly, the median element has passed from invention to innovation. An enterprise is founded for the diffusion of inventions and seeks to encourage technical progress. By hiring Watt, it subordinates invention to innovation. The role of science is minimal in the sense that it is above all a matter of employing a pre-existing stock of inventions. Coordination of the whole is the important factor now, as well as the whole coherence of all the innovations which often create yet more new inventions. On the whole, at least during the ancient epochs, this necessity for new inventions was never really appreciated.

Strictly speaking then, innovation becomes less important than diffusion, and particular and partial inventions become less important than global innovation. Collective efforts take on more importance, and temporary monopolies disappear. The influence of small groups also increases in importance. State intervention manifests itself; private investment must be accompanied by public funds (vital both for the constitution of an economic infrastructure, above all in the domain of communication and transport means, and for the development of education to raise the level of general scientific culture). Here the horizontal propagation is more rapid than the vertical spread of technology. This is illustrated by those countries which do not initiate technical progress themselves but try to adopt a more advanced foreign technical system in an overall manner. There are excellent examples of European countries in the first half of the 19th century seeking to borrow other foreign technical systems.

The general adjustment has passed from the simple technical stage as demonstrated by the preceding pattern. Indeed, it is at this level that coherence is most difficult to achieve between the other economic, social, institutional and political systems.

Thus, one may notice that the two models of technical progress (a. scientific progress-invention-innovation, and b. invention-innovation-economic growth) are clearly different from each other, but without being mutually exclusive. In fact, the most intense pressures for technical progress come from the two extremes common to both models, science on the one side and economic growth on the other. So the multiple relationships may be explained in terms of a certain scientific level on one side and by the needs of economic growth on the other.

Above all these two models apply to different stages of development. The first applies to the preparatory stage where the major components of the new technical system appear independently of each other. Against this, the second applies to the period of take-off, after the system has already attained a minimum of coherence, and when the search for coherence has been extended to other systems.

History is obviously concerned with the progess of evolutions. As one comes closer to contemporary epochs, one observes that there are important changes in the concepts which have been defined above, and some seem to progressively disappear altogether. Two phenomena seem rather evident:

A) First of all, and this fact needs little explanation, scientific progress and technical progress henceforth become more closely linked. Today there is no more possibility of major invention without extensive scientific knowledge. This raises fundamental problems which are continually echoed in the daily press, that is, should the goal of science be purely scientific progress or should it encourage more and more improved techniques? The question still stands.

B) Similarly, innovation and economic progress can no longer advance one without the other. Innovation has become one of the essential components of economic progress. And, to the same extent, collective action becomes more important than individual effort. The take-off of technical research, which requires ever greater fixed costs, has curbed individual research.

So the distinctions which we have established between the different concepts grow increasingly fainter.

Invention itself is disappearing as a distinct entity: it pales before the growing importance of the two elements on either side of it. When the need for a new technique arises and is designed for immediate use, it can no longer be invented without the aid of science. The barriers between science and technology are thus crossed, and just as scientific progress takes place in laboratories, so technical progress also stems from work in laboratories, a fact which implies identical intellectual structures for science and technology. It would be fascinating to study the history of the development of factory laboratories, their origin, advance, the way in which research was carried out, and on another level, the exact politics of business research. Although we know something of their main policies today, very little of their beginnings is known. For the iron and steel industry, we known that it was the Holtzer firm, which, in 1869, with Boussingault and Brustlein, created the first factory laboratory for reasearch on specialized steel. In 1880, Fayol hired scientists to study the same problem of specialized steel in a laboratory at his factory at Imphy. Successive monographs have enabled us to understand the precise way in which these factory laboratories dealt with such problems. With the aid of these monographs we may finally reach a more general and more global analysis.

Formerly, invention had to wait for the favourable technical, economic and social conditions to arise in order to be applied. Innovation follows invention. Henceforth it will be the desire for innovation that brings forth invention: the schema is completely reversed. An enterprise can begin research in its own labs "as soon as it discerns the possibility of profitable application." Dr. Holst, creator and first director of the Philips laboratories, launched a study into electrical discharges through gases just after the discoveries of Franck and Hertz, into solid state research after the fundamental work of Peierls, Bloch, Brillouin, and Heisenberg, and into nuclear physics and instrumentation after the discoveries of Joliot-Curie and Fermi. Applied research even sometimes advances ahead of fundamental research, in which case industrial laboratories establish scientific results from empirical data. But as the economy grows more impatient with the hazards of scientific progress, it tends increasingly to institutionalize fundamental research in its own laboratories. A direct link is thus established between economic and scientific progress.

So an evolution has gradually taken place. In the first metallurgical laboratories, there was a great deal of trial and error. Using scientific methods of observation, successive proportions of alloys were tried in order to determine the best one. For a long time industrial research remained at this stage. Two stages can be seen in the history of the Philips laboratories. The first consisted of using the results from fundamental research in an industrial and systematic way: or to put it another way, the method of experimentation was reversed as researchers began with basic scientific knowledge and sought to apply it industrially. The method reversed itself again in the last stage. In order to invent something new for which the entrepreneur discerned a future market, they needed to solve a given technical problem, whereupon they carried out the fundamental research necessary to achieve that level and solve the problem. This was the way in which fundamental research was undertaken on the properties of tungsten in the Philips laboratory.

But after this stage, the distinctions in these methods become increasingly vague. Invention can no longer be situated in relation to innovation. In a word, only the two extreme elements of scientific progress and economic growth remain.

The new relation among the elements of technical progress have raised several difficulties. Some remain strictly on the level which we are concerned with here. The high cost of research, either basic or applied, necessarily creates a choice among many alternatives. Consequently, the first choice is whether basic research should be given priority in the hopes that technical progress will result from it sooner or later. Or, on the other hand, since it is becoming more and more possible, should technical research be given priority while fundamental research receives only the funds necessary for it to support technical research? Apart from this, there is another series of questions which are as important. If today we are moving towards a new technical system, as we will try to demonstrate in the rest of this work, both a stable internal coherence and a coherence with other systems must be established. Although one is generally aware of this fact, the notion does not appear to have been seriously applied. The authors of the Fifth Plan of the National Planning Commission of France justified its work as follows:

Since research by its very nature involves a fundamental risk, it is the duty of a government in formulating a science policy to assess this risk in the interest of the public and assume explicit responsibility for attaining the cultural, social, economic and military objectives of the nation... The first reflex of a nation which wishes to preserve its independence is to seek to understand the relations between scientific research and the economy, and to determine the criteria that make some kinds of research more fruitful for economic growth than others.

This exemplifies just how important this first question is, but other considerations must be taken into account.

Obviously the increasing cost of research makes any individual initiative more and more impossible. So it has been necessary to construct either private or public collective institutions. The factory laboratory is consequently relegated to the simple role of controlling the production process. It has not, it can no longer have, the means of inventive research. We need hardly stress this aspect of the question as it is well known of everyone.

The integration of technical progress into a prospective aspect is more vague. Henceforth the term planning must be used. It is no longer acceptable for us to submit to the hazards of technical progress in which we have no choice, and we can no longer accept willy nilly what may happen in the field of technology and then try to make the necessary adaptions as best we can. In every field, the economic as well as the military, we must organize the future at the national level or at the level of enterprise, and accordingly we must also organize technical progress. In other words, to use the terminology which we have employed above, invention, in so far as it exists as a distinct entity, is necessarily specified and rationalized: it cannot be otherwise. The only difficulty is in what way this should be implemented.

Let us forget about the hopes of the modern counterparts of Jules Verne. We will return to the Delphi programme, the inadequacies of which will be emphasized. This only involves determining the actions made posssible by technical progress in a given time. The domestication of intelligent animals destined to perform basic tasks, to take but one example among many, does not constitute technological expectation, but rather one of the great hopes similar to the imaginary designs of some Renaissance men, and which was only possible when all the necessary conditions for the existence of a technical complex were realized.

Nevertheless, technological forecasting is conceivable, but it must be the result of precise and valid analyses, and today such analyses are largely lacking. New institutions of which we spoke earlier, diverse agencies, and even ministries of research and technology, as they exist in certain countries, have been created to meet this end. But one must not be mislead into thinking that the problem is solved because such institutions have been set up and charged with solving it.

As technical progress accelerates and grows increasingly more complicated, we should remember one last but vital question. As technical progress develops via uncertain, or seemingly uncertain, means, the adjustments of the new technical systems with the other systems have always taken place in a makeshift fashion, by the play of a number of freely acting forces, and with all the errors and reversals that were necessary until a satisfactory equilibrium came about. But henceforth, if technical progress is to become something rationally planned, that is to say the result of a methodical programme, simultaneously in reality, space and time, then programming must extend to all the necessary compatibilities in every field of society, in the economy (which is most often mentioned), society and culture, etc. In the absence of such research, it would be useless to try to impose any technical progress which did not meet the indispensable conditions required for general equilibrium and harmony.

DOCUMENTARY SOURCES

Where history is concerned, the importance of sources cannot be denied. But an idea of the form in which they appear is as important for their study as for their use. Lastly, it is impossible to assess the way in which methodology varies according to the type of sources to which it corresponds. Like any other subject, the history of techniques applies quite naturally to various different sources. Each period in this subject has its own characteristics. These remarks are just so many obvious truths.

Texts

Texts have long formed and still form the most important part of historical documentation, so particular attention will be given to these. Emphasis should also be given to the fact that these texts are often very diverse in nature, and a great deal of cross references should be made between these and other sources.

Among such texts, there are some which directly and others which indirectly, inform us of the techniques used at different epochs. We shall review them very briefly.

Technical literature Technical literature comes immediately to mind, as it probably provides the most direct source of knowledge about earlier techniques. It is a truism to say that it has evolved as much in its conception as in its presentation. The history of this technical literature, which has yet to be completed, presents numerous problems. Very recent studies have tackled this particular aspect and one hopes that a whole work on this subject will soon be undertaken.

When the subject of technology has finally lost the magical and religious character which it had originally, we may begin to a certain extent to codify and teach it, although how far this may be done has still to be determined. While the helping hand plays an essential role in the operation of a technique, it is impossible to give a written description of the work it does. All one can do is to list, on the one hand, the quality of the raw materials, and on the other, describe the instruments and the tools required for a particular product. If, however, the techniques in practice demand any kind of thought process, be it even partial, the so-called "technical treatise" becomes impossible to compile. But for a long time the technical treatise must have remained a compromise, even within some techniques, between passages which demonstrated reasoning and those which were the product of unorganised, empirical knowledge, in other words, they did not reveal any real methods of experimentation.

The Greeks seem to have been the first to try and create a form of technical literature. We will study this in context in more detail later. For the moment we will just point out that this literature was strictly limited to a description of the techniques defined above, that is techniques involving machines, certain elements of which could provide material for the basis of a theory, in fact the very theory that affects the five kinematic chains. Treatises on lifting apparatus were of this nature, as were those on war machines, and machines for lifting heavy weights. Techniques from which certain principles were drawn were similar, and treatises on fortification are good examples of this. Finally, accounts of certain combinations and the existence of a programme in a specific time led to the development of treatises on automation. Although our knowledge of Greek literature is incomplete, we do have almost everything produced by this civilization. There do not, however, seem to have been any treatises on architecture or agriculture.

The Roman period does not add much to our present knowledge. Their greatest contribution lies precisely in architecture and agriculture, which will be discussed later. They barely changed the concept of the technical treatise and were most probably content with what the Greeks had achieved in these well-defined areas. The only point which should be made here, is that some works exist which demonstrate the extent of the Roman contribution to the field of organization: Vegetius' treatise on military art and Frontinus' tract on aquaducts are not so much technical treatises, in the strict sense of the term, as administrative manuals which were themselves important. The late epoch, and here we are thinking of the period from the Byzantine age to the beginning of the Middle Ages, was content to accept all the classical literature. In the 5th century, the Byzantines may have tried to take up antique "syntaxes," but still using the same model into which they would integrate the few improvements realized since.

The Middle Ages undoubtedly marks a step back in the field of technical literature. In the same way that science was divided into a limited number of precise problems, technology was confined to works, the bulk of which were merely compendiums of unorganized facts. One might call this medieval literature a recipe book, usually containing disparate elements which were never systematized. We have numerous examples of various types of these. Many of these recipe books do not concentrate on a defined technology, but cover a wide number of subjects. The best example is that of the monk, Theophilus, who wrote on the subject of artistic techniques, or the chemical formulae recently studied by Pierre Cézard. Similarly another excellent example is provided by Guy de Vigevano's treatise on machines of war, compiled for the French King's departure for the Crussades. This was rapidly to lead to what the Renaissance termed "theatres of machines." These so-called "theatres of machines" took their definitive form from engineers' notebooks which were originally not intended for wide circulation. Each engineer noted down anything of interest or curiosity about his craft in a book. Although Leonardo da Vinci's notebooks are generally known, we forget that these were preceded by a great number of others, which various historians have tried to list.

The Middle Ages, however, did not ignore treatises centred around a given activity. Once again, these contain accumulated formulae but they are better organized and facts are linked together. Many examples of various subjects have been preserved. This is the case of agricultural treatises which increase in number from the works of the 13th century Anglo-Normans, notably Walter of Henley, to the great treatise by Pietro dei Crescenti of the 14th century. The treatises on the art of the farriery should also be mentioned, as well as those on small animals or hunting which were written in much the same vein. The treatise by Jean de Brie on shepherding is also similar.

All this literature was to inspire the work of the Renaissance, for it was to follow the example offered by the preceding period, although several modifications were introduced. It is thanks to Leonardo da Vinci that we know that engineers' notebooks were to enjoy the same success as they had done earlier. Similarly we know that the "theatres of machines" were very popular. The tradition remains the same from the manuscripts of the treatise by the German writer, Kyeser, to the magnificent editions by Ramelli (1588), Della Porta (1601), Zonca (1607), and Branca (1629). Such works were to continue throughout the 18th century with Leupold (1724) and König (1752). The latter is in fact a presentation of machines, based on engraving, frequently accompanied by brief descriptions. Undoubtedly the "descriptions" written in the 18th century are the most elaborate and organized form of this category.

But at the same time, another formula appeared which had already been introduced in the Middle Ages, but which was now more developed. These works always concentrated on a given technique. Writers tried to collect all the information concerning a particular production process, but this time the approach to the subject was more scientific, or at least more critical, which it had not been before. The subject of agriculture was certainly to profit greatly from this improved method, while the subjects of small animal hunting or big game hunting were to become rather vague. But books were to appear on almost every subject, some of which remained great classics for some time. Mining and metallurgy were particularly favoured, perhaps only because we know more of these: Agricola and Biringuccio, among others, are always remembered. Treatises were also written on the subject of guns, revealing the first stumbling attempts made in ballistics; there were also tracts on distilling, dyeing, and we could go on to cite many other examples. Leonardo da Vinci had envisaged several such treatises, among them a treatise on hydraulics. Need we mention the numerous tracts on architecture and town planning written during this period? In certain sectors therefore, technology was well-ordered in this way.

The whole state of mind or attitude had changed. This can be seen in the very titles of works, as from the end of the 16th century the formulae gave way to "reasons." In reality the scientific level reached at this epoch and the weak links between science and techniques, despite some hopes and successful results, blocked the development of a perfectly reasoned form of technology.

Written "descriptions" were developed at the end of the 17th century. We know that Colbert, probably more concerned with a stationary economy than progress itself, had enlisted the *Académie des sciences* to draw up some sort of balance sheet illustrating all the techniques in use. The best methods were to be chosen, improved upon if necessary – as the title of the collection was to be *Description et Perfection* –, so that they would achieve recognition in themselves. It was therefore a work compiled by technicians and checked by the highest scientific authority of the age. At the same time, moreover, the *Académie des sciences* began to publish a work entitled *Machines approuvées*, in other words a work on new inventions which had been submitted. Although some of these works were ready at the end of the 17th century, the first publication did not appear until 1762, which perfectly reflects the picture of classical techniques.

The *Encyclopédie* compiled by Diderot and d'Alembert followed exactly the same path. Some authors were involved with the *Encyclopédie* and the *Descriptions*, and both publications have stolen plates. But the greatest value of the *Encyclopédie* lies in its attempt to systematize enterprise and link techniques with other knowledge. But d'Alembert's Prospectus shows that the ideas on the constitution of a system of technology had hardly progressed.

On the whole, it is the workhand which creates the artist and there is nothing in these works which instructs one of the operations involved. The artist will only find in our work views which he may not have had before and observations made after many years of research. We offer to the studious reader imformation which he could have learnt from watching a craftsman at work out of curiosity, and to the artist we offer information which he could have learnt from philosophy in order to achieve perfection.

By making the volumes of the *Encyclopédie méthodique* more specific, Panckoucke gave his work greater depth, but in doing so he dissociated techniques from the general culture.

The technical treatises, which as we have seen were reintroduced at the end of the 15th century, continued to be produced in moderation during the 17th century, but were more abundant in most countries during the 18th century. The first great treatises on agriculture appeared, succeeding the books on "the country estate" which had been such a success from the mid 16th century: Jethro Tull provides a very good example. There was hardly any technical field which did not have one if not several works written on it, and virtually every country in Western Europe participated in this movement.

The 19th century finally abandoned the theatres of machines and descriptions in favour of the technical treatise. This resulted from progress made in the sciences, and the increasingly marked alliance between science and techniques. Naturally it is impossible to make a complete catalogue of these and we will return to this aspect of the question later.

From the end of the 18th century this technical literature was to be completed by technical journals. In this way, the public could be given an exact idea of the technical progress being made without having to recourse to technical treatises. Moreover, these journals are interesting as they reveal a number of important facts, such as definitive clarification, and the adaptation of different natural resources. In France some of these journals were originally published by the government, but in England they were produced by purely private sectors.

The *Transactions of the Society for Encouragement of Arts* was first published in London in 1783. France began to produce the *Journal*, in 1794, then the *Annales des Mines* in 1795, and the *Journal des Arts et Manufactures*. From 1801 the publication of the *Bulletin de la Société d'encouragement pour l'industrie nationale* marked the beginning of "industrial societies" which were to increase after this. Systematic perusal of these 19th century journals certainly provides the history of techniques with some valuable elements. Notice also that most of these journals did not strictly limit their curiosity to techniques, but economic problems which arose from technical progress were often mentioned.

A great effort must be made with regard to the whole of this technical literature. Naturally the first step would be to compile a critical list of all the publications, examining both the presentation and the content of these books and journals. What has been attempted on boat building, iron and steel metallurgy, and locksmithing illustrate all that may be learnt. Similarly, an attempt to publish or republish these works should be made, an idea which has been undertaken for some manuscript treatises. Apart from Leonardo da Vinci's notebooks, published in the last few years, works by Francesco di Giorgio, Kyeser and Taccola have also been published. The republishing of old works, which are often difficult to find, proves to be equally useful: from the end of the 19th century some works were reproduced in this way. Republishing still continue but more cautiously: the Italians have reproduced a work by Branca. The critical reeditions are far more interesting, as they provided notes on the original work which often contained passages which were difficult to understand or interpret. The United States began to undertake this task for some important works. Between the two world wars, President Hoover published an English translation of *De re metallica* by Agricola. More recently more systematic works have been undertaken. Publications concerned with metallurgy, some produced by C.S.Smith in America (from Biringuccio to Réaumur, thanks to The American Institute of Mining), and some by the Academy of Freiberg, which similarly deals with mining techniques, are also worth mentioning. There is still much to be done. Some Greek technical treatises have also been published, often in the original text, and occasionally with the necessary translations. A large number are still unpublished but medieval manuscripts and those of the first Renaissance are now better known.

Direct sources Sources which are directly useful are generally recent ones. They only actually date from the point at which administrative archives were really set up and organized. There are two types of direct sources: public administrative archives and company archives.

When large administrations were really formed, a certain type of document was introduced, and from this period onwards, that is the end of the 17th century, this documentation becomes more and more rich with the development of the mercantile age in which technology was the primary consideration of government concerns. The documentation is often important, diverse, and of varying value but no less negligible. It stems from three principle roles of the state: information, supervision, and protection of the rights of the individual.

Information and supervision are closely linked. From the end of the 17th century the state was interested in almost every sector of technical progress made both at home and abroad where techniques may be different or more improved. This interest in the latter was both for the smooth operation of its own services and the encouragement of the economic development of the country. A whole set of technical documentation was therefore created on the governmental level. In France, to take but one example, this effort was made in several directions. Inside the country itself, quality controls and extensive regulation was observed, which stimulated inquiries mainly concerned with technical problems. Similarly, missions were sent abroad, notably to England, but also Germany to observe in particular improvements and new ideas wrought in mining and metallurgical techniques. The archives of economic departments such as Agriculture, Industry and Public Works, which at the beginning of the 19th century replaced the old *Contrôle général*, contain many interesting files regarding this subject. The archives of departments which by their very nature used techniques, we are thinking here of the Ministry of War and the Navy, are also rich in information.

The state was also intended to serve official institutions which were charged with the task of advising and circulating technical knowledge. The *Académies des Sciences*, created at the end of the 17th and at the beginning of the 18th century, were technical as well as scientific in inspiration. Societies of agriculture, created from 1758 onwards in France, were similar as they were instructed to promote technical reform in farming.

The patent, to give it its modern name, had a dual aim. On the one hand, it was a necessity, for it seemed essential to protect the interests of the inventor, if only to allow inventive minds to develop. On the other hand, it was important that the invention should not become a factory secret, as the economists of the early 19th century suggested, primarily because the monopoly of a product is always a dangerous thing, and secondly, there was a natural agreement between the two standpoints, as the diffusion of inventions is beneficial for the economy. Interesting research has been done on the history of the patent, which varied in form according to the country and the epoch. The patent was first employed in the mercantile period, and it seemed to be general economic policy rather than intended for the protection of the individual. For a long time it did not protect inventions strictly speaking, but rather the importation of foreign techniques. It favoured less technical progress than the take-off of new industries within a country, using techniques already in practice in other countries.

An older text exists which also illustrates the age of the mercantile spirit, at least on the subject of techniques. In 1236, the King of England gave an inhabitant of Bordeaux the exclusive right for fifteen years to manufacture woollen cloth in the Flemish, French or English tradition. From this period onwards, some characteristics of the patent were established, including limited monopoly, and the guarantee of public power. The institution seems to have been systematized in Venice from the end of the 15th century, and it was already perfectly laid down in the English edict of 1623. Henceforth the composition and the law pertaining to the patent certainly evolved. Although the English or Dutch patent resembles the modern form, the French "privilège" has departed from the earlier form. It was often both limited in time and place, as it was under Colbert. Almost everywhere, research for precedence, closely linked to the present day patent did not exist. France inaugurated a new notion, with the control of the *Académie des sciences*, which has not been preserved, concerning the usefulness of a new process and a new machine: this fact was well noted in the statutes of the Académie of 1699.

In the 19th century, the institution was defined and improved. Public bodies were formed in almost every sector. While they were charged to protect the rights of the inventor, they were also appointed to circulate technical knowledge. Patents fairly rapidly became regular publications: in France in 1811, with recall of those granted since 1791, in

England in 1853, with recall of those granted since 1671. These publications thus formed an important source: a recent Parisian exhibition demonstrated the significance of these sources. A publication in Holland has retraced the country's "patents" from the 16th to the 18th century, but it would be useful to extend this research to other countries.

This documentation gradually becomes less abundant and more vague, apart from that which deals with the industrial functions of the State. Henceforth this technical knowledge was to spread by other means, in particular by the introduction of technical literature. Missions abroad and written works were thus less useful, although after the Second World War, many missions went to the United States to observe the technical progress initiated by the war. Larger enterprises also created their own documentation.

Company archives provide an indispensable source for the history of techniques from a certain period onwards. One would hope that this documentation does not disappear because it is thought out-dated. It is all the more interesting because it concerns both invention and innovation in their contexts.

Once again we will end, as we did for the subject of technical literature, by expressing a desire. In some countries research guides have been published about various fields of historical research. It would be useful, in some sectors this is even an urgent requirement, if such guides were given over to the history of techniques. It would not be impossible to complete this work in the field of direct sources. We believe that international coordination would be rather easy to formulate on this question.

Indirect sources It is difficult to present indirect sources in a logical manner as they are very varied. But these images and objects are all that remain for the reconstruction of the techniques in practice during early epochs.

Historical accounts, often literature of the time, may be useful sources, although they are frequently tricky to interpret. Need one quote Archimedes' technical experiments as recounted by Plutarch? The first example of the hydraulic saw appears in the poem *Mosella* by Ausonius, but we could cite many other examples. The greatest obstacle in any research of this nature is the time required to study these sources. Historical accounts may also provide us with some facts.

Sometimes the exceptional events, wonderful adventures, and astonishing occurrences portrayed by chroniclers and historians may offer descriptions and explanations from which the historian of techniques may benefit. There is one particular field where this literature is of exceptional value, and that is military art. Here again, the use of this documentary material requires close attention.

Diplomatic medieval sources form an important source of material for which Antiquity unfortunately has no equivalent. Once again, some kind of classification would be useful, even if it were to oversimplify the problem.

Acts of donation, in general to monastic establishments which are rather vaguely known as charters, represent, up to a certain epoch, a semi unique documentation. It is true that this documentation is often imprecise, although sometimes important facts may be revealed, particularly for the introduction and diffusion of some instruments. By studying these, we may certainly trace the spread and diversification of hydraulic energy before the 13th century. The publication of cartularies or even a simple bibliography of these could provide the first step in this research. It would be useful if editions of such documents included an index of subjects as this would considerably facilitate research: even a few tables show the advantage to be gained from this.

It would also be interesting if these documents' inventories concentrated on one industry, or a type of industry, thus assembling the whole of the documentation on a given subject. Some essays have been attempted on the fulling mills in England, and in France a compendium on the subject of the iron and steel industry has been started. All these documents assembled in this way provide diverse but partial elements, the accumulation of which permits the basis for hypotheses which no other form of documentation allows.

To be honest, statutory sources may be considered direct sources. They are incomplete in the sense that instructions in some cases, and bans in others, only provide half the picture of the techniques used.

Corporation regulations are by far the best known. In some towns they were united and codified: Etienne Boileau's book on the crafts of Paris is such an example which is often cited. Some collections have been formed on particular products in well-defined regions, offering invaluable variations to the history of techniques. Collections on the production of cloth in the north of France and Belgium provide examples of this. Similarly, many collections have been published in

Italy, often in topographical order, and remarkable monographs on the history of techniques have resulted from these texts. On the whole, the regulations were lists of bans rather than positive instructions. This proves, moreover, that these bans were imposed upon techniques which were considered inferior, but already in use. As more regulations were imposed, so the spinning wheel, certain ingredients used in dyeing, and methods of preparing textile materials became widespread.

Mining codes were much the same, the oldest of which undoubtedly represents the Roman law used in Spanish mines, formerly studied by Ardaillon. Unlike guild regulations, these mining codes were constructive, in the sense that a number of rules were applied for different reasons, such as relationships with neighbours, and safety measures. Some of the documents from these notebooks have been studied in detail, and we will mention in particular the studies undertaken on the codes dealing with mining methods in present day Yugoslavia.

The successive drafting of regulations and mining codes clearly conveys the evolution of techniques. The lifting of certain bans, such as those previously imposed on the spinning wheel, on carding, or on the use of certain materials employed in the process of dyeing, are symptomatic of an evolution which many considered during the period to constitute a decline in technical progress rather than an advance.

Finally, notarized acts may constitute a last important source. We should emphasize the fact that notaries were only first conceived in the north of France in the 15th century, but existed in the south in the 12th century. We need not define in detail at this point the types of acts passed by notaries, which may be interesting for the historian of techniques, such as inventories, estimates, and building contracts. Inventories often provide whole lists of tool types. Similarly, we possess very detailed contracts for the construction of ships, but this has rarely been consulted, apart from in some exceptional instances. It is also true that these sources demand long and careful perusal.

The interpretation of such texts raises serious difficulties, for literary texts in particular and some administrative documents are frequently inaccurate. One of the main difficulties results from the ambiguity of the language employed. Apart from a few rare exceptions, technical language has hardly been studied, although the development of some words would certainly be very revealing. A study of the mining codes in operation in the Italian mines of Massa, and research in the mining codes of present day Yugoslavia, have demonstrated that most mining terms in use in Europe were of German origin. In England, almost all the vocabulary used in iron and steel techniques are French, and in France a good number of naval terms are derived from Holland. The birth of a technical vocabulary presents some awkward problems: it is natural that in Greek the wheel and pulley should be associated, but surprisingly many Greek machines are associated with animals and so bear their names; while this might be natural for war machines, it is perhaps more surprising that lifting and building equipment should take their names from animals. (While this can be seen in the English language, for example, monkey wrench and crane, it is more obvious in the French where "chèvre" means both a goat and a hoist, 'mouton' a sheep and a ram, and "grue" means a crane, both the bird and the machine). Consequently there is a double problem, that of neologisms as techniques become more improved, and translations as tools and machines appear. All sorts of situations are possible, for a new instrument can be given an old name, or the tool and the name may be borrowed from abroad. We will return to this important question later.

Any appreciation of these texts naturally raises the problem of the methods of historical criticism common to all texts, such as dating, origin, influences, etc. The historian of techniques must add anything which he may deduce from his own subject to complete the analyses, and make comparisons with other texts. He must judge if the technique quoted corresponds to the norm, if dates match with what we know of the history of techniques, and in brief, he must add any supplementary information provided by specific study to the traditional critique.

Iconographic Sources

The inaccuracy of some texts, and the absence of old objects, render any image, whatever form it may take, of prime importance for the history of techniques. Blümmer, writing on the techniques of Antiquity, and Jal, writing on ship building, have both demonstrated the importance of iconographic sources. More recently, Captain Lefèbvre des Noëttes based his study of the harness mainly on iconographic documentation.

The problem is extensive in many respects, and varies depending upon the period and the technical sectors involved. We must therefore place the information at our disposal into some sort of order. Firstly, iconographic sources must be distinguished, for there are "images," in the widest sense of the term, in which the historian of techniques may find interesting elements, although this was not the artist's main objective. The cart and the ship in Breughel's *Fall of Icarus* furnishes such an example. On the other hand, we should consider the technical drawing proper, which began to appear with the introduction of the first technical treatises.

What one might call general iconography covers a very wide area of limitless variety. Hardly a single technique has not been portrayed by an artist, in one form or another. This type of documentation naturally evolved as the style of art changed.

Paintings, sculptures, vases, and the whole graphic decor have been used for the study of classical Antiquity, one of the most abundant and extensive sources of the history of techniques being found in pictorial representation. The corpus of Greek vases has doubtlessly been thoroughly examined, and the mosaics of Ostia provide remarkable images of Roman ships.

The Middle Ages used the same art forms, replacing vases with more paintings, manuscript miniatures, stained glass windows and seals. The documentation for this period is therefore extremely rich but very dispersed.

Documentation for the following centuries certainly diminishes in quantity but gains in quality and precision. There are yet more paintings and engravings, some of which are still very useful, despite the fact that technical drawing had become more advanced. Many views of factories and workshops were depicted by English, French and German painters of the Realist school, and numerous examples may be cited.

This iconographic documentation must be interpreted, and any attempt to do so should be made in several ways: collecting documentation and its critique, interpretation and use. In all these cases, especially in the first, research should be collectively organized on the international level.

The most important, but most demanding task lies in assembling all these images. Should one classify the images in sectors, or should one engage in a general but geographically limited perusal? In a systematic framework, the corpus of Greek vases already provides us with a type of pictorial representation which constitutes extremely important documentary material. This is the only general index we possess for Antiquity. Some photographic collections are particularly useful for the Middle Ages, such as the Marburg and Princeton collections, but less so for Antiquity. The Bibliothèque Nationale in Paris already possesses a sizeable photographic collection of manuscripts' miniatures. It would be fairly easy to draw up a list of the tools generally used, starting with centres which already exist and by using the work in progress, the result of which could only be known in a few years time: we know, for example, that France has also begun to assemble the corpus of the country's stained glass windows.

Another, less logical method, as it would require several systematic perusals of certain collections, would involve gathering all the information available on one particular technique. In certain fields, this would be facilitated by the existence of themes which have already been studied elsewhere on artistic and religious levels; such as the mystic press, the mystic mill, and the workshop of Saint Joseph for the examination of the carpenter's tools, etc. We have already mentioned Captain Lefèbvre des Noëttes' work on the problem of the harness. A similar study has recently been attempted by Mrs. Van Tyghem on the construction of buildings. Another series has already been started on watermills, windmills, working equipment, and tools used by the blacksmith, particularly for the medieval period. There is an important collection of seals depicting types of ships, providing one of the principle sources for the Middle Ages, now at Greenwich, in London.

Any critique of iconographic documentation is therefore very problematic. Datings, origins and influences all make the use of this material very difficult. First of all, one must be selective, for there are absurd representations which are a waste of time. Many other contemporaries have warned historians to take great care over these. Images of the plough which appear in the *Bible moralisée* of the 13th century, formerly published by Laborde, are purely fanciful. Similarly, a 14th century miniaturist, whose manuscript is conserved at the Bodleian Library places the ploughshare in front of the axle unit. Elsewhere oxen are illustrated pushing the plough rather than pulling it.

We shall only briefly touch upon the problem of dating: it is up to specialists in the history of art to provide all the elements necessary in this respect. The problems of localization are similar. Although, in some instances, the localization of a fresco or a mosaic is relatively easy, in other cases, particularly for manuscripts, the problem is more complex. It is useful for the historian of techniques to know if the so-called Tours' *Pentateuque* comes from North Africa. Note that the technician is often liable in these two cases to enlist the aid of the art historian.

Analysis of content must often remain hypothetical. If the true identity of the artist is known, then the problem is simplified but not completely solved. But many vital questions arise concerning the image itself which are frequently impossible to answer.

A) The first concerns the fidelity of the image. In many cases the artist was not quite sure what he wanted to depict. The object was sometimes relatively simple to represent, such as spinning with a distaff. The situation was rapidly complicated, however, when the technical instrument to be represented became more complex itself, as in the case of the plough and the ship. It was much later that artists took the trouble to depict everything down to the last detail, which is good news for historians of techniques.

B) Artists travelled far and wide. An artist from Italy travelling in Flanders, may have wished to depict an object from his native country. It is therefore necessary to establish this point, as in some cases the historian of techniques can discover, better than the art historian, what one could call an anomaly.

C) Lastly, we should consider reciprocating influences, which may be perfectly observed from one manuscript to another, where elements have been borrowed resulting in various deformations. In some instances one may even establish concrete relationships between them.

Iconographic documentation is therefore very rich and valuable because it provides information often absent in texts, but it always requires very delicate handling for all the reasons stated above. We should also look at the so-called technical drawing, of which we only have examples from after the Middle Ages. The earliest examples are provided by Greek Byzantine manuscripts of the 10th century, which contain certain treatises by mechanics of the school of Alexandria and technical Byzantine compilations. In Europe, we have the notebook by Villart de Honnecourt, an architect of Picardy in the second half of the 13th century. Following this, technical treatises such as engineers' notebooks were covered in drawings. The technical drawing was really introduced at this point and continued to develop to the present day.

Thus the need was soon felt to bypass the necessarily formal explanations of a written text by providing a graphic representation which could give the idea of a tool or a machine at first glance. In order to arrive at a satisfactory result, a number of difficulties had to be overcome. The first was the problem of conceiving a type of drawing which was capable of reconstructing the object, and not just making it part of the background decor, as had been the case in general iconographic works. We know that a simple picture, the reflection of a view, is necessarily almost always incomplete, as parts will inevitably be obscured. The second difficulty is identical to the first. If rapid notes could not be taken, as they were in the engineer's notebook, then a rough sketch could suffice, but when it came to the construction of the object by another person, all the surfaces had to be depicted, and accompanied by the relevant dimensions and measurements.

The technical drawing evolved in this way. At the beginning, and for a considerable time afterwards, only one drawing was provided, but in order to fulfil the requirements which we have pointed out, all the parts of a machine, to take the most complex example, had to be represented in a plan where they could be best seen. Carrara's seal, though it is not strictly speaking a technical drawing, shows a four-wheeled chariot, each wheel being represented by a circle. The war machines in Guy de Vigevano's military treatise, of the beginning of the 14th century, are represented in this way.

The discovery of perspective fairly rapidly corrected this somewhat primitive method of drawing and design. Thus the wheel could be ellipsed, and one needed only to place the object in a position so that the whole could be reproduced in a drawing. This method was to be used until the end of the 18th century, although it was gradually improved. The most important modification consisted of completing the drawing of the whole machine, to use the same example, with drawings of its various parts, which were assembled in a small composition, indeed even accompanied by a little setting. One of the most perfect examples of this system of drawing is provided by Agricola's *De re metallica* of the mid 16th century. The *Description des arts* by the *Académie des Sciences*, and the *Encyclopédie* which appeared in the middle of the 18th century, refined and completed this method of representation: in particular we find for a given loom all the necessary tools which accompany it, and if a machine is depicted, the representation is on different plans, cross-sections and profiles, and as in

Agricola's work, certain parts are drawn in detail. After the first technical plates of the 18th century, it became possible to construct models.

Another technique of graphic representation began to appear around the same time which was to fulfil the second requirement, that being quotation. In the middle of the 15th century, Ghiberti had already provided drawings illustrating bell casting, and these to some extent replaced the table. Similar drawings for boat frames were conceived by the English builder, Matthew Baker. From this point on, as far as the techiques of drawing lent themselves, technical designs were to be developed.

From the 18th century onwards, in some cases even from the end of the 17th century, the real industrial drawing was to emerge. Architects had already arrived at this stage. Cross sections, plans and profiles provided an exact method of estimating the amount of materials needed, and subsequently the way in which they could be reproduced materially.

Objects

Of all the testaments to the evolution of techniques, objects which have survived to the present day must constitute the most accurate and valuable sources. It is annoying that these have only recently been used to retrace the history of techniques, as many losses have resulted. In a good number of countries, these souvenirs of a material civilization are still rarely considered, and some historians and curators are often angrily obliged to relinquish frequently prestigious remains. In 1949, France refused England the chance of salvaging one of the old ships constructed in 1800 and captured in the Battle of Trafalgar.

Two types of objects must obviously be distinguished. On the one hand, there are the tools or machines used in production, and on the other hand, there are the manufactured products. For a long time an attempt has been made to trace the history of techniques by examining the development of production instruments, and research has been undertaken, at least in certain sectors, in the products made, in order to try to determine the techniques employed. Studies of these material sources, invaluable lists, and the analyses of which they must be the subject, are, to tell the truth, not available in most cases, and are only wishful thinking. Fortunately, exemplary attempts exist in various fields which promise hope for the future.

There are many more examples of the products made than the machines used in production. Naturally they are scattered far and wide and in many cases their artistic character renders them inaccessible to certain analyses. Or, to put it a better way, curators of archaeological museums or art schools have not yet grasped that some analyses could be done without incurring any damage to the object, and could provide irreplaceable information about the riches which they preserve.

As the material to be considered is extensive, we should first mention a few major difficulties, two of which seem indisputable, before approaching certain areas of this research: origin and dating. These difficulties increase by virtue of the fact that even for relatively recent periods, the objects which we possess are often exceptional works, and this is precisely the reason why they have been preserved. Leaving aside the Vix vase (the origin of which is still the object of hypotheses and suppositions), the existence of Byzantine or Byzantine type materials found in a tomb from the high Middle Ages, suggests two theories: it may be the pure and simple importation of a commodity, or an importation of Byzantine techniques and therefore an imitation. The numerous distant influences in art have been abundantly discussed, but few historians have considered the technical influences which may similarly be produced. Let us look at a recent example: the great chancel gates which exist in so many pilgrimage churches bear striking similarities. This is true, in particular, of the chancel gates in central France (Conques, Billom) and of those in Spain (Iguacel, Pampeluna etc.). Here again, several hypotheses may be put forward regarding the similarity of the techniques used in casting and assembling the gates: were the works themselves, the workers or the actual techniques spread abroad? In some cases we have the answers to these questions. It would be interesting, for example, to study Cistercian techniques in certain areas in order to discover how the diffusion of highly advanced techniques and their modification came about, and how these techniques themselves came about.

The problems involved in dating are no less awkward, but here we are joining the preoccupations of numerous other disciplines. Let us return to the previous example: was the simply constructed Romanesque gate a translation of an old product, a certain archaism, or the unskilled work of the village smith? Historians still argue over the introduction of the horseshoe, but it has been demonstrated that in some places the iron shoe could sink down into the earth over the years and appear on archaeological levels which do not correspond in reality to their true period. We can only arrive at a more or less approximate dating by making continual cross-checks, either with other objects or with texts or images.

The couple "instrument of production-object produced" is certainly the most interesting, but unfortunately also the most rare. Only a few elements may be associated with it. The Poles and Czechs have found traces of primitive reducing furnaces in recent excavations. By comparing these furnaces, the minerals used, the slag, and charcoal remains, archaeologists have been able to attempt to reconstruct an idea of what elements were employed, thus providing a rich harvest of information for the history of iron and steel making techniques. If these results are compared to ancient texts on the subject of mining in these estates and to the mining regulations which we have already mentioned, it is possible to suggest a social explanation for production.

Objects are therefore extremely important sources for the history of techniques, and research in this subject should be undertaken in two directions. On the one hand, it must deal with the study of techniques which have been preserved, and on the other, it should try to discover new evidence of techniques which have disappeared.

Two types of manufacturing instruments have been preserved. The first are almost always old tools, generally discovered in excavation sites. Such as the tools found at Pompeii, the ploughshares or coulters which can be seen in many museums, and the tools discovered in old mine galleries, such as those of Massa, in Italy, or those of Wieliezka, in Poland. As for machines, the earliest have undoubtedly disappeared for ever. The mills at Pompeii are probably the oldest known machines. The hand mill of Valère, at Sion, in Switzerland, dates from the 14th century, and lathes dating from the 16th century are conserved at the Brewers' House at Antwerp and in the Munich museum. More instruments, dating from the 18th century, used in the textile industry have been preserved. From this period onwards, moreover, public collections were better supplied, as is attested by the impressive collection of old steam-driven machines preserved in the Science Museum in London, and the very old rolling mill which appears in a private collection in Liège. Some period buildings can give us an impression of what old industrial plants were like, although this is often difficult to assess. We need only look at examples of certain

Cistercian buildings, such as the ironworks at Fontenay, and the works building of Royaumont, in France.

A few years ago, it might well have been possible to save from destruction material dating back to the 19th century, which has too almost completely disappeared. During the last few years, however, several countries have endeavoured to preserve old workshops and their tools. In Poland, an old nail factory containing a very interesting rolling mill was transformed into a museum. Similarly, in Sweden, an old lamination workshop was entirely preserved. A movement seems to have grown up today concerned with safeguarding the rare evidence of techniques which are not particularly old. [Transl. Note: the Ironbridge Museum at Telford (Great Britain) is also a case in point.]

The second area of study involves old techniques, modified to a greater or lesser degree, which have remained in use up until the present day. This material is undoubtedly both better preserved and studied due to the development of ethnological research. Thus collections of so-called "traditional" tools, have been assembled, indeed a few period workshops have been classified as historic monuments. But even more important are the precise studies made of certain crafts, such as the research undertaken by the French *Musée des Arts et Traditions populaires*, on iron tilt hammers, on the work of the clog maker or the stove maker of Villedieu-les-Pôeles, in Normandy. Collected works on instruments, tools and machines have been compiled virtually everywhere. The historian of techniques only needs to be able to distinguish from this important contribution, old techniques and those which may have resulted from relatively recent technical improvements.

Apart from the tools and machinery themselves, small-scale models are also interesting. Renaissance architecture had already discovered the use of the scale model for the construction of buildings. Similarly, the notion of making scale models naturally evolved, particularly for machines, as did the fashion for setting up exhibitions or collections with the aim of demonstrating and spreading technical progress. Descartes first had this idea. In 1683, the first exhibition of model machines was organized in Paris, a catalogue of which was made and which fortunately still exists today. This idea was taken up again by the great Swedish engineer, Christoffer Polhem, who made a great number of models of machines used in the mines: this collection may still be seen today. In the second half of the 18th century, model collections really became popular. Under the direction of the Périer brothers, acting on the instructions of Madame de Genlis, the mechanic, Calla, made small models from some of the illustrations from the *Encyclopédie*, for the future Louis-Philippe. At almost the same period, Vaucanson began to assemble a vast collection of machines and models at the *hotel de Mortagne*, bought by the king in 1783, which was to be the beginning of the *Conservatoire des Arts et Métiers*. There were other famous collections, such as that of the great Duhamel du Monceau, which was broken up only very recently.

The use of the model as an educational instrument declined with the development of the technical press, and the introduction of industrial design, but they retained their popularity when technical progress entered certain fields. In many instances models were complementary to patents, and it is in this way that the collections of the Smithsonian Institution, in Washington, came into being. In 1881, the Science Museum in London similarly took over the collections of the patent service. But private collections also existed: in 1900, the Science Museum was entrusted with the Maudslay Collection, originally started by this great English mechanic, and in 1903, the museum also acquired the Bennet Woodcroft Collection.

There is certainly a greater number of objects manufactured, from architectural remains upon which we will not dwell, to objects in the literal meaning of the word. Some are preserved *in situ*, and others are assembled in public or private collections. Numerous points should be clarified here, even with regard to architecture. In this domain, Roland Martin has studied the marks made on the stones of some Greek temples which prove that levering and lifting apparatus was used. This encompasses the whole problem of erecting Gothic cathedrals and the techniques which were employed. The history of art has often neglected to consider the history of techniques in this field.

Apart from objects found in archaeological digs, many museums have some quite exceptional pieces, rich or magnificent objects which are better identified on the chronological level regarding their origins or localizations, and are therefore equally valuable from the technical point of view.

Archaeologists were not interested in techniques for quite some time. It seems that it was only after the First World War that excavations were made specifically to bring to light manufacturing instruments and standard objects. Such efforts have usually been directed in particular areas. Research was first undertaken in Germany on reducing furnaces. Since the end of the Second World War, these digs have been systematized throughout Germany, England, Czechoslovakia, Hungary, Poland, and the U.S.S.R. These excavations have contributed extremely important information to our knowledge of iron and steel making techniques.

Ships discovered in Viking tombs have enabled us to partly reconstruct an evolution which otherwise would not have been known. The Nydam ship in Denmark, probably dating from the 3rd century AD, the 7th century Sutton Hoo ship in England, and the Gokstad and Oseberg ships in Norway, dating around the 9th or 10th centuries, perfectly mark out stages in the history of the ship. All these discoveries were made beteween 1863 and 1903. From 1950 onwards, other important finds have been made. Under-water research has contributed to our knowledge of antique vessels, as cargo ships have been located in the Mediterranean, brought to the surface, and even reconstructed. In Bremen, a 14th century ship was brought up from the mud, and in Stockholm a 17th century ship was refloated. (A more recent English example is provided by the successful attempts to raise the Mary Rose, now in Portsmouth.) Such prestigious remains are well worth the expense involved, as they provide accurate information about periods which other types of documentation cannot.

Although these chance discoveries have contributed many interesting facts to our knowledge of material life, no systematic research has yet been undertaken. Excavations of buried villages in Poland, and more recently in other countries, have enabled an accumulation of first-rate documentary material. In Czechoslovakia, systematic excavations have revealed numerous plough shares, with which it is possible to date the change from symetrical to asymetrical ploughing using the heavy plough.

Thus our knowledge of the material world and of techniques is gradually increasing. But it is still true to say that greater efforts must be made both in exploiting the existing material now preserved, and in gathering other evidence. An international collaboration should be organized to this effect, as international boundaries are of no concern in the study of the history of techniques. Links certainly already exist, even on the international level, such as the association of agricultural museums, regular symposiums of maritime history have been organized, and an international group has been formed for the study of ancient metals. We must, however, emphasize two other aspects of research concerned with objects.

The first concerns manufactured objects which are often dispersed throughout numerous collections. It would be useful if detailed list of analysis were established for each type of object. A recent example of this was proposed by the *Centre d'étude des textiles anciens*, recently established in Lyons. The example index included a large number of questions, the answers to which are vital for the history of techniques. In another field analyses of the old process of casting metal, undertaken by the *Société générale des conduites d'eau*, at Liège, has similarly demonstrated the interest brought forth by detailed research.

The second point is equally important. It would be interesting to organize research on the international level. Research was recently conducted on iron tilt hammers, not only in France, but also within countries which still use such tools, and there are many other examples of this sort.

AVAILABLE RESOURCES

We do not intend to make a complete list of all the various resources available at this point, but to provide some guidelines on the avenues already explored and those for future exploration.

Centres of Research

Research centres have been set up virtually all over the world, the sole or principal aim of which is to study techniques. It will often be necessary to establish a certain harmonization between them. Note also that the links established between these centres are still too much on a personal level.

Germany: the Union of German engineers, the seat of which is at Düsseldorf, has created a research group on the history of techniques.

Austria: the Austrian Institute of the history of techniques, created in 1931, was the first organization of its type.

France: several centres exist, some of which are strictly specialized. The *Institut d'histoire des sciences et des techniques* is one such centre, attached to the University of Paris, and created in 1932. The *Centre de recherches d'histoire des sciences et des techniques* is dependent upon the *Ecole des Hautes Etudes en Sciences Sociales. The Conservatoire national des arts et métiers* houses a documentation centre on the history of techniques. The *Centre de recherches d'histoire de la sidérurgie* is attached to the *Musée du fer* at Jarville, near Nancy. The *Centre international d'étude des textiles anciens*, founded in 1954, is associated with the *Musée historique des tissus* at Lyons.

Great Britain: the Newcomen Society for the Study of the History of Engineering is a well established centre and has many publications to its credit.

Italy: the Italian Institute for the history of techniques was created in 1961 and was established at Milan in the Museum of the history of sciences and techniques.

Eastern countries: there are Institutes of the history of sciences and techniques in almost every country (we shall cite those in Poland and the U.S.S.R.). Centres of research on material culture have similarly been established.

We have mentioned the Committee for the ancient iron industry, a product of the international Union of Prehistoric and Protohistoric sciences whose seat is at the Institute of Archeology at Prague. We can not list here all the centres of ethnographical research which are also of primary concern for historians of techniques: we have also already cited the French laboratory of ethonography which is attached to the *Musée des arts et traditions populaires*.

Two conclusions may be drawn from this brief survey. In the field of the history of techniques, it would be more judicious if research centres and museums were associated, as the material upon which a large part of the research rests should be at hand. This short list above, proves, moreover, that these centres were created rather haphazardly. Some centres have a dual function and other centres are totally lacking, such as those on farming and marine techniques. Each important technique should, indeed must have, a reseach centre. International cooperation could fulfil this necessity, and the regretful dispersal and uneven distribution of financial means could be avoided.

Museums

We have discussed the importance of objects and images as sources for the history of techniques. Objects and images are brought together, at least in part, in museums and collections, and the historian of techniques should not neglect this highly important aspect of documentation.

It is difficult to classify museums according to the information which they provide for the history of techniques. Let us say merely that there are some which are of secondary use in this field, and others whose contribution is more important. Other museums concentrate exclusively on just one technique. We will not discuss the first museums mentioned, as we would have to go into too much detail. A few concrete examples should be enough to provide some generally valuable information.

Art museums have large collections of images and objects. Drawings or paintings, from the very earliest to the Realist work of the 19th and 20th centuries, often form an important source of documentation, as René Evrard's book on *les Artistes et les usines à fer* proves. Here again catalogues have yet to be compiled and reproductions collected together, which has been done in some technical fields.

The same remarks may be made with regard to archaeological museums, both for their iconographic resources which are akin to preceding preoccupations, and for all the objects found in excavations – the tools and the manufactured objects. Some museum collections form genuine technological collections. Thus their catalogues become first-rate reference works: one need only look at some of the catalogues of the museum of Saint-Germain, near Paris. It is now certainly possible to make lists of equipment, the value of which is unquestionable.

Local museums are often extremely rich, be they museums which cover virtually every aspect of local life in art, history and archaeology, as is often the case in France, or purely historical museums, as is the case in Germany. Their collections range from archaeological finds to more recent objects collected with the aim of studying local ethnography.

Among the more specialized museums, we should first point out ethnographical museums. The importance of techniques in ethnographical research is well known. Georges-Henri Rivière has perfectly defined these museums and their possible roles in a number of domains in a volume in the *Pléiade* Series. This is also true of museums of general ethnography, of which the *Musée de l'Homme*, in Paris, is one of the most illustrious representatives. Even heterogeneous collections can sometimes contribute valuable information: here we are thinking particularly of the Missions Museum of Saint John Lateran in Rome, whose riches are quite remarkable.

At this point we can only provide a very brief list of all these museums. A comprehensive guide would, however, be most useful in directing research in areas which have not yet been examined, due probably to the absence of a vital thread. We must direct our attention to museums which concentrate principally on techniques. Our list does not claim to be exhaustive, but we will cite the most important ones and give a selection which may be of use to the researcher. General museums often possess more comprehensive lists, and inventories have already been undertaken which will be added to progressively.

Some museums concentrate on the whole technical ensemble. They are often associated with science museums, but here, we will indicate some of the most important centres.

Germany: *Deutsches Museum von Meisterwerken der Naturwissenschaft und der Technik*. This was founded in 1903, in Munich.

Austria: *Technisches Museum für Industrie und Gewerbe*. Created in 1908 and established in Vienna.

Denmark: *Tekniske Museum*.

United States of America: Smithsonian Institution, in Washington.

France: *Musée du Conservatoire national des arts et métiers*. We have already mentioned that the origin of this museum goes back to the model collection set up by Vaucanson. These objects, formerly instruments of documentation and demonstration, have gradually become museum pieces.

Great Britain: Science Museum, in London.

Italy: *Museo Nazionale della Scienza e della Tecnica*, in Milan.

Norway: Museum of techniques established in Oslo.

Sweden: *Tekniska Museet*, in Stockholm.

Czechoslovakia: Museum of techniques, in Prague.

U.S.S.R.: Museum of technology, in Moscow.

Most of these various museums have published catalogues, which often represent well-defined techniques. The catalogues published by the Science Museum, in London, do not just provide descriptions of the collections but also include information on the historical background of the techniques considered. Founded in 1857, it is undoubtedly the most experienced museum.

Many museums are given over to a single technique or a technical ensemble. These are without doubt the richest and consequently the most interesting museums. Because they have restricted themselves to one subject of study, the presentation of that technique is far more comprehensive. Larger museums, on the other hand, are often composed of collections or objects placed side by side without any general theme of presentation. Similarly, large museums often try to gather together the most exceptional pieces, but more specialized museums have another advantage, as the techniques are generally presented in their historical, social and economic context. We will give just a few examples, mentioning only the best stocked museums, and leaving aside the specialized sections in general museums.

Agriculture: apart from the museum of agriculture in Budapest, there are no large museums on the history of farming so a lamentable gap exists in this subject.

Mining: a mining museum exists at Bochum, in Germany, and another is presently in formation at Saint-Etienne, in France. Old mining sites are often converted into museums, such as those in Sweden, the most important of which is in Falun, and we should again mention the marvellous salt mines at Wieliezka in Poland, and there are several in Wales and in the North of England.

Military art: Army and War museums have been created almost everywhere. Firmly centred on the history of the army, these museums frequently preserve rich military material, at least from the 15th century onwards, which enables much work to be done on military techniques. The art of fortification has not been covered in so much depth although in truth this is considerably more difficult to present, but there are certainly some excellent examples of military architecture available for use.

Transport: virtually nothing exists on the transport system as a whole, although some sectors of it are very well represented. Almost every country which enjoyed maritime vocations now have maritime museums. Sometimes there are several such museums in one country. Although these museums are generally given over to the history of the military navy, the merchant navy is also given a noticeable place, such as the museum of Albenga which concentrates on under-water archaeological finds. Some museums, in particular those of Paris and Greenwich, are very rich in models and documentation of all kinds. Land transports are less well covered. There are railway museums at Utrecht, in Holland, and at Storhamar, in Norway, which was founded in 1896. Others are planned, as in Mulhouse, France. There is no one large automobile museum, but private collections exist, such as those of Ford or Fiat which are the richest, and there are many small, limited collections which together could form a very large and useful museum. Britain, however, boasts the London Transport Museum. Similarly, no large aviation museum exists as one might have expected, although the Battle of Britain Museum covers aircraft and weapons of that period. Rich Parisian collections are still waiting to be housed before they can be presented.

Agricultural Industries: museums on particular farming activities exist, although they are somewhat unequally represented. The most famous activity represented by museums is of course wine making, and many such museums exist within a single country. The most famous are at Beaune, in France, at Speyer, in Germany, and Krems, in Austria. Other fields of agriculture are less well represented by specialized museums, but in France there is a Tobacco museum at Bergerac, and an Olive museum at Cagnes-sur-Mer. Recently a cheese museum has been established in Auvergne. There is almost nothing, however, on fishing, forestry, and animal husbandry.

Diverse industries: these are perhaps better, but again only partially represented, and those that are do not have enough material.

The textile industry is represented by the Textile museum at Lyons, one of the most important in this field, the Hoisery museum at Troyes, France, and a Knitting museum in Denmark.

The iron industry is represented by the Iron museum at Jarville, near Nancy, which has clearly tried to show techniques in their human, historical and economic context. We should also mention the Cutlery museums at Sollingen, in Germany and at Langres, France.

Clock making is well represented at Besançon, France, and naturally in Switzerland.

For the glass industry we must cite the Glass museum at Murano, in Italy, and the large collection at the Corning Glass Centre in the state of New York.

The shoe industry is exhibited in several rooms in Fougères and Romans, in France.

For photography, the large collections of Eastmann-Kodak in Rochester in the United States are available, and a rather more modest museum exists in the suburbs of Paris, as well as one in Chalon-sur-Saône.

We have only given a brief selection here, and the list could easily be completed. As we see, there are many industries which are not represented, among them the most important. Although some information on industries, such as the chemical industry, is difficult to exhibit in museum collections, it is regretful that other industries do not have their own museums. Must we await their arrival?

Libraries and Archives

Similar comments and distinctions may be applied to libraries and archives. General institutions, national libraries and national archives exist which are naturally the first places of reference. But here, specialized centres are far less numerous.

For libraries, we must first point out those brought together by research centres, which have been discussed above. This is true of libraries which form specialized museums, some of which are a rich source of material. Specialized libraries concentrating on a particular subject are very rare. The Iron library at Schaffhausen, in Switzerland, is an exception, however, as it has a well-defined aim and extensive documentation.

Technical libraries exist, generally annexed to technical teaching establishments: we could put forward the library of the *Conservatoire national des arts et métiers*, in Paris, as a model example. The older these libraries are, the greater the chronological limits of their collections. Libraries belonging to certain associations could equally be cited, such as those of the association of engineers, and the association of manufacturers. In France there is the rich library of the *Société des ingénieurs civils* (Society of civil engineers) and that of the *Association nationale pour le développement de l'industrie* (National Association for the development of industry), a wealthy collection which unfortunately has suffered the consequences of regretful neglect. Those responsible for these institutions should not dispose of these invaluable documents under the pretext of getting rid of out-dated material.

The same observations may be made for archive material. We should first of all distinguish public and private archives. The former are naturally fascinating, in so far as public establishments are concerned with questions of techniques, because they use these techniques themselves, either with the aim of supervision, or to try to initiate innovation in order to sustain economic expansion. Military archives, those of Bridges and Roads, and institutions whose aim is the issue of patents, may be cited in the first category. The second category may be represented by ministry archives in the Departments of Mining, Industry, and Agriculture. The last category includes almost all the administrations of the state. We need not worry about the preservation of these archives, which have provided a large part of the documentation for historians of techniques, but an indexing system is required, however, as are study guides, which have been compiled in other domains.

The problem of private archives is more difficult to solve, and some cases are agonizing. By private archives we mean mainly business archives. Company technological archives may even relate to documents which do not concern technology in the first place: such is the case of accounting archives. But this documentation may concern application, adaptation, and the improvement of techniques used by the enterprise. Equally important, and more recent, are the archives for research purposes which appeared at the end of 19th century. Here, technology is less isolated than elsewhere as in these archives one finds the whole human environment with the labour force which commands it, and in the complete commercial, economic, and financial context etc. It would be a grave error to reduce the conservation of these technical archives. Herein lies the whole problem of company archives which must be resolved. Since the end of the Second World War interesting attempts in this direction have been made in several countries, but they have yet to be generalized and systematized. Similarly archives belonging to the association of manufacturers must be considered, some of which are very old. Thus technical problems are occasionally considerable.

TECHNIQUES AND CIVILIZATIONS

THE ORIGINS OF TECHNIQUES

There are innumerable works, scholarly and popular, on the origins of techniques. Prehistorians, philosophers and technicians have all considered this problem; their conclusions are varied and sometimes even contradict one another, which can leave the reader confused and uncertain. I would therefore like to reassure readers that this chapter will provide no new explanations. My aim here has been simply to classify and establish a certain order with regard to the facts, in an attempt to discover the first outlines of an evolutionary logic of techniques. This aim seems all the more worthwhile now that some very recent discoveries have completely shaken our ideas about the origins of humanity.

SOLUTIONS BORN OF IGNORANCE

This is not a question of so-called false solutions. To the extent that those who have dealt with the question of the origins of techniques were unable to detect any main theme, they fell back on explanations which placed these origins outside of the human domain. We should not be mistaken into thinking that the knowledge of different technical civilisations lasts for ever. Take for example the basic instrument of all technical activity, the tool. Certainly, tools evolve according to tendencies which can be easily studied. But there are tools which disappeared: the double-faced cut flint no longer exists in modern civilisation, the bow drill was removed about twenty years ago from the catalogue of this classic of mail-order companies, the *Manufacture d'Armes et de Cycles de Saint-Etienne*. Certain trades have almost completely disappeared, for instance there are no more blacksmiths in our villages.

The restoration of techniques used in the early ages of manking began only a century ago. One can understand, therefore, that the lack of a coherent body of knowledge should have led the first of those to write about the history of techniques to look for the earliest forms of evidence in exterior worlds. Within all these words, there were two privileged areas: the gift from the gods was the first of these, and this was succeeded much later by the imitation of nature.

Gifts from the Gods

There is a rich mythology of techniques, but it is unequally distributed. It exists in polytheistic religions, but rarely in monotheistic ones. The Greeks actually have such a mythology, the Romans far less so, and the Hebrews have virtually no mythology of techniques at all. The Greeks attributed the working of metals to a god, Hephaestus; the equivalent in Hebrew mythology was a man descended from Cain, known as Tubal Cain. We shall deal only with Greek mythology in this short part. It is the richest of its kind and the most widely studied.

I would like to point out first, as others have done, the use and profound significance of certain words, some of them close in meaning. *Tekhne* designates above all, practical activity, both material and manual. In Homer, this work evokes the idea of a gift from the heavens. P.-H. Michel even arrives at the conclusion that before the 6th century BC, science was indistinguishable from techniques and that the latter (a divine entity) bore witness to its religious origins. A second example is the word *metis*, which designates a form of practical intelligence. Remember, too, that Metis is the mother of Athene. Thus we have the outline of a filiation which permits the intervention of gods, demi-gods and heroes and heroines who are almost human.

In the first group, two closely-linked gods, Athene and Hephaestus, play a major role in the transmission of certain techniques to men.

We have just mentioned that Athene was the daughter of Metis; this connects her with practical activities which require a certain type of intelligence. Jealous of Demeter, who is reputed to have invented wheat, Athene is believed to have invented the swing plough which she gave to Atticus. But according to others, it was a woman, Myrmex, who stole the idea – one of the numerous examples of "stealing" from the gods. According to J.-P. Vernant, Athene is seen as a divine power devoted to work on the land, or more precisely, to plowing and its fecundating action. In other words, one could say that, from Demeter to Athene there has been a movement from gathering to agriculture. "Athene intervenes as a power endowed with *sollertia*, or manual dexterity and practical intelligence: she makes the instrument, the technical object which allows Demeter's wheat to be harvested more easily." Athene is certainly a technical power.

She is also the goddess of textiles. She not only made her beautiful peplos herself, but she also taught Pandora, the creature borne of Haphaestus and Athene, the art of making cloth. It should be easy to describe the different types of Athene's activities, all of which make her a polyvalent power, and indeed, this has already been done. For in fact she is also a warrior goddess, and familiar with everything concerned with weapons. The same goes for the art of carpentry, which along with cloth-making and the working of wood, leads us naturally to ship-building, from which Ulysses was to profit. Finally, was it not she who invented the bit, and thus enabled Bellerophon to tame Pegasus? Throughout all this a pattern emerges, techniques link together and become complementary. Take the horse, for example, which was used for agricultural purposes, but was equally important as a creature of war. Athene is well on the way to have conceived of and entrusted to men, whether voluntarily or not, a real technical system.

And now we come to Hephaestus, the technical god *par excellence*, who invented the working of metal by using fire. It is well-known that for a long time forging seemed a mysterious activity, verging on sorcery. Even in the 13th century, Biscournet, the locksmith who is reputed to have forged the hinges of Paris Notre-Dame is said to have been ready to sell his soul to the devil in order to acquire the techniques necessary for his work. Primitive, ancient and traditional civilisations have always attributed the smith with having supernatural powers: fire and the transubstantiation of all kinds of materials from stone to metal, remained incomprehensible for a long time.

Hephaestus worked all metals, and the works attributed to him are firm proof of this. Hera's earrings and Pandora's crown reveal his expertise as a goldsmith. Achilles' shield, Diomedes' and Heracles' breastplates prove him to be an effective maker of weapons and a skilful worker of bronze. He also worked with wood and leather: he made both chariots of war and harnesses for horses.

In a given technical system, therefore, Hephaestus appears to be the complementary power to Athene. These two represent virtually all the technical activities of the early era. But there is still more to say about Hephaestus which brings us almost to the time of Archytas. Hephaestus created veritable automata which were in fact more than mere robots, since they were actually alive. These creations were often made with Athene's help, and included the three-legged tables which could make their own way (*automatoi*) to an assembly of the gods, or the golden mechanical women who helped the crippled god to move. And then there was the creation of Pandora. Hephaestus was a true demiurge, therefore, giving life to everything he made. He also created Talos, the man of bronze who was appointed by Minos as the guard of Crete which he regularly patrolled at an unerring pace. This was the supreme warrior robot. Hephaestus was also familiar with pottery: Pandora was made with terra cotta before being brought to life. Similarly, there is a vase in the Berlin Museum which depicts Athene modelling a horse out of clay. Forging, smelting and casting bronze, working with precious metals and using the art of the potter: these were the main elements of Hephaestus' technical power, which to a certain extent overlaps with that of Athene's. Together, these two created woman, just as Hephaestus hepled Prometheus to create man.

With a little effort, it would be possible to establish when this mythology began to evolve. There are some techniques which actually go back as far as the VI millenium BC (agriculture and pottery for example), others are slightly more recent (the metallurgy of copper and bronze), and these are followed by even later ones (chariots of war). But generally speaking, they were all in existence by the beginning of the III millenium. During the 6th century BC, the new upsurge in technical development was in a sense the continuation of a long historical process, but adapted to an era when techniques were of prime importance.

With Prometheus we are breaking new ground, in that the first signs of a humanisation of techniques are becoming apparent. P.-H. Michel notes:

(It) remains a gift from the gods, but there is a poem by Homer which glorifies Hermes for it, who is not so much an inferior god, as a more human one in that he does not possess all skills, but rather he discovers them by intelligent means. Like a man of genius, he invents them and then passes the knowledge onto man. The guardian of the "secret" then teaches himself and he too acquires a knowledge which he is capable of passing on. Better still, skills are no longer considered as gifts from gods, but as the result of *a theft at the expense of the gods*. The myth of Prometheus is the most legendary expression of this belief which in the works of Hesiod and Pindar especially, has given new meaning to the word *tekhne*. Hesiod believes, for example, that alongside the skills of divine origin, there are purely human skills, such as the art of the caster.

In fact the conflict which broke out between Prometheus and Zeus first appeared in a late text by Diodorus. According to J.-P. Vernant, the promethean myth as represented by Hesiod is quite coherent, and already raises the problem of the function of techniques: labour appears to be consequence of the conflict between Zeus and Prometheus. The stealing of fire must be paid for and henceforth, labour was the condition of all wealth. From this stemmed the opposition between fertility and labour, between that which was supplied by nature through the intermediary of the gods, and techniques, which given to the accursed, or stolen by him, or which simply issued forth from his imagination. Demeter gave us wheat, but that was not enough; the gods gave us the wind, but that was not enough either. Athene gave us the plough, and ships and sails, and Prometheus gave us fire. In Plato, Prometheus represents the elaborate idea of techniques as a social function.

This, according to Vernant, was how the myth of Prometheus evolved. Labour and techniques were activities imposed upon us, but they were linked to religious thought. But in Plato, labour and techniques fulfil a social function which is entirely human and material, and they are therefore considered as human activities. Similarly, Aeschylus attempted to integrate labour and technical activities into the human world.

We must note, too, that the techniques associated with Prometheus are those which involve using fire: they concern metallurgists, potters and no doubt cooks as well. He is not connected with any of the other techniques. But these connections are relatively vague if one considers them from a purely technical point of view. In the latter sense, Prometheus emulated, if not competed with Hephaestus. For this hero, there was no question of weapons or metal implements. He was the symbol, quite simply, of all sources of heat and by virtues of this, he represented all those descendants of the devil and the sorcerors, the select guilds concerned with skills using fire. Furthermore, these were the first artisan guilds which developed outside of the domestic environment and whose members were part of the first specialised trades. Spinning and weaving could be done in the home, as could the working of leather and wood, but forging and pottery-firing was impossible. And of course the main unit of those working the land was the family. But activities concerned with fire, except cooking and baking, had to be done outside the house, and they were not practised within the family community, but by outsiders.

From our point of view, which is concerned strictly with the origins of techniques, the role of Prometheus, like that of his slow-witted brother Epimetheus, is particularly confusing. The creation of man and the stealing of fire are complementary, as man cannot live without fire, and this distinguishes him from animals. This seems to show a necessary link between the appearance of man and the birth of techniques. But henceforth there was to be an indispensable intermediary between the gods and men. Unable to accept the human origin of techniques, or at least of the most essential, man invented Prometheus.

Daedalus leads us to a third definitive stage, which is a kind of secularisation of techniques. But there is an important point to be made beforehand. Until this point, we were concerned only with broadly surveying the basic techniques which appear to have been given to man, such as metals, pottery, textiles, and ships. With Daedalus, we shall enter into the details of techniques, and in particular, its most important aspect, the tool.

But there is more. We are leaving behind divine origins. Daedalus is a man; he even has the makings of an historical context. He can be situated in time and space, and genealogically. He no longer receives, he "invents."

Chronologically, Daedalus was a Cretan sculptor in the 7th century BC. Many people, including Diodorus, believe he was an Athenian and of royal descent. His father was probably Metion, which would connect him with the *metis*, which we have discussed above. According to Pausanias, "Daedalus was famous throughout the world for his skills, as well as for his wanderings and misfortunes." But it is possible to be more precise about this. The Daedalus legends, or at least the oldest of them were brought to us by Diodorus which would place them in the first century BC. Prometheus was much older than this. During the period between Prometheus and Daedalus, the first major development of Greek techniques (whichever century one chooses to place him in – although it was most likely between the 6th and 4th centuries

BC) was an exclusively human achievement and was the first to be acknowledged as such. From here onwards, the secularisation of techniques was inevitable. No gods will feature in the following discussion, therefore, only men, and these shall have an historical reality. This includes Daedalus and his legend. We are no longer dealing with the hazy memories of the beginnings of agriculture, metallurgy and pottery where only gods and demi-gods played a part.

Françoise Frontisi-Ducroux has provided us with an excellent study of Daedalus and one of the things she deals with is the hero's name. "Daidalon" denotes anything which can be done in the material sense, anything that can be made: weapons and armour, jewels, cloths and fabrics, all wooden objects, chariots and the framework of ships. Moreover, the gods of *daidalon* are in fact Athene and Hephaestus. When a list was made of the objects coming under this generic term, it was noticed that out of forty-nine quotations, forty were concerned with metal objects, five with textiles and four with wood.

It is possible to see the formation of a system at this point. We have a hero whose name in everyday vocabulary denotes every object of craftsmanship, a hero who belongs to the royal dynasty of Athens, which itself goes back to Hephaestus and Athene. Daedalus seems to be the prototype of the artist and the artisan, and indeed, of the artist and artisan fused within the same person. He created the first divine images, but he also invented essential technical implements. Furthermore, he was also an architect and an engineer. He is the very embodiment of inventive genius and artistic talent. He is the true representative of *tekhne*, a term which denotes the skill of the doctor, the dexterity of the cobbler, the knowledge of the architect and the artistry of the musician, indeed everything which concerns material creation.

Daedalus was also a sculptor. He is reputed to have invented statues of gods made out of wood or wood covered with a layer of metal. The dedalic style of sculpting is characteristic: with their eyes open, legs apart, and arms stretched out in from away from the body, the statues seem almost alive. and like Athene and Hephaestus, Daedalus is supposed to have succeeded in giving them life, hence the legends of chained statues. This is an example of *tekhne*, the creation of illusion and the imitation of life which is such a passionate desire of all men from Daedalus to Descartes to Vaucanson. We are also nearing the Archaic period of Athenian art, with all the suggestions that these early statues could present to the eyes and minds if the inhabitants of Attica and Crete. We can see these statues in our museums. It seems then, that there has been some kind of a metamorphosis. That which was attributed to the gods Athene and Hephaestus in the ancient myths is now seen as the work of man. It is not impossible that Archytas' flying dove is also a symbol of the humanisation of the most advanced techniques.

Daedalus was a craftsman and a metallurgist, which means that he is not easily distinguished from the gods we have already mentioned who also worked metal. Daedalus, however, is important because he is supposed to have invented a large number of implements. These attributions vary from one author to another, and the differences depend largely on how much credit is awarded to Daedalus' nephew, Talos, his sister's son, who was his apprentice. Daedalus was so jealous of his success that he killed him by hurling him off the top of the Acropolis. This form of murder can be found time and time agan in Greek religious legends.

According to Pliny, Daedalus invented the saw, the hatchet, the plumbline, birdlime and fishglue. It is noteworthy that all these instruments are used for working wood. Birdlime and fishglue are both used in assembling. The plumbline is used by carpenters and architects. The hatchet, or adze, is used to trim wood and is extremely useful to carpenters, joiners and sculptors alike. There is no need to explain the saw. Also cited are the gimlet or drill, which Homer includes in the category of carpenters' tools. All these implements can be seen on a dish in the Berlin Museum which represents the workshop of a sculptor. An Etruscan golden bull, hammered and engraved anew during the 54th century BC, depicts Daedalus holding a saw in one hand and a broad axe in the other. This association is still apparent on a much later piece of engraved Roman glass, which depicts a ship being constructed apparently under the guidance of Athene.

Other legends involve Talos in all these "inventions," as a precociously gifted young man who is thereby doomed to die. As Frontisi-Ducroux has noted, the emphasis, as far as Talos' inventions are concerned is on the notion of circularity. According to Diodorus, we have him to thank for the wheel, the potter's wheel, and the compass, all of which are used for making or marking out circles. It is possible that a fishbone or a snake's jaw gave Talos the idea for the metal saw used for working with wood, and it was this saw which was to arouse Daedalus' anger.

Following the murder, Daedalus sought refuge in Crete, with Minos. Daedalus appears as a sculptor and woodworker in the legend of Pasiphae, but there is perhaps more, because *tekhne*, the creator of illusion and the imitator of life appears to be at work once more, just as it was with the living statues. In this sense, techniques were becoming a substitute for nature. On the other hand, Daedalus is not presented as being the creator of the labyrinth. It seems that he stayed in Egypt and was prompted by seeing King Mese's tomb which in fact was a converted fortified palace. It is not an architectural work but an enclosed space, with no ornamentation. However, it is as an architect that Daedalus went to Sicily after his escape from Crete. Diodorus writes of the construction of a dam near Megara, and fortifications at the Camicos citadel. This then, is the real hero: the architect and the "engineer." He also installed a water-cure establishment at Selinonte, and the temple of Aphrodite at Erice on an almost inaccessible rock. "He built", notes Diodorus, "a wall on the actual precipice, enlarging the vertically overhanging platform in an extraordinary fashion." It has been demonstrated that these works raised all the problems of controlling rivers and draining marshes (and it was the same in Greece) that were encountered by Sicilian engineers during the historical and archaic periods. And many of these works are built in suspension between the land and the heavens, as if they are somehow a link between the artistry of the gods, and the ingenuity of man.

The escape form Crete poses two problems. According to some, Daedalus escaped by sea, while Icarus was drowned and washed up onto the island which now bears his name. Daedalus is supposed to have been a carpenter and a ship-builder, and is reputed to have invented the sail: we can see again here the transfer of activities from the gods to man. According to the majority of writers, however, Daedalus escaped from Minos, the King of Crete, with his son Icarus, by sticking wings to his body with wax and flying away. It is well known that Icarus flew too close to the sun, which caused the wax to melt and thence his downfall.

Daedalus best represents the technical developments which began during the 6th century BC. For those writing much later – Diodorus was around in the 1st century BC and Pausanias in the 2nd century –

the history of the origins of more developed techniques is completely unknown. It was necessary to conceive of Daedalus in order to understand it. This demonstrates, with all due respect to Plato, the serious consideration given to techniques and the "consciousness of the role of technical development in the cultural progress of humanity."

Other characters can be found in this fantastic world. They are of a lesser calibre, but they still stand out in a group within which it is so difficult to tell the legendary beings from the historical ones. Let us take two examples.

Palamedes was a prolific inventor. He, too was a representative of *metis*. Acording to some interpretations, he invented certain letters of the alphabet, dice games, draughts and knucklebones, etc. Palamedes' aims seems rather surprising, and they are the object of a certain amount of ridicule in Plato's *Republic*.

Rhoicos of Samos and his son Theodoros have also been mentioned – both of whom were living during the 6th century BC. They are reputed to have created the Hera temple at Samos, and the labyrinth at Lemnos. They were also goldsmiths and sculptors and may have invented the wax casting method for making statues. Finally, we might have to thank them for the level and the compass. They are clearly still part of the same lineage.

Technical knowledge, and especially the notion of invention, considerably aroused the imagination of the Greeks. There had to be an explanation for it, its origins had to be told. And even if some have managed to provide a worthwhile analysis to explain the appearance of fire, be it either lightning striking a tree, or rubbing two very dry sticks together, they still found it necessary to place the responsibility for technical activity on a select few, who came first from the heavens, then from an intermediary position halfway between the heavens and earth, and finally, from Earth itself.

Nature's Example

This area is less well-defined. Ever since Antiquity, men have looked for the origins of tools and techniques in nature. There are examples of this from all periods.

We noticed above that Daedalus or Talos are supposed to have invented the saw by imitating either a fishbone or a snake's jaw. This is an example of the direct influence of observing nature on the making of the tools. The number of animals' names given to machines by the Greeks which have remained in use is surprising – the ram, the sawhorse, the crane, the grab [Transl. Note: *louve* in the french, meaning she-wolf], the crayfish and the slug (which is the name given to the archimedian screw) – and of course the same goes for weaponry: the battering ram, the onager and the scorpion, etc.

Of course, this is no doubt due to a certain resemblance, but of what sort? Did the animal inspire man? Or did the machine remind man of the animal simply on account of its general shape? Note that *cochlea*, like the Greek *cochlias*, is used in technology to designate the snail's shell, the tower with spiral staircase, the screw-press and the archimedian screw, all of which terms are derived from the word *cochlos*, which means snail. We shall return to this discussion later.

We hardly need dwell on Leonardo da Vinci's research into the flight of birds when working on his flying machine.

In a more contemporary vein, it is worth citing Charles Frémont, a notable historian of implements. Prompted by the sort of vocabulary we have just discussed, Frémont, like many others, sought the origins of the screw in natural objects, in particular helical shells, such as the snail. Reuleaux rejected this explanation, as did Fremont for the basic type of screw, however both of them found it acceptable to explain the screw and nut. Moreover, Frémont attributed different origins to the spiral staircase, the archimedian screw and the screw press. For Frémont, the relative movement created by the snail when it is taken out of its shell to be eaten is an exact and absolute representation of the screw and its nut – not to tighten adjoining parts, but to create a thrust, as in the case of the beam of a screw press. In order to extract the whole snail, one had to perform a giratory movement. Readers may think of this explanation as they wish.

There is a third type of explanation, which concerns the pebble-tools found some time ago in East Africa. In the words of Leroi-Gourhan:

A collection of eoliths taken from random samples has one striking characteristic: its is impossible to detect any kind of morphological constancy, and the distribution of forms seems to be entirely arbitrary. The only form of constancy which could be said to exist is of a purely mechanical order. This relates to the accidental thinness of a flint kidney, it affects the protruding parts indiscriminately, or result, on rather long flint kidneys, from fractures by bending.

And the conclusion which precedes this quotation:

Which means that if the percussion bulb indicates the likelihood of human intervention, it is also possible to detect a certain number of fragments which are the result of natural causes.

A flint which fell off a cliff and then shattered might well have been the first tool.

We should try to classify the above: the facts are diverse and they are the result of very different types of observation. They also presuppose, however, an imaginative effort which is relatively easy to define. When faced with a technical need, man has only to look to nature to find the solution, either by the simple discovery of a suitable object – a shattered flint – or by observing animal organs such as a fishbone or a snake's jaw. So it is no longer a question of a simple transposition from the natural object to the tool. And here we can distinguish between processes which are possible, and those which are not. To reproduce the shattered flint presupposes the notion of percussion. And the passage from a fishbone to a saw is much more complex. It seems to us, therefore, that within the natural universe, which represented for early man an enormous field of experiment, three different types of observation are discernible: (*a*) the effects of nature, which are clearly the most perceptible, even by primitive mae; (*b*) the general shape of beings, which certainly require more sustained attention and which could only date from relatively recent periods as we shall see; (*c*) the activities of animals, which are associated in part with conformation, but which are in fact an extension of it, the conformation being associated with a specific "task."

We shall not stay long with the first category. We have already mentioned the case of pebble tools which could have been the product of natural causes. There are other examp'.:s, one of the most typical of which is the tree trunk floating down a water course which probably provided the inspiration for the first one-man craft. Another perfect example of this is the case of Newton and the apple. Newton "understood" the fall of the apple because he was already thinking in those terms. The transition from a floating trunk to a craft is of the same nature, if not of the same degree. In other words, and this is the same in other cases, the problem necessarily preceded observation. Certainly, monkeys use sticks to bring down fruit, or as levers. In which case, it is best look for the differences between them and man:

the conservation of the tool for example. The transition from the wheel which turns according to a water current, to a mill is infinitely more difficult to explain. Initially, they must have looked for materials which were already prepared by nature, and which could be used immediately, on their own or with other objects.

Nature is unquestionably the mother of energy, and it has supplied the three essential forces on which man has survived for thousands of years, running water, wind and fire. But none of these forces leads directly to utilisation: we have indicated this already with regard to a piece of floating wood and sails, and lightning and rubbing together two pieces of wood. We saw above that these tranpositions were difficult to understand for the Ancients, who made up for this by appealing to the gods or heroes. A technical act is one which places man outside of natural phenomena, and which sometimes even goes against them.

The shape of objects demands long and pertinent observation, which, as we have already stated, must necessarily be preceded by the recognition of a problem to be solved: efficient observation is born of a need to be fulfilled. And this observation must be conducted according to certain rules which have gradually been put into place.

The first example is relatively unimportant. Around 1881, a glove manufacturer invented the press stud, which was to replace the impractical button and buttonhole. By relying on mechanical principles, he attempted to find the most suitable forms for a solid fastening. He gradually conceived of this object, outlined a design and made a prototype which he then tried out. Now the press stud exists naturally within certain animals: crabs, cephalopods, decapods and aquatic insects, etc. It is exactly the same shape and performs exactly the same function, holding together two seperate regions of the body. But it is certain that there is no link between the man-made and the natural object.

There is another, more significant example. Towards the end of 1789, Volta built his first battery, creating electricity by forcing contact between two different metals. On the 20th March 1800, he wrote an explanatory letter to Sir Joseph Banks, vice-president of the Royal Society of London. "As I shall demonstrate, this apparatus is fundamentally similar, in substance and in form, to the electrical organ of the torpedo, and of the eel, much more so than the Leyden jar and other known electric batteries, and I would therefore like to call it the artificial electric organ." The ancient voltaic battery was composed of three vertical rods, supported by a wooden platform; there is a raised column between the three rods formed by superposing in the same order a disc of zinc, then one of copper, and one of felt soaked in slightly acidic water. This structure is similar to the organ found in the torpedo. The liquid used in the Volta element is similar to that found in the intralamelliara, the metallic conducters being represented by the intermediary tissues.

In the first case, there is no possible link between invention and observation of nature. In the case of the snail and the screw, the link is very doubtful. For Leonardo da Vinci, the relationship is real, but nothing concrete emerged from his ideas. In the last case, was it a matter of genuine influence or retrospective comparison? It is very likely that it was the latter. There is a certan resemblance but no similarity, and it would have been difficult to conceive of the metal based battery from the organ found in the torpedo. Besides, those types of observation which provide ideas or proof of similarity, are relatively recent and imply a specialised knowledge of the natural world.

Some people have spoken of the "tools of living beings" and have cited examples which appear somewhat disquieting. "In spite of their diversity, cleaning apparatuses in animals are always composed of bristles or spines arranged in either one or several rows and suggesting thus the comb or brush used by man; these are either used together or separately." It is the same with coupling or fixing structures: hooks, fishhooks and suction pads. Whatever the function of the latter, it must always keep the subject firmly fixed to its support. This common feature requires two processes: on the one hand the isolation of a certain surface on which the edges of the suction pad are hermetically sealed, and on the other, a break in the balance of pressure surrounding and within the suction pad. Even before rubber was available, which was to enable man to create the perfect suction pad, he was familiar with the moistened washer secured with a cord. I am forced to admit my ignorance here, because we do not know when the suction pad was first used, or the nature of its original function. Consequently, we have no means of comparison.

The same goes for all types of pliers. These grasping devices are built from two main models, scissor pliers or pincers composed of two parts which are joined or crossed, or the sharp-edged, folding pliers, formed by the withdrawal of one or more of the terminal parts (jaws) into a thicker part immediately behind it (handle). The former are most often found in crustaceans, and the latter in insects. Again, it is difficult to know whether nature was a genuine influence. For it is undeniable that man has created a much more varied range of pliers than can be found in nature. These pliers have a commmon characteristic: the jaws or rough surfaces found on the contacting segments which serve to hold the object.

A. Tétry has noted the resemblances between nature and diving equipment. Man uses the supra-aquatic atmosphere for making such equipment, which makes oxygen available in greater or lesser amounts. This means that they are restricted in terms of time (apparatuses containing oxygen) or in space (feeding by a tube). Carbonic gas is eliminated either by the expulsion of the vitiated air or by chemical absorbtion. Human diving apparatuses correspond with the most specialised types which have been observed in nature. Tétry seems to indicate that one day it will be possible to create a device which will enable us to breath under water, a device which so far has been found only in nature. In any case, human exploits under water, real or legendary, have been taking place for centuries. Similarly, man made floats rest on the same principle, the structures are analogous – cork holds an important place – and they fulfil the same function as animal or vegetable floats.

I shall end on a more precise note. We are not yet at the stage of the luminous, cold organs of animals. Nature created the prototypes of certain stringed instruments (violin, guitar), and percussions instruments (tambourine, cymbals, xylophone) and even noise instruments, (grasshopper). But their aim is completely different to the artificial instruments used by man. Animals' instruments are used to call, to frighten, to attract the male or female, and perhaps to communicate. Man's musical instruments are altogether different and are used to supply ordered sounds, arranged according to a scale. The difference between the musical organs of animals and man-made instruments is so fundamental that it is impossible that any links should exist between them. "The grasshopper which rubs the rasp of its thigh on a rib from its outer wing is comparable to the violinist who slips his bow accross the strings of his violin." The two techniques are perhaps similar in terms of their overall structure, that is, they both make a suitable material vibrate by rubbing it with some kind of rasp. But the violin, even without including its organisation on the scale, is another matter: the different types of string, the bow, the resonance chamber, the range of possible sounds, etc., render it a completely different type of object.

Certainly, there are certain parallels to be drawn between animal and human implements. It could hardly be otherwise, since our needs are often the same, and there are not a thousand and one interchangeable means of action in any given circumstance. Yet resemblances, analogies or comparisons are not absolute. Certain animal tools are inefficient. Crustaceans' claws are an example of this: the arrangement of the contracting muscle, its length, and finally, the size of the pincer's two jaws reveal that the basic principle of all human pincers, the reduction ratio of levers, is absent. An animal must expend considerably more energy to produce the same result. But on the other hand, humans are familiar with only one type of elementary suction pad, while the same device found in animals can be very complex.

A fascinating film has been made about birds' beaks. Of course, we know that there are a great variety of these, from pointed beaks to spoonbills, whose differences depend on the nature of the bird's food and habitat. But again, man would have needed a large selection of samples and observation techniques which have only recently evolved.

Let us consider a few points.

First of all, the transformation from the animal tool, which is an integral part of the animal's physiology, to the human tool, can only take place after correct observation: of the purpose of the animal tool, its mode of operation, and its structure. In other words, they would have needed equipment, like magnifying glasses and microscopes, to observe. They would also have had to understand the meaning of this observation and its link with an eventual need, and finally, a fairly advanced knowledge of physiology would have been necessary. Research is being conducted today into certain types of animal behaviour, such as the dolphins' for example, in order to discover new techniques, but this would still have been out of the question not so very long ago.

Next, we should bring together both the animal tool and the need to be met by the human tool, and also the necessary change from a natural conformation to something which is very probably made of other materials, requiring further mechanical tranformations. The passage from a crab's claws to pincers completely transforms nature. Does taking a snail out of its shell really suggest to us the idea of the screw and the nut? This is hardly likely. Only Archimedes perhaps, in his bath (according to the legend) might have related a natural phenomenon with a scientific demonstration, the technical effects of which are well-known.

Ultimately, one can hardly assert that nature has contributed significantly to the development of techniques. Certainly, eating snails could bring to mind wine and the wine-press, but the relationship is a tenuous one. A crab's claws could not possibly have given rise to the fulcrum of pincers. It seems that the analogies were only discovered *a posteriori*.

The study of animal activities should certainly lead us to a similar conclusion. From the integrated tool, one can move on to the work it performs. Here again, there are numerous examples, most of which have been taken from Tétry's work.

The first example is rather a curious one, which concerns packing down the earth covering certain fossorial hymenopteras' burrow-holes. This is a common enough practice. Sometimes this is done with a part of the body, but there are several European and American species of the *Ammophila* type which use a small pebble held in the mandibles and handled like a rammer to conceal the burrow's hole. This seems to be evidence of exterior tool being used. Weaver ants are another example. They build their nests by sewing leaves together (often from mango trees) using threads secreted by ant larvae. Thus, they are not only using a tool, but they are working together as a logically organised team. It appears that octopuses use their suction pads to deposit small stones between open valves of mollusc shells in order to prevent the shell closing. The animal can then be easily extracted and eaten.

The utilisation of one animal by another occurs fairly frequently, especially in hunting, as is the case with humans. Parasites are evidence of this, but there are more complex examples: certain types of crab brandish a small sea-anenome between their claws, whose poisonous sting paralyses their prey.

"Constructions" have altogether a different meaning. Generally, they are used for habitation. It is known that certain types of birds are thereby able to make perfectly constructed nests, because their young are born still in the embryonic stage, and they remain in the nest for a long time. Sparrows' nests are remarkable for their intricate network of interwoven twigs which surround a deep bowl lined with the most delicate of materials. Some have even managed to make closed nests. As above in the case of the ant, there is a warbler which makes its nest by sewing leaves together. We have included some drawings which show the different forms of knots and ties used by the weavers to make their nests (fig. 1).

We need only remind readers of the number of burrows which often contain many long galleries.

No animal has perhaps affected man more by its work than the beaver. There are beavers which live in burrows whose opening is above the water level, and there are others which build lodges out of branches and mud, which are completely surrounded and immersed in water. If a brook does not provide a sufficiently big stretch of water, they build their famous dams, thereby creating artificial lakes. Two beavers were once observed over two months, during which time they used dead wood and boughs from freshly-cut aspen and alder trees to build a dam thirty metres long and one metre high. The upper side was covered with a mixture of mud and stones. The leaf-fall in Autumn served to seal off the numerous holes. The animals also accumulated stocks of winter food. In their surroundings, beavers dig canals measuring up to hundreds of metres so that they can float the logs they have cut down to their lakes. Does this represent a series of associated instincts or foresighted intelligence? In fact, there is no intelligence involved. Beavers have been observed looking after man-made stone dams for an entire summer, as if they thought repairs were needed. Furthermore, although they live in colonies, beavers work individually. When they cut trees down, some of which are very large, the beavers are forced to leave large parts behind, because the trees have fallen in such a way that they cannot be removed.

Also relevant here are the devices used by a number of animals. Monkeys, and especially chimpanzees use tools in a way most similar to man. These animals have prehensile hands and are thus able to throw stones and tree-branches which they have broken off, either in order to defend themselves or to get fruit down from trees. In captivity, they use a stick to obtain an object which is out of reach. Does this not represent a kind of training?

At first sight, the imitation of nature appears more logical than the idea of gifts from gods. In fact, however, it seems that at no time in history has nature inspired man in this way. When methods of observation were gradually developed, some curious minds were there to find

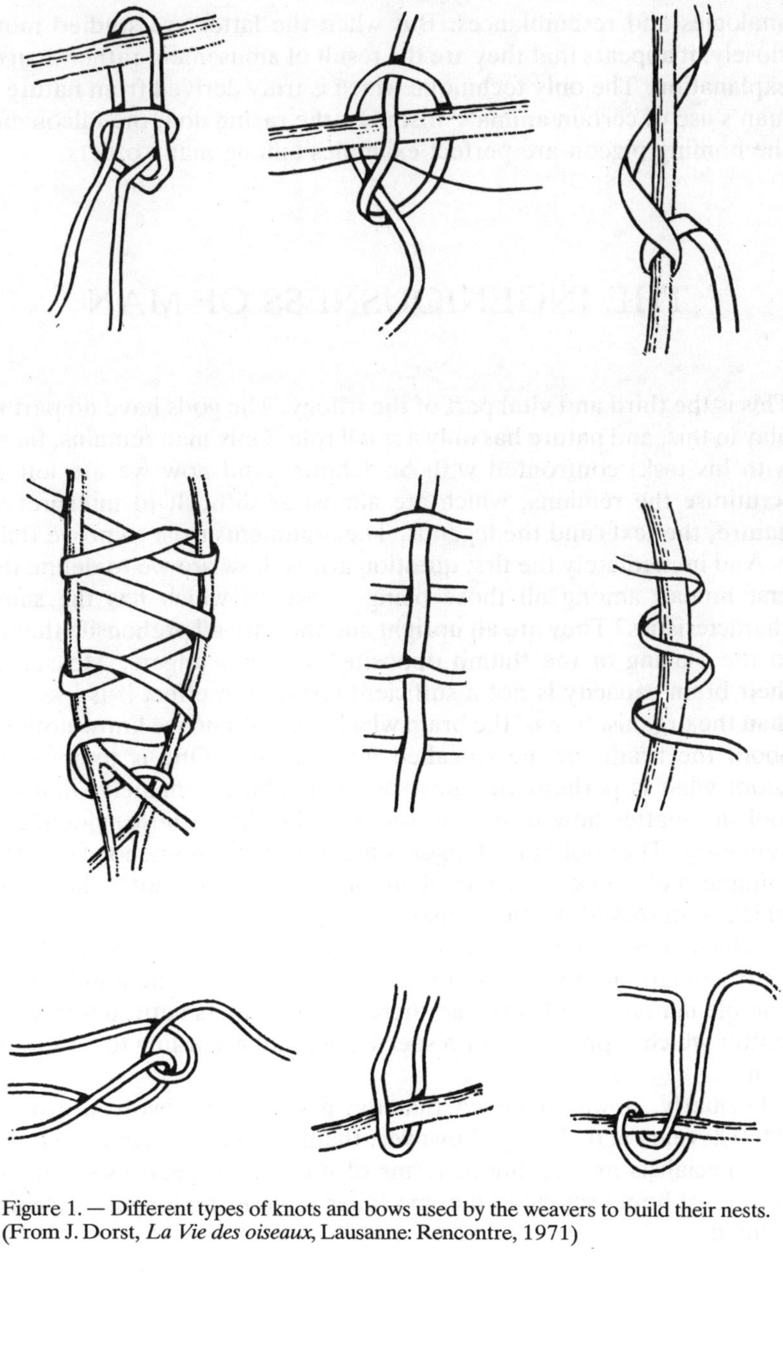

Figure 1. — Different types of knots and bows used by the weavers to build their nests. (From J. Dorst, *La Vie des oiseaux*, Lausanne: Rencontre, 1971)

analogies and resemblances. But when the latter are studied more closely, it appears that they are the result of amusement rather than of explanation. The only technique which is truly derived from nature is man's use of certain animals' instincts: the racing dog, the falcon and the homing pigeon are perfect examples among many others.

THE INGENIOUSNESS OF MAN

This is the third and vital part of the trilogy. The gods have no part to play in this, and nature has only a small role. Only man remains, faced with his task, confronted with his labour. And now we are left to scrutinise the remains, which are almost as difficult to interpret as nature, the texts and the legends. The arguments exist to prove this.

And immediately the first question arises: how are we to define the first human among all those beings, each of which has the same characteristics? They are all upright and they are all prehensile thanks to the placing of the thumb opposite the other fingers. Measuring their brain capacity is not a sufficient test and in effect it is less that than the organisation of the brain which counts, and we know nothing about the brains of the so-called first humans. One is tempted to adopt what is perhaps the most rational solution, namely, that the tool, no matter how primitive, seems to be the earliest evidence of humanity. The tool is no longer a hereditary phenomenon, like the chimpanzee's stick which is abandoned after use, but a heritage, which is preserved. Both the materials, and the knowledge of the way in which it is made, are handed down to the next generation. It represents in fact the birth of a culture. The tool is not simply some suitable matter found here and there and used in its natural form, but matter which is prepared for a specific use and according to a rational form.

Formerly, it was believed that the first humans evolved around 500,000 years ago. Today, however, thanks to fairly recent discoveries, specialists are talking in terms of millions of years, so that the notions of homo sapien and homo faber are now no longer so clearly defined.

Lucy, or the First Outlines

Certainly, "human" remains do exist. The borders of Kenya and Abyssinia are valuable areas for research. From Olduvai in Tanzania to the river Omo in Ethiopia which supplies Lake Rudolph, an impressive number of sites have been providing us for the last twenty-five years or so with evidence which has completely changed the human genealogical table. This area seems to have been the cradle of the human race: the bones of hominidae have been found, and some of the world's oldest fossils providing evidence of abundant wildlife, and sometimes even of primitive forms of industry. These remains come from the entire Pleistocene period, which ranges according to interpretation from 2.6 to 1.5 millions of years. It was in the Omo valley that the earliest, almost complete skeleton was found, a young girl baptised by prehistorians as Lucy.

East Africa thus appears to have been the cradle from which the human species gradually emerged as a race distinct from that of the common "trunk" shared by primates and hominids. A tooth, discovered at N'Gorora, near Lake Baringo and thought to be 11.5 million years old, heralds quite clearly the hominid line. The discoveries made at Zothagam, which are about 5.5 to 5 million years old, help to mark out this evolution. Then, at the end of 1972, discoveries made near Lake Rudolph revealed that hominids and Australopithecines were living alongside one another 2.6 million years ago. The hominid's brain had an interior capacity of 880 cubic centimetres compared to that of 500 for a robust Australopithecine, 620 for a gracile Australopithecine and 750 for a Pithecanthrope a million years ago.

The numerous sites excavated in this region have provided us with some important information. It is now thought that there was a "clumsy" Australopithecine, which was herbivorous. It was unable to walk on its hind legs and no doubt moved around with the help of its finger joints (in the same way as gorillas and orangutans). The other type of hominid, much more gracile, was omniverous and bipedal, and thus more highly evolved.

The oldest implements known today were found on several sites in the Omo valley and on the eastern banks of Lake Rudolph. From about 2.5 million years ago, hominids were using shattered stones, sharpened bones and naturally pointed teeth (mainly hippopotamus' and pigs' canines). By the middle of 1972, the remains of 26 individuals and more than 300 tools had been excavated.

Until recently, prehistorians were agreed that the first implements were pebble tools or choppers as they are sometimes known. One or two fragments taken from a stone smaller than a fist was sufficient to give the latter a cutting edge which was fairly rough but sufficient nonetheless to be suitable for cutting or scraping. The double-faced flint was derived from this extremely primitive tool, and was characteristic of numerous paleolithic industries, with its precise cutting edge giving a triangular shape. Such was the classical theory, almost universally accepted until 1972 when all this was suddenly called into question (fig. 2).

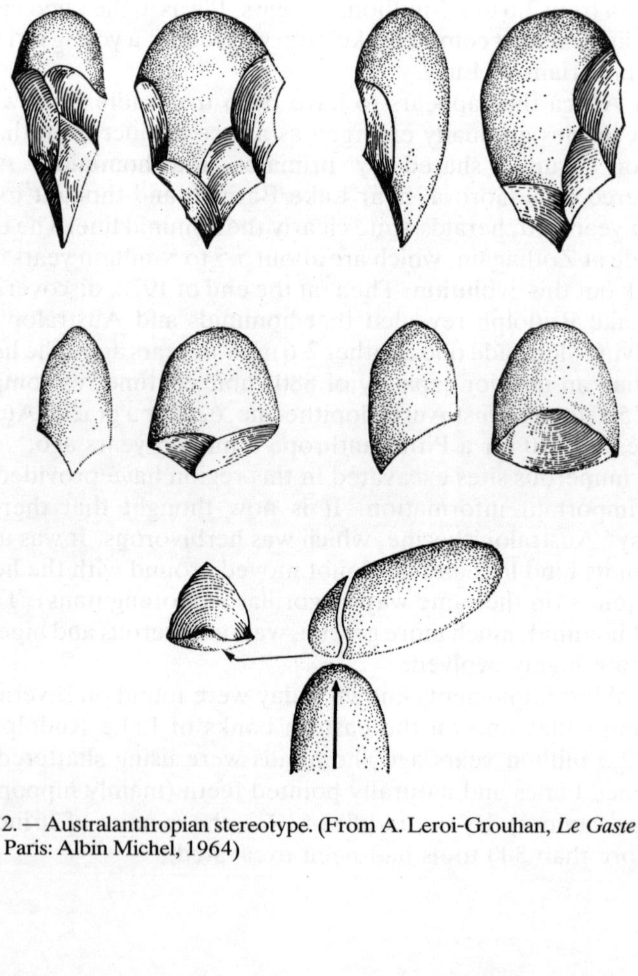

Figure 2. — Australanthropian stereotype. (From A. Leroi-Grouhan, *Le Gaste et la parole*, Paris: Albin Michel, 1964)

Many of these primitive tools were not actually found in situ, but rolled about and carried along by streams and rivers they had been broken up by the process of erosion. It is difficult to date such fragments, but a quartz pebble tool, discovered in 1969 by the French expedition to the Omo valley, has been dated by the gangue of concretions which covered it as being some 2.2 to 2.1 million years old: until new finds are made, this would appear to be in all likelihood the oldest known tool in existence. In the absence of tools made of any other materials, some authorities have even suggested that this must have been the only tool in use, that is to say, the universal tool. Flakes have also been found in place, in layers of gravel over 2 million years old, with some rare pebble tools and teeth belonging to Australopithecine man. Nicks and furrows on the stones suggest that the larger flakes were used as cutting edges. One must therefore re-evaluate everything that has been put forward. Such pebble tools could in fact be the remains of the raw material used to obtain the flakes, rather than tools themselves. Thus the history of the most primitive tool has been completely turned about, but is this not a simple hypothesis of work?

In 1959, Mary Leakey discovered pebble tools linked with the very first hominids in the Olduvai Gorge, thus dating them between 1.8 million and 800,000 years old in this region. Before this discovery only a few pebble tools in Morocco and South Africa, thought to be about a million years old, had been found. The pebble tools "made" in the Omo Valley date back as far as 2.2 to 2.1 million years, and recently discovered flakes go back nearly 3 million years in time.

A provisionary genealogy was thus possible, and we have schematized this below.

clumsy Australopithecine	Homo habilis gracile Australopithecine	Pithecanthropus
extinct	Homo sapiens	extinct

To quote André Leroi-Gourhan, writing before the latest discoveries in the Omo valley:

The African pebble industry really corresponds to what one may imagine as the first distinct shape of the rough stone. Identification of these first products of human industry is not easy, and has been the preoccupation of prehistorians from the beginning of the

60's of the last century onwards. Although it is easy to recognize tools from the point at which secondary adjustments gave them a definite shape, it is more difficult to do so for cut stones which may only be untreated fragments.

Let us continue to look at this quotation, as it presents precisely the problem of conscious human intervention. "The blow required to determine the shape of the flake tool must be applied with a direction and force which presupposes conscious intervention, but of the thousands of blows produced on pebbles by undertow or water falls, chance determines a certain number of flakes which appear to have been retouched by humans." Apart from the possibility of natural causes, the tool is thus a real product of the body and the brain of the Anthropiens. "It is logical in this case to give to an artificial instrument the norms of natural instruments: it must correspond with these constant forms, and to a genuine stereotype." And for the prehistorians, the pebble tool corresponds to a stereotype as attested by millions of objects.

Obviously two pebbles were necessary, as one was used for striking the other.

(the blow is) delivered to one of the sides, perpendicular to the surface, and removes a flake from the pebble leaving a sharp edge; two or three such blows would make a longer and more sinuous edge. Delivered to one side, the operation gave birth to the "chopper," and applied to two sides it became a "chopping tool." One must state that the operation implied only one type of action, that being the most simple: hitting the side of the pebble at an angle of 90 degrees. No further identification other than that of *one action* producing *one* cutting edge is possible.

To make his tool the Australopithecine employed the simplest action used when he wished to break up bones (and there are many examples of broken bones), crack nuts or kill animals. This technical nature corresponds to the little that we know about his brain. It is human and "appears coherent with the organism of the being which it completed. It implies a genuine state of technical awareness."

It is true that archaeology is still only at the start of its investigation. "It would be difficult to go further back than the Australopithecine in the study of the origins of the tool. Moreover I regret this deficiency," admits Leroi-Gourhan, "as the Australopithecine is certainly not the initiator of manual operations. Obviously we have not yet found Adam and Eve, and we cannot say if we are even close. It is only recently by going further still, about 600 kilometres north of Addis Ababa, that other discoveries have upset hypotheses already made. And here again we see that the Australopithecines and man lived together.

Was this first mankind limited to the zones of East Africa which have been defined above? It appears that some remains, previously ignored in other countries, now attract attention. At Chilhac, not far from Le Puy, in France, three or four pebble tools have been found in torrential deposits combined with lakeside sediment and volcanic material. Research undertaken to date them has suggested that they are 1.8 million years old. Around 1 million years ago, hominids lived on the Côte d'Azur as they did on the banks of the Somme river, in the north of France, and pebble tools have also been found in Spain, Italy, Germany, and Yugoslavia.

Subsequently, this wealth of human remains and the "tools" which accompany them have enabled the prehistorians to date the first man as far back as 3 million years. Life was difficult for our ancestors living in the Vallonnet cave, near Roquebrune-Cap-d'Ail, on the French Mediterranean coast. The cold transformed the poplar forests into grasslands, or even into steppes. The cave was not so much a dwelling as a lair. Food consisted of land animals, such as rhinoceros, deer or cattle, most of them old, and even whales and seals, whose vertebrae and jaws have been discovered. They were mostly old animals which were easier to hunt, and half-rotting beached whales which were easier to cut up with the pebble tools. Naturally, all this did not constitute any great advance.

Stages of Progress

Progress was most certainly slow in the beginning. An interesting exerise is to scale down the progress of techniques to the duration of a single year. By this time-scale the first tool would have appeared on the Ist January, fire arrived between the Ist and the 16th October, the first funeral rites on 22nd December, the birth of art on the 28th December, agriculture and the Neolithic revolution on the 30th December at 5 pm, the steam driven engine on the 31st December at 11.20 pm, and nuclear energy at 11.54 and 35 seconds. Thus we can see the enormous temporal distance which separates the discovery of the first technique from the Neolithic revolution of the next chapter.

All too frequently, at least until a few years ago, the work of prehistorians lacked any systematization. Everything was diluted in terms which were difficult to relate to each other. And this problem was compounded by discrepancies in studies of various regions. The most precise effort of clarification to date is to be found in the various works of Leroi-Gourhan. So the reader will not be surprised if we follow him step by step.

Classifications In order to trace the stages of progress in the domain of the tools cut from stone, one must first establish as strict a classification as possible, which is exactly what the pioneers of natural sciences have tried to do. In the subject which concerns us here, it is not sufficient just to classify, or, in other words, identify clearly defined characteristics, but also to deduce the actions involved, both in the actual manufacture of the stone tool, and if possible its use. Once this list has been made, the researcher has a solid foundation on which to base his definitions of the technical civilization which he is studying.

The problem should be considered in terms of progress and systems. It will be difficult, as far as this period is concerned, to distinguish saturations and imbalances, given that only a limited number of tools are available, mainly of stone – wood and bone tools having disappeared – and partly because there are few remaining examples of the objects made, which are necessary counterparts to the tools. Having said all this, it is nevertheless essential to provide descriptions and classifications, and even more important to possess an overall view of the morphology of these tools. Once we have acquired the necessary vocabulary, and we cannot overemphasize the importance of this problem, we may then determine the typology of a period, place it upstream or downstream in the global technical concatenation and also to relate it to the technical civilizations which exist side by side. Most interesting comparisons may be made between technical civilizations which are very distant from each other: the overall time-lags, concomitants and the purely general laws of technical progress of each category should be noted, as well as the the relevance of the grouping within types.

And immediately, essential differences may be observed. In a word, the earliest form of tool was an instrument for cutting, albeit a very basic one. This may have been a pebble already formed by nature which needed only to be adapted to perform a certain type of work, or it could have been a rough, unworked block prior to shaping used by man for the appropriate cutting job, known as the core, and which had to be given the shape and the proportions suitable for certain types of cutting. In the first instance, the raw material, or the original pebble, was prepared for use, and in the second, the raw material itself was first worked in order to provide, technically speaking, usable flakes.

It is obvious that the nature of the materials used was an important consideration in the methods employed for the manufacture of tools. But whatever the case, the method of preparation determined the exact shape of the quarried product. In other words, man knew the type of tool he required and from the start began to take the steps necessary to obtain it.

The shaping of the core-produced flakes may be explained in various ways. One may regard them as mistakes or scraps, and sites reveal many hundreds of flakes discarded in this way. Similarly, these flakes could be used to make new tools. This question has already arisen elsewhere with reference to the first human tools. These middens, or large amounts of discards, indicate either abandonment or the beginning of a new industry, depending upon particular cultures and local resources. This leads to the problem of assessing how it was that these two ideas could eventually be combined.

One quickly arrives at the question of modules for cutting, whose form was strictly dictated by a given material either because of its resistance or because of its size (fig. 3). It is quite likely that men, at different stages of their civilizations had always been aware of reasons for saving raw materials. And here we find one of man's characteristics: we might even say, up to a point, that in primitive civilizations, economy, such as the use of flakes produced by cutting, marks progress.

Apart from cutting, the fashioning or flaking of the stone tool should also be examined. Flaking was performed by the removal of very small parts from the core, a process known as retouching, work which demanded delicacy and precision. There are many types of retouching, unifacial or bifacial, or sucessive retouching by a sort of slow approach to the definitive form, which suggests the extent of the retouches made (fig. 4).

Thus the stages in the production of stone tools may be defined. It will always be difficult, if not quite impossible, to explain the development of the methods of producing these tools. And here the question arises as to how a tool is born. Of course a tool never comes into being by itself but was designed to fulfil an inevitably pre-existing need. Thus one must start with an existing tool and transform it in order to obtain a suitable one; it might even in a sense be necessary to invent one. It is essential to study, for example, the progress from the stone tool to the metal tool, which presupposes totally different techniques. Some authorities have noted that a copy of a stone tool made in metal and vice versa works equally well. We will return to this later.

The tables reproduced here, which are the work of Leroi-Gourhan, aptly illustrate the evolution of techniques (fig. 5). The first concerns the subject discussed above. The length of the cutting-edge obtained in relationship to a given weight of material used provides important data. Clearly the use of waste flakes, once they were put to use, considerably increased the proportions of the length of the cutting-edge, right up to the fashioning of microliths. The curve expresses an apparently characteristic movement. After a relatively slow take-off period, indicated by the flatness of the line at the level of the biface (a phenomenon of saturation), the use of the waste flakes permits a redress, in the form of a parabola until the appearance of the microliths. At this point there should be a new flattening of the curve: the knapped stone industry is in its turn saturated and apart from a few rare improvements, it cannot be developed any further.

The second diagram is equally revealing (fig.6). The number of types increases more rapidly than the length of the cutting edge obtained from a given mass of raw material. The two curves increase less rapidly at the beginning than the curve illustrating the growth of cranial capacity. The latter slows down at the same time as the two other curves rise. At this period, cranial capacity reaches its peak, although this is not the only point in question, as the texture of the brain is certainly far more important.

Apart from techniques involved in manufacture, the other important element is constituted by tool types, the two being indissolubly linked. Here, morphology and typology merge. One may note that in general the number of tool types does not cease to rise, and that tool systems are developed, revealing by that very fact the same chronological stages. On the other hand we witness some disappearances which may in fact be no more than transformations: the "chopper" thus becomes the biface. But tools rapidly become integrated, the transformed tool, intended for some other use, does not eliminate the earlier tool which had its own use. We should finally

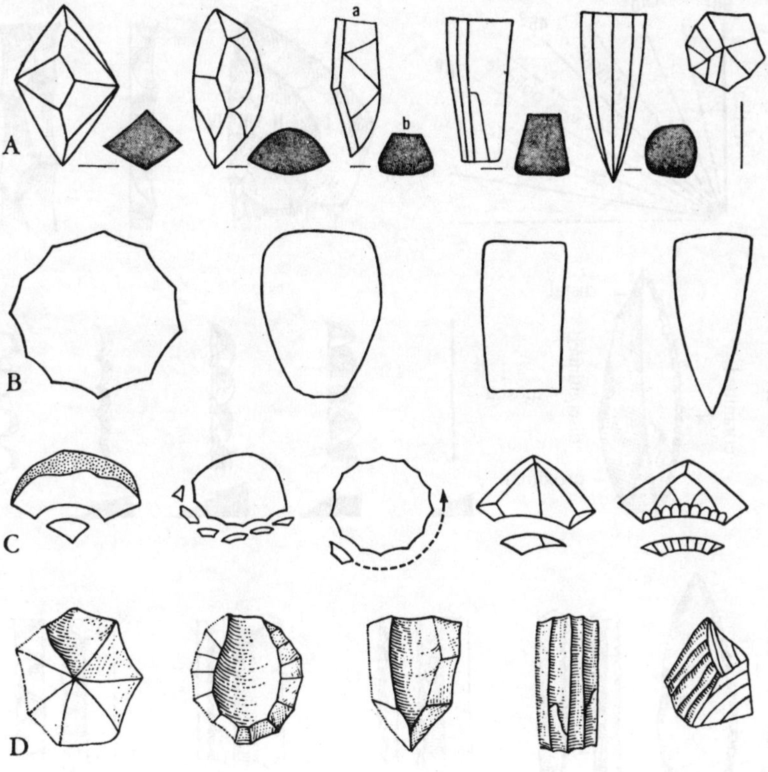

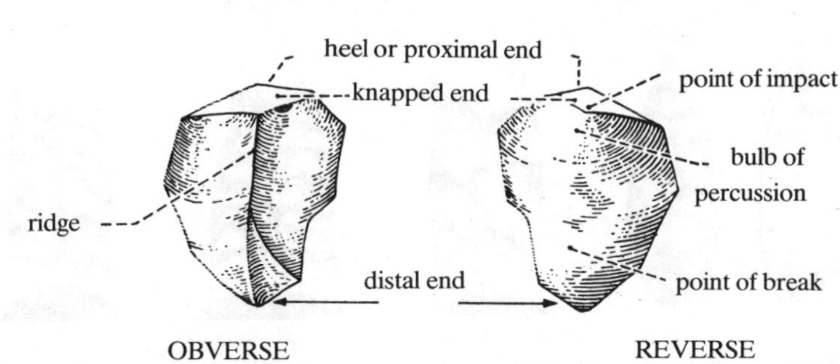

Figure 3. — The nucleus and the way in which it is cut. (From A. Leroi-Gourhan, *La Préhistoire*, Paris: P.U.F., 1968)

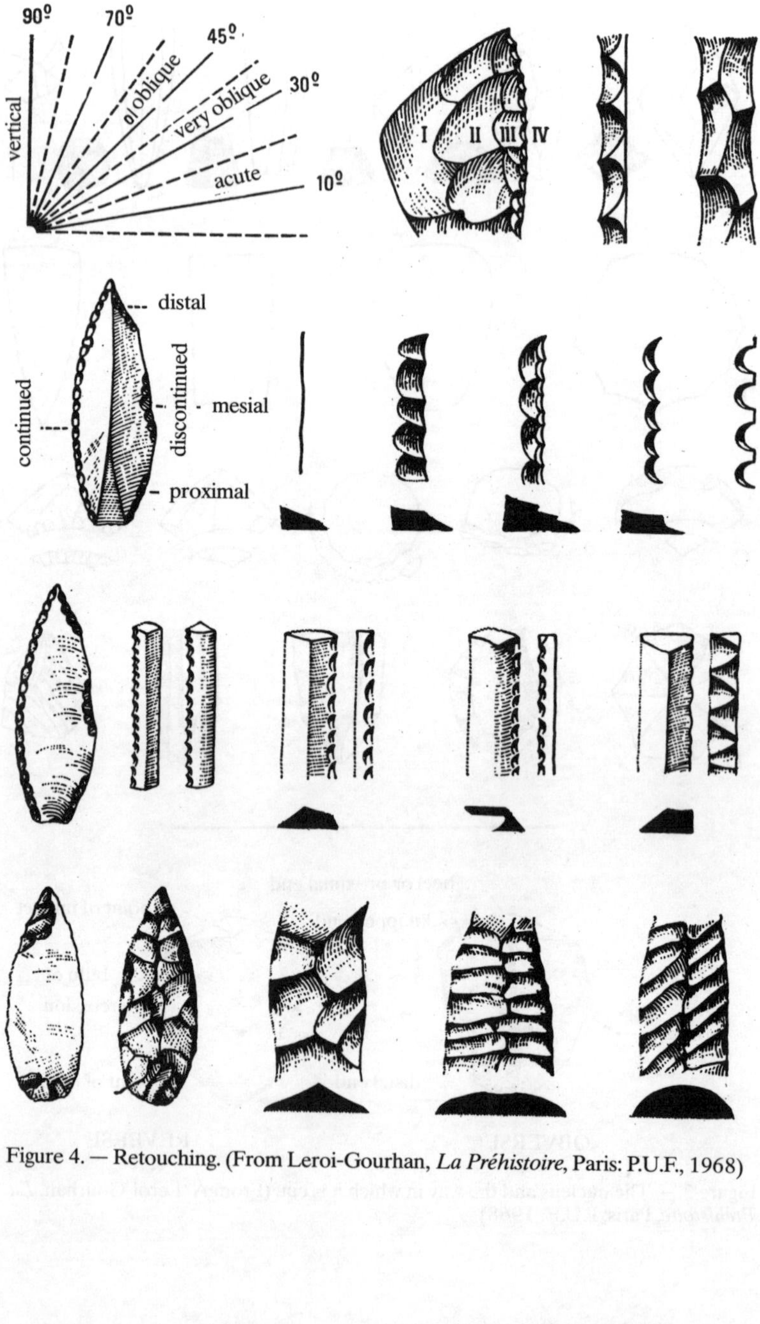

Figure 4. — Retouching. (From Leroi-Gourhan, *La Préhistoire*, Paris: P.U.F., 1968)

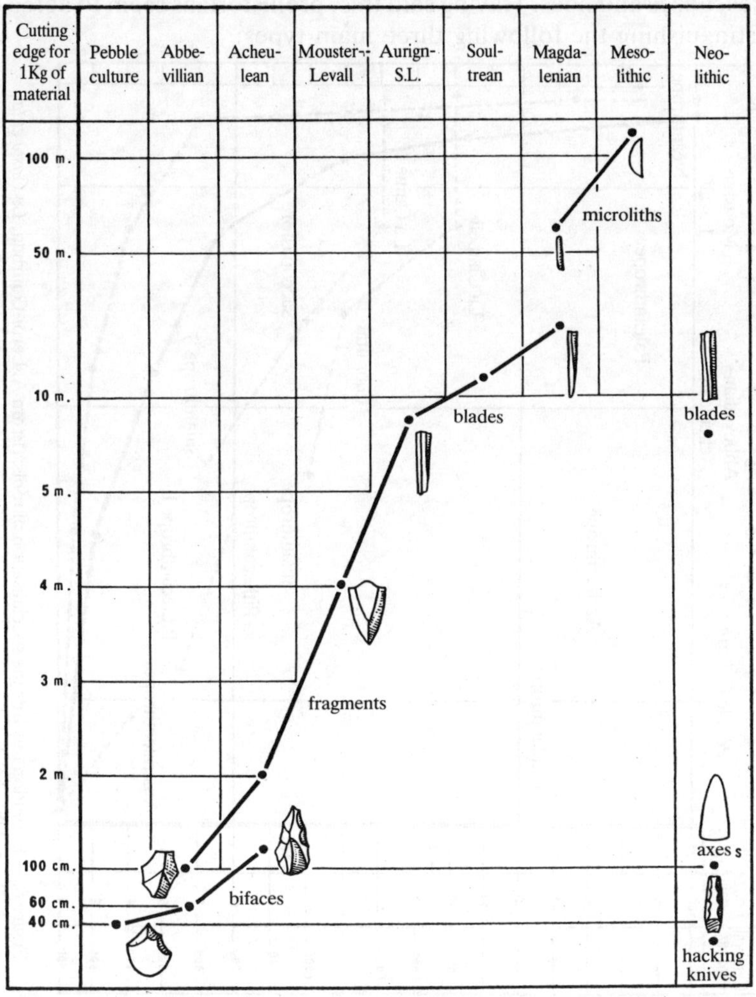

Figure 5.— Relationship between the length of the cutting edge obtained and the weight of the material used. (From A. Leroi-Gourhan, *Le Gaste et la parole*, Paris: Albin Michel, 1964)

point out that the genealogy of tools is still very vague. The "chopper" is similar to the biface, while axes and knives are closely related to axes, adzes and hoes. Having said this, prehistorians seem to agree on distinguishing the following three main types:

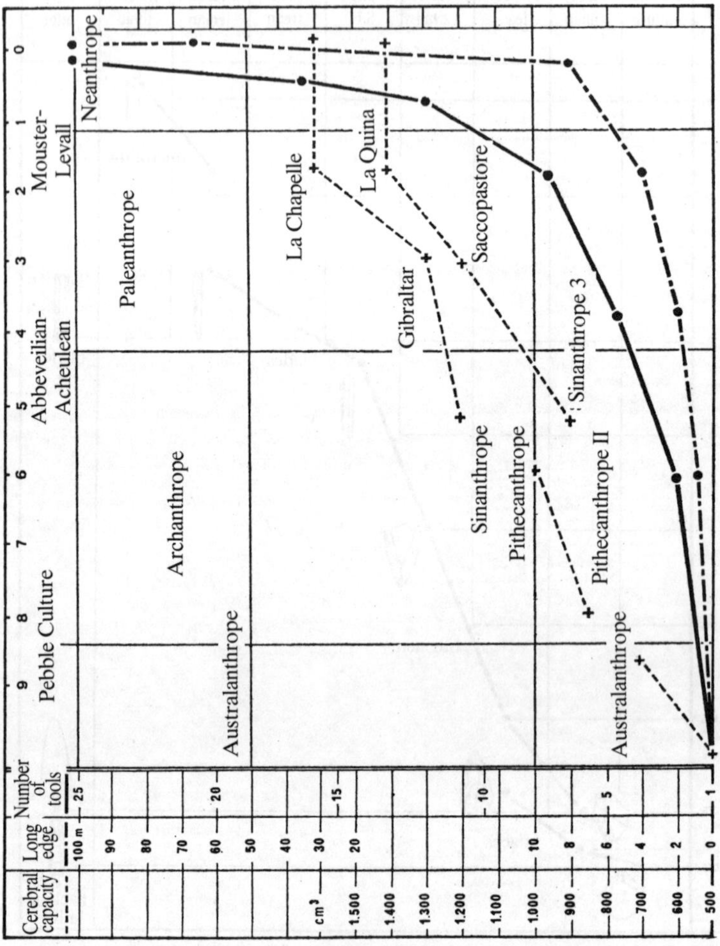

Figure 6. — Evolution towards the specialisation of tools. (From A. Leroi-Gourhan, *Le Gaste et la parole*, Paris: Albin Michel, 1964)

1) tools with distal cutting edges, from the "chopper," thus the most primitive tool, with unifacial then bifacial retouching ("chopping tool"), to the axe, the bevel of which forms the cutting edge and exists prior to flaking, and finally the knife which also possessed a pre-existing bevel which is either retouched or fashioned;

2) axes, adzes, and hoes: objects of polished stone, of animal or vegetable material, comprising important morphological distinctions (fig. 7);

3) chisels, tools with straight cutting edges obtained by the removal of a blade known as "chisel cut." Here again there is a considerable number of morphological varieties (fig. 8);

4) bifaces, scrapers and points "together with the "chopper" constitute one single family, characterized by a generally oval form, the sides of which are most often dissymmetrical in outline, or by the nature of the retouching." The variability of these tools is significant (fig. 9);

5) objects with worn and jagged edges, geometric tools: "a large morphological family is formed by blades or bladelets, or more rarely flake tools, on which sheer or vertical, direct or inverse retouching ensures the creation of varied dissymmetrical forms following their outline or the nature of the alteration." Geometric implements are most often represented by microliths;

6) foliated pieces, either bi- or unifacial, made from stone or bone material, are almost limitless in the variety of forms;

7) scrapers in particular have been the object of studies: their purpose is to grate bone or wood material. "Technical determinism thus plays a large part in the form and function of the cutting-edge." Scrapers are classified into three main large categories, according to their use: those for the removal of thin material, those for the removal of whole strips, and "ridged" scrapers (fig. 10);

8) perforating implements and borers: there are innumerable varieties of these. The borers either had a straight or slanting axis, or were angled. The notches were either on an oblique or double axial end (fig. 11);

9) points made of bone material display pronounced similarities in every civilization. Although they were often used to serve as spear or arrow-heads, they could also have been used for other purposes which are difficult to determine.

Thus we possess a long list which indicates relations, families and varieties. Only by studying such instruments designed for work may we define the civilizations involved, and then draw up chronologies. It would, however, be going too far to base any system of techniques upon descriptions of groups of tools alone. Two considerations must

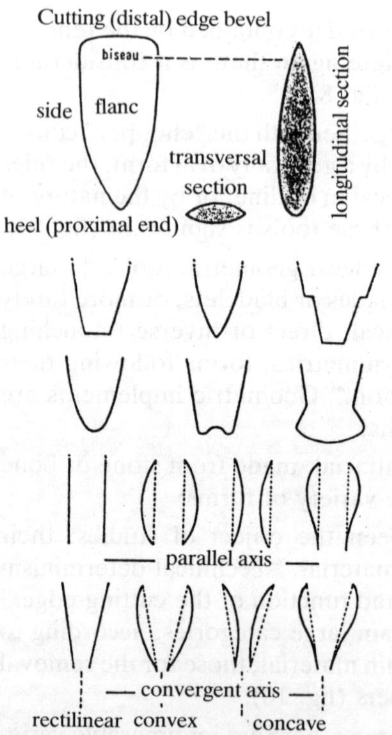

Figure 7. — Axes, Adzes and Hoes. (From A. Leroi-Gourhan, *La Préhistoire*, Paris: P.U.F., 1968)

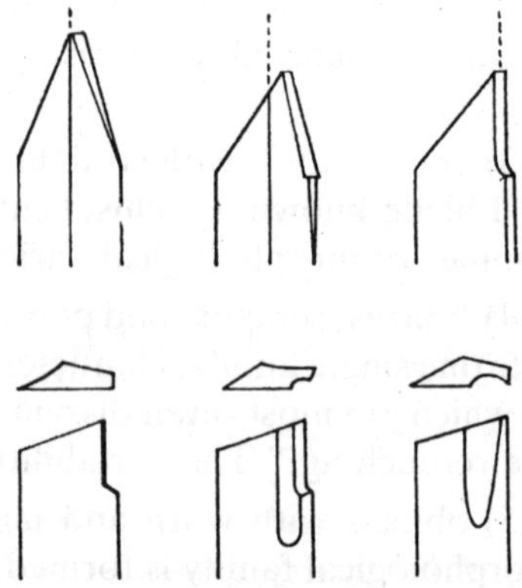

Figure 8. — Chisels. (From A. Leroi-Gourhan, *La Préhistoire*, Paris: P.U.F., 1968)

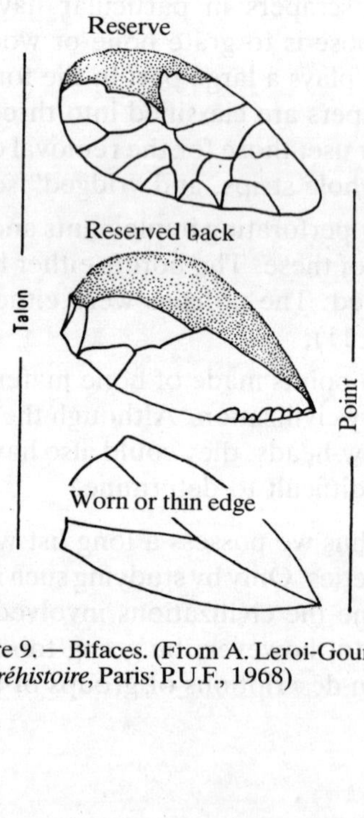

Figure 9. — Bifaces. (From A. Leroi-Gourhan, *La Préhistoire*, Paris: P.U.F., 1968)

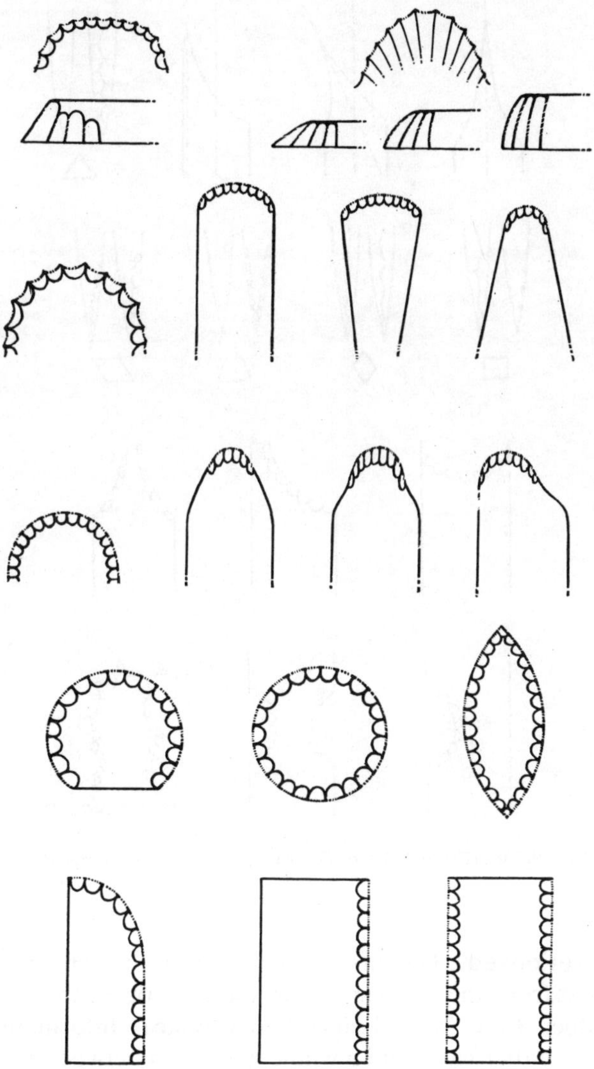

Figure 10. — Scrapers. (From A. Leroi-Gourhan, *La Préhistoire*, Paris: P.U.F., 1968)

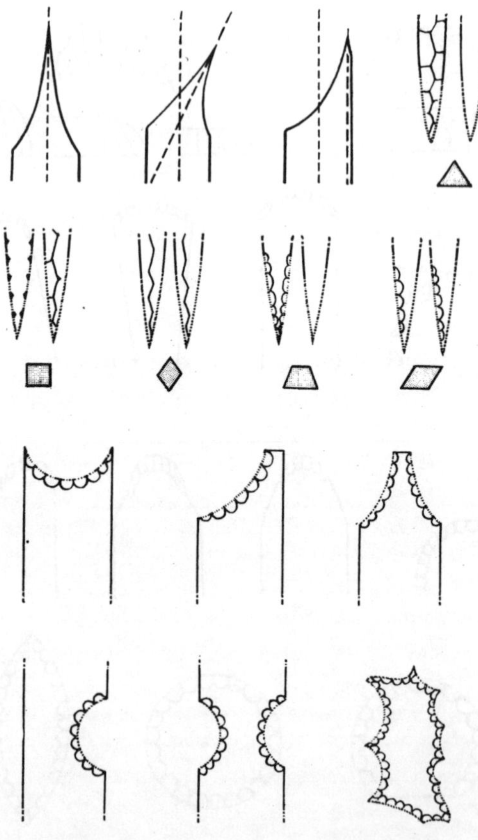

Figure 11. — Points. (From A. Leroi-Gourhan, *La Préhistoire*, Paris: P.U.F., 1968)

not be overlooked. The first concerns tools, which are naturally inseparable from the objects, or, in the case of weapons, from the effects which they were designed to produce. Information is not sufficient in either case, and a whole gap exists between the tool and the object which is extremely difficult to fill. Although, to a certain degree, an abundance of tools suggests an abundance of objects, this hypothesis is not immediately valid, for civilizations which are very rich in objects and relatively poor in tools are on record, and the reverse is equally true. This question has certainly not been neglected, but a general absence of objects produced does not permit broad conclusions to be made. Insofar as a system of tools and a system of objects are indeed closely interdependent, we lack here an essential explanation.

The second consideration is no less important. Tools, or systems of tools, only represent one aspect of techniques but there are many others of which we are almost totally ignorant, such as fire, which should easily be detected in prehistoric deposits. According to our present knowledge, apparently the oldest trace of fire in the world is that found in the Escale cave, at Saint-Estève-Janson, near Aix-en-Provence, France: this has been dated back some 700,000 or 600,000 years. More recent are the fires of Terra-Amata (near Nice, France), Verstèszöllös (Hungary) and those of Chou Kou-Tien (China). And we could cite many other techniques here: clothing, gathering, habitation, hunting or fishing, and trapping... We often imagine them by a far-from-convincing analogy with the technical practices of so-called primitive peoples. All that remains today is the stone or bone tool and, once again, we are not really sure how or for what purpose it was used.

Stages of development We will continue to plunder shamelessly Leroi-Gourhan's exemplary works. Indeed his books provide a global view of the evolution of techniques, which pleases and reassures the mind in a field in which the accumulation of details often leads to confusion. He distinguishes several chronological stages which may, moreover, be placed in the context of temporal space at levels often removed from each other.

Archanthropian From the very beginning of the 20th century, the hominids who are known today as Archanthropines, were hailed as being the first of the human race. Now it seems that they are merely one link in an extensive chain of history which goes back long before their appearance. "In their time, the Archanthropines already had very distant human predecessors."

The Pithecanthropines and Sinanthropines are virtually the same. The Sinanthropines knew how to make fire, the Atlanthropines had already developed tools, and the Pithecanthropines had a femur similar to ours. Their face is enormous and their cerebral capacity is about two thirds the size of modern man. Their forehead is like that of the Australanthropines, with a very sizeable orbital circle. They evolved during the ancient Quaternary period. There was, however, some progress, as the cerebral capacity of the Archanthropines is almost double that of their predecessors.

All of these stages in man's development come in what is usually known as the Lower Paleolithic period, the best known of which is the Acheulean culture. Let us locate them geographically: Pithecanthropines were discovered in Java, Sinanthropines in China, Atlanthropines in North Africa, and Australanthropines in East Africa.

Only the Sinanthropines have been found in their habitat, and this area has revealed millions of tools worked in a material which was hardly suitable for the functions they were designed to perform. The Atlanthropines display an industry which may be attributed to a very primitive Acheulean tradition. From their bifaces and axes one can estimate that the others had formed a similar technical civilization.

Using the tools found in the area inhabited by the Atlanthropines, it is possible to reconstruct the features of the industrial stereotype of these populations which are distributed over a large part of the earth (fig.12).

Primitive methods remained in use, in particular the method of flaking by perpendicular percussion. Bifaces tended to replace "choppers," but a second series of actions seems to have been introduced "by which the core of the stone is struck tangentially, rather than perpendicularly to its axis, producing flakes which are much longer, sharper, and already very similar to the flakes used by the Paleanthropines." Although the type of equipment remained very limited, flakes were used directly and "core tools" continued to be fashioned, so we must therefore emphasize the number of additional new actions. But as Leroi-Gourhan notes, "this acquisition corresponds to something more than just a simple addition, as it already implies on the individual level, a higher rate of planning in the progress of technical operations." This differs fundamentally from the process used by the Australanthropine. The latter conceived his tool well and chose the pebble accordingly: the possibilities were enormous and the imagination of the creator of the tool very limited. In the case of the Archanthropine, the processes were virtually the opposite. He possessed the raw material and had to choose, if for example he wished to make an axe, the point at which a large fragment would break off, the cutting edge of which would constitute the working edge of a future tool, as well as

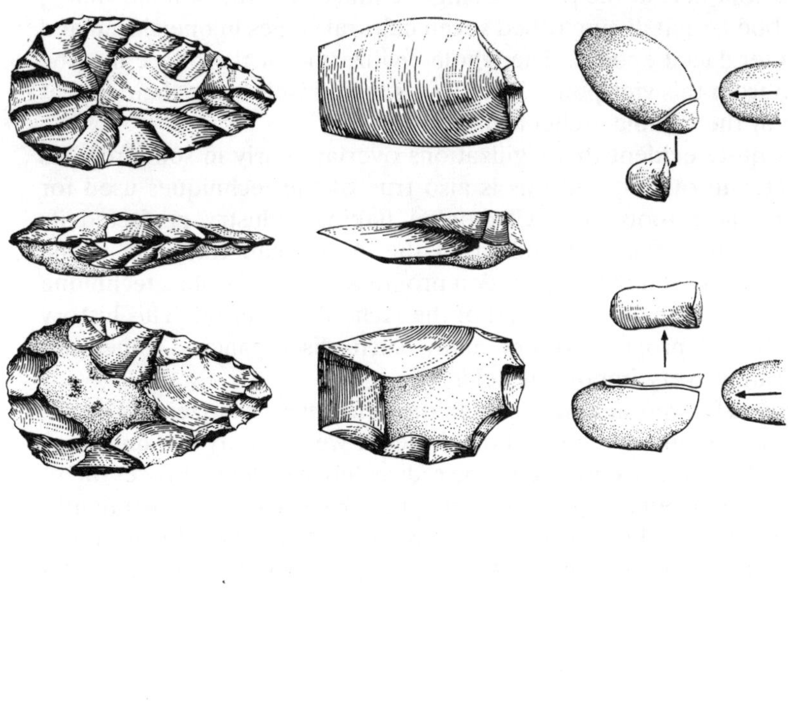

Figure 12. — Archanthropian stereotype. (From A. Leroi-Gourhan, *Le Geste et la parole*, Paris: Albin Michel, 1964)

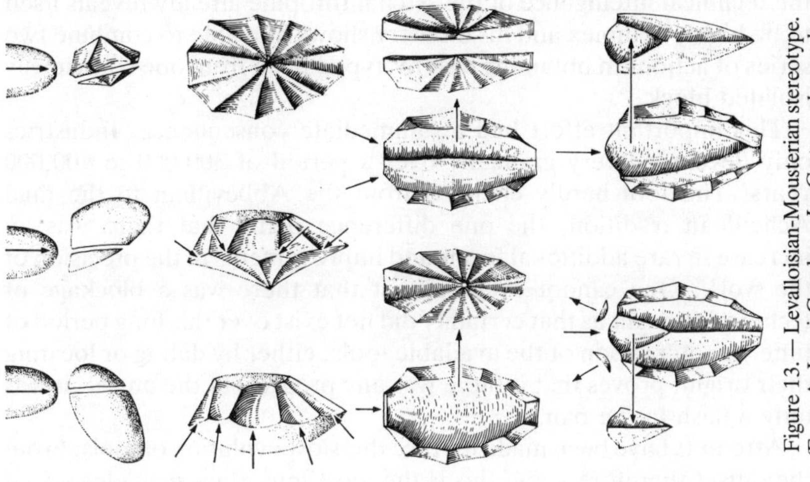

Figure 13. — Levalloisian-Mousterian stereotype. (From A. Leroi-Gourhan, *Le Geste et la parole*, Paris: Albin Michel, 1964)

deciding upon any secondary improvements to be made. A similar but rather less deliberate method was used for making the biface. "Thus the technical intelligence of the Archanthropine already reveals itself to be highly complex and his industry shows him able to combine two series of actions in obtaining a stereotyped form from one consciously isolated block."

This important effort had no immediate consequence. Industries only developed very gradually over a period of 300,000 to 400,000 years. The tool hardly changed from the Abbevillian to the final Acheulean tradition, the one difference being that there was an increase in rare additional forms and improvements in the precision of the work. One cannot even suggest that there was a blockage of technical thought as that certainly did not exist over this long period of time. A comparison of the available tools, either by dating or locating their origin, proves that if there was any progress at the outset, it was only a flash in the pan.

Attempts have been made to trace this slow evolution of tools. From the outset therefore – and this is the most important new idea – tool parts were trimmed with a hammer and finished with a wooden tool. Some bifaces, however, were entirely fashioned with a stone hammer struck obliquely at the piece. Using the finds from the Somme valley, the Abbé Breuil distinguished seven cultural stages in one glacial and two interglacial epochs. The whole civilization of the biface may be traced from this viewpoint. Thus it was that chisels made their appearance in the Middle Acheulean period.

It is quite evident that civilizations overlap, early in some regions and later in others, and this is also true of the techniques used for making flake tools. The Clactonian flaking industry perfected its methods from the Abbevillian to the Acheulean tradition. Abbé Breuil considered that the sudden progress in the Levallois technique could have occurred at the end of the Acheulean period. This history of technical progress offers many such discrepancies sometimes occurring in very important fields.

Obviously even summary explanations for these changes and developments are out of the question. Indeed, we can only ask questions and our knowledge may need to be redirected in order to answer them. How and why did one pass from the primitive stage (the Australanthropines are still known as Chelleans or Abbevillians) to the more developed techniques of the Acheulean tradition? Can one talk of a revolution, or is it merely a question of a very gradual evolution by slow stages? Works on prehistory are rather misleading in this respect, as they swamp the reader with avalanches of discoveries and evidence, mainly tools, without (apart from exceptional cases) supplying us with diagrams in a chronological framework, albeit hypothetical, which could provide us with the exact direction of the evolution, even though we may never know the reasons to explain it. Consequently, seeing where technical "progress" came and trying to comprehend this phenomenon becomes an almost impossible task.

Paleanthropian This stage represents a period of transition between the Lower Paleolithic and the Middle Paleolithic. It is therefore a relatively vague and imprecise notion, "which is normal if one conceives evolution as a progressive phenomenon." Beings from this period have been discovered virtually all over the world, together with their tools and often in their habitat. Their age corresponds to the second part of the penultimate interglacial period and to the first part of the last glacial period, one which is relatively short in comparison to the preceding one, and would provide evidence of the acceleration of the progress of techniques (fig. 13).

The final culmination in human evolution would be Neanderthal man. Despite the eye socket which remains the same, the cerebral capacity of this man approaches that of our own. "One must admit that apart from a detail, the exiguity of the prefrontal areas, but nonetheless an important one, the brain of Neanderthal man corresponds approximately to ours, in that he possesses the same cell endowment, for the middle cortex area in particular."

We will now witness what might be called the Levallois "revolution." In the words of Leroi-Gourhan:

> During the Middle Paleolithic age, a very important evolution came about in the lithic tool. The Archanthropines of the preceding period for a large part still followed primitive tradition, and their tools, bifaces and hand-axes, were still hewed from a block in the same way that the "chopper" had been by the Australopithecines. From this block came by-products in the form of flakes, the cutting edge of which could or could not be used. During the Acheulean period, the thinning of the edges of the biface by tangential percussion detached many wide and thin fragments from the matrix, which were consequently used as cutting tools. From this development of cutting the biface arose the technique described by prehistorians as the Levallois technique. The block which was originally destined to become one amygdaloid shaped tool thus became the material for predetermined shaped fragments which were to serve as tools themselves. In order to arrive at this result, the core was first worked like a rough biface, then prepared for the extraction of a flake, and sharpened for successive extractions until it was used up. The

preparation might reach the point at which the pin came out of the nucleus with a single blow, either in a triangular point, a subcircular-shaped fragment, or a long, narrow blade.

And in conclusion: "the Levallois technique represents the most advanced method for the making of flint tools created by man." Our scientists are aided by the considerable number of tools at their disposal, and by the discovery of real workshops where evidence of every stage of tool making is available.

It is difficult, if not impossible, to define the period and the place where this new civilization was born. One notes, moreover, that it presents two aspects. The Moustiers (in the Dordogne, France) was still a cave civilization while Levallois was an open-air one. The profound similarities are so numerous that many have suggested that there was a winter and summer civilization. The Levallois encircles the Mediterranean in a ring more than a thousand kilometres wide: in Africa, it extends as far down as Kenya. One assumes that these techniques were established over a period of some 5000 years. Although this might seem rapid in comparison with others, the spread of more developed techniques was still very slow.

The diversification of the tool is now more pronounced than before, but even so one can still detect a certain sameness. The original block, which was previously made into the tool, is henceforth the source from which tools are made. Production was mainly in cutting tools, and apart from the old tools, knives, scrapers, points and notches could be obtained. This is clear evidence of an already advanced technical intelligence. A bone and wood industry certainly also existed, but unfortunately very few relics remain, the rest having disappeared. "For the bone industry," notes Leroi-Gourhan, "it seems that the situation had hardly changed since the Australopithecines." The absence of any tools made out of bone and the remarkable proliferation of flint flake tools on which marks indicate that they were used either to work bone or wood, suggest that working in wood was already a very important occupation.

So far we have been concerned with the tool and ignored everything else, in particular indications of habitat. Without doubt the idea of the winter cave dwelling must be abandoned. But man had always inhabited caves during this period: can we not learn something from their existence? We have not yet reached the period when man began to settle: it is therefore more accurate to speak of provisional camps, all the more destructible precisely because of their temporary nature. There are only a few rare examples, two or three in the West and in the U.S.S.R. Let us look in particular at the deer hunters' camp found at Pincevent (near Montereau in Seine-et-Marne, France). "An open air settlement was found, with more or less circular shaped areas indicating the remains of ancient huts. Domestic organisation did not appear to be very structured. Man lived in a circle, a few metres in diametre, outside of which he threw the debris of comsumption." Here again, there seems to have been little progress of any note from the Archanthropine period.

We are reduced to making hypothetical observations about other techniques. Traces observed on animal bones seem to indicate that they were skinned, so that their fur or skin could provide clothing, or could be used for some sort of bedding. The climates in which the Neanderthal man lived were fairly varied so that it is absolutely impossible to generalize on this subject.

Once more we return to the same question: how and why did this new civilization arise? Imagination alone is not sufficient to explain this. Research has been undertaken in one direction, the only one which might have provided the material for a valid explanation. As has been pointed out above, we are now approaching the last stage in the development of the brain, at least in its capacity, if not in its structure. Cerebral development would therefore have been the factor which led man to seek the greater improvement of techniques. We have still to ascertain why the human brain, whose capacity had barely evolved from that of the Chapelle-au-Saint man, could henceforth achieve such incredible progress in the field of techniques. But by limiting ourselves to our remote period, the logical step would be to search for other traces of intelligence in Neanderthal man which did not exist in his predecessors. Are there any burial places, for example? And has any evidence of graphic signs been preserved?

Replies to these questions are hesitant. Excavations, particularly those made in earlier times, have overturned sites before accurate observations could be made, and the few remaining undisturbed examples are too slight, too fragile to allow any useful conclusions to be made for our purpose. The area in which skeletons have been discovered is not convincing. A few rare cases might suggest that cannibalism was practised: above all there are abandoned bodies, whose limbs were probably dispersed by animals in search of food. The discovery of the Neanderthalian skull at Mont Circé in 1939 is very feeble evidence. It was found lying on the floor of the cave, surrounded by some stones and a pile of animal bones, which seemed to have been placed there intentionally. The lack of tools proves that it was not a habitat which had been lived in for long. Was it really a burial place? The notion of a cult of skulls, bones, bears and foxes appears to be the result of misinterpretation. Consequently there is very little more one can safely say on this subject.

At several sites, indications of ochre have been found in Mousterian strata. To deduce from this that these people practised a form of art, would be pure speculation. Various discussions and interpretations have surrounded the discovery made in 1968, in the Pech-de-Lazé cave, near Sarlat, in the Dordogne, of an animal's bone, on which engravings made by a flint were observed, but it is very difficult to date this bovine rib. These engravings would seem to have been deliberate, as the traces are parallel and the V-shaped drawing are so clear that they cannot just be marks from scraping, resulting for example from cutting off meat. The author of this find, François Bordes, "did not dare suggest the significance of these engravings." He simply recognized that this engraved bone is by far the oldest known to date, the oldest until then having been discovered in Aurignacien or Perigourdin deposits of the late Paleolithic period. An American researcher, Alexander Marshack, on the other hand, has attempted to distinguish the earliest engraved marks and the tools used to incise them. He has even gone as far as to suggest that he had recognized a lunar calendar. So, on the one hand, Mr. Bordes estimates that it is the work of "an idle hunter amusing himself", and on the other, Marshack detects "the mark of a prescriptive stage made with the aim of using and reusing this image in a ritual or ceremonial way." He goes on to conclude that the Acheulean hunters or their immediate predecessors already commanded a complex language. The historian subsequently finds himself in the awkward position of having to choose between these theories.

Neanthropian We enter a very different and considerably more advanced world during the Upper Paleolithic period. "The development of habitation structures, the importance of bone material used to create weapons, tools, or jewellery, the use of colouring materials, and the appearance of the first works of descriptive art" bear witness to the profound evolution of mankind. From now on we are in *Homo sapiens*' age, one of the most famous representatives of which is Cro-Magnon man. This civilization covered the period from 30,000 BC to 8000 BC.

There are some quite considerable changes. Man generally no longer lived in caves, but instead built huts and tents. He made clothes out of skins, skillfully sewn together, and jewellery was introduced. As a furrier, he was a hunter and an assegai user at that, and as a basket maker, he would undoubtedly have used wood and bark.

The "chopper" had been replaced by the biface, a tool kept in use for thousands of years without undergoing any radical development. During the Levalloisian era, the biface itself became the nucleus for the creation of flakes which in turn were worked. "The Mousterians brought about the most important technical revolution, perhaps the greatest in the whole history of mankind, by providing the solution to the flaked core tool with a pre-determined shape." Subsequently, and in the period which concerns us here, "the evolution continued through minor improvements in quarrying devices." They succeeded in making narrower and thinner blades. But the flakes were also used according to their shape: henceforth the tool certainly became more and more specified and, at the same time, less was wasted. It has been possible to calculate that towards 12,000 BC, during the Magdalenian period, man could make several hundred tools out of two or three kilos of broken flint. This trend towards microlithism led to an extremely widespread use of tools, even in areas where no raw material was available.

Figure 14 represents a table, borrowed once again from Leroi-Gourhan, which clearly demonstrates the development of tools, and thus the diversification of techniques (fig.14).

> The first three stages proceed following a single order, by the accumulation of new forms derived from old ones, but without totally abandoning the latter (...) The third stage is already continuing in a new direction (...) There is a complete contrast by the fourth stage. Due to a transition which rapidly developed over the years between 35,000 and 30,000 BC, in Western Europe, one is faced with not only three times as many tool types, but also tools and objects which are reflected in present primitive cultures.

And in conclusion: "thus yet another technical world is in the making, ours." The making of bone objects, both for tools and jewellery, seems like a real achievement, so much so that in some places attempts have been made to imitate the bone points in stone.

Material	First stage	Second stage	Third stage	Fourth stage
	Perpendicular percussion Core-based tool	Perpendicular percussion Tangential percussion Core-based tool	Perpendicular percussion Tangential percussion Fashioned core Fragment-based tool	Perpendicular percussion Tangential percussion Fashioned core Fragment-based tool Blade tool
Stone	Chopper Stone fragment	Chopper Biface Stone fragment Flaky fragment Hatchet (Scraper)	Chopper Biface Stone fragment Flaky fragment Levalloisian flake Hatchet Scraper Levalloisian point Nock (Blade with back) (Chisel) (Scraper)	Flaky fragment Blade and flake Blade with back Blade with nock Foliaceous point Point with notch Geometrical pieces Nock Scraper Chisel Gimlet
Bone Material			(Awl)	Awl Needle Assegai Harpoon Propeller Pierced stick Spatula Half-round twig Wedge Pickaxe
Various	Polyhedrons	Polyhedrons ?	Polyhedrons (Dyes) (Fossils) (Huts) (Tombs)	Dyes Fossils Costume Lamp Huts Tombs Figurative art

Figure 14. — Evolution of tools. (From A. Leroi-Gourhan, *La Préhistoire*, Paris: Albin Michel, 1968)

It may therefore be possible to distinguish types of civilizations, even though certain tools, such as Solutrian points, are widely distributed. The historian can now make distinctions and establish classifications for approximately twenty tools and identify more than two hundred variants. And we can already observe the extent to which the notion of a technical system may enable greater undertanding of the subject. "The development of the bone industry is directly linked to the increase of scrapers and chisels." It is not only a matter of multiplication but, above all, of diversification. The knife and the point appear as tools, and as major acquisitions.

Little more remains to be said on the subject. Today, detailed studies are being made which, with the discovery of new sites, will certainly contribute important elements to an understanding of this evolution of techniques.

These marked stages are undeniably revealed and defined by different technical systems. From the birth of the tool, which probably soon ceased to be unique to the Levallois "revolution," and then to the "revolution" of the Upper Paleolithic age. Of course evolution is not so clear-cut, or so chronologically distinct. At the end of each period, transformations are difficult to detect, preparations are slow, and technical civilizations overlap.

We lack, and will most probably continue to do so, models of explanation. They certainly exist, for instance the capacity of the brain, and the evolution of climates which influence the whole environment. But the brain and the hand, and successive glaciations cannot explain everything. Societies, beliefs, and many other factors must also have played an important part, the significance of which we are almost incapable of assessing.

THE FIRST GREAT TECHNICAL CIVILIZATIONS

Some facts are self explanatory. It took Homo Sapiens some thirty thousand years to develop farming. In three thousand years, man acquired agricultural know-how, animal domestication, pottery and metals, and two thousand years later he developed the written word. In the words of André Leroi-Gourhan:

> Farming was barely consolidated around 6000 BC, pottery appeared to be already very advanced, and in approximately 3500 BC, metal and writing began to dawn; all this must lead one to consider that 2500 years of farming were sufficient for Western society to acquire the technical and economical foundations upon which the whole structure of mankind is based.

The reasons for the sudden acceleration of this evolution are still rather obscure and were, in any case, all suggested by specialists: evolution of the human race, and climatic changes. Firstly, we should note that the cerebral capacity of Neolithic man was no greater than that of the Paleanthropines of La Chapelle-Aux-Saints or of La Quina: we are now dealing with a brain volume of between 1,300 and 1,500 cubic centimetres, the same as modern man. As for the question of the climate, at this moment in time, research is still at an early stage of its investigation.

One may discourse interminably about the "inventions" of this period. Their genesis will always remain difficult to explain. Two examples will serve to demonstrate the limits of our understanding. Throughout the world there was a variety of clays which made possible the development of pottery, and man had known the secret of fire for some time. Despite this, however, it is not until the dawn of the sixth millennium BC that one sees the beginnings of ceramics. The second example is almost the reverse of the first. For any type or any nature of metallury to appear, at least the reduction of a mineral was required. One had to use the necessary apparatus, and an adequate temperature had to be maintained: this might have been how ceramics were produced, but the genesis of the potter's kiln still remains to be explained. Furthermore, one must recognize the mineral types and imagine the melting agents required, and subsequently, the method of working metal, which is fundamentally different from that of stone. Unfortunately we cannot provide a satisfactory explanation for all this. The historian of techniques is condemned to suggesting only chronological landmarks which have been broadly clarified in the last few decades.

Documentary material is thus now abundant, although there are still many areas of obscurity. Only explanations are lacking, and we must assume that they will always be so. Although it is relatively easy, when all is said and done, to find out how man at this epoch proceeded in obtaining well-defined products, it is still difficult to establish the means by which he arrived at these methods. Moreover, the Ancients had clearly understood the problem, as we have seen in the preceding chapter. Attributing various inventions to the gods or heros was not only an easy solution, but also demonstrated the recognition of their ignorance, and at the same time, emphasized their astonishment at having created these processes.

The stages are clearly defined. At the beginning, there was the "Neolithic revolution," a term which should be clarified. This took place between the eighth millennium BC and the beginning of the third, and it was during this period that the first great technical civilizations appeared in Mesopotamia and Egypt. Progress then spread to neighbouring Mycenaean and Hittite regions, and step by step, but visibly slower, to successive concentric areas. Both developed and underdeveloped countries existed at this epoch, as they do today, whose frontiers and chronology are difficult to determine: one must establish the whole difference between Vercingetorix and Cleopatra, between the Germans and the school of Alexandria, without considering the still more uncultivated civilizations of Africa or America, at least in less advanced regions. We lack especially synoptic tables listing precisely the material means which were available to man.

THE NEOLITHIC REVOLUTION

We will once again quote Leroi-Gourhan:

> At the end of the Paleolithic epoch, in societies surrounding the Mediterranean a radical techno-economic conversion took place. Between 8000 and 5000 years before our era, the techno-economic mechanism based upon agriculture and animal husbandry appeared, and societies adopted a completely different form from their original one. On a geological scale, there is only an instant between the last hunter of aurochs and the Mesopotamian scribe, and the accession to new economies is explosive.

This statement encompasses the whole problem.

Childe was the first to talk of a Neolithic "revolution." This archaeologist considers this to be one of the most significant and fundamental transformations experienced by humanity. Indeed, the first advanced technical system seems to have appeared at this epoch. Population settlement, agriculture, animal domestication, and pottery are the major acquisitions of the Neolithic period, and these were undoubtedly far more important developments than stone polishing.

And all this appears to have happened very quickly, in reality from the beginning of the eighth millennium BC to the fourth millennium BC. Between 6500 and 5700 BC, Çatal Hüyük seems already to have possessed an advanced technical civilization: one need only admire the mirror of obsidian which was produced there to be convinced of this fact.

Did these important discoveries, the term "discovery" being more applicable in this instance than invention, arise spontaneously, or were they the result of progressive circulation among localized populations? It will certainly be impossible to say. It is possible, however, in the light of present knowledge, which is rapidly increasing with the course of excavations, to define a region; a long band stretching from Thessaly and Greek Macedonia to Northern Iran, with continuations towards the South, in Syria and Palestine, beginning in the eighth and seventh millennia BC (fig. 1). In Iraq and Palestine, we may note in the ninth and eighth millennia BC the growing importance of harvesting grasses, executed with the aid of skillfully adapted equipment: sickles, grindstones, bowls and pestles. The domestication of sheep and goats is well on the way. Can this really be called true farming, even if silos appeared in Iraq during this period? In any case, specialists now favour

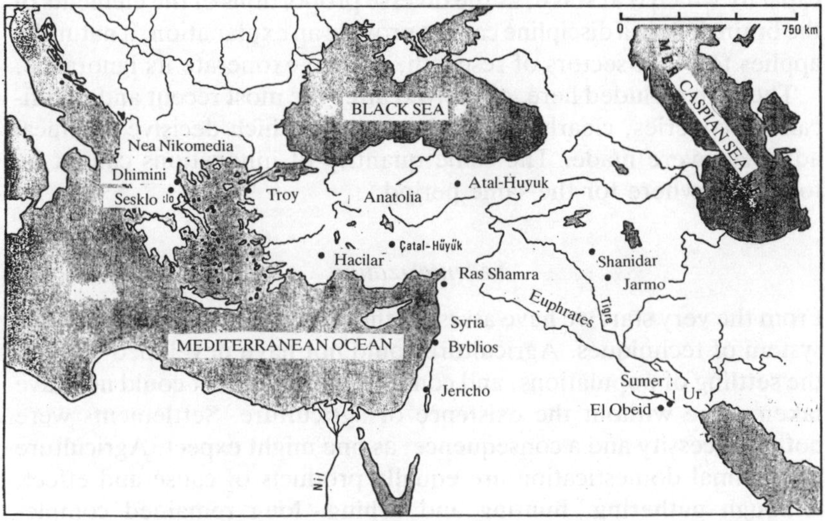

Figure 1. — The Eastern Mediterranean during the neolithic age.

this privileged zone. In this area, ceramics were developed after agriculture and animal domestication.

Reasons to explain this genuine technical revolution have naturally been sought, and many have been found, none of which are really convincing. Is this due to changes in the climate, or better natural conditions? The arguments for this do not seem to hold up. Indeed, the change of climate could equally have reached Italy or Spain, or any region situated on the same latitude. Climatic and natural differences in this zone, in particular in the soils, are so numerous that one searches in vain for any unity. As Leroi-Gourhan so correctly points out, tropical areas did not need to keep stocks and consequently had no need for agriculture: the nature of the products harvested, and the distribution of ripening times were sufficient to support a relatively large population. Why then the existence of Nordic regions and their restrictions? The only valid explanation, again a theoretical one, is that it was man himself who reached technical maturity in this area. And we return to the problem presented above: it is not so much the cerebral capacity which is at stake, as the diverse proportions of the elements of the brain. While a discipline cannot provide an explanation, it naturally applies to other sectors of research, as if to exonerate its ignorance.

The map included here, conceived after the most recent and significant discoveries, clearly shows the zone in which decisive technical advances were made. The same quantity of innovations cannot be found elsewhere for the same period.

Sedentarization

From the very start we have assessed the importance of the notion of a system of techniques. Agriculture could not have developed without the settling of populations, and conversely, settlement could not have taken place without the existence of agriculture. Settlements were both a necessity and a consequence, as one might expect. Agriculture and animal domestication are equally products of cause and effect, although gathering, hunting and fishing, long remained complementary activities. If man settles, livestock will immediately appear. Leroi-Gourhan attributes this situation to the restrictions of settling around food depots.

The following stages have been cited for the first permanent population settlements: around 7000 BC in Northern Iraq; 6000 BC in Mesopotamia, Syria, Lebanon, Anatolia, Macedonia, and Thessaly; 5000 BC in Egypt; and 4000 BC from the Sudan to Baloutchistan.

Dwelling places, originally consisting of a single room, were built in groups. They were constructed of material intended to last for a relatively long period: generally mud-brick which was dried in the sun. Buildings effectively became permanent in Jericho, Ras Shamra, in Kurdistan, and Haçilar, during the eighth and sixth millennia BC. Conglomerations exist where dwellings were made of clay walls, and supported on stone foundations. We have reached the upper stage of the proto-Neolithic period. During this last stage, dwellings also consisted of one room, either square or rectangular, possibly with a covered porch. But they were no more than simple huts, built of posts driven into the ground, supporting a wall made of grids of reeds and mud-plastered branches. The floor consisted of a layer of pebbles or a hard-packed surface. Dwellings were of this kind at Sesklo (Thessaly), and Nea Nikomedia (in Greek Macedonia). Despite the fragility and light weight of these constructions, they were, as other elements prove, stable buildings.

Some authorities have even suggested that towns existed. Today, we seem to have gone too far in interpreting archaeological data. Childe did not envisage an urban "revolution" until the third millennium BC. The tenants of Neolithic towns cite two exceptional cases: Jericho A pre-pottery, at the beginning of the eighth millennium, and Çatal Hüyük in the seventh and sixth millennia BC.

Obviously, everything depends on what one understands by the term "town." For some a town is a fortified conglomeration, such as the square tower of Jericho, a sort of rempart formed by the houses leaning against each other and without an opening towards the outside, which encircled the conglomeration of Çatal Hüyük. The facts themselves are very dubious. In Mesopotamia, the remparts were only generalized from the third millennium BC onwards. And the unquestionable towns of Egypt, or even those of the Late Roman Empire, never had fortified walls. Others quote a few agricultural activities and a developed cottage industry. But archaeological digs do not provide very fruitful material. The obsidian mirror of Çatal Hüyük is feeble proof because it remains unique. No potters' quarters have been found in any of these conglomerations. We know that the numeric criteria of the population is not an obvious indication. While Jericho probably had a population of 3000, we know from other sources that large villages had populations of 10,000 predominantly farming inhabitants. If one bases a theory on the criterion of urbanism, it is in the Mesopotamian and Syrian towns of the late fourth millennium BC that one finds this concern in an evolution involving highways. We might even say, and with good reason, that these two "towns" in fact represent a pre-urban stage.

On the other hand, there is evidence of early fortresses, a sign of an advanced society organized within a hierarchy. The remains of the sixth millennium BC towns of Sesklo and Dimini are striking. The acropolis of Dimini is surrounded by three concentric walls, foreshadowing Mycenaean palaces (fig. 2). It was to be the same at Troy, founded around 3500 or 3200 BC. But these cannot be called towns either. These fortress-palaces, however, witness the settling of peoples and the beginning of political power.

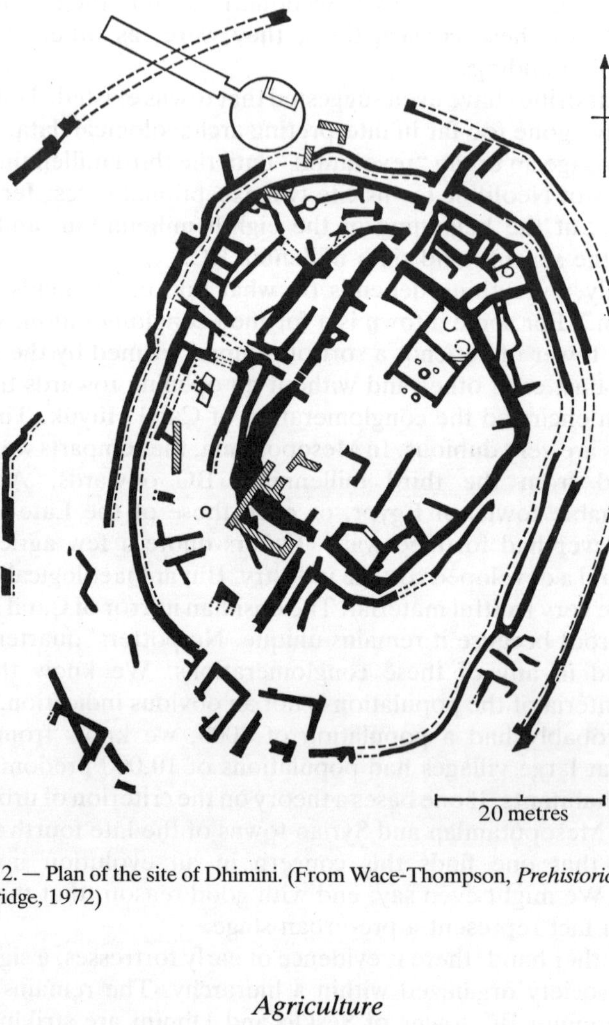

Figure 2. — Plan of the site of Dhimini. (From Wace-Thompson, *Prehistoric Thessaly*, Cambridge, 1972)

Agriculture

Agriculture appeared at the end of the Mesolithic epoch. Truth to tell, it is sometimes difficult to establish the facts concerning this, and our sources of documentation are often ambiguous. Seeds found in centres could equally have come from wild plants, in the beginning at least, as well as cultivated ones. This is similar for certain instruments which could be used either for gathering or cultivation, as is exemplified in the case of the sickle. The dawn of agriculture necessitated the recognition and selection of profitable varieties of plants consumed before, but as yet not cultivated, then improvements by hybridization. It is a patent fact that in regions where farming appeared for the first time, agriculture was always linked with animal domestication, an obvious indication of the existence of a technical system. In preceding epochs, moreover, man had always practised a mixed food economy, finding a balance in the complementary exploitation of the animal and vegetable world. The change to a more settled way of life would therefore suggest two activities. Leroi-Gourhan estimates, however, that some farmers may have complemented their resources by hunting, and some shepherds complemented theirs by gathering wild plants.

Zoological and botanical conditions must have been equally favourable for the simultaneous appearance of animal rearing and agriculture in the same regions. Plants with consumable seeds were particularly developed throughout the temperate band, especially in the southern zone, Africa, to the north of the tropic zone, the Middle East, and Central Asia. Again in the words of Leroi-Gourhan:

> Before the present desication, during the epoch when man changed to proto-agriculture, the periodic exploitation of herbaceous plants and seeds certainly constituted an essential part in the search for food. Grasses themselves played an important part among these plants: despite the small size of their seeds, they represented a high quality nutritional food, and could be preserved for a long time. It has recently been suggested that the regions of the Near East, in particular the north of Iraq possessed grasses with large seeds, ancestors of present day cereals, at least from the middle of the last glacial period. The basic conditions for the progressively agricultural exploitation of wild wheat were consequently developed in the same regions where goat herding was first manifested.

Agriculture was certainly not "invented," but was the result of a process of progressive change from one situation to another, sometimes without any links common to farming and animal rearing.

It is not clear which centuries witnessed the birth of agriculture, and scientists rarely agree with each other. According to some, cultivated cereals appeared with the Mesolithic Natufians of Palestine around 7800 BC, and in Jarmo and Jericho around 7000 BC. According to others, the growing importance of seeds in both Iraq and Palestine, in the ninth and eighth millennia BC, is attested by the abundance of certain tools, such as sickles, grindstones, bowls and pestles made of stone. Decriptions of silos in Zawi-Chemi and Shanidar in Iraq have even been cited.

During the seventh millennium BC, the cultivation of wheat and barley seems to have been developed in a great number of regions in the Eastern basin of the Mediterranean. The proto-Neolithic period in Sesklo, in Thessaly, has revealed the cultivation of hard wheat (*Triticum dicoctum*), einkorn (*Triticum monococcum*), barley (*Hordeum vulgare*), and millet (*Panicum miliaceum*). While millet is European, all the other cereals in Thessaly originated in Asia. The wild barley region stretches from Anatolia to Northern Iran. Hard wheat seems to have originated in the area of Palestine and Northern Iraq, and Asia Minor seems to have known another primitive variety of wheat (*Triticum aegilopoides*).

We are thus concerned with a system of agriculture which is already passably diversified, perhaps also already adapted to different soils. This diversity is attested by the existence of other cultivated species: the broad bean and the pea were known to ancient Neolithic man, and probably even to proto-Neolithic civilizations, and the lentil was also cultivated at this stage.

At Nea Nikomedia, in Greek Macedonia, virtually at the same period as at Sesklo, one finds wheat and barley, as well as peas and lentils. But nearby, the populations of Northern Greece lived on additional wild cereals and plants gathered in the area.

During the upper Neolithic culture, agriculture was definitively accepted in Dimini. New relationships between man and the natural environment were established. Ground was cleared to provide artificial environments which could be worked more easily. It is even possible that irrigation and manuring were introduced. Certainly, the same cereals were cultivated as those during the pre-pottery period. Of those leguminous plants which can be traced, the lentil, pea, and broad bean are extremely common. It is more difficult to say whether wild fruits were picked or whether fruit trees had already been cultivated. In any case, the olive, pear, almond and fig have been pointed out.

It has been noted that towards the end of this epoch the vine may have been introduced, in evidence at the beginning of the ancient Bronze age. It is well known that botanists are not in agreement about the origin of the vine. Nevertheless, it would seem that the greatest number of wild species were to be found in the vast zone stretching from the Black Sea to the Indus valley. The vine spread extensively during the period when temperatures rose, which took place between the last two glacial periods, as certain discoveries in the Neckar valley have shown.

Archaeological discoveries made in the last twenty years have thus clarified the transition from gathering to cultivation. The area of transition was previously thought to have been located between the Mediterranean and the Caspian. Today, the north of Iraq, Syria, Libya, Palestine, Turkey and some parts of Greece, have been explored, revealing sites with traces of fairly diverse cultures, dating between 8000 and 6000 years BC: thus, sites have been found at Jarmo, Shanidar, Zawi-Chemi and at Çatal Hüyük. "This transition took place almost imperceptibly; sickles, as we have said, were present before agriculture, and only statistics show that goats ceased to be hunted. The example of Iraq corresponds to an ideal demonstration, as within the space of a few centuries a transformation had taken place, without entailing any upheavals which could jeopardize cultural survival."

Animal Rearing

The domestication of animals poses other problems which are just as complex and similar in nature. The material available is difficult to interprete. Moreover, animal rearing did not eliminate hunting. It is almost impossible to trace the genealogy of some animals, indeed even to distinguish between them at these remote epochs, and this is particularly so in the case of goats and sheep.

The development of animal domestication, which may have advanced with hunting, requires favourable environmental conditions, as it presupposes that the hunter and the hunted maintain some sort of personal relationship. Excluded from this fact are the great migrant herbivores, whose herds pass once or twice a year within weapon range; also excluded are the large and dangerous or fast herbivores, such as the ox, the bison, and the horse, animals covering a large area who are difficult to approach and impossible to contain. When one analyses the factors contributing to the possible transition towards animal domestication, one realizes that the conditions of the physical environment are even more imperative than bio-zoological conditions, and that the chances of animal husbandry being established in the Steppes of Africa or Central Asia were very slim.

We have seen that conditions were favourable in Northern Iraq for the domestication of the goat. From this one may conclude that proto-rearing most probably originated in the mountains.

To be honest, it is difficult to say whether the goat or the sheep was domesticated first. This probably varies according to the region. Dates suggested for this are as vague as those put forward for agriculture. The sheep appears to have been domesticated around 9000 BC in Northern Iraq, according to the dates established for remains found on sites at Zawi-Chemi and Shanidar. It has been said that at Jarmo and Jericho, an important interval existed between the domestication of the first animal and the second: in this area the goat appears to have been domesticated approximately 1500 years after the sheep.

In the seventh millennium BC, goats and sheep, the first domesticated animals, were present throughout the zone which has been defined above for the beginning of farming. As has been noted, these two activities of man match almost exactly, both geographically and chronologically.

The transition from mountain goat or sheep herding to the rearing of larger herbivores and pigs has not yet been elucidated. It seems, however, to be linked to the first impetus given by proto-rearing of the bezoar, as it developed a little later, within a radius of the original centre. Between 6000 and 3000 years before our era, the sheep, ox, pig, donkey and horse were domesticated, then towards the Indus valley, the buffalo, zebu and elephant were reared and spread to the Near East, Asia, Europe and Africa. In this movement, the inital momentum is the only point to be called into question, as, apart from the American llama, all animal rearing constitutes a coherent historical block. The principle once accepted, its application to new species, offers fewer difficulties than the transition of ceramics to metallurgy. It is interesting to note that, apart from the reindeer whose food conditions are very particular, all the domesticated herbivores are actually grass eaters (cattle, sheep, horses and camels), animals grouped in dense societies on a continuous vegetable carpet and whose flight behaviour is one of grouping. The beating action adopted by shepherds working with dogs applies equally to these animals. Leaf eaters (cervidaes, living in small hordes under cover) whose flight behaviour is one of dispersion, remained undomesticated.

The same difficulties of dating still exist. Pigs and cattle existed in the proto-Neolithic period at Sesklo, in Thessaly. At Nea Nikomedia, around 6200 BC, goats and sheep lived together with cattle and pigs, and at Jarmo, pigs appeared towards 6500 BC, and oxen around 5000 BC.

By-products from animal rearing were very quickly used. Wool was produced at Çatal Hüyük in approximately 6000 BC, and cheese was made at Dimini during the recent Neolithic period.

Hunting, however, still remained a privileged method of providing supplies. At Dimini, in Thessaly, auroch, stag, deer, boar and hare were all hunted. Fishing also continued in this large sea-side village.

But this type of food was very limited compared to what was provided by cultivation and animal rearing.

Let us note finally, and we will return to the subject of Egypt later, that there was an attempt, albeit much later, to rear certain species which were eventually abandoned, such as the hyena and the crane. Broadly speaking, we do not know, during this period, of the rearing of poultry.

A specialist has tried to clarify these points:

> When one sees the most perfect and universal of our domestic animals suddenly appear together, in what one could call the prehistoric history of civilisations, does one not wonder if this achievement was the venture of a special human race, endowed with faculties which his predecessors lacked, a race which was the result of sudden change rather than slow evolution, which suddenly created a new being.

We will now return to the ideas outlined above: the perfect concordance and complementarity between agriculture and domestication. "The processes started during the Mesolithic period in the Near East in around 8000, had already completely transformed the structure of societies in around 5000, from Mesopotamia to Turkey, Greece and Egypt. A basic economy is formed even before the appearance of pottery, by the association of wheat or barley with sheep, goats and pigs, and the first permanent villages appeared."

We can easily imagine the influence of this fundamental technical change on forms of societies, even if, at the beginning, life styles were mixed before they became combined. Gatherers, hunters, farmers and shepherds probably lived side by side, definitively integrated in complex societies, which may have already recognized a certain division of labour.

Ceramics

The reader will not be astonished to learn that the date or the area in which pottery was developed is not known. Specialists agree that it did not appear before 7000 BC. In fact, the material existed almost everywhere, and no concrete technical obstacle had blocked its development since the Paleolithic period. To be honest, nomadism is hardly favourable for ceramic activity: drying is a slow process and frequent movements are not suitable for a material which, by definition, is relatively fragile. Sedentarization and agriculture were to favour its development: harvests, the handling and stocking of products, and gradually refined methods of preparing food required numerous and varied receptacles, to complement the more ancient practice of basket work. We should also note that clay was employed in building as well.

Truth to tell, opinions differ. In this area, as in others recently discussed, every archaeologist claims to have discovered the oldest known document, usually accompanying it with astonishing explanations, all of which must be verified.

Before the discoveries made at Sesklo and Nea Nikomedia, pottery from Çatal Hüyük dating from the seventh millennium BC or from the beginning of the sixth millennium, was thought to be an exception to the general rule. As far as carbon dating can be accurate, pottery existed at Sesklo, in Thessaly, in around 6500 BC, and at Nea Nikomedia, in Greek Macedonia, in approximately 6200 BC. These dates would appear to be the oldest in comparison with others which have been suggested. According to some, this ceramic industry arose from the difficulty which these populations found in trying to make stone vases. Others estimate that pottery was also known in Iraq, Syria, Libya, Palestine and Anatolia at this time. Different regional forms were quickly developed, a fact which eliminates the notion of a single centre of invention. In any case, in approximately 6000 BC, pottery appeared from Thrace to Iran, to Çatal Hüyük in Anatolia, and Ras Shamra in Northern Syria.

Between 6000 and 5000 BC, the development of ceramics reached neighbouring regions: Cyprus in approximately 5250 BC, Haçilar around 5500 BC, and Jarmo towards 4600 BC. In most of these regions, pottery was developed after agriculture and animal husbandry: in the northern regions of Europe it is the reverse, for pottery preceded farming and animal domestication. This is an important fact, as it completely destroys any idea of the logical development of a material culture.

The problem of fire similarly arises. Apart from its culinary use, in around 3500 BC fire made dyeing possible with ferruginous ochres: rock paintings provide magnificent examples of this. The firing of clay, however, was accidentally produced in the fireplaces of cave dwellers. One can only imagine the potter's kiln, used no doubt, for the creation of modelled figurines. Leroi-Gourhan suggests that plaster made its appearance at the same time as pottery, and that "from Mesopotamia to the Mediterranean, the reduction by fire of gypsum in plaster, provided a coating on the floor and walls of the kilns." Pottery and plaster indicate the existence of an already proved skill in obtaining temperatures between 500 and 700 degrees, and the possibility of exceeding 1000 degrees for very limited and adequately ventilated parts of the hearth. We will return later to the subject of the progressive acquisition of higher temperatures which play a predominant role in the preparation of a large number of products (fig. 3).

Metallurgy

Once again, we are dealing with new mysteries which are very obscure. The notion that the use of native metals was relatively simple to conceive, already poses problems. It is not only the idea of metal which is involved, but also all the processes required to produce a usable material: it has to be heated and forged, thus entailing operations which bore no similarity to earlier techniques. Even the glitter of gold or silver is not sufficient to explain the appearance of metallurgy.

It is quite different for the reduction of minerals. Man must know how to extract the metal ore from the rock, how to choose the mineral and how to construct a furnace, which presupposes that heat is the essential vector of the operation; understanding that in some cases the mineral must first be roasted, and that melting agents must often be added, how to create charcoal and knowing that it is vital for reduction, presupposes workings of the mind which are quite astonishing and which would undoubtedly be impossible to reconstruct. According to some, one is left with the theory that all this resulted by chance, but this is difficult to accept: a potter's kiln built of stones containing a mineral, with lime or plaster as the reducing agent, and which, in certain places, was of a sufficient temperature could not have come about by mere accident. Let us simply admit our ignorance and agree with Leroi-Gourhan by saying that the production of pottery implies some knowledge of working with fire.

The region would also have had to have been rich in minerals. Here again, cultural, technical, geographical and geological interactions are necessary. We find the same notion of environment as was suggested for the birth of agriculture or animal husbandry. Access to exploitable deposits and the level of techno-economic development are indispensable preconditions. It is unthinkable that metallurgy could have been developed during the Acheulean epoch. Social conditions must also be

Figure 3. — Neolithic Greek sites and pottery. 1, outline of a dwelling of the Chalcolithic period; 2, statuette; 3 to 5, ceramic; 6 to 12, late Neolithic; 6, outline of dwelling; 7 and 8, arrowheads; 9, axe; 10, statuette; 11 and 12, ceramic; 13 to 20, Middle Neolithic (Sesklo); 13, outline of dwelling; 14 and 15, axe and adze; 16, statuette; 17, seal; 18 to 20, ceramic; 21 to 25, pre-ceramic Neolithic; 21 and 22, blades; 23, ear disc; 24, sling ball; 25, hook; 26 and 27, early Neolithic: ceramic. (From A. Leroi-Gourhan, *La Préhistoire*, Paris, 1968)

reunited: ceramics may have been a family activity, but metallurgy could not have. Originally, the latter seems to have been the product of an industry of specialists, who worked for a market open to a fairly large number of people. It is a remarkable fact that the consumers, at least in early times, were often not the producers: either one bought outside the industry, or one seized the metal by means of war. These mineral riches are not all included in the zones which we have defined for agriculture and animal rearing, but are often close to their borders.

In the south-east of the Mediterranean, an experimental study has given some idea of the possible uses of fire as the transforming agent of matter. So that after pottery, metal, lime, and glass appeared at almost the same time. Notice also that the east Mediterranean had developed charcoal and a method of constructing a kiln which could reach very high temperatures. And as we have previously stated, this lime could already act as a flux.

In the last few years, the time at which metals might have been known, in one form or another, has been placed at perceptively earlier dates. Çatal Hüyük may have possessed lead and more abundantly copper in about 6600–6300 BC, that is to say at the same time as pottery was developed, and Haçilar towards 5400 BC. Around the beginning of the fifth millennium, the process of casting this metal was accomplished. It is very difficult to determine the content of this metal and its uses, but these examples seem to be exceptional isolated cases, due to the presence of sizeable deposits. It is certain that the Chalcolithic civilization, taken as a whole, was much later.

The same may be said of bronze, which gave birth to certain empires. Bronze poses the same difficult problems. Was it a natural alloy, or an intentional one? We will probably never know the answer, or which one preceded the other.

Briefly, and we will return to this later, we should mention the presence of iron in the royal tombs of Çatal Hüyük, constructed in the third millennium. There is such a quantity of this material that we must suggest that it was undoubtedly in this area of Anatolia that iron metallurgy was invented.

Stone

We should not forget, however, that stone was the basic tool material. Even when metal was introduced, the metal tool was often made in stone where metal deposits were not available.

Lithic industries are well differentiated throughout the zone we have defined, reaching a peak in the seventh millennium BC. All the sites already mentioned have provided an abundance of evidence testifying to this side of man's activity: Jarmo in Iraq, Ras Shamra in Syria, Haçilar in Anatolia, Nahal Oren on the Palestinian coast, Jericho in Palestine near the Jordan, and Seyl Aqlat in Southern Palestine. This industry is most often microlithic. On the other hand, there are few geometric microliths accompanying an industry of blades, bladelets and fine perforating implements in Iraq. In Northern Syria and Anatolia, where the mesolithic substratum is poor, there was no microlith tool industry.

A tendancy for more diversified tools followed. The production of objects in bone, introduced in the recent Paleolithic period, was equally developed.

Neolithic tool material in the Near East seems in fact to have followed ancient tendancies, but in using this material, this culture resoundingly manifested imagination lacked by its precedessors.

THE TECHNIQUES OF THE FIRST GREAT KINGDOMS

History really begins at the end of the fourth and the beginning of the third millennium BC. Thus the Near East already possessed agriculture, metallurgy, and domesticated animals: it was very quickly to acquire the written word, towns, and political and military power.

The Egyptian civilization began around the years 3500 BC. Between 3500 and 3000 BC, the two kingdoms of North and South asserted themselves, and writing was introduced. Around 3000 BC, Narmer finally unified the country. In approximately 2780 BC, Djozer founded Memphis and built the first pyramid at Saqqara, thus inaugurating the first stone architecture. From then on a new civilization was born.

It was around the year 2800 BC that the dynasties of Uruk and Ur were founded, and the royal tombs of Ur appeared. Here again, almost at the same time, civilization advanced rapidly and ceaselessly affirmed itself in the following centuries. This simultaneous emergence is extraordinary, even if no apparent influence was originally exerted by one region on the other.

Egypt

Apart from some enigmas which have still to be solved and certain influences exerted by neighbouring civilizations, the technical system of the Egyptians poses some difficult problems.

When the first kingdoms emerged, men had been living for a long time on the lateral terraces of the Nile, hunters and fishermen who used stone tools. The advance of the desert had overtaken the swamps inhabited by other men. They then joined the first inhabitants of the valley and the oasis, whereupon they all began to gradually cultivate the land. Lithic industry was refined, and the first pottery was introduced. So in around 3000 BC, Narmer unified the country, and established and wore the two crowns of Upper and Lower Egypt. He also introduced the first hieroglyphics, and it has been suggested that some of the drawings of this epoch recall the art of Mesopotamia. The pre-dynastic era saw the arrival of some metals: gold, silver, and lead were introduced at the beginning of the third millennium BC. Buildings were constructed out of sun-dried mud-bricks. I and II Dynasty tombs have been found at Saqqara, Balla, Negada, and Helowan. Furniture consisted mainly of pottery, alabaster and hard stone vases, and schist objects.

Suddenly – and herein lies the first problem – from the years 2778 to 2423 BC, the Old Kingdom became an advanced technical civilization. From the III to the V Dynasty, we see the building of pyramids and this whole civilization which is illustrated in engraved and painted frescoes. These frescoes, such as the Ti mastabas (dating around 2560) and the Mereruka mastabas (dating around 2420), reveal as much on the suject of agriculture as on the subject of the arts and crafts which possessed already definitively formed equipment and produced very elaborate products. It was at this epoch that stone architecture made its appearance. Imhotep, architect, doctor, and chief minister to King Djoser, built the first pyramid at Saqqara, and a whole complex of palaces and shrines, where for the first time one saw fluted columns which were to make Karnak and Luxor famous. This dates back to 2780 BC. The earliest inscriptions, those from the palette of Narmer, present the image of a still primitive language and script.

It is difficult to account for the sudden appearance of an advanced technical system, and it is frequently awkward to determine the material of this equipment, almost certainly still Lithic in tradition, and possibly, in certain cases, made out of copper. Obviously our ignorance of the exact details of the Neolithic techniques which we have just mentioned, precludes the measurement of the differences between the two cultures. One cannot honestly say then if this was a sudden process or a slow change. May one talk of an "industrial revolution"? As G. Posener notes, "And then, the invention of a new tool suddenly opened up a whole crop of possibilities: the use of the hand hammer or the miner's pick which increased the power of the hand, probably inaugurated the era of using building stone, and of working quarries and mines. More inventions were to follow." According to the same author, the gestation period seems to have begun in the pre-dynastic age and ceased with the Thinite kings, probably in the region of the years of the fifth millenium BC to about 3300 BC. "And from the first dynasties onwards, the Egyptians had developed all their building methods." Undoubtedly this view is rather too schematic and, as we have said above, the suddenness and the extent of this "revolution" must be qualified.

At this epoch, the evolution of the first dynasties conquered by Macedonia appears at first sight to have been limited, according to all the illustrations of the Egyptian civilization. New ideas or implements are rare, such as the bellows, or the Archimedes' screw. Similarly, some evolution in tools may be observed, due probably to the introduction of metals; bronze in the Middle Kingdom (2160 to 1580 BC), and iron during the epoch of the XXV Dynasty (712 to 663 BC). But, on the whole, the technical system remained the same until the Macedonian conquest in 332 BC. Only then were more advanced Greek techniques put into practice, particularly the use of machines.

Once again this statement needs qualifying. Our documentation is mostly pictorial in origin. In the main, these images date from most of the mastabas of Saqqara, that is towards the middle of the second millennium BC, then from the tombs of the Valley of the Kings, around the middle of the first millennium, and finally from the Scrolls of the Dead which are more recent. In fact, if exact comparisons were made – they have not been as yet -, one would realize that this stability is perhaps more apparent than real. We have already mentioned the bellows and the Archimedes' screw, but other examples may be cited;

vertical weaving looms of the Middle Kingdom, the horse, the wheel and the wagon, and glass under the New Kingdom. And this is the same for tools where evolution and enrichment are manifest.

This would appear to be the classical evolution of a technical system, improvements of which are to some extent foreseeable from the outset. Also, is it not somewhat hazardous to suggest a blockage of Egyptian techniques, as some have done? The Egyptian technical system seemed to reach its perfection at the same time as the Greek system. Let us say that evolution is slow and that the history of an independent Egypt largely unfolds under the same technical system. Undoubtedly several reasons may account for this.

The civilization of Ancient Egypt was a purely rural one. The country was separated by the sea and by the marshes of the Nile delta which were difficult to cross. The Egyptians were never sailors although they were boatmen. One may refer to economical phenomena, linked to the lie of the land and the nature of the soil and its hydrography. Others have considered that this relative stagnation was the result of a fixed society. For our part, we believe that all these factors could have been involved.

The beginnings Since around seventy years our knowledge of Neolithic Egypt has increased considerably. So much so, that the notion of the sudden appearance of an advanced civilization must be rejected in favour of an evolution, perhaps a more rapid one than elsewhere, but nevertheless a continuous evolution which suddenly appeared revolutionary when the technical system became coherent.

It is true that the region was favourable. Leaving aside for a moment the image of the Nile floods, one must remember that the valleys were lush with grasses, and that the fauna was particularly rich with wild asses, Barbary sheep, urus, antelopes, gazelles, and giraffes, lions and leopards. This fauna was still hunted in the Middle Kingdom. In the valley itself were elephants, kudus, two species of pig, hippopotami, crocodiles and wild boars.

The hunters of the plains had visited the valley during the Lower Paleolithic period: their tools have be found on the high terraces of the two banks, testifying to a Levallois tradition of microlithic forms. As the drought periods became more frequent, and the need for water more pressing, nomads took refuge in well-watered regions. They "were to be found, therefore, in conditions which forced them to abandon a parasitic life and lead a productive one." Some historians, in particular the American, Perry, have vehemently supported the theory of the birth of agriculture in Egypt: new life was produced without human labour, in the silt of the annual floods where plants, ancestors of wheat and barley, reproduced themselves.

The Badarian civilization, established on the barren spurs which swept down to the marshy valley below, had already cultivated emmer wheat and barley, as has been proved by grains found in their more or less permanent encampments. Hand mills, flag stones with concave surfaces, and grinders are manifest proof, and we have already mentioned the discovery of goat and sheep bones. But this is only the beginning of a new civilization: sedentarization does not appear to have taken place as yet, and hunting and fishing were still important activities. They worked with wood, pottery existed, and remains of cloth have been found. They probably also painted their faces and eyes.

Sites at the Fayum present an image of already very advanced techniques. The inhabitants of this area had managed to cultivate emmer and barley, but had also produced flax. Harvests were preserved in silos sunk in the ground. Pigs, cattle, sheep and goats had been domesticated, and hunting and fishing with bows and arrows continued. The presence of scrapers suggests that animal skins were used for making clothes.

The site at Merimde gives a similar picture. There is evidence that the sling was used, and textile production is attested by spindle-whorls. It is the same at Debono, near Helwan. From the Tasians to the Badarians, the same evolution towards new techniques can be found. Further south, near Khartoum, one leaves these technically advanced tribes to rediscover a Neolithic culture, certainly one which is relatively advanced, but which had not developed agriculture, even though an excellent well-fired ceramic industry existed, and goats were reared.

In Upper Egypt, the Badarian culture was followed by the Amratian, and in all likelihood, proceeded from it. The inhabitants of this region embarked upon a systematic cultivation of the plains irrigated by the floods. Farming and animal rearing were henceforth balanced, and large cattle were reared for their milk, meat and labour, even though the complementary activities of hunting, fishing and gathering were still required. Rock paintings give us the impression of a new advance in the field of traditional techniques. Sedentarization gave birth to large and permanent villages. The flint industry was improved and it is posssible that the ass was already domesticated. The Amratians navigated the Nile, but in crafts without sails. Pottery, on the other hand, made no notable progress and if anything ceramics were not as advanced as those of the Badarians.

It is not until the Gerzian epoch that a new stage begins. At this point agriculture became an essential resource and henceforth may have been founded, to a certain extent, on artificial irrigation. Pressure flaking reached a point of excellence in flint production. "The most significant development was the comprehension and application of the principle of a well-thought-out metallurgy." Tools and weapons made of cast metal were used, although they were relatively rare, and earthenware was developed although stone vessels were still numerous and were made of harder rock. All these products reveal a modified society, nourishing its specialists on a better developed agriculture. Dwellings became more solid and more durable in construction. It is possible, if not highly probable, that the whole of this civilization was produced by outside influence.

Thus, we have defined a stage where new techniques were acquired and a new technical system was developed. There is no sudden transformation, but an evolution in which one may recognize peak periods and domains where old traditions continue or sometimes even decrease. But from now on everything had reached a stage where the first Pharaohs' dynasties could begin to initiate change: henceforth, agriculture is perfectly adapted, involving all the necessary secondary operations; metallurgy had already been tried and tested, writing was to be born, and the whole technical ensemble was to reach a level of equilibrium.

Exploitation techniques Agriculture had therefore existed in a certain number of regions in the Near East when the first dynasties began. We will not repeat here what has been widely discussed about the Nile floods and their relationship with farming operations. The many agricultural scenes in Egyptian tombs and imagery provide abundant documentation on this subject.

Cereal production was the basis of most cultivation. Barley, corn, emmer, and from the beginning of the New Kingdom, durra were the main crops. Leguminous plants were already numerous; lentils, beans, onions, chick-peas, cucumbers and watermelon were richly represented. The Egyptians also grew a variety of fruits: pomegranates, figs, jujubes, olives, carobs and dates. It is more difficult to date the arrival of the vine. Sycamore provided both wood and fruit. This list demonstrates that a wide variety of crops had already been developed, which were far from primitive species. To be honest, we have little information about the species cultivated by the Ancient Egyptians. We must also add a new product to the above, flax which rapidly gained in importance as it was lifted and not cut for harvest.

Farming material was very limited. The hoe, used equally well for gardening as well as for making irrigation channels, was a special type as we have illustrated. It appears in this form on a Theban temple of the XVIII Dynasty (1580 to 1314 BC), now in the Louvre. On the other hand, more elaborate, modern-looking instruments with handles may also be found in the Ti mastaba (2500 BC) at Saqqara, and in the Menna tomb (1412 BC) near Thebes (fig. 4).

The instrument used for ploughing should not present any problems. This tool, the ard is virtually identical except for a few details to early representations of the V Dynasty (the Ti mastaba at Saqqara, dating around 2500 BC), that of the XVIII Dynasty (1580 to 1314 BC), and to the Deir el-Bahari papyrus of the XXI Dynasty (1060 to 950 BC), and we might safely say without too much risk of error that it resembles the plough used today (fig. 5). The absence of any modification and the continued use of this tool is absolutely remarkable. This ard was quite different from the ard with a share-beam used on the northern border of the Mediterranean. This sort of plough is illustrated many times, both in tombs and on papyrus. It appears as an ard, (a primitive plough) with two stilts separated by one or more cross-bars, up to four in the Nakht tomb near Thebes (1415 BC). A link, usually made of rope, supported the coulter and shaft. The team always consisted of a pair of oxen. The shaft appears – drawings are sometimes very ambiguous in this respect – to have been linked to a light yoke placed in front of the animals' horns. But there are also illustrations of oxen yoked around the neck, like horses. Ploughing is generally performed by two men, one to hold the ard, and the other to lead the team. On a fresco in the tomb of Sennedjem, near Thebes, just one man is pictured performing both these operations. Lastly, an ard drawn by four men is represented on the same fresco in the Louvre which we have already mentioned. The Egyptian ard has a slim frame, and only scratches the surface of a loose, light soil.

Harvesting was done using a sickle (fig. 6). This implement may have originally been equipped with small flint blades slotted into a wooden groove. Once it was made in metal, it became an instrument with a large curvature. This tool was also hardly modified between the VI Dynasty, from which date the earliest representations, and the time of Alexander the Great. The method of harvesting allowed the worker to stand completely upright, as the ears of corn were held close to the grain and thus cut high on the stalk.

The ears of corn were then transported in large baskets, carried by two men by means of a bar placed on their shoulders. The grain was placed in a pile, then spread in a circle around the threshing floor. The only technique used at this epoch consisted of threshing under the hooves of either oxen or cattle. The grain then had to be cleaned, in order to separate it from the broken straw, which was done using spades, and finally it was winnowed. The last two operations were performed by women.

One realizes then that equipment was limited to only a few implements, even for one of the most essential cultures. Apart from a small number of modifications to tools arising from the introduction of metal, the equipment used in farming remained much the same during the many centuries of Ancient Egyptian history.

As a matter of fact, we know very little about types of cultivations, and we possess few illustrations of them. There is only a limited number of images illustrating the vine (from XVIII Dynasty Thebian tombs of Nakht and Sennedjem). They show high growing vines, not our modern low vines. Representations show the grape picking, pressing by foot, and then the way in which the liquid was stored in amphores. The method of grape pressing was certainly not adequate, for the grape really needs to be crushed. This was done by placing the grapes in sacks between two poles. In the Middle Kingdom, at Beni-Hassan, this process was perfected. "The sack is placed in a stall and fixed to one side of a wall. On the other was a braid, through which three men passed a rigid shaft, enabling them to exert greater pressure on the sack. This system did not require either the complicated gymnastics of the Ancient Kingdom, or very great effort." So there was some evolution from the Ti mastaba (approximately 2500 BC) to the tombs of the XVIII Dynasty (middle of the second millennium). Unfortunately we do not have any more images after this date.

Figure 4. — Sickle, adze, knife and hoe. (Old Kingdom)

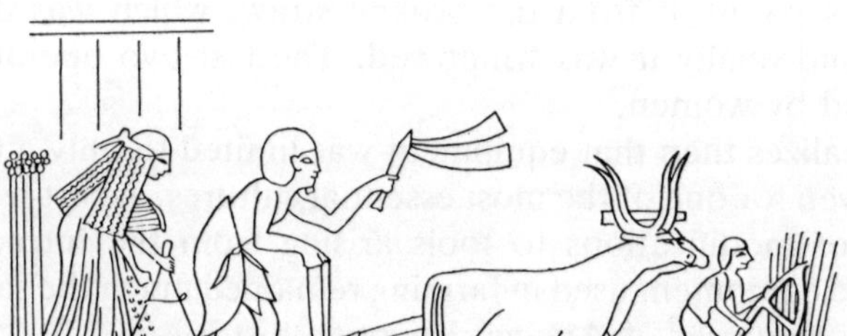

Figure 5. — Ard. (XVIII Dynasty)

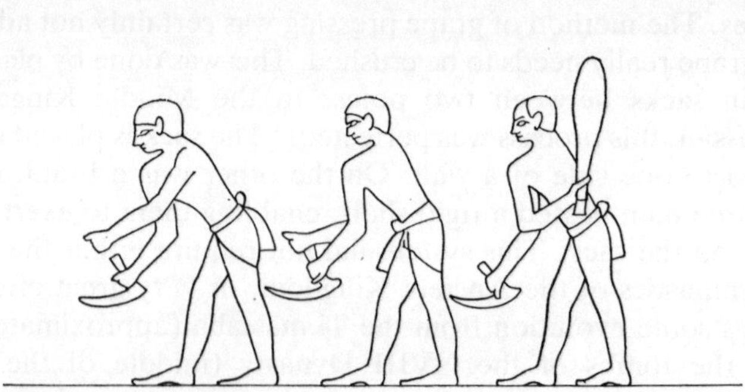

Figure 6. — Harvesting with a sickle.

As G. Posener has pointed out, there was a certain degree of experimentation in animal rearing and in the species which man attempted to tame. The early domestication of some animals appears to have been abandoned later on, such as the hyena, the crane and a few other species, including the gazelle, oryx, bubale, addax and ibex. Stabling was fairly standard, at least for some animals. Let us return to our Neolithic chronology.

Sheep appear to have been domesticated before goats, the latter being a mountain animal. Sheep rearing is only traced from the V Dynasty onwards, but we are dependent on images for this information. Cattle, on the other hand, were domesticated from the time of the early dynasties. According to a specialist, at this epoch, that is to say the middle of the third millennium BC, cattle breeding had not been a long precedence. Indeed, under the Ancient Kingdom one still sees cows tied by the legs while they were milked, a fact which suggests that they were still nearly wild (fig. 7). Two breeds of cattle can be distinguished for this period; one is large and strong with long horns, and the other is small without horns. Pigs had also been domesticated, but we do not know very much about this.

The Egyptian poultry was similarly well provided. The goose existed from the time of the early dynasties, as can be seen from the famous

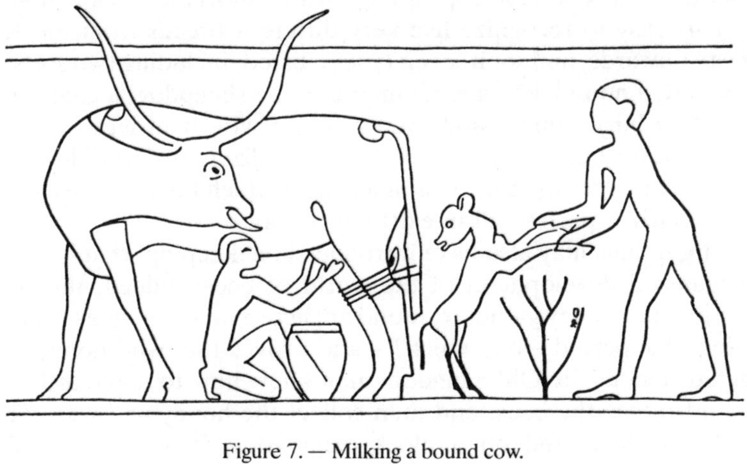

Figure 7. — Milking a bound cow.

and very beautiful representation at Meidum, dating around 2720 BC. Cranes were also well represented at Saqqara around 2500 BC. Geese and cranes were probably fattened for their meat rather than for the liver (fig. 8). Ducks were also a source of food, although they did not appear anywhere in the poultry: moreover, rearing ducks presented problems for the Romans who appear to have been the first to succeed in domesticating them. The pigeon, on the other hand, appeared as early as the very first dynasties: it was used for carrying news, but was probably also popular on the table.

There is some debate surrounding the subject of other animals. Despite the opinions of other authors, the ass or donkey seems to have been around for a very long time: it threshed corn, worked or carried loads. "It was, until the XVIII Dynasty, the only animal which could transport people and lead caravans across the desert to the Red Sea." Illustrations of this date from the XII Dynasty (fig. 9). The hen, most probably Persian in origin, only appears in sketchy representations during the Greek epoch. A mutilated passage from the epoch of Tuthmosis III speaks of a bird which laid an egg per day. Could this be the hen? The horse only appeared in the New Kingdom, to be more exact after 1300 BC, and the camel was only domesticated in the last century before the beginning of our era.

From the IV Dynasty onwards, dogs were kept on leads, moreover with hyenas. Some of these dogs were genuine greyhounds, with svelte bodies, straight ears, thin flanks and fawn coats. One can also see a smaller dog, more solid in appearance, with a short muzzel and floppy ears. It is easy to recognize five very different breeds from the XII Dynasty onwards, and soon seven types existed, including two sorts of greyhound, a hound with a medium muzzel, a sheepdog, a guard dog similar to a huge hound, and even a basset with bent legs. The cat appeared at the beginning of the Egyptian civilization. From the start, there are two types of feline distinguishable which bear similarities to the wild native species still in existence today.

The Egyptians may also have introduced beekeeping, that is to say the organized development of the world of bees: indeed, the rock painting of the honey gatherer, found at Bicorp in Spain, goes back to the Neolithic period – but which? – and proves that wild honey was collected. Under the Old Kingdom men knew how to make artificial hives, introduce the bees, and then collect the honey and wax. This is perfectly illustrated on a block originating from Abousir. All

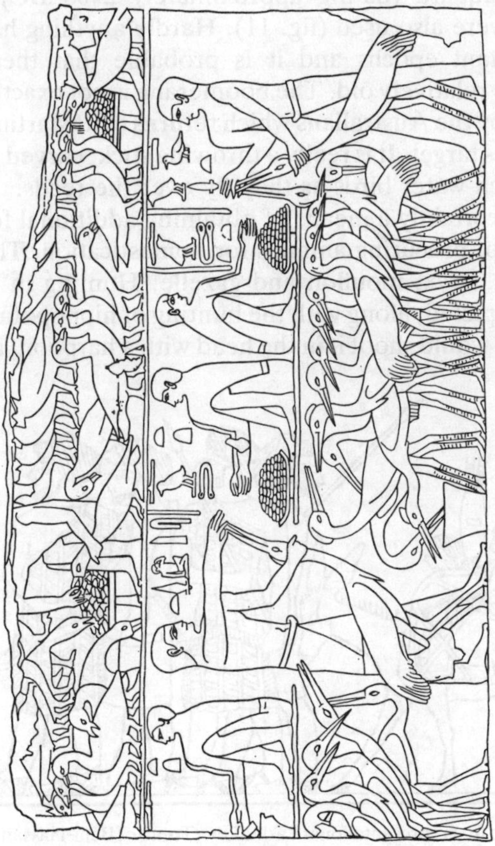

Figure 8. — Force-feeding cranes.

these operations are represented in the tomb of Rekhmire, near Thebes (XVIII Dynasty; fig. 10). The technique seems to have been perfected since the V dynasty. Beehives were smoked and honey poured into large jars, and it is possible that gingerbread was made. Honey and wax were very important in the Egyptian way of life.

Hunting and fishing were abundant and ancient practices. Many scenes of these activities are illustrated, represented on funerary monuments and temples.

The bow, lasso and boomerang were used in hunting, traps were laid, and the falcon was employed, as can be seen on the mastaba at

Mereruka at Saqqara (dating approximately 2400 BC). Nets for catching birds were also used (fig. 11). Hardly anything has changed since these distant epochs and it is probable that these hunting techniques were relatively old. The boomerang is not exactly the same as the one used by the Australians which returns to its starting point if it does not reach its target. It is really a throwing stick, curved at one end, designed to catch water birds as they fly over the reeds.

Hunting was certainly a means of obtaining additional food. But it was also a way of procuring animals for domestication. They hunted wild ox, ostrich, oryx, mouflon and gazelle. Hunting in thickets of reeds is often depicted, along with the hunting of hippopotami in small boats. First they would shoot it in the head with a harpoon, and then as

Figure 9. — One of the first representations of a donkey. (Tomb of Beni-Hassan, XII Dynasty)

Figure 10. — Removing honeycombs from Hives. (Tomb of Rekhmire at Thebes, XVIII Dynasty)

Figure 11. — Quails caught in nets in cornfields. (Old Kingdom)

the beast turned round with its mouth open, they would kill it with a lance.

Fishing was done with nets, hoop nets, harpoons and fish spears, all of which have been in existence since the Old Kingdom and have hardly changed. Of course, this type of fishing was done only in rivers or lakes. Two boats were necessary when using hoop nets, as one can see in the Ti mastabah.

The production of metals in Ancient Egypt poses several problems, not all of which have been resolved. Or, to put it more accurately, all kinds of interpretations have been put forward, some of them somewhat thoughtlessly. Although there was an abundance of certain deposits, like the iron deposits in the Aswan region, for example, Egypt was not rich in any of the other metals, except for some copper mines in Sinai. The Sinai mines were most probably developed around 2700 BC, but we know nothing about Egyptian mining techniques. We have found no trace of exploitation, nor any figurative representations. Some think that they used pillars, while their solution to the problem of lighting remains a mystery.

We are uncertain about the chronology of metals. Gold, silver and lead were already in existence during the predynastic period, while copper appeared during the Badarian period. The Old Kingdom has bequeathed us several tools made of copper (scissors, cold chisels and nails). Metal was hardened by hammering. Bronze was in common use by 1160 BC, and probably came from abroad: it was known in Ur by 3500–3200 BC. There has been a great deal of discussion about iron. Some people have spoken of ancient iron beads, but in fact this is probably meteoric iron. Others believe that iron first appeared during the VI Dynasty (2423–2263 BC), or the XII Dynasty (2000–1785 BC). However, it is generally accepted today that iron was commonly used by the XXV Dynasty, despite the discovery of an iron dagger in Tutankhamun's tomb, which was most probably imported.

The processes used to reduce minerals were probably the same for all of Europe and the Near East in the same period. It is impossible to say more, since no metallurgical furnace has been discovered in Egypt. However, we possess some pictures depicting the working of metal. The main difficulty appears to have been heating the metal to a sufficiently high temperature to be able to work it. This raises the problem of the bellows used, of which there were three types. There was probably no evolution involved here, since each one was adapted for a particular operation (fig. 12). The first of these was the blowpipe, which was most likely used for precision work (it is depicted on a XVIII Dynasty tomb in relation to the making of enamel). Then there is the sarbacane, the first representation of which is found in the Mereuka mastabah (around 2400 BC): here it is being used to make jewelry. Finally, there are the goatskin foot bellows (Rekhmara's tomb, XVIII Dynasty). The heel is used to keep the central opening of the diaphragm closed during compression. The diaphragm is raised with a rope, with the foot at the point of the bellows, and the heel raised to keep the opening free.

Anyone who has seen the Egyptian monuments, at least those built after the III Dynasty, is quite justified in thinking that the Egyptians were skilled quarriers. Some of these quarries are still in existence, and furthermore, in quite good condition, which is hardly surprising sincesubsequent generations were content to dismember the ancient monuments to get the stone they needed for new buildings. There are quarries near Memphis and Aswan, the pink granite found in the latter region being used in abundance.

The techniques employed to cut stone are virtually unknown. Limestone was not difficult to quarry, but it was completely otherwise

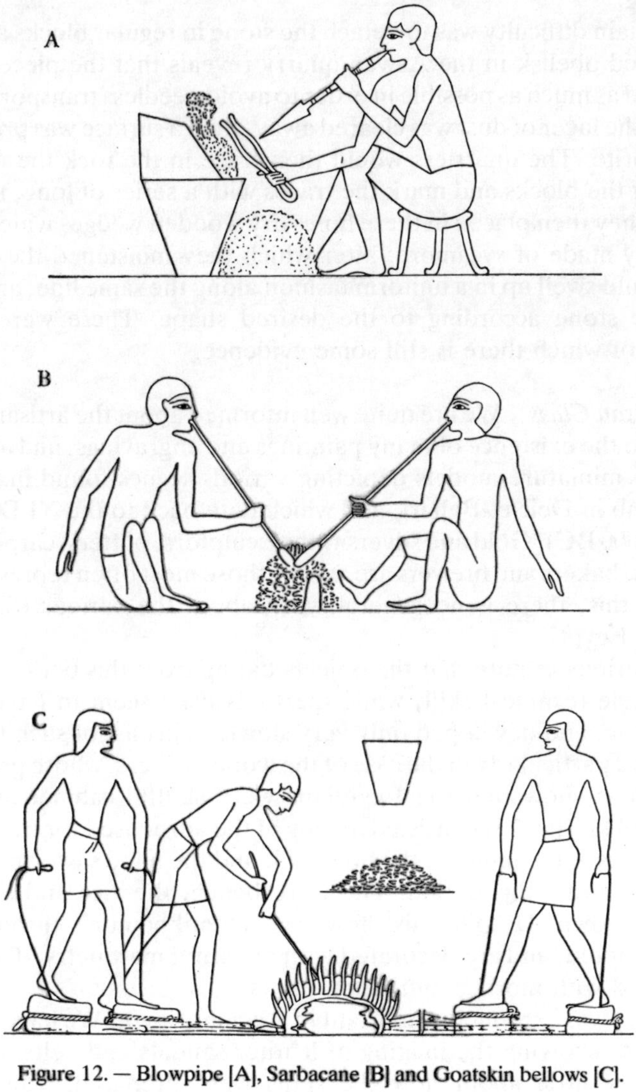

Figure 12. — Blowpipe [A], Sarbacane [B] and Goatskin bellows [C].

with granite and diorite. Large constructions were made with granite and limestone, while diorite was hardly used at all, except for sculpting.

The main difficulty was to detach the stone in regular blocks and the unfinished obelisk in the Aswan quarry reveals that the pieces were prepared as much as possible in order to avoid needless transportation. Firstly, the layer of dust was cleared away, then a surface was prepared with diorite. The quarriers would then trace in the rock the desired shape of the blocks and mark the traces with a series of long, narrow holes. They then placed in these holes dry wooden wedges which were generally made of sycamore, after which they moistened the wood. This would swell up in a uniform fashion along the same line, and thus split the stone according to the desired shape. There were some failures of which there is still some evidence.

The Artisan Class We are quite well informed about the artisan class, thanks to the existence of many paintings and engravings, and also the valuable miniature models depicting various scenes found in Meketre's tomb in Deir-el-Bahari, and which date back to the XI Dynasty (2160–2000 BC). Gold and silversmiths, sculptors, potters, carpenters, cobblers, bakers and brewers are among those most often represented. Despite this, there is no general work about the various trades of Ancient Egypt.

It is curious to note that the objects dating from this period reveal remarkable technical skill, while the tools used seem to have been fairly basic, and developed only very slowly. This list of such tools is short, and particularly in the case of the woodworkers, whose products nonetheless show that the Egyptians were skillful cabinet makers. Their tool kit (fig. 13) consisted mainly of the adze, used for trimming, and scissors and hammers, the bow drill, and the hand saw, which was invented during this period. The work bench, the vice and the nail were unknown. Despite this, however, Tutankhamun's thrones are finely sculpted, and are decorated with beautiful marquetry. They are assembled with mortice and tenon joints.

Leather was extremely important in Ancient Egypt. We have plenty of pictures showing the making of leather sandals and belts, but we know very little about how the skins were prepared. Tanin was produced from acacia pods, and chamois were made with oil which was pressed into the skins with a hammer, after which the leather was

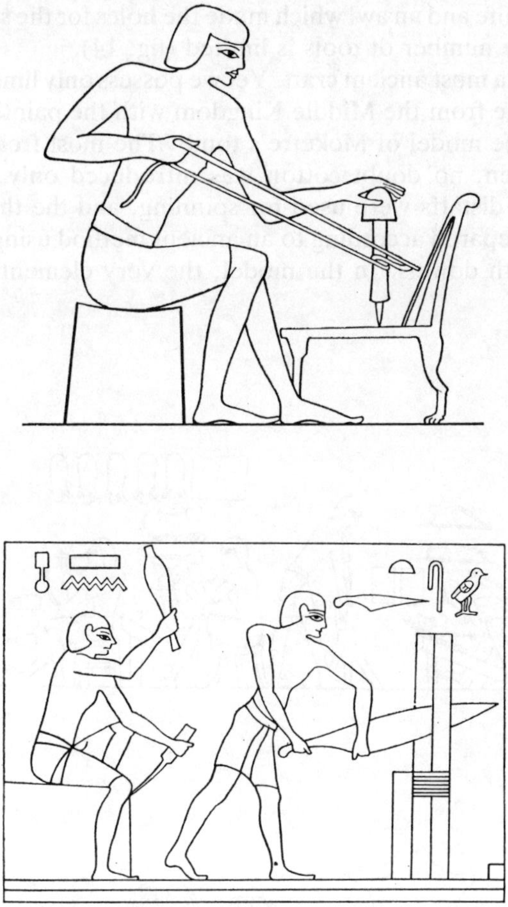

Figure 13. — Egyptian woodworker drilling with bow [above]; hand saw, shears (Cutters) and hammer [below].

stretched to make it soft. Tawing was done with alum. The oldest depiction of this industry comes from the tomb of Ti (c. 2500 BC). The curriers are pictured stretching the skins. Leather belt makers can also be seen on Rekhmire's tomb (XVIII Dynasty). One of the workers is cutting out round thongs from a large skin with a leather knife which closely resembles the same tool used today. Sandal makers also used the leather knife and an awl which made the holes for the straps. In this case, too, the number of tools is limited (fig. 14).

Weaving is a most ancient craft. Yet we possess only limited sources, and these date from the Middle Kingdom with the paintings by Beni Hasan and the model of Meketre's tomb. The most frequently used fibre was linen, no doubt cotton was introduced only much later. Spindles and distaffs were used for spinning, and the threads of the warp were prepared according to an ancient method using the surface of a wall, with dowels. In the model, the very elementary weaving

Figure 14. — Making sandals [above] and making leather straps in Ancient Egypt [below].

loom is placed flat. There was no comb, and the threads were tightened by use of a stick. From the time of the Middle Kingdom, large, upright weaving looms were used, and these were no doubt a fairly recent development. The remains of Djer were wrapped in material which had 60 threads on the warp, and 48 on the weft, which was already enough to produce a beautiful piece of fabric. Materials have been found dating back to the XVIII Dynasty which were 138 x 40 and 128 x 56. From this period onwards, the Egyptians became past masters in the art of cloth-making. The cloths were often dyed, using primary colours.

Pottery became important during the Old Kingdom, and was practised alongside the making of stone vases. Alabaster and schists were easily worked, but this was not so with diorite or granite. There are some pictures of workers making stone vases, but drills with weights on their upper end, chisels and hammers are not sufficient to enable us to understand exactly how large pieces were made. Terra cotta vases were made using a potter's wheel (fig. 15). Egyptian pottery were very ordinary, and there is no evidence that they used kilns. However, enamelling was very popular for making small figurines, and was done using silica and soda, with a copper-based colourant. These figurines were made using moulds, which explains their abundance. The oldest dated piece of glass is an imitation turquoise in the shape of a blue glass eye which was made during the reign of Amenophis I (1558–1530 BC). The use of glass became popular after this, but glass-blowing was never practised by the Egyptians.

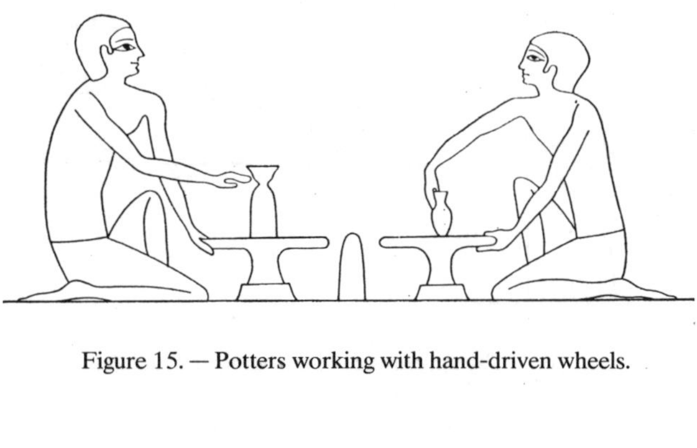

Figure 15. — Potters working with hand-driven wheels.

The jewels in our museums, and Tutankhamun's treasure are more than sufficient evidence of the mastery of Egyptian jewellers when working with metals and stones. The paintings of jewellers at work in Mereruka's mastabah in Sakkara (c. 2400 BC) show that they too had a limited number of tools, without counting the blowpipes used for melting gold. The method used to make certain types of beads of stone is still a mystery for the historian of techniques.

The study of the ancient Egyptian artisans' techniques still remains to be done. It would be helpful, too, to have precise lists of tools, like the one compiled by Blümner for classical Antiquity. And this should be accompanied by precise research into the objects which we do possess: the study of certain types of woodwork and certain cloths would be just as revealing as that of tomb paintings. But at this point a problem arises which is rarely discussed: the relationships between manual skills which can be passed on through apprenticeship, and the paucity of tools, and all this combined with an advanced knowledge of the materials being used. Certainly, there have been some important

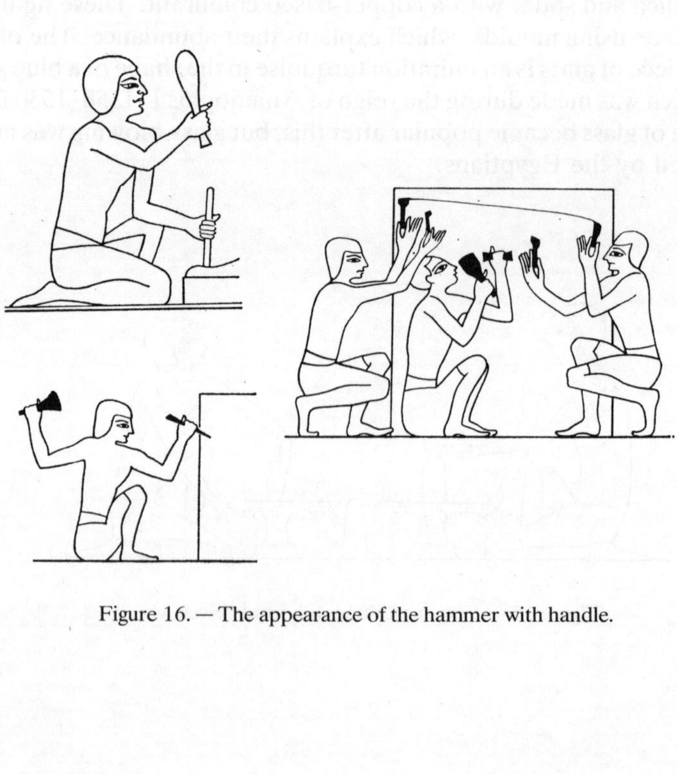

Figure 16. — The appearance of the hammer with handle.

stages: the change from stone implements, which were used during the very first dynasties, to metal ones is a patent example (fig. 16). As for working with stone, it seems that they used strikers made of diorite, and abrasives – usually a quartz powder – were used for sawing and polishing. Finally, metal chisels were introduced. However, the drilling of carnelian pearls still remains a mystery.

Egyptian artisans can be credited with some spectacular achievements, but they also neglected certain areas: one of these was pottery, which was very badly fired. It is true that a civilization will achieve more in one field than it will in another. And one must not confuse the exceptional results achieved by a few, for a few, with everyday techniques, which in a sense are the yardstick of a technical system.

Construction The architectural achievements of the ancient Egyptians still cause wonder and surprise, and one cannot deny that the pyramids at Sakkara and Gizeh, and the great temples of Karnak and Luxor still arouse our admiration. But we should remember that we have seen only those buildings which were exceptional: temples and funeral monuments. Almost nothing remains of the palaces which must have existed, and absolutely nothing of ordinary architecture. This is an important point, which tends to prove that as in Greece – but unlike Rome a,d the Roman Empire – monumental architecture was quite different to common architecture. This becomes more meaningful when one thinks that we are familiar with the houses of Çatal Hüyük and Jericho built during the neolithic period, but nothing is left of those in Thebes or Memphis, except a few famous remains. This is because these towns were built using clay bricks, an extremely perishable material. This is a whole side of construction techniques which eludes us. Ordinary construction, then, was done with clay bricks, and this was so for all villages and most towns. There are paintings which show these bricks being made, usually with a mixture of sand and straw to prevent shrinkage. They were sometimes moulded, but in any case, they were made according to standard sizes, and then dried in the sun. This material was entirely suitable in a hot country such as this with very little rain. However, it could not escape the erosion of time, and the dwellings were constantly having to be rebuilt. This explains why nothing is left of these lightweight constructions.

The large stone constructions were an entirely different matter. The Egyptians built their heavy constructions in the same way as their clay-brick houses. Just as the bricks had to be piled up, so did the heavy elements of the stone monuments. They had to take a great deal more care over the foundations, however.

The stone was assembled using dry joints; mortar was never used by the ancient Egyptians. The building of uneven foundations, which strengthened the solidity of the whole, was widely practiced from the Old to the Middle Kingdom. It seems that dovetails with bronze reinforcements were used in Abydos during the XIX Dynasty (1314-1200 BC), and subsequently became very popular.

One problem which has aroused a great deal of curiosity and polemic concerns how they managed to lift giant slabs of stone, sometimes weighing several tons, to considerable heights. The dimensions of the pyramids are well-known. The largest obelisk in Egypt is that of Queen Hatshepsut's in Karnak: it is 29.50 metres high and weighs 374 tons. The unfinished obelisk from the Aswan quarry would have measured 41.75 metres high and weighed about 1200 tons. Before the Macedonian conquest, the Egyptians had no form of lifting apparatus so they used two types of simple machines which were the oldest known: the lever and the ramp (fig. 17). For transport, they used sledges and rollers (fig. 18). These great monuments, still standing today, were built with very primitive materials therefore, and using only the driving force of humans and animals.

In fact, the Egyptians must have embedded the construction into the sand as it got higher (the advancing desert embedding everything in its wake must have given them the idea for this), thus creating ramps on which they could mount the pieces into place. Obelisks were erected in the same fashion. Once the work was finished, the monument was levelled out. There does not seem to be any other satisfactory explanation.

All this naturally implies long periods of time. The pyramid of Chephren was built in thirty years, and required a large work-force. The temple at Edfu, built during the Ptolemaic period, took one hundred and eighty years to complete. Did time and the numbers of men compensate for the real lack of technological progress? It is almost certainly true that other factors were important as well. Furthermore, we still have the same problems about the building of our great cathedrals.

The pyramids unleashed the imaginations of our contemporaries, who saw in them a reflection of all the esoteric sciences. Several

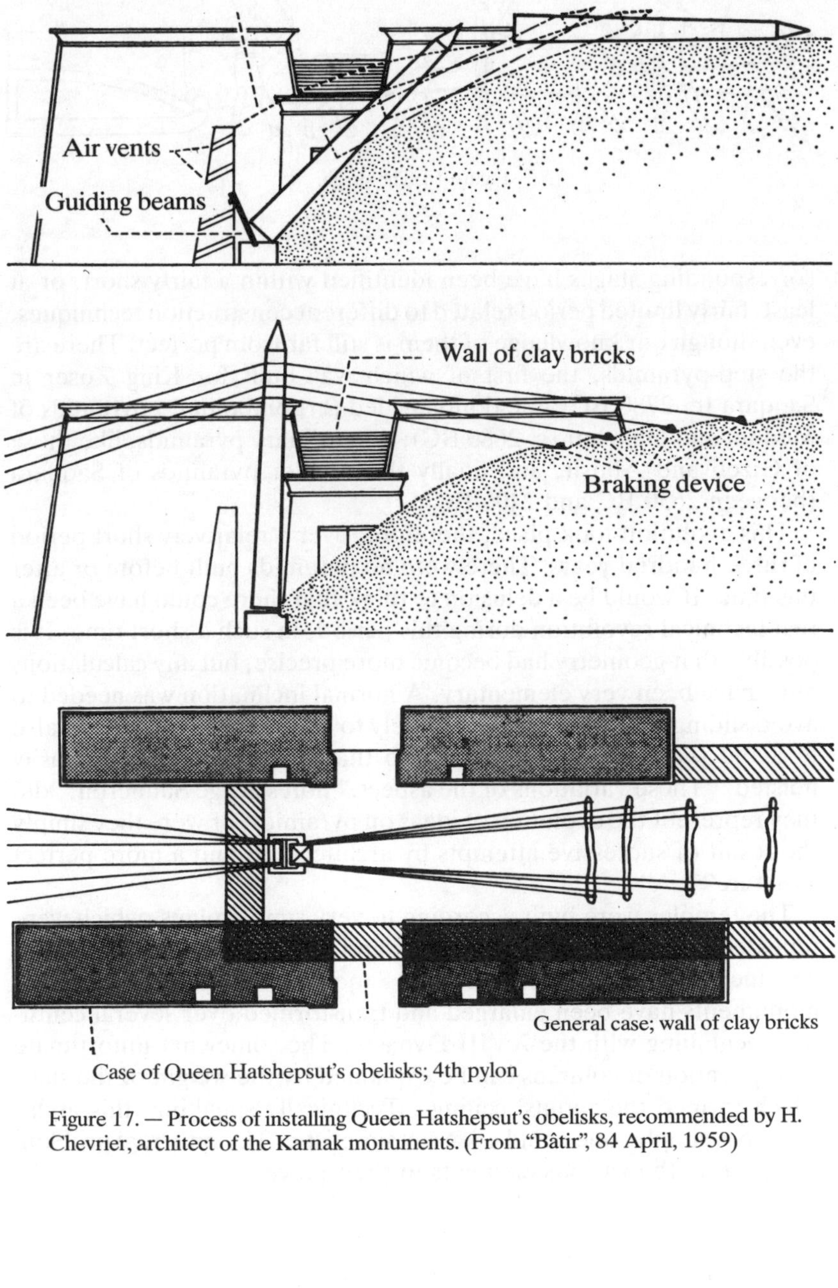

Figure 17. — Process of installing Queen Hatshepsut's obelisks, recommended by H. Chevrier, architect of the Karnak monuments. (From "Bâtir", 84 April, 1959)

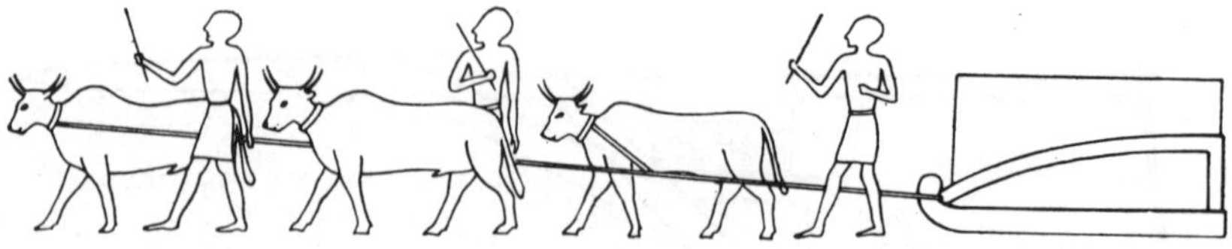

Figure 18. The sleigh.

corresponding stages have been identified within a fairly short, or at least, fairly limited period related to different construction techniques, even though our knowledge of them is still far from perfect. There are the step-pyramids, the first of which was built for King Zoser in Saqqara (c. 2780 BC); the double-sided ("rhomboidal") pyramids of Medum and Dahchur (c. 2680 BC); the ordinary pyramids, like those of Gizeh and Abusir, and finally the smaller pyramids of Saqqara (between 2650 BC and 2400 BC).

This evolution took place, therefore, over a relatively short period of three hundred years. There were no pyramids built before or after this time. It would be a delusion to think that there could have been a real technical revolution during this period, in such a short time. It is possible that geometry had become more precise, but any calculations must have been very elementary. A normal inclination was needed to avoid sliding, but this was hardly likely to happen with stone, and also to avoid too much cutting work, so that the stone could be easily hoisted. "These variations of the aspect," notes Serge Sauneron, "did they represent the evolution of ideas on pyramids, or were they simply the result of successive attempts by architects to find a more perfect structure?"

The temples were built according to very simple plans, which were repeated endlessly. Those of Karnak and Luxor are the most ancient, not including the restorations at Saqqara. But these two great monuments have been enlarged and transformed over several centuries, beginning with the XVIII Dynasty. The somewhat unfortunate multiplication of columns can be explained by the weight of the slabs which formed the rooms' ceilings. Technically speaking, this architecture is simple, one could even say simplistic. The only real problem was putting the various elements in their place.

The vault was known, but seldom used. The ribbed vault, made with mud bricks and without arches, on a mounting of oblique layers, dates back to the first Dynasty (3300–3000 BC). After the VI Dynasty, stone vaults with full arches demanded a more careful attention to form. Corbelled vaults were also in use.

These were then, fairly crude architectural techniques which did not entail any great practical problems. This was grandeur without risks. The temples of Deir el-Bahari, on the other hand, reveal a sense of balance and perspective verging on perfection.

Space The organisation of space is dominated in Egypt by its geography. Egypt is a long, narrow stretch of land with a river running through the centre. Major floods occur regularly, with various consequences.

There was no need for any roads since the Nile was a natural route for transport. In any case, they would have been difficult to trace in the silt or the sand. Consequently, Egyptians did not use wagons, either. The chariot was introduced much later on, but was used only for war. The limited size of the oasis bordering the river meant that all large buildings were built near the water course. Sometimes, as in Aswan for example, canals were dug in order to reach the quarries.

Figure 19. — Construction of a boat using very small pieces of wood. (Middle Kingdom)

Humans were usually used for transporting goods – pictures often show those great baskets suspended from a pole and carried on the shoulders of two men. This is how the harvests were usually brought in. Animals, particularly donkeys, were also used for transport from the XII Dynasty. In other paintings we can see rollers and sledges being used to move large objects. The whole thing would be pulled either by men or by animals.

There were various types of small crafts used on the Nile. First of all, there was the small skiff used for hunting in marshy areas. There were also cargo boats, often depicted in later eras (fig. 19). They were of various sizes, but the prow and stern were always raised and the draught was slight. They carry a mast with a square sail, and are manoeuvred by two lateral rudders. They were powered either by sails or by oars. There is a painting from Deir el-Bahari depicting a large transport barge, of a more rounded form, carrying obelisks. It was most likely pulled by other ships (fig 20).

It is particularly difficult to reconstitute large pieces of work. There is a singular lack of paintings and written sources, and it is possible that some contemporary historians may have let their imaginations run away with them.

There were some towns in ancient Egypt, Memphis and Thebes being perhaps the most well-known. All of the buildings in these towns, except for the temples, were built using clay bricks with the result that they have almost completely disappeared. The only remaining dwellings today are from the temporary camps which housed those working on the temples or tombs, in Gizeh, Kahun or Deir el-Bahari, for example. During the earliest times, the houses seem to have had elements which are still popular today: a reception room, a storage part and a private section. A chicane shaped entrance would open out on to the street. In Kahun, workers' dwellings had six to seven rooms (measuring about 100 square metres). Bigger houses with a large interior portico sometimes measured as much as 2400 square metres. The "workers town" of Tel el-Amarna was composed of four-room houses, all of them identical. There were some storeyed houses in the towns, probably because the land was almost certainly harder to come by and more expensive. We have pictures and models of these. The work was done on the ground floor, the reception rooms were on the first and the private section of the house was on the second floor with a terrace on the top. These working towns were built

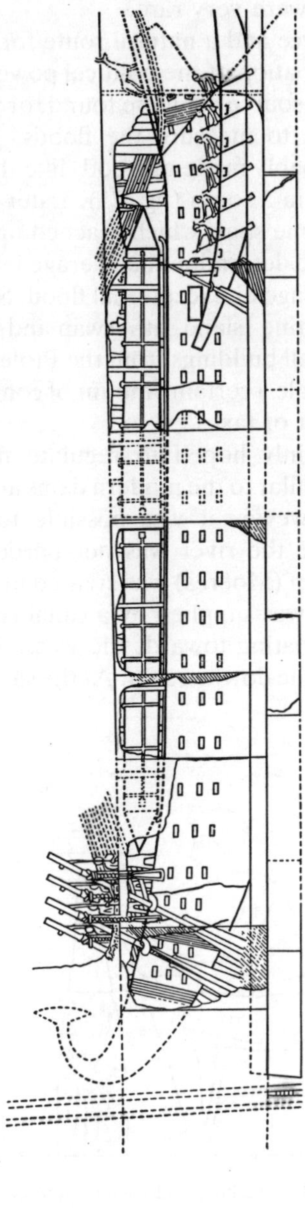

Figure 20. — Barge transporting obelisks.

according to standard plans, but there were no plans for the large cities. Perimeter walls were very rare.

As both a water source and a natural route for transport, the Nile was a constant preoccupation of the political powers. They knew very little about it, where its source could be found, or why the floods took place. They were able to measure the floods, however. The first "nilometers" were probably made of wood, like those depicted in the process of being made in a tomb (fig. 21). Later they used masonry structures, a type of stone shaft which reached up to the level of the water: they were able to determine the average level with markers or steps, and then the average or exceptional flood. Some of these shafts still remain, at Elephantine Island, at Aswan and near the temples of Kom Ombo and Edfu, all buildings from the Ptolemaic period. These calculations made possible a certain amount of control work and also a basis for the assessment of taxes.

The Egyptians certainly hoped to regulate the floods, but the construction of dams similar to the modern dams at Aswan would have been unthinkable. However, it was possible to create diversions instead, at areas where the river was not bordered by cliffs. It is possible that Lake Karun (Moëres) was created in a depression with a similar aim in mind. It was supplied by a canal running first parallel with the Nile, then deviating towards the west. Two parallel works then linked it with the Nile downstream. At the same time, the oasis at

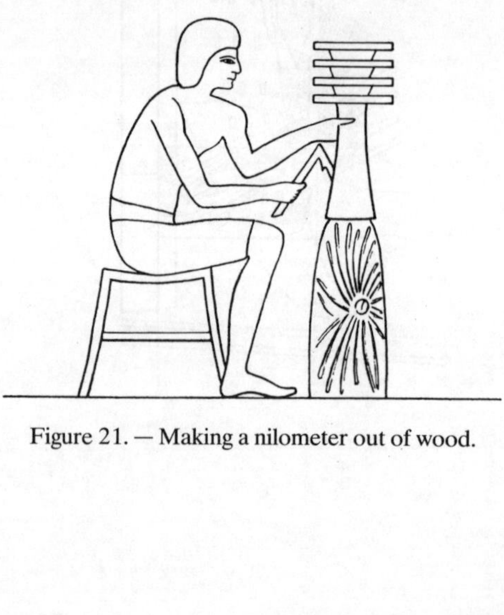

Figure 21. — Making a nilometer out of wood.

Fayum was developed. It is difficult to say at which period these works were executed: the most ancient monuments of this region date back to the XII Dynasty (2000-1785 BC).

All the irrigation systems have completely disappeared. These were very lightweight works, and were probably remade after each flood. In the Nakht's tomb (around 1415 BC), workers can be seen in the process of digging a canal: one of them is cutting down a tree with an axe, while another is digging with a hoe. Like the Dutch, they had to raise the water from one level to another in order to use it. For this, they used two instruments, the shadouf, a very ancient implement, and the Archimedes' screw, which no doubt existed before the time of the famous scholar whose name it bears (fig. 22).

It was hardly likely that the Egyptians would not have thought of developing the Nile into a major channel of communication, but to actually do this was an entirely different matter. There has been a lot of discussion about the so-called canal of the two seas, the idea of which was taken up again in the 19th century, in competition with the Suez canal. A diversion was created downstream at Bubastis (Zagazig),

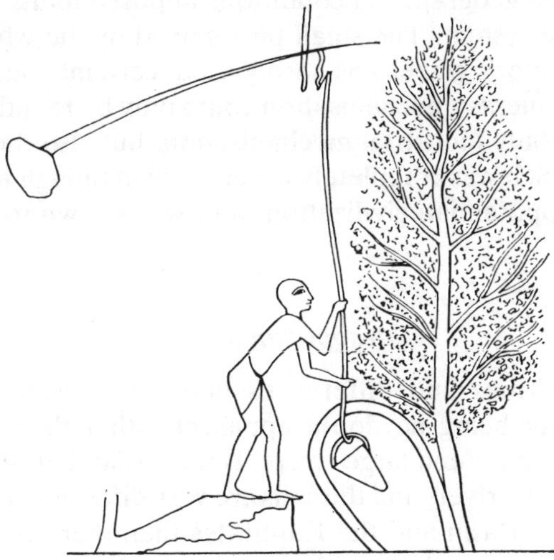

Figure 22. — A shaduf.

from the pelusic branch of the Nile, and finishing up at Ismailiya. From here, a canal was dug from the Bitter lakes to end up at what is now Suez. It is claimed that its breadth was wide enough to take three triremes abreast. Some attribute this giant project to the VI Dynasty (2423-2263 BC), but others, perhaps more accurately, date it at the end of the 6th century (around 520 BC).

There were no doubt many small utilitarian diversions, many of which were only used in case of flooding. We have already mentioned the Aswan quarries. The remains of a dam can still be seen at Helwan, dating back to the fourth millenium BC and which supplied water to the alabaster quarries. Some canals were built from the Nile to supply the towns and villages with water – as at Neuser-Re in Abussir, under the V Dynasty (2563-2423 BC), for example. The piping was made of copper with plaster joints. At Tanis, pottery cones were used.

Research is still needed to complete the basic picture which we have just sketched: we can only hopes that someone will choose to do it soon.

It appears therefore, that the Egyptians did indeed possess a true coherent technical system, even if its means appear to have been limited. Some of their projects still arouse surprise, even if one considers their equivalents elsewhere (the pyramids in Mesopotamia and Mexico). The geographical conditions imposed limits which perhaps blocked the system. The small part played by the wheel, which except for the potter's wheel was hardly used, certainly put the brake on some developments: mechanisation could only be found outside of Egypt then, not only industrial mechanisation, but also the mechanisation of war, which was completely absent – even though it existed in the nearby Mesopotamian civilization, and was known to the Egyptians.

Mesopotamia

There are similarities and differences between Mesopotamia and Egypt, the former being no doubt apparent rather than real. Both regions are centred around large rivers. But beyond that, the climate, and the nature of the rivers and the soils are very different. There is less unity in the Tigris Basin and the Euphrates than there is in the Nile valley. Towards the south, Babylon is closely dependent upon the rivers whose flooding and irrigation are its life-source. Further north, a plain of low ground is crossed by several water-courses, each one of which is an economic entity in itself. In Assyria and Syria, rains from the west make agriculture possible without irrigation.

It would be impossible for two climates to be the same over such a long distance from north to south. Winter is mild in the southern regions in the Sumer countries, and this provides fertile ground for the date palm. In the Akkad countries in the north, however, winters are cold. Snow occasionaly falls in Assyria and Syria, and here the date palm naturally gives way to vines and fruit trees.

Natural resources are limited. Building stone is of a mediocre quality, and flints are rare. On the other hand, the abundance of minerals in Anatolia explains the early use of metal in this region.

It is only natural that these conditions should have influenced the evolution of techniques. It is also obvious that they imposed necessities of a different type than those which are found in Egypt. This is in spite of chronological coincidences which are largely the result of contact between the inhabitants of the Nile banks and those from the banks of the Tigris and the Euphrates.

A breaking off We have already spoken about certain sites which reveal an advanced Neolithic civilization, especially in the northern and western regions of this vast geographical entity. During the fourth millenium BC the entire country was able to profit from a fairly advanced technical civilization. The contribution of various different races – as in Egypt – must have enhanced the elaboration of a rapidly perfected technical system. The first stages are marked by a series of successive discoveries which appear to have displaced the centre of this civilization from the north towards the south. Jarmo has already been mentioned. The hamlet, or rather the superimposed hamlets, are situated on a spur above the valley of a tributary from the Tigris. It would seem that the beginnings of a farming industry led to the permanent settlement of part of the population. Grains of wheat and barley, millstones and sickle-teeth are in evidence, along with the bones of domestic animals (sheep or goats, cows, pigs and dogs). The houses are made of pressed clay, and pise, with reed floors. This site was no doubt formed by primitive farmers, and dates back to around 5000 BC. The techniques used at this point were still simplistic.

The settlement at Hassuna, in Assyria on the western banks of the Tigris, marks a new stage of progress. A steady farming industry was assured by two permanent water-courses and regular rain-fall. Some consider this to be the first farming industry in the world. Although there are almost no remains of any houses (fig. 23), large jars of coarse pottery have been found, mixed with chopped straw and imperfectly fired. Large hand-mills and tools which are generally agreed to be hoes, and the bones of large cattle, sheep and goats are evidence of farming and rearing. Slings were used for hunting, with bullets of baked clay. No arrowheads were used. Wood was almost certainly worked with axes and adzes. Subsequently, sedentary peasant villages were established near the first settlement, on six different levels. Only a few curvilinear walls remain of the oldest village. Level II already boasts rectangular houses. In villages III to V, the houses had three or four rooms, and were probably grouped around a court. The wooden doors were actually able to pivot on socketed stones. There were bread

Figure 23. — Isometric view of Hassuna 'standard' farm. (From Lloyd and Safar, *Journal of near eastern Studies*, 1945)

ovens and silos made of unbaked mud jars mixed with straw and externally coated with bitumen.

Farming tools were already growing more sophisticated. Their pottery consisted of a few different types of well-fired painted vases, decorated with motifs similar to those used in basketry. These small tribes imported obsidian and amethysts, which they knew how to perforate. They do not appear to have used any copper. Childe believes that the now well-established farming industry contributed to a population growth, which would explain the expansion of this civilization.

The pottery of level VI in Hassuna marks a break with the past. This new culture was christened "Halafian," after the Tel Halaf site, and stretches from the Iranian hills to the east of the Tigris to the Mediterranean. This stage was clearly more highly developed.

Emmer and barley were cultivated. They reared two types of cattle, sheep, goats and pigs, although hunting (still with slings) supplied much of their food. There is evidence of some form of a textile industry but we do not know if they used flax or wool. Copper had made its appearance in the form of small beads which probably came from local deposits. Obsidian and chert were still used to make knives and sickle-teeth, and adzes were still made using celts of polished stone mounted in knee-shafts.

An Halafian vase has led some to believe that the wheeled-vehicles existed at this time, but this interpretation is very unlikely to be correct. The Halafians had not heard of the saw, which is essential for wheel-making, and they had not even heard of the potter's wheel. In fact, we do not know of any spoked wheels dating back further than 2000 BC.

The paste used in pottery was very delicate and covered with a layer of slip, which was sometimes polished. It was already possible to heat the kilns to quite considerable temperatures, sometimes as high as 1200 degrees.

Stone-working was still excellent, even with the hardest of stones. Vases, pearls and even amulets provide evidence of this.

If one accepts that a public highway is evidence of urbanisation, then it should be said that the roads in Arpachiya were paved with shingle. Furthermore, there were constructions which have been called *tholoi*, from which primitive temples were made, or lofts, or shops, but which all signify the life of an organised collective. The Halafian civilization, which was also fairly widespread in all this region, was supplanted by the sumerian civilization of al'Ubaid, on the delta site. The original Sumerians, according to Childe, must have come from an older country, perhaps from the steppes in the North-West, or from the mountains in the East. Here, urials, mouflons and goats ran wild, and cereals grew naturally. In any case, this delta country was extraordinarily fertile, and was re-fertilised each year by alluvium deposits. The lagoons abounded in fish and prey of all kinds. This may have been a natural paradise, but it needed intensive work "and the organised cooperation of large human collectives." They had to create areas for cultivation, drain the marshes, and control the irrigation and flooding. At the end of this chapter we shall consider the social and political conditions, also to be found in Egypt in this region, and which are perhaps at the basis of these technical innovations.

"It may be inferred that the Ubaid farmers of Sumer had devised a rural economy, based on irrigation, which was able to produce sufficient food to support a growing rural population, and also yielding a social surplus which could be used for non-productive works and trade."

There was no stone in the region, so it was necessary to maximise the resources they did have, which were essentially clay and reeds. Sickles were made with hard-baked clay but a new element was brought into construction, that of moulded clay, or the brick. The clay brick – destined for a long future – was first used in Eridu. Stone-tools were imported and were used in much the same way as during the Halafian period. Copper, which was cheaper, more durable, and could be sharpened, was hardly used. The imitations of metal tools made out of clay or stone are evidence that copper was known, however.

Pottery was hand-made, and the number of different types of pottery designs increased considerably. The surface was left unblazed but decorated with geometric designs using paint with a light sheen.

By now, the temple had become an extremely important factor in settlements. For these rural civilizations, which were based on a strict organisation of irrigation, the temple was not simply a religious sanctuary, but also a meeting-place. The insignificant remains and the lack of written sources prevents us from pursuing this point. "In any case, all known Ubaid settlements remain villages; their limited size and lack of functional differences tell us that these local groups of the population did not yet merit the name of cities. It was only during the Uruk period that some of them were to achieve this distinction."

Archaeologists have defined the Uruk period by the appearance of unpainted red-slipped and grey pottery. The earlier stages of the Uruk civilization coincide with the final Ubaid period (c. 3900–3500 BC). It is thought that this new civilization was the result of people coming from the West or the North-West, who would have been Semites. During this period, both technical and social structures were transformed. Childe notes:

> Even the earliest Uruk pottery, from Erech -IV and Eridu -V includes wheel-made vases. Admittedly, the potter's wheel was not only a significant discovery in itself, but it also implied the existence of a more articulated economy, the use of metal tools and therefore the existence of professional smiths, and is nearly always associated with wheeled vehicles.

In other words, a new technical system was evolving. And furthermore:

> The development of the Tigris-Euphrates delta would be calculated to produce just those social and economic conditions in which metallurgy and wheeled vehicles would be serviceable and under which the unit of settlement could and must be relatively large and the social surplus sufficient to the support of professional potters, smiths and other full-time specialists.

The growth in population was certainly a direct consequence of this.

The temple, a religious monument, a meeting place and a centre for stocking part of the farming produce, continued to evolve. At Eridu, the temples of ancient Uruk were made according to obeidian traditions, but on a larger scale. Temple -III was built on a base covered with limestone blocks. The first pillar temple with columns was built on level -IV B at Erech, the columns being made of brick. Some of the surface area was reinforced and decorated by a mosaic of baked clay cones hammered into the mud plaster when it was still soft.

The seals and tablets produced during the last part of the Uruk period are incised with conventional numerals and characters: these documents – the oldest written ones extant – represent the temple's bookkeeping. The numeration is sexagesimal, whilst the writing is pictographical and probably ideographical.

Civilization has taken there a great step forward. Writing was practised and conglomerations were being formed – Eridu, Erech, Ur, Lagash, Ugair – which could almost be called towns. Temples were getting bigger. All this conveys an unprecendented degree of social organisation and technical progress, two inextricably linked factors. The plough – or more accurately speaking the swing-plough – appeared at this time, as did probably the metal scythe. Henceforth, farming was organised work. Cows provided milk, and goats were raised along with two types of sheep, one hairy, like the Egyptian sheep, and the other with curly fleece. Bows were used for hunting, and lances were tipped with metal points. Large and small rivers accomodated boats with high prow and stern, whilst sledges and wheeled carts were used on land. The traction animal most widely used was the ox, and it is possible that the Asiatic ass, the onager, had already been domesticated.

This group of advanced techniques supplied the necessary supplements to the ever-expanding group of craftsmen: there were already many workers of metal and leather, sculptors, carpenters and potters. They had learnt to alloy copper and lead, thus lowering the melting point of the former, and they could use wax casting. In contrast, tin bronze was unknown and most metals were imported from Anatolia. The art of pottery became more widespread and seems to have been industrialised during the later Uruk period.

We are now at the end of the fourht, and at the beginning of the third millenium BC. An entire technical system has thus evolved, a system which is coherent, with diverse and complementary activities. This technical system is very similar to that of Egypt during the thinite dynasties of the same epoque. The processes of development were identical, even if there were marked differences in detail. The clothes worn by lion-hunters were exactly the same, an Uruk ship was exactly the same as the "foreign" ships of the late Gerzean period. The jar with bent spout from the late Uruk period could almost be mistaken for the one depicted on the Narmer palette. Some characteristically Sumerian traits seem to indicate that in some respects, credit should go to that region.

Several hypotheses have been formed about this surprising synchronism, even given the relative proximity of the two regions. The major factor seems to be the role played by Arab, Semitic peoples coming from the North of Arabia and spreading out to the west and east of their original territory, thereby transmitting to various areas discoveries made on the borders of the Nile, the Tigris and the Euphrates. Some Mesopotamian techniques, like various forms of art for instance, would thus have been passed on to Egypt, while the bow and self-coloured pottery would have come from the East to Sumer.

We enter the historical period at around 3000 BC. Unfortunately, as far as techniques are concerned, the historians of ancient Mesopotamia seem to be entirely unconcerned about establishing a strict chronology. As in the case of Egypt, not enough is made of the evolution of a technical system, which certainly did not remain fixed throughout the entire period and whose evolution was inevitable from the start. The technical system, in fact, had to be developed to its limits, and one can see this process beginning from the time of Jemdet Nasr. Of course stone and baked clay were still largely used, but metal were becoming more and more popular, for making fish hooks, chisels, axes with handles, and copper, lead and silver vessels. They had also learnt to separate silver from lead. At the same time, Susa I A was still only a fairly densely populated village, but the materials in evidence closely resemble those of Sumer. It was with Susa C during the late Uruk period, that a real civilization was born: writing and the establishment of proper towns bear witness to this fact. It is thought that the innovations of Susa B were due to the infiltration of elements which passed on identical techniques to Sumer during the Uruk period. Thus, it was between 3000 and 1500 BC that the technical system was finally perfected, and once more this echoes the developments in Egypt.

Primary activities Agriculture held a place of prime importance in the economy at the beginning of the third millenium BC, which naturally made farming techniques very important. As with all types of agriculture, evolution was relatively slow, but it was progressive. Even the invasions and other upheavals suffered by these regions during this period do not seem to have made any great difference.

The plants grown were more or less the same as those cultivated during the advanced neolithic period, and those grown by the Egyptians at the same time. Barley was the staple cereal, with emmer and millet. We shall not dwell on the point made by certain archaeologists, that Mesopotamia was the birthplace of cereals. It was only later that gardens were producing a complementary diet which was both more varied and superior to the leguminous plants produced in the neolithic period. We do have a list of these, a relatively late one dating only from the 8th century BC: garlic, onions, leeks, cabbages, lettuces, fennel, beet, rape and radishes. Several herbs were also grown : three types of mint, basil, saffron, coriander, rue, thyme, and pistachio. Persian ferrula was also popular and there were several varieties of poppy. The present state of documentation on this period makes it very difficult to know from which date these various vegetables were first cultivated. There is a similar lack of information about fruit trees. Vines were grown from the Sumerian period onwards. Fruit trees, the date palm first of all, became widespread rapidly in the more northern regions. During the Babylonian period, virtually all fruits were grown: pomegranates, medlars, apricots, peaches and figs.

A very few paintings depict farming implements. We saw that at the end of the archaic period, the swing plough seems to have complemented the hoe which had long been in use. This presents a problem, which has not yet been resolved. In Egypt, the farmer used an ard; here, on the other hand a characteristic tooth plough was used (fig. 24). The difference is notable and deserves serious research. If the interpretation is accurate, it was often surmounted by a seed coffer. However, this addition seems to have been a later development (figs. 25 and 26). By the end of the Uruk period, the wooden sickle with small cutting stones had probably already given way to the sickle made of metal.

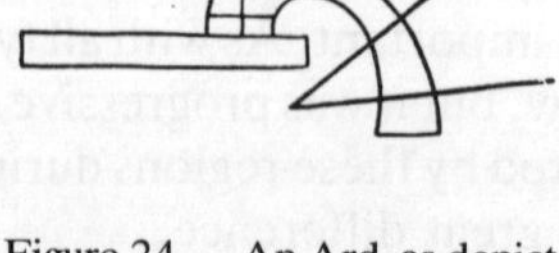

Figure 24. — An Ard, as depicted on an Uruk pictographical sign.

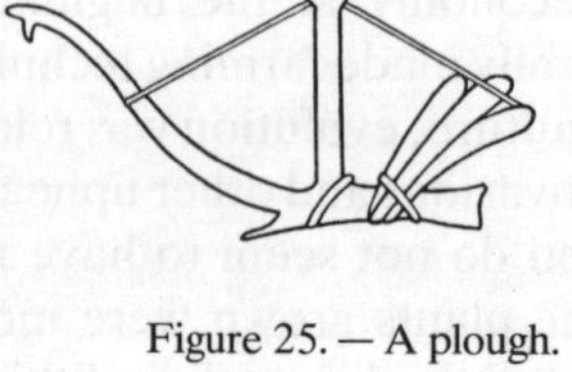

Figure 25. — A plough.

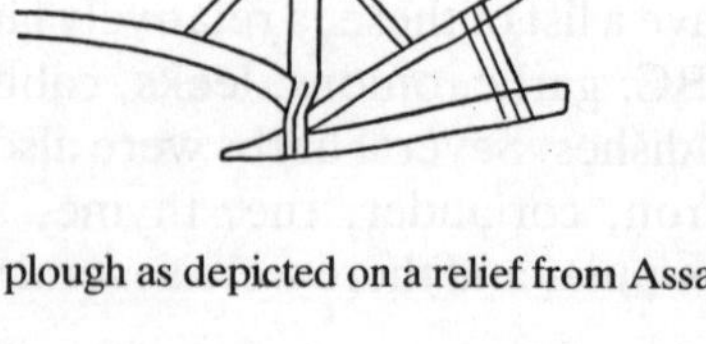

Figure 26. — A plough as depicted on a relief from Assarhadon.

Grain was usually kept in silos whose evolution was slow and limited. These techniques hardly differed from those of the archaic period.

Stock rearing rapidly developed: cattle, pigs, donkeys, goats and especially sheep were the first domestic animals. For a long time, since metal tools were lacking, the wool was ripped from the animal's back instead of being cut. From the Ubaid period, as illustrated in a famous bas-relief, milk, butter and cheese were being produced. Milking took place in cowsheds made of reeds. The milk was poured into a large jar, which was kept constantly moving, constituting the churn. The butter was filtered through a funnel which kept back all the lumps – these were then packed down into another jar.

Poultry quickly made their appearance: first geese and ducks, then much later the chicken.

It seems that hunting came to be practised largely by the ruling classes, for pleasure rather than for obtaining food. Hunting scenes are often represented, mainly that of lion hunting from the earlier periods. Bows and lances were used. Boars and partridges were also hunted, and falcons were added to this list from the first millenium BC. It is also noteworthy that locusts were often eaten on brochettes.

Fishing was also widely practised, thanks to the many lagoons and canals. This was an important supplementary resource, and one tablet mentions eighteen different types of fish actually caught. Fishponds soon came into use. Rods and fish hooks (which were soon made of metal) were used for fishing, and also nets and harpoons for larger fish.

Mesopotamia had few resources of wood, so they were forced to import it. The same phenomenon was apparent in Egypt and we saw that it set limits to the development of certain techniques. The most important tree was the date palm, for its wood and fibres which were used for making rope, and for the fruit-stones which were used for fuel or crushed to make animal fodder.

There were no metalliferous deposits anywhere in this region. Metals came either from the rich mines of Anatolia, or from the Persian Gulf. Copper appeared at the end of the Ubaid period, but was still rare. Mirrors were found in Susa made of plaques of polished copper, along with axes and other instruments made of the same metal. Analyses of the objects coming from the Susa necropolis reveal 92.12% pure metal, with no iron, sulphur, lead, zinc or manganese, but with traces of nickel. So at the dawn of the third millenium BC, copper was still used as a pure metal. It came from Oman, the Iranian plateau, and Anatolia.

Real bronze, which was not simply an alloy of lead and copper, appeared only during the third millenium. This was tin bronze or antimony bronze. A tablet from the III Ur Dynasty gives the composition of bronze as the following: 80.05% lead, and 5.84% antimony and the rest is unidentified.

The appearance of iron will no doubt remain a mystery for a long time. Here and there iron objects have been found which date back to declining periods, but these finds are isolated. In a predynastic tomb in El Gerzeh in Egypt, some elements of iron oxide were found in a necklace, and some tools of soft iron in the Cheops pyramid. In Mesopotamia, from around 2700 BC, which is about the same time, some fragments of iron were found at Tell-Asmar, and among some objects found from the 15th century BC was the handle of a bronze dagger which still contained some particles of thc iron blade. Opinions are divided as to whether this was iron from deposits or meteoric iron, but the dagger's blade found at Khafaje was smelted from terrestrial minerals. Some, like Childe, believe it was imported, probably from Armenia, where later events point to the fact that a primitive tribe had found an economic process for reducing the minerals which they managed to keep secret until the end of the second millenium BC.

It is almost certain that as in Egypt, iron was not an innovation of these two ancient civilisations, and furthermore, Egypt had some deposits which were totally lacking in Mesopotamia. It is supposed that the Hittites supplied most of the iron. In the 18th century BC, iron was found in places, but still in relatively small quantities. It was only at the end of the 12th century BC that its use became more widespread. We have seen that the real age of iron came even later in Egypt. It seems that iron was used originally for producing weapons, while most tools were still made using copper and bronze.

One material was adopted quite rapidly by the Mesopotamians, and that was glass. It was probably around the beginning of the second millenium BC that these peoples discovered not only ordinary glass, but also coloured glass.

In the field of primary activities, then, it seems that the Mesopotamians were somewhat in advance of the Egyptian civilization, especially where the production of materials was concerned. The inhabitants of the Nile valley had the advantage of much larger, and more varied natural resources. The Mesopotamians depended largely on supplies from elsewhere, and there were periods when they lacked certain materials. At the very end of the Archaic Dynasty II, the Sumerian copper workers were using bronze which contained 6 to 10% tin – we do not know where this tin came from. There were periods when tin was completely lacking.

Secondary activities In a general sense, we have very little information about the techniques of Mesopotamia's secondary sector. We have hardly any written sources, and figurative representations, like those found in Egypt, are rare. And it is difficult to define the methods of fabrication simply using those objects which can still be found today. This lack of information means that the following discussion will only skim the surface of the issue, and this all the more so because archaeologists have hardly dealt with these problems.

Textiles are usually dealt with first in these discussions, and this one is no exception. The main textile fibres were used successively, the first of which, wool, dominated the industry for a long time. For a long period, the wool was torn from the sheep and not sheared, which required large herds. The wool was first sheared at the beginning of the first millenium BC, when iron was finally introduced into the country. Bleaching the wool – even though they knew how to marry white with black wools – was done with soap of various kinds (soda, potassium from ashes, alum and resin).

Linen was produced relatively early in Palestine. Once the wool had been torn off, it was dried in the sun on the roof of houses. After the retting, it was dried out again in kilns, after which it was hammered down and combed. In Mesopotamia, flax was unable to compete with wool. We should also mention hemp here, and cotton was also cultivated from time to time. Around 700 BC, Sennacherib attempted to increase the amounts grown in Assyria, with plants probably from the Indus valley. The fall of Nineveh in 612 BC halted this expansion.

The oldest textiles known to us are those discovered at Susa, which date back to the end of the Ubaid period. These materials range from untreated fabric, like swaddling cloth, to delicate cambrics. They were made of flax, with threads twisted at both ends for the weft. The instruments used produced poor results. There is one representation of a spinner from Susa, who as to be expected, is using a spindle and a distaff. One seal depicts a workshop for weaving (fig. 27). The cloth,

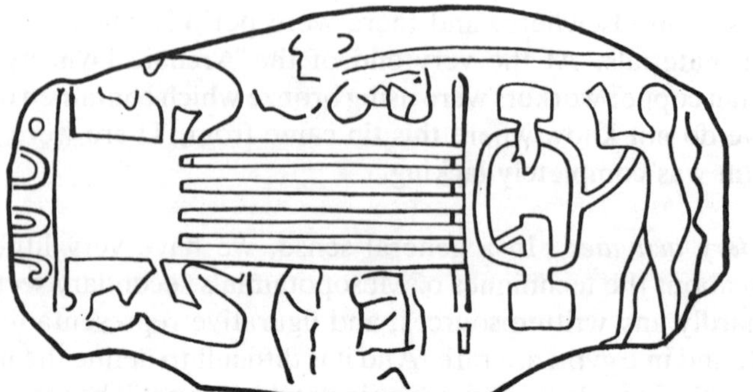

Figure 27. — Scene depicting weaving. (From Contenau, *la Civilisation d'Assur et de Babylone*, Paris, 1937)

half-completed, is stretched by two people probably on a horizontal loom. In Palestine, the vertical loom was adopted quite early. The shuttle appears only in a later document (Job, VII, 6).

Originally, clothes were almost certainly made using sheepskin. The designs used were then imitated using wool, then draping, which was finished with embroidery. During the first millenium BC, draping was abandoned in favour of the long tunic. Working men wore short tunics with a belt. Embroidery was very popular with the upper classes.

These cloths were probably dyed during the later periods. The primary colours were provided by indigo, saffron and cochineal, and other colours using mixtures of these. Carpets were probably used during the Assyrian period, evidenced by the step of a door from the palace of Khorsabad (8th century BC), which has a sculpture of an imitation carpet with fringes. It is not impossible that the Egyptian textile industry was, from a certain time onwards, technically superior to that of Mesopotamia.

Like the Egyptians, the Mesopotamians used leather a great deal, but its use was gradually limited to make shoes and military equipment. The techniques used were probably identical to those used in Egypt, but since there are no figurative representation of techniques using leather this cannot be proved. The skin, closed like a sack, was immersed in a tank filled with tanin. It was used to make shields, scabbards, quivers, baldrics and girths. But the Mesopotamians went further and used it to make waterbags, sacks and parasols. Even door-hinges and boats were made of sewn leather. The growth of Assyrian power prompted the substitution of leather protection for coats of mail and metal helmets.

The Mesopotamians were certainly not as skilled wood workers as the Egyptians. There is a major reason for this: the lack of raw material since the use of the date palm was restricted. This is evidenced by contracts for selling property, which reveal that vendors used to take all their precious wooden fencing with them. Probably the only construction to have frameworks made of wood were public buildings. Woodworkers and cabinet-makers – the number of whom was surpisingly high – made these frames, and handles for tools, weapons, and chests, which were perhaps the only type of furniture known.

We noted above that the Sumerians used fossilised sea-shells from the lagoons for glyptography and inlaying. These were of an ivory shade. Later, actual ivory was used: elephants could still be found in Syria in the second half of the Second millenium. A few years ago, some interesting specimens of ivory-work were found in Assyria, Syria and Palestine.

Like everywhere else, fire was used a great deal in varied techniques which required different sorts of kilns. It seems that techniques using fire rapidly developed and kilns were built to reach temperatures of up to 1200°C. Fire was used primarily for preparing food and drink, and among the latter was beer, which was also popular in Egypt. Fermented barley, and very rarely, wheat were used, the grains of which were contained in large jars kept at a warm temperature. The grain was then dried in the sun and roasted in kilns to stop fermentation and conserve the malt sugar. After this, it was ground with a pestle and the husk was removed using riddles. Kept in jars, the malt was cooked in rounds with spices. From Ur III, it seems that pulp was preferred. The malt bread was soaked in water, the fermentation being carefully controlled. Several types of yeast were used. The basic tool kit consisted of stone mortar boards, kilns and jars. Pottery, an ancient art, was still being perfected, and the appearance of the wheel from the fourth millenium BC which succeeded the lathe was a major step forward. We still have a painting of a potter's kiln on a Susa cylinder. It has an elongated form and has two low domes at both ends which maximise the amount of heat retained. The humidity produced by firing escapes through two rows of ventilation holes. Other types of circular kiln were also used. Charcoal was obtained from the date palm, from undergrowth, dried scrub and date-stones.

Brick-making slowly developed. The rectangular bricks in the shape of tiles used in Uruk constructions were eventually replaced by small, plano-convex bricks of which one side was flat and the other "cushion shaped." Originally, these bricks were mud bricks, dried in the sun in wooden moulds. Little by little they were relegated to the role of filling, and the facings were made of fired-brick and enamelled from the 15th century BC onwards. The total lack of stone forced Mesopotamian civilizations to use brick alone while the Egyptians used an infinitely more solid material, at least for building their great monuments. Before the introduction of fired and enamelled brick, it is obvious that rebuilding must have been very frequent, even in a country where atmospheric precipitation was rare, which is the case in the southern regions.

Although the Mesopotamians had very few minerals, they nevertheless worked much with metal. There were many blacksmiths, silver and goldsmiths, and the objects which remain for our scrutiny testify that they were extremely skilled craftsmen. Once again, we know little about the techniques employed: we know of no well-preserved metallurgic furnace, or of the techniques used for moulding and casting, and little of the processes for forging the metal. The Mesopotamians cast the metal in moulds, then using a file and chisel. They also developed a method for stamping metal. For metal statuary, they used the methods employed by the Egyptians where thin sheets of metal were fixed with nails to a wooden model.

The tools used by the Mesopotamians were as limited as those used by the Egyptians during the same period, so the following list is relatively short:

Flat chisels with a pointed heel; saws with a tang projecting from the middle of the butt; small pliers formed of two metal bands soldered together; harpoons or hooks with two or three barbs; axes for hewing inclined towards the base relative to the socket; straight adzes or broad axes; oval daggers reinforced with a median rib; crescentic axes with scalloped backs; spear-heads with a projecting heel, the blade of which is leaf or poker shaped with an octagonal base from which a quadrangular tang tapers so that it can be inserted into a reed handle;

double-barbed arrow-heads; single-barbed spear-heads; arrow and javelin heels, forked and fixed in a reed shaft.

These are quite limited tools which, like weapons, evolved very little despite the transition from one metal to another.

As was the case in Egypt, machines were virtually unknown in Mesopotamia. We should point out, however, that rams were used in military concerns, ancient representations of which are available. The lance and bow, together with the chariot, which will be discussed later, were the basic weapons.

Space Here again, we encounter the same problems, and apart from a few exceptions, the same solutions which were developed by the Egyptians were practised by the Mesopotamians.

Transport was an absolute necessity, particularly for resupplying certain materials which were not available in the region. Land transport followed fairly stable routes. Pack animals were undoubtedly widely used for carrying loads, as they were in the West up until the Middle Ages.

Historians do not entirely agree on the date at which the first vehicle appeared, generally known as wagons or chariots. The existence of the wagon from the fourth millennium BC onwards is attested by archaic tablets from Uruq (level VI), the semi-pictorial system of which illustrates a certain number of recognizable objects (fig. 29). According to Paul Garelli, one of these might represent a four-wheeled wagon, obviously derived from a sort of sledge. The military use of the chariot from the years 2800 BC is only attested by a vase painting from Hafagi (fig. 30). Elena Cassin believes that the Vautours stele and the so-called Standard of Ur mosaic dating around the third millennium BC, show four-wheeled chariots with extensive bodywork, but the wheels are still solid. Moreover, in certain tombs at Ur and Kisch, of the same epoch, illustrations of four-wheeled wagons pulled by oxen, donkeys and onagers can be found.

Thus an evolution has taken place, a branch of which leads to the development of the chariot of war. The first modification was made to the wheels. They were hollowed out to form four spokes. Acccording to Garelli, the Mesopotamians followed the Mitanian example by adopting the six or eight spoked-wheel, considerably reinforcing it from the 15th century BC. The chassis was also appreciably lightened.

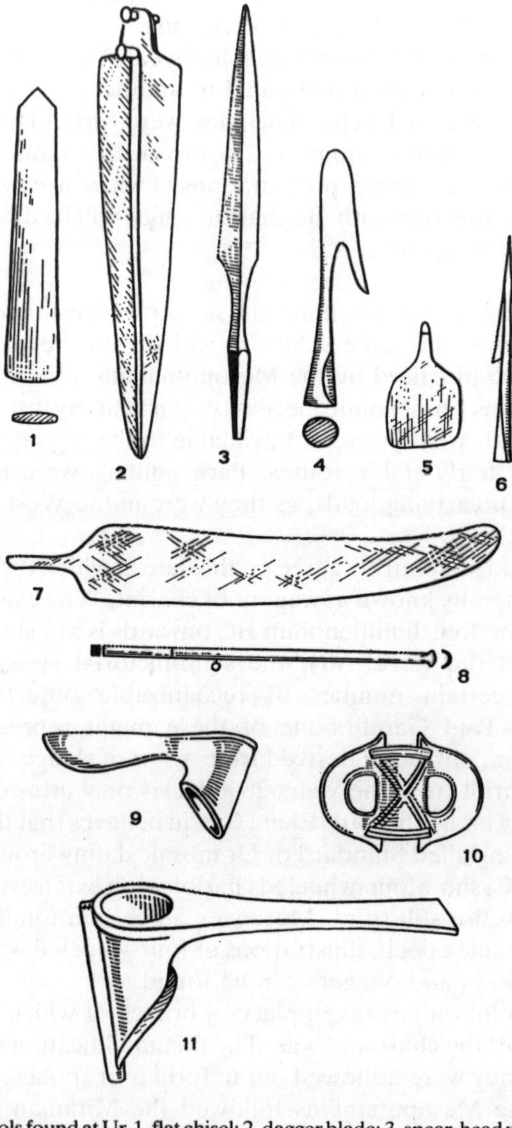

Figure 28. — Tools found at Ur. 1, flat chisel; 2, dagger blade; 3, spear-head with poker end; 4, socketed harpoon-head; 5, razor; 6, double-barbed spear-head; 7, saw; 8, gouge (drilled); 9 and 10, axes; 11, cross-axe. (From V.G. Childe, *New lights on the Ancient Middle East*, London, 1958)

Figure 29. — Chariot from a pictorial sign from tablets at Uruk.

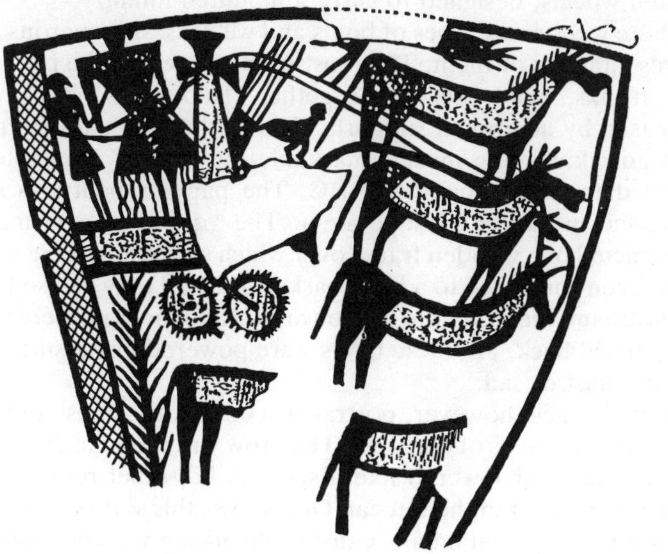

Figure 30. — A military chariot. Funeral vase in scarlet pottery originating from Hafagi (British Museum)

But above all, and this fact is complementary to technical changes, the use of the horse most certainly revolutionized transport conditions. The horse seems to have been used from early Antiquity in countries where it had been domesticated. Georges Contenau believes that it was gradually introduced into Mesopotamia, but that it was probably the Hyksos invasion, which had introduced the chariot of war led by teams of horses, that generalized the employment of the horse in countries which had only just begun to make use of it. The horse had occasionally been used around 2800 BC, is cited positively around 2000 BC, and was widely used from the 17th century BC onwards.

According to Arnas Salonen, the quadriga first appeared between the end of the 9th century BC and the beginning of the 7th century BC. The earliest representation of this is to be found on a fresco from Till Barsif, dating from the years 744/727 BC. The Sargonids had indeed created a chariot, which was heavier, larger and taller, with eight-spoked wheels, designed to carry a mounted infantry.

There were many types of boats and we possess numerous pictorial representations of them. The most simple type was the raft, made of tree trunks, firmly secured together. Buoyancy may have been increased by additional supports. The "kelek" was also a raft fitted with goatskins. The most remarkable illustrations available of this vessel date from around 700 BC. The papyrus boat was similarly witnessed in Lower Mesopotamia. The "quffa" is a round vessel, constructed of a wooden frame over which sewn skins were stretched: it has been compared to a large basket. It could only be used to travel downstream: the wood was sold on arrival and the skins were folded up and carried back. All these boats were powered by oarsmen, as they had no mast or sail.

Some images, however, portray warships with a mast and sail, but also using two rows of oarsmen. The prow was raised in the form of an animal's head or lowered like a spur. A bas-relief representing the Sennacherib fleet in the Persian Gulf shows this sort of vessel. It may be that the Mesopotamians enlisted Phenician workers, whose boat building skills were famous, in order to create this advanced navy (fig. 31 to 35).

Of constructing, we only know that of palaces and temples. Houses were frequently built in mud-brick, but, as in Egypt, they have completely disappeared. The palace consisted of the cell type house, multiplied around a series of courtyards: Mari and Khorsabad are

Figure 31. — Warship. (From Contenau)

good examples of this. To avoid floods, these edifices were built on banked up and leveled terraces of land.

We have indicated the materials which were most frequently used, that is mud-brick. From the proto-historic period, bitumen mixed with clay was used as mortar, or as a putty and watertight covering. Treated in this way, bitumen was also widely employed for sea walls, quays, and in actual construction work, as well as for calking boats.

Roofs were terraced, but the short span of the palm trunks must be considered, as they dictated the construction of long, narrow rooms, and thick windowless walls to combat the materials' friability. The Mesopotamians do not seem to have known about vaulting, but the absence of stone probably accounts for this. Contenau, however, cites some examples: a bas-relief portrays houses with mitre-shaped roofs (fig. 36), and the rectangular King's chamber at Ur has an ellipsoidal vault. The sewer system of the palace at Khorsabad is composed of long, narrow corridors with corbelled vaulting, but this construction is much more recent, dating around the 8th century BC. The Sumerians used the column, which consisted of enormous square mud-brick pillars. The column, as we can figure it, is made of a stone socle and a wooden shaft. Here again, the absence of stone prevented the use of

Figure 32. — God of vegetation navigating a swamp where reeds are growing on board a boat known today as a belem. (From Contenau)

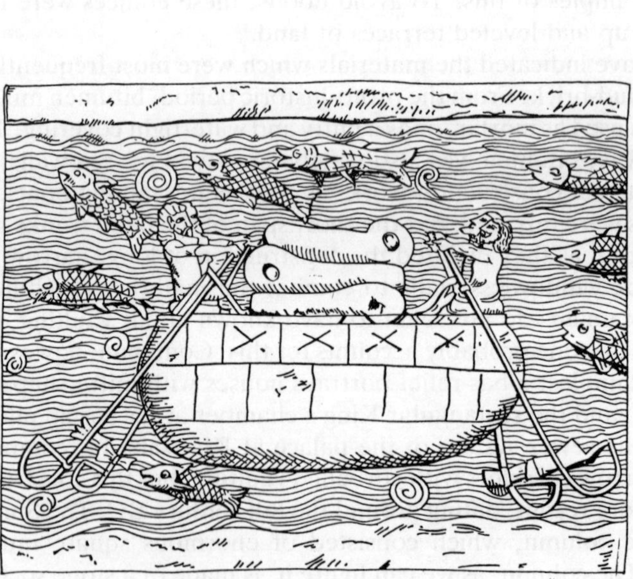

Figure 33. — Couffe type boat. (From Contenau)

Figure 34. — Raft or kelek. (From Contenau)

the column in Mesopotamian countries. In the Neo-Babylonian Empire, façades were covered in fired and enamelled bricks to produce reliefs of subjects.

The buildings which have attracted the most attention are the storied towers or "ziqqurat" which accompany temples, and are rather like the campaniles in Italy (fig. 37). Contenau has pointed out two types of tower. The Sumer type is composed of several superimposed terraces, quadrangular rather than square in shape, of unequal sizes. On the last platform is a small temple. Steps at the side of the terraces allow climbing from one storey to the next. The ziqqurat at Ur is designed in this way, and dates back to the III Dynasty of Ur (12th-11th centuries BC). The northern type of tower is examplified by the storied tower at Khorsabad. This has seven terraces, rather than five found in the Sumer ziqqurat, which are square in shape, and the top is reached by mounting a gentle ramp which runs around the edifice. The purpose of these edifices, prototypes of the Babel tower, and different both in conception and construction to the pyramids of Ancient Egypt, has been extensively discussed. They represent an already advanced technology.

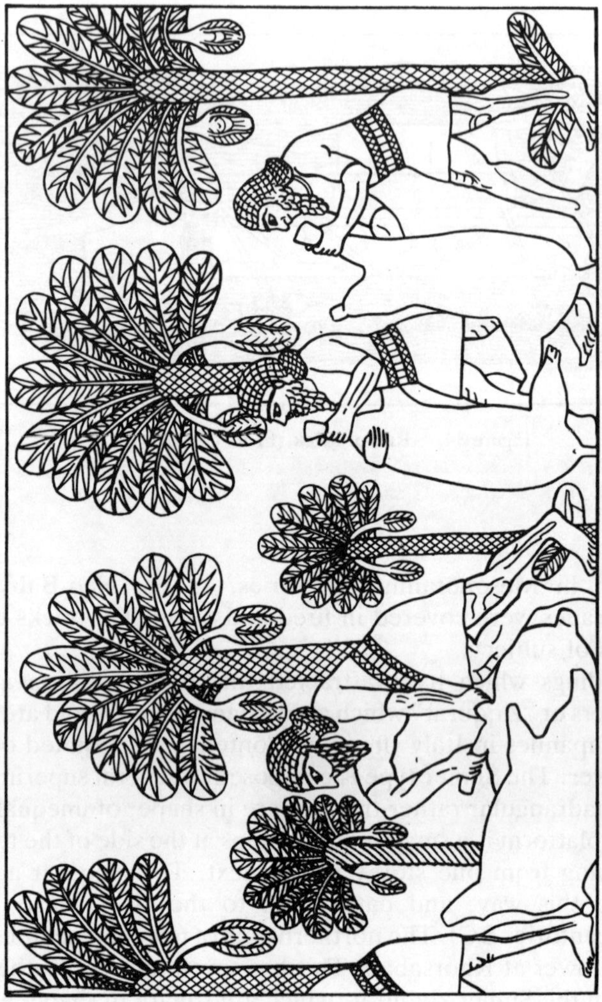

Figure 35. — Inflating goatskins for a river crossing. (From Contenau)

Childe places the urban revolution of Mesopotamia in the third millennium. Standard plans were quickly used. In the square plan, perpendicular routes delimit the various quarters. But within these,

Figure 36. — Roofs in Assyria.

free ranging imagination recovered its possession. A Babylonian tablet, dating from the Neo-Babylonian monarchy of the 6th-5th centuries BC, listed that the town had 103 temples, 900 chapels, and 180 altars dedicated to the goddess Hishtar.

Figure 37. — A Ziqqurat or step pyramid.

We know very little about the equipment of Mesopotamian towns. Roads were only occasionally paved owing to the lack of stone. An aqueduct was found three hundred metres above a small valley, which supplied drinking water from the mountains to Nineveh, some fifty kilometres away. Stone slabs were cemented with bitumen to ensure they were watertight.

Towns were also fortified, which was a new idea in itself, for as we have seen fortification was rarely practised by the Egyptians. The surrounding walls were often doubled and they appears to have been flanked by many raised towers (fig. 38).

Agriculture naturally depended upon a water-supply. The Mesopotamians used the same techniques as the Egyptians, with successive canals and "shadoofs," often built in batteries. All these constructions required close attention.

The technical systems of Mesopotamia and Egypt were thus close parallels, although there were most certainly reciprocal influences. The differences between the two civilizations mainly stem from the natural conditions and probably from certain earlier traditions. Stone

Figure 38. — Storming a town.

constructions are frequent in Egypt but rare in Mesopotamia. The use of metals, and the introduction of the wagon and the horse were certainly earlier on the banks of the Euphrates and the Tigris. Other techniques were first put into practice by the Mesopotamians. But the similarity of these two systems remains no less important. Until the Macedonian conquest, Egypt and Mesopotamia existed on technical systems which were relatively advanced, rapidly established and slowly improved.

THE EXPANSION OF THE NEW TECHNICAL SYSTEM

The technical system was definitively established in the region which stretches from the banks of the Nile to those of the Tigris and the Euphrates, and after promising beginnings in Thessaly, Thrace, and Anatolia, it quite naturally spread to regions which had remained technically backward. This technological transfer, as it is known today, obviously first affected closer regions, and then progressively reached more distant areas. Very little is known at present of the history of this diffusion, of its chronology or the nature of its form. Consequently, the table we have included is of necessity rather schematic and summary (fig. 39).

Our ignorance of the expansion of a relatively advanced technical civilization is due to two main reasons. The first stems from the lack of documentation, particularly graphic documentation. In this respect, we are no longer in the realm of the riches of the Egyptian tombs. On the other hand, we are concerned with epochs and civilizations for whom art, in the general sense of the term, takes precedence over techniques, according to archaeologists and historians. Strictly speaking we are entering the historical period where researchers, at least up until a few decades ago, were less concerned with purely material problems.

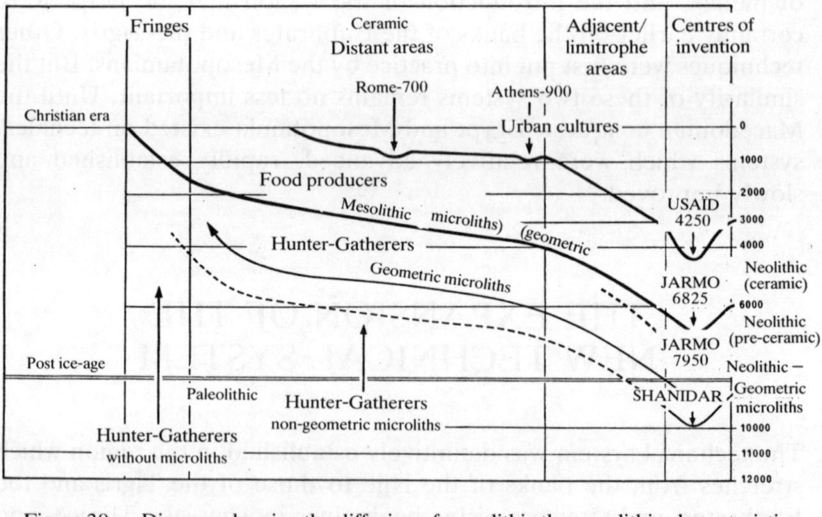

Figure 39. — Diagram showing the diffusion of mesolithic then neolithic characteristics from the Near-East towards Europe. (From A. Leroi-Gourhan)

The Near Regions

These regions were obviously the first to be affected. Within these bordering areas were regions which were naturally destined to benefit rapidly from the technical progress of the Egyptians and the Mesopotamians. To some extent these regions occupied an intermediary zone, which was true of both Syria and Palestine. In the Lower valleys of Ras Shamra, Jericho and Elkhiam, agriculture developed very quickly, and towards the third millennium BC, groups of inhabitants gave birth to towns. Cereals were the basis of all food produce. Pottery appeared around 6000 BC in Northern Syria and at the same time in Thrace (Nea Nikomedia), in Anatolia (Çatal Hüyül), and in Iran (Tepe Asiab). Copper was produced in all these regions from approximately 5000 BC, and during the fourth millennium, the whole of the Near East had developed metal casting techniques.

When Mesopotamia swung towards a new technical civilization, at the end of the fourth millennium BC, all these regions, including Egypt, benefitted almost immediately. The intermixing of peoples, moreover, facilitated this diffusion. Byblos and Ugarit had already reached a high level of technical culture.

Another problem confronts us, still more controversial and more difficult to explain. To the east of all these favoured regions, a long way to the east as the Lower valleys of the Indus and its tributaries are involved, a chronological synchronism exists with Egyptian and Mesopotamian technical developments. It is true that these regions bear similar characteristics to Egypt and Babylonia, as they are based on alluvial plains upon which any sedentary agriculture relies on a natural or artificial irrigation system owing to the lack of sufficient rain. A coordinated, organized effort is required in some areas, which inevitably leads to a remarkable cultural unity.

Urban civilization was already well under way during the third millennium BC, with developed walls and imposing citadels, protected by ramparts of mud-brick (Harappa and Mohenjo-daro). Capitals evolved, already housing both leaders and administrations, and serving as large grain depots. Towns were well built according to carefully studied plans and equipped with improved sewerage systems.

In the valleys, everything depended upon irrigated cultivation: wheat was grown (*Triticum compactum*), barley, peas, sesame, and almost certainly rice, and dates were gathered. Among the animals reared were Indinan hump-backed cattle, another breed of cattle without humps, buffalo, goats, and sheep, but poultry and elephants had also been domesticated. We are less certain whether pigs were domesticated at this stage.

Buildings were built in kiln-fired bricks. Often mud-bricks and fired-bricks were used alternately. Towers were frequently reinforced with beams. Stone was rare, but wood was relatively abundant.

The principal metals used by these civilizations were known to be gold and silver, lead and copper. Tin bronze was produced, as were alloys of copper and arsenic, with 3.4 to 4.4% for this last element. The peoples of the Indus knew how to cast iron using the wax casting method.

While the art of ceramics was more advanced in Sumer, weapons and tools were more primitive. On the one hand, there are slings and clubs, and on the other, there is an array of tools including flat chisels, enlarged and flattened towards the handle, flat axes with curved cutting edges, small saws, daggers, spear and arrow-heads, knives and razors, billhooks and sickles. The method of fitting handles onto all these tools differs from that used in Sumer. Carts and small boats are similar to those which were still in use in the country.

Childe's conclusions merit being reproduced in this discussion:

> Enough has been said to show that India confronts Egypt and Babylonia by the third millennium BC with a thoroughly individual and independent civilization of her own, technically the peer of the rest. And plainly it is deeply rooted in Indian soil. The Indus civilization represents a very perfect adjustment of human life to a specific environment, that can only have resulted from years of patient effort. And it has endured; it is already specifically Indian and forms the basis of modern Indian culture. In architecture and industry, still more in dress and religion, Mohenjo-daro reveals features that have always been characteristic of historical India.

A technical civilization which thus remained independent. But how does one reconcile the developments of the technical civilizations of the East and West? Could it have been that for some the point of departure was situated on the borders of the Iranian plateau, for exchanges across this plateau would not have been difficult? This is where the process leading to material civilizations at a high level of development would have begun. The parallelism of these developments is undeniable, and similarities between certain products are striking. Cereals grew in the wild around the Iranian plateau, as did fruit trees, apricot, peach, and possibly the vine, and wild sheep wandered around there. The migrations from the valleys to the mountains turned this area into a region favourable for the "Neolithic revolution." While Persia remained without the written word until the first millennium BC, the civilizations of Western Iran had developed a mixed agricultural economy, which possessed sickles with flint teeth, had domesticated cattle and two breeds of sheep from which milk and wool could be used. They used adzes, hoes from shaped stone fragments, ceramics and metallurgy, and the potter's wheel, the same potter's wheel which was used in Mesopotamia *x* centuries before 3000 BC and in the Indus basin *y* centuries before 2500 BC.

Once again let us return to the words of Vere Gordon Childe:

> The use of the potter's wheel or metal tools does not depend simply on technical knowledge. Mass production of fragile pots is only worthwhile when the numbers living together in the same village have reached a substantial, but as yet indeterminate, figure. A community can only use metal tools when it is producing an effective social surplus. In each case the threshold is determined by economic and social factors – the efficiency of the rural economy and the concentration, or at least fluidity, of the social surplus. Hand-made pots and stone tools may be indices of the smallness and poverty of a community rather than of its antiquity.

There we are dealing with the relations between a technical system and an economic, social and subsequently political system. But let us continue to study Childe:

> In the second place, metallurgy and the manufacture of pots on the wheel and very likely wagon-building too, were crafts, exercised by full-time specialists. These were, at least economically, released from kinship bonds; their mysterious skills could earn them a livelihood wherever the requisite social surplus was available. After all, a normal farming village requires and can afford the services of only one or two smiths and potters. The younger children of such a village artisan had better seek employment and sustenance in some neighbouring settlement not yet provided. Accordingly at this stage the diffusion of ceramic techniques and even ceramic forms need not imply the migration of a whole people, but only the immigration of specialist potters.

Thus we may observe similarities in techniques and tools between the artisans of Mesopotamia and of the Indus basin.

Let us reconsider the dates above. Metallurgy and the use of the potter's wheel were both well-established in Mesopotamia during the Uruk period, *x* centuries before 3000 BC. They reached the Indus basin during the Amri period *y* centuries before 2500 BC. In the case of pottery, "there is little evidence of divergent technical developments in these two extreme regions. But the mounting of axes by a shaft hole through the head is, however, more than a local trick, and its supremacy has been demonstrated by its universal adoption." The fact that this process found its way to India only later in the third millenium BC even though metallurgy was already known, means that this process could not have been diffused along with the art of metallurgy itself. Childe thus concludes that the technique of using the axe with handle "must be the result of a secondary diffusion eastwards. Sumer was not the birthplace of metallurgy, but simply a secondary centre where the shaft-hole axe was invented.

Between the "horns of the fertile crescent" were countries which were singularly well-placed for benefiting from the perfection of technical developments achieved as much by the Egyptians as the Mesopotamians: this was especially so for Palestine and Syria. The civilizations of Ugarit and Byblos rapidly attained a higher level of techniques because of this. The potters wheel was being used in Byblos during the time of the first dynasty in Egypt and Byblos IV is surrounded by a thick wall, made of dressed sandstone blocks. The first temple was built during Byblos V, and a second during Byblos IV, around the time of the Egyptian II Dynasty. Metallurgy was established after this, evidenced by silver ornaments and weapons of bronze.

In Palestine, neolithic Jericho was already a permanent settlement. Stock rearing was practised.

> The villagers built houses with walls of clay, plastered inside with lime and sometimes painted, with wooden posts helping to support the roofs. As wood-working tools polished celts of fine-grained rock were used besides flaked flint celts. Pottery was apparently unknown, but bowls were carved out of calcite and mortars from basalt and limestone blocks.

In the bronze age, fruit trees, vines and almond trees were cultivated along with wheat, barley millet and lentils. It was also at this time that the settlements grew into small walled towns: Jericho VII covered an area of 280 to 320 acres, and was encircled by a wall of mud bricks on a stone base. Ay and the Megiddo citadel had stone ramparts which were 6.5 and 4.5 metres thick.

Metal was known, but rare. Childe believes that metal work was done by itinerant smiths rather than resident professionals. Implements were usually made of stone, but there were copper adzes, and the latter was also used to make weapons, daggers, arrow-tips, and spears. Pottery was still largely hand-made during the old bronze age, but the wheel was used in some areas.

It is possible that some elements coming from Egypt during the early dynasties were transplanted in what was already an advanced autonomous civilization.

We must advance in time to examine the next stage. It was the first half of the second millenium BC that saw the flourishing of two civilizations already familiar with fairly advanced techniques. One of these, the Hittite civilization in Anatolia was slightly earlier than the second, the Cretan Minoans and the Mycenaean civilization from the Greek world. These two civilizations were probably formed by the slow, progressive invasions of peoples who already had a technical history and who settled in regions which had already experienced the "neolithic revolution." This represents an aggregate of civilizations therefore, who were also in contact with advanced techniques via Egypt, Palestine and Syria.

The Hittites arrived in Anatolia at the beginning of the second millenium BC; they were the first historical Indo-Europeans. Their empire rapidly expanded with the seizure of Alep and Babylon at the end of the 19th century BC, and the Hyksos invasion of Egypt during the course of the 18th century BC. By the 9th century BC, the Hittite empire had disappeared. On arriving in Anatolia, the Hittites encountered the already advanced civilization of the Semite peoples. This is evidenced by the capadocian tablets, particularly those of Kultepe, which tell of advanced commercial exchanges. The Hittites adapted these practises to their own traditions, but they certainly borrowed a great deal from the peoples who preceded them.

The Hittites have been credited with bringing about two essential developments. It is well-known that metallurgy was one of the great riches of Asia Minor, but the innovation was the use of iron. It was used in the great Hittite empire during the 20th and 18th century BC. Its use became widespread by 1100 BC. Iron has been traced back to earlier civilizations, but these pieces were rare, often unique and almost certainly imported – we noticed the same for ancient Egypt. An iron amulet was discovered in Byblos, dating back to the 20th century BC, and an iron ingot was found in a tomb in Knossos built at the end of the 19th century. It was the Hittites which transmitted this iron in appreciable quantities both to Egypt and Mesopotamia.

The second major contribution was the horse. It is often difficult to date the arrival of the horse in ancient civilizations, but it is certain that the domestication of the donkey was centuries old. The four-wheeled chariot which figures on a Capadocian cylinder, seems to have been pulled by donkeys (fig. 40). The Sumerians designated the donkey with an ideogram, and when the horse was used eventually, it was designated by a periphrase, the mountain donkey , just as the camel was called the donkey from the South. The oldest royal tombs of Ur, which date back to 3000 BC, depict only donkeys. The earliest date at which there is unequivocal reference to the horse in Mesopotamia is only 2000 BC. Between this time, it seems that they used the hemione and no doubt also the onager, which came from Iran. At this time, the horse was becoming domesticated in Elam and Cappadocia. The Hyksos invasion, the success of which was no doubt due to the fleets of chariots drawn by horses, was probably the first great expansion of the use of this animal. The Cappadocian horse will always be renowned in Antiquity. A squire named Kikkuli, of the court of Boghaz-Keui in ancient Hattusa, wrote a treatise on horse training in the 14th century BC: each day of this process, which lasted about six months is described in the smallest detail.

We have very little information about other techniques used by the Hittites. Judging by the amount of contact between them and Mesopotamia and Egypt, the level of development was probably very similar. The figurative representations which are most often of religious ceremonies, do reveal some rare precise elements: bows and spears were used for hunting and in war, the two-horse chariot was the essential part of any invasion (the Egyptians were to profit from this.) Helmets with crests and greaves which are represented on the armour of the Greek hoplite (armoured infantrymen) were found in Zendjirli, dating back to the New Empire.

We have virtually no information about the Hittite art of building or their organisation of space. The towns were planned in the same way as those in Mesopotamia and Syria. At Boghaz-Keui, the ramparts were made of earth with a stone facing, and in some places there was a double façade with an interior filling (fig. 41). The exterior towers no doubt served to consolidate the wall rather than to facilitate defence. The gates are slightly different from those seen in areas near the banks of the Tigris and the Euphrates (fig. 42). The temples were probably decorated with wooden columns mounted on stone plinths.

Figure 40. — Cappadocian cylinder from the de Clercq collection.

The facts are similar concerning the greek world, both on the continent and on the islands and in Crete, until the end of the Mycenaean period. A more accurate contemporary knowledge of demographic movements indicates that the first people were Asiatic, followed by Indo-Europeans.

Farming was the main activity in this region, despite the marked development of trade, but farming techniques were among the most stable. The same plants cultivated were the same as those grown at the end of the Neolithic period : cereals and leguminous vegetables. Olive trees, vines, fig, pear and apple trees became widespread. The carob tree appears to have been imported from the Levant. Wine was probably introduced during the middle Minoan period in Crete, i.e., at the beginning of the second millenium BC. They drank milk, beer, and infusions. The main fat used was oil. Stock rearing was also important, of long and short-horned beef cattle, sheep, goats and pigs. Dogs were added to this group later on.

The problem of the horse now seems to have been solved. The wagon with wheels finally appeared at the beginning of the second

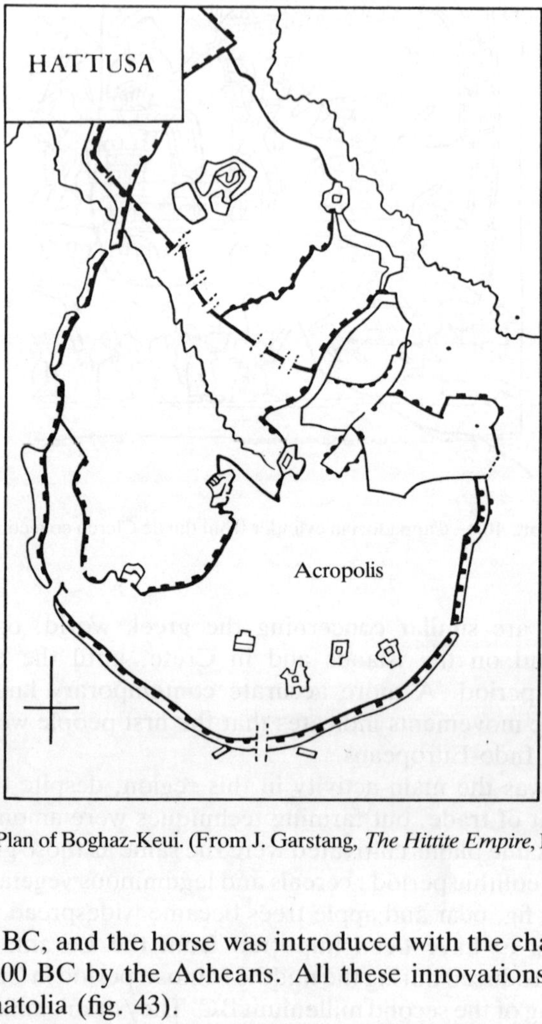

Figure 41. — Plan of Boghaz-Keui. (From J. Garstang, *The Hittite Empire*, London, 1929)

millenium BC, and the horse was introduced with the chariot around 1400 to 1300 BC by the Acheans. All these innovations came from nearby Anatolia (fig. 43).

Gold and copper were known from the Dimini period. Bronze first appeared at the end of the third and at the beginning of the second millenium BC. Greece possessed some of the rare stanniferous deposits which were found in the Near-East, near Kirrha, between Delphi and Corinth. Greece and its islands were also rich in mineral resources:

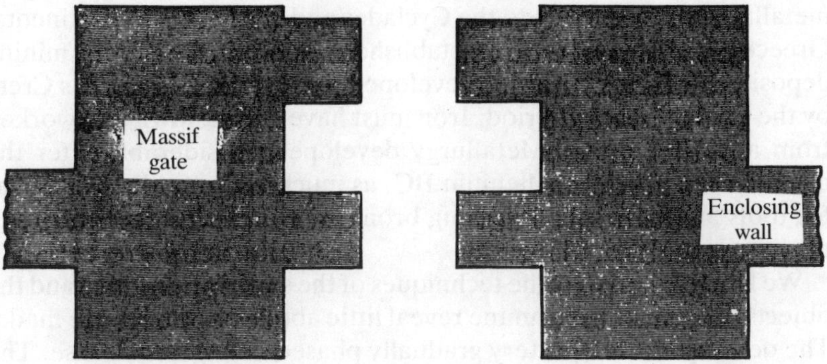

Figure 42. — Zendjirli Gate. (From J. Garstang, *The Hittite Empire*, London, 1929)

Figure 43. — Mycenian geometric chariots. (From Garlan, *la Guerre dans l'Antiquité*, Paris, 1972)

copper from Euboea, lead and silver from Laurion and gold, silver, copper and lead from the Cyclades. The populations of Asia minor came during the second half of the third millenium BC, bringing metallurgical techniques to the Cyclades and then towards continental Greece. They appear to have established trading posts near the mining deposits, then a colonisation developed which reached as far as Crete by the early Minoan I period. Iron must have been known and worked from about 1200 BC. Metallurgy developed considerably after the middle of the second millenium BC, as much for everyday objects as for tools and weapons. The long bronze sword was perhaps a Mycenaean innovation.

We know little about the techniques of the secondary sector, and the objects which we can examine reveal little about how they were made. The development of pottery gradually phased out the stone vase. The potter's wheel was used from the first quarter of the second millenium BC, coming either from Anatolia, or from Crete. Thc widespread use of the turning wheel followed closely.

We still have some elements of weaponry which reveal that materials used for offensive and defensive purposes were getting progressively lighter. In the 14th century BC, breastplates were made of linen and reinforced with metallic scales. The Dendra metallic breastplate was probably the first or one of the first of these types. At the end of the 13th century BC, linen and leather were used with metal scales to make coats of mail which were thus much lighter. Shields were also lighter, and were now circular. Bronze pointed spears, swords, daggers, slings and bows completed the military armoury.

Excavations have revealed the methods and systems of the construction and organisation of towns. The technique of building walls on a foundation of herring-bone masonry almost certainly came from Anatolia. One can find buildings made in this way in Eutresis, and Beotia, as well as in Haghios Kosmas in Attica, or in Lerne in Peloponnesia. This technique was used for a long time. Houses were smaller in the beginning of the Second millennium, and apsidal houses were first built at this time. No doubt these also came from Asia Minor – they can be seen in Troy – to the Cyclades first and then to continental Greece.

The most important development was the palatial revolution during the middle bronze period in Crete. It is certainly linked to a political revolution – the emergence of the first monarchies. We know little about those first palaces, those of Knossos, Mallia, and Phaistos, becauses they were constantly being rebuilt. They were erected on hillsides, the land being admirably suited to terracing. They were built around a vast central courtyard. The workshops and shops were situated on the ground floor, while the living rooms were on the first. During this period, an aquaduct was built in Knossos to tap the water from fifteen kilometres away. The first half of the second millenium BC was the great period for building palaces. The one at Knossos forms a rectangle of 150 by 100 metres, and as was customary, arranged around a courtyard with porticos (fig. 44).

We know very little about the towns, which were built using light materials so that they have almost completely disappeared. Excavations in Gournia have revealed the absence of any actual plan for the town. The roads are winding, and the only element of any geometric symmetry is the palace around which they are centred (fig. 45). Gournia was a town of craftsmen and potters. All Minoan cities in Crete were completely open, they therefore counted on their insularity and a powerful flotilla to protect them from invasion. Troy was fortified, but it appears to have suffered as much damage from earth tremors as from wars.

The only noticeable innovation was the tombs with cupola, which in Crete date back to 1600–1550 BC, and which rapidly became popular. It seems that this was the only use made of the cupola (fig. 46).

The Mycenaeans did not make any significant changes to the techniques used in construction, but they did introduce a few innovations, and they helped popularise the tombs with cupola. Greek territory was invaded by megaron style houses, enormous water tanks and agora porticos, but there were further changes. Palaces were being built very high up, and with the acropoles, they were the final part of an extremely well-developed system of defence. The outer walls were of a cyclopean type, with very few openings, and the town was dominated by the royal residential citadel. Houses had one or two floors, and were still built on stone bases. Wooden boards were inserted into the walls, which according to some made them more flexible – this would have been important in this area which was subject to earth tremors. The tiling was exceptional and the roofs were usually terraced. Water was supplied from wells or kept in water tanks. The walls were often decorated with paintings or frescoes.

We are prevented on several counts from giving an accurate assessment of Greece in the bronze age. There is nothing like the wealth of

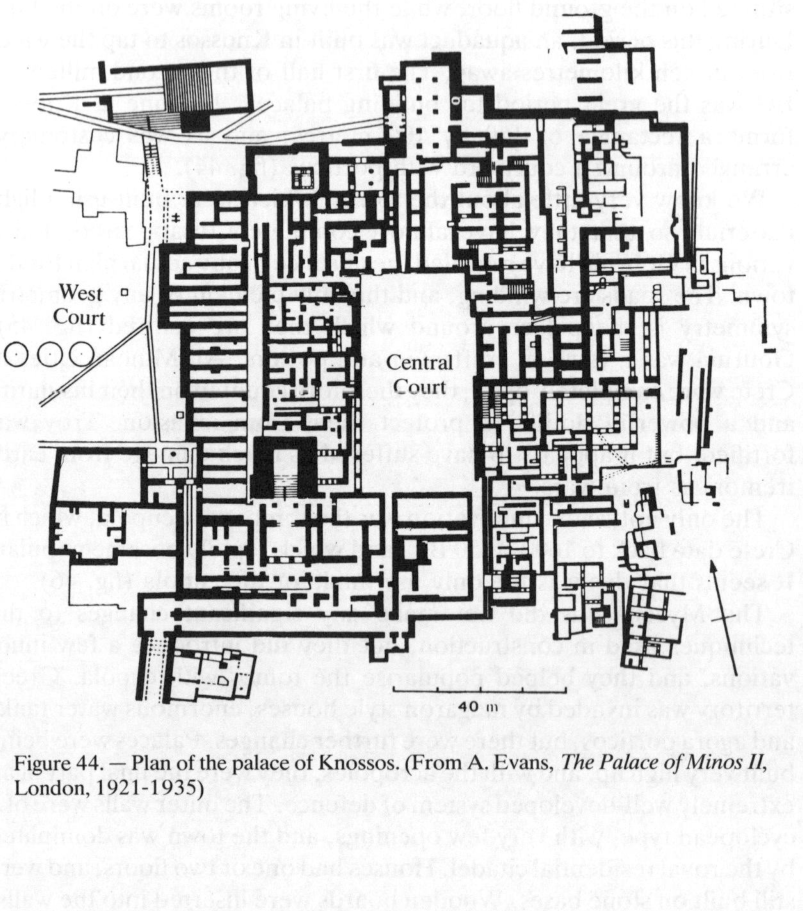

Figure 44. — Plan of the palace of Knossos. (From A. Evans, *The Palace of Minos II*, London, 1921-1935)

figurative representations from Egypt or the number of texts from Mesopotamia for the same period. To take but one example, we know nothing about ploughing implements used in Crete or during the Mycenian empire. Our knowledge is very vague concerning metallurgical techniques. Only one thing is certain: in most cases, technical progress came from the outside, principaly from Asia Minor: advanced metallurgical techniques and consequently the introduction

Figure 45. — Plan of Gournia. (From H. Boyd-Harves, *Gournia*, London, 1951)

of iron, construction techniques, the horse and the chariot were actually all imported from Anatolia, and no doubt so were many more aspects of technology. This is a perfect example of the diffusion of a technically advanced civilization. And even when the Mycenaeans began to use the techniques of fortification, it is highly probably that they borrowed the ideas from Asia Minor. Hence, it is difficult to establish which innovations should be attributed to the people of

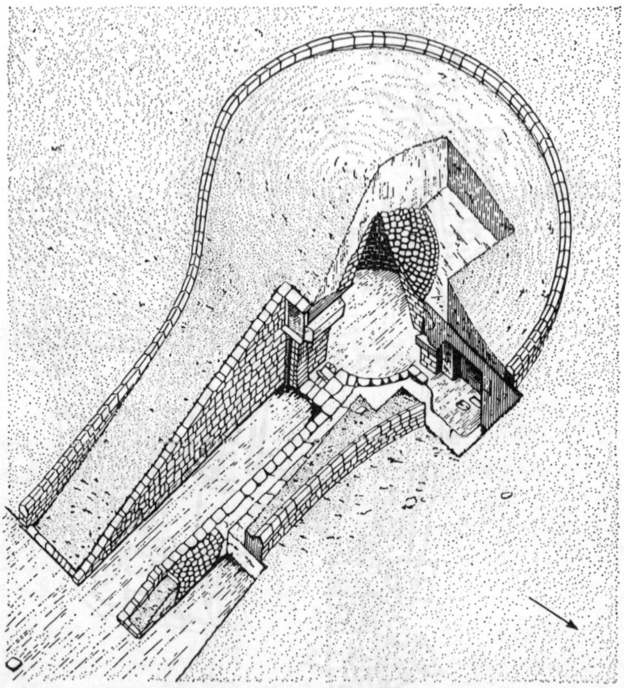

Figure 46. — Isometric view of "The Treasure of Atrea". (From Hood, cited by Rachet, *Archéologie de la Grèce préhistorique*, Verviers, 1969)

ancient Greece: it seems that they are found more in art forms than in the techniques of production.

Distant Areas

The problem of these distant areas is difficult to solve. There is often a break between prehistory and history which it is not easy to close. There is often a lack of documentation and certain aspects, even in recent works, are rarely discussed: the development of agriculture, the emergence of the potter's wheel, and the domestication of animals are examples of this. It is also virtually impossible to make any comparisons because of the lack of any chronological precision, even at the most basic level.

Certain reservations should be made clear from the start. The successive stages of progress made in the material world appear to have been a general rule: it was thus for the transition from the Neolithic to the Chalcolithic period – which took place only in certain areas – and from thence to the bronze and iron ages. The speed of transition from one stage to another varied considerably, each region having its own evolutionary rhythm. The speed of progress seems to have depended largely on the state of natural resources. Where there were none of these, and before widespread trade enabled distant regions to benefit from the knowledge of advanced technical civilizations, the significant amount of borrowing seems to indicate that a knowledge of bronze could be acquired even where there was no local industry. Thus, in certain regions, stone implements were direct imitations of those made of bronze, even to the point of reproducing the flaws in the metal. In other sectors technical developments were reversed. Agriculture was known in the Near-East before pottery, but the whole of northern Eurasia was familiar with pottery some time before agriculture. These regions have a wealth of forests and water, where fishing and hunting were practised long before land cultivation. It is difficult, therefore, to produce a synoptic picture of technical progress and the acquisition of new techniques. Population movements were often at the base of these technical transfers.

The Balkans, or rather the whole of south-east Europe no doubt soon benefited from the technical progress made in the eastern Mediterranean. The development of south-east Europe was slightly behind that of south-west Asia but the difference is marginal. The lack of natural resources, and in particular the lack of animals which could be domesticated could have been one of the reasons for this. The sites in Nea Nikomedeia correspond to those in Bulgaria and Yugoslavia in the Starčevo civilization, which, allowing for regional variations, stretched from Yugoslavia to the Ukraine. The differences will become sharper during the fourth millenium BC. In the third millenium BC, bronze appeared, and the copper age was born in Hungary.

An homogeneous civilization was in existence during the second half of the fifth millenium BC, which stretched from the Rhine to the western Ukraine. Only the most fertile land was developed, the famous loess. This farming was still nomadic, or at least semi-nomadic, and the clearing was done by essarting. The cultivation of barley and wheat was of prime importance, more so than animal domestication, although all the species known during the Neolithic period were reared: cows, sheep, goats and pigs. Neither hunting or fishing were particularly important. As well as adzes made of hard rock, sickles were used, along with scrapers and arrows. The villages were sparsely populated. The houses were rectangular and had walls with wooden posts. They had double-sided roofs.

These differences point to the third millenium BC. Stock rearing was becoming more important, and the dwellings were being built higher up, or were surrounded by ditches. Weapons were also increasing in number.

A study of the central and western Mediterranean is not an easy undertaking. Although navigation had hardly developed by the third millenium BC, numerous changes must have taken place (fig. 47). The emergence of the Montserratian civilization along the Spanish, French and Ligurian coasts from the sixth millenium BC, is now attributed to acculturation, and not as was once thought, foreign imports. The rearing of small livestock was more important than farming, and hunting and fishing were widely practised. It was not until the third millenium BC that the growth of navigation brought about the development of the chalcolithic civilizations of Sicily, Sardinia, Malta, and then Spain and France. There is no denying the influence of the

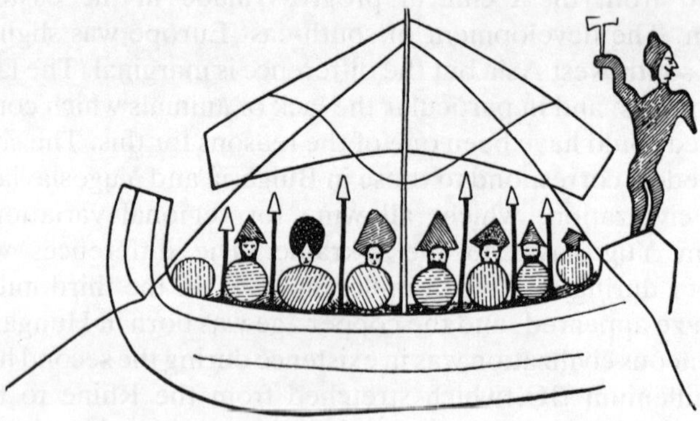

Figure 47. — Warship.

Near-East. Relatively few metals were in use before the end of the third millenium BC.

We now know more about northern Africa, and it is certain that the coasts, as well as the Sudanian regions beginning with those nearest Egypt, underwent technical development. Fishing and goat-rearing were the livelihood of the Sudan during the second half of the second millenium BC, and by the third millenium BC, it was the turn of the bovines. Farming developed relatively late in these regions, but it was known earlier on in northern Africa due to the influence of eastern populations. But here, as in Sicily, Spain, Sardinia and especially in Tunisia, the real bonds were forged in Carthage by the Phoenician settlements, during the first half of the fourth millenium BC. It was this advanced civilization of Ugarit and Byblos which disembarked from the eastern ships.

Let us go back to the first half of the second millenium BC. All of Europe and the African shores of the Mediterranean were familiar with farming, rearing and no doubt the use of bronze in most areas. The Phoenician region was clearly more advanced. As for continental Europe, we can glean some information from the striking rock engravings from the Val Camonica in northern Italy. We are able to reconstruct the village with the plan of Bedolina, the houses below, and higher up the mosaic of fields where the different types of cultivation are marked by a series of dots. A river winds its way through the countryside. Other lines mark with precision the separating walls, the furrows and the canals: all this reveals an already well-organised agriculture (figs 48 and 49). It is difficult to discern exactly the type of swing plough used, but it seems most likely to have been the toothed-plough, which was used on all the northern banks of the Mediterannean, and one of the plougher's feet seems to be resting on the heel of the share (fig. 50). And what about the animals harnessed to this apparatus? At this date it is hardly likely to have been a horse. Equids? One of the animals seems to have a pair of horns, but this is only supposition. The only thing which is certain is that no horse has been depicted on any rock engraving, and the wagons which we shall see shortly were all pulled by oxen, and a donkey also seems to be distinguishable. Horses have been depicted on more recent engravings.

The picture of the smith is also unclear (fig. 51). Bronze, like copper was certainly forged, but iron smelting would have been very surpris-

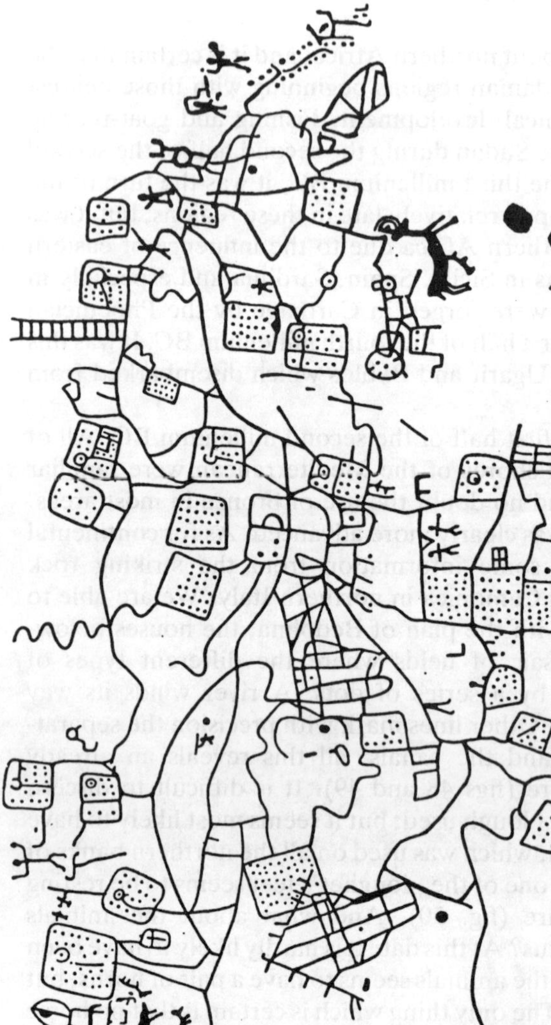

Figure 48. – Plan of the village of Camonica Valley.

Figure 49. — A house at Camonica Valley.

ing at this date. Finally, we have included pictures of two vehicles, one with two wheels (fig. 52), and the other with four, the latter being pulled by oxen, as we indicated above (fig. 53). Spears were frequently used as weapons and for hunting, and the animals most widely hunted were cervidae. Although the dating is sometimes a bit uncertain, and even if they are not always as perfect as the Egyptian paintings, the engravings of Val Camonica are exceptionally valuable to the historian concerned with material life.

We are about to see a fundamental difference between two regions very distant from the advanced civilizations which we have just discussed, and very distant from each other. China offered a field of possibilities similar to those of Mesopotamia and the Nile and Indus valleys. The neolithic period of northern China appears complete to some, in full possession of all its means, which makes it difficult to consider it as "invention" in seclusion. The dating is uncertain: was it the third millenium? It does seem to have had certain characteristics peculiar to itself, however. The plants grown and the animals reared were indigeneous: these were millet, then later rice, and dogs, cows, pigs, sheep, poultry and silkworms. The pottery round first appeared during the second millenium BC. The shift towards southern China brought with it some changes, notably the predominance of rice over millet. Furthermore, bronze was first used at this time.

Figure 50. — The Ard (Camonica Valley).

Central America constitutes a second autonomous area of technical progress. At the end of the seventh and during the sixth millenium BC, the inhabitants of the Tehuacan valley used to collect wild plants, but

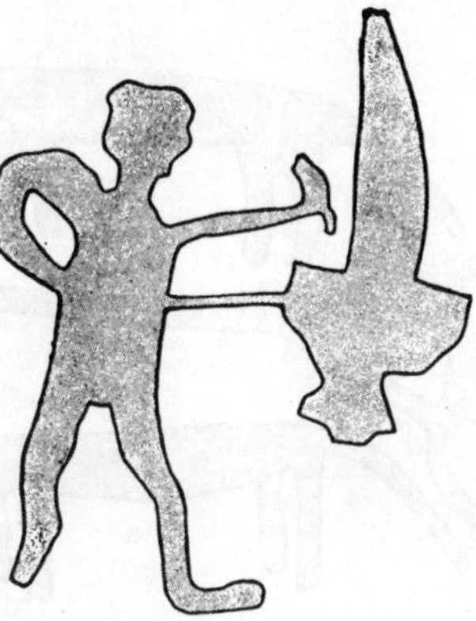

Figure 51. — The smith (Camonica Valley).

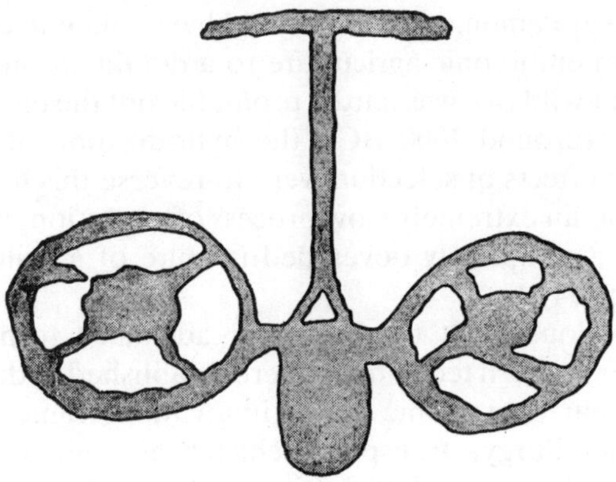

Figure 52. — Barrow with two wheels (Camonica Valley).

they were also beginning to cultivate some of these, such as marrows and pimentoes. Implements were becoming more diverse and some remains of nets, cloths and baskets have been found. The number of plants grown for food also increased until the middle of the fourth millenium BC: maize, a type of haricot, and gourds, which were used

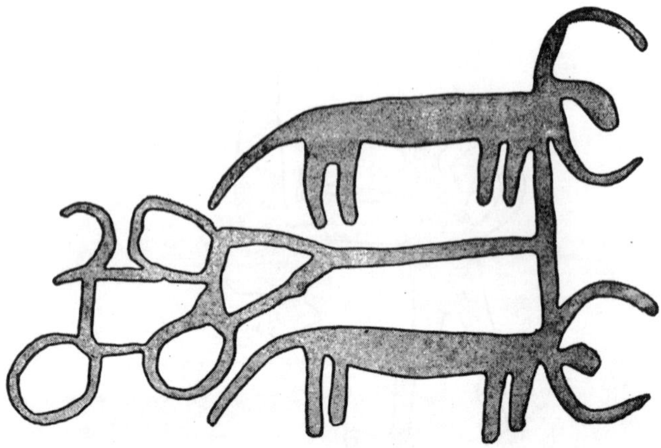

Figure 53. — Wagon with four wheels (Camonica Valley).

as recipients, were among these. It would have taken five milleniums to pass from an embryonic agriculture to a dominant one. Botanists have shown that wild rice was hardly profitable but the mutation of the pod-corn gene (around 2000 BC), the hybridization of maize with teosint and the effects of selection were to reverse this tendency. We have here, then, an extremely slow process of evolution, which, as we shall see, never completely developed in spite of a brilliant artistic civilization (fig. 54).

There is no doubt that the bases for advanced techniques, the ancestors of our modern techniques, were established in the privileged areas of the Near-East. Some major innovations came from there: agriculture, metallurgy, transport vehicles as well as writing all emerged during the first half of the first millenium BC with the great ancient civilizations. The change is still difficult to explain, as are the methods of transference towards less advanced civilizations, and advances and delays are still surrounded by mystery. These forms of progress should be linked with social transformations and economic development, which themselves depend in part on the development of contacts. The Hittites, Egyptians, Mesopotamians, Phoenicians and Syrians all benefited from their constant relationships with other civilizations, and the contact with other techniques, processes and

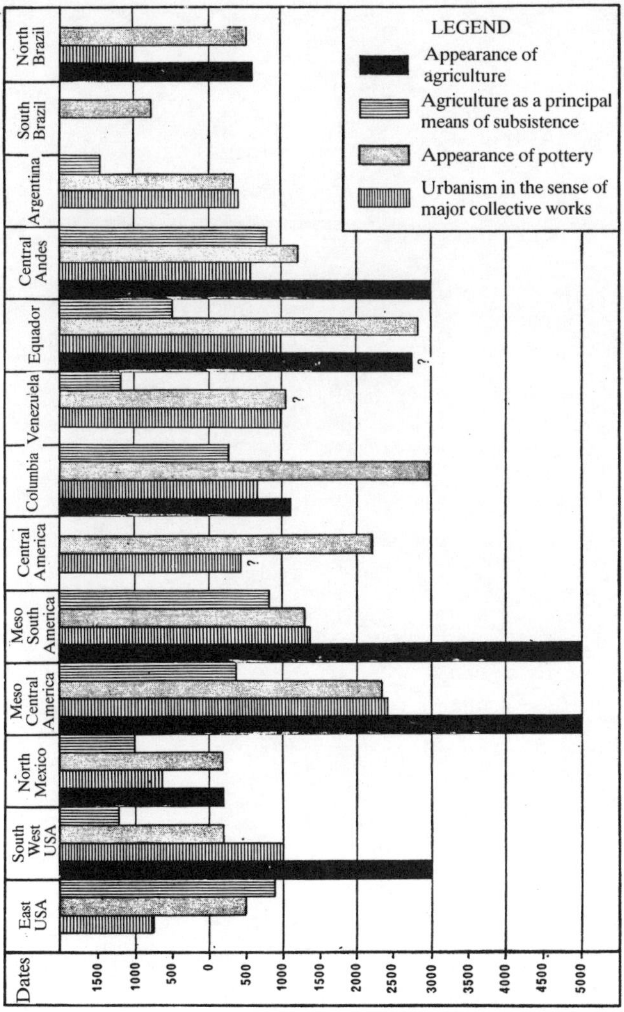

Figure 54. — Emergence of agriculture, pottery and urbanism. (From A. Leroi-Gourhan)

products. One major point emerging from this is the fact that closed civilizations are less innovative than open ones.

THE GREEK TECHNICAL SYSTEM

From the period of the Creto-Mycenaeans onwards, primitive Greek civilizations were already familiar with advanced techniques, some of which they had developed themselves, while others had been borrowed from their Egyptian or Mesopotamian neighbours. It seems that it was only from the 6th century BC onwards that a real technical system slowly began to develop, which was both different from and more sophisticated than earlier systems.

We should briefly emphasize some of the conditions under which this Greek technical system was formed, which may subsequently help to answer some of the many questions asked, often inadequately, about this subject.

Firstly, the general poverty of the country of Greece must be indicated. Plains are relatively rare, often marshy, although having said this some very lush plains do exist. Mountains are dry, arid, and rocky, and there are only a few waterways which flow irregularly. Forests are poor and regenerate with difficulty. There are almost no mining resources, apart from argentiferous lead. The wealth of the country therefore lies in dry farming lands on the one hand, and the sea on the other. The most important communications were enabled by the sea, which provided a large part of the food requirements, either in produce from the sea itself, or by facilitating the importation of various types of merchandize. These conditions explain why colonies were established in the richest areas in all respects.

Politically, this barren land was divided between a multitude of rival cities. Neither internal political quarrels nor civil wars were favourable for technical expansion. Small city-state economies, and perpetually unhealthy finances naturally limited any progress, if the latter was at all possible. The arrival of the Macedonians from rich provinces was necessary to encourage the building of the first great empires. It was exactly at this moment that an interest in advanced techniques was felt, although this may have been a little late in the day.

Our knowledge of Greek economies is still rather fragmentary. In any case, the two points which we have just emphasized – economic division and the necessity for massive imports – were the only two factors which hindered technical development, and the importance of these cannot be stressed enough. Our knowledge of Greek demography is no better, and we cannot say whether this also constituted an essential factor and what influence it may have exerted.

At the end of the chapter, we will return to the question of the contempt which the Greeks displayed with regard to the material world, although this characteristic of their nature may have been exaggerated.

TRADITIONAL SECTORS AND PROGRESSIVE SECTORS

It is self evident that the technical development of the Greek civilization was uneven, but these uneven advances were also typical of other civilizations. Some techniques obviously demand a slow process of evolution, while others are quickly introduced or improved. First and foremost, we should point out that from the beginning of the 6th century BC onwards, progress had been generalized in all sectors of material life to a greater or lesser degree, and this fact has undoubtedly not yet been fully developed. It would seem that "leading sectors" to some extent influenced more traditional ones. The formation of a coherent technical system required, moreover, a certain compatibility between the diverse techniques in use. We will return to this problem at the end of the chapter in order to make some conclusions about the blockage of technical thought.

Poor soil hardly encourages the development of agricultural techniques, except perhaps in some of the more fertile plains. The Mesogee plain in Attica with fertile red clays, the countryside of Thria and Eleusis, the Boeotian valley of Cephisus, the black lands with the chalky clays of Thessaly, Argolis and Mantinea, the valley of Pamisos, the rich alluvial deposits of Eurotas were as famous as the wealth of Laconia. These areas did not produce enough, however, and the rest of Greece was occupied by mountains and rocky zones. Greece has consequently always depended upon large imports of foodstuffs from Sicily, Egypt, and Thrace, or from the banks of the Black Sea. Soil improvement must have been very limited, as the lack of water did not enable a sizeable system of irrigation, and land was rarely drained. Soil enrichment was virtually impossible due to the absence of essential raw materials. Manure was rare as animal rearing was not yet widespread, and stalling was unecessary throughout most of the country because of the climatic conditions. Marl was also rarely available.

We have only vague ideas about the types of crops which were cultivated. In fact, Greek or Sicilian wheat were members of fairly diverse species. Paul Cloché notes that the most widespread crop was a sort of durum, an autumn species with a long bearded ear and reddish in colour, which was produced in the countryside of Metapontum, Syracuse and Segesta. *Poulard* wheat or duck-bill, depicted on an obol from Orchomenus in Boeotia, had large ears of corn with bulging grain. During later periods, the Greeks cultivated clad corns which were grown less during the classical period, in favour of the cultivation of wheat. The most widespread type of wheat was einkorn, a long pointed grain which remained in the husk during shelling. Barley appeared to be the most popular cereal of the Greek world, as it was one of the first to be used for human consumption and formed an important crop, even after the introduction and large scale cultivation of wheat. Aristophanes notes that during the 5th century BC, milk gruel and barley flour were still eaten. Barley survived the dry summer periods of much better than other cereals. The most widespread type of barley cultivated was that with six rows. Greek botanists thought of oats as mere weeds, and rye was completely unknown.

Cultivation methods seem to have remained fairly primitive. There were three ploughing times, spring, summer, and autumn. Ploughing was very superficial, as the shallowness of the plant earth made deep ploughing impossible, except in the very fertile areas of the plains. Although numerous pictorial representations illustrate that the swing-plough was used in Egypt, the few representations which we have of Greece demonstrate that the dental swing-plough was used (fig. 1). The primitive swing-plough may have consisted of a single piece, as a

Figure 1. — Greek ard.

text by Hesiod indicates. The image which Paul Cloché gives us is more fanciful and contrary to what he says, such an instrument would have been difficult to make out of one piece. A terracotta statuette from Tanagra, dating from the 7th century BC, a cup of Nikosthenes from the 6th century BC (Berlin Museum), and an Attic cup from the 6th century BC (the Louvre), all depict identical instruments. Thc latter even illustrates the characteristic position of the foot on the plough-tail. Hesiod, who also mentions this composite plough, writes that the beam was made of laurel or elm wood, the sole was made of oak, and the beam of ilex or holm oak. There was only one silt and for a long time the ploughshare lacked an iron support. At one time it was reinforced with frets or iron strips. During the 4th century BC, iron ploughshares were introduced. Other traditional instruments were used apart from the plough, including the pickaxe and the two-pronged hoe. A hoe for weeding was also used, as well as the three-pronged fork for turning soil. The plough was pulled by a team of oxen who were yoked either around the horns or the withers.

Cereals were "sawn" using sickles. A certain time lapse was thought necessary between harvesting and threshing, as long as the ears of corn were bound in sheaves. Threshing was done under animals' hooves. Then the whole lot was pitched high with shovels so that the straw and chaff could be separated from the grain. The grain was then stored in earthenware jars for transportation, as can be seen on a vase in the Louvre in Paris.

There were many flourishing olive plantations, as the olive tree was perfectly suited to the soil in a great many parts of Greece. This plant was well protected, moreover, even supported by the authorities, despite its delicacy and expense. The main fat element of Greek material life, olive oil was abundantly used in many ways. The olive could also be harvested straight from the tree, and at different times, for the fruits could be picked before they were ripe, half ripe or fully ripe depending upon the uses one wanted to make of it.

The grape vine could be successfully cultivated with the olive tree. The Attic vines of Diacria and Corinthia created marvellous vineyards. The vineyards of the Greek islands and those on the shores of the Aegean Sea were famous, but vineyards could be found almost everywhere. The Greeks cultivated two types of grape vine, one being a trailing type, and the other a climbing vine which ran around a network of trees and was supported by a latticework of poles. Like olive picking, the grape harvest was one of the most popular occupations in the Attic countryside.

Cereals, olives and grapes were not the only widespread crops. Fruit trees were also cultivated on a large scale, and figs were particularly popular.

The development of a certain mechanization, which we will study in more depth later, was to facilitate change in the whole process of agricultural production. A 6th century BC crater, now in Leningrad, shows grain being crushed in a cone-shaped mortar. As Hesiod indicated, this mortar was made of wood, but those made of stone (basalt or granite) have also been found. The most widely used millstone up to the Hellenistic period, was the so-called "coming and going" millstone, driven by hand on a stone platfrom around a pivot exterior to the centre stone. The grinder, similarly in stone, had a parallelepiped form: this was a sort of hollow funnel, with a long narrow opening at the base, and like the platform it was lined with herring bone ridges. A long wooden handle, linked to the pivot, was fixed in mortises on the upper side.

Two types of rotating treadmills were probably introduced at the end of the Hellenistic period. An example of the first, however, used for grinding minerals in the lead mines of the Laurion, may be found in the Classical period (fig. 2). The giant millstone is conical in shape at the top, and the running stone resembles an hour glass, the lower half of which covers the conical top of the fixed stone. The upper part of the hour glass serves as the funnel. A slight gap is kept between the two stones so that the moving stone can pivot around a wooden axle sunk into the fixed one; the latter's suspension on this axle enables the gap between the two stones to be maintained. This type of treadmill which could be driven by animals walking round the axle in an never-ending circle, was to enjoy a remarkable expansion in the Roman world.

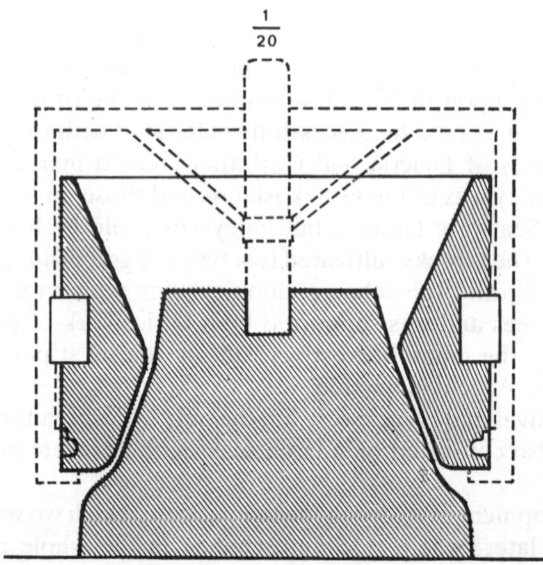

Figure 2. — Grindstone for crushing.

The production of olive oil was more complicated than that of flour. The first operation consisted of separating the stone from the pulp and extracting from the latter, firstly, the *amorge* – a bitter liquid which could be used as a fertilizer or a material for treating wood or leather – and secondly, the thick juice from the oil. Originally, the olives were undoubtedly crushed by treading, the feet being protected with clogs, and the *amorge* was collected by a pipe. It is even possible that pestles were used. At first, this pulp was crushed in a basket, using a heavy stone: such a scene is depicted on a Greek vase from Delos where the olives are shown piled into a large mortar. It is probably much later that the oil mill was introduced, which the Romans called a *trapetum*. This consisted of two plano-convex millstones, placed vertically, supported by a horizontal axle turning on a vertical pivot. The recumbent millstone consisted of a stone vat, the sides of which fitted the external profile of the two running stones. Here again, a space had to be maintained between the two stones. Excavations at Olynthus have provided examples of this, dating from around the 5th century BC. The pulp was then pressed, and a 6th century BC Greek vase depicts a still very primitive instrument, consisting of a sort of basket in which the pulp was pressed with a stone, as it was in the first stage of the operation. Presses similar to those used for making wine were then employed. The use of the lever balanced with enormous counterweights also attested by a vase from the 6th century BC, now preserved in Boston. The fruits to be pressed were placed in quadrangular envelopes mounted on a board, on which the lever was placed. The process of making wine was very similar to that of oil. The grapes were first trodden underfoot in transportable, wooden vats. The weighted lever was then used, and although it is difficult to specify when this was introduced, it must go back to the same period when the press for making oil was used. Finally, according to Heron of Alexandria, a system of numerous winches and pullies was employed instead of the feet.

Cloché states, but not precisely, that during the 4th century BC, methods of market gardening were greatly improved in some regions, such as those practised on the outskirts of Athens.

The nature of the soil necessarily dictated animal husbandry. There were a few rich areas in Greece where cattle and horses existed, such as in Thessaly, Epirus, Boeotia, Messene, and Euboea. Almost everywhere else donkeys and mules were used, or small livestock, such as sheep and goats were reared. The Athenians had to import most of the horses and cattle they needed. Pigs were very common, and two breeds of sheep were reared: the long-tailed sheep, like small Algerian sheep, and the broad-tailed sheep, endowed with a fatty growth which invaded the whole of the caudal appendix. Sheep were reared especially for their wool.

Hunting still continued and some animals existed in such a large quantity that the Greeks had to control their numbers by literally waging war on them. This was particularly true in the case of the hare which overran the islands. Fishing was obviously an important resource. At the time of Homer, the great have tended to scorn fish as a source of food. Fishing began in the Archaic period and fish gradually became plentiful on the markets. The Greeks knew and used every different fishing technique: fishing using a floating line, a sinking line, fly fishing, casting long nets, harpoon fishing and using a hoopnet, were probably methods which had been employed for a very long time. Anchovies, sardines and tunas were the most common catches.

Thus, despite an environment which was not very favourable, at least in most of the Greek countryside, the Hellenistic civilization managed to develop a more advanced system of agriculture than that of other, richer regions. Whatever types of plants were cultivated, a certain number of improvements were implemented in the methods used to grow them, and more particularly in preparing produce from the land. It is possible that agrarian problems, which existed within many Greek societies, may have contributed to the blockage of the development of farming techniques. It appears, however, that natural conditions and the technical environment played a more important role. Greek agriculture did not seem to have developed the means to progress beyond the stage at which it had arrived after considerable efforts.

Development of the sub-soil also raised problems which were difficult to overcome. We have said that Greece was not rich in mineral deposits. Copper was found in Euboea and Cyprus, iron in the Cyclades, some gold in Boeotia and Euboea, and silver at Thasos and Siphnos. During the 7th and 6th centuries BC, and for this very reason these two towns were the major flourishing centres of silver production. But all this mining activity was to be overshadowed by the Laurion mines, where lead was concentrated in contact zones between schists and limestone.

It is likely that the Laurion was exploited very early on. There were various gallery types, square, trapezoidal, and irregular, but they were never more than a metre in height, often only as high as 60 centimetres, and 60 to 90 centimetres in width, thus forming veritable passages which did not demand excessive labour so that the veins could be reached quickly. If it was too costly to reach the veins by digging passages, either square or rectangular vertical wells were sunk, measuring a surface area of approximately 2 square metres. The facings were perfectly smooth, and mortises for the ladders were installed at various intervals giving access to the shafts. Thus they could dig to depths of about 120 metres. The tools used included the flat headed hammer and a spike. The former was used for striking the dresser, a rod measuring 2 to 3 centimetres in diameter and 25 to 30 centimetres in length, the end of which was bevelled with 2 or 4 edges. The pick consisted of a flat, thick blade, pointed at one end, and folded at the other in a socket shape so that it could take a fairly strong handle. The mattock was a sort of spade, the socket of which was curved like that of a modern spade. The first three tools were made of hammered and quenched iron. Lamps were usually made of terracotta, but occasionally of lead. Finally, the mineral was transported in bags or baskets (fig. 3).

The discovery of mineral deposits was relatively easy for those who had considered the location of metaliferous masses and realized that they usually lay in pockets. Depending upon where the gallery hit the pocket, the mineral could be extracted either by beginning at the top or at the bottom. Fire was only rarely used to break up the rock, if at all. Ventilation was provided by waving cloths, but carefully caliphated vertical wood partitions were also used in the shafts, which produced a sort of siphon effect. As was the case in the Laurion mines, the passages through the schists and limestone did not need supporting works. A system of pillars was used in the deposit itself.

Considerable progress seems to have been made during the 5th and 4th centuries BC, and it is at this point that we arrive at the third contact. During this period, vertical shafts were improved and the knowledge of where the deposits lay and how to extract them increased. We know that production in the Laurion mines grew

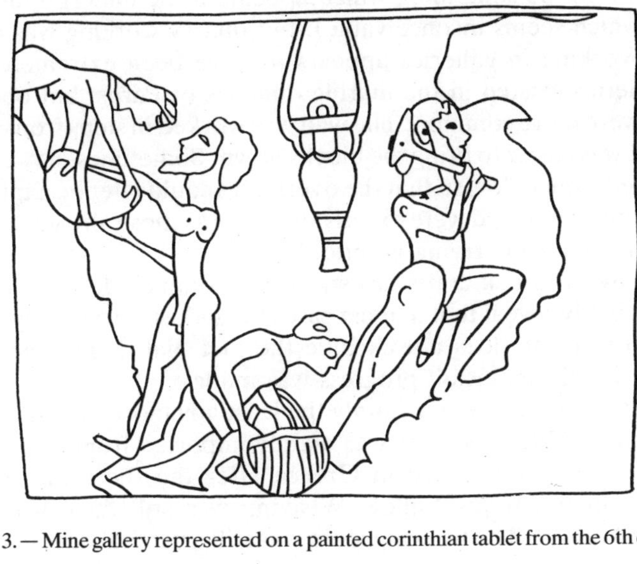

Figure 3. — Mine gallery represented on a painted corinthian tablet from the 6th century BC.

considerably under the archonship of Nikomedes (484/483 BC). Similarly, we know of the remarkable work carried out at the hill of Cephalari, near lake Copais, under the direction of the specialist engineer, Crates of Chalcis during the 4th century BC. All this would seem to lead to the conclusion, albeit not a concrete one, that noticeable progress was made in the art of mining at the end of the 6th and the beginning of the 5th century BC.

The mineral was first sorted and then crushed, originally in mortars, and later by the millstones with two conical parts which have already been described. The mineral was then washed in a simple tank, the floor of which was inclined. Water was probably collected in water tanks during propitious seasons, for there lay a major problem. The mineral was then roasted and reduced by methods of which we know very little. It seems that this operation was performed by artificially blowing air over the mineral. Finally, the lead and silver were separated by the process of cupellation. Our knowledge of these operations is still too sketchy to say whether any real progress was made during this period. It seems likely that the growth in the production of minerals prompted not only an increase in processing plants, but also some degree of improvement in roasting and reducing methods.

Digging out clay did not present any great difficulties (an image on a Greek vase shows something which appears to be underground clay mining, which seems inconceivable), but quarry working was not so simple. Working in galleries appears to have been extremely rare: some galleries existed in the marble quarries of Paros, but open air quarries were more common. The wall was worked in steps from above because it was easier to bring the blocks down. Sometimes, however, it was worked from below so that the overhangs could later be exploited. The methods employed were exactly the same as those already used by the Egyptians for quarrying granite.

The study of Greek crafts has still to be completed. Such a study would probably reveal that in most cases the industry was fairly static, but this conclusion does not exclude the fact that there were some sectors in which significant progress was made.

Textile techniques were certainly the least advanced. Wool was the most common fibre used, flax was mainly imported from Egypt, and cotton had been introduced in Greece after the the campaigns of Alexander in the Indus Valley. Weaving was still done using the spindle and distaff. The weaving loom was still a vertical one, the warp of which hung from a crosswise bar supported on two posts and stretched by weights attached at each end of the threads (fig. 4). A moving bar separated the even or odd threads or groups of threads, and the shuttle looked like a flat wooden needle. The cloth was woven from the top to the bottom and a card tightened the cross threads of the web. The length of the cloth was determined by the height of the loom.

Woodworking was important in a great many domains, such as construction, boat building (fig. 5), and furnishings, etc. In most cases the tools used were those which had been employed by preceding civilizations, except that the metal and the forms may have been improved. Originally the saw, axe and adze were used, followed by planes, augers, gimlets, bow-drills, bow-trepans, chisels, scissors, and polishers. The Greeks had invented two indispensable instruments, probably around the 6th century BC. The first was the curved saw which was easier to handle. The second and more important innovation was the lathe, which was introduced by Theodore of Samos, bronze-smith, engineer, sculptor and architect, one of the men of the Greek Belle Epoque similar to the "engineers" of the Renaissance. Although the lathe did not operate on a connecting rod and crank system, the use of this instrument was to enable great improvement in woodworking and carpentry.

From the time of Homer, pot makers used the potter's wheel. A second person was needed to turn the wheel so that the potter could have both hands free to work the clay. Very little progress seems to have been made in this craft. After the vase had been dried, it was polished, but we will not dicuss at length techniques which have been studied in depth, such as the methods of firing and painting, and the colours used. Two points concerning the baked clay, however, must be clarified. The bricks, usually graded in three sizes, remained untreated for a long time, and were merely dried in the sun. We have examples of the moulds which were used in order to obtain an exact gauge. Some rare evidence of the processes employed before the appearance of Roman techniques has been found from baked bricks. It is possible that this type of brick was used right up until the very moment when Roman influence was exerted. On the other hand, terracotta tiles were extensively used as it was impossible to use mud tiles. There were two types of tiles, each of which had variations: the Laconian tile which was curved in shape with a curved lap-joint and the Corinthian tile which was straight with two edges and an angled lap-joint. According to

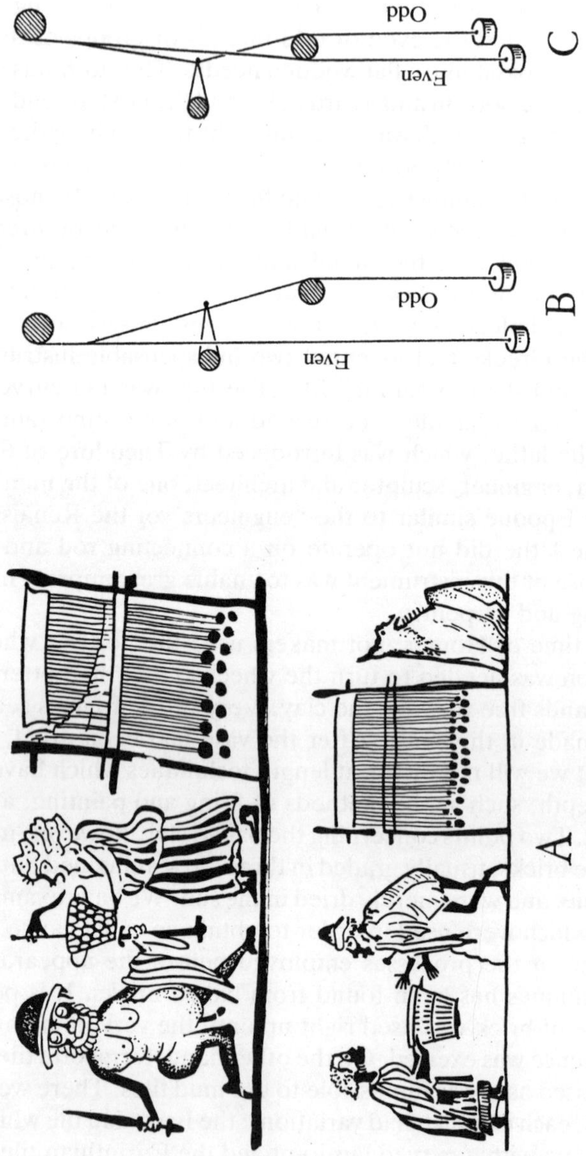

Figure 4. – Weaving looms. Cabirionese vases [A] and successive positions of the warp's threads on the loom [B and C]

Figure 5. — Greek boat. Painting on an Attic vase (circa 500/475 BC)

Pliny, Corinthian tiles, which were also cast from moulds, date from the 6th century BC.

We know very little about the art of metalwork, but the Greeks appear to have been extremely skilled smiths. The images left to us, however, must be reinterpreted in a less fanciful way. The wine pitcher from the 6th century, now in the British Museum, depicts a forge and not a foundry workshop. The central image clearly demonstrates that iron rather than copper is being worked. The furnace is therefore merely a forge; indeed pig iron was not known at this epoch. Bellows were made out of two twin goat skins. Equipment was already very diversified, but there are no new tools at this stage.

Bronze was frequently used for making statues. Casting methods do not appear to have been very advanced as yet, as small and large statues were cast in several pieces and then assembled. Air bubbles and vitreous scoria appeared in the metal owing to the use of sand in moulds.

The early techniques which the Greeks visibly began to improve upon were those used in construction. Columns, pillars and architraves in early temples, as well as in domestic architecture, were all made of wood. This technique was continued until the middle of the 3rd century, as the buildings at Samos or at the asclepieion on Kos illustrate. Virtually all the available types of wood were used, including oak, fir, rotproof cypress and ceder, walnut, olive wood and even vine. Bricks were used for walls, but huge walls were also built out of clay. The temple of Patrai was built of brick. Similarly brick was used for the construction of surrounding walls: during the 5th century for the wall of Themistocles at Athens, the walls of Platea, Thebes, and Tegea, in the 4th century for the walls of Apollonia and Mantinea, and even in the walls of Sparta. These walls were often mounted on stone socles and were reinforced with wooden beams, according to a technique widely used by the Creto-Mycenaeans. In the 3rd century BC, Philo of Byzantium recommended that wooden chains be used for brick fortifications. They were also used for repairing parts of the brick wall of Athens in 307/306 BC.

Major revolution came about in the substitution of wood for other materials. This substitution began, tentatively as yet, around the middle of the 7th century BC, and the operation was undoubtedly progressive in the beginning: rotten wood columns were replaced in stone. This was the case at the Temple of Thermos at Heraion at Olympia. Then, apart from reconstructions, this progressed to the replacement of the whole structure, as is demonstrated at Marmara, Delphi, Corfu and Calydon. Nevertheless, entablatures were still made of wood and covered with terracotta. The use of stone was generalized during the 6th century and entablatures were "petrified," as can be seen on the Temple of Apollo at Corinth constructed in the second half of the 6th century BC. It is difficult to say whether these early columns were monolithic or not: this could have been possible with techniques which had been tried and tested in Egypt. In any case, the drum column rapidly appeared, and it was already in use in the Temple of Hera at Olympia. One advantage certainly lay in the transportation of the materials required, as the column could only be mounted vertically, which presupposes that lifting equipment had already been developed. Although the Temple of Apollo at Corinth, erected around 540 BC, still possessed monolithic columns, it seems that this was probably a late and exceptional example.

Finally, the substitution of stone for other early materials resulted to a large extent from the development of transport and lifting instruments. The Greeks employed Egyptian methods of transportation, as well as more improved equipment, such as rollers, carriages like the

Eleusis' wheels for moving logs, and round packages of wood (fig. 6, 7 and 8). Roads were an obvious necessity, and examples of these have been found near quarries and temples. Lifting operations were more difficult: a drum column could weigh up to five tons, and a piece of entablature might be six to nine metres long and weigh as much as thirty tons. Temporary inclined levels may have been used for moving very large loads, as was so for the Parthenon entablatures. But these costly and clumsy procedures were replaced by lifting apparatus, based on a system of pullies, blocks and winches. Signs of such operations have been found at Delphi and at Olympia, dating from the second half of the 6th century BC. Chersiphron and Metagenes, who undertook the building of the temples of Ephesus, invented mechanisms for transporting and lifting the materials to the desired location. Some centuries later, Vitruvius wrote of the admiration which these constructions inspired in the building world, as they still do today. Vitruvius also mentions the work which these two architects wrote on the new machines they had created, and which have been passed down through the centuries by generations of architects, without requiring any major improvements. Hoists and pullies were used in the construction of the Erechtheion, built at the end of the 5th century BC. At Didyma, two machines were coupled together in order to hoist columns and pieces of entablature which measured 7.5 metres long and 2.20 metres thick. Heron was to make a classification of these various machines, in particular, for hoists with 1, 2, 3, or 4 beams forming a trestle.

The examination of archaeological remains has enabled a greater understanding of the construction methods used during this period. Foundations were poor, and they were even content to level the rock when possible. A solid base was therefore essential. The walls of large temples were built with regularly drafted blocks, assembled with even courses without mortar. Finishings were done when the construction work had been completed, in other words, the columns were fluted after they had been erected. The outside surface of the wall was decorated after the whole wall had been built, although surfaces were marked for this operation before being erected. The most difficult problem was that of ensuring the solidity of this stack. The column drums were linked together with cedar pins, placed in a square cavity in the middle of the drum, and embedded with lead. The blocks of the walls were linked together, horizontally and vertically, with double bracket dovetail-shaped metal clamps. Doubting, quite wrongly, the solidity of marble, Greek architects introduced veritable iron frameworks to reinforce their buildings. Thus, they not only strengthened the walls, but also lintels, corniches, and architraves, to prevent collapsing. The entablatures of the Propylaea at Delphi were reinforced in the 4th century BC, and we may observe the strength of the irons used.

A study of the marks on these stones illustrates the whole technology developed in the field of construction. Grooves, protuberances, and V- or U-shaped channels are manifest proof that apparatus, ropes, nippers and lewises were used from the 6th century BC onwards (fig.9 and 10). Marks made from assembling the buildings demonstrate that the work was logically organized. Some progress may even be detected, such as the use of the swallow-tail clamp for reinforcement at the end of the 6th to the beginning of the 5th century BC, and then the introduction of single or double brackets. Securing the drums was perfected around the middle of the 4th century BC by attaching central

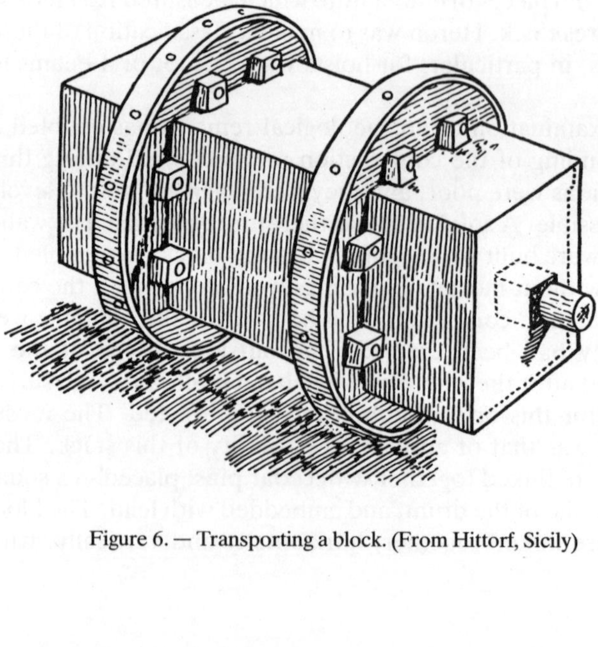

Figure 6. — Transporting a block. (From Hittorf, Sicily)

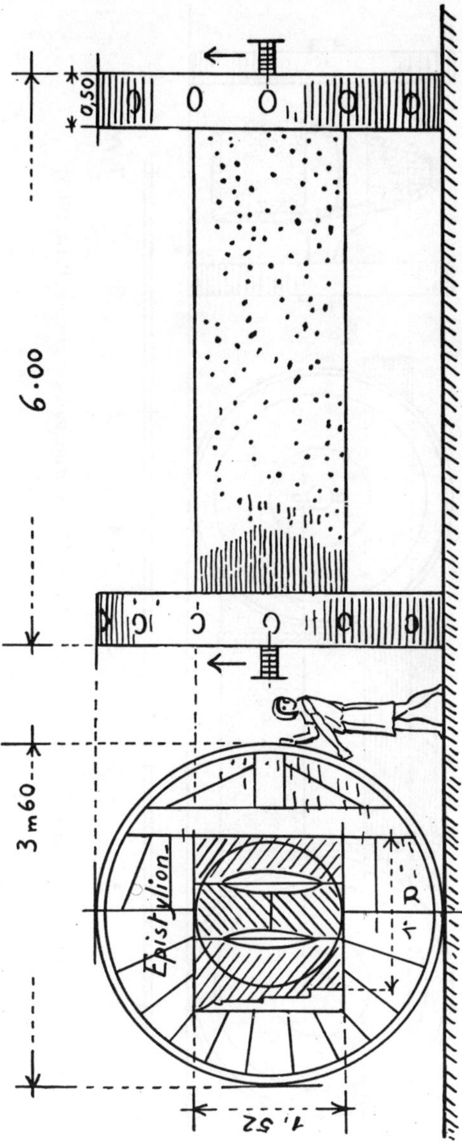

Figure 7. — System of transporting blocks at Ephesius. (From Vitruvius)

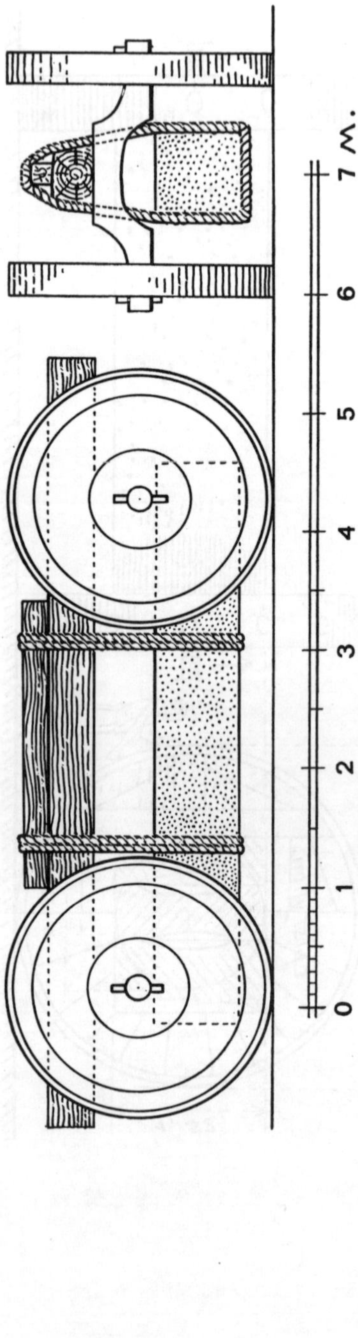

Figure 8. — Reconstruction of Eleusis' wheels. (From R. Martin, *Manuel d'archéologie grecque*, Paris, 1965)

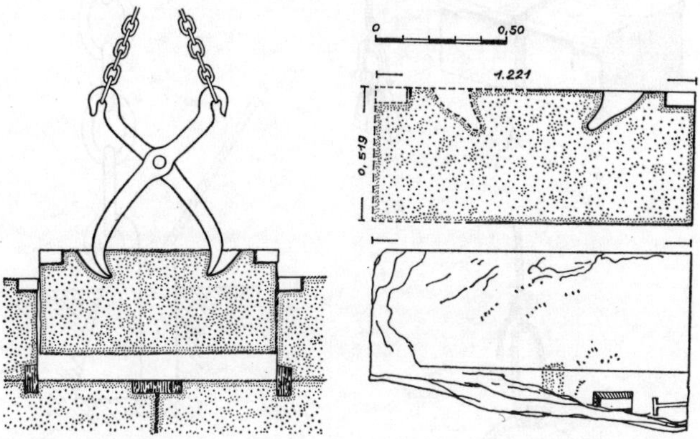

Figure 9. — Lowering a block into place. The cavities made for the tongs (Parthenon). (From R. Martin)

pins to tenons placed at the edges. It is likely that the architects worked from scale models made beforehand, and also used these to present their projects before the proper authorities. This procedure was still used by Renaissance architects at the beginning of the 16th century. From the end of the 5th century BC, they benefited from all that the science of perspective and the study of shade could contribute. Presentations of plan, section and elevation could thus clarify important points presented by the scale models. We cannot overemphasize the major improvements made to the building of large stone edifices during the height of the Classical period.

The art of roofing buildings, on the other hand, does not seem to have made very much progress. The framework merely consisted of a simple stack of wood, as is illustrated in the arsenal of Piraeus. This process was very expensive and meant that the timber frame was extremely heavy. Even if, according to some authorities, science could not contribute a better solution, the Greeks demonstrated great skill in other fields which required woodwork and which could have enabled them to modify roofing frameworks. This could also have resulted from the quality of timber which was used.

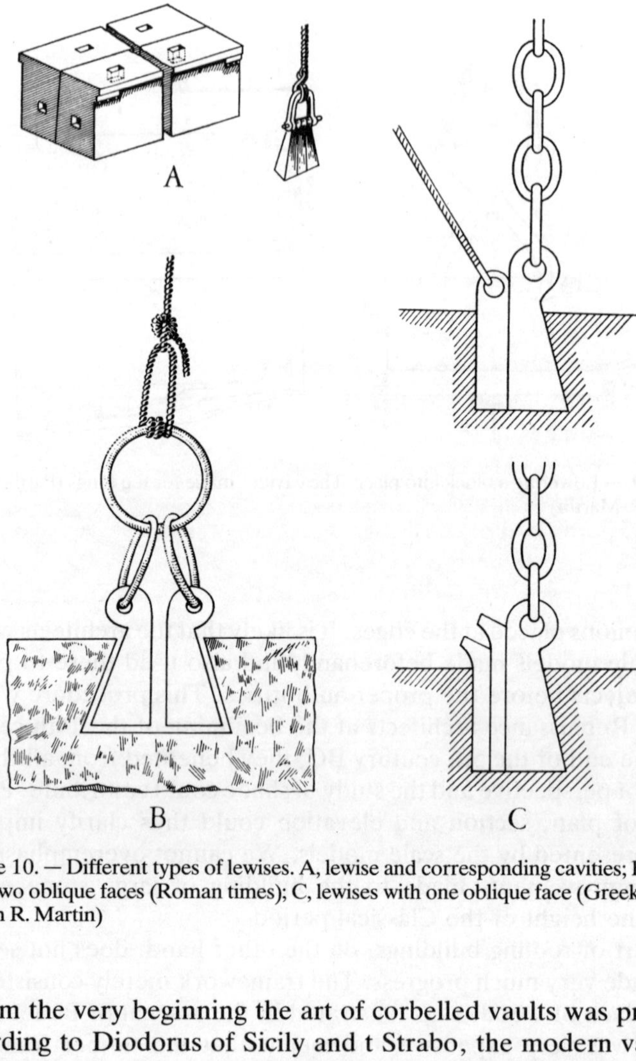

Figure 10. — Different types of lewises. A, lewise and corresponding cavities; B, lewise with two oblique faces (Roman times); C, lewises with one oblique face (Greek times). (From R. Martin)

From the very beginning the art of corbelled vaults was practised. According to Diodorus of Sicily and t Strabo, the modern vault had been invented by Democritus of Abdera, around 440 BC. The earliest examples we know of are thought to date from the 5th or 4th century BC.

The delicate problem of transport now remains to be discussed. It is difficult to dissociate the different elements involved. We have already emphasized the extent to which the notions of structure and system are essential in order to arrive at worthwhile conclusions. If these elements were separated, we would end up with completely valueless judgements. Transport and roadways, transport and harnessing, and transport and cattle domestication are inseparable elements, as are ships and ports. Transport and economy, in other words, weight and bulk, must also be considered.

Land transport involves both roadways and all the different sizes of materials to be transported, whereas roadways are determined by the size of the country, its geographic layout, political organization, and financial status. Apart from these, the available materials and the appropriate techniques are also factors which determine the progress of transport. Crete had developed a road network from the second millennium BC, whereas there were very few Greek roads during the Classical period, and those that did exist were badly maintained and were often no more than rough tracks. Bridges were virtually nonexistant, although the technology to build them existed, as we know of advanced paths and paved sacred ways which were not very long but which nevertheless went straight to their destination. The roadways at Delphi, Eleusis, and Olympia were paved in this fashion and had raised pavements. In the towns, roads were more often than not reduced to mires. It was only from the 5th century BC onwards that urban routes were paved and maintained. Political division and the poor financial status of the towns were equally powerful blockages to any progress, as was the lack of technical initiative.

We should also stress the fact that goods were not widely circulated. Trade was relatively limited and was only concerned with the quantity of goods imported, in particular with grain and certain materials. Up until very recently, at least until the end of the Middle Ages, beasts of burden could suffice if need be, particularly for transporting foodstuffs. The transportation of certain large loads could certainly be difficult, which may account for the limitation of some construction work. These loads were carried by donkeys and mules so that tracks were the only type of roadway required.

The Greek chariot was generally two-wheeled. An Attic crater dating from the 8th century BC, now in the National Museum of Athens, may be interpreted as representing a four-wheeled vehicle. But these are probably funeral chariots which only had to make short journeys. The four-wheeled vehicle, which did in fact exist, did not possess a front axle assembly unit and was consequently almost impossible to manoeuvre (fig. 11). The system of wheels and axles was still very primitive. Antique harnessing is generally known to have been poorly adapted for haulage, as the harness strangled the animal around the throat and more or less prevented it from breathing. The horseshoe was also unknown. Loads transported in such a way could not exceed five hundred kilogrammes per vehicle. But the question of transport also raises another aspect; that of animal domestication. Pastures were poor in most areas, and we have already seen that oats were virtually unknown. In such conditions, whatever improvement could be made to vehicles, no significant progress could be made in transport methods.

The horse probably first served as a mount from the 11th century BC. The texts of Xenophon seem to suggest that the saddle existed from the beginning of the 4th century BC, but spurs and stirrups were not known to the Greeks.

In a country bordered by coasts, maritime vessels were naturally the most developed form of transport. The most wealthy and powerful cities, at least those that remained so, were situated near the sea. Three factors are involved in naval techniques: the ship, the port and navigation, although the latter was undoubtedly the most primitive, and we know very little about it. In fact, the Greeks had illustrious

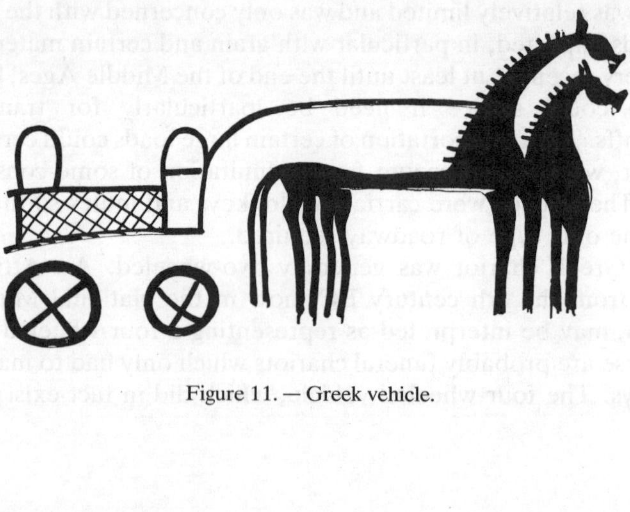

Figure 11. — Greek vehicle.

Phoenician predecessors, from whom they had certainly borrowed the best part of their techniques. Coastal navigation was obviously practised, but ships could also sail straight across certain axes, such as those which lead to Egypt. Lack of observation means, and the absence of adequate maps undoubtedly restricted navigation to mere guesswork.

Many suggestions have been put forward on the subject of ships, few of which probably reflect the true nature of things. There is very little change, or at least hardly any significant transformation, from one image to another. Obviously merchant ships should be distinguished from war ships, the latter being subject to stricter strains, particularly where speed and manageability are concerned. War ships, which have been more often the subject of representations, were more streamlined than merchant ships and raised at the bow and the stern, as is exemplified on the Nikosthenes cup, now in the Louvre. Merchant ships were small, with a capacity of about forty to sixty tons, rounded and low in the water. The bridges in these ships were very reduced: simple shelters, situated at the bow and the stern.

Propulsion was provided by both sails and oars. All the sails were quadrangular in shape and fixed to a single mast by means of one yard perpendicular to the top of the mast. The sails were made out of leather or canvas, but we have little information regarding their fabrication. The single sail could only be propelled by wind from behind, or at the very least allow supporting trim. It was therefore naturally impossible to tack or to hazard more exact manoeuvres. The oar was consequently indispensable, and the tiered rows of oarsmen have often been remarked upon; triremes and quadriremes would have been boats with three or four rows of oarsmen. This is a purely theoretical view, however, as ships would have had to have been very tall but these were unstable unless insufferable ballasting was used. According to this view, the ship of Ptolemy Soter would therefore have carried twelve rows of oarsmen which is absolutely unthinkable. It is very likely that the terms used applied to the number of oarsmen per oar. Were there no ships which had several superimposed rows of oarsmen? It is difficult to answer this as iconographic interpretation is often difficult.

From the most remote antiquity the ship's rudder had always been a lateral one. It has been demonstrated that no great importance should be attached to this fact, especially for ships with oars which in fact did not need a rudder, as the teams of oarsmen were well trained to row together.

The port was perhaps the most remarkable invention of the Greek Classical period. Formerly, as has been shown in the writings of Homer, landing beaches were used, a fact which entails significant consequences. The Mediterranean has no tides, so it was therefore necessary to pull the boats ashore. This also meant that keels could not be very deep. The introduction of the port must have enabled some progress in the development of ships, as vessels could subsequently take a larger keel. The Greeks created every type of port, ranging from inlet ports which were easy to build due to the rocky coastline, to artificial ports constructed with dikes, quays, and moles.

Spectacular examples of these ports remain today. There was a double port at Cnidus, the building of which was enabled by the construction of jetties. We should also mention the port of Alexandria, where the island of Pharos was linked at the end of the 4th century BC to the mainland by a dike some one thousand three hundred metres long with two openings by which the two docks could communicate. In the port of Antioch, at Seleucia in Pietria, the dock was entirely dug in an unsheltered coast, and was linked to the sea by a canal. Finally, the port of Piraeus should also be mentioned as this was built in a remarkably propituous site.

It is curious that the Greeks attempted to cut through the Isthmus of Corinth. This work had been conceived around 600 BC by the tyrant Periander, and was taken up by Demetrius Poliorcetes at the end of the 4th or the beginning of the 3rd century BC. This scheme was later abandoned because the engineers were convinced that there was a difference in the levels of the two coasts. This was the very problem which for some time prevented the building of the Suez Canal.

It was in 283 BC, under the reign of Ptolemy Soter, that the world's first lighthouse was built. The architect, Sostratus of Chidus, erected a tower nearly 85 metres in height, the levels of which were set back above each other. The light itself was created by a large furnace in front of a huge mirror which reflected light for more than 55 kilometres out to sea. Unfortunately it disappeared in the 14th century in an earthquake.

With construction and its lifting apparatus, with transport techniques, and the whole geographical organization which they presuppose, we have reached a turning point. Diverse requirements exist for the creation of a technical system, taking into account relationships between the various technical, economic and social systems. It would be unfair to reproach the Greeks with the slowness of the evolution of their agricultural techniques, as climatic and pedagogical conditions, the absence of a comprehensive botanic science, and possibly the weakness of demographic movements are extremely restricting factors. And one can easily understand that despite the cost of transport, the importation of foodstuffs was preferable to expensive hydraulic installations, to the search for effective enrichment of the land, and to the development of new agricultural techniques. On the other hand, some techniques had to be renewed for various reasons, and we will now turn our attention to these.

The "Greek miracle" of the 5th and the beginning of the 4th century BC was to affect technical progress. As far as we can tell from rare existing texts, it was really during this epoch that the Greek, and for that matter the Antique technical system, was formed. Comparative studies with Mesopotamian or Egyptian technical civilizations are not yet detailed enough to enable us to establish very clear lines of demarcation and, in certain cases, doubts still remain as to the exact extent of the technical contribution made by the Greeks.

As most contemporary authors have liked to point out, technical progress of this period presents several aspects which should be emphasized. There is no global progress to speak of, but on the contrary the so-called "leading technology" may be observed to take place alongside traditional artisan techniques which evolve almost imperceptibly. We have only to look at the techniques of the textile industry which, as we have already seen, used distaff spinning and the vertical weaving loom for some time. The construction of standard buildings, rather than those which were to profit from the latest techniques, was not to change very radically. There is therefore a visible difference between stable technical sectors and progressive sectors.

As we will see, not only progressive techniques arise haphazardly, but rational techniques also occur, which require that science gradually develop principles at the same time and which similarly provide the latter with some of the elements which it previously lacked. Science may be an independant entity only when technology does not require collections of particular cases in order to resolve the problems which arise, but becomes the application of general principles. This link between science and technology was to find fortune during the Renaissance. Although at the end of the 15th century, it was possible to bring this remarkable conjunction of two previously separate fields of research to light, this is much more difficult for the period which concerns us here. The history of science is known in its entirety, but that of techniques is less well known, and above all, there are few historians who have attempted to synthesize science and technology, which was one of the central preoccupations of the engineers of Alexandria.

And almost immediately an exact chronological concordance may be noted. This technological take-off, which began somewhat timidly during the 6th century BC, took shape in the following century and exploded in an extensive form at the end of the 4th and the beginning of the 3rd century BC. It was similarly from the 6th to the 4th century BC that Hellenistic science, as Tannery terms it, developed from the emergence of the Ionian school, possessing similar techniques, to the sudden expansion of the Greek world after the conquests of Alexander. Reciprocal interventions continued throughout the whole of this period. We find men who simultaneously lend their attention to both science and techniques. This tradition begins with Thales of Miletus, physician, mathematician and technician, to Archytas, one of the greatest scientists of his time and undoubtedly a very great technician, to whom one attributes perhaps too much, and continues with Archimedes. Thus the links between science and technology do not flag, on the contrary they grow closer.

It is true, however, that physics remained rather hesitant, concerned as this subject is with abstract notions and which future developments will sweep clean. Technicians could hardly use physics to aid them, and scientists could learn nothing from a technique whose problems only gradually arose within a system. We will judge later the extent of this separation between physics and technology, and the extent to which it was felt, with Strato of Lampsacus, who was not as celebrated as many of the others. "Physics for entertainment" with which Greek engineers have often been mocked, may only have been a reaction against a purely abstract science.

Progress made in mathematics during the same period was infinitely superior. Arithmetic, in which the Pythagoreans were passionately interested, continued to develop. This is also true of geometry which, indisputably born of practical problems, was raised to the rank of science, entailing all the latter's requirements. One has passed from intuition or guesswork and from technical demonstration, which was subsequently ceaselessly dismissed to the rank of impure unessentials, to discursive virtue. Geometry quickly took the lead over other scientific disciplines, as it lent itself better than any other science to purely intellectual thought, and because it also had a visual value of demonstration.

Geometry, however, remained closely linked to technology, and although it succeeded in becoming a very advanced body of doctrine, it provided the material arts with all the services which the latter had lent it for its own development. The most concrete link between science and technology is undoubtedly music. The range of notes had been perceived by the ear and later translated into a mathematical form. The translation from a range of notes to a range of numbers came about by constant confrontations between art and hypotheses. The study of technical vocabulary, which to our knowledge has never been attempted, would contribute extremely important elements to this research. Architects, who were also technicians, and military "engineers" were to quickly make use of the science of numbers.

As scientists began to organize their knowledge by forming mathematical formulae, technicians similarly started to compile the first technical treatises. Thus, a form of technology began to appear, precisely in the domains where changes were most pronounced.

Before going into detail on this issue, it would seem useful to clarify certain important changes which largely conditioned the evolution of a great number of techniques. The best example is provided by the definitive clarification of the classical kinematic chains. Their attribution and dating is somewhat hazardous. Archytas is credited with the screw. It is probably around the 5th and 4th centuries BC that the screw, pulley, cogwheel, and their meshing together contributed an appreciable number of fundamental new ideas to the technology of mechanics. This was also true of the complementary assembly of various kinematic chains; one was to progress from the weighted lever press to the press with winches and pullies.

From this moment on, all the problems of elevation, both in large constructions and in the mines, were partially resolved. And what about the use of the pulley in sailing ships? If energy sources remained the same, the means of transmitting forces and their reduction, enabled by cogwheels and pullies combined to form block and tackle, led to the spread and diversification of the use of energy.

We are extremely badly informed on the subject of materials. Even at the height of their expansion, Greece was not rich in minerals, and we will return to this subject later. No noticeable progress seems to have been made in reducing furnaces, although work in the mines had been facilitated by elevation methods, indeed by the use of the Archimedes' screw for raising or supporting mine levels. Furnaces discovered at Agros Sosti on the island of Sifnos, probably dating from the 6th century BC, offer the earliest examples of chimney furnaces, with a clay covering, an air vent, and a casting hole. While some progress was made in primitive furnaces, there does not seem to have been any later significant progress. It is highly likely that the Greeks used imported metals, the only exceptions being lead and silver which were produced by the famous Laurion mines.

Military techniques were probably no more spectacular, although more is known about these as Greek historiography has provided many facts on this subject, despite, on the whole, some late and often inaccurate compilations. Thucydides in particular, writes several passages on various machines, but apart from the case of battering rams, he also uses vague terms, the value of which the historian of techniques has many problems in recognizing. It is therefore very difficult to draw an evolutive table of the military art of ancient Greece. It would also be indispensable to extend these studies to the whole of the Near East, from whom the Greeks often had to borrow techniques.

A few years ago, Andre Aymard referred to the long delay in the organization of military techniques, both for attacking and defending fortified towns. "It was only from the beginning of the 4th century BC that a profound evolution began to take place. This evolution was very slow until the stunning campaigns of Philip of Macedonia and his son Alexander opened everybody's eyes." Formerly, the techniques of poliorcetics and war machines were totally rudimentary, even if the machines mentioned by our sources were employed. It is likely that only battering rams were used. Does not historians' uncertainty in this subject stem from the ambiguity of the language used? We have already seen this confusion in other areas.

Names are extremely rare before the 4th century BC. We may mention Mandrokles, who threw a floating bridge over the Bosporus Straits for Darius' armies (522/485 BC), "executing exactly what king

Darius had envisaged." He was so delighted with the bridge that he commissioned a large picture of it. We should add moreover that Mandrokles came from Samos.

Of course we would like to know more about Artemon of Clazomenes (around 469/429 BC), whom some authorities have hailed as the "founder of the new art of siegecraft." He was an "engineer" and a builder of machines. He demonstrated his genius in both these fields at the siege of Samos (430/429 BC). As Diodorus of Sicily notes, "Pericles was the first to use machines known as battering rams and testudos, built to the designs of Artemon of Clazmenes." According to Plutarch, the statesman admired these new machines, which may not have been in current use.

According to the writings of Ephorus it was the first time that battering engines had been used to knock down large walls, and Pericles thought this a wonderful new idea. The engineer who had invented them was carried about in a chair, in order to direct and hasten the work because he was crippled in one leg and for this reason he was called Periphoretos.

Plutarch gives other accounts of this engineer which do not paint a very flattering picture of him.

He was so fearful of things that most of the time he didn't move from his house, where he sat attended by two servants at his side who held a copper shield over his head so that nothing would fall on him.

One can imagine the unfortunate inventor of these war machines absolutely terrified, although Plutarch may have exaggerated his story a little. Unless, of course, Artemon of Clazomenes may have seen himself as a moral victim of his instruments of death and felt the same feelings of regret which some of today's scientists experience.

Aymard notes that historians have long recognized the fact that a genuine military revolution took place in Greece from the 4th century BC onwards, even before the intervention of Philip of Macedonia. There was certainly a revolution in the infantry's armament, and in the use of mercenaries, but the role and the improvement of fortifications also grew during the course of the first half of the 4th century BC. Cities built higher stone walls, dug ditches in order to ward off attacking machines, which would seem to prove that machines had become more powerful and effective. They also adapted plans to follow the relief of the land, and facilitated flanking by recessing walls, building jutting angles, and increasing the number of towers.

The power of the new artillery – and we will encounter this phenomenon on several occasions – thus compelled an evolution in fortification. It is possible that at this epoch, the mechanical genius of the Greeks was considerably advanced. It would not be surprising if military art, in particular the machines used, in fact profited from certain works which combined pure and applied research. Once again we come across Archytas of Tarentum, born around 430 BC, and who died in a shipwreck on the coast of Apulia in approximately 348 BC. This great individual was both a statesman and a man of science, but was also interested in techniques. A disciple of Philolaus and a member of the school of Pythagoras, he wrote many works of which only minute fragments remain. He was interested in mathematics and physics, and proved himself to be a scientist equal if not superior to Democritus or Pythagoras.

But on the other hand, Archytas was presented as a veritable inventor. The screw, pulley, rattle, and kite were all attributed to him. He constructed a flying dove, which the whole of Antiquity related with admiration. Vitruvius quotes Archytas as having written tracts on applied mechanics. "Archytas, who excelled in the fabrication of machines, wished to incorporate geometry and speculation in everyday uses, having applied it in all sorts of ways." Again Plutarch writes of him:

> The art of the inventor, of organizing instruments and machines which is termed mechanics or organics, greatly valued and admired by all sorts of people, was first put into practice by Archytas and Eudoxus of Cnidus, partly to embellish somewhat the science of geometry by this favour, and partly to support and strengthen some geometric propositions by visible and material examples of instruments, intellectual examples of which could only be found by concrete and necessary reasoning.

And Plutarch continues to cite the solution to the problem provided by this instrument which was called the mesolabe.

Although Plutarch may be an excellent and faithful portraitist, there is a more interesting element to be found in this short passage. All the evidence illustrates that Archytas was a prodigious inventor and mechanic, but he was never mentioned among the mechanics of the school of Alexander. But more especially, this passage enables us to grasp the link between science and technology to which we referred earlier. Like arithmetic, geometry was to henceforth serve the technician. Arithmetic was to aid the scientist in arriving at certain solutions which could not be achieved by applying the process of reason: thus special instruments were necessary, the most famous of which was the mesolabe. We will also return to this form of mechanics which was "greatly valued and admired by all sorts of people," a statement which would appear to contradict other opinions.

As far as we may judge, Archytas appeared to play an essential role in introducing a more rational system of applied mechanics. He represents an irreplaceable stage, and it is most regrettable that his work is not better known. Let us say rather, in order not to favour an author to whom much has been attributed because he is not well known, that the end of the 5th century and the beginning of the 4th century BC certainly marked a turning point in the composition of infinitely more refined, advanced, researched, and not to say more scientific techniques. They did not have to wait long for the results of the evolution translated by this composition.

The considerable expansion of the art of poliorcetics may be observed in the first half of the 4th century BC. Henceforth, techniques were fairly well incorporated into a system for which complete expositions, if not a well established theory, could be made. These accounts are probably the first technical treatises which we possess. Unfortunately our sources are always extremely discreet and tenuous.

During this time, the Greeks certainly borrowed some of their war machines from other civilizations. The Orientals were past masters in this art of poliorcetics. Xenophon attributed his hero with a keen concern for the machines, which he did not use moreover, aimed at bringing down the walls of Babylon. Was this a personal experience taken from the first half of the 4th century? Or the result of a slow but sure evolution which took place in Greece? Whatever the case, extremely old Assyrian bas-reliefs depict battering rams mounted on a kind of tower on rollers.

It is possible that ideas were also borrowed through Sicily. The Carthaginians seem to have been the first to create a large collection of machines for use in sieges. To besiege Selinus, the Carthaginians took and erected a considerable number of such machines. Diodorus relates that the towers were taller than any others built so far, and that iron-tipped battering rams were also used. All these machines terrified the people of Selinus, and the walls of the city effectively gave way under their force.

Stirring accounts written some years later illustrate the preparations made by the tyrant Dionysius the Elder and his allies against the

Carthaginians. Interest in machines had spread, and leaders undertook to develop and increase their number. In 399 BC, Dionysius called upon the workers of all the villages he had conquered to build his armament. But workers also came from Italy, Greece and even from the Carthaginian states, drawn by the promise of high wages and substantial rewards. The rationalization of this production should also be pointed out, as numerous types of machines had been adopted so that each troop on the pay role could find weapons used in his native country in this arsenal. Moreover, inventions were encouraged and were appropriately rewarded. According to Diodorus and Aelian but later contested by Pliny, the catapult was developed in this way. Diodorus also states that quadrireme and pentaric ships were first constructed on this occasion. A considerable armament had therefore been formed bearing, one may suppose, valuable comparisons.

This whole armament was used against the town of Motya. It is interesting to note that when Dionysius arrived before the enemy city, he examined the defenses "with his architects." The siege did not only depend upon bravery, speed and attack, but also required technical understanding. The machines were positioned both against the fortress and an eventual rescue fleet. When the latter arrived, under the command of Himlico, catapults were launched against it, new missiles which had recently been invented "and which were used for the first time", striking terror in the heart of the enemy. The catapults pushed back the defenders of the walls which could then be battered down with the rams. Elsewhere mobile wooden towers, with six floors, moved on wheels and equipped with movable bridges which were proped against the top of the curtains, were also used.

It was therefore the first time that a modern siege had taken place using an important and extensive armament in a rational and calculated way. A new technique had truly been created, and it is not surprising that shortly after this the first treatise on siegecraft could be compiled.

Aeneas lived at the beginning of the reign of Philip of Macedonia. According to Polybius, Suidas and even some texts by Aeneas which have come down to us, this author had compiled a number of military works: *Signals by fire*; *Stratagems*; *Preparations for war*; *Finances*; *Of castrametation*.

We have only the treatise on the defence of fortified towns, and that is merely an abridged version which, according to Aelian, was compiled several years later by Cyneas of Thessaly, general and confident to Pyrrhus, the king of Epirus. We are fairly certain of the period when Aeneas lived. According to the most recent episode mentioned in his work and in the absence of any mention of the third holy war or of the campaigns of Philip, we can date the work between 380 and 360 BC, in other words the first half of the 4th century BC. There is no evidence to suppose that any author preceded him, and it is worth noting that his successors quote only him and appear to consider him the first and foremost military author whom they know.

Information on machines is very rare in this work on the defense of fortified towns. It is possible that Aeneas may have included them in his work on the preparations for war, or they were recorded in a specific treatise of which no trace remains. The remaining work does not deal either with fortifications, which formed the subject of a specific work by this author and has also disappeared. The subjects which do survive in the remaining work are notes on incendiary materials made from pitch, sulphur, oakum and grains of incense, and notes on methods of protecting oneself from them, particularly by using vinegar which the Ancients believed had great power. Aeneas also writes of countermines and the method of using bronze shields to detect the underground approach of the assailants.

The compilation of this technical work was in keeping with the taste for contemporary thought and didactic literature devoted to military subjects, as is attested by many of the works by Xenophon. Our knowledge of this early author remains very limited, for we know nothing of where he came from or of his life. We cannot say whether he merely compiled facts or whether he actually participated in these military campaigns.

It is highly likely that the need was felt for a sort of new codification of military rules and techniques following all the inventions which had revolutionized the art of war from Artemon of Clazomenes to the Cathaginians and the engineers of Dionysius. In the first third of the 4th century BC, the Greek world thus possessed a new and relatively complete system of mechanized war machines, which needed only to be perfected. Against this, a new art of fortification was to arise, of which we are vaguely aware by observing the achievements of this epoch. Simultaneously, engineers, in the sense that they used engines, or architects existed – although it is difficult to name them – to whom powerful leaders, tyrants or monarchs, accorded favours in order to procure their services. Artemon of Clazomenes was the first of these, and although we do not have the names of those who served Dionysius the Elder, it is already possible that these individuals had travelled throughout the Greek world and learnt their trade from their elders.

The campaigns of Philip and Alexander were to enable the whole of this partly new mechanization to be put into a concrete and more perfect form, as certain historians have already pointed out. Apart from this, schools also seem to have been formed during this period, and the most famous engineers were soon surrounded by pupils who disagreed with the teaching of their masters and dreamt of their own successes.

Polyidos of Thessaly was engineer to Philip of Macedonia. His followers depicted him as a talented man, a fact which does not deny the victories of his monarch. In particular he regularized the application of the battering ram, used with great success at the siege of Byzantium. Athenodorus and Vitruvius quote him for these same reasons and speak very highly of him.

We are better informed about his pupils, Diades and Charias, engineers to Alexander the Great, both of whom had learnt their trade from Polyidos. In the 1st century BC, Athenodorus still knew of Diades' treatise on war engines, which we unfortunately no longer possess. In this treatise, Diades credited himself with the invention of transportable towers and engines known by the names of trepans, "crows" and hanging bridges. He also employed the battering ram mounted on wheels, or at least he gives a description of it. It is only Athenodorus who is content to summarize his predecessor, and he is the source of all our information on Diades' machines.

The height of the smallest movable tower was usually 60 cubits (27.72 m) and with a square base of 17 cubits (7.90 m) on the side. They had ten storeys, the highest of which was a smaller square, measuring a fifth of the base surface area. Other towers were 90 cubits high (41.60 m) and had fifteen storeys, and lastly towers measuring 124 cubits (about 57.30 m) were constructed, which had twenty storeys with a side base of 23 1/2 cubits (10.86 m). Each tower storey was encircled by a covered way three cubits wide (0.39 m), to facilitate aid in case of fire. The towers were mounted on six or eight wheels, owing to their considerable mass. The height of the storeys decreased towards the top, as the size of the planes got smaller. Care was taken in the construction of these contraptions, to conserve the same proportions, dimensions, and materials in towers of different heights.

This was also true of the testudo which carried the battering ram (fig. 12 and 13). The largest was 30 cubits in width (13.86 m) and 40 cubits in length (18.48 m), with a height of 13 cubits (6.00 m), without counting the attic or roof trussing which could be as much as 16 cubits (7.39 m). The attic had a covered way which encircled the apparatus. Diades constructed a three storeyed turret on the ground, which housed a catapult above and a water supply below in case of fire.

The trepan had a similar testudo. A channel was incorporated on the base, very like that in euthytonan catapults, and like these machines was equipped with a cross winch. Two pullies were fixed at the other end of the channel, used to fire the steel tipped beam which advanced on rollers. The beam was then pulled back by capstans. "With such a well constructed machine," notes Athenodorus, "an engineer could achieve glory."

Athenodorus did not describe the crow and Diades himself had not included a description of the hanging bridge. One notes, nevertheless, with Alexander's engineer, a singular growth in the types and numbers of Macedonian armed machines. Diades had also created war engines for naval combat, but we have no precise information on this subject.

Historians' accounts show us all this material at work, and we know the admiration excited by the skill with which Alexander's armies lay hold of towns. The famous sieges of Perinthus and Tyre were of special note. One must admit that this success was undoubtedly due as much to the extensive use of the machines as to the sustained violence of the attack.

Names of engineers proliferate at the end of the 4th century BC. The Athenian, Epimachos, was engineer to Alexander's nephew, Demetrius Poliorcetes. At the siege of Rhodes, he created the helepolis. This was a machine 83 cubits in height (41.58 m) and 50 cubits wide (23 m), in the form of a tower which could withstand a stone of three talents (78.53 kg), a feat which terrified the enemy and secured victory for the king of Macedonia. His distant followers continued to speak of it with great admiration and the machine was described in treatise after treatise.

Philo of Athens lived at the end of the 4th century BC. According to Vitruvius he worked on the temples of Ceres and Proserpine at Eleusis, and he would have written a treatise on the proportions of the temples. Similarly, he should also be credited with the arsenal of Piraeus, later destroyed by Sulla, and on which he wrote another tract

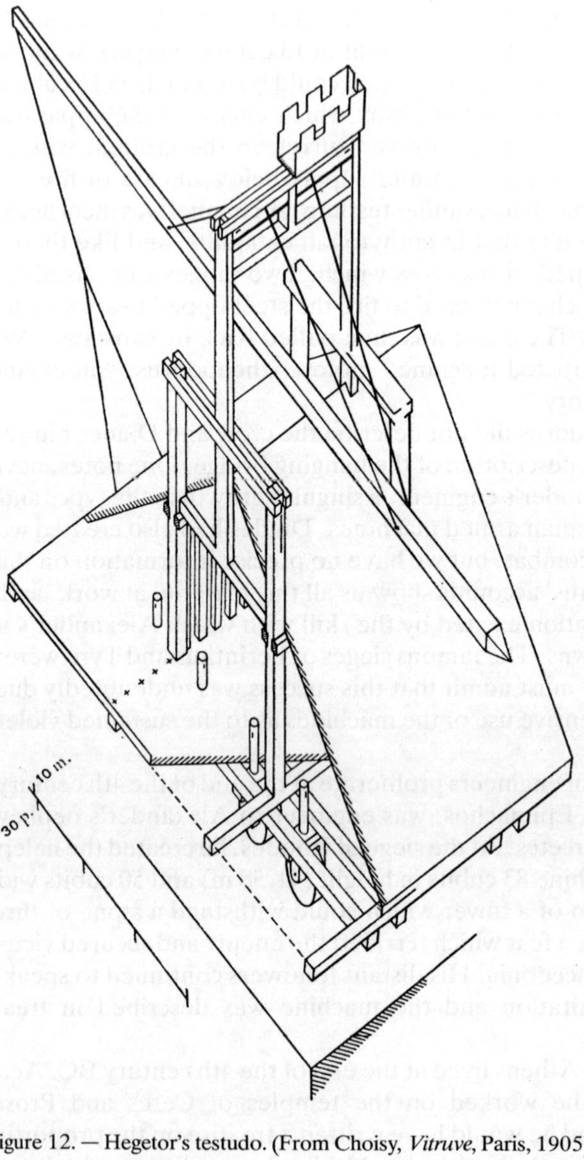

Figure 12. — Hegetor's testudo. (From Choisy, *Vitruve*, Paris, 1905)

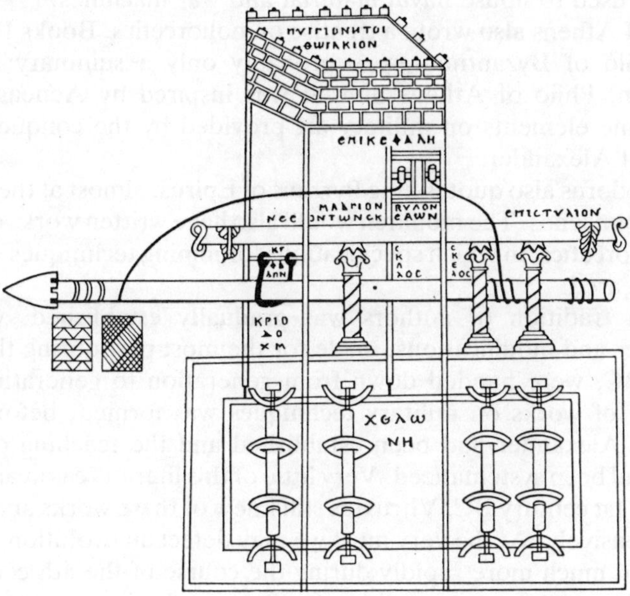

Figure 13. — Plan and elevation of Hegetor's testudo. (From a 10th century Greek manuscript)

which was still considered the definitive work at the time of Vitruvius. Plutarch said that this arsenal was a unanimously admired edifice. An inscription engraved on a marble flag has since been preserved. The work was started in 346 BC and completed in 328 BC. The building incorporated both a public walk and an arsenal, situated between the agora and the ships' berths. The interior was long and narrow in the form of a gallery, divided into three naves by two rows of slender pillars, with a central nave reserved for people to walk down. Loopholes were built high in the walls and there were two large doors at either end. The walls were built without foundations, in layers of superimposed stones, without mortar but held together with iron fitments and sealed with lead. The framework consisted of supporting wooden structures, so the idea of roof timbers is absent. It consisted mainly of a pile of wood beams which rested against each other, the weight of which acted vertically without causing tension. These premises were used to house naval material and war machines.

Philo of Athens also wrote a treatise on poliorcetics. Books IV and V by Philo of Byzantium were probably only a summary or an adaptation. Philo of Athens himself was inspired by Aeneas, and added some elements on military art provided by the conquests of Philip and Alexander.

Athenodorus also quotes king Pyrrhus of Epirus, almost at the same date as Epimachos. The monarch would also have written works on the art of poliorcetics, in which specifications for mining techniques would have appeared.

Thus a tradition of authors was gradually established, whose discoveries and modifications, made for the most part during the 4th century BC, were handed down from generation to generation. A collection of works on military techniques was formed, before the school of Alexander had been established and the teaching of this subject had been systematized. Very little of this literature is available, but in the 1st century BC, Vitruvius still knew of these works and used them extensively. At the very most we may detect an evolution which progressed much more rapidly during the course of the 4th century. Nevertheless, one must stress this attempt to synthesize, to put technical knowledge affecting many domains into a didactic form. The introduction of a form of technical literature is an extremely important fact.

It is certainly true that our authors are far more meagre when it comes to other techniques. Indeed, Greece very rapidly undertook large scale works, whereby the skill and science of the engineer could be displayed. In many cases, this did not involve the simple work of the entrepreneur, but an obvious concern for the application of scientific theories was demonstrated, indeed even shown to be of paramount importance.

Information about Hippodamus, who lived in the second half of the 5th century BC, is rather more vague. According to some Antique authorities, he would have instigated a veritable revolution by constructing several large towns on a regular plan, with wide roads intersecting at right angles: Piraeus, Thurii in 443 BC, and Rhodes in 408/407 BC. This town planner would have had a profound influence. Aristotle said that he was skilled in the science of nature.

In fact, the Ionians seem to have developed towns built to regular plans before Hippodamus, in particular colonies, or certain colonies of

Miletus. Miletus was rebuilt on regular plan, moreover, in 479 BC, after the partial destruction of the town. When the Miton of Aristophanes designed his town of Nephelococcygia, he did it according to a perfectly regular plan. In any case, most recent historians doubt that Hippodamus could have constructed both Piraeus, built in 479 BC, and Rhodes. Hippodamus would have been born in Miletus in around 500 BC, and he would have arrived in Athens under Themistocles. His work is difficult to identify precisely. But Hippodamus, or others, certainly developed towns on a regular plan, be they circular, square or rectangular, in the beginning of the 5th century BC.

Herodotes elaborated on the Samians because it was in their town he could see the three most famous works of Antiquity. The first was the temple, constructed by Rhoicos, son of Philo, of Samos. The second was the breakwater, actually built in the sea to protect the port. It was constructed at a depth of twenty orgyies and was the length of two stadiums. Lastly, there was the tunnel which enabled water conveyance to Samos. This had been built in the 6th century BC, under the direction of Eupalinus, son of Naustrophus, born at Megara. The underground conduit was 835 metres long and the tunnel which carried it across a hilly terrain was 1100 metres long. Twenty vertical wells provided access holes so that the conduit could be cleaned and maintained. The height and length of the construction was eight feet. The adduction canal, twenty cubits deep and three feet wide, ran alongside one of the walls, and the channel was made of terracotta. On the other side was a path where a man could walk round. The whole construction was faced and vaulted. The completion of this work presupposes the aquisition of major techniques: some defects show, however, that the technique or techniques employed were not absolutely perfect. Building had been started simultaneously at both ends, so that a process of levelling had to be put into operation, possibly using methods which we will find once more with Hero of Alexandria. Nevertheless, errors were made and at 425 metres from the south exit, the tunnel must have deverted, as it was designed to be straight. The slope was gentle, so gentle in fact that it proved to be inadequate. The end of the tunnel subsequently had to be dug deeper and deeper towards the north exit. After the tunnel, 400 metres of underground channel was laid to reach the town. Despite the faults which were later corrected, the Samos tunnel represents a significant achievement, requiring already advanced techniques.

The techniques used in building aqueducts were both simpler and older. We should mention the aqueduct built by Pisistrates at Athens in the 6th century BC. This went from the Hymettus to the theatre of Herodes Atticus, and had two branches. It incorporated the old distribution network by a system of water tanks. Water flowed regularly and freely, and manholes were built every thirty to forty metres along its length.

Plato rated Thales of Miletus among the men who "were learned in the mechanical arts." According to a tradition reported by Herodotus, in 558 BC Thales had diverted the course of the river Halys, in order to enable Creosus' army to cross to the other side. At least he may have made it possible to ford by digging a diversion canal. There are still doubts about the origin of the siphon. The first example is that of the Lycia aqueduct, but unfortunately we do not know its date. In any case, it crossed a valley on a wall at the base of which there were two corbelled openings. Pergamum's aqueduct, built in the 2nd century BC, is the most interesting. The two end points were situated at 375 metres and 332 metres. But it had to cross two valleys whih were 172 metres and 195 metres deep. Pressure at the bottom of the siphons was 20 and 17 atmospheres respectively. The pipes were made of bronze or lead, as terracotta could not support such pressures. There are not many examples of these but we can cite a few significant ones, which show that the Greeks had achieved a degree of skill in this field.

We cannot simply say that Greek techniques were blocked, to the point that some authorities have suggested, that the technical contribution of the Greeks was relatively poor. It is therefore necessary to make an appraisal, and only after this, may we discover why Greek techniques did not undergo any other developments. Chronologies should not be forgotten either. We will see in a moment that the Greeks themselves attempted this appraisal with the school of Alexandria. Was it possible to continue the effort required in technical imagination? We have already reached the age of the explosion of Alexander's empire, and the beginning of the first Roman enterprises. The political situation was hardly favourable for this development, but contrary to the suggestions made by a number of historians, Greek science was not capable of providing the Hellenistic civilization with techniques, as these same historians have suggested. We will therefore very briefly consider these two problems.

THE SCHOOL OF ALEXANDRIA AND ITS POTENTIAL

At the end of the 4th century, Greek technology found its development blocked, at the same time as institutions and societies of the Classical period had just disappeared under the influence of the first great empires. But we could also suggest that the new masters of the Antique world, smitten with power and concerned for efficiency, were to attach greater importance to the possession of a material world, or at the very least a material world which could be of service to them. As we have pointed out, Philip and Alexander had already manifested an interest with regard to this. Most astonishing is the work of the various Ptolemies of Alexandria, who in many ways foreshadowed the efforts of the enlightened Renaissance princes. The extraordinary convergence between science and techniques, both striving for fulfilment if not progress, should be emphasized.

Ptolemy Sôtêr, who took the throne of Egypt in 305 BC, was certainly not very cultivated, but this is not so surprising considering the origins of the man. Nevertheless, he demonstrated great taste in patronizing and gathering together the greatest scientists and technicians of his time. Yet again, the Renaissance provides similar examples. The Greek expansion had, to some extent, scattered these attempts, and the respective isolation of the various schools seemed to have slowed down any progress. It therefore seemed good politics, in every sense of the term, to reunite in one place the most distinguished men and books of the Greek world. On the succession of Ptolemy Philadelphus, in 283 BC, for whom virtually everything had been completed, the school of Alexandria was already enjoying the height of its fame.

Zenodotus, the chief head of the library, relentlessly collected the most famous works, bought them from many countries, and even requisitioned books found on ships. The expense involved and the fame of the library undoubtedly enabled a rapid increase in collections of works. Neleus, curator of the books of Aristotle and Theophrastus, sold those under his charge. During the epoch of Ptolemy Euergetes (246/221 BC), Callimachus states that the library contained 400,000 unordered scrolls, containing several works, and 90,000 properly arranged scrolls. A secondary library, containing 42,800 scrolls, was established at the Sarapeum. Even if some works were duplicated, the members of the school could certainly find there almost everything which Greece had provided.

Demetrius of Phaleron, who had established the peripatetic school at Athens, also founded the museum of Alexandria on the same model. Apart from a library, it had a walk, a large dinning room for communial meals, and sleeping quarters. But some completely new elements may be observed, such as the introduction of dissecting rooms, observatories, and a zoological garden.

Obviously we do not know exactly what Ptolemy Sôtêr demanded of the people he sent for, or much about the people who often came to see the treasures collected at Alexandria. It was probably less to do with imagination than with acquiring knowledge, less concerned with progress than with an appraisal. The founders of the school of Alexandria wished above all to collect together all the information avaliable, and to make a sort of encyclopedia. We should certainly stress two points which may or may not have been known at the outset. Insofar as an assessment of the whole was undertaken, knowledge was accumulated and organized, that is to say the arrangement of scientific systems was to be initiated, which could in turn stimulate new discoveries. Apart from this – and we will become increasingly aware of this phenomenon – they were to arrive at a confrontation of the various kinds of knowledge. Thus, they did not stop at the noble sciences, sciences which had become pure, but introduced into research not only medicine, already elevated by Hippocrates, but also techniques; techniques which had fervent followers, and which, as we have seen, were the subjects of written works. Perhaps these pure sciences were found to be less pure than one had wished, and that the techniques had become more reasoned than had previously been thought. It would therefore seem interesting to make a very rapid survey of the whole range of works.

Euclid was one of these scientists who attempted to make such an appraisal: there were, moreover, earlier examples in certain subjects. He must have arrived at Alexandria around 300 BC. He was both a mathematician and a physicist, but unfortunately we only possess a fragment of his work. Apart from the *Elements* which made him famous, he wrote three other books on *Data* (definitions), two books on *Places on the surface*, four books on *Conics*, which replaced the treatises by Apollonius of Perga, and one work on *Paralogisms* in geometry. On the subject of physics, he is credited with a work on optics, and astronomical phenomena. Finally, he probably compiled a work on mechanics, *De levi et ponderoso*. What is particularly interesting in this work, is the attempt on the part of the author to generalize a geometric system. If Euclid followed the Pythagorean theory of rays emitted by the eye, he only made a geometric hypothesis: his ray of light is in fact a straight line. By the way, he did reveal some facts which technicians may have known already, which they certainly used: shadows to calculate heights, laws of reflection on flat surfaces and the possibility of lighting a fire by using a mirror. He may have used identical methods for anything mechanical.

Strato of Lampsacus had followed the lessons of Theophrastus before coming to Alexandria where he was entrusted with the task of educating Ptolemy Sôtêr's son, Philadelphus. He arrived in Alexandria in about 299 BC, and left in 285 BC to succeed to Theophrastus. In the assessment he made of the physics of his age, and of which we only possess a few fragments, Strato gave experimentation a greater place, wishing to make physics a more positive activity, and thus more available to technicians. But in necessarily tackling specific questions, his form of physics could not be established in a system. His reflections on gravity and space are interesting but they could not be explanatory. It was almost as a technician that he recognized the proportionality of gravity to mass, and the phenomenon of increasing acceleration in natural movements.

Herophilus was probably the founder of the medical school of Alexandria. He also appeared to be the true founder of the science of anatomy, on which he wrote a book which has since disappeared. With him began the reign of empiricism. He may have been the first to practise dissection, as he was interested in the vascular system and in certain organs. He studied the nervous system, the centre of which he located in the brain, and sketched an outline of the respiratory circulation. He did not seek to provide general explanations, or abstract relationships. His therapeutics were also based on concrete observation.

Aristarchus of Samos had been Strato's pupil, and was interested in physics and the problem of light, but he was especially fascinated by astronomy, and made an observation of the summer solstice in either 281 BC or 280 BC. He wrote a treatise dealing with the dimensions and distance of the sun and the moon, which has fortunately been preserved. His method is purely geometric and founded on the measurement of angles and on a problem of trigonometry. The results are deceptive, mainly because he lacked suitable instruments. His compatriot, Conon, was also interested in astronomy and geometry. He wrote seven books on astronomy, particularly on eclipses. Unfortunately he died too young to demonstrate all his ability.

Eratosthenes appeared in the second generation of the Alexandria school. Born at Cyrene, he went to Athens before arriving at Alexandria, around 244 BC. There, he became head of the library. Two of his works are particularly important, the geography, and the treatise on measures, only parts of which are extant. He also attempted to construct a geometry of space: Aristarchus had tried to prove that the sun was the centre of the system, and Eratosthenes declared his belief that the earth was round. He managed to make a fairly accurate assessment of the distance between the earth and the sun, but his use of the famous method of the moon's dichotomy was not so successful, for his calculation of the distance beteen the moon and earth was wildly inaccurate.

Erasistratus, a doctor, was a contemporary of Eratosthenes, and it seems he was known chiefly for his works on physiology and pathological anatomy. He also completed Herophilus' works on nerves and the brain. He studied the heart, but encountered several difficulties when it came to the circulation of the blood.

So within two generations, Alexandrine science had managed to make some important contributions. It is possible that the first of the Ptolomies, whose education had been somewhat limited, had wanted to express his awareness of the culture in which he lived. This is difficult to prove of course, but if one considers this in the light of the Renaissance, it is less easily dismissed. He therefore gave his son the best education possible, and this explains the almost gluttonous desire to bring together the greatest minds, works and scientific knowledge in this period of Greek history. It is also possible that he discouraged fruitless abstract discussion, and promoted the ideas of practical assessment, concrete results and clear explanation. All this was practised within the limited confines of their scientific knowledge consisting of geometry and arithmetics. Similar efforts were to be made by the Sforza, Montefeltro and Malatesta families. The techniques of observation and experimentation were gradually acquired in this way: the improvement of techniques was the main object, and all the more so because this area of research above all others was closest to the Sovereign's heart.

This preoccupation is evident in the problems of pseudo-Aristotelian *Mechanica*, now considered to date from 280 BC, and which was written very much in the spirit of the Alexandrian school. It is a theoretical work, but nonetheless it still deals at length with practical aspects. All the problems of mechanics are explained with reference to the principle of the lever, which was itself derived from the circle. Distinctions are made between the natural aspects of this technique, and the human elements, which are often contrary to nature: by means of a lever, a light weight can lift a much heavier one. This work shows a good understanding of statics, but the treatment of dynamics is not nearly so successful. Hero of Alexandria also dealt with a large number of these problems, and conducted many concrete experiments besides. He experimented with the transport of heavy loads by means of rollers and carts and with the chock wedge, capable of holding enormous loads in place, simply by opposing the latter's weight with the weight of the earth. He also experimented with pincers and finally, with groups of pulleys in the aim of reducing the amount of power necessary to lift a weight – this became known as the blade and tackle. One of the notable characteristics of Alexandrian scholars was their refusal to consider any technical problem that could not be explained in scientific terms. Thus, for example, they were able to explain bodies in equilibrium, but not in movement.

Archimedes is a good symbol of the technico-scientific ideal of the Alexandrian school. He lived in Alexandria when he was still fairly young, rubbing shoulders with the scientists we discussed above, and reaping all the benefits afforded by the exceptional library and museum. He strengthened the growing awareness of the close relationship between science and techniques as a result of his research into practical statics, and his declaration that every massive body has a centre of gravity. He then used a minimum of postulates – whose technical basis cannot be disputed – to establish a logical series of laws. In fact, it was techniques which were the real basis of his theoretical work. The theory of floating bodies, which is the base of hydrostatics, is a similar case.

A scientist of genius, Archimedes also enjoyed the reputation of being a skilful technician, so much so that the Ancients, especially during the later period, gave him the credit for a considerable number of inventions. Of course, geniuses are apt to receive the credit for everything, and it is noteworthy that the technical curiosities usually associated with him are also found related to all the other technicians from the School. He does not seem to have written anything technical, which according to Plutarch and those after him represented a disdain for material things but which in fact was not at all in the spirit of the Alexandrian school. The catalogue of Archimedes' technical works, such as it has been passed down to us today, is hardly comprehensive. Archimedes' "inventions" can be divided into two groups, those concerning the siege of Syracuse and the others. Ballistas used for throwing stones, and arrows were already known at this time, and furthermore, Marcellus used a machine which was so enormous that it took eight galleys joined together to carry it. It is possible that the Archimedes' machines were larger and better equipped, they were capable of throwing stones weighing ten talents – or about 260 kg – but the principles of these machine were no different to those already in use. We know that Euclid was familiar with the convergent mirror, which he had described in geometrical terms, but it is possible that he was unable to put theory into practice, because of the inadequacy of current polishing techniques. As far as fortification was concerned, Archimedes confined himself to building some loopholes, and fortifying the lower parts of the town. And after all, was not the town finally captured?

As for the rest, the so-called archimedean screw had been introduced into the Nile valley long before the time of Archimedes, and it was mentioned in the pseudo-Aristotelian *Mechanica*. Hydraulic organs were known to Ctesibios, an older contemporary of Archimedes, whom we shall discuss later. The apparatuses for raising ships were perhaps more original – then again this was a problem that interested all technicians during this period.

There is a great deal of evidence to show that Archimedes was certainly interested in machines, and it is not impossible that with a more extensive knowledge of science, he could have improved the basic principles of some of these, and perhaps even have expounded the theories on which they were based. This brings to mind a term coined by Plutarch, "simple games of geometry." And once more we find the one great autonomous science (or which had become autonomous), being extended to science and techniques. The desire to reduce techniques to an exercise in geometry began at this point and continued right up until the Renaissance. But unfortunately, mechanics does not consist merely of statics and geometry, and the failure by the scientists and technicians to grasp the theory of dynamics necessarily limited their ambitions. With all due respect to Plutarch, it is very likely that Archimedes did not write anything on techniques, because since he did not invent anything of note, he had nothing particularly new or original to say. It is far easier to imagine Archimedes rationalising known techniques, pointing out mistakes of assembly or dimensions, and in particular, attempting to induce general laws from the confused tangle of empirical results obtained from experiments.

The Alexandrian school was the centre then, of a remarkable convergence of all sorts of intellectual endeavours. The preference for concrete experiment and observation necessarily prompted an interest in techniques. Some works on techniques had already been written, but these were fragmented and incomplete, and an assessment was needed. The range of techniques considered in these assessments was relatively small. Military techniques, recent inventions, the techniques which could be "geometrised" and finally an element which has been rather scornfully called, "the physics for amusement," formed the main basis of such works. The works dating from this period reflect the limited scope of their concerns which did not begin to include architecture, artisan techniques, ships, and many other areas. But it is possible of course that some of the works may have disappeared.

We know very little about the apparent founder of the Alexandrian school, Ctesibios, and all that we do know has been gleaned from the rare allusions made to him in works written by his immediate and distant successors. He is mentioned along with Archimedes in Vitruvius' first book, as being one of main authorities for the training of engineers, and as being himself one of the most famous mechanicians in Antiquity. His written work was obviously still in existence at this time. He was a mysterious character, who lived during the first half of the 3rd century BC, which made him a contemporary of Archimedes. He came from relatively humble origins, the son of a barber established in a suburb of Alexandria. He appears to have manifested a remarkable gift for mechanics from a very early age, and his work on the subject was apparently a major one, which regrettably has entirely disappeared. This means that we are limited to discussing only those "inventions" which are usually attributed to him.

It was Ctesibios who invented the water organ, which has been described by Philo of Byzantium, Hero, and Vitruvius (fig. 14). As the diagrams show, the organ was powered by hand-bellows. There was a tank of water whose pressure controlled and prolonged the air-flow which the bellows forced back into the apparatus. There were three sets of bellows arranged in a triangle. The air accumulated in goatskin bags, and was then passed into the apparatus, from whence it escaped at a regular pressure into the sound pipes. Philo and Vitruvius have given us some good descriptions of the force-pump, which is so well-known that its importance does not need to be emphasized here (fig. 15). According to Vitruvius, the pistons were polished with a lathe, rubbed down and oiled. We know nothing about how the cylinders were made, but this was certainly a more complex operation. The difficulties involved in constructing these machines meant that only relatively few were built.

Ctesibios is also supposed to have invented the water clock, whose basic principle is fairly simple, but Vitruvius did not conceal the fact that this clock was not entirely reliable. Philo also mentions a military apparatus which was apparently invented by Ctesibios, but he did not expand on this. Athenaios mentions a machine which was capable of penetrating enemy walls, consisting of a large pole which pivoted on spindles. There is mention of a machine called a "bronze-spring engine" (*chalkeutone*) which is essentially a ballista, but equipped with a metal spring instead of a rope. The spring could only have been made of steel, however, which was hardly known at this time. The final machine in this list was known as an air-spring engine, or *aerotonon*, which was in fact a machine using compressed air acting on a spring, so that it merely operated as a trigger. It was made of pounded bronze, shaped on a lathe with flat regular surfaces, and was equipped with pistons.

It seems that unlike his contemporaries, Ctesibios did not attempt to provide any assessments. As far as we can gather, he simply confined himself to expounding various ideas based on observation, such as the elasticity of air, for example, and the incompressibility of water. He also considered what could be taken from these ideas to create machines, even though the latter were in fact technically unrealisable. At any event, he is certainly accepted as being the founder of this school of mechanicians, and his mechanical skills were clearly recognised, as were his numerous inventions.

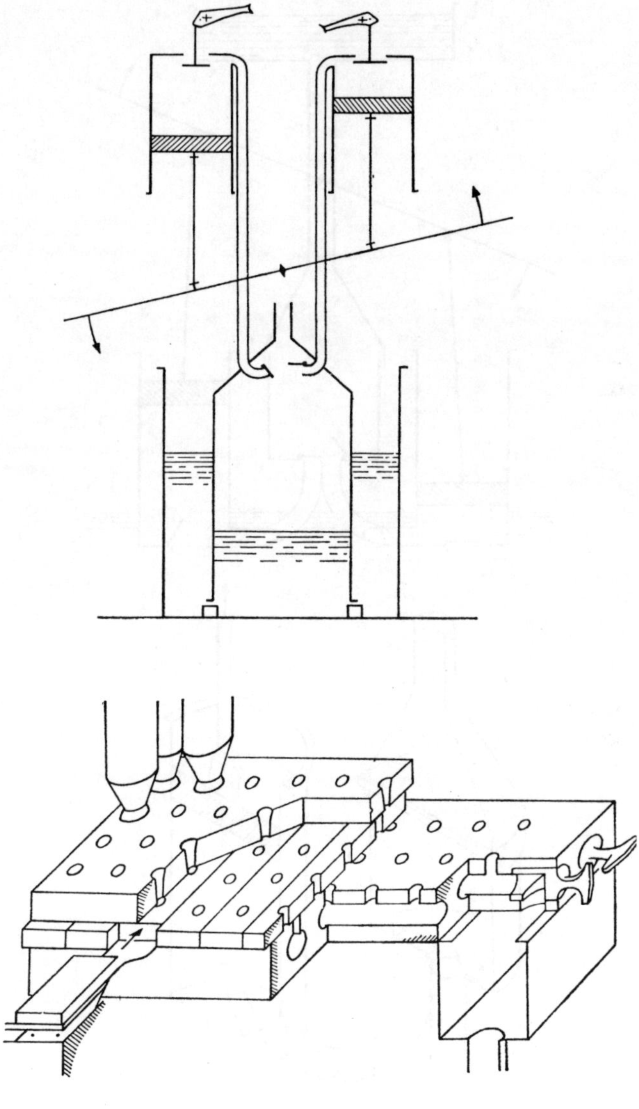

Figure 14. — Water organs. (From Choisy)

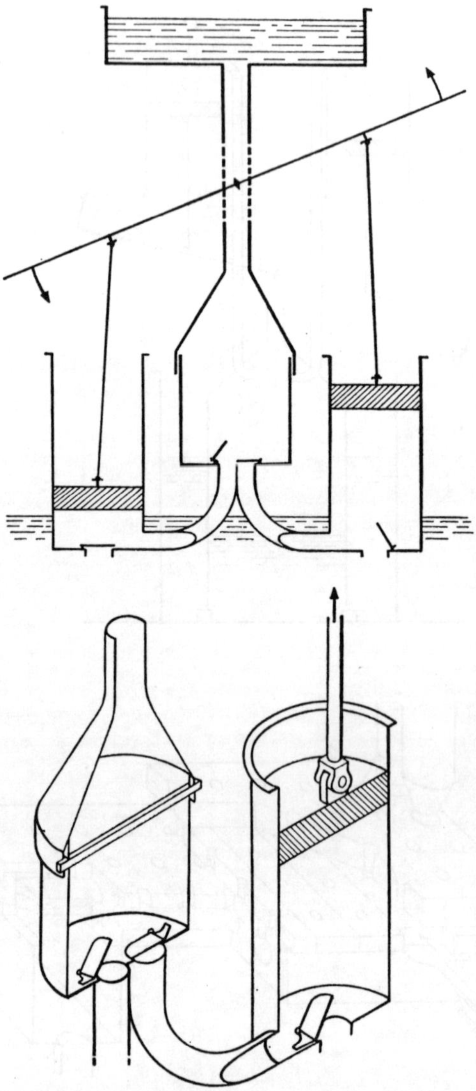

Figure 15. — Suction pump by Ctesibios. (From Choisy)

Vitruvius, for his part, writes about him:

> Furthermore, this was not the only inspirational idea of Ctesibios, for he also discovered another principle whose use was manifold: this was that of liquid forced under pressure to produce impulses found in nature. For example, blackbird song could be produced by the action of water and divers, and the same water could produce moving figures and many other things, whose effects charm the eye and delight the ear.

Ctesibios must also have attempted, therefore, the difficult art of constructing automatons, which was to give rise to many mechanisms so familiar to us today.

It is not inconceivable that a part of Ctesibios' work was included in that of his successor's, Philo of Byzantium, even though in certain passages the latter took great care to stress his originality with regard to the former. In any event, Philo's work is the first one by a mechanician which is still largely intact. This takes us to the programme of mechanical "syntax," whose importance lasted until the 10th century. The titles themselves of the programme make interesting reading (the extant works are preceded by the "+" sign): *Treatise on levers*; (+) *Pneumatics*; *Treatise on automatons*; *Treatise on wonderful instruments (organs and pipes)*; *Treatise on the traction of heavy burdens (the Baroulkos)*; (+) *Treatise on clepsydras*; *Treatise on the building of sea ports*; (+) *Treatise on the building of fortresses*; (+) *Treatise on war engines*; (+) *Treatise on self-propelled wheels*. It is possible that these various treatises were preceded by a general introduction on mathematics and general physics. This collection was completed by Hero. The progamme was concerned only with new techniques, not with traditional ones. Again, it was mainly an assessment, so that not everything should be accredited to the authors.

The treatise on *Pneumatics* is known through an arabic translation: it is well written, succinct, and comprehensive. It begins with a series of theories on the nature of air and its relationship and junction with water, and on vacuums. In other words, it deals with the same problems as those already considered by Ctesibios. The physics is fairly elementary, founded on a number of simple experiments and resulting in the idea that no vacuum can possibly exist. Philo also establishes the compressibility of air, the equilibrium of fluids in communicating vessels, and the principle of the siphon. The rest of the work is devoted to a presentation of the "marvellous containers", or "magic jars" which use these various properties: like the intermittent fountains, vessels with two liquids, and the wine-thieving cup. This section ends with a presentation of some automatons, such as the horse and trough, or the bird in its nest, flapping its wings and frightening away a snake which jumps up from the ground. A description of multi-channel taps follows this, along with whistling waterwheels and singing animals (compressed water, with air escaping at a controlled rate). The work ends with more descriptions of more utilitarian objects, like the noria with bucket line chains, Ctesibios' air pump, various elevating apparatuses with balance beams, winches, wheels with squirrel cages and camshafts. After one general essay, the *Pneumatics* rapidly turns into what was later to be called a "theatre of machines."

The work on clepsydras was probably written by Philo, although there is no definite proof of this. We know of this work only through the arab version, which is somewhat truncated and poorly written. The dimensions of clepsydras are carefully noted in the account of the main mechanism. According to this particular description, the clock is regulated by an elbow-pipe, the perpendicular part of which turned within a semi-circle divided up into the signs of the zodiac. A pointer is connected to this pipe, which controls the flow. Time was calculated in terms of temporal hours; i.e., hours which varied according to the seasons and climates. This was an ancient and widely used formula. The apparatus could be equipped with various automatons, with bells and different characters, for example, such as the flute player which performed by means of a horn, a siphon and compressed air.

Philo's other treatises all dealt with the art of war. We still have a treatise on poliorcetics and a treatise on war engines. The treatise on fortification is the earliest complete work of this kind in our possession. Philo was part of a long line of tradition, however, and his fourth and fifth books are merely taking up the ideas of a treatise by Philo of Athens. The work lacks logic, although it does contain all the major principles of classical fortification in this period. There is less innovation than codification, and all the references are there, in particular those refering to Rhodian engineers. It is in fact a scientific mixture of formulas and empiricism. "The location of all ramparts, as well as their slope, curvature and extensions, should always be determined by the nature of the area." This is the first rule: adapt according to the terrain. It is only at the end of the work that Philo talks about the great systems of fortification built by his contempories, and the traces they adopted according to the type of environment: "maiander traces" or meander systems in fields, semi circular or saw-tooth traces for undulating areas, double systems for pronounced re-entrants, and traces with oblique curtains for triangular salients. Lateral flanking was evidently one of the main principles of these traces. Towers and posterns were essential elements, outflanking on the outside, and aligned on the inside. Notes about the principle of covering the stonework with earth "face-masking" the priority of thickness before height, the maximum tilt of flanking parts, and the need for an easy passage around the walls are accompanied by notations about the problems of providing supplies and the modernisation processes of ancient fortifications. The subject is dealt with thoroughly, therefore, and its treatment reveals a complete mastery of some techniques. The work ends with a discussion on attacks. Apart from a few psychological or tactical considerations, it represents in embryonic form a treatise on war machines: scaling instruments, testudos, and armoured moveable towers. It seems that the author has acknowledged the lessons learnt from the last great campaigns of Philip of Macedonia and Alexander the Great.

The treatise on missile-throwing machines is extremely important in that it enables us to take a much closer, more direct look at the way in which these mechanicians worked. According to Philo, previous authors – which proves that his was not the first treatise – differed on the essential principles, particularly about how to establish the proportions of the smaller parts. Philo believed that a module should be used so that the machines could be constructed according to clearly defined proportions. This module, the *τὸνοζ*, represents the diameter of the rope of sinews which serves as the spring of the arm. Naturally, it is proportional to the weight of the projectile: by increasing the cubic root of the projectile's weight – calculated in dactyls – by a tenth, one can calculate the the diameter of the rope in inches. The extra tenth represents the approximation of the cubic root. The formula goes thus:

$$d = 1{,}1\sqrt[3]{p}$$

They arrived at this formula by a process of trial and error, and by drawing up tables, some examples of the most common of which are provided by Philo. There was also the problem of measuring the projectiles, although Philo does not mention this. The dimensions of the various pieces of the machine were all drawn from this module: the spring box, the thickness of the boss, the support, the arm and the length of the spring, which had to be twice that of the arm. And Philo provides us with all the various applicable proportions. In order to assess certain sizes, he employed the method of doubling the cube. The method he used for this was already known, but he took great pleasure in using it himself. Existing machines could be made smaller or larger using similar methods. This rationality in the construction of war engines was not a new phenomenon in Philo's time: "No-one dared depart from the formula," he notes. This was not technology, i.e., a technique founded on a rational basis of scientific explanations, but there was something new in drawing up tables of repeated and organised observations – and the scientists of the Middle Ages and the Renaissance were to do the same. Philo's reflections were particularly interesting with regard to a technique which would have welcomed a more rational basis, but which lacked the necessary scientific information. Only the practitioner, armed with practical knowledge could obtain his objective, but this was invariably dependent upon the whims of chance.

Many practitioners have constructed machines of an equal size and identical composition to those built by the ancients, they have used similar pieces of wood, and equal quantities of iron. However, although they used different types of loading systems, the range of their machines was far longer, and the effects much greater. They were also a lot more powerful. If they were ever asked how they produced such results, however, they were unable to explain.

Philo considered this lack of rational justification as very serious. However, there was also the negative side to the problem. "Some aspects cannot be understood by reason and mechancial processes alone. And many discoveries are the result of experiment." This was acknowledging the limits of contemporary science. The most important thing in Philo's opinion was that experiments should be organised, and not totally uncontrolled.

Philo's principle tenet was that technics could no longer be such an uncertain activity. A scientific technology was in sight, but still difficult to understand. The limits of science and practice ultimately blocked possible progress: but one wonders whether it was possible at all, or whether it was the techniques which had simply reached their limits. Philo was quite correct in noting that the art of machinery was difficult when so many necessary conditions came into play. "This is what happens to our art when so many calculations are required to obtain the final product, or to complete and perfect a work. The slightest deviation from the correct process can lead to faults in the final result." Talking about one piece of a machine, he specifies: "It should not be considered haphazardly, but with a certain method." Even though there should be no misunderstanding about the "many calculations", Philo perceived – better than his teacher Ctesibios, that science was just one of the many tools indispensable to techniques. Moreover, he clearly stated that he would base the descriptions of his machines on "explanations based as much on mechanical demonstration as on natural reasoning." He never neglected the opportunity to show off his scientific, geometrical or mechanical skills. This attitude of a technician goes beyond that of simply describing machines.

Philo also emphasised the "wedge machine," i.e., the machine with penetrating arrows, stretched with wedges, of his own invention. The description of the mechanism is somewhat confusing, and the manuscripts have no illustrations for reference. The author remarked that the ancient machines were "tiresome and costly, and virtually worthless since they could not withstand prolonged tension." The new machine had a much longer range, it was easier to construct and dismantle, and it was cheaper. The aspects of dismantling and cost are mentioned here for the first time. Philo also insists on the quality of the materials. The iron used to make the springs had to be "extremely pure." They should not be subjected to brutal forging, but rather hammered when cold so that only the outside hardens while the inside remains supple. Metal, therefore, was coming into play; these machines were no longer simply neuro-ballistic.

After having described Ctesibios' machine, Philo goes on to talk about the machine called "polybole." This was probably invented by an Alexandrian "engineer" named Dionysus. The arrangement of this machine was quite unusual and very complicated, but often ingenious. It required careful attention, however, being difficult to drive and to build. The final machine dealt with by Philo was Ctesibios' air-spring engine, or aerotonon.

This work by Philo of Byzantium, albeit narrowly conceived, is indicative of what we consider to be a state of technical and scientific progress in comparison with the past, even though we know so little about the latter. The most important point is that thanks to all the many scientists and important works under its roofs, the Alexandria

School enabled links to be established between science and techniques which hitherto had been somewhat ephemeral. Even if Philo was only a compiler, and even if there were other engineers whose names he mentions, nevertheless he conveys an attitude which appears to be highly evolved. Philo's "mechanical syntax" was to be widely studied, plagiarised and developed, and more publicised each time a new discovery was made.

No doubt parallel trends were developing elsewhere. Take the Rhodes School, for example, which we mentioned above. Apollonius of Perga, who was living in the first half of the 3rd century BC, spent a long time in Alexandria, where he studied geometry under the disciples of Euclid. Apart from *Conics*, which quite rightly proved to be his greatest claim to fame, he also practised applied science. The Arabs have attributed him certain treatises on mechanics, which are now lost. Vitruvius mentions him alongside Archytas and Archimedes and others too, who "with the help of calculations and the knowledge of Nature's secrets, have made great discoveries in mechanics and gnomics, and have left expert treatises to their successors." We are still dealing with the same kind of channels here, then. According to Proclus and Pappus, Apollonius also wrote about the screw. History still has a place for a man called Agesistratos, who, according to Athenaeus, wrote a treatise on machines. This is confirmed by Vitruvius who also suggests that his reputation was a major one. Athenaeus' following statement is fairly representative:

> It is obvious that if someone is placed in charge of a town's defence, they should be able to trace working drawings in order to provide defence against attacking machines. They should also be able to construct attacking engines capable of resisting defence. For a beginner, this would be extremely difficult, and only those who have made a careful study of this art, who have examined all the relevant research, and who are aware of the precise nature of all the experts' writings and the latest facts regarding this science, will find this task within their capabilities. We need to use worthwhile inventions, and not simply innovate for the sake of it, unless of course we wish to be like the people who seek to deceive those in ignorance, and who prefer the appearance of truth to truth itself.

The science of engines took on its definitive aspect; henceforth, in fact, it was blocked.

Hero of Alexandria is still the object of much discussion: as Dain quite rightly notes, "Hero was for applied sciences what Euclid was for geometry, and judging by the number of fragments which have been preserved under the name of this writer, he could be connected with every single treatise on applied science." Hero as a person is an elusive topic of study, but the halo surrounding his works makes them an equally awkward subject. He was probably a man of humble origins, like Ctesibios, and may have started off as a cobbler. The majority of those who have studied Hero's life believe that he featured during the end of the 2nd century BC, which is noticeably later than the other scientists we have already discussed. Athenaeus, Vitruvius and Pliny make no mention of him, but they do talk about Ctesibios and Philo. On the other hand, Pappos, Eutocius and Hero of Byzantium, who were all writing later, mention Hero of Alexandria but not Ctesibios. Apart from this chronological quarrel – which is not without its own interest – Hero quite clearly belongs to the lineage of the Alexandrian mechanicians.

He seems to have produced about the same amount of works as Philo, a large part of which, and perhaps the most important part, is still with us today. Thanks both to Hero himself and certain writers of a later period, we are able to list these works here:

Firstly, two works of a scientific nature: *The Metrics*; *Definitions and Commentary on Euclid's Elements*.

Four works mix scientific preoccupations with practical applications: (+) *General Mechanics and the Baroulkos*; (+) *Pneumatics*; *Catoptrics*; (+) *Dioptra*.

Finally, four works of applied mechanics: *Water clocks*; (+) *War Engines*; (+) *Cheiroballistra*; (+) *Automatons*.

Those works preceded by a "+" are extant, and it is evident that most of these are particularly interesting. It is also worth noting that this list is slightly different to the one we drew up for Philo. Fortification problems do not feature here, but is this not perhaps an indication that Hero was living at a time when the Roman conquest had made fortification largely futile? It is also possible that Hero chose not to deal with the techniques which had not made any marked progress, and which did not therefore require re-consideration.

We possess only fragments of the *Metrics*, with numerous interpolations. This work is a kind of elementary manual of mathematics. The first part contains definitions of arithmetical terms with examples, like the approximate extraction of the square roots of numbers which are not perfect squares. The second part is devoted to geometrical terms. Finally, there is a practical treatise on plane geometry. The *Definitions and Commentary on Euclid's Elements* was to be a succinct account of

Euclid's geometry. It ended up in fact as the technicians' mathematical ABC.

The same was to be true of *Mechanics*. This treatise owes a great deal to Aristotle and Archimedes – the latter is quoted nine times regarding the problems of equilibrium and the distribution of weights on supports. The stamp of peripatetic thought is evident in this work with the emphasis on causal explanations and the reduction of mechanical phenomena to basic principles. Hero reduces "powers" to a unique nature, the simplest concrete expression of which is to be found in the lever, and the abstract, and therefore symbolic expression of which is to be found in the circle. But the expression of the reduction of a wedges angle, and of the screw to the lever is less clear than in Aristotle. The work is not without finesse, in its examination of the effect of percussion on wedges, and of the more or less favourable conditions in which screws could be applied. Nor does this work lack skill, which is attested by Hero's method of slowing down machines whose reduction ratio is greater. Hero devoted a great deal of time to studying circular movement, whether concentric or not. His conception of gravity was not totally Aristotelian: he clearly states that heavier bodies fall more quickly than lighter ones, and that the shape of these bodies is also significant.

This work is situated midway between science and techniques. Hero may have admitted that experimentation was the best teacher, which was thereby to admit that as a whole, science was unable to allow for technical reality, but he declared elsewhere: "It is quite clearly necessary that those who are learning about mechanical arts should understand the nature of the centre of gravity and gravity itself." Consequently, the treatise is laced with practical problems. The information on gearing gathers most of the propositions known at that time: two wheels on the same axle turn in the same direction; geared wheels turn in opposite directions; if two circles turn around the same axle, the larger ones will move more quickly than the smaller ones; if two circles turn on two different axles and are geared to one another, then the smaller ones will turn faster than the larger ones. All the problems concerning ratio and reduction are considered, reduced for the most part to the problem of levers. This book concludes with the problem of supporting heavy burdens, which was particularly relevant for architects (columns and lintels) and which itself was reduced to that of the equilibrium of a balance beam.

The second part broaches more theoretical questions, relating to five simple machines, or the five kinematic chains: the windlass, the lever, the pulley, the wedge and the screw. The study of the windlass in summary: Hero establishes the law of equilibrium between power and weight, and provides elementary calculations of ratios. His treatment of levers is taken from Archimedes, and there is nothing original in the study of pulleys and mouflons. Hero emphasises the difficulties of constructing and using screws. His definition of the screw, which he likens to a curved wedge, is somewhat curious. The most obvious effect of all these machines is moving considerable weights with moderate forces.

Ultimately, this work may be merely one of compilation, but we should recognise that everything is clearly presented and logically ordered. As for the simple machines, the author notes: "for the most part we were content to agree with those who have dealt with the subject in the past." The practical demonstration of all these principles was provided by means of the baroulkos, a machine used for moving heavy weights, precisely with moderate forces. This machine was known before Hero's time, and was dealt with by Philo. The problem was to move a given burden with a given force, by means of a gearing system. In this case, a force of five talents was to be used to lift a burden of a thousand talents. With their knowledge of gearing, this was hardly a difficult task.

The third part is devoted to a number of composite machines and is the most practical part of the work. The first of these machines were the lifting engines: the mast crane and pulley, the engine with two posts and mouflon, and the three or four-legged hoist. Hero suggested that it was better to avoid using nails which would weaken the resistance of the wood, and to use ropes instead, which are still used today. After this he deals with crabs, which pick up stones, and the work concludes with a discussion on various types of press: the screw press; the press with central screw; the press with screw and lever, the lever press with winch. According to Pliny, some of these apparatuses were new to Hero's time. This part is fairly similar to the final book in Vitruvius' work.

There are some interesting comparisons to be made between Hero's *Pneumatics* and those of his predecessors. The basis of his work is the same as Philo's, but there are some notable differences, however. A. G. Drachmann has drawn up some comparative tables of the two works. In Hero's work, forty-four of the apparatuses featured can be found in Philo's Pneumatics, twenty-three are completely new, and eight provide different solutions. On the other hand, entire groups of instruments dealt with by Philo are neglected by Hero. Hero's originality – and this has sometimes led to his fame being somewhat exaggerated – stems from the machines using the force of steam, like the famous aeolipile, for example.

The preambule of the treatise recalls certain basic principles, and these are accompanied by discussions on opinions which differ to Hero's. Throughout the work, and this is the same with Philo's *Pneumatics*, there are the most ingenious applications of all the Alexandrian scientists' knowledge concerning the elastic and motor force of steam and gases under the action of heat and pressure, and especially on the action of this compressed and dilated steam and gas on the equilibrium and movement of liquids. It would be tedious to relate all the experiments concerning suction pads, syringes, lamps, siphons, suction pumps and water organs (which are no longer the object of a single treatise). The fountain and aeolipile, the small whirling device driven by steam which escapes by means of an ingenious mechanism composed of two bent pipes, are well known. In this work, Hero reveals a greater mastery of techniques than his predecessor.

Optical science was made up of three main parts: optics; catoptrics (mirrors, theory and application); and dioptrics (dioptra, theory and application). Science and techniques are closely linked here. Unfortunately, as far as Hero is concerned, the extant texts are inaccurate, incomplete and interpolated. The fundamental elements of optics were to have been dealt with at the beginning of the section on catoptrics. As Ronchi has shown, Hero made the mistake of extending the law of the minimum way to mean refraction, which meant that he was led into making appalling generalisations about all phenomena of light. The first part of catoptrics contains two optical theorems and six theorems on flat and convex mirrors. In the second part, there are nine theories concerning the construction and use of various apparatuses composed of flat, convex and concave mirrors. As with the mechanics and pneumatics, Hero provides further evidence of his broad knowledge and contributes some interesting complements.

The *Treatise on dioptrics* consists of a description of the dioptra, followed by several examples of how it could be applied. The dioptra is a surveying instrument. Hipparchos' dioptra had served to measure the visible diameter of the sun and the moon. Hero's dioptra consists principally of a mobile water-level on a tripod; but this level can be removed and replaced either by a simple mobile alidade placed horizontally or vertically, or by a circular plate divided up into degrees and which could be fixed according to any oblique plane (fig. 16). In some ways, this apparatus resembles the modern theodolite. The second part of this work deals with a number of measuring problems, several of which resemble the same ones experienced by the Roman agrimensors. Moreover, several solutions are obtained by a process of trial and error and trying different approaches: such as the tracing of a straight line between two points so that neither point can be seen from the other, or simply measuring a field, for example. Other problems concern civil or military operations. We should note in particular the aspects of underground works, such as tracing a straight line through a mountain to join together two points on the opposite flancs, or digging shafts which drop perpendicularly onto an excavation or a gallery. Hero was especially interested in the problem of measuring distances which were out of range, such as the length of a river, the distance between two far away points, or the height of an inaccessible point or the depth of a ditch, for example. He also considered the question of the best distance at which to place an artillery engine. Practical and agrarian geometry was also well represented, with measuring and parcelling up, etc. Finally, he also dealt with astronomical problems and the art of measuring angular distances in the heavens: he attempted to measure the angular distance between two stars, and to determine the distance between two places in different climates by observing an eclipse of the sun in two different areas – this was done in places whose latitude was apparently known. Hero also provides us with a description of the hodometer, a fairly ingenious counter adapted for wagons and consisting of an endless screw and cog-wheels. The same apparatus, with the addition of a paddle wheel, makes it possible to measure the travel of a ship. Finally, he assesses the asterix, a very similar instrument to the Roman groma.

So in fact this is a treatise on practical geometry, and represents as such the precious remains of a long tradition. Hero may not have made a significantly original contribution to this tradition, but he at least proposed some new and sometimes elegant solutions here and there to old problems, such as calculating the area of a triangle according to its

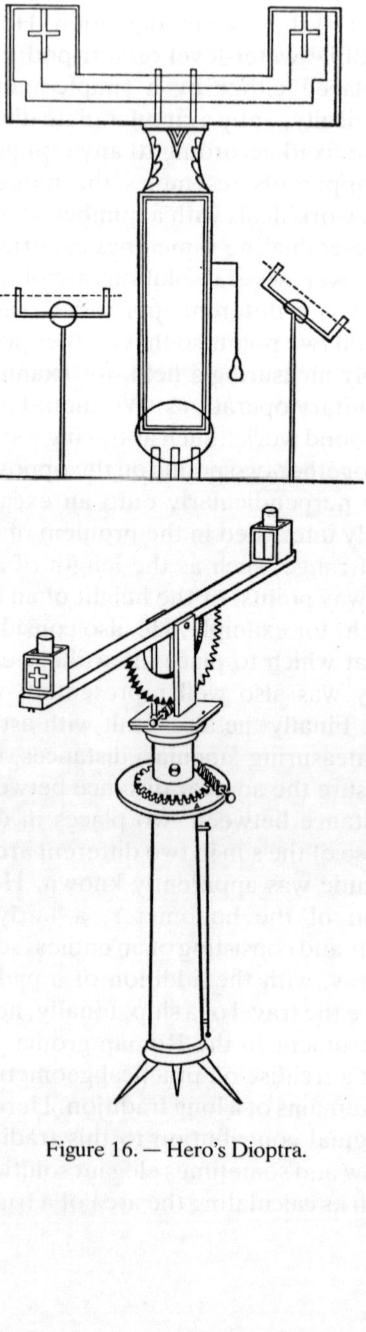

Figure 16. — Hero's Dioptra.

three sides. The most interesting aspect for us is the ensemble of technical means available to the Ancients for solving these problems.

Hero's treatise on throwing machines is less comprehensive than Philo's, however it is more detailed:

It is true that our predecessors wrote a great deal on ballistics, giving the dimensions and designs of machines. However, not one of them gives any proper indication as to how these should be built, or how they should be used, it being their habit to write only for those who were already in possession of the necessary details.

Hero effectively describes every piece of every machine, and gives their dimensions. But all these details do not make our understanding of these machines any more precise. It is noteworthy that Hero's text is very close to that of Vitruvius, who, in chapter 12 of Book X dealt with the same subjects.

Hero made a classification, and a somewhat artificial one, of throwing machines: (*A*) euthytone machines which cast only arrows. These are sometimes called scorpions; (*B*) palintone machines which can throw not just arrows but stones as well – the latter machines are called "lithobolos."

The principles of all these machines are the same. A bow string is stretched between two flexible arms or held with a flexible material. The string is stretched by means of a winch into a cross piece which is manoeuvred by a spring. This is the classic throwing machine of Antiquity. "The aim of ballistics" Hero emphasises, "is to throw an arrow in such a way as to hit a long-distance object in a given direction." He reminds us that ever since history began, people had sought to throw heavier and heavier projectiles for even greater distances. This had led to the time when the bow string could no longer be held by hand, giving rise to the introduction of winches.

The machine had to be completely mobile, and therefore dismantable and portable. The dimensions had to be completely normalised so that each used or broken piece could be easily replaced. Philo was already convinced that this was possible, and as far as the dimensions and proportions of the machines were concerned, Hero used the same method as that of his predecessor: determining a module and formulas from which all the dimensions could be drawn. All the formulas are identical. "The diameter of the stretcher's opening is the point of departure: it is therefore the stretcher which is the principle and foundation for all the rest." Hero adds to this the formula for the arrow throwing machine, which did not feature in Philo. The module, which again was the diamater of the stretcher, had to be equal to a ninth of the arrow's length, or:

$$d = \frac{L}{9}$$

So for an arrow measuring 3 cubits, the module would be 8 inches. Here again, Hero's attitude is very interesting. "It must be certain that the figures for the dimensions have been determined by experiment." So this was not a matter of reasoned formulas, as certain contemporary historians have claimed. Hero emphasises that it took a long time to arrive at these results: "Little by little, they succeeded in constructing machines which were as logically ordered as they were powerful." Nothing had changed since Philo's time, then. We said above that the science of war engines was blocked: there is no better proof than this. It was only with Biton that the formulas were not uniquely the result of accumulated experiment. Vitruvius continued Hero's tradition.

There was already a long tradition of treatises on automatons by Hero's time. And no doubt emulation was probably at its height in this field. The imagination was of little use in the field of surveying instruments or war engines, but automatons were quite the opposite, giving full rein to inspiration. Everyone wanted to demonstrate their skill in this field. "We wanted to write something more original about fixed assembly automatons" noted Hero, "something better than our predecessors, and at the same time more instructive; and as far as this subject is concerned, we shall not include any of the writings of Philo of Byzantium." We should note straight away that the field of automatons has a particular significance for the history of techniques, with its combinations of the most diverse mechanisms, the invention of new types of transmission, and what is more important, of progamming. These were to constitute an exceptional support for certain advanced techniques in the centuries to come.

The treatise is divided into two main parts: automatons on moving stages, and those on fixed stages. The former operated in effect on a kind of rolling chest which contained and therefore concealed the motor mechanism, which served not only to drive the theatrical automatons, but also to manoeuvre the stage around. As for the fixed automatons, the chest remained stationary, and was used as a veritable miniature theatre, where elegant puppets performed plays of several acts, with intervals and scene-changes. In both systems, the motor consisted of a descending counterweight. There were two types of theatrical representations. Hero devoted the first book of his treatise on mobile theatres to a description of Bacchus' apotheosis. There was only one scene, all the characters were arranged at the god's feet and they executed their roles in the open-air. There were no scene-changes, and the whole thing took place in the middle of the audience which was seated in a circle around the stage. The second book, on the other hand, describes a performance of the myth of Nauplius, which was actually a tragedy in five acts, with intervals and scene-changes, and which took place on a fixed stage. Hero writes in remarkably precise detail about how the various processes should be executed. His obvious preference for fixed theatres is apparent in this system which is technically superior, not only by virtue of the fluent and dependable machanisms, but also because of the variety of theatrical combinations.

The mechanism of the moving theatre is extremely simple. It consists of a three-wheeled carriage on rails. Movement is generated by a counterweight fixed to the end of a coiled string on the axle of the motor wheels. They were aware of the principle of accelerated movement in falling bodies, so the movement was controlled by lodging the counterweight in a container and resting it on a bed of grains of millet and mustard seeds. These were both light and slippery of course, and could escape through an appropriately-sized hole which opened and closed by means of a small gate. The weight was calculated according to the various resistances, one of which was the ascent of another counterweight, or in other words, the string winding around the other axle, which thereby produced movement in the opposite direction. The length of the strings was also determined. The same counterweight system applied to the fixed theatre, but a whole host of other mechanisms were added to this to bring the characters to life. Since the stage no longer moved, the counterweight could be arranged more effectively. The seed was replaced with dry sand, which was more economical and durable. There were several different types of mechanism brought into play: levers, counterweights and ratchet wheels serving as cams. The main counterweight was linked all the various movements, and the mobile parts were supported by a dowel until this was released by the counterweight.

Use was therefore made of mechanisms which could also be introduced in other machines. All the kinematic chain systems are represented here, often adapted to produce a particular effect, like the cam, for example, which is effectively a lever defined in time. All the principles to be found in the *Pneumatics* are expressed here. Ultimately, the automatons appear to represent the use of scientifically described phenomena which have been actually constructed to produce their successive effects in time and in an ordered fashion. But the very fact that these mechanisms were used for amusement inhibited some from transposing their assemblies for use in the more general material world. But this would be to forget that small operations cannot necessarily be effected on much larger scales. This was the main issue, transposing the model into reality, especially where war engines were concerned, and this was quite clearly perceived by the mechanics of this time. Vaucanson and Jacquet Droz's automatons were separated by a similar distance from certain modern automatic machines. Apart from its popular and even literary interest, the automatic theatre is significant for its role in the research into mechanical combinations. It is difficult for us to measure Hero's progress in relation to Philo, because we do not have the latter's treatise. But nonetheless it would be unjust to ignore the incredible skill of this Alexandrian mechanician in the very area where he wanted to contribute something new. This tradition abated somewhat after Hero's time, only to be resumed with the Arab mechanics . The Renaissance "engineers" occupied themselves with large automatons destined to colour princely fetes.

Hero clearly marks the end of the Alexandrian School. He may have refined some of his precursors' positions, but ultimately he worked with the same attitudes, the same methods, and not surprisingly, he achieved the same results. Of course it is difficult for us to appreciate the work done by the Alexandrians in terms of the general evolution of Greek techniques because of the lack of earlier treatises, of which there was a considerable number. It is possible, however, that the confrontation of disciplines was greater in Alexandria than in any of the earlier schools, with the exception perhaps of Rhodes which we know very little about, except that some of its members were at one time scholars at the Alexandrian School.

One can understand why certain techniques, notably those concerned with crafts, did not figure in Alexandrian research, but it is not so clear why certain others were omitted as well. This is particularly true of architecture, which could have stimulated some interesting developments, notably in terms of perspective and proportions. But perhaps we are not aware of all the works produced by the School. And it is also possible that the existing treatises were felt to be adequate, and that no further revisions were needed. Hero did not actually write anything on this subject, but Aenaes' treatise was revised by Philo.

It hardly seems necessary to state that all this needs to be discussed in more precise terms. These texts are difficult to edit and translate and some of them have not been published since the 16th or 17th centuries. This is one of the areas where collaboration between specialists of Greek language, technicians and historians would be particularly fruitful: the recent editions of Archimedes' works have shown us the way.

BLOCKAGE OR LIMITS

The historian of Greek techniques could not possibly avoid dealing with one particular problem which has and still is, causing a great deal of ink to flow. Apart from a few minor differences, all the solutions found thus far, except for a few variations, are largely in concordance. André Aymard posed the question in clear terms:

> At the heart of Antiquity, during the time of the Hellenic civilization, the stage was set for a steady yet radical transformation of the conditions of everyday life. But this transformation did not take place. The ingenuity brought to bear in using the properties of matter, and taming the great forces of nature, in order to fight off the enemy or entertain the innocent and the curious, gave way to an attitude verging on indifference when it came to using the same qualities to alleviate the misery and physical pain of man (...) so it was not through ignorance that the Ancients sinned then, but through their own refusal. A refusal which had its exceptions, but which is apparent in too many cases to bear witness to anything other than a principled attitude. A mental or social attitude, or both perhaps: the question cannot be ignored.

This means that there was probably a blockage in technical thought, even though all the elements of a major evolution were available to hand. This blockage may have been due either to a reaction against manual work and the components of material life, or it could have been because of the existence of a particular social system one of the elements of which – and this is the essential point – was slavery. Contemporary writers adhering to the first opinion have based their beliefs on texts, while those holding the second opinion base themselves merely on deduction. It is worth evaluating both sides of the argument.

Before anything else, we should say that these texts are relatively well known, and are as short as they are few in number. It is indisputable that slavery brought discredit to manual labour, even though the latter was practised largely by free men. In any case, both Plato (*Laws*, VIII, 846) and Aristotle (*Politics*, III, 5) proposed that in their ideal cities no manual worker could be a citizen. Everything concerned with craft and handiwork brings shame and deforms the soul as well as the body. In the *Gorgias*, Plato lays into "engineers," the type of engineers known to us and who emerged during the very same period as this great philosopher. "You pour scorn on them and their art, you call them makers of (war) machines when you wish to insult them, and you would never give your daughter to their sons, or marry their daughters yourself."

The other texts are later ones, being attributed to Plutarch. The first (*Marcellus*, XXI) actually deals with some of Plato's attitudes. We shall now quote Amyot:

> But since then, Plato became angry with them, maintaining that they were corrupting and spoiling the dignity and excellence of geometry by bringing it down from the realm of the intellectual and incorporeal to the lowly world of sensible and material things, and by using it with corporeal matter, which entails the vile and base practice of working with one's hands; since then, I say, mechanics, or the art of the engineer, has been separated from geometry, and having long been held in contempt by philosophers, became a military art.

In a way this is an exegesis of the texts cited above. The second quotation, taken from the life of Marcellus, deals with Archimedes' attitudes (*Marcellus*, XXVI):

> Archimede's elevated soul, his profound and searching mind, and his wealth of geometrical theories were such that he wanted nothing left in writing about the construction of the machines which had brought him such glory (...) regarding mechanics, and indeed any art which fulfils mankind's needs as obscure and vile.

And the other text, also well known, where Archimedes assesses the machines devised by Appius, Marcellus' lieutenant in the army which besieged Syracuse:

Archimedes did not rate any of these machines very highly, which in fact were nothing in comparison to his own; not that he would have held his own inventions to be worth very much, he regarded them as mere geometrical games which had helped to fill some moments of leisure, and most of which were requested by King Hieron, who was constantly persuading him to deflect his skills from purely intellectual things towards sensible objects, and to make his reasonings accessible to the sense and palpable to the common man by applying them to everyday objects.

The conclusion is obvious:

And repudiating this entire science of inventing and building machines, and generally any art which serves some use, vile, base and mercenary, he turned his mind and studies instead to writing about things whose beauty and subtlety were in no way linked with necessity.

Pierre-Maxime Schulhl deduced quite rightly, with Aristotele, (*Nicomachean Ethics* X, 7) that for all philosophers the contemplative life was far superior to even the most higly rated of its practical elements. He estimated that "Even the greatest of the engineers from Antiquity, Archimedes, did not succeed in convincing himself of the legitimacy of his works on mechanics." He notes that this school of thought began in Iona, particularly with Archytas and Eudoxos. It advocated the creation of a scientific technology but was eventually thwarted by Archimedes and by the platonist influence. It is thus possible to see the opposition of servility and freedom prolonged by that of technology and science. "Researching for practical application is to demean oneself, to lower oneself, and can only be admitted to as a form of diversion or amusement." The Romans had much the same attitude:

For Seneca, contemporary inventions – the use of transparent glass, heating stoves (...) are the works of the vilest slaves, of experienced minds, penetrating perhaps, but not great, not elevated, just like everything else which can only be done with the body bent down, and the spirit turned towards earth. They are the products of reason, but not of the right reason: all these superflous inventions subject the soul to the body, as a master turned slave.

Aymard used a different approach, but arrived at the same conclusions. His portrait of Archimedes conforms to the generally offered picture, but it is furnished with more detail which we shall consider later. "He steeped himself in aristocratic ideals, in the pride of the ancient, hereditary nobility, he heaped scorn on the acquisition of wealth which was disrupting the caste system, and on commerce which was producing the nouveau-riche whose insolence he found appalling." This ideal would have contributed to a decline in the status of techniques. It contained in effect a hierarchy of activities necessary to the life of a collective and the lowest of these were allocated to the most inferior ranks of society. This, then, represented both intellectual and social discredit: working with materials necessarily produces the soul of a slave.

We have already said that the texts were few in number and that they were inevitably ambiguous. We should first put aside an opposition to be found in the first of Plutarch's texts. There is no doubt that Plato disapproved of certain demonstration techniques: he felt that resolving the problem of doubling the cube using a mesolabe was not, and should not be seen by scientists as a rational process, as the process of a normal mind. Geometrical demonstration should be done in quite a different way. The engineers at the Alexandrian School obviously understood this: they struggled to use several exercises of pure sciencc to show that they possessed a perfect understanding of the virtues of discursive demonstration. It was correct to have a clear separation between two disciplines which although sometimes related and which originally had been treated as one, did in fact derive from two different types of logic. No doubt this was no longer true of physics: Archimedes obviously clearly understood this, and the link between pure science and technical experimentation could still not be completly severed, even though it had been earlier with geometry. Plato firmly adhered to the tradition of ancient techniques which allowed only for the ruler and compasses: these were tools whose operations had logical results, results which could be seen and understood immediately by means of a rational intuition. The mesolabe was something quite different: it gave the solution without making clear the reasoning process. It was here that the fault lay. The formation of a system of geometry such as that devised by Euclid – a global and homogenous construction – required a logical development, not mechanical construction. Archytas was the last to pursue this line of thought. The necessary separation between science and technics and between mechanics and geometry was self-evident. As soon as science was no longer an element of techniques, as soon as it had its own particular function, it could not be otherwise. According to Eudemos, Pythagoras' greatest merit was to have made mathematics into a liberal discipline by studying it from an immaterial and rational point of view. Arithmetics and geometry were only able to assert their independence and become recognized as pure sciences by separating themselves from their founding father, technics.

The second problem is linked both to the preceding and the following ones. It concerns the scorn that the Greeks may have had for manual labour. In fact, for the upper classes – and as Aymard has noted, scientists and philosophers belonged to these ranks – this may have been an intellectual attitude, instilled by the practise of a pure, logical and demonstrative science, as opposed to empirical activities which are the work of the body, not of the mind. It is also possible, as we noted earlier, that the existence of slavery brought material operations into disrepute, since these were performed only, – or nearly only – by slaves.

One point springs immediately to mind. The rejection by intellectuals of a certain kind of material world is not a purely antique attitude. It has existed in other periods and other places, without provoking a blockage in technical thought.

The second remark is perhaps more important. We are dealing with three testimonies, two of which come from philosophers, the other from a scientist who was also a great technician. These three testimonies may be very prestigious, but do they reflect the general opinion? Historians should have searched for evidence to the contrary, particularly with regard to the facts. In opposition to Aristotle and Plato, one can also cite one of Solon's laws which required all Athenians to teach their sons a skill, and this included manual ones. There was another law in Athens which threatened to punish any citizen criticising another with regard to their profession. Glotz has noted that Athens and Corinth were the only towns where manual work was not frowned upon. This disdain – of a varying nature and origin which it would be worthwhile discussing in more precise terms – would have been prevalent in areas like Thebes, Thespies and Sparta, where technical progress was far less marked. Of course it could be objected that this leglislation was prior to Aristotle and Plato, and that it was promulgated precisely at a time when technical progress was still evident.

There is more to this, however. We emphasised earlier these words of Aymard's: "The acquisition of wealth which was disrupting and the role of trade which was creating the rise of the nouveaux riches whose insolence was found appalling." There was, perhaps, then, an attitude held by intellectuals in the face of a world which concerned itself precisely with material techniques. In a sense, it represented a reaction against current trends. Similarly, one of Plutarch's phrases mentioned earlier is also noteworthy here: mechanics "so highly valued and loved by all sorts of people." And if Plato refused to let his daughter marry the son of an engineer, was it not because such marriages actually took place? Our information on this subject is too scanty and imprecise to allow for anything other than hypotheses. It is undeniable that those people who had the difficult responsiblity of moulding the material world tended to concern themselves with the sort of problems which arose in their own lives every day, and that they favoured those who could bring them power and prosperity, that is, the technicians. On the other hand, it was no less natural that the creators of "a better world" should have sought refuge in immaterial principles: again, this was an attitude which existed in other periods and in other places.

And so finally to slavery. Two opposite strands of reasoning can be used here. The insufficient means available through techniques forced the perpetuation of slavery. Since there were neither sufficient sources of natural energy or mechanisation the ancient Greeks, like the ancient Romans after them, were forced to recruit the services of a servile work-force. This argument was also upheld by Lefèbvre des Noëttes. But the reverse is also logical: the existence of slavery blocks technical progress because if a servile work-force is available, there is no need for improved techniques to reduce the amount of work.

Both these arguments have been presented by Aymard. He begins the discussion with Aristotle's well known reflection. At the beginning of his *Politics*, Aristotle indeed declares that slavery would cease to exist if shuttles and plectrums were able to start up by themselves. He did not say that slavery was actually preventing the shuttles and plectrums from starting up by themselves.

Aymard continues:

Sensing the links between the two facts, the Ancients saw the cause in one, the absence of machines, and the effect in the other, slavery. It is even possible, following Aristotle's example, that the most foresighted among them and also the most tormented, found in this cause-effect relationship something to appease their conciences. Because it enabled them to conceive of slavery, unjustifiable in principle, as an inescapable necessity in the life of a collective.

But the historian of Greek Antiquity believed this traditional explanation to be opposed to "the very conduct of the Ancients" with regard to what we call technical progress.

Far from being unaware of scientific knowledge or of practical developments which would have enabled them to initiate and develop [technical progress], it was actually by design that they refrained from using them to reduce the work load by a greater, faster rate of production. The absence of machines does not, therefore, constitute a first fact, from which one is entitled to explain the existence of slavery.

It was necessary then to find a cause, or causes for this deliberate refusal, and one of these was quite possibly the contempt held for manual work. After having acknowledged Aristotle's position, then, Aymard arrives at the antithetical explanation: "slavery should not be considered as a consequence of an absence of machines: on the contrary, the latter presents itself to the historian as a consequence of slavery."

The problem is difficult to resolve. Was it really necessary to follow this line of reasoning? Aristotle quite clearly said that slavery was the result of technical insufficiencies. Why reverse his explanation under the pretext that it does not concord with the attitudes of the Ancients, and conclude that the technical blockage was a result of slavery? Our knowledge of slavery in Antiquity is still too fragmentary for us to draw any definite conclusions. Only reasoning a priori can set us on a more fruitful path which it would be worthwhile exploring in more detail. Jacques Ellul noted quite rightly – and the comparisons are very enlightening – that technical progress was far greater in civilisations which practised slavery, like Egypt for example, than in those where it was virtually unknown, such as Israel. There was far more technical progress during the slavery period of Roman history than during the time of the great emancipations. And the freeing of slaves during the invasions did not lead to any marked technical progress.

The economic point of view is also fairly instructive. Even in difficult areas of work like mining, for example, the extensive use of slaves due to a lack of mechanisation made for a far less profitable concern than if techniques had been manpower-saving. This would appear self-evident. And even in the craft industries, a great deal of capital which could have been used in other ways was needed to provide a professionally worthwhile slave. It is in their interests, continues Ellul, "not to waste [this capital], not to use it here, there and everywhere, and if it is possible to make the work easier and more efficient, then it is in the masters' interests to make sure that this is done, just as shown by Cato the Elder."

As for the idea that there was a technical blockage because of the existence of slavery, then, we gladly reiterate Ellul's judgement: "In reality, we are faced here with one of those facile, sensationalist and completely non-historical explanations, habitually propounded by those whose sole purpose is to justify theories."

It was therefore necessary to put the problem another way, and specify some of its aspects. Contemporary historians are still divided about the actual development of techniques during certain periods, the boundaries of which are often unclear. Vernant believes that the Greeks borrowed their technical knowledge very early on from the East, and that this knowledge was not profoundly changed by any new discoveries. "The innovations or improvements which they introduced in certain fields did not extend beyond the framework of the technological system which appeared to be already fixed by the classical period." There is some chronological confusion on this matter. True enough, as we noted above, advanced techniques from the East and Egypt were effectively available to the Greeks at the beginning of their era. After the 6th century BC, however, it is undeniable that some profound changes occured in clearly defined fields, which represented major developments of the techniques they had acquired from elsewhere: *invention* – in the true sense of the term – is clearly what happened then. It is no less obvious that the classical technical system was structured at the beginning of the 3rd century BC. The Alexandrians did codify, indeed they eventually perfected several techniques, but there were no more major transformations. In our opinion, between the 6th and the 3rd century BC, the Greeks contributed a great deal to the field of techniques. The blockage probably occurred after the efforts to synthesise made by the Alexandrian School.

The second point is equally important. If there was actually a refusal of technical progress, it would be as well to define just what exactly was the object of this refusal. It is the nature of this blockage, then, which is in question here. There are two different standpoints with regard to this, different in spirit, certainly, but to a degree they are concordant, and one could almost say complementary. The idea of a technical blockage, according to many historians, emerged from the idea that the Greeks possessed all the necessary elements for major technical progress to take place. The following passage is again taken from Aymard:

The Greeks were enlivened by a true scientific spirit, and all that was needed was for them to apply in practice the principles discovered through their reasoning (...). Even more important, they did actually began to apply them, either to satisfy royal fantasies or to defend Syracuse against the Romans (...). It was not through ignorance that the people of Antiquity sinned, then, but through their own refusal.

Contemporary historians still believe that "the development of mechanisation would have been logically possible." One should ask, therefore, whether or not the Greeks had the necessary intellectual tools to promote technical progress. It is futile making the point – even though it is still done today – that Hero of Alexandria's aeolipile could not logically lead to the steam engine. The latter was born out of an understanding of vacuums, condensation and atmospheric pressure, notions which were unknown to the Greeks. And this could also reflect an inaccurate interpretation of texts. The same is true of the crank and connecting rod system, unknown in Antiquity, and which was the basis of all advanced mechanisation: there was no slavery in the Middle Ages, but it was not discovered then, either. It is also somewhat futile to ask oneself why a mechanism so simple in principle was not invented earlier. The confusion between rotary and linear motion was first discussed only during the Renaissance. Moreover, the absence of a dynamic in the Greek scientific system considerably hindered the development of mechanisation. Levers, pullies, winches, cog wheels, reduction: all the potential of Greek physics had been realised. This had happened during the 6th and 3rd centuries BC and it represented a great deal of progress. Equally, the camshaft, used by Hero of Alexandria, could not be put to practical use. The possible applications of Greek science should be specified. Apart from those which were effectively made, those aspects which could be applied were no doubt very few.

There were other difficulties of a purely material nature. Resources were limited in the Mediterranean countries, particularly Greece. The Greeks were dependent for metal on foreign imports; these were sometimes from very distant regions and very often the routes were cut off. It was the same for wood; there was very little of it, it grew again only with difficulty, and it was often knotted and twisted. This shortage of wood caused a lot of problems, not only in terms of materials, but also of heat energy. Besides this, it is well known – as the history of techniques from the 15th to the 18th centuries clearly shows – that real mechanisation, at least advanced mechanisation, was conceivable only using metal. The lack of water precluded the widespread use of hydraulic power. As for harnessing, this was dependent on both the development of animal domestication, which was more advanced in some areas than others, and on the existence of an expansive road network which the divisions of the States and the financial mediocrity of the cities made impossible. Furthermore, transport by animal porterage continued to be used until the Middle Ages: the absence of the front-axle unit, a much more recent invention, made heavy transport impossible, at least over long distances.

After the scientific impossibilities and material difficulties, J.-P. Vernant adds a third aspect to his explanation. It is true that by the 6th century BC, the secularisation of techniques was an accepted fact. They were freed from magic and religion. According to Espinas, the beginning of the 5th century BC was a watershed in technical history, when the unconsious practising of techniques gradually evolved into an actual technology. He also believed the sophist movement to have been the first to attempt to define and assert technical thought. Two movements grew from this: the theoretical elucidation of certain technical problems, found in the pseudo-Aristotelian work and a movement of codifying and teaching recipes which was expressed in "manuals" written by the mechanicians of the Alexandrian School. It was possible to develop the second movement, at least in those areas where techniques could be codified – this did not include the techniques of artisans, for example, where the work was done by hand – but it was not so easy for the first movement to base itself on a science still incomplete and difficult to use. We have already underlined the extent to which the absence of a dynamic, perhaps blocked by the arguments of Zeno of Elea, constituted a powerful break on the technical system. Philo of Byzantium insisted that technical research could not be conducted by pure reasoning and rigourous demonstration. Vernant has suggested that it was perhaps because of this and not through any aristocratic concerns, that Archimedes refrained from writing anything on techniques. The explanation of techniques would inevitably be a compromise between theory and experiment, each of them badly articulated. Vernant also believes that in the thought of Hero of Alexandria, the five kinematic chains formed a coherent system, closed in on itself, and excluding innovation and progress. The very rationality of the system implied its limitation and completion. J.-P. Vernant's conclusions are formal:

(Greek technics) did not yet have what we consider to be the characteristics of technical intelligence and which are the basis of its dynamism. It was rarely – or badly – linked to science. Experimental thought was not considered. And the failure to elaborate the notions of natural law, physical mechanisms and technical devices meant that there was no conceptual framework to assure its progress.

Denying the technical thought of the sophists, he concludes thus: "The technical stagnation of the Greeks went hand in hand with the absence of any real technical thought."

Is it surprising, even without alluding to intellectual attitudes, or allowing for the existence of slave societies, that at a certain stage the Greeks should find themselves in a technical impasse, and for reasons purely internal to techniques? It is also necessary to know if other systems and structures were able to bring the limits of this technical system to light, thereby generating essential transformations. We have seen that Alexander's creation of a great empire and the subsequent constitution of strong and extensive monarchies did facilitate the development, if not the transformation, of acquired techniques, in the nature of systemisation and improvements. A mediocre economy and a relatively stable population was hardly favourable to technical progress. The general conditions for a technical mutation were far from being united.

THE ROMANS AND THEIR SUCCESSORS

It seems that generally speaking, the technical system perfected by the Greeks remained as it was for several centuries. There were a few changes, but these were only minor ones and they had little effect. Even the relatively quick disappearance of slavery and the civilization's shift to the North to areas which were far richer in resources did not profoundly modify the situation. The stability of tradition is testimony of this.

As far as techniques were concerned, the Romans were not innovators. P.-M. Duval has made a list of Roman "inventions," which is relatively short and limited in its range. It is also debateable whether or not they did actually invent all the objects listed: the vaulted cupola or dome, the aqueduct, and the glass pane did definitely appear during the Roman period, however. And what is more important, perhaps, they also invented several tools, which will be considered later. It is less clear whether they did actually invent or not balances for weighing, direct action screw presses, and bows with catches. They also used water mills, chemical fertilisers, wax candles, metal imposts, codices and stenography. This list is certainly not insignificant, but it in no way represents a mutation of the system established by the Greeks and their eastern and Egyptian neighbours. Finally, it is important to remember that some of these inventions were not simply the the work of the Romans, but of their subject peoples as well.

The achievements of the Roman Empire are generally accepted. Many historians, sometimes the same ones who emphasise the blockages inherent in the Greek technical system, find them a fascinating object of study. Much of their "success" lies with the visible remains of their civilization: the carefully planned towns, the magnificent monuments ranging from the Pantheon in Rome to the *Pont du Gard*, the triomphal arches, still arouse admiration and rightly so, because they are evidence of their almost perfect mastery of the material world. Perhaps the Romans' greatest talent was their organisation of space, which underpinned the Empire's greatness and still forms the basis of what has been called "the Roman genius."

CIRCUMSTANCES AND SURROUNDINGS

Since we intend to examine the principle elements of the Roman technical system and its differences with that of the Greeks, it would be useful first to outline briefly its circumstances and its context.

Good Pupils and Conscientious Observers

Up until a fairly late period, the Romans, who were essentially a farming people, were familiar with all the agricultural techniques used in the Mediterranean world. It was their successive conquests which brought them into contact with civilizations technically more advanced than their own. From the first century BC, these conquests also brought them into contact with agricultural areas where conditions were quite different from those in the mediterranean basin, with heavier soils, larger forests, water courses, and more pronounced rainy climates. They became aware of the Hellenic world through their conquest of Sicily, and Carthage opened up similar but more varied paths. Later it was to be Spain, Gaul and then a part of Germany as well. Such that the Roman civilisation became a kind of melting pot which brought together very diverse civilizations, although their respective elements could not necessarily be applied in all areas throughout the Roman world.

The Romans were certainly good pupils and conscientious observers, but it was obviously in their interests to assimilate the techniques unknown before their expansion, and which would permit them to exploit to the full their new-found immense empire. They probably not only assimilated these techniques, but improved them as well.

As good executors rather than innovators, the Romans were not prolific writers. There are only a few people who exercised their talent in this area, and none of them attempted to write one of the great encylopaedias so prized by the Greeks. Generally, the authors of latin technical works do acknowledge the debt owed to the writers of other civilizations of which theirs was essentially a tributary, and as mediocre technicians, they seem to have felt it just as important to rely on others as it was to show their erudition. This lends a particular characteristic to their works. They represent not so much actual technology as an insertion of technical norms into a general pattern.

The number of works on agronomy is hardly surprising and they have been well studied. These works seem to have several origins, the first of which seems to have been Carthaginian. We know from Columella that Carthage counted among its people several agronomists, of whom the most famous was Magon. His work was translated in Rome and was the most popular book of its kind. Magon dealt with agriculture on a very general level and was not confined to any one branch of interest, and this was probably the reason why his work became the farming bible for the Romans.

Columella cites about forty Greek agronomists, Varro about fifty, so that it is not surprising that Greek works on agriculture were more than abundant. However, apart from a few famous works like those by Hesiod and Xenophon, very few of these are still in existence. Of course it is likely that they would have cited any author who had even the most tenuous links with agriculture just to show that they had read everything, such was the case with Aristotle, for example. It is probable that the Romans learned of Hellenic agronomy through the Sicilian writers, of which there were about four.

The very first agronomic treatises date back to the 2nd century BC, and were written by Cato and the two Sarsenas, a father and son. The latter texts no longer extant, but no doubt they were very close to the former. Roger Martin has noted that these were not agronomic treatises as such, but more like these "maisons rustiques" which were so successful in the 16th century. These had no scientific pretensions, but were anthologies of all sorts of formulas for solving the hundred and one problems encountered every day by landowners. Cato's success was chiefly due to his personality, and Varro seems to have had little respect for his work. The latter is in fact an *Economics*, in the sense meant by Xenophon.

We shall not deal much with Virgil, more a poet than an agronomist. Yet his work is significant in that the descriptions of the Po valley reveal a richer, technically more advanced agriculture. After devoting a large part of his treatise to the rural economy, Varro deals with agriculture itself, in terms of its technical aspects. Book II is concerned with rearing, and Book III with specialised areas of breeding, (beekeeping, raising fish and game). The part concerned with actual techniques is very short. One of the greatest Latin agronomists was probably Cneius Tremelius Scrofa, whose work unfortunately no longer exists, but which was highly praised by Varro. According to the latter, he had brought a new spirit to the subject "which one could go so far as to call rational and scientific," which brought him closer to Hellenic techniques than Cato. Agriculture was for him an art i.e., a technique and a science which implies both precise knowledge and specialised research. He anticipated Columella, and was responsible for the introduction of the actual *science* of agronomy into Rome. Scrofa's conception of farming techniques made his work a major turning point.

Columella's work truly was a national manifesto, written in an attempt to save an ailing Latin agriculture. He believed that they should fight against the growing sterility of Italian soil, against the technical recession due to insufficient manuring, and finally against economic and social recession. Amateur dabbling in farming was not enough, landowners should be adequately trained even though there was no instruction in agriculture provided in Rome. Columella also advocated large scale exploitation as the basis of all good organisation and techniques.

It is worthwhile considering in a little more detail what can be learnt on a technical level from the Latin agronomists. We need to know more about their predecessors in order to assess their originality but this is easier said than done since we know very little about them. But it seems – and this is mere hypothesis – that most Latin agronomists were compilers, and contented themselves with developing and perfecting existing techniques. It is very unlikely that these changes altered the preceding technical system to any great extent.

Pliny the Elder's *Natural History* is a work in a category all of its own. He did not want it to be a technical treatise; it is rather a collection of observations, which is very rarely prescriptive. It is less valuable therefore, for the history of technical thought, than for the history of techniques. Other than these differences, his method is like that of all the other Latin agronomists. "The perusal of about 2000 volumes very few of which, owing to the obscurity of their contents, are ever handled by scholars; 20,000 noteworthy facts obtained from one hundred authors that we have explored, and with a great number of other facts in addition that were either ignored by our predecessors or have been discovered by subsequent experience." This extract from the preface clearly reflects the characteristics of most Latin writers: an almost absurd erudition, and good observation which combine to result in what was essentially a collection rather than a work of systematic knowledge.

Vitruvius, the author of *De Architectura*, lived in Rome during the period of Augustus. Like Pliny, he wanted to compile a sort of *compendium* but in a much smaller area, that of architecture. and only one particular aspect of architecture at that. In one work he dealt with matters which had been treated separately by his forerunners, in a shortened, more accessible form, so that they would be useful not only to architects and professionals, but also to anyone who was interested in the subject. Like the writers above, Vitruvius drew from the best Greek sources, and like them also, he was honest and generous in his acknowledgments so that thanks to him we know of works otherwise extant. His last books, particularly Book X which concerns civil and military machines, in fact consist solely of resumés of many books which we do still possess. This book is very similar to Athenaeus' treatise, who was probably slightly older than Vitruvius. Once more, we find this characteristic habit, shared by all these Latin writers, of collecting all that they could of their ancient – mainly Hellenic – heritage. The content of Vitruvius' work is surprising: the part on architecture is by no means complete, and it even leaves out some of the most notable aspects of Roman architecture, such as the theatres and the triomphal arches, etc. By contrast, most of the last book is concerned with military machines, which have very little to do with architecture, and the same goes for hydraulics, the subject of Book VIII, in which aqueducts are barely mentioned.

Books I to VII deal with public and private buildings, considered separately or in groups, in terms of construction, decoration and fittings. There is a rather summary treatment of hydraulics in Book VIII, and then the discussions on sun dials and clepsydras provide an excuse for him to talk about geometry and astronomy in Book IX.

Book X of course is concerned with military and civil machinery. Mortet has attempted to explain this surprising number of gaps – including as mentioned above the absence of certain monuments, and very cursory allusions to certain construction processes, arcades, vaults and arches for example. He believed that Vitruvius' treatise was meant to provide technical knowledge for the functionaries of the council of public works. But why all these gaps, and why a book on military machines? In our opinion, Vitruvius simply borrowed wholesale from his predecessors' works, in particular the Greeks, and for this reason wrote very little about Roman innovations, with the exception of the water mill which was also of Greek origin anyway, as its name in Vitruvius' work implies. Virtually everything in this work is Greek, from the terminology to the architectural decoration.

Nothing is more representative of the methodical Latin approach to "technical treatises" than the works of Frontinus and Vegetius. The treatise on Roman aqueducts was written at the very end of the 1st century AD, and is essentially an administrative work, composed with information taken from the water service archives. Consequently, the technical problems it deals with are only those encountered by this small administration: problems with slopes, basins and especially with measuring the flow. One could say, and the same thing has been said in specialist critiques of this work, that this treatise is really only an examination of the administration of water in Rome, and of the solutions to the various related problems.

Vegetius *De re militari*, written at the end of the 4th century AD, can hardly be called a technical work, dealing as it does with the organisation and administration of an army. One only has to set what he says about military machines, against the texts by Vitruvius or the Greek mechanicians of Alexandria: there is no comparison.

We shall end this rapid survey of Roman technical science with a brief look at the many agrimensors. The growth of the Empire and the subsequent growth of the administration considerably increased the task of those who were originally simply land surveyors. The widespread use of cadastres had necessitated a review of the techniques used by the Alexandria School, which were already fairly advanced. The replacement of the dioptra with the *groma* caused problems which are in evidence in all the technical treatises. Epaphroditus, Vitruvius Rufus and Balbus, attempted to synthesise their knowledge on the subject and show that the change to the *groma* had provided a satisfactory solution to the political and adminstrative problems. The problems of measuring land were thus overcome, as were the same sort of problems in construction: a new concept appeared for the first time in written form in these treatises, that of the "measurement."

The Romans did not produce any outstanding scientists, and probably for the same reasons they had no outstanding technicians either. There is striking confirmation of this in their written works. They knew what should be done, and they were sometimes able to add their own elements. But it was rare that they rationalised their actions. Their successes constituted a run of individual achievements, whose real causes were never examined. The mechanicians from Alexandria were effective in practice, but their mental attitude did not match this. Of course, there is always a certain discrepancy between technical literature – one of our major sources – and technical practice. Cicero pointed out quite rightly through one of his characters, that very few farmers, landowners and developers had ever read any agronomists. And many agronomists display an empiricism which could only have been based on elementary practices. It is clear that they simply "collected" information, which of course is the worst way to go about forming a rationalised set of techniques. Finally, it is worth noting that this system was very highly organised, so that any new technique inevitably caused problems. A well-ordered organisation necessarily needs fixed techniques.

Geography and Techniques

It is undeniable that techniques in any epoch are closely linked to local natural resources. Large deposits of metal ores, an abundance of energy resources, natural motor forces, humid climates favouring agriculture and pasturage, rich soils, all these are factors which should be taken into account. Since the time of Bodin and Montchrestien, however, this has not been the case, except perhaps during the recent events concerning petrol.

The Mediterranean Basin and the outlying area, i.e., the Near East, was not particularly rich in such resources. There were few water courses, and those that did exist were usually extremely difficult to manage, with the exception of the Po valley in northern Italy, and after the Roman conquest, certain areas of Spain and Gaul. There were few minerals, and the forests were meagre and slow-growing. On the whole, the situation was not a favourable one. It was exacerbated when supply routes were cut off and the effects were sometimes disastrous. Columella deplored the decline of food production in Italy, and the fact that the Romans had to use supplies from elsewhere, like Egypt, for example.

Not enough research has gone into the dependence of ancient civilizations on others for their supplies, and the effect this had on their technical development, but there is at least an increased awareness of this subject's importance today. Thus, such reliance on external support meant that some technical developments were impossible, or at least very difficult.

The Romans' conquests brought them some important resources which did not exist in the Mediterranean, and this prompted some profound changes. The Romans found the iron they needed and which had been so lacking in Antiquity, in Spain, Gaul, Styria and Carinthia. Spain also provided them with copper and lead. It was not only metallurgical techniques which changed as a result of this, but other techniques using these metals, such as iron working, for example. Finally, it was the abundance of water courses that made possible the widespread use of the water mill.

On the other hand, as they moved further north, the Romans actually adapted some of their techniques, such as grape-cultivation, for example, which was changed to suit climates quite different from those in the Mediterranean.

There is another important point to be made here, namely that the so-called Barbarian peoples subjugated by the Romans all had their own techniques, which were the fruit of their natural resources and long-established traditions. It is very likely that their conquerors adopted many of these techniques where it was possible and in their interests to do so. "It is generally accepted," notes P.-M. Duval, "that from the 2nd and 3rd centuries, the Romans adopted the long, welded and damasked sword, which they had seen at the hands of their Barbarian cavalry auxiliaries, and which had come from workshops in the Rhino-Danube frontiers." Pliny mentions hard iron from the Noric and Carinthia provinces. There has been a lot of discussion about the techniques borrowed from Gaul and no doubt these will continue: take the forecarriage ploughing instrument, for example, the famous gallic harvester, and the barrel.

It is possible that the demands made on the Empire by the Romans would have caused techniques to change even in those areas beyond the famous *limes*. Excavations in Poland, in the Kielce region and therefore beyond the frontier, have provided evidence of the extensive exploitation of mineral resources between the 1st and 4th centuries. Polish experts can think of only one good reason for such widespread exploitation: exporting to the Romans. Furthermore, it was at this time that the furnaces for reducing minerals were somewhat developed and improved.

Organisation

The immensity of the Roman Empire demanded a highly structured organisation. The Romans were not only perfectly aware of this, but they also knew exactly how to put this into practise. Such an organisation could only exist on a basis of advanced techniques from many diverse sectors.

The first of these was the institutional sector. The Roman preoccupation with law, both public and private, is well-known and indeed, law was the fundamental base of this organisation. Some of these laws concerned major technical problems, of which two examples will suffice here. There was a limit on the weight of wagon-loads which could circulate on the roads, rather like the laws we have today. There was strict control of mine exploitation, – state authorisation was required and the inmost depths were state property, – and there were strict laws on extraction methods. There are many more examples of this.

The adminstration of such a large territory required a great deal of technical support. Again, two examples are sufficient to illustrate this.

I. – Water conveyance had been widely practised by the ancient Greeks, and as we mentioned above, this had required some large building projects. But Roman towns were even more extensive, so that water conveyance caused many more unprecedented problems. They had to go further afield to find water sources, and this itself entailed more complicated technques, such as the building of aqueducts – like the ones in Rome with which we are so familiar thanks to Frontinus – and also aqueduct bridges, like the famous ones in Gard in France, and Segovia in Spain. Construction work on the first aqueduct (the Appian) began in 312 BC. After this the system gradually expanded as Rome grew bigger, until it was finally reorganised and extended under Claudius and Nero (fig. 1).

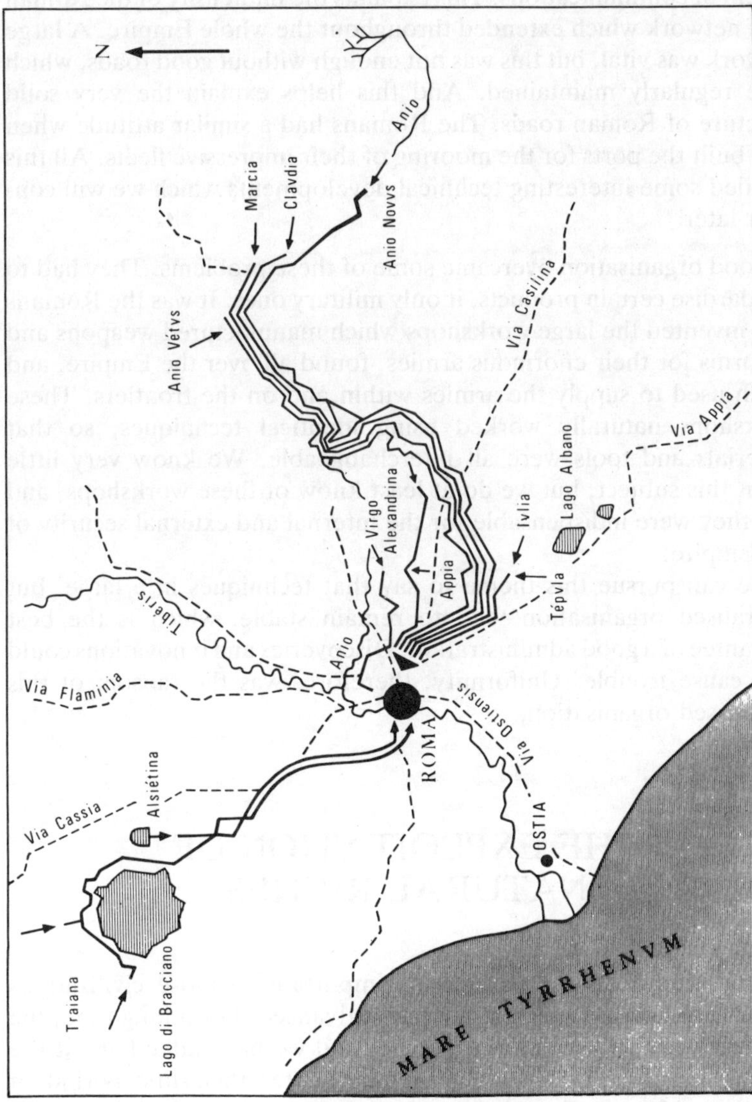

Figure 1. — The Roman aqueducts system.

II. – The management of the Empire required a well-organised system of communications. This explains the immensity of the Roman road network which extended throughout the whole Empire. A large network was vital, but this was not enough without good roads, which were regularly maintained. And this helps explain the very solid structure of Roman roads. The Romans had a similar attitude when they built the ports for the mooring of their impressive fleets. All this entailed some interesting technical developments which we will consider later.

Good organisation overcame some of these problems. They had to standardise certain products, if only military ones. It was the Romans who invented the large workshops which manufactured weapons and uniforms for their enormous armies, found all over the Empire, and which used to supply the armies within and on the frontiers. These workshops naturally worked using identical techniques, so that materials and tools were all interchangeable. We know very little about this subject, but we do at least know of these workshops, and that they were indispensable for the internal and external security of the Empire.

We can pursue this theme to say that techniques in a large, but centralised organisation tend to remain stable, which is the best guarantee of a good administration. Discoveries and innovations could only cause trouble. Uniformity, therefore, was the ransom of this centralised organisation.

THE EXPLOITATION OF NATURAL RICHES

Natural resources were extremely important to those civilizations whose technical systems are not very advanced. For a long time, the Romans were an agricultural people, and we have talked about the abundance of literature on agronomy. After their first period of expansion, they continued also to benefit from the wealth beneath the soil. As farmers and miners the Romans always remained primarily interested in these two areas of techniques.

Agriculture

Technically speaking, until the beginning of the modern period, Roman agriculture hardly differed from the other types of agriculture surrounding Mediterranean regions, from the Greek to the Carthaginian. However, it did benefit from some of their developments, in the categories of cultivations, in the general organisation of labour, and in the types of tools used.

As far as we can tell from Pliny and other agronomists, the botanic and animal worlds underwent little change. The Romans even contributed to the expansion of certain plant species which previously existed only in the mediterranean basin, such as the chestnut, peach and apricot trees. Perhaps their greatest contribution was to have cultivated vines which could stand the oceanic climate (Spain and Bordelais in Aquitaine), the more northern climates stretching from the Rhône and Saône up to the Loire, the banks of the Moselle and finally up to Scandinavia.

The Latin writers do not fail to mention all the types of species grown, and how they differed according to climate and soil. There is evidence of more than six types of wheat, which Columella however reduced to three, soft wheat, durum wheat and "poulard" wheat, and that does not include spelt. Barley, millet and leguminous plants (vetch, lupin, and chick-peas) constituted with wheat the staple products. Cato cites 7 varieties of vine, Columella 58, Pliny 91 (and he estimates that this figure was possibly as high as 400). The olive tree gradually found its way into Italy: 10 to 15 varieties have been cited. The large number of varieties is testimony of a sound knowledge of arboriculture, but this is rarely alluded to in the treatises, however. Progress was certainly made in the techniques of species breeding, in cutting, layering and especially grafting, but unfortunately, we know very little about this. This is more evidence of the Romans' concern to find plants and species suitable for all the regions of the Empire.

As for stock rearing, progress was slow, but constant. The Romans made very little contribution to the techniques of selection, which is hardly surprising since the main agronomists assumed the female had the dominant role in selection, which is completely wrong. Bovines were reared, but usually they lived on older types of cattle, such as the tall Epirus breed, selected during the Pyhrric period. It was the Romans who began the practice of temporary stabling. All equine techniques had been developed exclusively by the Greeks.

Four types of sheep were known during the Roman period: Etruscan, or Bergamask, those from Magna Graecia (southern Italy), the fat-tailed Steatopygous and Merinos. The Steatopygus was reared after the conquest of Carthage, and the Merinos came from Betica after Hannibal's defeat. The stabling of sheep was common.

Pigs and goats were reared in many areas, as they had been by the Greeks, but the donkey was a more recent acquisition. They reared the same poultry as the Greeks and only the rabbit had been recently imported. Aristotle did not seem to know of its existence, and the historians only allude to it from the beginning of this period, and even then it is considered a relatively exotic species. According to Pliny and Catullus it came from Spain; Strabo believed it came from the Baleares. It was probably still confined to certain areas of the Mediterranean and some regions of North Africa. The Greeks knew of gallinaceans only towards the end of this period, and it was only with the dawn of the last century before the following period that the Romans had several different races of this species, about half a dozen in fact. Their use became more widespread and the Romans also practised artificial incubation. The rearing of ducks was also a fairly recent practice when Columella was writing, and he suggests that precautions should be taken to prevent them from escaping, demonstrating perhaps that these were still relatively wild species. Neither the guinea-fowl nor the turkey were reared by the Romans.

The history of cultivation methods and tools is a familiar subject, although the agronomists, including Pliny, were rather vague in their treatment of it. The first and most important of these instruments was the one used for ploughing. The type most often used was the swing plough, or ard, used for symetric ploughing. Of course, there were regional varieties of these. Pliny distinguishes between ploughs used for the denser soil around Rome, and those in Campania, where the soil was more porous. There has been a lot of discussion about the type of plough actually used. Some think that the Chambige plough was used in the Po valley and Tuscany, but it is generally accepted that the swing plough with share-beam was the most popular and widely used. The ploughshare is widely depicted in ancient representations of ploughs: it is arranged vertically, but there is no space between the base of the beam and the share. This was so in both the Arezzo and Talamone ploughs depicted on a Sicilian coin dating back, at the earliest, to about 258 BC.

In more recent documents, the plough's beam and share are curved, each quite distinct, and fitted together within the share beam. All the colonial coins circulated during the two centuries before and after the birth of the Christian era depict the swing plough with share-beam, consequently, it is generally believed that this instrument was put into widespread use by the Romans, or even imposed by them.

It is not sufficient to simply know the type of instrument used, however. We should also understand how it worked. We know from Columella that the Romans practised assymetric ploughing with a symetric instrument. All they needed to do was to tip the plough to one side. In the 1st century AD, Varro recommended attaching two planks to either side of the share during the third ploughing so that the seeds on the wings of the furrow could be covered. Virgil talked of "wings" as if it were a commonly used term. Palladius mentions not only the *aratra simplicia*, but also the *aratra aurista* which was designed specifically for flat land. It is possible that eventually one wing was used instead of two, and this became the mouldboard.

Pliny states: "The coulter is the part which cuts the hard earth before it is broken up, marking out the tracks for the future furrows with incisions which the share sloping backwards is to bite out in the process of ploughing". This statement has provoked a lot of argument. Some people believe that Pliny was referring to the plough with both mouldboard and coulter. According to more recent interpretations, however, Pliny was referring to the *coutrier*, or subsoil plough, which was quite independent of the plough. Nevertheless, this was a major step forward.

Pliny also distinguished several types of ploughshares, the most common of which was "a lever with a pointed beak" This sort of "ard-point" was fairly common at this time. He mentions another type of plough which would correspond to the socketed ploughshare, placed on the extreme end of the share-beam. An arrow-shaped ploughshare cut the soil horizontally. Pliny also noted the innovatory use of the wheeled forecarriage in Rhaetia and Gaul. Far from being an ordinary plough as it has often mistakenly been called, this was a swing plough, a symetric implement, which was fixed to a small carriage.

The harrow also seems to have undergone some changes. Varro says that its use was limited, to pulling up weeds, and covering the seeds. It was a crude instrument, consisting of twisted riddles in a wooden box with bundles of thorns, which by all accounts was virtually useless. The harrow described by Pliny had metal teeth and cut much deeper. This meant that some of the weeding was unnecessary.

Sickles were used for reaping and Pliny mentions a reaping machine with wheels. There were a number of hypotheses made about this before a picture on a somewhat dilapidated bas-relief was found in Arlon in Belgium. "An enormous box" writes Duval, "on two wheels, equipped with teeth, and pushed across the field by an ox, used for carrying the ears of corn pulled up by the teeth." R. Martin, after the descriptions of Pliny and Palladius, put this machine back in its proper place: it was used for pulling up corn ears, but the straw which the animals had trampled on obviously could not be used. This meant that part of the corn was inevitably lost because of the unequal length of the ears. Such an instrument would have been useless in areas where intensive cultivation was practised (fig. 2).

The Romans used to thresh the grain by using animals, sledges, rollers or *tribula* or corn drags, and of course, the flail.

The organisation of cultivation was another important question, and techniques for improving the land were widely practised. All the agronomists insist on the importance of fertilisers, and using the ash from stubble-burning was an age old process. Farm manure was more plentiful by now, as a result of the increase in stock rearing, especially that of bovines. In Gaul, marling was also used for this purpose. Columella and Palladius estimated that between 16 and 21 tonnes of manure per acre was needed for growing wheat, providing that it was grown only for one year if crops were rotated.

Cato and Varro suggest that biennial rotation was the usual practice: according to Virgil and Pliny, plants alternated with wheat, while Columella suggests that wheat alternated with beans. Columella advocated leaving the land fallow for one year. Triennial rotation was much more rare, but it was practised nevertheless:

according to Columella:	turnip – wheat – beans
	millet – spelt – barley
according to Pliny:	millet – rape – wheat

Long root crops were alternated with short rooted ones, and the Romans were obviously familiar with the beneficial effect of plants on the soil, particulary that of vetches and lupins. They also advocated that fallow land should be ploughed to prevent wild plants from

Figure 2. – Gallic harvesters. (From M. Daumas, *Histoire générale des techniques* t.I Paris, 1962)

weakening the earth. It seems that there was considerable progress made in all these areas from the time of the Republic to the Empire.

Western Europe will always be indebted to the Romans for extending the areas used for vine-growing. The vine was actually introduced into Gaul by the Greeks, but grapes were grown only in restricted areas near the Mediterranean coast. It was only much later that the large Narbonensis' vineyard was established in that province. It seems that the Romans originally wanted to protect the Italian vineyards: only those with the rights of the Roman city were authorised to plant vines in Gaul. One of the first "Nordic" vineyards was set up in Gaillac, Then Hermitage and the Côe Rôtie were established, along the Rhône. New types of vine had been required for this, and the pinot grape became very popular as a result. The vines grown in Bordeaux had no doubt been imported from Spain. The conquest advanced rapidly: by the 3rd century AD, the Burgundy had become a major vine-growing region, and even Trier as well – Ausonius talks of Moselle wines in the 4th century. Three and a half centuries after Strabo, vine-growing was practised extensively throughout France. There were many different varieties according to the varying geographical conditions. In 71 BC, Pliny noted that a new type of grape was being grown in Vienne, in France. This was no doubt in the face of numerous obstacles, such as the edict passed by the Emperor Domitian, which was passed in an attempt to stop expansion in order to protect the vineyards in Italy.

It is not surprising, therefore, that vines were carefully cultivated. They were frequently earthed up, either with a hoe, or a light plough. The agronomists have left us some clear descriptions of pruning techniques. Columella distinguishes several ways of leading the vines: creeping vines; trellised low vines; trellised vines without yoke; trellised vines with simple yoke; trellised vines with a yoke on four poles. All the large Roman vineyards, of which the Falernian was perhaps the most prized, were arranged on trees.

The agronomists wrote in very precise terms about the work calendar. This seems to be part of a tradition going back to Hesiod's *Works and Days*. As always, chronology is all important. Almost all the agronomists advocated multiple ploughing: Pliny for example, informs us that it was done five times in areas of hard earth, and up to nine times in Tuscany. Virgil believed four times was sufficient and most of the agronomists reduced this to three. Both Varro and Xenophon thought two ploughings were sufficient for vines (in early spring and mid-summer). Later, Columella and Palladius raised this to three ploughings. It is evident from all this that the number of ploughings depended very much on the nature of the soil and the type of cultivation.

The same was true of harvesting times, which were closely determined by weather conditions and the degree to which the grain was ripened. From the technical point of view, Columella distinguished between three different harvesting methods: the grain was cut either close to the ground, half-way up or at the very end of the stem.

Assessing the profitability of Roman farming is not easy, and the figures which we do have are certainly at the extreme ends of the spectrum. Furthermore, the units of measurement are often different. Cicero talks of a yield of 140 to 180 gallons per acre, while Varro claims that in Etruria one acre produced 90 to 135 gallons, which hardly seems possible.

Our basic lack of knowledge about Greek agriculture, particularly that of the early period, prevents us from making any comparisons with the Roman period, or from accurately assessing the progress made during this time. We are probably fairly safe in saying that there was no revolution, however. There was a great deal of progress made in some areas however, and the success of the vine is a striking example of this, as is rearing and the introduction of some agricultural implements. It is not surprising that such a fundamentally rural people should seek to perfect the agricultural techniques of others rather than other types of techniques – this does not include their concern with urban projects, which we shall consider later. So perhaps the techniques of Roman agriculture are best described as a synthesis of all the most refined techniques. Virgil, and Hesiod both wax lyrical on this subject, and the differences between the two are rather revealing. However, it is no less true that the agrarian and argricultural technical system did not change, and that the Romans merely developed the potential of the preceding one. Unless of course, more specialised research eventually invalidates this interpretation.

The Mining Industry

It is difficult to assess the progress made in this industry, because we know so little about the older Greek techniques, and because our information on the Roman industry is so fragmentary and sometimes even suspect.

One aspect is of prime importance: there were relatively few areas mined in Ancient Greece, but these expanded enormously within the Roman Empire in the wake of the conquests. The Spanish mines, in particular, provided an excellent test-bed for the conquerors to try out new improved techniques. In many other areas – and we are thinking particularly of iron ores here – the many banks of alluvial deposits needed only opencast mining. Furthermore, and this is important from the technical point of view, there were private mines alongside state-owned ones, exploited either directly or in concession. In a system as diverse as the Roman mining industry, certain rules were needed, and record of these has been found on bronze tablets discovered in the mining region of Ajustrel, in Spain. These rules concerned both private and public rights, but they also dealt with exploitation and purely technical questions. The *lex metallis dicta* is present almost in its totality in the mining rule books of the Middle Ages.

Our sources on the details of exploitation are fairly thin on the ground, and apart from those we have just cited, they come mainly from the finds of archaeological excavations, particularly in Spain. According to these, the Romans do not seem to have perfected Greek techniques in any significant way.

There is still one major problem, which does not appear to have been satisfactorily resolved. The Romans probably invented the wheel pump which raised water from the mines. We have two main pieces of evidence concerning this question: the Venafro wheel, which is now in the Museum of Naples, and the reconstructed assembly of machines from the Tartessus mines in southern Spain. (fig. 3) According to the available recreations, these wheels were driven by water, itself brought by successive flights of norias (fig. 4). Now it seems that friction and wastage – such as the loss of energy – made it impossible to pump all the water powering the wheels, and that consequently, the mines were easily flooded. They could not have used Ctesibios' force pump since they were unaware of the crankshaft system.

Some limited, but not unimportant progress was made in metallurgy, which was largely due to the greater use of metals. In Noricum, for example, veins of "natural steel" – no doubt an alloy of iron and manganese – were exploited. The technique of steel-making by cementation was an age-old method, but it produced only superficial

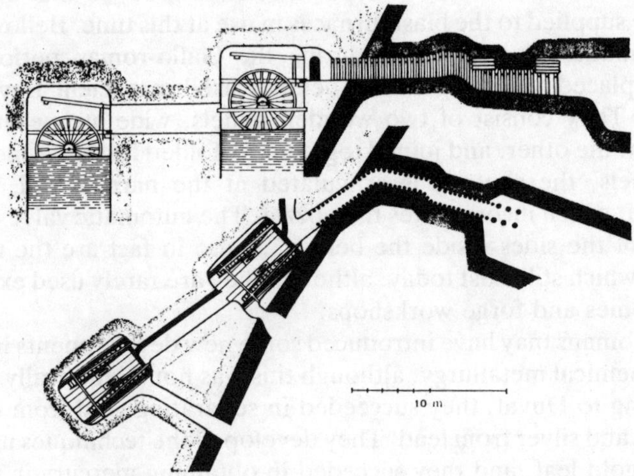

Figure 3. — The whole ensemble of wheels at Tharsis. (From J.-C. Baroja, *Norias, Azudas, Aceñas,* Madrid)

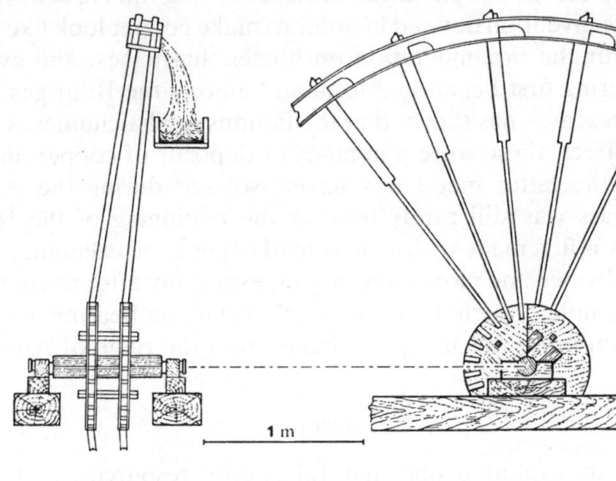

Figure 4. — One of the Tharsis wheels. (From J.-C. Baroja)

plating. According to Duval, it was very likely the Romans who developed and perfected the type of bellows used, so that extra air could be supplied to the blast furnaces in use at this time. Bellows with an automatic valve appeared during the Gallo-roman period, and these replaced the old goatskin bellows used in ancient Egypt and Greece. They consist of two wooden panels, wide at one end and narrow at the other. and joined together by folded leather. One of the two panels, the shutter, is articulated at the narrow end, or the headstall, which incorporates the nozzle. The automatic valve is fixed on one of the sides inside the bellow. These in fact are the type of bellows which still exist today, although they are rarely used except in some homes and forge workshops.

The Romans may have introduced some new developments into the use of chemical metallurgy, although this was not scientifically based. According to Duval, they succeeded in separating gold from copper mineral, and silver from lead. They developed the techniques used for making gold leaf, and they succeded in obtaining mercury in the 1st century from the famous mines of Almaden in Spain. One of their more specialised techniques was the tinning of copper and silver, which according to Pliny was done with "white lead," i.e. tin. He believed this was a Gallic invention devised in order to make copper look like silver, and used for the ornamentation on horses' harnesses, and even on carts. Silvering first began in Alesia and among the Bituriges.

Recent research has shown that the famous orichalchum was in fact brass. In effect, there were a number of deposits of copper and zinc alloy, but the latter metal was never isolated during the Ancient period. Brass was still rarely used at the beginning of the Roman period. It was first made with a mixture of copper and calamine, and its use gradually became more widespread, especially after the conquest of the Republic which had such rich calamine-bearing deposits. Orichalchum-brass was used in coinage from the reign of Augustus.

Energy

The Ancients exploited only natural energy resources, and if one considers technical development in terms of the amount of energy being used, the Romans performed no better than the Greeks. Power was provided mostly by animals and humans. Wind power was used only by ships.

The appearance of the water mill did not in fact constitute a revolution. The weak currents in some of the rivers of the Mediterranean Basin, the irregular flow patterns, the strength of some mountain currents and the sheer size of some of the larger rivers such as the Po, were all major obstacles against its effective use. Besides which it seems that for some time its use was limited simply to milling grain. "There should be no illusions about this," writes Marc Bloch, "the water mill is an ancient invention, which was age-old by the time it actually became widespread" (fig. 5).

Bloch has painstakingly sought out all the texts and evidence which mark the history of the water mill, a task which so far has almost not been followed up. There is evidence of a water mill in 18 BC, in

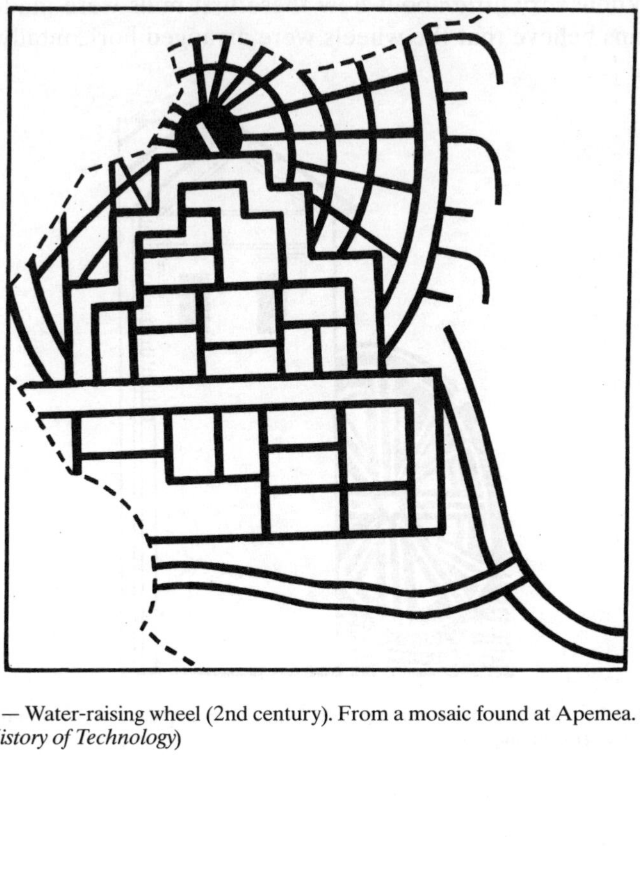

Figure 5. — Water-raising wheel (2nd century). From a mosaic found at Apemea. (From Singer, *History of Technology*)

Cabeira in Pontus among the outbuildings of a palace erected by Mithridates between 120 and 63 BC. A Greek epigram, probably dating from the period of Augustus, suggests that the water mill was a recent development. Vitruvius describes a water mill in his *De architectura*, under its Greek name, *hydraletes*, and some time later, Pliny talks about water mills on the Italian rivers. It seems, then, that the water mill was first used to a significant extent during the last century BC, in the mediterranean East. The mill was certainly not imported from the Chinese, although they were also using it at about this time.

There is very little evidence of these mills after this period. Ausonius mentions a water mill situated in the Moselle in the 3rd century, and in the 4th century, there is evidence of mills in the Janiculum supplied with water by an aqueduct. During the 6th century, there were no more than ten to be found in Gaul, despite the infinitely more favourable natural conditions (fig. 6).

We know very little about how these first mills were made. Most historians believe that the wheels were arranged horizontally, which

Figure 6. — A water mill. From a mosaic from the great palace of Byzantium (5th century). (From Singer)

helped to avoid more complicated ways of transmitting the movement. But the earliest, most precise description of a water mill provided by Vitruvius indicates the opposite, that the wheel was vertical. This is supported by further evidence of vertical wheel mills in Barbegal in Provence, and a mosaic found at Apamea in the second century, together with the Byzantine mosaic from the first half of the fifth century consolidates this. Pliny states that the mills were established by the rivers in Italy, but we know that some were also powered by tributaries and, as in the case of the Janiculum mills, by aqueducts. The group of mills in Barbegal in southern France were built in the 3rd or 4th centuries. They were powered by an aqueduct with a distributing basin divided into two channels which supplied eight different chambers and which had a fall in level of 18.60 metres. The Janiculum mills or the one at the agora of Athens, which all operated in the same way, date from the 5th century. There were also some boat mills, like those installed by Belisarius on the Tiber when supplies from the Janiculum mills were cut off by the siege of Rome.

Most of these mills, as we have said, were used for grinding corn, but the Mosellan mill described by Ausonius, may also have been used for sawing stone. The Apameus mosaic depicts a lifting machine, like the one described by Vitruvius (fig. 7). There may also have been some of these in the Tartessus mines in Portugal, but there is some doubt about this (fig 8).

IMPLEMENTS

The subject of implements has rarely attracted the attention of historians, even historians of techniques. The only exception to this is where prehistory is concerned, because apart from the remains of a few skeletons. tools are often the only form of evidence. For the sake of clarity, we shall divide this discussion into two categories: simple tools, and complex ones, or machines.

Simple Tools

The historian of implements has only two main works at his disposal, by Blümner and Frémond. The first has a much wider

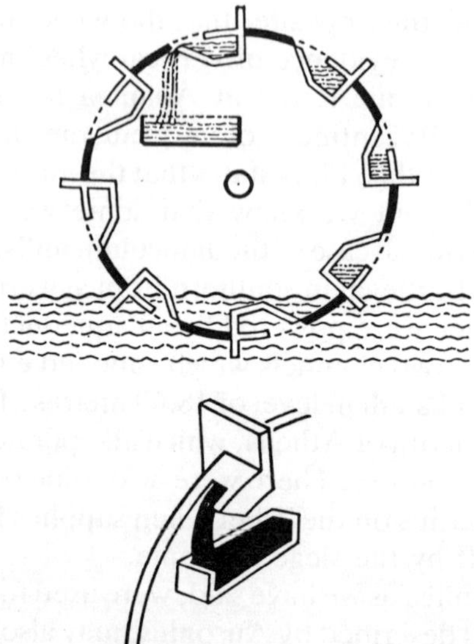

Figure 7. — Reconstituted water wheels. (From Choisy, *Vitruve*)

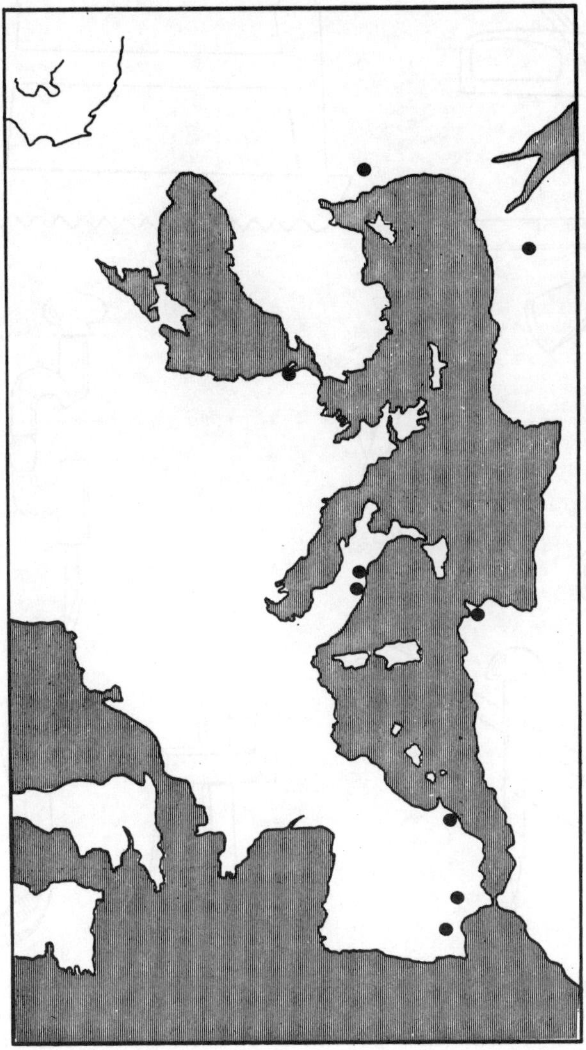

Figure 8. — Map indicating location of water-raising wheels during the Roman period. (From J.-C. Baroja)

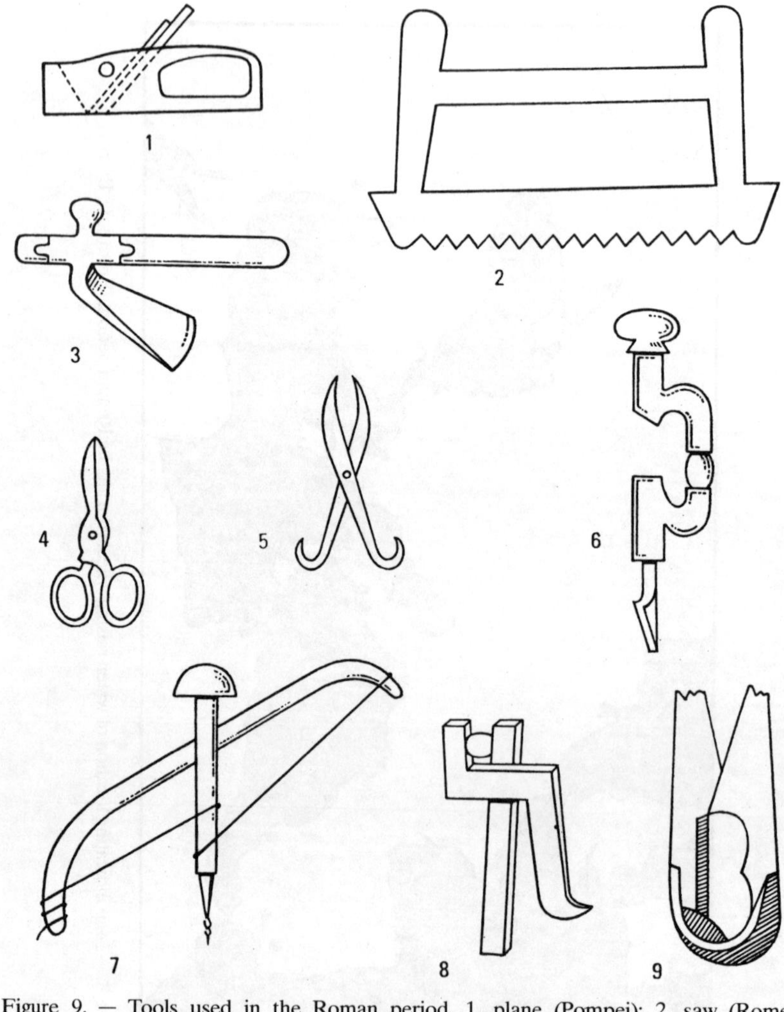

Figure 9. — Tools used in the Roman period. 1, plane (Pompei); 2, saw (Rome); 3, hammer-adze (Gaul); 4, scissors (Rondsen); 5, scissors (Priene); 6, Bit-brace (Naples); 7, anger (Thebes); 8, Punch (Autun); 9, shears with interchangeable blades (Pompei). (From M. Daumas)

approach, but the second contains more refined analyses of certain tools.

We are more or less certain that implements underwent considerable development during the roman period (fig. 9). It is difficult to see why, because there were few changes in techniques as a whole. In part, this phenomenom can be explained by the availability of more metal, a greater diversity in the type of consumer products in demand, and the effect of new techniques introduced to the Romans by subject peoples. This, again, has hardly been studied.

There are various different types of sources for this subject. There are the figurative representations, which were used especially by Blümner. Then there are archaeological objects, either the tools themselves which are the most valuable, or the objects made by these tools, which of course immediately precludes most of their wood work. Our ignorance about Greek implements or those of the so-called Barbarian peoples also hinders a full appreciation of the Roman contribution in this field.

We are therefore limited to discussing only those areas which have been most fully researched. But we would hope that eventually, an overall assessment will be possible.

The Romans were quite innovative when it came to cutting implements. The first pivoted scissors, like the ones we use today, were first produced at this time, replacing the old spring scissors. We have them to thank for the frame saw, while older civilizations had known only the small hand saw. The blade was fixed firmly onto the frame. Stone saws have also been found, without teeth, which were probably used with sand. The Romans are also reputed to have invented the punch, the gimlet, the milling-cutter and the spiral gimlet with three points. They are also thought to have invented the bit-brace, the auger probably coming from the Gallic civilization (the word *taratrum* seems to be of celtic origin).

The Romans were familiar with files: they used the *scobina*, or the wood rasp, and the *lima*, or metal file. The oldest ones seem to originate from the Gallo-roman period. They are of various shapes, flat, half-round and square. There are some with two crossed cutting edges, and others with just one, such as the adjusting file, for example.

There was one important development in wood working during the Roman period, which was the introduction of the plane, unknown in preceding civilizations. This made wooden structures far more solid. It is thought that other tools related to the plane followed rapidly: the jack-plane, and the ordinary plane, longer than the one used today, the rabbet plane and planes used for preparing wood for use in assemblies. The working of metal and other substances with a vice probably dates from the Gallo-roman period.

If one compares the figurative representations of tools from the Roman period, and those from the Egyptian period, a quite remarkable path of development is evident. The type of tools used was to remain virtually unchanged from the later period until the 13th century. The changes wrought were mainly those of diversification and increased specialisation of tools: the tool became more and more specifically designed for one clearly defined operation. There were several types of hammer used during the Gallo-roman period, for example, each one designed for use in a different trade. There were seperate hammers for the coppersmith, gold and silversmiths, coiners, cobblers, carpenters and woodworkers, stonecutters, and even hammers with nail claws, so that they could extract nails. There were many other types as well which we have not yet managed to identify. Scissors underwent the same type of evolution.

It is perhaps easiest to distinguish between the different types of operation in woodworking, and one very good example of this is the nature of the saw. The following group of saws used in the Roman period remained the same until the 14th century: (*a*) short saw (or semi-long) with a solid handle on one side, gripping the saw with a half-length of wood: this is a push saw which is the ancester of our small hand saw; (*b*) the long stone saw, held in the middle of a large rectangular frame, like the saw used by pitsawyers; (*c*) the ordinary frame saw; (*d*) the stone saw for cutting crosswise (with two small handles): the equivalent of our backed saw; (*e*) "harnessed" saw, with one handle.

Blümner attempted to establish an inventory of tools used by ancient peoples, particularly the Romans. This work, completed at the end of the last century is of immense value. However, it still needs to be reassessed and completed in the light of new archaeological discoveries. A good comprehensive inventory of these tools would provide directions for important fruitful research. This can only be achieved by searching all museums for their collections, many of which are probably lying forgotten in their storerooms.

Machines

The machine did not develop – except in certain specific sectors – as much as the basic implements. Vitruvius' work in this field demonstrates this clearly, as it is based almost entirely on Greek sources. Some machines evolved as a result of other innovations, such as the noria, derived from the water mill. There was progress made in agricultural machinery: of those used in cultivation (ploughs, harvesters), and in the preparation of agricultural products (especially presses).

The mechanisation of war, which had played such a large role in the initial growth of technology, remained largely dependent on the Greek system and heritage. Thus, artillery was still largely neuro-ballistic, the principle types of which were frequently used: the ballista; the catapult; the scorpion or the onager inspired by the principle of the sling.

The only new idea lay in the primitive crossbow, an image of which is provided by two bas-reliefs at the *Musée du Puy*, in France. It illustrates a hunting bow with catch. But could this not be a derivative of the chiroballista devised by Hero of Alexandria? The term *arcuballista* can be found in the 4th century with Vegetius. So we agree with P.-M. Duval when he suggests that the improvement of these machines had taken on the characteristics of an invention. We are certainly witnessing a process of development, and this process was relatively limited.

As for lifting appliances, Vitruvius' texts are virtually all Greek in origin. Lifting apparatus with squirrel cage wheels however, may be of Roman origin (fig. 10). In any case these are the only pictorial representations we have for the whole of Antique civilization (fig. 11).

The use of the screw as an instrument of pressure in a machine must have been initiated much later. The only example we have of it, at least in its early form, is the screw press, used for pressing wine or cloth, the latter only attested by a wall painting at Pompeii. The cloth press was used to clean the material of all the liquid products which had been impregnated during the process of fabrication, and the instrument itself consisted of a frame with two wooden screws.

Similarly, the improvement of the bellows should also be mentioned. Up until the Roman age, a certain number of instruments had been used, the effects or results of which were identical but whose

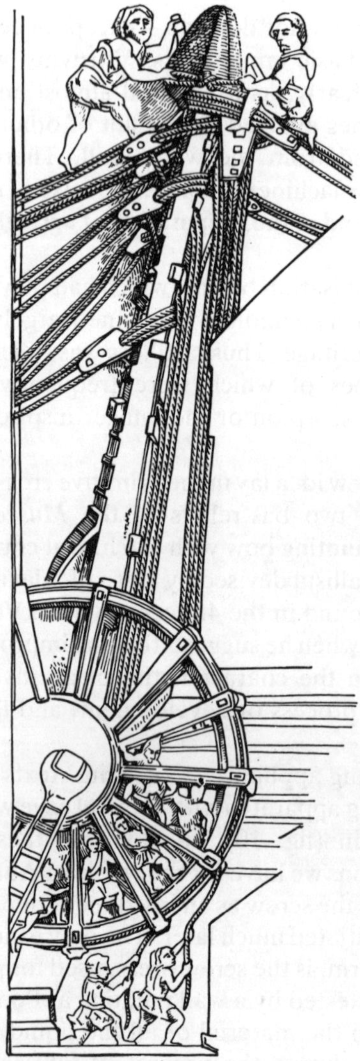

Figure 10. — Lifting machine with squirrel cage (Bas-relief from Latran). (From Daremberg and Saglio)

Figure 11. — Lifting machine with squirrel cage (Bas-relief from Capoua). (From Daremberg and Saglio)

Figure 12. — Winch and hoist (from Vitruvius of the Vatican). (From Daremberg and Saglio)

structures were completely different: blow pipes, goatskin bellows, diaphragm bellows, cylindrical and pump bellows. It seems that our modern fire bellows used in the home, which are angular deplacement bellows with a valve, a wooden framework and leather sides, may be of Roman origin.

The same is true of Roman tools. Detailed and extensive inventories of all the machines used by Antiquity should be made according to different criterion: practical use, which implies the range of crafts, and the types of mechanisms employed, which implies the structural range in use, linked also to the type of energy harnessed (fig. 12). The long evolution of machines, in other words, of the technology which we consider today to be the most advanced, may be classed within strictly defined stages. This obviously first presupposes a classification of machines, which is still lacking despite some of the commendable efforts which have been made.

SOME TRADITIONAL TECHNIQUES

This is undoubtedly the most static sector in the domain of technology, and this was indeed the case for quite some time. We will quickly review the different aspects of this. Here we will observe a combination of traditional techniques taken from the Greeks and the Barbarians, the origin of which is very difficult to discern.

There is nothing new, that is to say fundamentally new, in the textile industry, as spinning, weaving and fulling cloth remained completely unchanged. Only the weaver's card, used to separate threads, was introduced in the late Empire. As Duval has pointed out, changes were made in dress and costume. But here again, both the necessities of a more severe climate and the influence of the conquered peoples must have led to new ideas in clothing. The Gauls introduced clothes which were cut, close-fitted and sewn, as well as breeches (an early form of trousers), socks, coats, the use of furs, and probably buckles. We have very little information on dyeing techniques which most probably evolved in the sense that a wider range of materials was used.

Ceramics must have been a highly saturated technique, as according to Duval, the only Roman contribution lay in the systematic use of the hollow or smooth decorated mould for the mass production of huge quantities of pottery pieces, which were often decorated in relief. But here again we find that the generalization of known methods, made possible by the new political organisation of geographical space, does not necessarily constitute change in the fullest sense of the term. It is perhaps useful to point out a new fact: the large scale organisation of production in regions which were highly concentrated, geographically speaking, such as Arezzo in Italy, Lezoux and La Gaufresenque in France. We will consider this later. Much larger kilns were required to produce a greater amount of pottery. The kiln consisted of three parts on two levels: the hearth was below, usually under the ground, then an adjoining hearth, known as the hearth of the hovel, which was partitioned into two by a wall which helped to support the floor of the firing chamber on the third level. From an inscription at La Gaufresenque we know of kilns which could each hold 3000 vases.

The Romans were certainly more innovatory in glass-making techniques. Cicero tells us that blown glass appeared in Rome in the year 20 AD, originating from Syria. Before the introduction of blown glass, the only other known forms of glass were drawn and pressed glass. Henceforth the Romans were no longer limited by the imposition of one obligatory form, so that production could take on a greater freedom. In the 2nd century, translucent glass resembling crystal was already known, as were the window, the mirror on a metal backing, and the magnifying glass in the form of a water-filled ball. Once again, it is difficult to gauge what progress was made, as our ignorance of the beginnings of glass-making precludes any theories. Nevertheless, significant statistics exist to show the presence of glass in different archaeological levels.

We lack any detailed study on the subject of carpentry and furniture making. Roman furniture was undoubtedly produced in the Mediterranean Basin but we do not know much more about this area. Here again, techniques used by conquered peoples, particularly those of the Celts, must have been adopted and adapted. The Gallic wicker chair with a high, round back has already been mentioned. The dining table with four legs and covered with a fringed table-cloth, around which one sat to eat, was of the same origin. Duval notes that "the furniture industry advanced considerably during this period." But this is only a very general opinion, as details of the industry are unfortunately unavailable.

In the chemical industries, one should note the appearance of the wax candle, with a wick made of vegetable fibre, papyrus, oakum or off rush cores. It seems that soap (derived from the Gallic *sapo*) was first used to dye hair, and was only introduced in Rome as a hygiene product in the 4th century. It was originally made from a mixture of tallow, ashes, and probably red herb juice which had curative properties. It was later used to clean the skin and to wash clothes.

The *codex* succeeded the cumbersome scroll (*volumen*). Pages were first made from papyrus, and then from parchment at the end of the 2nd century. Ink was based on a thinned down soot mixture: it was that which came from Pergamum (which gave its name to parchment).

BUILDING

The Romans have long been considered past masters in the art of construction. The abundance and size of some of their monuments, generally better preserved than those of earlier periods, provide a picture of their indisputable skill.

In fact, urban programmes and building styles seem to have been handed down by the Greeks. Two problems should be considered here: building types and the techniques used to construct them.

Types of Buildings

In reality, research in this area falls within the subject of the history of architecture. But we should note here some of the forms for which Roman architects and engineers became famous: the triumphal arch, the amphitheatre, the aqueduct, the bridge, the trophy, the pantheon, the votive column or decorated funerary pile, and the villa.

This list is not without ambiguity. As we have seen, the Greeks did actually construct aqueducts, and the lines and profiles of their water conduits were to be envied (fig. 13). It is more difficult to say whether or not non-Roman Antiquity had known bridge building. On the contrary, it seems that we must credit the Romans with this work, but only insofar as it fitted into a vast communications system which Greece had not developed or had not needed to develop.

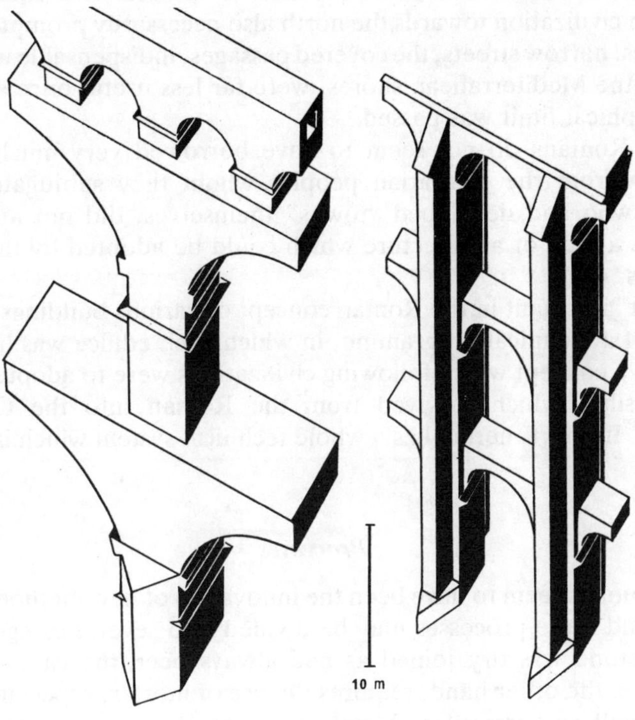

Figure 13. — Arch from the Pont-du-Gard [left] and arch from the Segovia aqueduct [right]. (From A. Choisy)

All this represents a phenomenon which depended upon the art of techniques and politics. Art in a technical form because certain edifices demanded new solutions, and in a political from insofar as these constructions were one of the ways in which the power and glory of Rome could be manifested.

These buildings were the result of a certain organisation, either a political or social, rather than an economic one. A new concept of town planning also emerged, and political bodies existed which no longer enjoyed the same significance as they had previously had in Greek civilizations, as from a particular period onwards the people were no longer masters of the city. From this ensued a profound reorganisation of the structure of the town. The extension of the Roman civilization towards the north also necessarily prompted some changes: narrow streets, the covered passages, indispensable in the hot sun of the Mediterranean shores, were far less useful once a certain geographical limit was passed.

The Romans do not seem to have borrowed very much in this domain from the Barbarian peoples whom they subjugated. The latter, who had developed "towns" themselves, did not appear to possess a type of architecture which could be adopted by their conquerors.

Most important in the Roman concept of various buildings was the unity of a technical programme, in which each edifice was linked to others, a concept which following civilizations were to adopt (such as the basilica which survived from the Roman into the Christian world). In short, herein lies a whole technical system which is closely united.

Processes

The Romans seem to have been the innovators of new methods in this field, and these processes may be divided into several categories.

Freestone was dry joined as had always been the case. Smaller stone, on the other hand, requires the use of mortar, especially if it is very small and irregular. Mortar was therefore very important in construction work of this type. Its composition had been determined for some time, as it was used in Hellenistic palaces, principally at Pergamum, in a lime, sand and clay form. These mortars hardened over a period of time to form a conglomerate with the stones which they were to bind. It was easy to make and set fairly quickly. The quality of the mixture naturally depended upon that of its components, and the way in which it was used could prevent air bubbles from forming and remaining when it hardened. It was compressed by the use of a pestle working between vertical boards. Analysis of Roman mortar has shown that fine grains of sand, lime mixed with charcoal dust, clay, and if needed, gravel and crushed brick were used. Throughout the Empire the proportions of the mixture were generally the same, from one century to another: one measure of lime to three of sand, or two to five , depending on what was required. Sand and water were mixed together with a binding agent of lime and clay. It was for spreading these mortars that the Romans invented the trowel, the shape of which was very similar to the trowel used today.

Care had to be taken to ensure that the wall did not warp during the drying process; also to ensure perfect cohesion of the facing itself and to prevent cracks. Pieces of scaffolding were left in the wall cavities when the supporting scaffolding was sawn off on either side. Brick level fittings were incorporated across the wall mass which re-established a horizontal plane every time the work was restarted, and ensured the cohesion of the facing to the other surface for the total height of the structure. It is quite possible that these were similar in design to the latticework of wooden beams used by the Gauls. Courses were constructed diagonally for the wall facings, arranged in herring bone or ears of corn patterns. It was possible to build brick chains, increase the number of discharging arches, and apply coatings, owing to branches and geometric incisions made in the surface of the rubble stone wall facings.

Rome adopted the Greek tile, which we now call the Roman or round tile. Similarly, sheets of stone or marble were also used. Window glass was incorporated in the walls from the 3rd century onwards, and coloured stained-glass windows were also introduced during this period.

Innovations, on the other hand, were implemented in the way in which doors were constructed and used. The Greeks had not developed the system of hanging doors on hinges as we do today. The Romans appear to have used the split hinge, which consisted of alternate locks on either side of the hinge which then fitted together and were held by a peg. Valuable examples of these can be found in the museum of excavations in the villa of Catullus at Sirmio, on the shore of Lake Garda (fig. 14). The villa was built and extended between the 1st century BC and the 1st century AD. This was a visible improvement upon the old hinge system which consisted of manoeuvring pegs in the sockets at the top and the bottom of the door.

Similarly, the Romans must be credited with another important technique: the web key and the modern lock. Before this invention, the Romans probably only used the wooden latch, the reliability of which was questionable. It is difficult to give an exact date as to when this type of modern lock first appeared, but it may have been during the 3rd century; Frémont's research on this matter has largely been

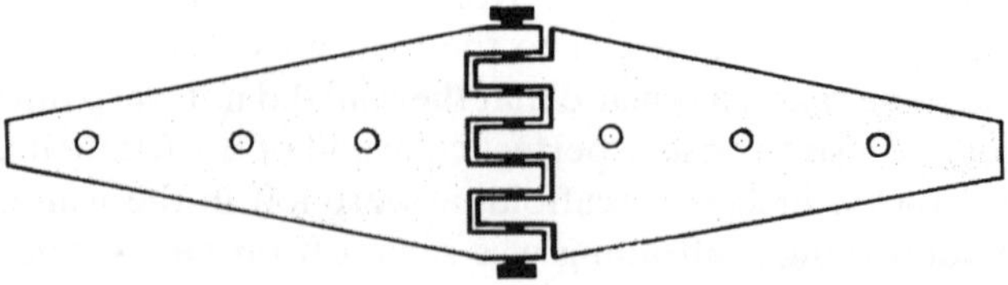

Figure 14. — Split hinge (Sirmio).

outdated by archaeological finds made since the publication of his work on locksmithing.

The vault is a more controversial subject and there are many doubts surrounding this architectural feature. It is agreed, however, that it was of Greek origin, a fact which is forgotten in the face of the profusion of Roman vaults. The Romans, however, managed to build larger spans than those constructed in earlier periods. The increase in the number of storeys, stairs, bridges, tiers, and amphitheatres meant that the use of the vault became extremely generalized, and led to marked improvements in the way in which it was built (fig. 15). It provided the ideal cover for many public buildings. This architectural feature required, moreover, a particular size and cut of stone, and certain specific assembling techniques. It was certainly the acquisition of these techniques which enabled the remarkable spread of this method of construct-ion. Did stereotomy, the daughter of geometry, exist in the Greek civilization?

On the other hand, it is quite possible that the Romans invented the dome, a specialised form of vaulting (fig. 16). Whether they did or not, Roman architects used two basic types of dome; the dome mounted on squinches, and the dome on pendentives. Once again, construction methods were the essential consideration, and the Romans quickly realized that the dome could be built without using scaffolding. The Pantheon dome, in Rome, is one of the most magnificent examples of this.

The Romans also made great progress in the art of framework construction, whereas the Greeks had long remained at a semi primi-tive stage. We might suggest that the building of immense scaffolding, necessitated by huge constructions, enabled Roman architects to resolve certain problems which had until then remained inadequately

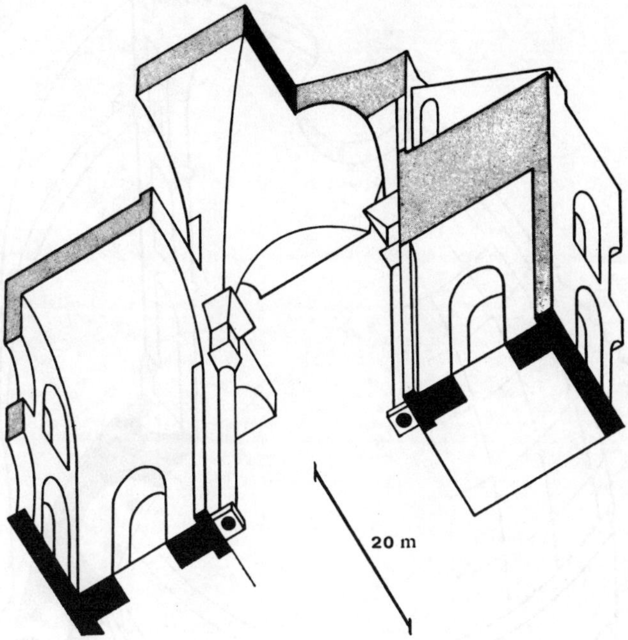

Figure 15. — The Basilica of Maxence. (From A. Choisy)

dealt with. The great glory of the age of Roman Antiquity subsequently lies in the invention of roof trussing and triangular structures which more than favourably replaced the timber piles built before. It was the beginning of modern framework construction as we know it today.

The arched bridge was undoubtedly a Roman invention. The generalization of the vault favoured this construction and ensured its sucess. As we have already said, it is unlikely that the Greeks knew the bridge, which may explain why they had virtually no road network. The development of land communications and the necessity for efficient administration, to ensure that couriers and troops could travel easily to their destinations, was to make the bridge an important architectural feature. Unfortunately we do not know how the piers of the bridges were based – and there are some which have a great many, often embedded in fast flowing rivers. The Alcantara bridge has arches

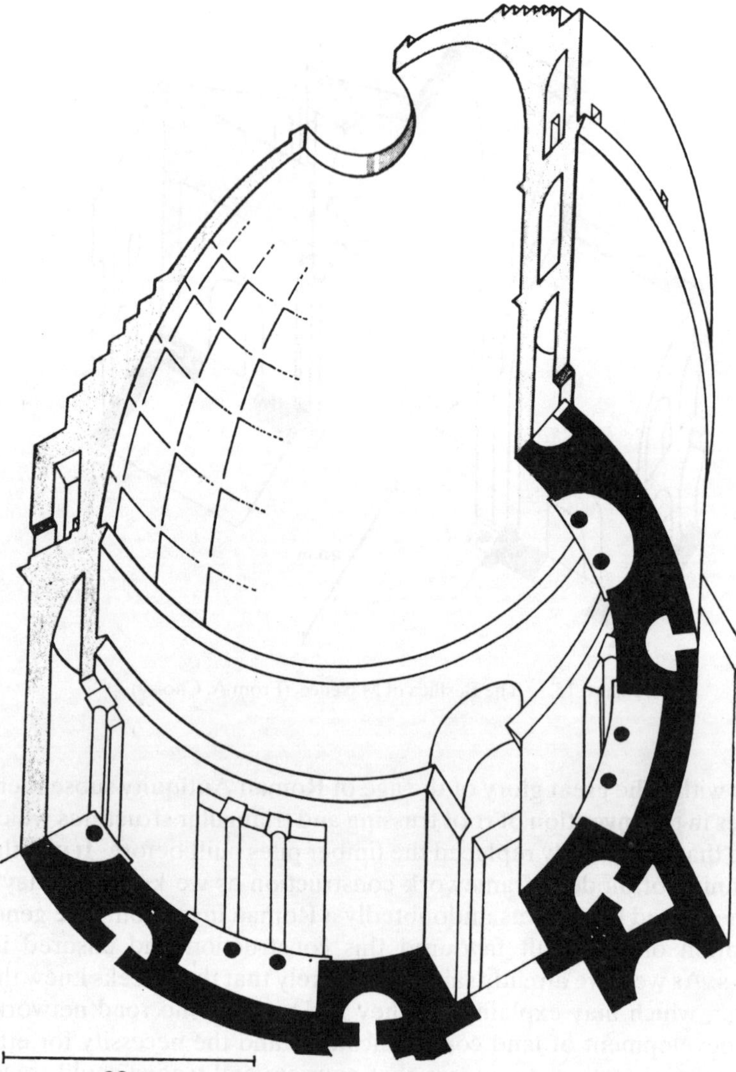

Figure 16. — The Pantheon. (From A. Choisy)

with a 27 metre span, and the arches of the Narni bridge had a 32 metre span. The piers of the Saint Angelo bridge in Rome were sunk some five metres in the bed of the River Tiber. More detailed and systematic studies are required in order to understand the methods employed by Roman engineers for the construction of such bridges. Even the list we have of the bridges that existed is hardly comprehensive.

Heating systems may also have been a Roman invention (fig. 17). Heating was closely linked to the edifice itself, principally in the Thermae. The most popular system was the hypocaust, which consisted of a chamber of warm air situated under the rooms to be heated.

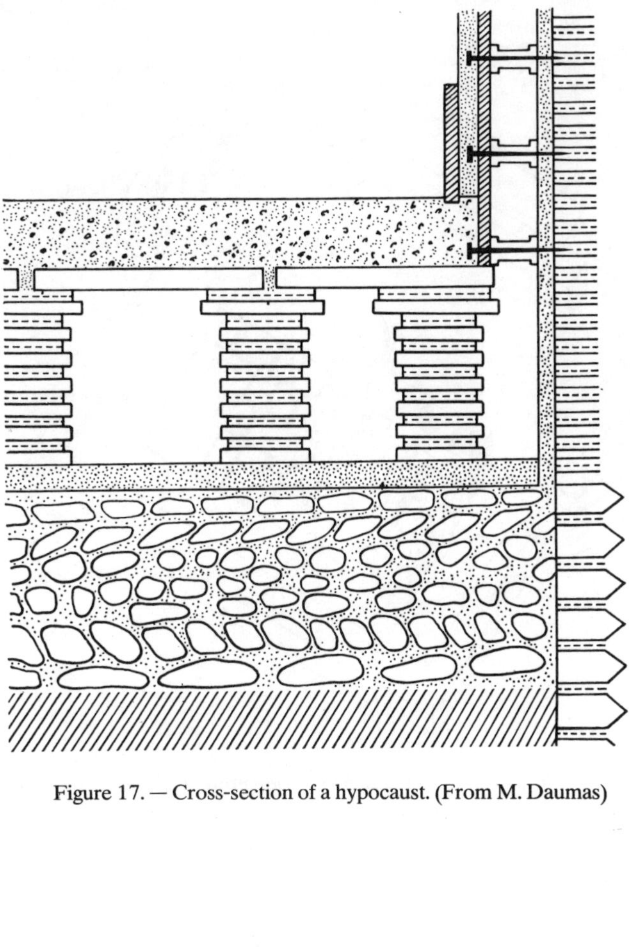

Figure 17. — Cross-section of a hypocaust. (From M. Daumas)

The process was certainly an old one, as this system had been in existence in the civilizations of the Indus from the time of the third millennium BC onwards. Systematization of this process necessarily entailed improvement, improvement which was aided by the use of brick consructions (fig. 18).

On the whole, the Romans adapted and developed the maximum from the techniques they had inherited. They had adopted the Greek techniques of working stone, and had extensively developed all the techniques of making and building with brick, which was to become the major material used in Roman Antiquity. They undoubtedly learnt from the Barbarians a certain number of techniques for working with wood, which was largely used in regions where resources of this

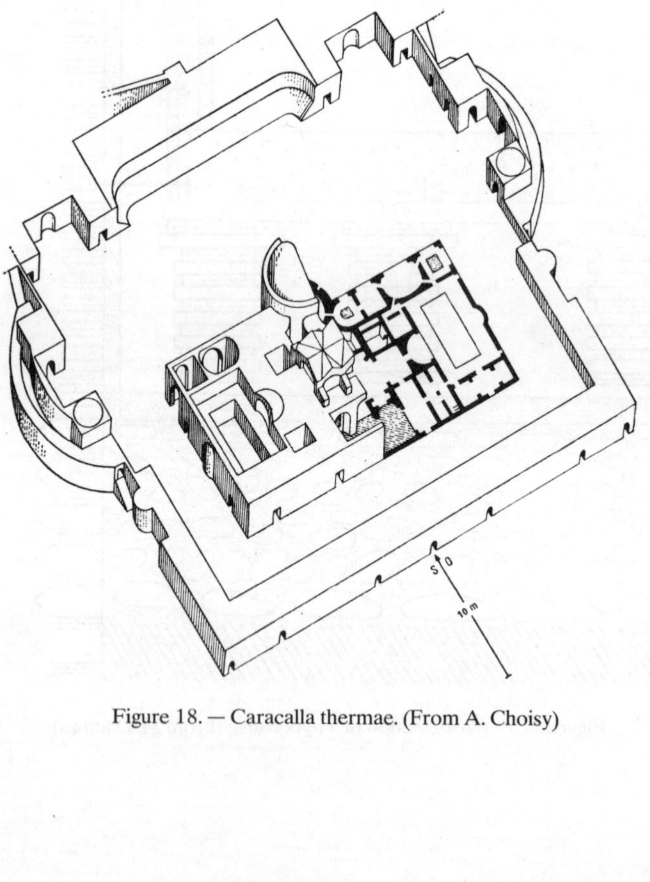

Figure 18. — Caracalla thermae. (From A. Choisy)

material were infinitely superior to those known by the Mediterranean world. Their genius lay in their ability to combine all these techniques and to exploit them to their greatest potential, and it is above all this quality which enabled them to develop the expertise which we still admire today.

GEOGRAPHICAL SPACE: A NEW SUBJECT OF TECHNICAL ORGANISATION

Very large countries which could promote technical change did not really appear until the reign of Alexander. Ancient Egypt only consisted of a narrow strip of land which stretched along the banks of the Nile. The originality of the Roman Empire lay not only in its extraordinary geographical extension, but also in the fact that it spread to regions which were very diverse both in climate and natural resources. We have already stressed the importance of this fact. Geographical space had to be organised, and as we have already seen, the techniques involved in this organisation were particularly pursued by the Romans.

Means of Transport

Methods of transport concern both transport techniques themselves as well as all the resulting indispensable infrastructures. Our knowledge in this field is very unequal: the famous Roman roads have been the subject of thorough investigation. For several years, thanks to a fashion – to be sure, a very profitable fashion – naval archaeological research has increased with the exploration of wrecks. Apart from these favoured sectors, there are other fields which have hardly attracted historians' attention.

Need we remind the reader that the latin names of land vehicles were almost all borrowed from the Celtic language: *carpentum*, the chariot built on a framework, *carrus*, the cart, *benna*, the truck, the four-wheeled *reda* and *petorritum*. Most of the vehicles used by the Romans seem to have been two-wheeled. Four-wheeled vehicles might have had a mobile front-axle unit, for without this they would have been very difficult to manoeuvre. Iconographic sources depicting these do exist, but the use of such vehicles must have been extremely limited.

It is virtually unanimously agreed that the horseshoe was not yet used at this time. Similarly modern harness with a shoulder collar was not yet known either, and only the Antique form of harness, which went round the throat of the animal, was used, meaning that only light loads could be pulled. Land transport would thus have had to have been carried out by porterage rather than by vehicles, and this was to remain so until the late Middle Ages. Moreover, an edict by Diocletian limited the loads allowed for two-wheeled vehicles on important roads.

Navigable waterways were extensively used, and we possess numerous references to this. Once again, it would be useful to have a map showing the navigation routes followed in the Roman epoch, and this could certainly be done without too much difficulty. Some archaeological discoveries have brought to light the type of craft used. The Utrecht Museum preserves a vessel which was probably used on the Rhine or the Moselle. The boat is relatively flat and raised at both ends which are absolutely parallel. The transport capacity of such a craft was fairly limited, if indeed it was used for transporting merchandise.

Maritime navigation must have made considerable progress. Here again, the Romans had inherited traditions from the Greeks and almost certainly also profited from the techniques of other peoples, particularly the maritime tribes of Gaul (we know that Ceasar greatly admired the ships of the Veneti). Veneti ships had iron tackles, oak hulls fitted together with long nails, and semi-rigid sails of skins sewn together. According to Lucian, the boat which brought corn from Egypt during the 2nd century, had sails made of skins. This system had its drawbacks, however, particularly when it came to manoeuvring the craft (fig. 19).

Underwater excavations have revealed a great deal of information on the Roman navy, whereas the historian formerly had to resort to studying pictorial representations which were often difficult to interpret. Research has been carried out on more than twenty wrecks, ranging in date from the 2nd century BC to the 4th century AD. These are usually keeled vessels consisting of an internal structure formed of continuous or discontinuous frames, and planks which are joined longitudinally, (stringers and wales) and transversely (frames and beams). The planking, which may have been single or double, was always freeboard, held by mortice and tenon pin joints. It was carefully calked and sometimes covered with an exterior lead sheathing. The pieces were mainly joined to each other with pegs and nailed-down copper plates. The wood used for the construction of such vessels was fairly varied and was chosen according to the type and function of the ship to be built. Already curved wood was used, or it could be bent in the heat. Planking seemed to be joined on the keel, and the frames were inserted after the planking had been assembled. All these techniques were directly derived from Egyptian methods.

Figure 19. — Roman ship from Sausse. (From M. Daumas)

Some evolution may be observed from the Greek type of wreck of the "Grand Congloué" of the 2nd century BC to those of the 4th century AD. After the middle of the 1st century BC, most ships were constructed in the single planking method, thus forming a lighter hull, which presupposes that the rib was reinforced. The practice of using lead sheathing was abandoned between the 1st and the end of the 2nd century AD, for reasons which are difficult to assertain. This evolution was to make the hull much lighter and to reinforce the internal framework (more nailing was used). This evolution was a continuation of the Mediterranean tradition initiated by the Egyptians. The last stage in this development of ship building is illustrated by the vessel found in 1975 at Marseille. Also noteworthy is the appearance of second mast which inclines towards the bow.

The standard capacity of the ship in the Hellenistic epoch was approximately 130 tons. During the Roman period, imperial transport reached a capacity of 340 tons, and 1300 tons could be carried by the ship built at Ostia, designed to transport corn from Egypt. We should lastly point out that there was a large number of different ships.

The Infrastructure of Transport

Infrastructure techniques involved in transports were just as important as the means of transport themselves. In this area techniques were necessarily linked to each other, as well as being linked to other techniques of a different type.

It has often been suggested that the Romans did not develop rivers and waterways, but this affirmation does not seem to hold true. We have pointed out the existence of the canal built by Corbulo, between the Rhine and the Meuse, the Fossa Drusiana, crossing the dangerous delta, and the Marius canal, between Arles and Fos. This sector of Roman engineering has been dreadfully neglected.

The Roman road, on the other hand, has been the object of high praise and has rightly remained famous (fig. 20). It represents an extremely important innovation which was probably developed at a relatively late date, as it was virtually non existant in Ancient Greece. Its structure is well known: it mainly consists of a deep roadway set in a vast trench, flanked by drainage ditches and gentle embankments, so we might almost call it a sunken wall. The real innovation lay in the superposition of layers of large flat rubble stone, gravel, diagonally

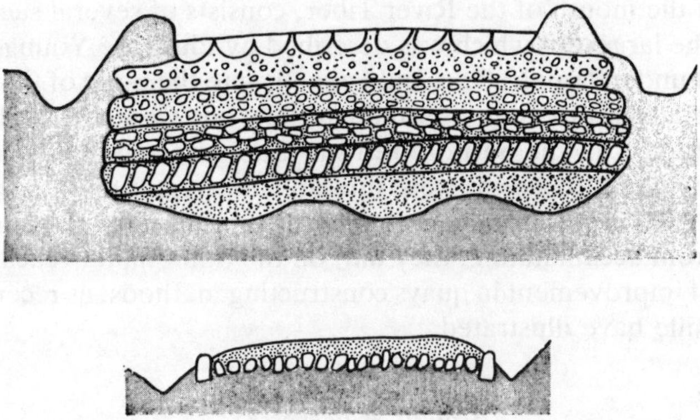

Figure 20. — Cross sections of a Roman road. (From M. Daumas)

placed slabs, dry rubble, and sometimes layers of mortar. The road was steeply cambered and raised slightly above the surrounding terrain. It possessed a sort of elasticity which contributed to its solidity. The only drawback was frost, particularly in the north of the Empire, as water could penetrate this veritable stonework and would, as it began to ice and expand, dislocate the stone construction.

Pliny recommended that the road be 18 feet (5.35 metres) wide, but main roads were often double this width. The problems of the course of the road have been the subject of many discussions but we should obviously qualify all this: the frequently used straight route was not as systematic in construction as was thought and could only be used in areas which were not hilly. The other idea which is generally accepted is the permanence of these roads. The Romans indisputably adapted their road network to traffic, and to the inevitable modifications in the latter. Whatever the case may be, the Romans established a vast road network: it has been estimated that 90,000 kilometres of major roads were built, and 200,000 kilometres of secondary roads existed.

The Western Mediterranean was appreciably poorer in natural ports than the rest of the area. This explains why the Romans had to create more artificial harbours than the Greeks. The Greeks, however, had shown themselves to be quite capable of building ports in the Hellenistic epoch, as is demonstrated by the port of Alexandria. Remarkable examples of Roman work in this field have remained. The port of Ostia, at the mouth of the River Tiber, consists of several successive basins, the largest of which was described by Pliny the Younger, and was built under Trajan. One could equally cite the ports of Civitavecchia, Terracina, Pozzuoli, Misenum in Italy, and Cherchell and Leptis Magna in Africa. The port of Fréjus in Gaul was dug out of the solid ground by hand; its basins were probably used right up into the 17th century. Although the Romans did not perfect any method of preventing silt from accumulating, they did, on the other hand, reach a high degree of improvement in quays constructing methods, as recent finds at Marseille have illustrated.

Roman Waterworks

The extent of Roman waterworks and water systems is astonishing, but it was only partly innovatory. The Samos tunnel, the aqueduct of Pisistratus, and the techniques for channelling water developed in Mesopotamia and Persia, are overwhelming proof that mankind had made great attempts to advance in this domain. Once again, the sheer scale of the work astonishes us today.

Extensive lead production, principally from Spanish mines, undoubtedly made possible the construction of the innovatory aqueduct bridge for crossing small valleys, the development of the principle of the inverted siphon and enabled greater water distribution in towns and cities. The Romans, masters of an advanced technology, had solved the problems of levelling, and applied tried and tested systems of organisation, (a perfect example of which is provided by Frontinus), and so did not hesitate to use water supplies which were situated 33, 50, or even 100 kilometres away. All the aqueducts were constructed in such a way as to ensure that the slope was both adequate and regular enough to transport the water. If the slope was too steep, then it was broken by a system of cascades. The twenty-four cascades between the Morvan mountains and the town of Autun (France) thus made possible the transportation of a water supply to the town without any difficulty.

Tunnel passages demonstrate some of the mistakes which were made, as exemplified in Bougie around the middle of the 2nd century. On the other hand, springs were harnessed very successfully, and reservoirs, manholes, and distribution towers were also successfully constructed. Slopes were usually slight, 0.2 in 1000. We need only look at the magnificent achievements of the Pont du Gard and Segovia aqueducts. Lyons' inverted siphon required some 26 kilometres of lead piping, and lead conduits also crossed from the Rhône to Arles.

The Arabs have often been credited for the extensive irrigation systems in both North Africa and Spain. In fact the Arabs often only reused and reactivated Roman installations. But again we must admit that we lack precise studies on these works, some of which were still in use at the end of the 16th century.

In all these fields, the extent of these achievements presented new problems. It was because of their geographical expansion that the Romans could exploit to saturation point the techniques which they had inherited from the Greeks and the countries of the Near East. The Greeks had been prevented from making further progress by the exiguity of their cities' territory. The Alexandrian Empire had lacked the time to develop techniques because it was involved in conquering other lands, but the Romans had both the space and the passage of many centuries in which to concentrate on technical development.

It is difficult to judge globaly accurately the technical level achieved by the Romans, and errors have undoubtedly been made on this account. We may summarize by suggesting that the part played by the Romans was of stabilizing technical development which had almost been constant up until then.

We may detect only a very few new ideas. As we have emphasized several times, the Romans were the ideal inheritors of Greek techniques in almost every sector and, via the latter, of all the techniques of the Eastern basin of the Mediterrranean.

Obviously their conquests and the geographical expansion of their civilization, meant that the Romans could profit from techniques which had been unknown in Mediterrranean regions, as has been exemplified by the use of Gallic or Germanic techniques as well as about Pergamum. It was in this way that the Mediterranean technical system could be completed, but without entailing any profound upheavals. The so-called barbarian techniques were perfectly integrated into the pre-existing system.

Some sectors of technology remained in which improvements were possible and desirable, as we have quite clearly demonstrated for tools. But we should not forget that the Romans had at their disposal an infinitely greater quantity of certain materials than had their precursors, their teachers: lead and copper from Spain, and iron provided by all the regions north of Italy. This supply was invaluable, as an abundance of materials necessarily leads not only to an abundance of technical processes but also to their diversification. This we can easily notice nowadays.

Finally, let us review the conclusions which will terminate this discussion on the organisation of geographical space. The growth of the latter indubitably encourages unnecessary or needless improvements and adaptations in restricted areas. Even today, do we not talk of the technological power of great empires? Although the Romans may not have been innovators as such, they were, on the other hand, obliged to be great realizers of technical systems, and the transition from small scale systems to large scale works could not have been achieved without initiating some changes.

Another problem has yet to be discussed which historians have virtually ignored, no doubt because it confuses their arguments. We have looked at the blockage which halted the development of Greek tech-niques: let us look again at the texts which have been frequently cited. They could equally apply to the Roman epoch which did not show any signs of either important innovation or of developing an advanced system of mechanization, even during the period when the reasons put forward for this, such as the institution of slavery or the contempt of manual labour, were less convincing or no longer existed. And this was despite the improvement of certain material conditions (natural riches were far more abundant).

We must therefore find other reasons for this phenomenon, but they are more difficult to define. One explanation springs immediately to mind: the Romans made absolutely no progress in the field of science, or at least in sciences upon which technical progress could depend. Neither their physics nor their knowledge of materials were superior to those which the Greeks could have known, notwithstanding the geographical growth of their civilization. Research in this domain is singularly limited, and we may be correct in suggesting that there was indeed a blockage, similar to the Chinese technological civilization which was also blocked from a certain point onwards. The perception of this halt, as this term is better than that of blockage, will always be as difficult to explain as the new advance which took place in the middle of the 12th century. We are asking a question which is impossible to answer, as it will almost always be impossible to account logically for any negative movements which occur within a civilization.

BYZANTIUM

The same characteristics and problems are to be found in this area, as Hellenistic Greece, Rome, and Byzantium shared the same technical system. Only narrow-minded specialists claim otherwise. The brilliance of the Byzantine civilization has often obscured its lack of technical imagination. Similarly, one may speak of the brilliance of the Hellenistic and Roman civilizations.

Chronologically speaking, the history of Byzantium is very extensive, as it develops over ten centuries, from around 500 AD to the middle of the 15th century. From the 11th century onwards we have rightly included Byzantium in the Western European world. The "take-off" of Western Europe from the 12th century encompasses the whole of the Christian civilization, and so Byzantium too. From the 6th to the 11th century, Byzantium was a faithful reflection of earlier civilizations.

Technical Thought

The blockage of technical thought is well illustrated by the appearance of technical literature which merely repeated old Hellenistic treatises. Alphonse Dain has demonstrated how the relationship of ideas stemming from the Alexandrian school, or sometimes even earlier civilizations, were followed until the end of the 10th century. This is one of the most remarkable features of technical stabilization, of which there are few other examples. An attitude characteristic of a blockage of technical thought is exemplified by the fact that treatises dating back some twelve or thirteen centuries were thought to contain new ideas by mankind in the 10th century.

To a large extent works on military techniques were handed down from man to man, but unfortunately we know very little about these people, and are even uncertain of their names. Apollodorus of Damascus was in fact very romanized and indeed worked for Trajan in

Rome. He wrote a treatise on siegecraft, in which there are few innovations. On the whole, he tends to repeat Philo of Byzantium on the subject of fortification, and Athenodorus on the subject of war machines. The tactics of Onesandrus in the 1st century recalls those of Ennius writing in the 4th century BC. Following in the same tradition were Aelian, Arrian and Polyaenus: only their historic examples change and demonstrate the stability of the techniques employed.

A certain awakening seems to have come about during the period of Justinian in the 6th century AD, but what had been written often several centuries earlier was still repeated. Possibly the most original works were those written by Urbicius, the strategies and a treatise on hunting, a tract on mechanical studies by Anthemius of Tralles who began the construction of Saint Sophia, and two anonymous works on strategies and siegecraft. But on the whole such treatises remained indisputably immersed in the tradition of the ancient texts, contributing only details. Thus the technical spirit of the age tended to degrade that of the Alexandrine epoch.

A similar revival took place in the 10th century, with Leo VI the Wise and Constantine VII Porphyrogenitus. Dain has listed seven compilations from this period, by which we can tell the subjects of the great texts of the Alexandrine school: compilations on machines, siegecraft, strategies and tactics, as well as compilations on naval war, which may be new as we do not have any evidence of earlier works on this subject. They broadly repeat all the known collections, and updating of older texts appear to be rare and fairly insignificant. Nicephorus Ouranos, who was in a way the culmination of this movement, demonstrated little more originality. The *Geoponica*, the only collection of texts made in the same period on the subject of agriculture, really brought together everything contributed up until then by Antique agronomy.

One of the most celebrated authors of this epoch was unquestionably Hero of Byzantium. Both this author's name and work prove everything that has already been said on this subject. The name is significant as there is no evidence to show that such a person even existed. All the works on applied mechanics were probably attributed to one Hero. His *Geodesy* and his *Treatise on machines* seem to be a semi-complete reproduction of works from the Alexandrine period. Of all these treatises only a scientific edition could obviously demonstrate the stability of the Byzantine technical world.

The Principal Techniques of Byzantium

Although this area still requires more detailed study, major Byzantine techniques only contributed a few innovations. Like the Romans, the Byzantines profited from some of the improvements made by neighbouring peoples, and advances undoubtedly resulted from local conditions rather than from inventive genius, which was extremely limited in any event.

No changes may be observed in exploitation techniques. On the subject of farming, to take just one example, the swing plough, an everyday tool according to pictorial representations, was an obvious legacy from Classical Antiquity. Cultivation techniques, other farming tools, and the crop types, remained the same. The *Geoponica* do not reveal any major transformations.

No new techniques were introduced either in mining or metal working. The only interesting fact was the exceptional skill displayed in smelting and casting bronze. The great bronze gates, which for many years opened out upon Western Europe, are proof of this expertise: the gates of San Zeno in Verona, those of Saint Sabina in Rome, and possibly the gate-posts of Aix-la-Chapelle are the most favoured examples of this. Until the introduction of iron hinges, these bronze doors were certainly the best fences devised to enclose edifices which were high in quality and rich in materials. The Byzantine Empire was too poor in iron for any progress to be made in iron- and steel-making techniques.

Similarly, little progress was made in the domain of energy. Hydraulic energy does not seem to have been particularly developed, even in areas of the Empire which were more favourable for its advance. But here again, we lack detailed studies to substantiate this claim. Nor may we base any opinions on iconographic sources, such as the Apamea mill, without being certain that this evidence is true to the overall situation.

Transport methods are identical to those used in Classical Antiquity. in fact, harnessing methods had changed significantly at the same time in Western Europe as in Byzantium, from the beginning of the 12th century, although the iron horseshoe and the shoulder collar had been introduced some decades earlier.

Recent studies have demonstrated that Byzantine ships were directly derived from Greek vessels, apart from the method of caulking which had been borrowed from the Arabs. Despite this, some evolution may be observed in boat building.

Before the Arab invasions, ships were exactly the same as those used by the Greeks. We need only look at the dromon, which had a single row of oarsmen, and the liburnian, which was simply a light version of the dromon.

The period from the 7th to the 12th century marks the peak of the imperial Byzantine navy. A dromon, with two rows of oarsmen and which could carry up to a hundred men, was developed, but apart from this, the ships were identical to those used in preceding civilizations.

Although the Byzantines were undoubtedly skilled builders, they did not contribute any important innovations to the techniques handed down by either the Sassanids or more particularly by the Romans. They mainly used brick according to the methods inspired by the former, but they used the Roman method of building vaults without curves or bends, known as groin vaulting. They also employed Roman mortar in brick constructions. Most of the achievements of the Byzantine Empire, Saint Sophia (6th century, fig.21) and the church of Saint Mark in Venice (11th century, fig.22), as well as aqueducts and water tanks, did not show any new technical innovations.

This is also true of town planning where again the Byzantines followed the tradition laid down by the Romans. At the end of the 5th century, edicts issued by Zeno specified road widths and projecting balconies according to the same norms which Ancient Rome had established some time before.

The grandeur of Byzantine techniques lay more in the perfection of the execution and decoration of their work than in the methods employed. We will illustrate this fact with just a few examples. Weaving techniques had been borrowed from Egypt. A still very simple form of the simple loom seems to have been introduced during the 5th century. For some time, silk factories, which were controlled by the state, continued to use Sassanid techniques. The beauty and the variety of the colours and designs of Byzantine textiles, some of which were Persian in inspiration, were their most striking qualities, and they spread throughout the Western world. The same applies to their silver ware, and more especially to their cloisonné enamels which were particularly popular in Byzantium. Byzantine mosaics were also far more superior in quality to the work of the Romans, as the wealth of the materials, and the variety of colours, especially the gold

Figure 21. — Saint Sophia. (From A. Choisy)

backgrounds (small flakes of gold set between two glass sheets), are extremely impressive on first sight.

Byzantine techniques had therefore been borrowed from various conflicting civilizations which had gradually become integrated. Byzantium was in fact the natural continuation of the Roman tradition of assimilation. The Mediterranean and southern position of the Byzantine Empire explains certain choices. And we can attempt a list of the contributing countries, with the techniques they "gave": (*a*) from Greece came the techniques of agriculture, hunting, fishing, and boat building; (*b*) from Rome, all the techniques of construction;

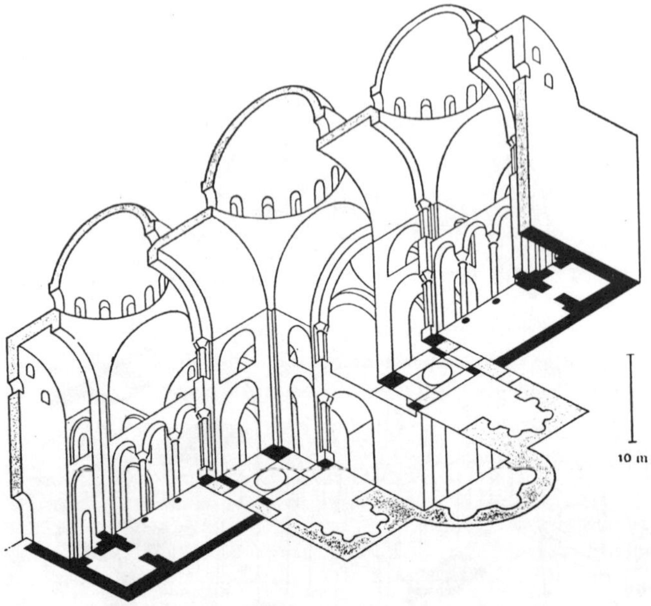

Figure 22. — Saint Mark. (From A. Choisy)

(*c*) from the Orient stemmed Sassanid brick, Arabic damascening and calking, and Egyptian fabrics.

We should seek to find out why the Byzantine Empire preserved the technical system which it had inherited from its predecessors or its neighbours, as we have done for the Roman Empire. If historians appear not have asked this question, it is in fact because the technical reality was to some extent eclipsed by an incredibly brilliant civilization. The artistic wealth of Byzantium actually hid a perfectly stable technology. Moreover, we should clarify the extent to which the Byzantine Empire could or could not profit from new techniques introduced in Western Europe from the 12th century onwards. The ever limited geographical space which Arab and Turkish invasions left the Empire of the East, could no longer provide the raw materials indispensable for more advanced techniques. It is astonishing that the Latin Empire of the Orient, arising moreover from the technical civilization of Western Europe, should have contributed so little in this area, other than better fortifications, and the windmill. Lack of water, and the absence of metalliferous deposits literally blocked the development of techniques in this part of the Christian world. It could hardly do otherwise. The end of the Oriental Empire was not merely the result of a wave of invasions, but was also produced by the gradual disappearance of natural resources indispensable for the survival of a nation.

Greek Fire as an Example

For a long time Byzantium has been credited with the use of what was known during the period as Greek fire, of which there may have been several types. Recent research has shown that this was not a Byzantine idea, and that naphtha, as well as a one particular powder, had been known for some time.

During the Persian siege of Athens in 480 BC, Herodotus mentioned the use of arrows tipped with oakum which was lit before the arrow was shot into the air. The oakum must therefore have been impregnated with an inflammable material. During the sieges of Plataea in 479 BC and Delium in 424 BC, Thucydides cites the use of bellows for fanning the flames of a fire, the basic material of which was bitumen. Aeneas the Tactician, describes in some detail the use of incendary missiles in chapters 33 to 35 of his *Tactical Theory*. There are many such quotations up to the Byzantine period. The use of "Greek fire" was first mentioned in 678 when it was employed against the Arab fleet. As we have just seen the name was new but the weapon itself was already quite old at this epoch.

The material used was most probably asphalt oil, either by itself or possibly mixed with sulphur and pitch. Athenodorus, and Julius the African revealed the first mixtures of spontaneous ignition which were mainly composed of hydrocarbons, with quick lime to raise the temperature, and once again saltpetre, or "salt dust" as it was described in a Greek formula.

Although incendiary weapons were only sporadically used in Classical Antiquity, they became more standardized during the Byzantine period, and it is very likely that their application was considerably improved. Parallel to the unlimited progress and great effectiveness of these incendiary weapons, was the introduction of a new concept, which consisted of confining a fuse product in a relatively enclosed and sufficiently resistant space, like a tube for example. These were steps towards the rocket, jet propelled by the escape of combustion gases. Reeds were often used for the tube, and although the mixture remained a state secret we know that it was formed of a hydocarbon, sulphur and an inflammable material which could only have been saltpetre.

Finally, tubes were required to launch the rockets. "It consisted of a reed pipe in which certain materials were packed; in order to use it, the reed pipe was placed in a bronze tube, lit at one end, and then after a great deal of thunder and smoke, it rose into the air like a burning meteor and reached its intended target." This suggests that fixed tubes must have been used, like those which can be seen on the bows of boats illustrated in some manuscript miniatures, as well as mobile tubes, of which we have no pictorial representations. Maurice Mercier, a French historian who specializes in all these machines, suggests that the Byzantines were on the way to developing the canon and gun, where the rocket itself was to become the cartridge.

A miniature in a Vatican manuscript shows a "siphonal" boat, as they were termed by Leo VI the Wise. The tube was necessarily oblique in order to draw out the inflammable liquid. This liquid was drawn out by the aid of a suction and force pump, and launched ablaze towards the opponent. The tube was fixed rather like the mast of the bowsprit on the sterm of the ship, so it was useless in rough sea.

To conclude, Leo VI recommended that "the pots filled with fireworks would also set fire to their boats when they broke up." Another text tells us that "these closed pots contained fire which suddenly burst in flashes and set fire to the objects which they hit." For a long time we have been unsure of the composition of the spontaneous ignition product. The chemist, Forbes, a good historian on the subject of oil, believes that if an exact combination of oil and quick lime was exposed to humidity, the heat developed by the contact of the lime with humidity might be sufficient to light the mixture spontaneously, particularly if it contained fine particles.

The Byzantines therefore appear to have contributed major improvements to extremely old techniques in this field, although some points have yet to be clarified.

The whole of this period, covering twelve centuries from the 2nd century BC to the 10th century AD, a history rich in events, reveals a remarkable technical stability. In some instances, however, there was even some deterioration in the techniques which had already been acquired, as is demonstrated by the formulae applied to war machines by Philo of Byzantium and Hero of Alexandria. These disappear after Vitruvius. We should certainly not ignore some of the interesting improvements made by both the Romans and the Byzantines, although these improvements do not constitute radical upheaval.

BLOCKED TECHNICAL SYSTEMS

To be honest, the title of this chapter is rather ambiguous, all the more so because it applies to three civilizations which are very removed from each other: pre-Colombian America, China and the Islamic world. We have already seen other blocked systems, such as that of the Egyptians, for example, and the same has been said of the Greek technical system. Similarly, what of the so-called primitive peoples who remained on the technical level of stone tools?

From the 12th century onwards, only Western Europe saw successive changes in its technical system, in the industrial revolution of the Middle Ages (to take the title of a recent work), in the technical revolution of the Renaissance, the technical revolution of the 18th century, that of the second half of the 19th century, and in yet another which is changing the world today. The real problem lies in the fact that other systems seem to have been blocked at the end of the Middle Ages.

We hope, however, that a brief survey of the three civilizations mentioned above will help to reveal the crux of the problem.

CHINESE TECHNIQUES

The study of Chinese techniques presents many, various difficulties. Firstly, there is an internal problem, as the chronology of Chinese technical discoveries is uncertain. One particular authority, reputed moreover for his knowledge of material life in China, attributes one invention to epochs which are separated by several centuries and all in the space of a few pages. There is yet another problem. For some time China has played an important part in the history of science and techniques. We may enviously recall the early beginnings of the Chinese civilization, and its sudden halt in the 15th or 16th century. Many tables have been drawn up to demonstrate the technical intelligence of the Chinese during very early periods, and praises have been sung about inventions which mainly preceded discoveries made in the West. And we should cite all the contributions made by the Chinese civilization during its technical expansion to European techniques in order to illustrate the extent of the progress made by the former.

But a redress was necessary: The relatively recent discovery of the fact that China remained fixed within an out-moded technical system has to some extent discredited the Far East. Suddenly we witness a complete reversal. This was also true of the Middle Ages in Western Europe which had long remained in the dark and then was suddenly brought to light with some exaggerations. Should we need point out, moreover, that the division between these disciplines is such today that a sinologist is completely unaware of the civilization of Ancient Egypt and vice versa. There is a total lack of chronological tables relating to Chinese technical progress, and it would be desirable if one could re-establish the correct balances and furthermore ask some essential questions. An exact chronology is therefore indispensable. The water mill is a very revealing example: Tsouei Leang (386/334 BC) has been cited for building water mills in the middle of the 4th century BC, but on the other hand, the construction of the first water mill has been dated in the year 30 BC. In this last case, it would have been exactly contemporary with the introduction of the Western mill. Without attempting to find forerunners, would it not be logical to suggest that technical acquisitions could have taken place at the same date in civilizations which were very removed from each other, but which were aware of similar developments?

The second question relates to the early beginnings of technical progress, at least in some areas of technology, and we should be cautious in our investigation here. During the Neolithic period, China was at exactly the same stage as the rest of Western Europe. Major developments were virtually contemporary with those made on the borders of the Oriental basin of the Mediterranean, but later than those of Mesopotamia and Egypt. Chinese techniques seem to have began to advance over a technically fixed Classical civilization around the Christian era. What were the reasons for this, and what medium did technical progress in China take? We must admit that there is still a great deal to be learnt about the ancient economic and demographic evolution which took place in China. Our knowledge of the history of science is far more comprehensive.

Finally, there was a sudden halt in technical progress which should be taken into account and must also be explained. It is universally agreed that the Chinese technical system was blocked from a certain time onwards, but the causes and significance of this are still a subject for endless discussion.

The last problem concerns the transference of a number of Chinese techniques to Western Europe. Perhaps the contribution made by China to a Western world which was trying to disengage itself from a Greco-Roman technical environment has been over-emphasized. The following elements are always cited: the voyage of Marco Polo, the compass, paper, gunpowder, and the modern harness. Paper undoubtedly originated in the Far East. Despite the linguistic efforts made by André-Georges Haudricourt, harnessing still remains a rather vague subject. The idea that gunpowder was a Chinese contribution has recently been challenged, and it is very unlikely that the same phenomenon occurred for the techniques of smelting, unthinkable as it was for the water mill. We may suggest the parallel development of concomitant inventions, which were the result of identical technical reasoning: the water mill is one such example. All things considered, if we look at the mid 13th century, we would notice a brilliant technical civilization: should this exclude the remarkable cathedrals, irrigation systems of Spain, and the Scandinavian ships which crossed the Atlantic?

Techniques and Technology

One of the features of technically advanced civilizations is the development of their own systems of technology. It is extremely instructive to define the domains covered by all these ancient treatises and, if the case arises, their chronological sequence. They exactly define the limits of techniques which had arrived at a certain maturity and which were consequently suitable material for dogmatic accounts. It is quite natural that China should have arrived at this stage. But here again, we should define chronological limits. Although a military treatise existed from the beginning of the 5th century BC, we may observe that most of these written works, in fact almost all, date from the 10th and 11th centuries, with only a few dating before this (one from the 6th century). This should be clarified in comparison to Western technical literature which was particularly prolific during the period from the 4th to the 1st centuries BC.

Some of these works testify that Chinese technology, at least certain Chinese techniques were not totally empirical, but that technical knowledge was organized in some areas. The great clock tower of Kaifeng, built by Su Song in 1088, was preceeded by a specific theoretical treatise, developed by his pupil, Han Gong Lian, who studied gear systems and the basic principles of general mechanics. This treatise should be compared to some of the very early works of the Alexandrian school. Such a comparison would enable us to ask a question which has rarely been considered: which techniques did China borrow from the Western world?

Let us look at this type of literature in more detail, and attempt to classify such works according to their subject matter.

I – Agricultural literature was certainly very important. Six treatises have been attributed to the Han epoch (in other words approximately two centuries before and two centuries after our era). The first was written by Fang Cheng-tche in the 1st century AD but unfortunately these have all disappeared today. The first complete treatise available for our examination was written by Kia Sseu-Sie (533-546), and we have another by Tchou-Fou (1101-1103). Both works reflect a traditional farming system, the rules of which were no doubt established and laid down many years before. Some more specialized treatises were also compiled, for example works written on horticulture during the Song epoch (960-1279), eight treatises on veterinary art under the Souei dynasty (581-617), and the great tract on agricultural mechanization by Wang Chen, the preface of which dates back to 1313. The same dogmatic spirit can be found in all these works, together with the same ensemble of formulae naturally transposed into a slightly different environment, which may also be observed in the works of Latin agronomists. The spirit in which these works were written is identical.

II – Literature on the subject of building is much less prolific. There is, however, the treatise on architecture by Li Jie (circa 1100), and a manual on carpentry by Yu Hao (10th century). The second work represents something new, as did the tract on farming machines mentioned above. It is significant that new areas were explored relatively late in comparison to Western technical literature.

III – The first large compendium of military techniques was compiled by Zeng Gong Liang aound 500 BC. Literature on the subject of military techniques does not appear to have undergone a very great development in China, although this was also true of Europe, apart perhaps from works written on poliorcetics and treatises on war engines. In this field, China seems to have been behind the techniques developed by Western Europe.

IV – On the subject of mechanics, Chinese literature was on the whole well behind that of the West. We have already mentioned, however, an 11th century treatise on gear systems. Also, around 1030 Yan Su compiled a work on clock making which was far more advanced than the many treatises on water-clocks of Classical Antiquity. China does actually seem to have been ahead of Europe in this particular field, as it was only in the 14th century that an equivalent was produced in Europe with a tract by Dondi.

To the above list we may add a treatise on sugar by Weng Chao (960-1026), a work on wine by Sou Che (1036-1101), and a few descriptions on silk making. This virtually completes the survey of Chinese technology.

Several remarks may be made ragarding this very brief survey of Chinese technical literature. The first point concerns the extent of this technology. It does not appear to differ radically from what we have seen in Greece and Rome, or elsewhere in the Middle Ages in Europe. Many large collections dealt with agriculture and the military arts, and some important treatises concentrated on what is known today as the "latest technology." We have said before that we should think of this from a chronological viewpoint, but we unfortunately lack comprehensive editions in European languages, by which we could comparé these two traditions and examine the Arab tradition which has long been suspected as having provided an intermediary between the Far East and Europe. The excellent drawings which are usually reproduced are extracts from the famous work on *The Exploitation of Works of Nature*, written in 1637, at a time when

Western techniques were already more advanced than those of China.

Exploitation of Natural Resources

Chinese agriculture as a whole seems to have known all the stages of other agricultural methods. The processes of technical development are virtually the same everywhere. Naturally variations of climatic and hydrologic regimes, and vegetable and biological differences play a part. It is upon these differences that we must concentrate, as the rest concerns a common technical system.

Like everywhere else, crop types tended to multiply, and this growth seems to have been very rapid. Under the Zhou dynasty, around the beginning of the first millennium BC, the main cereals of the country were already extensively cultivated: panic grass, different varieties of rice, and barley. Later, spikelet millet, hemp, sesame, and corn were grown. But rice very quickly became the most important crop. From the beginning of the Christian era to the first half of the first millennium, so from the Han to the Song dynasty, new plants appeared which came from either the steppes or tropical zones. Vine is thought to have been introduced at the end of the 6th or the beginning of the 7th century. At the end of the Tang dynasty, that is to say around the year 1000 AD, water rice was introduced, which was to spread from Southern to Northern China. Kapok was also extensively cultivated throughout the country between the 10th and 13th centuries.

If technical treatises are to be believed, then around the year 1000, Chinese agriculture was already based on a scientific method, as selection and hybridizations were common practices, and already seven types of rice were known.

Fruit trees were cultivated very early on; first of all the peach, apricot, guinee, and the jujube were introduced, and later the plum and pear tree. Grafting was often practised, and market gardening was equally widespread. At the beginning of the 12th century, there was a great variety of leguminous plants including the bean and the pea, and some specialized plants were to make their appearance. Egyptian experts may have arrived to improve sugar cane producing methods. Here again, Chinese agronomic science had almost nothing to envy in the Arab gardens of Southern Spain. At the beginning of the 12th century there were eleven varieties of apricot, eighteen types of bean, and eight species of pea.

The ancient practice of horticulture saw major development between the 10th and 13th centuries. The peony, chrysanthemum, lotus, orchid, day lily, lily, magnolia, and begonia all featured in ornamental gardens. The peony and the narcissus were borrowed from the Arabs.

Farming tools and cultivation methods were certainly developed very quickly. The Chinese had a good knowledge of soils, and the standard practice of using manure is attested in the 13th century. But China was not so fortunate as Western Europe which had an abundance of manure. Chinese farmers consequently ploughed in nitrate fertilizers instead of using manure for crop cultivation. They also made a type of compost consisting of lime, cob from old walls, and mud from ditches. We will later examine how drainage and irrigation systems were extensively employed.

Farming tools undoubtedly profited from a highly developed metallurgy. The swing plough evolved during the period between the Han and the Song dynasties, in other words in the first half of the first millennium AD. The sole-ard was used, different to the ard of Classical Antiquity, but similar to the one which spread throughout the northern border of the Mediterranean during the same period. Other instruments differed little to those used by other civilizations.

Agricultural treatises, like governmental orders issued during epochs when agriculure was controlled by the state, established strict programmes for any farm work. A treatise dating from around the 10th century proves that crop rotation was practised, and that the alternation of millet and beans was advocated.

Thus Chinese farming techniques showed not only some degree of evolution but were also at a fairly high level of development. But these techniques were barely more advanced than those of the Western world at the same period, except perhaps for those of horticulture. This is also true of the principles involved in other techniques for the exploitation of nature. Neither animal domestication, hunting nor fishing demonstrates any fundamental differences.

Mining techniques appear to have been fairly well developed, but truth to tell we lack precise data and subsequently any comprehensive chronology for these techniques. We will later examine the dates proposed for the appearance of some of the various metals. We do know that the Chinese quickly exploited salt deposits, as the industry goes back to the 1st century BC. A work dating from the year 1334 AD provides more precise information on this subject. The same techniques were used by both China and the West: water was sent into the mine, and then the brine was brought up and reduced in boilers. In 1080, a well was sunk to a depth of 3000 feet. The Chinese were fortunate in having a choice material available for such a task, as bamboo was abundant and could be made into considerable quantities of very long pipes, by fitting one bamboo pole into another. We know very little, however, about other Chinese methods of mining coal and oil.

The vitality of a technical system and the power of an economy are often measured in relation to the quantity of energy available within a country. Energy problems are therefore primary considerations. On the one hand, there is natural energy, and China and Western Europe were well provided in this domain, far more so than Mediterranean countries, and on the other hand, there are energy converters.

Like everywhere else, the first type of energy to be exploited was animal energy. The wheel was certainly borrowed from Western civilizations. We know that harnessing methods were very important. We can say very little on the subject of yokes, but the horse involves more innovations. The saddle, stirrup, and bit were probably invented in Central Asia, according to some authorities as early as the 3rd century BC. Haudricourt has shown that a very clear separation exists on either side of a line drawn from Central Vietnam to the Baltic. Below this line was the chariot with a shaft attachment and pulled by two animals, and above is the chariot pulled by a single animal in shafts. Harnessing methods could not have been the same in both cases. Representations dating from the Han dynasty show an already advanced harnessing system. The two shafts were curved and joined together at one end, where a small yoke was attached which formed the main support. Traction was maintained by a breast strap, similar to our breast harness. Later, the yoke disappeared and the shafts were shortened, whereupon the support became a backing which was placed on the horse's back. The breast harness and the breech band remained fixed to the shafts, rather than to the chariot itself. This evolution more or less ended around the 8th and 9th centuries AD. The shoulder collar, fixed to the vehicle by traces, the swingle-tree, and the horse shoe were later imports of the 15th century, so China was well behind the West in this field.

Dates quoted for the water mill do not match exactly. We have mentioned one mill builder who would have lived around the 4th century BC. But on the other hand, the first reliable evidence of a water mill dates in the year 30 BC which coincides exactly with its appearance in the Western basin of the Mediterranean (fig 1). A text by Du Shi, prefect of Nanyang, writes that in the year 31 AD water mills were used to power the bellows in metallurgic furnaces. We have a 14th century image of an iron tilt hammer (the oldest Western example of this instrument dates from the end of the 12th century). Windmills were used for irrigation and drainage in Asia and China in around the 10th century (the 7th or 8th century in Iran).

Any developed system of mechanization required not only something to produce energy, but also basic mechanisms to transmit or reduce and transform movements. In this domain, China seems to have been greatly outdistanced by Western Europe and we will return to this fact later. Unfortunately documentation is very poor for this subject. The school of Alexandria with its already advanced theories of camshafts and cogwheels, most certainly outshone Chinese techniques. It has been suggested that cogwheels were known during the Han dynasty, in other words in the period of four centuries before the Christian era, but texts regarding this are not really very precise. China does not appear either to have developed the connecting rod and crank system.

Preparation of Raw Materials

Once again we lack precise information regarding this subject. We should point out, however, the dates at which new techniques were introduced and what improvements were made to them. The Chinese appear to have been the initiators of several of these techniques in this field, notably in the domain of metallurgy.

Below are the most important dates in metallurgy:Copper was introduced in approximately 3500 BC (circa 5000 BC in Turkey, Syria, Iraq and Iran).Bronze circa 1400 BC (the third millennium BC in Turkey, Greece and the Balkans).Iron in the 6th century BC (15th century BC in Anatolia).

It is quite obvious, as we have already indicated, that technical principles were developed on the continent in a region which lay approximately in the oriental basin of the Mediterranean, in the Near

Figure 1. — Waterwheel. (From J. Needham, *Science and Civilisation in China*, Cambridge University Press, 1954-1967)

East, from which these techniques could spread both to the east and to the west. Due to the absence of Carbon 14 dating, the exact date of Neolithic beginnings remains very inaccurate: perhaps the third millennium BC?

It was in Northern China that bronze first appeared, which was a higly developed metallurgy from the beginning. Chinese bronze casters were extremely skilled, as has been stated in many works which have been handed down to us. Over the years the proportion of copper used was replaced in favour of tin. In Southern China and in the bordering countries, bronze was only developed in the first millennium BC. The spread of metallurgic techniques was thus particularly slow.

The introduction of iron was also relatively late in comparison to Near Eastern civilizations. The imperial code of the T'sin state is reproduced on an iron cooking pot in 513 BC. The most remarkable phenomenon is the appearance of cast iron, which some authorities may have exaggerated by dating its appearance in the 4th century BC. Cast iron could only have been made by using relatively powerful bellows, and was only produced in Western Europe with the aid of hydraulic bellows. In fact, the latter were only developed in China at the beginning of the Christian era. Contemporary historians only use vague and often absurd terms, and we only possess a few objects made of cast iron which are difficult to date. The oldest cast iron pagoda in the province of Hubei dates from 1061 AD. During the period when kingdoms were at war, in other words from the 5th to the 3rd centuries BC, iron moulds were used to cast hoes and axes. It is difficult to believe that the metal cast could have been iron, as certain authors have maintained. Should one attribute the discovery of cast iron to the use of coal with a higher calorific value than that of charcoal? But coke would also have had to had been discovered for this process. And should the discovery of cylinders and pistons be attributed to the development of the bellows? The mystery remains virtually impenetrable (fig. 2).

The problem of the introduction of steel is just as difficult to resolve. It is possible that steel was produced by "cofusion," according to a method which Reaumur clearly defined, despite the absence of precise scientific knowledge in the beginning of the 18th century. The process mainly involved the "cofusion" of iron and cast iron. The art of welding soft steels was probably known during the 3rd century AD, and would

Figure 2. — Metallurgical blowing-engine worked by water-power. (From J. Needham)

have lead to the development of damask steel used in Japanese sabres in the 11th century.

Other metals quickly followed suit; antimony at the end of the 3rd century AD, zinc at the beginning of the Christian era, and brass around the 5th century AD. White copper, which appears to have been nickel silver, was mentioned by a 4th century author. Lead by-products were quickly used for make-up. Some authorities have accepted the hypothesis that this by-product was used in Rome at almost the same date, causing lead poisoning which might have contributed to the demographic decline of the Roman Empire.

Fire-based arts certainly made progress possible in metallurgic techniques. Glass has been found on prehistoric sites in China, Japan and Korea. Confusion and the absence of exact details often makes us doubt suggested theories and the reality of the facts cited. It may be that during much later periods, glass was imported from the West. Glasses, the existence of which is attested by a 14th century text, had an Arabic name indicating their origin.

A strict chronology of Chinese pottery is required. There is no doubt that ceramics were developed during the Neolithic period, but at exactly which epoch did this happen? What form, moreover, did the evolution of this technique take? Slender glazed vases have been attributed to the 15th century BC. During the Zhou dynasty, of the first half of the first millennnium BC, the Chinese used powerful kilns by which they could obtain stoneware with an overall glazed paste. Under the Han dynasty, potters could obtain a temperature of 130-0æoæ C. Proto-porcelains were developed before 500 AD with the introduction of jade-green ceramics. This technique was very rapidly diversified, and real porcelain appeared in the form of almost translucent glazed stoneware, requiring temperatures of 1450°C. Ceramics reached their peak with the Ming vases during the period of the same name at the end of the 14th century.

The technique of lacquering, using the milky sap from *Rhus vernicifera*, would have come from Central and Southern Asia, during the Chang dynasty. This quickly became a highly developed technique. In fact the scum or lather of the lacquer was used, often coloured with mineral dyes. This medium required a long preparation before it could be applied. For obvious reasons this technique was limited to the Far East.

Paper was undoubtedly one of the major discoveries of China. This fact is unquestionable despite all the obscurities which surround this invention. From the 3rd century BC, paper was produced on a small scale in Asia, using extremely diversified materials. Paper made from the sap of the mulbery tree dates from the 2nd century BC. Silk, bamboo, flax, straw or corn were employed with equal success. This method of paper making survived into the Middle Ages, owing to Chinese workers who were taken prisonner by the Irano-Arabs during the battle of Talas (751).

We are less certain whether China should be credited with the invention of gunpowder. According to some authorities, China appears to have developed a mixture similar to gunpowder from the 1st century BC, but it was only much later that it was used in military arts. Other sources are more precise and date the earliest use of gunpowder in the year AD 85. The Chinese were most probably the first to use sulphur, and colour the fire it produced with the aid of metal oxides, thus creating the first fireworks. We even know the formula they used: 75.7% of nitre, 14.4% of coal, and 9.9% of sulphur. It is unlikely that they used it for "fire-arms."

Other preparations were more specific. The sugar refining process most certainly came from the West, around the end of the 6th or the beginning of the 7th century. Every civilization, at some point in their history, developed a method of alcoholic fermentation. Cereal fermentation was practised in China from the Zhou dynasty onwards. Treatises written in the 11th and 12th centuries provide ample information on these techniques which differ very little from those of Western Europe. Many forms of alcohol were based on cereals (millet and rice). Wine made from grapes was only developed a short time before the beginning of the Christian era.

The Crafts

China naturally practised all the crafts which may be observed in other civilizations, some with great perfection which still compel our admiration today.

The textile industry took off in a big way. Of the textiles used, wool was never very important apart from in the north west of the country. Felt was introduced in Central Asia and it would have arrived in China in the 4th century BC. Cotton only gradually spread towards the East from its region of origin, but the Chinese first bought it from India or Java in the 4th century AD, and from Turkistan at the beginning of the 8th century AD. From the latter came hemp and flax. As we have seen the arrival of kapok was much later.

Silk, which established the fame of the Chinese textile industry, was obtained from the Eurasian region, and from either wild or domesticated lepidopterans. The Chinese may have been the first to make silk material as they possessed the best silk producing insect, the *Bombyx Mori*. We have examples of silk material dating from the Jin dynasty. Hellenistic Greece particularly appreciated these Chinese products and tried to implant mulberry trees and silk worms in the Mediterranean basin (most probably these imports came from the oases of Central Asia).

We know very little about spinning techniques (fig. 3). The spinning wheel, which in all likelihood was powered by hand, was known to the Chinese at the beginning of the Christian era. The first pictorial representation available dates from 1210. Similarly, a machine for unwinding silk existed in 1090. The cocoons were placed in a bath of warm water, the silk came out through small guiding rings and was then wound smoothly onto a large bobin by a backwards and forwards motion (fig. 4).

We know more about the types of weaving loom used. The *Book of agriculture and weaving*, compiled in around 1210, provides images and descriptions of these, particularly for silk weaving. The superiority of Chinese weaving looms has already been mentioned. Two types of loom were used which left the weaver's arms free to work: the traction loom which may have originated in China and Europe, and the pedal loom which is attributed to the Chinese. Representations of these appear on 12th and 13th century manuscripts.

Upon close examination, these looms display many similarities to European looms of the same period. It has been said that damask silk cloth existed from the time of the Jin dynasty, and during the 2nd century BC, four heddle looms were built, which henceforth enabled the weaving of silk brocade. Egyptian cloth, which was also very high in quality, should be compared to Byzantine cloth which stemmed from Egyptian techniques, and strict chronologies could be proposed in order to determine the extent of some of the contributions made: these contributions may have been reciprocal.

Printing was one of the major discoveries attributed to the Far East. It seems very difficult to accept, as some have done, the theory that printing was derived from the seal, an object which was long known by other civilizations which did not develop printing. We should also try to establish a strict order of the stages contributing to the development of this process, which is illustrated by texts and objects.

Xylography was introduced from 770 AD onwards for the reproduction of Buddhist texts on paper rolls. In 1907, Sir Aurel Stein discovered the *Diamond Sûtra*, printed in 868 on a paper scroll, in the caves of Tan-Huang. This must therefore be the first example of a printed "book." Some recent authors, however, deny that xylography could have been an ancestor of typography. Moreover, raised stamps, used for the capital letters of manuscripts, had long been known in the West.

Whatever the cause, popular Buddhist works spread to the west and the east from the 9th century onwards. Was there an industrial stage for the production of relatively short texts rather than images? This question should be clarified. The translation in 982 of the *Tripitaka*, a Buddhist canon, has been cited, which would have required 30,000 plates.

Figure 3. — Silk spinning machine powered by a waterwheel (1313). (From J. Needham)

Figure 4. — Machine for unwinding silk (1090). (From J. Needham)

The invention of typography has been attributed to Pi Cheng (1041-1048) who must have therefore imagined the first mobile characters, which were sculpted separately out of viscous clay, hardened in fire and then assembled with a medium composed of a mixture of resin, wax and paper ash. As a matter of fact, this material does not seem to have been very strong. Another question arises to which there seems to be no answer: was there a press? Pi Cheng's invention does not appear to have had any repercussions as it was so quickly forgotten.

Typography was therefore re-invented and improved by Wang Tchen, magistrate of Tsing-tü and author of a classical treatise on agriculture. In this work he relates the invention and perfection of typography. He found the handling of xylographes tedious, so he engraved the characters on mobile blocks, and facilitated their manipulation by classifying them in files, arranged on a surface which revolved on a vertical axis. The characters seem to have been made in wood, but we know nothing about the presses or the inks used. To make an actual impression, the characters were merely rubbed on the surface of the paper. As the wood would have split under the weight and action of the press, the main benefit of this form of printing lay in the fact that characters could be re-used for a greater length of time.

Metal characters (lead or copper) have been attributed to Korea, and their introduction is precisely dated for the year 1403. Korean characters would have arrived in China at the end of the 15th century, but they only enjoyed a modest success as Chinese paper was very fine and could not take metal characters, particularly if both sides of the paper were printed. The problem of ink was equally important: in 1398 Chen Ki-souen wrote a treatise on this subject. The inks which were employed were therefore unsuitable for use with metal characters.

In fact, true printing was really invented in Europe around the middle of the 15th century. It consisted of metal type characters made from a special alloy, a method for assembling these characters, a suitable ink, and a press. There seems to have been no influence whatsoever from the Far East on the inventions attributed to Gütenberg. Only the printing techniques of China and Korea, which were only to be fully developed shortly before the middle of the 15th century, remain to be defined.

Geography

For much of its history, China was a unified country, and the Chinese Empire subsequently represented a geographical territory which only the Late Roman Empire rivalled in size. And that is naturally without taking into consideration outside exchanges. All large territories require many specialized and coherent techniques.

We have already discussed the problems of harnessing on the subject of land transports. The wheel was a Western invention. Vehicles were two-wheeled, and China did not use, at least up until very recently, the mobile front axle unit. Porterage was certainly very widespread, as was also the case in Western Europe during the Middle Ages. The wheel was as old an invention in China as in the West. But like the West, a logical road network could only result from a centralized administration. Standardized, imperial roads date from the Han dynasty. Although the Chinese road may have been a little earlier than the Roman road, which was created during the expansion of the Roman Empire, it was nevertheless very similar to its Latin counterpart, in that it was built as straight as possible, with organized stopping places and was a result of administrative and military concerns. In fact, it is especially this sort of road which may be seen in North China.

Certain aspects of the Chinese river network often led to highly original solutions. The flooding of large rivers and the abundance of swamps necessitated a range and scale of work which Mediterranean countries were not to develop. This was particularly so in the basin of the Yellow River where dykes and diversions were necessary in order to avoid disasters. From the end of the 4th century BC, Si Men-po ordered that twelve canals be dug the length of the Chang River, which flowed from the Yellow river, in order to reduce the effects of low water. The same was executed at the end of the 3rd century AD, for the Min River, flowing from the Blue River. This work also provided irrigation for important zones. It was also during the 4th century BC that canalizations united the Blue River with the Hai River, the beginning of which was to become the Great Canal under the Souei dynasty. The Great Canal network, systematization of various works undertaken during most periods, was to be completed in the 13th and 14th centuries.

Bridge building in China appears to have been fairly successful at a relatively early age. Wooden bridges, and flexible rattan or bamboo suspension bridges were very old inventions. During the 7th century, bridges with segmented arches and bridges suspended by iron chains were built. Some authorities maintain that the latter type were developed in the 11th century. Stone bridges also existed. The bridge at Lou kou kiao (end of the 12th century) was 350 paces long, 18 wide and had 11 arches.

We know relatively little about boat building in Ancient China (fig. 5). In fact, techniques differed greatly from north to south, and from the rivers to the sea. There were various types of sampans and junks and these were continually modified (fig. 6). The large sea-going junk is thought to have existed in the 9th century; if not at this date, then it was certainly in use in the 11th century. The Chinese fleet continued to develop, but it is difficult to trace the progression of these changes. The few images which are available tend to prove that this evolution was virtually identical to that of Western ships. The stern rudder only seems to have been developed by the Chinese well after it had been

Figure 5. — Three mast junk (engraving from 1757). (From J. Needham)

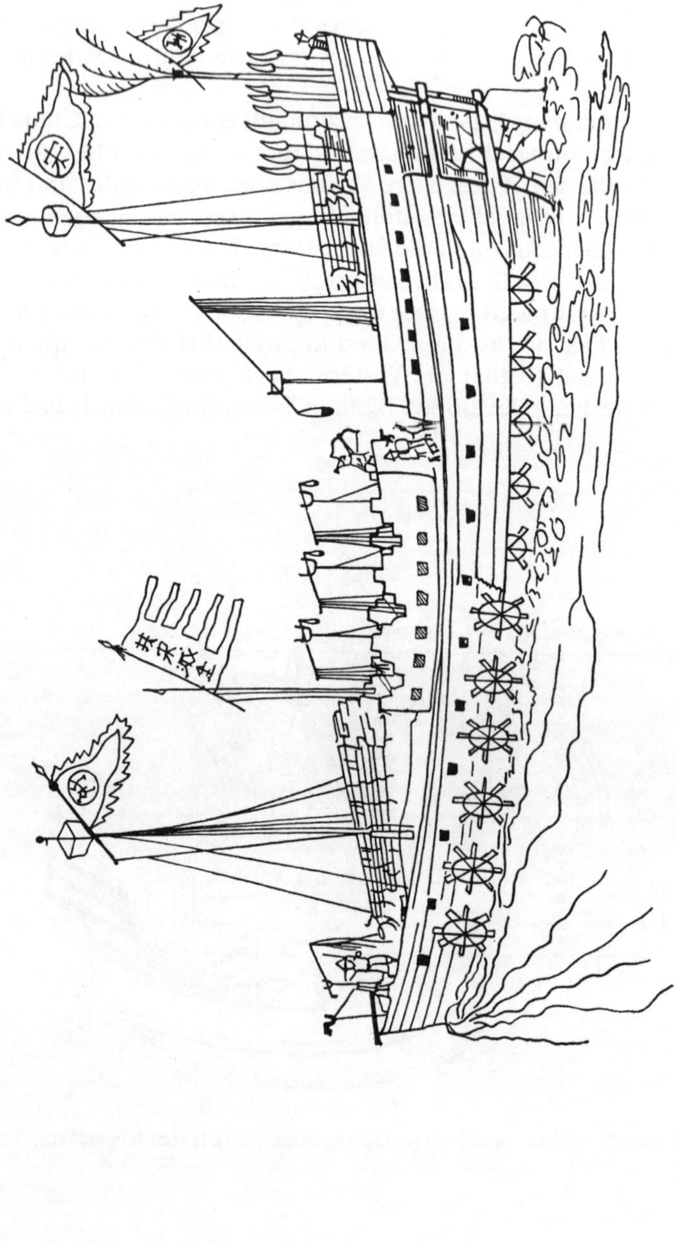

Figure 6. — Reconstruction of a multiple paddle-wheel warship (1130). (From J. Needham)

discovered in Western Europe, although an example of this rudder type has been cited for the 1st century BC.

Let us remind ourselves of the kite which was, however, produced by many civilizations at completely different epochs.

Very little has been written on Chinese towns and cities, or on fortifications which were probably built very quickly around the towns. Only the story of the Great Wall is known, which cannot be compared to the Roman *limes*. In approximately 220 BC, old discontinued works were linked together to form a rampart of earth, about nine metres high, paved on either side with stones or bricks, and watchtowers were built at regular intervals along the length of the wall.

Blockage

We have arrived at the main topic of this discussion, which may be summarized under two headings; the advance of Chinese techniques over Western methods, and the blockage of the Chinese technical system.

We indicated our position with regard to the first point at the beginning of this chapter. We still lack any exact chronological order of the appearance of Chinese techniques. We have been able to note concordances in many fields between West European and Chinese technical developments, a phenomenon which may be explained by their concomitant evolutions. Perhaps the practice of borrowing Western techniques by China has not been emphasized enough. Metallurgy, glass making and ceramics all originated from Western techniques, whereupon improvements were made within each technical zone. If we look at dates common to both China and Europe, it becomes evident that certain techniques were more advanced in one country, and that reciprocal influences within the other were definitively less widespread than we might have thought. It is more difficult to accept certain opinions in other domains, a phenomenon which we will illustrate with just two examples.

Taking as first problem the measurement of time and distances, it would seem that the West was not behind the East (fig. 7). The water-clock was developed in both areas at almost the same time, although it may possibly have appeared slightly earlier in the West. The Greeks of Alexandria had resolved the problem of the unequal length of days, whereas the Chinese had not come to any satisfactory

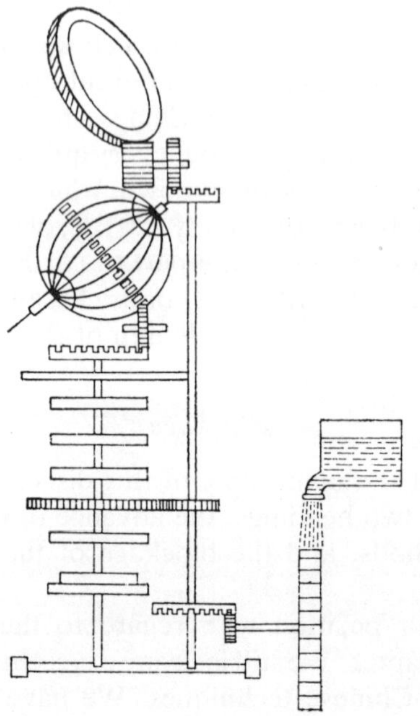

Figure 7. — Astronomical clock. (From J. Needham)

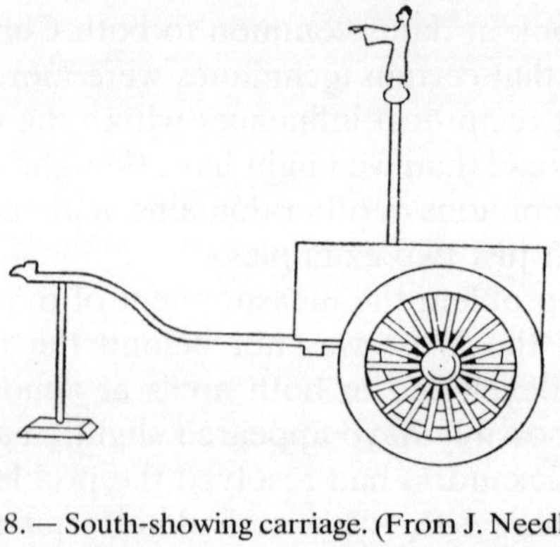

Figure 8. — South-showing carriage. (From J. Needham)

conclusion. It is highly unlikely that the Chinese developed the mechanical clock which the West had put into practice from the 14th century onwards.

As for the problem of measuring distances, in 1027, Lou Taolong presented the Emperor Jen Tsong with a chariot for measuring distances, while the Greeks of Alexandria had written of a similar method in the 3rd or 2nd century BC.

The second example which is often quoted is the compass. Let us ignore for the moment the "south-pointing carriage" which was a purely mechanical device (fig. 8). The "south-pointing" needle is mentioned for the first time in the 11th century, by Chen Koua. There is certainly much evidence which testifies to a knowledge of the existence of the magnet from the 3rd century BC onwards. But the problem of the Earth's magnetism emerged much later. The use of the compass in navigation techniques dates from the 12th century BC. It remains necessary to find out the exact use of the compass for a type of navigation which was on the whole coastal. Pierre de Maricourt was one of the first in the West to study the effects of magnetism, at virtually the same time as the Chinese began to question its uses.

From this point of view, it is important that specialized scientists studying these different questions unite, in order to establish an assessement which the lone researcher has great difficulty in achieving. In this case, the work to be undertaken notably exceeds the simple problem of priorities: in a word, it is a matter of establishing how certain techniques were spread abroad, and on the other hand, if virtually chronologically concomitant parallel developments were possible.

The second problem is that of a blockage of technical progress. We have seen that this was not peculiar to China alone at the end of the Medieval period, but was also experienced by the Eygptians, the Greeks, and many other civilizations. There are several sides to this question. Blockages may have arisen from internal factors, in other words they were produced by purely technical problems. This is conceivable only insofar as global technical systems are envisaged, as traditional technical structures may block changes in the system even if advance is visible in other sectors. Equally, the coherence of a technical system with other economic, social, and scientific systems must be taken into account. The problem must therefore be considered on a large scale, as indeed it has been by many specialists.

In fact, the reasons for this blockage have been sought in the domains of intellectual pursuit and society, in the same way as the blockage of the Greek technical system. If we say that the great inventions, those which marked China's advance over the West, such as printing, gunpowder, and the compass, only had a slight repercussion in the Chinese world, this manifests exactly the phenomenon we have just pointed out. Techniques which are today called the "leading" or the "latest technology" can only be incorporated into a global technical system if a certain coherence exists between those techniques. The delay or slow progress of other technical structures will naturally block the introduction of more advanced inventions. In the West, printing resulted from a prodigious intellectual movement, the compass only became an important instrument with the development of navigation techniques and the crossing of the Atlantic, and gunpowder was introduced with the development of a sizeable artillery, which presupposes hundreds of other techniques (besides, the metal cannon was a virtually contemporary development in the Far East and the West, at the end of the 13th and the beginning of the 14th century; fig. 9).

More pertinent reasons are required to explain the blockage of Chinese techniques while the West, on the contrary, was initiating a new technical revolution at almost the same period. Apart from reasons specific to technology itself, other circumstances must be considered.

Techniques always apply, in one way or another, to science. Needham has ably emphasized the stagnation of Chinese scientific thought during a time when the West was in the process of a scientific revolution which was to eventually culminate in the development of modern science. The product of both technicians and scientists, this modern science was to provide technology with new and irreplaceable elements. Or, to put it a better way, the union between technology and science was to provide the two domains, the scientific and the technological, with tremendous and enriching contributions. Nothing of this kind happened in China, where to some extent science gets bogged down in out-moded traditionalism. Here we must stress the blockage of thought, the significance of which is obvious.

The determining role, in every respect, of neo-Confucianism has been pointed out. The appearance of an ethic capable of resolving all nature of problems, scientific and technical, as well as economic,

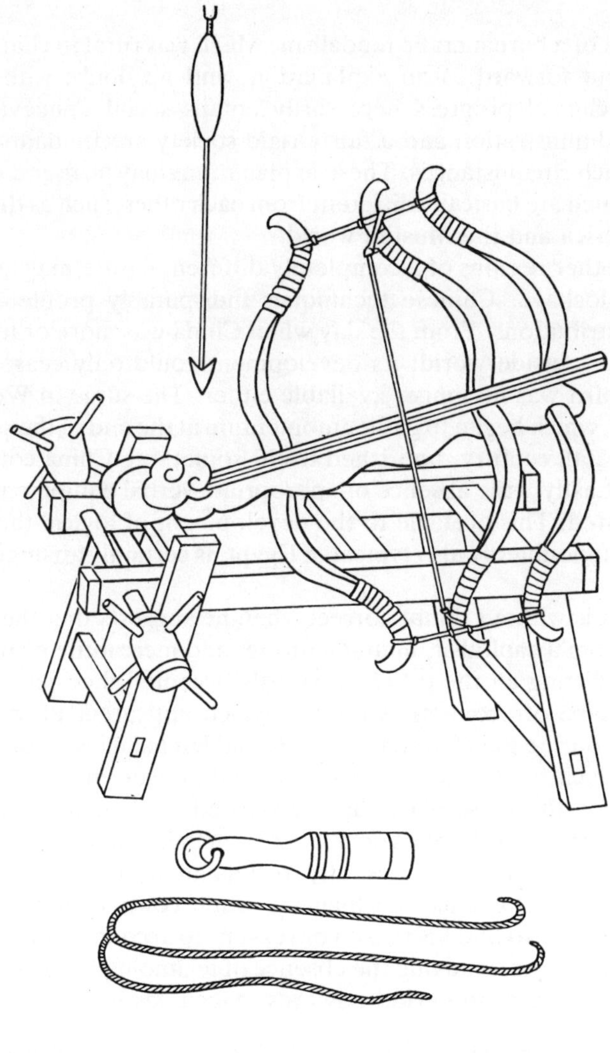

Figure 9. — Principle of Chinese crossbow, drawn to artillery's scale (1044). (The greek mechanics of Alexandria were more advanced.) (From J. Needham)

political and social questions, certainly contributed to the stagnation of the Chinese technical system. This agrees in part with what we have just stated.

The birth of a bureaucratic feudalism, which was rural in character, has been put forward as an explanation, and no doubt with good reason. Technical progress necessarily entails social upheavals. A powerful administration and a fairly rigid society are fundamentally hostile in such circumstances. These explanations may be found within societies which are basically different from each other, such as those of South America and the Muslim world.

Finally, other reasons of a completely different nature may explain technical blockage. Chinese techniques indisputably profited from outside contributions. From the day when China was more or less cut off from the outside world, its development could only cease, as a technical spirit was no longer available either. The surge in Western technology, which began to gather momentum at the end of the second half of the 15th century, nourished itself, something China could no longer do. Lastly, "the absence of appropriate verbal equipment" has been indicated. This obstacle to the development of logical thought, which was undoubtedly also typical of Egypt, is certainly an important factor.

Needham is without doubt correct when he suggests that the West benefited from a capitalist, manufacturing, and mercantile economy, thus contradicting all the other previously accepted theories. It was really this type of economic structure which appeared, after many beginnings, during the 15th century, and enabled technical progress in the West which the Chinese civilization could never enjoy.

In fact, for the most part, these proposed explanations are not wholly supported. Authorities do not agree on the existence of some of the phenomena, and this naturally results in a sort of confused catalogue, some elements of which are often contradictory. Every author would like to find his reason or reasons to account for technical blockage. Needham, studying the absence of technological innovation in China from a certain period onwards, concludes in another text:

> Tradesmen were always kept in their place and were not permitted to hold responsible positions within the State. Obviously there were guilds, but these were not as important as those in Europe. Perhaps here we can put our finger on the main reason to explain why the Chinese civilization could not create a modern technology, as it is universally accepted that the development of technology in Europe was closely linked to the growth in the power of the merchant class.

Rostow criticizes this view, and would seem to be quite justified in doing so. He maintains that the merchant class in China was more powerful and less despised than has been previously thought. In the last analysis, guilds were hardly, if at all, inclined towards technical innovation.

Mark Elvin rejects other explanations, such as inadequate capital, limited markets, political obstacles, and the incapacity of the Chinese to create large scale and long term enterprise. He suggests that in fact the Chinese had arrived at a stage where sudden and significant progress had to be made. He also adds to the above the factors of the low income per head caused by demographic pressure, the increased cost of raw materials, and the lack of internal cotton resources. But here again, Rostow objects to this view on the grounds that this impasse was also true of the West, and stresses "the scientific, philosophical, inventive and innovative ferment which was distinctive of Europe during the same period". And he quotes Elvin:

> Almost every element usually regarded by historians as a major contributory cause to the industrial revolution in north-western Europe was also present in China (...) Only the Galilean-Newtonian science was missing; but in the short run this was not important. Had the Chinese possessed, or developed, the 17th European mania for tinkering and improving, they could easily have made an efficient spinning machine out of the primitive model described by Wang Chen. A steam engine would have been more difficult; but it should not have posed insuperable difficulties to a people who had been building double-acting piston flame-throwers in the Sung dynasty. The crucial point is that nobody tried. In most fields, agriculture being the chief exception, Chinese technology stopped progressing well before the point at which a lack of basic scientific knowledge had become a serious obstacle.

Rostow insists on the absence of a "scientific revolution" to explain this technical blockage.

The debate remains open to discussion.

PRE-COLUMBIAN AMERICAN TECHNIQUES

The civilizations of pre-Columbian America present, more than any other, the image of a technical blockage at a fairly primitive level, which is astonishing to we Europeans. The existence today of an overdeveloped America, the recollection of a frequently marvellous art, and an often incredible form of architecture, have all helped to conceal some aspects of the problem. We are less surprized that some African, Melanesian, and Oceanian civilizations have remained at a more primitive stage. It would therefore seem illogical to conclude that more advanced pre-Columbian techniques were suddenly halted in their development. Many explanations for this have been given, none of which are really adequate, such as the isolation of one area in relation to other continents (Europe, Asia, and Africa which communicate easily between each other), the division of internal civilizations, and societies which are too rigidly structured. All, or virtually all, these explanations have been put forward. We are forced to fall back on rather vague formulae, which in truth only shift the emphasis of the problem: "the American man was not a technician by birth (...) Inventive stagnation, coupled with technical inertia which did not exclude manual skill."

We are undoubtedly obsessed by the idea of natural, logical and inescapable technical progress. From this obsession stems our astonishment at certain ignorances: the explanation, such at it is, therefore obliges us to look for exogeneous causes. We have already seen this in the techniques of Classical Greece. We must also add the many difficulties faced by the historian in the consideration of pre-Columbian techniques. Most peoples of pre-Columbian America did not have any system of writing, which is obviously an important technique for communication: we have no written evidence of the form it might have taken, and the only indications which we do possess date after the conquest, resulting either from natives but more often from conquerors who were careful to record all they had seen or noticed. Technical material made, as we will see, almost totally from organic material has almost completely disappeared. And this lack of detailed and comprehensive documentation relates equally to the whole technical environment, in particular to the demography and the economy, as well as science, which are so closely linked to technology. All this subsequently leaves the field open to often unwarranted and excessive suppositions.

The main point which should be remembered, and contemporary historians have all agreed on this, is constituted by a certain number of "missing" techniques essential for the development of advanced technical systems. This is to some extent the key of the blockage.

Apart from writing, which we have just mentioned, there were three large technical sectors which did not evolve in any way for some time.

I – It is not strictly true to say that pre-Columbian America did not domesticate animals. Some domestication did take place in various regions and for different purposes. The Aztecs domesticated both the turkey and the dog for food, and the Incas domesticated several types of camelidae, such as the alpaca and the vicuna, which provided the textile industry with wool from its thick fleece. It is true that the llama and the guanaco were poor carriers (approximately 30 kilogrammes each), but over short distances they could serve as beasts of burden. All these animals also provided meat, their skins could be used to make sandals, straps, and bags, and their bones were used to make needles and various other tools. Finally, their excrement was collected and used for fuel.

It is nevertheless essential to note that in many regions the absence of animal domestication led to important consequences. From the standpoint of providing food, there was no meat, milk or animal fats, and from the industrial standpoint, there was no leather, skins, or wool. It was probably in the domain of energy that these repercussions were the most serious, as there could be no porterage or traction, either for transport, agriculture, or power to drive grindstones or mill wheels, and of course no manure for cultivation.

Some questions must be asked which are rather awkward to answer. The domestication of animals, and more widely animal rearing, required a certain number of necessary conditions. The first requirement was the existence of animals which could be domesticated, and we know that America possessed neither the horse nor cattle, which were the most useful and valuable animals in the civilizations of the Ancient World. It is equally important to have the necessary food available to feed the animals reared, and again we know of the difficulty which Europe found with regard to this, even in regions which were relatively rich, whereas the great civilizations of America only possessed extremely poor pastures. A balance was therefore necessary, to be achieved at a later date, for all animal rearing peoples, of human and animal food-producing cultures, as an agro-pastoral equilibrium was difficult to establish in areas where agriculture was not as yet very advanced. All this presupposes suitable civilizations and societies. A conjunction of a number of vital elements, one of which was insufficient alone, was therefore required.

The available documentation is undoubtedly too limited to provide the solutions to these questions.

II – The problem of metals is virtually the same. Broadly speaking, the Americans before Christopher Columbus hardly used all the metals available in their original form: gold, silver and copper. Lead had been used on a small scale, and bronze had probably been discovered by accident. The greatest handicap to these American civilizations was the absence of iron: meteoric iron was only used by the Eskimos. On the whole, historians are rather confused on this subject, as although pre-Columbian Americans understood the procedures and tools need for the fusion of minerals, and we will return to this later, they were certainly far less skilled in the more complicated operation required for reducing them – complicated, that is, to such an extent that no one can clearly, or even vaguely, explain how reduction metallurgy was developed in the Old World. At the beginning they probably used native deposits as was the case in America, but there was no iron, apart from meteoric iron. It is absolutely impossible to determine the cause or causes of the following state of affairs: in some regions men discovered the method of reducing minerals, and in others they did not.

In certain areas of America, metallurgy had been particularly advanced for the metals we have just mentioned. Metallurgy was most probably developed somewhat later in the area between Panama and the Equator, present day Columbia, and we will return to this aspect of the question further on. In North America, there were large deposits of native copper. In Mexico, metallurgy was similarly fairly advanced.

The major deficiency, therefore, was that of iron. Neither copper nor bronze, and as one might expect gold or silver, were hard enough in quality for the fabrication of tools. This absence of a hard metal is well marked by the long persistence of stone tools, even though copper and bronze had been developed between the 8th and the 10th centuries AD. The isolation, or relative isolation, of America during this period of the Western Middle Ages, may explain on the one hand, the lack of importations from far-off countries, which had existed in Europe and Asia, and on the other, the transference of technology. It is obvious that in many domains, the ignorance of iron making was one of the main reasons for the blockage of a large number of techniques: there is little point in going into detail on this subject as the great drawback of this "lack" may easily be appreciated. It would therefore be useful to make a list of fire-based techniques, in order to understand better the blockage of American civilizations.

III – A third fact, still more curious and more difficult to explain, remains to be discussed, which has been attempted, so far unsuccessfully by many authors. Let us make it quite clear, however, that we do not intend to give either here, or further on, the definitive explanation, but to consider the elements of the problem. Before the 8th century, the children of Veracruz in Mexico played with toy dogs, made out of wood and mounted on four wheels which pivoted on two axles. Despite that, the wheel was never used in the civilizations of pre-Columbian America. Important consequences of this may be immediately observed: neither the pulley, trestle or lifting apparatus, nor the lathe or any type of land vehicle, or for that matter anything which could result from the wheel, could be developed, and subsequently the screw, flywheel, gears, and the reduction or exploitation of hydraulic or wind energy could not be initiated.

We should certainly not be mistaken on this account. The West developed the wheel much later in certain sectors, yet it is undeniably true that at the moment of conquest, Europe had developed virtually every possible use of wheel and consequently every form of continuous circular movement. We must emphasize that it is quite wrong to suppose the natural use of a given technique, and consequently to be surprised at its late introduction, or as the case is here, at its complete absence in some civilizations.

Let us summarize briefly what we have been able to observe in the development of techniques in the oldest civilizations of the Old World. The most primitive use of the wheel was undoubtedly in land transport. It was only after this development that the wheel could be employed for other jobs: the pulley and trestle, the potter's wheel, etc. One might almost say that it was thanks to land transports, to the vehicle's wheels, that the East could become aware of the possibilities offered by the wheel, or more broadly, by circular movement. So what happens in America? Land transport naturally presupposes the development of other techniques, particularly those of animal domestication (as vehicles pulled by humans could only have been of limited use), and the development of certain metallurgies used to reinforce the most exposed parts of the vehicle (axles, wheel hoops, etc.). America did not know of these associated techniques which gave birth to the development of the four-wheeled vehicle. The transition from the toy to the land vehicle was not as simple as some authorities would have us think. Once again, the idea of a technical system increases in value. We will not cite, as some have done, the wheel barrow, which the West only developed much later, probably in the 8th century. Thus it seems we may better appreciate, if not fully comprehend, the astonishing absence on first sight of the wheel in the civilizations under discussion here.

The conjunction of these three important "lacks" must necessarily limit the technological development of the pre-Columbian civilizations of America, and moreover prevent some techniques, particularly those involving the arts, from reaching a very high level. The absence of certain techniques blocked to some extent those techniques established at a relatively low level.

In the domain of agriculture, the distinction should be made between soil development, cultural techniques and the species cultivated. Soil development was certainly very limited. In some regions, in particular the Andes, there was systematic irrigation, with often highly developed canals – that of Chicoma was 120 kilometre long –, reservoirs, and certainly a whole organization which supposes the development of complete networks. On the other hand, there appear to have been no drainage systems, but crops were frequently grown on terraces with dry stone walls. Clearings were made by burning and felling trees, and the terrain, exhausted after a period of a few years, returned to fallow land. But no progress seems to have been made in crop rotation. This derived partly from cultivation methods, from the plants they cultivated, and especially, the lack of fertilisers. Fish manure was used in Peru, and the Aztecs used human compost. But American civilizations lacked natural forms of manure, because there was no large-scale animal rearing. Natural fertilisers were almost unknown.

The Americans also had no animal traction, which meant that the instruments used to work the land tended to be very primitive (fig. 10). The shortage of iron exacerbated this problem, although of course Western civilizations also used very little iron in cultivation. The earth was usually worked with a hoe, made of wood and sometimes with a stone blade. Digging sticks were also used, they were curved, with a foot rest (fig. 11). In the central Andes, clods of earth were broken up using clubs with stone heads. These were also used as weapons. Generally speaking, there was not a great deal of diversity in agricul-

Figure 10. — Aztec digging stick. (From M. Daumas, *Histoire générale des techniques*, P.U.F., Paris, 1962)

Figure 11. — Inca digging stick. (From M. Daumas)

Figure 12. — Inca weeding. (From M. Daumas)

tural tools. Tubers and roots were gathered up with a spade or a hoe (fig. 12), and cereals were usually harvested with stone sickles. The produce was usually stored in jars, baskets and silos.

The pre-Colombian farmers had very limited means at their disposal, but nevertheless, they knew how to exploit to the full the plants and crops which they did grow. This was achieved principally through selection and specialisation. The Andean civilizations are supposed to have cultivated more than forty different types of crop, each one corresponding to a particular ecological stage. We cannot list all the plants grown, but the following are some of the most important: potatoes, maize, haricots, peppers, sweet potatoes, gourds, cassava, peanuts, avocados, cotton, amaranth and sage. Another important crop was lakinoa, a kind of Andean rice which grows in coppices. Many tubers were cultivated.

There were still many difficulties to overcome in this area, particularly in the Andes, which has a mixed climate. The potato, for example, was grown on the level immediately below that of the steppe: this gave a greater yield, but there were several problems because of the variations in climate. Hence, it was important to keep stocks. The tubers, dehydrated by the sun and the frost, remained edible for several years. Maize and rice were grown on lower land, in areas exposed to the sun but sheltered from winds.

Figure 13. — Aztec metal smelting. (From M. Daumas)

Arboriculture was not widely practised: the only two fruit grown were pineapples and pawpaws. We know little about their use of forests, except that, like the West, wood was the material most widely used.

There was food in abundance, then, some of which was very rich. However, overall it was not very well balanced. There were two major deficiences, in animal proteins and mineral salts.

Very few animals were reared, but dog meat was eaten, particularly in Mexico. Turkey was also popular, especially in America with the Aztecs, who were to pass this on to the Europeans in the 16th century. Alpaca, vicuna , llama and guanaco were all reared in the Andes, but this was for their fleece and for traction rather than their meat. They also hunted and fished for their food: game was regularly hunted, as were rabbit, hare, venison, peccaries, and all types of bird. This was despite the restrictions of a somewhat limited range of hunting instruments. The equipment used for fishing by the Aztecs in Mexico was similarly restricted: they used nets and spears. Sage did contain some oil, but on the whole the food in pre-Columbian America had very little oil or fat. This meant that everything was either grilled, or as was more often the case, boiled. One of their main drinks was beer made from maize.

It seems that the American peoples knew exactly how to exploit the different types of tree growing in that region. But again, the tools they used for this were very crude. Apart from different types of finishing tools, such as knives, awls, scrapers, pincers and polishers, they also used adzes. The blade was made of polished stone, although the Eskimos used meteoric iron, and in Mexico and Peru, they were made of copper. Sometimes, the same instrument was used for several different operations. It is difficult to see how they managed to make assemblies, because they had no nails, even copper ones. They did use fish glue, however, but this inevitably shortened the life of the constructions. Some assemblies were made by sewing the materials together with the roots from fir trees, in the same way as the Vikings made their ships.

It is difficult to establish the history of pre-Columbian metallurgy. The metals used were the ones to be found in this region, mainly copper and gold, of which there were large deposits in North America. This was the position for a very long time, no doubt because they had not conceived of reducing minerals. There were very few mines originally, and even when they were forced to open up new ones because the surface deposits had been exhausted, they were fairly shallow: the greatest depth of a mine was seven metres. Real metallurgy, i.e., the reducing of minerals, began in the 8th and 10th century in Columbia and southern Peru. We are not sure of the processes of reduction: of the type of furnaces, how they achieved the air flow, how they melted the metals, or if the case arose, how they roasted sulpherous minerals.

The only metals smelted were those with a low melting point, and this was usually gold. There are some pictures of cooking pots made of fired clay (fig. 13), and figure 14 shows a blowpipe being used to sustain the fire. The use of alloys was quickly established in South America, although this was probably originally by accident, either prompted by the existence of two metals in one deposit, or by mixtures created in founding furnaces. The copper-gold alloy is known to have been used in the Andes from the 6th century, but it was in Columbia from the 10th century that the alloys were systematised. One alloy used more often than most was tumbaga, made of 82% gold, and 18% copper. It was first heated then placed in an acid bath, giving it a surface of gold. It was then refired and pounded, after which it had the same strength as bronze or soft steel. There were many binary alloys: gold-silver (with 35 to 50% silver); silver-copper (with 20% copper); and copper-arsenic (with 5% arsenic). Most of these alloys came from the southern coast of Peru, where a tertiary alloy had been developed, that of gold, silver and copper. The copper-tin alloy, i.e., bronze, appeared later, in Tiahuanaco in the 11th and 14th centuries, then in the northern coast of Peru in the 14th and 15th centuries. A small proportion of tin, usually about 12% or less, produced a bronze which could be pounded, and with which tools were usually made. But, because of its low tin content, this same type of bronze was not usually very strong. Bronze with a higher proportion of tin could not be pounded, but it was useful in smelting and casting.

The south-American ethnic groups had certainly achieved a certain amount of expertise in the use of certain metals, although we should make clear that these alloys were usually for artistic purposes. This meant that except for bronze, whose applications were limited, there were not many metals which could be used for technical purposes. One can understand, therefore, why polished cut stone should have been used for so long.

Figure 14. — Worker using a blowpipe. (From M. Daumas)

Metal working remained fairly crude, mainly because of the lack of the necessary tools. The metals, used mainly for making ornaments or jewellery, were usually pounded and beaten into thin sheets or sometimes even leaf. These pieces were then carved, worked in repousse or embossed, and then engraved. Gilding and silvering were invented between the 5th and 11th centuries, by the Mochica civilization from northern Peru. Heat working was usually done by annealing, lashing or welding with the hammer. This could be combined with quenching. But all these operations were unable to produce a metal as hard as iron or steel. Wax casting was the most widely used method. Different pieces of metal were joined either by riveting or soldering. The latter process was practiced in the central Andes between the 7th and the 11th centuries. These two different types of metal working, one with hammers, and the other with heat were used in conjunction. Gilding could be done in three different ways. The colouring was achieved in the way described above for the copper-gold alloy. Alternatively, they covered the mould with a fine gold leaf before filling it with another metal. Plate-gilding was done by engraving the silver or copper object with acid, then coating the engraving with ground gold and mercury. The mercury disappeared in the heating process, while the gold was soldered to the other metal. Plants or siliceous sands were used for polishing.

The remaining momuments built by some of the pre-Columbian peoples are a testament of their expertise at working with stone. Unfortunately, although the quarries and tools still exist, there are no written sources to tell us about the processes they used. Most of the tools were made of stone, but wood, reeds, bone, leather, and much later, copper, were also used. Miss Chita de la Calle has provided us with this excellent summary:

Stone tools, such as mallets, axes, wedges, scissors, scrapers, piercers and chisels enabled the craftsman to grind, break up, slit, crumble, sharpen, rasp, pierce, ream, engrave or polish. Those made of wood or bone were used to rasp, pierce, or alter by pressure. Reeds, bone and metal were used to perforate stone and split leather. Very hard abrasives were unknown, so the craftsman would use the same stone with which he was working, just as with diamonds. Sand and water were also used as abrasives.

The techniques of flint cutting were the same as those used in the rest of the world. Very hard stones which were difficult to work were used for ornamentation and jewellery, which were difficult to work: the Aztecs used rock crystal in the 15th century, with emeries and a tool of quenched copper. Amethyst, chalcedony, agate, sard, carnelian, jasper, jade and turquoise were also worked. Jade was cut into plates by means of leather straps with sand and water. Tools made of more unlikely materials, such as bone, leather, reed and copper, were in fact only the supports, either of claystone and the same stone which was being worked, or of silaceous sands. The real tools are the abrasives. They even went so far as to use ears of corn as gimlets, which in the 13th century were replaced by tubular piercers of copper which provided a regular supply of water and abrasives.

We know very little about how the large slabs of stone for construction – used mainly for great monuments – were quarried, worked, and dressed. The interior was made of piled up rubble-stones, or of flat volcanic stone. The coating was made of basalt, granite or sandstone, squared off or polished. No cement was used, but sometimes the rubble-stones were stuck together with mud. In Mexico, they used mortar made of sand and very hard lime.

Not everything was known about fire-using techniques during the pre-Columbian period, but nevertheless, they were widely practised throughout the continent. They were used in ceramics, in spite of the absence of the potter's wheel, which was so popular in other ancient civilizations. Of course this restricted the potters to sculpting. According to Favre, the paste was usually rolled up then twisted into a coil to make the sides of the desired object. The techniques of moulding were adopted fairly quickly, thereby facilitating production on a larger scale. It was probably invented in the valley of Mexico during the 6th century, in the Maya region during the 8th century, from where it was probably passed to the northern coast of Peru (although some believe that there is evidence of moulding from the southern coast of Peru dating back to the beginning of the Christian age). Moulding made possible the anthropromorphous and zoomorphic pottery whose realism is still so striking today. However, the use of moulding did not wipe out the technique of coiling used for large pieces, particularly jars for containing food. At any event, the abundance of clay beds all over America greatly stimulated the development of pottery industry.

Plastic clay was used for decoration purposes: cutting was done with a point or a stamp with die made of a hard material, usually fired clay or stone. Most pottery was painted, and archaelogical remains show the remarkable development of polychromy in certain regions, particularly in the northern coast of Peru. The pottery was fired in a kiln, which was usually open. Wood was piled up around the pieces to be fired, and the fire was kept at a constant heat of about 400° to 600°C for several hours. In some areas, the firing was done in a well, with the hearth at the top. In both cases, however, they used the reduction process by means of the air current. Fired in this way, the crockery was extremely tough, and it assumed either a brown, red or cream shade characteristic of the pottery of certain civilizations. Precolumbian America never used the horizontal corridor kiln so popular in the Old World. The kilns were not bricked up and tended to be vertical. They worked at a low temperature, and the air flow was fairly low, just enough to keep an oxidized atmosphere. The colouring was usually light as a result of this. It seems, therefore, that American pottery techniques were perfectly autonomous.

Bricks were also used in construction for building many great monuments. Usually, these bricks were simply dried in the sun. In the Andes, however, all constructions were made of stone.

Assembly techniques tend to relate to those sectors where there is no great need for implements, but in wood assemblies for example, the lack of tools did prevent the creation of any complicated assemblies. This meant that the elastic properties of wood were fully exploited: frames were made of poles bent into an arc, and trunks were made of thin planks which were softened by steam so that they could be bent to a right angle. Neither nails or dowels were used, and the planks were sewn together. Basketry was popular, and there were many different types: spiraled basketwork among the Eskimos, spirals of woven strands, especially in North America, but also in the Amazon valley, basketwork with twisted ropes, found along the West Coast, basketwork with woven strands, which was popular almost everywhere, and diagonal basketwork for mats.

The Americans were very skilful in all textile techniques. It seems that they developed rapidly first in Peru, then spread throughout the continent. The most popular fibre was cotton, grown on the coasts of South America, and again, especially in Peru. In a few regions, other vegetable fibres were used, notably agave and reed. They soon developed the use of animal fibres: rabbit hair, and even human hair, and also wool. The latter came from camelidae, llama and especially alpaca which were reared specifically for this, but also from wild animals, such as guanaco and vicuna. Wool from the latter was particularly fine and manageable, and was easily dyed. We do not know how these various fibres were prepared, but no doubt the techniques were very rudimentary. Similarly, we know virtually nothing about dyeing techniques, but the fibres were usually dyed before spinning, with vegetable and animal (shells) colourants; mineral dyes were also used: ochre, calcinated lime, chalk and manganese. Alum was used as the mordant. The craftsmen from the southern coasts of Peru therefore had more than one hundred and ninety colours at their disposal.

The fibres were spun with a spindle, made of wood or thorn, with little ceramic spindle-whorls. The Peruvians used to let the spindle hang freely from the thread, but almost everywhere else they threw the spindle onto a plate or onto the ground. After a while, the spindle would slow down, and stop. It then needed to be wound on once more. The type of distaff used is unclear, but the Incas of the 16th century used a forked stick to wind on the thread. As might be expected, the Americans never used the spinning wheel.

The techniques of weaving were similarly crude. Looms with straps were widespread, and are very like the ones depicted on Greek vases (fig. 15). The American loom was slightly more sophisticated: it had a beam, breastbeam, heddles and shafts, crossing rods, and loom slays with combs. But some elements were never used: the harness, treadles, or the shuttle with spool. The low warp loom was the most commonly used, the high warp loom was relatively rare. Brocade was a natural outcome of the invention of the warp. From the 7th century, tapestry, embroidery and gingham were all practised. Tapestry making was at its height from the 11th to the 12th centuries, with more

Figure 15. — Inca weaving loom. (From M. Daumas)

complicated fabrics, double cloths, braid, dyed cloth and velvet. Knitting was also practised during this period; and it appears that this technique was unknown in Western medieval civilizations. After this, the most developed techniques tended to decline, in particular tapestry and embroidery, but also knitting and brocade. Gauze was used by the Mexicans. Textile production in Peru was a large industry, largely due to very elaborate techniques organised on a wide scale by the "monasteries of the Sun."

In the field of architecture, we know more about urban techniques than anything else. Of course there was a non-urban type of architecture, with several different styles, ranging from the tent to half-buried houses with clay and wooden walls, and houses made of dried bricks and with thatched roofs. There were even some storeyed houses, with stairs of clay bricks. All these buildings had common characteristics: they had only a single room, no windows or floors, and they were ventilated through the roof or simply from above.

Architecture in the towns was certainly more sophisticated, but the lack of tools which were still at the neolithic stage meant that the development of techniques was particularly slow. Wood was used for the frame, lintel and roof, and stone for the fittings after which it was drafted. Pebbles and rubble-stones mixed with mud for the filling-in were also used, however. We have already mentioned that they used clay bricks; real bricks were very rare but there is evidence of these in the ancient Maya empire between the 4th and 10th century, and in the Mexico valley, from the 9th to the 12th century. Lime, mixed with sand and loose chippings, provided them with a particularly tough mortar. Again, a distinction should be made between ordinary dwellings – less well conserved and therefore less well known – and the large public buildings, fortifications, palaces, temples and especially the pyramids, which were all built using more sophisticated techniques. It is invariably this type of construction which interests historians.

The foundations were rudimentary: terracing for buildings on mountains, and platforms for those on level land. The walls followed the natural angle of the mountain or hill slope and therefore retained the rubble stone on the inside. The drafting was still fairly crude: this was because the walls were rarely left plain, but were generally faced with stucco, usually polychrome or with mosaics. There were few openings onto the exterior, and ventilation was normally provided by small holes in the partitions and at the top of the walls. The doors usually consisted of a lintel and inclined posts open at the bottom. They could not be closed, but there was usually a curtain or matting. Roofs were mostly terraced on a frame which rested on the actual walls, and which was covered by shingle, lime and mortar. They were probably supported by columns or pillars. From the 4th century, the Mayas were building false corbelled vaults, as well as arches. This technique also enabled them to build aquaducts, drains and triumphal arches. It is worth remembering here the slow development of the vault in Europe. The tools they possessed for this were much more sophisticated, and enabled them to cut with much more precision into harder stone which could withstand the pressures created by the vault. The pilaster was used in the Mexico valley before the 10th century, at which time serpentine pillars could also be found. The stone column was a tolteque invention (10th-12th century): it was made of several assembled pieces, adjoined by mortice and tenon joints.

Most public buildings, palaces or temples had only one storey, but in some regions they had two, and sometimes even four, with recessed fronts.

It should be clear by now that the pyramids are not necessarily proof of an advanced technology. In Egypt the pyramids date from a time when techniques were still fairly crude and the stones were probably still being hoisted on inclined planes. The civilizations which built pyramids did not use the wheel, but whether of not they used rollers is uncertain.

The pre-Columbian Americans also built fortifications. The fortresses of Sacsahuaman in Cuzco and of Ollantaytambo, in the valley of Urumbamba, are both good examples of the type of military construction built by the Incas. The fortress of Ollantaytambo is composed of two high walls, which, like the triple remparts of the former, are made of cyclopean bonded megaliths. These buildings were fixed to the side of a natural hill dominated by an enormous tower. On the other hand, their armaments were relatively primitive. Their round shields made of wood or reeds were covered with feathers, and the double-edged swords had obsidian blades. Bows and other types of missile projectors were used, with pointed arrows of stone as well as javelins. They wore wooden helmets and their armour was a simple tunic padded with cotton.

Urbanism had developed under the pre-Columbian civilizations, and there were some large cities. The Incas and the Aztecs were both expert at such building: the towns were usually carefully planned, according to a quadrangular arrangement with a square in the centre, and roads coming off it at right angles. The lie of the land was particularly awkward in the city of Mexico, so embankments were used which enabled them to build a large town with wide straight roads and canals which carried most of the traffic. These canals had bridges made of large beams. The surface of the roads was made of beaten down earth. The only inconvenience of this arrrangement was the fluctuations in the water level which either caused the canals to overflow onto the roads, or to run dry. They also had to make provisions for the supply of soft water to supplement other sources which proved to be insufficient. The aquaduct they built was five kilometres long and crossed the canals by means of hollow deep bridges. A second aquaduct eventually had to be built to supply the growing population. In 1449, dykes were built to protect certain areas of the town from flooding. The problem of household waste was solved by the availability of a large workforce to dispose of it.

Despite the lack of certain means, these civilizations had obtained a quite high level of expertise in urban techniques. Unfortunately, it is difficult to study these pre-Columbian cities, because they have almost completely disappeared.

We shall close this section with a discussion of the problem of communications, which may provide the answer to where the blockage of the pre-Columbian technical system lies. A technique can only be fully exploited if it is applied within a large area both in geographical and human terms. As soon as there are difficulties in communication, and as soon as exchange becomes almost impossible, the entire system finds itself more or less blocked. So what was this region like before the Europeans arrived?

We have already made the important point that at this time knowledge could only be communicated orally. All messages were transmitted in this way, and to help them, they used mnemonics such as designs or knotted cords. The absence of writing was already a severe handicap by this time.

The routes of communication were of course closely linked to the means of transport. Maritime transport was still very primitive, with canoes made of skin or bark mounted on frames being used in the North. These were also used in tropical regions. There were also one-man boats and pirogues, made either using fire or with adzes. In the islands, the sails were made of matting. Further south, various types of rafts were used. At any event, the pre-Columbian peoples were not a maritime civilization. They did not possess the necessary equipment to build ships: they could have built the hull and the sails, but they had no winches, hoists, or pullies. Consequently no ports were built either.

There were two main techniques connected with land traffic, porterage and traction, which was either animal or human. Animal porterage was rarely used, because the animals were not big enough. The peoples in the North mainly used traction. Long-distance exchanges were practically impossible, and the consequent fragmentation of economies was a prime cause of technical stagnation.

The roads were little more than paths, barely marked out, and these were used essentially to administer and control the country. Only the Incas managed to establish a relatively extensive road network. Some of their roads were even paved in the areas surrounding the towns, no doubt to facilitate the supply of provisions coming in from neighbouring areas. Some remains of these networks were found in Yucatan, dating back to the 7th century. The roads were 4.50 metres wide and were raised from about 0.60 to 2.50 metres above ground level. The base was made of heaped up stones, and the sides were protected by roughly shaped stones. The surface was made of metalling levelled out by a roller usually pushed by about fifteen men.

A civilization does not need very advanced techniques in order to stand out among the rest. Pre-Columbian civilizations were remarkable for their arts and this often leads to erroneous conlusions about the level of techniques as a whole. It may be difficult to explain the technical deficiencies of this time, but it is through such deficiencies that one can understand the technical blocks which were to afflict this area until the European conquest.

THE MOSLEM WORLD

Even though there is an abundance of information and evidence about Arab and Moslem civilizations, any study of this subject seems to consist almost entirely of problems which are difficult to resolve.

In the words of one recent author writing about these civilizations: "One has the impression, not of material progress, but rather of the stagnation of the Moslem world. It is difficult to see any consolidation of acquired knowledge, unless of course our perception of this period is distorted by discoveries made in more recent times, which has made us more severe in our judgement." This, then is the issue: the Moslem world is not seen to have made any significant contribution to the domain of techniques.

This is not then simply a case of a blocked technical system, but of the heritage of a blocked technical system, which proved impossible to change. In this sense, the Moslem world was almost regressing in comparison to the other civilizations which we have just discussed. This ties in well with the more general historical context. The Arabs were nomads, and any nomadic people necessarily has a limited range of techniques – limited in the sense that they have no permanent settlements which are the condition of technical development. On the other hand, these people were conquerors, who sought to maximise the profit afforded by their conquests. They had no desire to change the techniques they acquired and practiced, because such change was bound to bring about social and economic disorder. It was in their interests therefore, to maintain the existing level of techniques.

However there is another side to this. At the time of their conquests, the Arabs were not a technically advanced people, which caused them several problems. Their approach to these problems is proof of their extremely flexible attitude.

They were able not only to use the techniques of the occupied countries, but also to gradually re-establish the techniques which had slowly disappeared in the past, particularly in the field of hydraulics.

The second point relates to all peoples who conquered large domains. The Arab empire covered diverse civilizations, all at different stages of technical evolution. It was one of the Arabs' strong points that they, like the Romans, ensured that all of the areas they occupied benefitted from the most advanced of these techniques.

The third point is no less important. The Arabs were able to assimilate all the most advanced techniques that they found in the countries they conquered. We noted above that many of the works of the Alexandrian School are known to us through their arabic versions, and the same was true of many agricultural treatises. This was almost certainly the result of preocupations in western caliphates, notably those of southern Spain.

Beyond these technological transfers, it is very likely that the Arabs were not so much an innovative people, but one which perfected the techniques they inherited. Al Jazari's treatise on automatons and certain agricultural treatises bear witness to this.

The Arabs seemed to have fulfilled the same role as the Romans, then, but perhaps they did so even with more dynamism. They also held a strategic position between the Western and Asiatic worlds, thus permitting vital technical exchanges to take place, such as the importation into the West of certain types of steel from India, along with paper and the compass, to take but a few examples. This meant that both Western and Eastern people were able to profit from the same developments. A fairly high level of technical education was essential for this. Very little research has been conducted into the technical mind of the Arabs, largely because precise facts are lacking, as is a precise chronology of the history of the Moslem technical world which might help us to appreciate its full value.

This is all the more curious because despite the Arabs reluctance to innovate, they possessed a relatively large number of written technical works. This feature is evident in Damas and Cordoba, the two great intellectual centres of the Moslem world at its height.

Of course, not everything in these works was original. Far from having destroyed the Alexandria library, as some people have unjustly supposed, they in fact took great care to translate many of these Greek and Latin works. It would be helpful to have a proper inventory of these translations, since some of these were of Persian origin. Philon of Byzantium's work on pneumatic apparatuses is one of the works that we know of only through the Arab translation.

This Arab literature is just further proof that in the field of agriculture, Islam drew from Greco-Roman experience. The numerous technical treatises are in fact mainly compilations, and it has been shown that the general course of development did not change. In fact, neither the land nor the climate necessitated any great alterations in the technical system. In the 10th century, when the centre of the arts and sciences shifted from Baghdad to Cordoba. the same traditions of the classical agronomists were perpetuated. However, the importance of the Persian- and Arab-inspired *Agricultura nabatea* should be noted. The treatise of Ibn-al-Awan, written in the 12th century, reveals the same influences of nabatean agriculture born in the Chaldean region, possibly one of the cradles of agriculture itself. As well as the prescriptions which are to be found throughout the Mediterranean basin, there is an element of magic which is very important. In fact, in the geographical zone occupied by the Arabs, the nature of the soil, the scarcity of water and the hot climates were all difficult constraints to overcome, and were necessarily imposed on all technicians of agriculture.

In other fields, Arabian literature was no doubt more limited. We shall quote once more the famous treatise on automatons by Al Jazari from the 13th century. This author was familiar with the Greek mechanicians from Alexandria, but he tried to go beyond them. Some of his writings on mechanics were certainly new, but the basic substance remained the same. A treatise on armoury, written for Saladin (in the second half of the 12th century) describes a type of machinery used in war – the counterweight artillery – which was borrowed almost completely from the West. We shall return to this point later.

The Arabs were good mathematicians and astronomers, not to mention doctors and alchemists, and their scientific literature is evidence of this, with works by such famous scholars as Avicenne and Averroes. However, it seems that they were less interested in more technical questions.

Rather surprisingly, the Arabs do not seem to have transferred to certain parts of their empire those techniques which were improved in other areas. So while farming techniques in Egypt used the same processes, types of crop and instruments that had been in use for centuries, the techniques used in Southern Spain, on the other hand, underwent a marked development. Perhaps one of the reasons for this was the fragmentation of the Empire into independent caliphates. Having invented nothing themselves, the Arabs had absolutely no interest in anything that might have been invented by the people they had conquered, thus sealing off one important channel of progress.

Any history of techniques in the Moslem world is eventually plagued by the lack of written sources. There are very few technical treatises and virtually no figurative representations. All that is available are a few Persian miniatures dating from the 12th century, which can give us only a vague idea of the techniques used in that area; and of course one should remember that what was valid in Persia was not necessarily so in Egypt, North Africa or Spain.

The Arabs, and especially the Moslems, were not great innovators of cultivated species. At the very most, one can credit them with having extended the cultivation of certain crops, especially sugar cane, cotton and perhaps also rice. From the 13th century, these were grown in North Africa and throughout the southern regions of Europe: Spain, Italy and Sicily. One distinct contribution of the Arabs was the group of exceptional gardens in the south of Spain, during the time of the Cordoba caliphate in the 10th century: these botanical gardens were used for the same purpose as experimental fields of seeds, to acclimatise and develop the various species coming from the East. Most of these were medicinal plants, but experiments were also conducted on plants with a view to large scale cultivation.

There were no notable changes in agricultural implements, or in the techniques of tilling the land (fig. 16). Furthermore, it seems that the same regional varieties still persisted. The Persians used a swing plough with share-beam, for example, while the Egyptians were still using the heel-stilt plough (fig. 17).

The same phenomenon is apparent concerning the exploitation of the sub-soil. In Egypt, the Near-East, and Spain, the Arabs were content to simply work the old mines with old techniques (fig. 18). This is especially evident in the lead mines in southern Spain, and in the mercury mines at Almaden, further north. The Arabs were certainly

Figure 16. — Spade with footrest. (From M. Daumas)

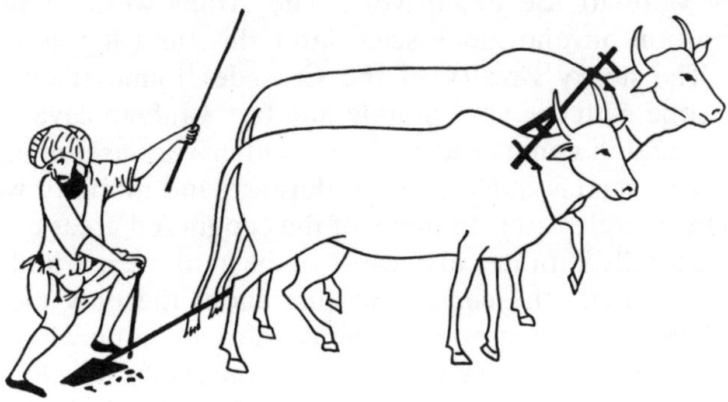

Figure 17. — Persian ard. (From M. Daumas)

short of iron, and the lack of fuels compounded the problems caused by the scarcity of deposits. Maurice Lombard has clearly shown how the Arab and Moslem worlds were dependent for wood and metals on the countries of Western Europe. In exchange, they supplied spices and gold.

Figure 18. — Miner with pick. (From M. Daumas)

Animal domestication did undergo some changes, the most striking of which was the widespread practise of breeding horses. For a long time, Asia Minor was the equestrian reservoir of the Eastern basin of the Mediterranean. It was probably the Persian Sassanid dynasty (226 to 632) which perfected the art of horse breeding, and it was here that the Arabs found what later became known as the "Arab" horse, and which they were to use extensively. The Arabs were to practice horse-rearing on an enormous scale after this, and it was to reap dividends. The heavy cavalry of the Crusades found it extremely difficult to cope with the very mobile and fast Arabian cavalry. The Arabs also found in conquered regions another very useful animal – the camel. The camels qualities of endurance and mobility were to make it an invaluable asset. In many of the conquered areas of Islam, wagons were totally impractical because of the muddy earth and desert sand. In some areas, the camel was just about the only means of transport. This also dispensed with the need to maintain what would have had to be an enormous road network. The Arabs developed the species of camel most suited to the kind of work required: the province of Oman in Southern Arabia was famous for its racing camels, for example. The Arabs may have found the horse which was to bear their name in Persia, but it seems they were actually responsible for developing the different camel species. Between 400 and 1300, the East was closely dependent on this animal, causing the virtual disappearance of the wagon.

The richness of the Iranian cuisine stands in great contrast to Arabian food, which had few variations, and was often badly prepared. Persian gastronomy was renowned even under the Sassanids. It was in the Moslem world that sugar cane first emerged, and its use spread eventually to the West and the Far East. We shall discuss a bit later the development of Morocco, which has been the subject of some recent studies. Moslem craftmanship is nearly always hailed as an exceptional achievement. However, it is noteworthy that the Arabs found many of these developed techniques among the peoples they subjugated: Egyptian cloths are one example. Furthermore, a number of these highly praised craftsmen were not Arabs, but again, came from the subjugated peoples. And nomadic industries are invariably relatively primitive. Apart from a few improvements and additions, the tools used to not appear to have developed beyond the level of the antique period. (fig. 19). It has even been said that for certain trades, these tools were actually diminished in number. As nomads, they possessed hardly any tools, or furniture: this only changed when palaces were built. The textile industry, which was flourishing in Hellenic Egypt, was the glory of ancient Persia and byzantine Syria. Persian carpets were already being acclaimed by Sidoine Apollinaire. Knotted carpets were already being made under the Sassanids. At the beginning of the 7th century, Huien-Tsiang from China expressed his surprise at the skills of Persian craftsmen and their expertise at weaving silk brocade, woollen cloths and carpets.

Figure 19. — Frame saw inherited from the Roman world. (From M. Daumas)

A Chinese, captured in the battle of Talas in 751, is reputed to have taught the Arabs the art of paper-making. Papyrus was becoming scarce and expensive, so this technical transfer was all the more important. The first paper factory was esatablished in Baghdad in 795, and by the 10th century paper had taken over from papyrus. "Hundreds" of paper mills are supposed to have existed in Fes from the 12th century. These were almost certainly hand mills.

Ceramics was an ancient occupation of the Arabs, and extremely fine pottery was made all over the Moslem world: it has been said that the pottery from Persia, Mesopotamia, Egypt and Syria easily rivalled the glazed earthenware of Tunis and Cordoba. The "tiles" from Malaga became the Spanish *azulejos*, and also important was the pottery enamelled with reflecting mineral substances. The only treatise on ceramics which we still have today was written at the end of the 13th century by a Persian workshop manager from Kashan. There was diversity in all aspects of ceramic processes, from the manufacturing techniques, the enamelling and the firing, to the forms and type of decoration.

Arab military techniques profited a great deal from western inventions. All their equipment used in war was identical to that used in the West, from the armour to the coats of mail, and from the swords to the lances. The manufacture of steel, described in the treatise on armoury written for Saladin, seems to be the product of a vivid imagination rather than sound technical knowledge. The crossbow was used by the Arabs comparitively late, from the 9th century (figs. 20 and 21).

Most of the original elements of their equipment used in siege warfare seem to have come from Sassanid Persia. It appears that the Arabs were not in fact the forerunners in the handling of counterweight artillery, which had been known in the West from the 9th century. It is possible however that some of the best machines were built by the Arabs, and better maintained, which would perhaps explain the surprise expressed by Joinville (fig. 22).

Both the Arabs and the Moslems were affected by the problems of space, which were very similar to those encountered by the Romans. Consequently, some of their solutions were the same. The first major problem was posed by traffic. We have already mentioned that the intensive use of pack camels reduced the roads' network to a minimum. It seems that there were very few civil engineering works created. In some countries, like North Africa and Spain, the Arabs

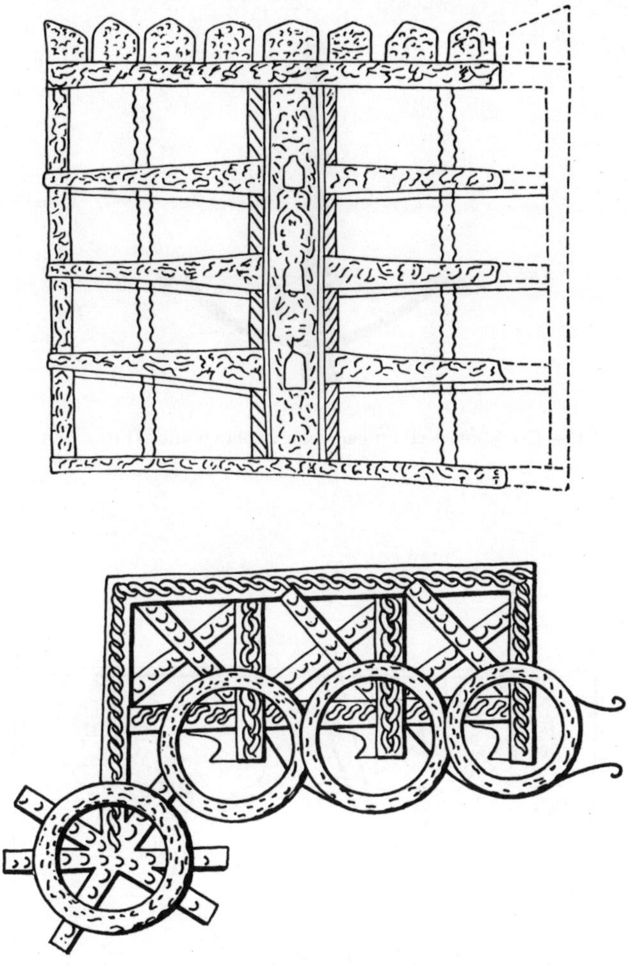

Figure 20. — Crossbow tower [above]; Triple crossbow [below]. (From M. Daumas)

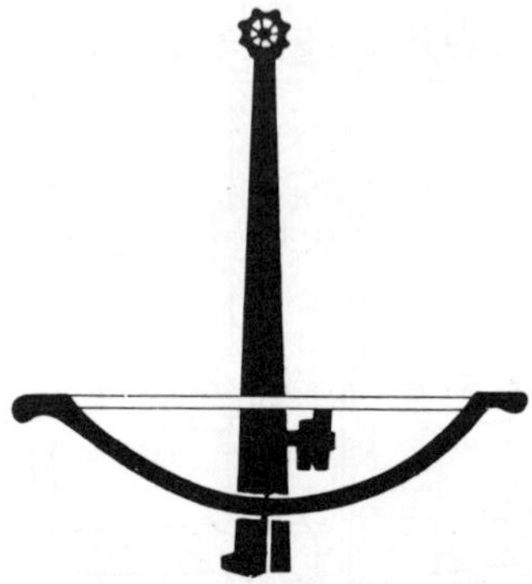

Figure 21. — Crossbow used for launching naphta bottles. (From M. Daumas)

Figure 22. — Trebuchet. (From M. Daumas)

discovered Roman works. It was not before the 15th century that stone bridges were built in Turkish Europe.

River traffic on the Nile and the Euphrates remained much as it had been in Antiquity (fig. 23). Originally, the Arabs were not a great sea-going nation, and the only forms of naval transport they possessed were very crude small decked crafts. Around the coast of Oman in the Red Sea, they used boats made of pieces of wood sewn together: in Syria, the planks were nailed. Wood from the coconut tree was used in the boats made of sewn wood, and the fibres were spun to make the masts. The leaves were used to make the sails and mooring ropes. (fig. 24). According to R. Lopez, it was the Arabs who made the first "lateen" sail. A Persian miniature painted in 1237 from the school of Baghdad depicts a stern-post rudder, which it seems was already used on the seas of northern Europe.

It has been clearly shown that the importance of religion in Islamic life was the main reason why the towns were established. Indeed, at least twenty of these urban creations were established in the countries under Islamic rule. These were also based on the elements of preceding civilizations: the plans of Arabian and Moslem towns were very similar to those of Greece and Rome. As in Antiquity, these towns were organised around the area of worship which were usually beside a market place. Some were military towns, built around a military centre used for administering occupied territories. These towns were often flanked by forteress-palaces, of which there are still some striking remains, from Grenada to Cairo. They were also often surrounded by high walls. There is no notable difference between this military architecture – unknown to the nomads but imposed on them by their conquests – and similar architecture in western Europe. The fortress of Cairo was built in the 11th century, and it was only after this that similar constructions were built in many other areas.

The building trade continued to use local techniques: they used bricks or stone depending on the area. The mosque of Ibn Tulun in Cairo which dates from the 9th century, and which is traditionally believed to have been built by a Christian, is made entirely of bricks. In Mesopotamia, Egypt and Persia, military constructions were often built using stones from antique monuments. The Arabs in Egypt were the first people to practice large scale demolition of the monuments built by the Pharaohs.

Figure 23. — Boat from the Euphrates. (From M. Daumas)

Figure 24. — Small craft with sails. (From M. Daumas)

The Moslems were very aware of the importance of hydraulics, and it was used to supply water to the many settlements that had been established before the conquest. Irrigation in these semi arid countries had been a vital necessity for centuries, so when the Arabs invaded, they were able to benefit from fairly advanced techniques, perfectly suited to the various needs of each region. They even had at their disposal the economic, social, judicial and political organisation which had evolved to make such works possible. In some cases, they were able to reap the benefits of these original efforts by systematizing and applying the same techniques throughout their Empire.

It was in Iran that they discovered the use of underground water levels built up by infiltrations – which tended to be large in areas where gravel and masses of fallen rocks accumulated at the bottom of mountain ranges – and by drainage tunnels called "quanats." H. Goblot, the main authority on this subject, has shown that this technique was relatively limited to Iran and the neighbouring regions. Before the time of the Islamic civilization, it spread to the banks of the Mediterranean and the Arabian peninsular, after which it underwent a remarkable expansion in North Africa, Sicily and Spain. It is possible of course that after developing this technique, the Romans might already have introduced it into North Africa.

Building dams was a second method of irrigation. On the larger rivers, these were often simply water catchments. The Romans were no doubt already familiar with gravity dams. The arched dam was apparently first built in Glanum in Roman Gaul, in the 2nd or 3rd century. The Kebar dam in Iran, probably built in the 14th century, is twenty-six metres high. The Saveh dam dates from the 13th century. At Kebar, the outlet at the bottom is situated about six metres below the first dam crest, and about ten metres after the extra elevation. This ensured a very large cut off wall. The entire length of the crest was supposed to serve as a spillway in the event of flooding. It has been claimed that by this time, the Sassanids had already installed arched dams at Aden.

Roman antiquity had conceived of *norias*, which curiously enough is an arabic term denoting a specifically Roman instrument. These instruments were powered by horse-whims. Were the water-raising wheels an Arab invention? It is difficult to say. The oldest extant ones are to be found in Hama, on the Oronte, and they date back to the 8th and 9th centuries. The Toledo wheels, the oldest in Europe, are cited by Al Ma'mun in 1043. Historians still disagree about the role of the Arabs in the transmission and development of these techniques.

The recent discoveries made by Bertier in Morocco have revealed the rather extraordinary installations built for irrigating sugar cane plantations and for powering hammer mills. It is difficult to date them, but they are most likely from the 13th century. Up until now, none like them had been found elsewhere, even in areas like Egypt for example, where sugar cane was extensively cultivated.

In general, the technical civilization of the Moslem world poses many problems. As in the case of China, we need an exact inventory and an exact chronology to study it properly, both of which are lacking. Many myths still abound about the inventions or the transmission of inventions attributed to the Arabs. Some geographical sectors need to be better studied. Many comprehensive works have been written about Spain, so historians would do well to turn their efforts towards Persia, particularly during the Sassanid period. Every step of thc way, we find ancient examples of the most advanced techniques in these regions – in textiles, pottery, the building of dams, wind mills with horizontal sails, stirrups, etc. Furthermore, it is through the Persian miniatures that we know about the Moslem technical world. Was Iran an important centre of technical innovation? Was it a meeting point for the two technical civilizations of the East and the West? These questions still remain to be answered. One main point is certain: the Arabs were not a particularly innovative people.

But let us leave the Arabs aside. They had no technical system as such, or more accurately perhaps, the techniques which they were using when still a nomadic people had hardly changed. They were able to profit from the advances made by the people they conquered, but they were content with simply this, and they did not exploit the situation nearly as much as they might have done. As a central focal point for several civilizations, their role was mainly one of transmission.

But the Chinese and the South-Americans did have technical systems: the former was advanced, the second was relatively limited. The progress of both, however, ended abruptly. Isolation cannnot be blamed for this, especially since China was never totally isolated, and Europe managed to advance without outside help. So the blockages must have been caused by something else. Was it because scientific knowledge ceased to progress? This is probably part of the reason, as, if scientific knowledge is not directly useful in terms of techniques, yet it can act as a stimulant. This bears a certain resemblance to the case of Greece during the hellenic period. In America, the absence of writing, and in China the complications of the writing system were no doubt also important factors. There was another point in common between these two civilizations: the rigidity of their social structures. This was probably one of the major causes of technical stagnation.

THE MEDIEVAL TECHNICAL SYSTEM

Contrary to a once widely accepted theory, we now know that the Middle Ages was a period of intense technical activity, which we are beginning to understand better. Systematic research still remains to be completed in many domains, however, in order to achieve an exact view of the movement of medieval techniques. In some cases we are ignorant of the precise significance and the chronological progress of this movement. Some attempts in this field are interesting but still debatable.

As we have already seen on several occasions, our first step must be to establish some sort of chronological order: naturally techniques must be considered together with other human activities and reciprocal effects should not be isolated. And as we are partly concerned with these temporal divisions in conventions, we should try to put "our" Middle Ages into perspective. Thus our period begins around the middle of the 12th century and ends in the early years of the 14th century. There was a marked rate of progress at the beginning and clear signs of crisis and depression at the end. At the outset, there were new elements and innovations and at the end, discord and war.

The 12th century has undoubtedly been less well studied in terms of its originality in comparison with preceding centuries. At least today, authorities stress a fairly rapid demographic growth, the increasing number of clearings which were made, and a form of feudalism which was henceforth organized. Cultural and technical exchanges increased and the Crusades opened wide the gates to the East and Asia. But already profound movements were taking shape, the importance of which was to grow: they were the natural consequence of demographic expansion, population intermixing and displacement and the movement of communal liberty.

Progress was constant in the second half of the 12th century and throughout the 13th century: the rapid boom of markets, urban development, the revival of striking gold coinage, large building works, and the strengthening of feudal monarchies, all seem to prove that growth was manifested in all walks of life. Obviously it is impossible to provide figures, or to measure the extent of the expansion of the medieval West, but we believe that the collection of certain data would be sufficient to provide an extremely clear impression of this phenomenon.

Tensions began to arise at the very end of the 13th and the beginning of the 14th centuries, as monarchies, economies, and established positions became vulnerable, although the exact causes of the great crises of the 14th century have not yet been accurately detected. Philip the Fair of France was the first to implement monetary changes, as poor harvests and famines appeared in the first decades of the 14th century, and social problems began to manifest themselves in various regions. The West quickly began to know great financial crashes, devastating epidemics, and interminable wars.

This is the accepted outline of events today. For a century and a half, the European West had seen real expansion. This expansion was naturally to create tensions, and structures adapted much more slowly. Is it not logical to suggest that in such a vast movement techniques might have played an important role, that they had a significant part in the processes of growth, as they had in those which produced these final tensions? Some authorities have suggested and maintained as much, but this may not yet be a general conviction. Inasmuch as a technical system and economic progress are closely linked, we should observe, if not a total technical revolution, then at least an ensemble of new technical ideas, or to be more precise significant technical innovations, largely taking place around the middle of the 12th century. In fact, although some earlier discoveries had been made, a coherent body of advanced techniques was formed during this period. Similarly the pursuit for economic growth must have been maintained by a pronounced technical dynamism which we call development today. Finally, we are able to observe a sort of decline in innovation at the end of the period thus defined, at the end of the 13th and the beginning of the 14th century. This could provide a model for studying the creation and evolution of the medieval technical system.

THE FRAMEWORK OF TECHNICAL PROGRESS

As we have already emphasized, techniques must be incorporated both within the history of mankind and an ensemble of material data. In order to understand this better, it seems therefore necessary to precisely define the nature of its context, such as the particular intellectual environment and the physical conditions involved, in brief, everything that concerns technology "upstream," which is generally less well known than the economic evolution situated "downstream."

Medieval mysticism has often been discussed, a spirit of contemplation and simultaneously a certain contempt for work, stemming from different motives, which would have prolonged an ancient attitude. We are dealing with the same difficulties here which arose in Antiquity, and it is probably for these same reasons that the problem has remained unsolved. Technology and the notion of work are closely linked. But unfortunately we lack any detailed and unbiased study on the second section. In a recent article, Maurice de Gandillac has demonstrated the value attached by some 13th century clerks to the "mechanical arts." He refers to Lulle for both educational and economic reasons. The latter actually restores the notion of work to favour "without reference to a theology (almost unknown during the Middle Ages) which put forward the example of Jesus the carpenter, and of Paul making tents, in order to glorify this *Formierung* by which raw material was technically improved and converted into a specifically human form of work." There is still cause to recall the mystics of the Rhine, such as Eckart and Tauler who, half a century later, criticized the excess of monastic *otium*, but without praising the material fruits of labour. It is obvious that fidelity was still shown to Aristotle's traditional classification of various ways of life, where technical activity was placed well below others. But, as was the case with Antiquity, we still need to .neasure the exact significance of this notion and in any case not attribute to it a general character. The favour shown by princes towards "engineers" and technicians for many and various reasons, the attention displayed by some towns towards technical progress, the protection of manufacturing secrets and all the technical regulations, are once again proof of an interest which refutes the intellectual state of mind too complacently cited. We are here in the realm of the history of ideas, certainly, but provided that one distinguishes between the history of exceptional ideas and the history of standard ideas. The latter remains to be studied.

We will pass over the doubts of the moralist and the theologian. Apart from these, there were scientists and practicians who, for whatever grounds, were committed to a material life. Even before the appearance of Vincent de Beauvais who is often cited, a genuine mystic, Hugo de Saint-Victor, demonstrated an intellectual spirit which is not commonly represented. The *Didascalicon* set out to instruct what one should read, the order in which one should read it, and to comment upon the text. Hugo de Saint-Victor saw four main sciences, which contained all the others: theory, which works towards the speculation of reality, practice, which is the discipline of morals, mechanics, which dispenses the necessary actions for life here below, and finally logic, which teaches one how to speak well and discuss with finesse. Theory was itself divided into theology, mathematics (or *quadrivium*) and an as yet imperfect system of physics. Mechanics comprised: (1) clothing (*lanificium*); (2) armament (*haec duas habet species: architectonicum et fabrilem architectonica, dividitur in coementariam quae ad latomos et coementarios; et in carpentariam quae ad carpentarios et tignarios pertinet aliosque hujus modi artifices. Fabrilis dividitur in mallestoriam (...) et in exclusioram*); (3) navigation; (4) agriculture; (5) hunting; (6) medicine; (7) the organization of entertainments (*theatrica scientia*).

Briefly then, these are the various aspects of material life. But the mechanical arts are only sciences to a certain degree: *Poteste namque idem actus, et ad philosophiam pertinere secundum rationem suam, et ab ea excludi secundum administrationem. Verbigratia, agriculturae rationem philosophi est, administrationem rustici.* Agronomy is a science, cultivation [is] the countryman's toil.

These texts are interesting for several reasons. Note briefly that this original classification of knowledge also incorporates techniques. And, in this last case, a form of technology, albeit still partial, undeniably exists.

At almost the same epoch, Domingo Gundisalvo added to al-Farabi's statistics of sciences new ideas which are in the same spirit as those of Hugo de Saint-Victor. The *De scientiis* was presented as a veritable treatise for careers' advising. What we today rather narrowly call science, formed in the classification of al-Farabi, *scientia doctrinalis*; arithmetic, geometry and optics (*scientia de aspectibus*), the science of weights (*scientia de ponderibus*, in other words statics and metrology), and lastly, the science of machines (*scientia de engeniis*) or technology. One should also note that speculative mathematics, popularized by Nicomachus and Boethius, which was paralleled with the arithmetic taught by treatises on abaci or algorithms (*arithmetica alia est pratica, alia est theoria*), found numerous practical applications. Similarly, geometry could be either active or speculative, the former being employed by surveyors, carpenters, smiths and builders.

The incorporation of the science of machines or engines in doctrinal knowledge was the major new idea. "The science of engines teaches us how to imagine and invent a way of adjusting natural bodies with *ad hoc* devices, conforming to a numerical calculation which may be applied to whatever purpose we wish." We may see the extent to which techniques, in the widest sense of the term, were integrated with didactic knowledge. The famous remark made by Leonardo da Vinci, "Mechanics is the paradise of mathematics," is at the end of the road.

We find the same ideas with Vincent de Beauvais in the 13th century (*Speculum doctrinale*). He praised architecture to the skies and drew many forms of knowledge from it, particularly geometry, arithmetic for estimates, music for acoustics, and astronomy for sundials. We may therefore observe the same viewpoint and aims.

In 1296, Raymond Lulle reviewed these various sciences in his *Arbor scientiae*. Apart from *trivium* and *quadrivium*, he cited successively: *de arte fabrili* (metallurgy); *de arte carpentaria seu aedificatoria* (building); *de arte sartoriae* (clothing); *de arte agricolae* (agriculture); *de arte mercimoniae* (commerce); *de arte nautarum* (navigation); *de arte militiae* (military art). Like his predecessors, Lulle emphasized the need for theoretical knowledge required by technicians. A certain tendancy is therefore extremely clear. A didactic technique existed which closely affected speculative science, where in some instances the technique itself became a science and in every case, its usefulness was a guarantee of its respectability.

Apart from this position of principle, other interesting elements may be observed. Scholars themselves maintained that science and technology were closely linked and that if technicians required theoretic knowledge, then science itself also required technology. Scholars' technical preoccupations were genuine, and from the 13th century they were to shatter the rigid barriers to which our manuals have made us accustomed. The *Liber de ratione ponderis* by Jordanus de Nemore, which was certainly written before 1200, deals with inclined planes and curved levers, and puts forwards a number of propositions, some of which were incorrect, which demonstrate, as Guy Beaujouan has shown, an undeniable interest for the work of engineers. This is also true of the 13th century work by Gerard of Brussels, *Liber de motu*, or of William of Moerbeke who had already translated the works of Archimedes in around 1276.

Roger Bacon has drawn the portrait of Pierre de Maricourt who, according to him, incarnates the ideal scientist:

> He knew by experiment the laws of nature, medicine and alchemy, as well as matters concerning the heavens and here below (...) He investigated the methods of metal casters, and what they do with gold, silver, other metals and all minerals; and he learned all about warfare, arms and hunting; he examined everything concerned with agriculture, land surveying, and the work of peasants; he even studied the doings of old witches, their divinations and incantations, and those of all the sorcerers, and likewise the illusions and tricks of all the jugglers (...) But this admirable man was not admired for his real worth, for had he wished to remain in the entourage of kings and princes, he would easily have found himself covered in honours and riches.

Although techniques and science had many and various links during the epoch which interests us here, this could not be otherwise in the field of combat. Moreover, the last phrase of Bacon's text which we have just cited proves this readily. The favours shown by the great clearly demonstrate the importance of material problems within a civilization which has been depicted rather too systematically as concerned with moral or religious problems. Certainly many princes honoured engineers and technicians, particularly military technicians who ensured their ruler's glory and power. Technicians were therefore received with honour at many courts. Roger II of Sicily called Arab mechanics to his court, and the Hohenstaufen, especially Frederick II, followed this taste. In the 13th century, Alphonse the Wise, in Spain, had great respect for the engineers whose talents he employed.

As is so today, distinctions were certainly established between various techniques. There were respected techniques and poor relations. As always military techniques were frequently the first consideration, and this was a substantial reason, albeit not always the major one, for progress. We should also mention the marvellous techniques made in automated toys and other things of amusement enjoyed by some princes. In both cases, such inventions were frequently beneficial to the whole technical system.

It is curious to note, moreover, the permanence of some sectors of technical research. Here we may bridge the gap between the work of those mechanics of Antiquity we have discussed, and the ideas of the Renaissance, all too often illustrated solely by Leonardo da Vinci. In the *Epistola de secretis geribus*, Roger Bacon prefigured the great Florentine and provides a detailed picture of the technical preoccupations of his age:

It's possible to make Engines to sail withall, as that either fresh or salt water vessels may be guided by the help of one man, and made sail with a greater swiftness, than others will which are full of men to help them.

It's possible to make a Chariot move with an inestimable swiftness (such as the "currum falcati" were; wherein our forefathers of old fought,) and this motion to be without the help of any living creature.

It's possible to invent an Engine of a little bulk, yet of great efficacy, either to the depressing or elevation of the very greatest weight, which would be of much consequence in several Accidents: For hereby a man may either ascend or descend any walls, delivering himself or comrads from prison; and this Engine is only three fingers high, and four broad.

A man may easily make an Instrument, whereby one man may in despight of all opposition, draw a thousand men to himself, or any other thing, which is tractable.

A man may make an Engine, whereby without any corporal danger, he may walk in the bottome of the Sea, or other water. These Alexander (as the Heathen Astronomer assures us) used to see the secrets of the deeps.

Such Engines as these were of old, and are made even in our dayes. These all of them (excepting only that instrument of flying, which I never saw or know any, who hath seen it, though I am exceedingly acquainted with a very prudent man, who hath invented the whole Artifice) with infinite such like inventions, Engines and devices are feasable, as making of Bridges over Rivers without pillars or supporters.

Does this not present the notion of a degree of technical progress? Does not this demonstrate the belief that all material advances were possible? This was, in any case, to be the programme which all technical research was to take up until the Renaissance, illustrated in the famous letter by Leonardo da Vinci which is virtually an exact copy word for word of the above text.

jewellery, alchemy, automata etc. Some of these techniques were covered very thoroughly, such as colourings, and to a large extent, metallurgy.

Two other types of document, concerning more particularly architecture and mechanics, stem from these books of formulae. Architecture is more personal in nature, as this involves the notebooks of lists which architects and engineers kept for their own use, noting down anything about their craft or which attracted their curiosity. The only work of this type preserved from this epoch is the album of Villard de Honnecourt which probably dates from the first half of the 13th century. He cannot have been the only one in this genre, which enjoyed great popularity during the Renaissance.

The second type of document represents a more advanced type of notebook. These are compendiums of machines, drawings and explanations, arranged around specific themes, many of which were military in nature. In 1268, the engineer Assaut wrote to ask for an audience with Alphonse de Poitiers, who was preparing for a crusade, in order to show him his collection of machines. In 1328, Guido da Vigevano made the same request to Philip VI, proposing to show him a treatise comprising entirely of a collection of machines designed for use in distant wars. We still have this manuscript. Here we have the first examples of what were to be termed, two centuries later, "theatres of machines" (fig. 1).

Techniques had therefore provided much food for thought during the Middle Ages, organized to a greater or lesser degree, and more or less systematic in nature depending upon the subjects considered. Certainly, the technological aspect was still slight, but not completely absent: empiricism on the whole remained dominant. So too were technical instructions which encompassed the status of corporations: this primarily concerned bans rather than actual instructions on the subject of the employment of a particular material or the use of a certain process. Yet there was nevertheless a desire to instruct the world in the best of the techncial knowledge of the age.

We should not forget either the intellectual entourage surrounding techniques. Very rapidly – for whatever reasons of wealth or power – authorities, scientists and philosophers leaned towards the problem of technology. While our lack of texts, particularly for more remote periods, should not be forgotten, we may nevertheless note that this preoccupation for techniques manifested itself around the middle of

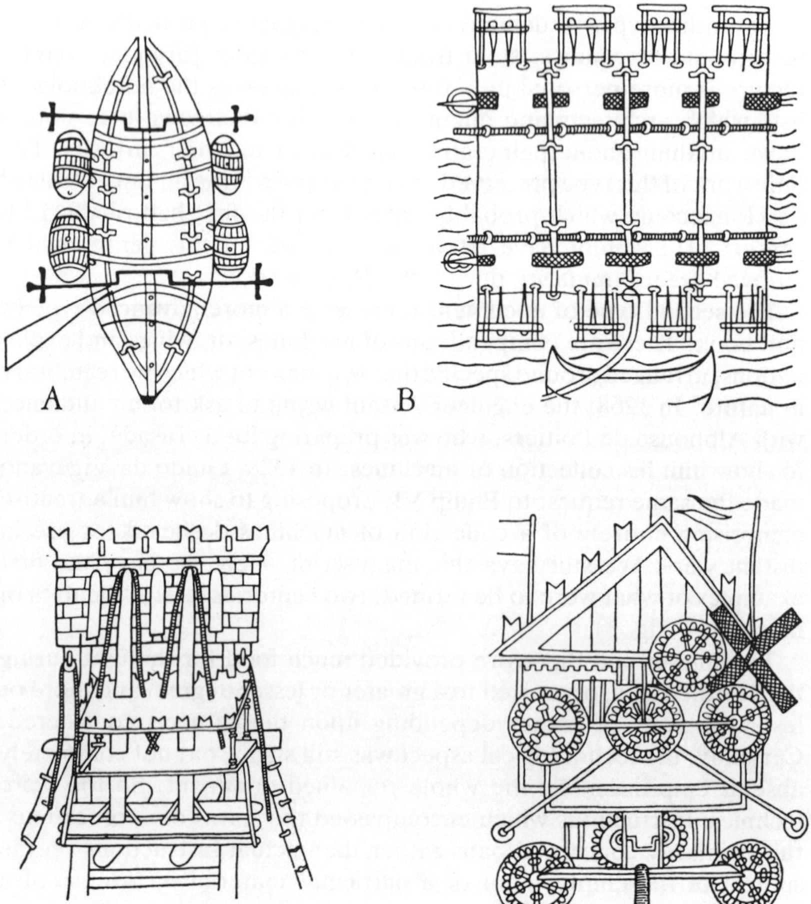

Figure 1. — Machines designed by Guido da Vigevano (circa 1327). A, crank-propelled paddle-boat; B, portable suspension bridge; C, movable tower; D, wind-driven fighting car.

the 12th century and continued to develop until the end of the 13th and the beginning of the 14th century. Here an additional concordance should be emphasized. Apart from a few rare examples, there is hardly any of the contempt shown for techniques which has been so frequently suggested. In order to apply this thesis, might one not suggest that at the very beginning of our period, it was the Cistercian monks who initiated the rebirth of iron and steel making, developed the use of hydraulic energy, and invented new building methods? The introduction of a working monasticism and the vigour which it gave to some techniques would seem to constitute unquestionable proof. Was it not precisely as a result of this new spirit that techniques could once again make progress? In order to reply to a developing economy and an increasing population, they attempted to gather together both fairly ancient techniques and different traditional techniques from those of Antiquity, and to create an ensemble of techniques at a level on which a new technical system could be formed. This necessitated the re-reading of certain texts and the collection of particular facts with this aim in mind: in such a way, matters could be clarified.

The question of a natural context is equally important. The centre of gravity of the medieval West was no longer the Mediterranean region, but had visibly shifted further north. As all technology is necessarily dependent upon natural conditions, and even more so in early epochs than nowadays, new or adapted techniques naturally appeared. As we will see, although Antiquity did indeed invent certain methods and machines, the natural conditions of its geographical space prevented or prohibited any further development. Aside from these natural forces, economic factors also played a part, the effects of which are quite obvious: the climate is one of the most restricting forces.

The natural environment of the Middle Ages in the West was therefore not exactly the same as that of classical Antiquity. The Mediterranean border of Europe had certainly not changed, and techniques which had barely evolved were still in use: biennial land rotation, the use of the swing-plough, the continuation of the kiln and many other elements are evidence of this. As one moves towards the North, so conditions change totally. Lands are lush, dense and rich, and require modified tools. The Gallic plough, very different from the Roman swing-plough, had already extended into the Po Valley from the time of Pliny the Elder. Different crops were also cultivated and there was a return to the widespread growth of certain Southern plants, in particular the vine. On the other hand, plants which could not be grown in the dry climates of these regions could now be cultivated. Forests had a greater variety of species, thus enabling the medieval West to become a civilization of wood, far more so than had been the case in Antiquity. These Northern forests regenerated themselves more easily, providing a great quantity of wood which Antique civilizations had never known: the Parisian middle classes of the 13th century could use a fireplace for generous heating, consuming a lot of wood, while the Romans in the time of Caesar could only use a brazier. This is also true of water systems, as rivers and waterways flowed more regularly and did not suffer the drought periods of Mediterranean rivers. The proliferation of the water mill, developed moreover in Late Antiquity, was the most immediate result of this different state of affairs. Metals were more abundant, especially iron ore, deposits of which were scattered throughout almost the whole of Northern Europe. Rome only enjoyed the exploitation of real mineral deposits after the conquest of Spain, and the Empire's iron came from Alpine or even more northern regions, and from still barbarian areas, even at the height of its expansion: recent discoveries in Poland seem to demonstrate that deposits between Warsaw and Cracow supported the Roman Empire, endowing them with considerable importance.

Exchanges – and who will deny the pressure of exchanges on production – also required modification. The Antique Mediterranean survived on maritime commerce, but medieval Europe was to survive above all on land commerce, although with much more difficulty. Antiquity could concentrate its food-producing crops, as the circulation of food stuffs was easy, and could build marble temples, for marble columns were transported by ship (as underwater excavations have proved). Difficulty in transportation implies on the contrary that centres of production were widely dispersed in the Middle Ages. This explains why the vine was cultivated in higher regions, why smaller deposits of iron were exploited, why an undifferentiated and extensive cultivation was practiced, why the number of fish ponds increased, and the reason why local material was used. It is only at the end of our period that maritime trade was to be revitalized.

Aside from these physical conditions which were essential for material life, other economic factors played an important part. Here again, astonishing chronological coincidences may be observed. The history of the climate, which should one day be included in our general history, shows a certain unity, characterized by a glacial thrust. Thus temperatures generally cooled down around 1200, perhaps even as early as 1150, to 1300, maybe even to 1350, according to the outline suggested by Le Roy-Ladurie. The climate seems to have been fairly mild during the four centuries from the Carolingians to the great clearings. At this point in time the vine began to advance towards the north, although in the 11th century the hazel tree had not won back its optimum positions in the north of Scandinavia. From the end of the 12th century onwards, failing rainfall and rapid evaporation gave way to a far more humid and cooler climate. Even if such conclusions seem hasty, and without exaggerating the differences in temperature between periods, this fact and its limits in time should be noted.

Demographic movement is the last of these great conjunctional movements which exerted a considerable influence on economic developments and subsequently on techniques. The rapid progress of demographic expansion is obvious, and virtually every contemporary author remarked upon it. It began around the middle of the 12th century and continued into the second half of the 13th century, perhaps even to the end of that century. Evidence available to prove this expansion emphasizes the significance of this phenomenon: extensive clearings could only be undertaken by a larger population, and it is precisely for reasons of increased food requirements, very clear urbanization, (at least in some regions), together with all the implications involved in the development of towns, (particularly in the domains of industry and commerce) as well as social changes, which are the elements which naturally exert pressure on the technical systems in force. A more detailed study of the beginnings and the rapid take-off of this demographic growth would be most useful: some regions were more favoured than others in this respect.

Contrary to this, at the other end of our period, the importance of demographic experience may have placed a brake on technical progress, while the interval between the two movements had become too considerable. It is a generally accepted fact that an extremely dense population tends to preserve traditional techniques, as is exemplified by China at the same period, but undoubtedly there is more to this than meets the eye. Conversely, technical stability in the face of population growth prompts imbalances which present serious drawbacks to the economy. We will see that certain authorities have emphasized the inability of farming techniques to reach the level required by the demographic density of the time, although clearing work had been undertaken. The transition from extensive to intensive agriculture could not be achieved, resulting in considerable food deficits. The great catastrophes of the 14th century, ranging from the first famines of 1316-1317 to the Great Plague of the middle of the 14th century, from the monetary changes of Philip IV the Fair to the great Italian crashes again during the middle of the 14th century, undoubtedly stem from these internal tensions in the medieval West in the centre of which were the problems posed by techniques.

In the light of the intellectual viewpoints discussed before, it would seem that the great technical movement of the Middle Ages should be situated within this whole natural complex, so that we might better comprehend this phenomenon. Let us remind ourselves of the two essential points in our demonstration. The first concerns chronological order and we have attempted to underline all the concordances involved. The second concerns the constitution of a new technical system which is largely different from Antique systems. We must therefore examine this new technical system.

THE INNOVATIVE FIELDS

We should certainly be cautious in our investigation here in order to avoid error. Any new technical system does not necessarily imply general progress and total upheaval of the old system. In some sectors the evolution need only be significant enough to initiate profound change in the conditions of economic life and to maintain some coherence between traditional sectors. If, at the end of the period in question, economic growth produces, in an unequal way, imbalances between the various advanced sectors, or between these and traditional sectors, a discrepancy will arise between technical structures and other structures, which will in turn lead to a state of crisis. It is with this example in mind that the progress of medieval techniques should be studied.

One cannot overemphasize the importance of distinguishing between the birth of an invention and its generalized use, in other words its innovation. Social, economic and natural conditions thus come into play. The history of techniques is therefore situated on two levels, but it is easier to exemplify this by looking at concrete examples. The water mill was used in Antiquity, modern harnessing was known in the East from the 2nd century AD, and agriculture was far less developed than iron and steel making techniques: these are accepted facts. On a higher level, one might suggest that for various reasons the Western Middle Ages had at their disposal the hydraulic energy necessary for the generalization of the use of the water mill, that from a particular time onwards a more progressive economy based on products from the land could benefit from modern harnessing, that the richness of the land, the climate and the possibility of extensive farming could dispense with technical changes in cultivation methods, and that any other conditions merely formed previous but uncirculated experience which could be coherently organised within a system from the middle of the 12th century. In this case, we are concerned not so much with listing inventions but eventually demonstrating that a new technical system emerged, necessarily heterogeneous in its origins, simultaneously composed of old inventions, some new techniques, and a degree of conservatism.

The importance of the quantity of available energy has always been an essential factor in the economy, even more so for economic growth. On the other hand, forms of this available energy or energies (as there are usually several types at the same time), play a significant role in the evolution of techniques which are closely dependent upon this factor. An increase in the availability of energy exerts a very clear pressure on all the techniques situated "downstream."

Antiquity already knew and used every form of energy employed by the Middle Ages in the West: animal energy had long been employed, wind power, particularly for ships, had been used for some time, and hydraulic energy was employed from the 1st century AD. The major achievement made in the Middle Ages lay in having considerably increased the quantity of available energy and using it more effectively. The major assets of our period are therefore of two types. On one side, there is an infinitely greater quantity of natural energy worldwide: water, wood, and animal rearing progressed as a result of natural grassland for grazing. And on the other side, undoubtedly owing to this abundant energy source, a diversification of its uses was possible with the employment of old or recent mechanisms in which there had been little previous interest, due to the lack of sufficient energy resources.

Truth to tell, very little may be said on the subject of animal energy. Nature undoubtedly had a greater hand in things than the adoption of new energy converters, the exact limits of which we will examine later. The abundance of the grasslands led to an increase in the number of animals and gave them greater strength and resistance. We will consider later the changes in harnessing and the introduction of the iron horseshoe which meant that the horse could be used better and more efficiently. The problem of breeds may possibly have entered into this, but our information on this subject is rather scant: new breeds were probably imported by the Barbarians, and breeds were improved by selection and cross-breeding. It was during this period that breeds from the Perche and the heavy draught-horse from the Boulogne region were introduced. We know that veritable stud-farms existed, particularly in Normandy, where these slow changes may have originated. We have already considered above the progress of veterinary medicine. Although the history of the horse is better known, cattle, on the contrary, which were essential for farming work and for some types of transport over a short distance, have not yet been the subject of any serious study.

The large scale use of hydraulic energy undoubtedly represents a substantial medieval innovation. Hydraulic energy was used in Antiquity, but for natural reasons it could not enjoy the expansion it warranted. Between the 1st and 5th centuries the use of hydraulic systems became more widespread. We may observe here the prodigious expansion of the use of this form of energy in more northern regions around the middle of the 12th century. Although statistics for this may be presented, we should point out, however, that we are necessarily dependent upon our documentation which is only really copious from the middle of the 12th century. Having said this, some interesting figures are available. At the end of the 11th century, the *Domesday Book* mentioned in its census 5624 water mills in England, which may represent approximately 16,000 horsepowers. In the Aube department in France, where all the texts have been reviewed, they numbered 14 in the 11th century, 60 in the 12th and over 200 in the 13th century. In the Forez, where all such texts have been similarly published, there are none counted for the 12th century, but there are

70 in the 13th century. According to all the studies undertaken in this field, it seems that the use of hydraulic energy must have already been manifest from the 9th century onwards, at least in some regions. Hydraulic energy gradually increased in importance and appears to have grown rapidly from the middle of the 12th century.

A few points may be clarified with the aid of figurative representations, a recent census of which has been almost completed (fig. 2). All the examples we have of these, from between the 9th and the 13th centuries, illustrate mills with vertical wheels, which was also the system used in Antiquity. The arrangement of the wheel in relation to the flow of water is equally important: the overshot wheel, powered by water falling on the wheel from paddles above, has the highest output, of approximately 75% against the 60% output of the breastshot wheel, where the wheel is placed at the side, and the 25% productivity of the undershot wheel, driven by a current moving against the paddles at the bottom of the wheel. The position of the wheel is closely linked to hydraulic conversion. The three wheel positions were all known in the Middle Ages as they had been in Antiquity, but our texts do not permit any definitive conclusion to be made on this subject.

The water mill was infinitely more efficient than traditional techniques. Most certainly, the continuation of hand-grinding, which was firmly defended in some regions, proves that the water mill was not economically vital. In order to maintain its use, it was necessary to implement a ban, but this was very rarely applied to other uses of the water mill. In other instances, on the contrary, there was considerable economic gain to be made and primitive techniques were no longer used, to such an extent that in some cases mills became collectively owned, such as fulling and tanning mills.

Let us look now at a last form of mill which used a particular type of energy, that is the tide mill, which operated with the ebb and flow of the tides in areas where holding dikes could be readily built. It is easy to understand why Antiquity had not thought of developing these. The tide mills at the port of Dover apparently existed from the time of William the Conqueror, and mills on the Adour River, in France, are mentioned around 1120-1125.

Apart from sailing, the use of wind energy was introduced in the West during the medieval age. Texts would seem to prove that windmills were used in the 7th century on the plateaux of Iran. Two

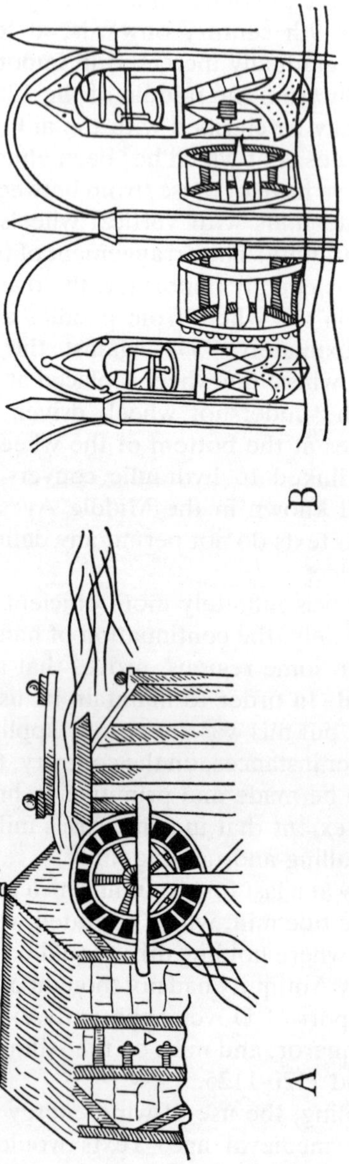

A

B

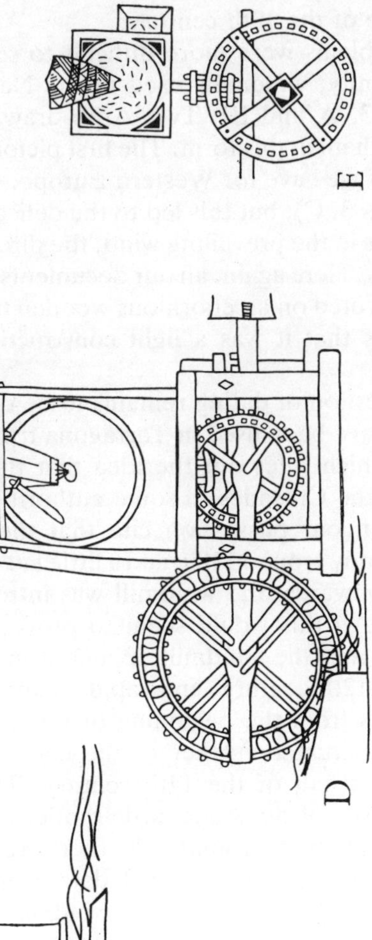

Figure 2. — The watermill. A, mill with overshot wheel (Luttrell psalter, circa 1338); B, floating mills or boat-mills (legend of Saint Denis, circa 1317); C, mill with undershot wheel (rentier of Audenarde, circa 1270); D, mill mechanism ("Hortus deliciarum", end of the 12th century); E, mill mechanism (fresco of San Isidro in Madrid, second half of the 13th century).

Persian geographers mention windmills in the province of Seistan around the middle of the 10th century.

Mechanical problems were more difficult to solve. It is probable that up until recently, windmills used in the Near East were on a vertical axis (fig. 3, A and B). Two Arab drawings from the 13th century still show them in this form. The first pictorial representations of windmills which we have for Western Europe, show the mills on a horizontal axis (fig. 3, C); but this led to the delicate problem of how to position the sails in the prevailing wind, the direction of which may change appreciably. Here again, all our documents tally: as a whole all these windmills pivoted on an enormous wooden tripod, or post-mill, which presupposes that it was a light construction always built in wood.

The difficult question of dating remains to be discussed. In Spain, windmills would have been used in Tarragona from the 10th century onwards, a fact which rules out the idea that the windmill was an importation from the Crusades as some authorities have suggested. The oldest texts in our possession cite that there was a working windmill in Normandy around 1180, and a little later in England. From the 13th century onwards, the windmill was introduced throughout Western Europe. All this would seem to prove, apart from a few exceptional cases, that the windmill appeared in our regions in the second half of the 12th century, and rapidly expanded to areas with regular wind forces from the beginning of the 13th century, finally reaching countries situated further to the east, such as Poland or Sweden, in the first half of the 14th century. The oldest pictorial representations, those of the Audenarde rentier and the Saint Vaast d'Arras psalter all date from around 1270: there are no representations in the terrier of the bishop of Lille, which dates from the same period. We are quite sure that at this date the windmill was only used for grinding grain.

We have already pointed out the exceptional abundance of wood in the more northern zones of Western and Central Europe. This therefore provided a considerable source of thermal energy which classical and Mediterranean Antiquity did not possess. The importance of this source of energy, for a great number of industries (in lime, glass and metal production), for heating, and for the fabrication of bricks and tiles, need hardly be emphasized. It stemmed from natural resources.

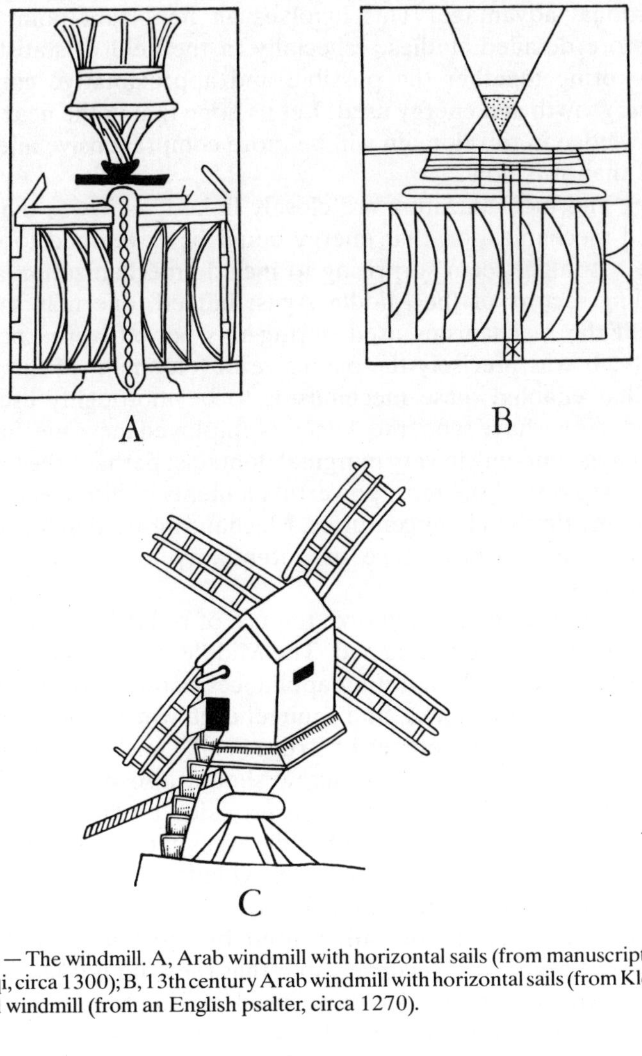

Figure 3. — The windmill. A, Arab windmill with horizontal sails (from manuscript from Al Dimashqi, circa 1300); B, 13th century Arab windmill with horizontal sails (from Klemm); C, medieval windmill (from an English psalter, circa 1270).

It therefore seems obvious that the availability of energy was far greater in the Middle Ages than it had been during Antiquity. Energy sources were more considerable and also more diversified, which was an additional advantage. This involves an important point which merits more detailed studies, especially in the field of statistics, in order to bring together the possible and approximative curves of economic growth and energy used. Let us hope that in the near future our knowledge in this domain will be more comprehensive and more detailed than it is now.

Energy and mechanization are closely linked, as it was thanks to advanced mechanization that energy could be developed to its full potential. It might seem surprising to include mechanization among the leading sectors of the Middle Ages: indeed, one may say that almost all the mechanisms used during this period were known in Antiquity. It was precisely the extensive development of the use of energy that enabled these mechanisms to be thoroughly used and improved. As we have seen, the Ancients employed these mechanisms in many cases, but only in very marginal domains: perhaps the taste for automata was one of the most powerful elements of discovery, as will be seen many times in later centuries. Mechanisms used in the Middle Ages progressed from a stage of potentiality to one of concrete technical reality.

There is little need to emphasize the use of reduction mechanisms which have already been discussed. The Middle Ages had in fact only developed the use of those lifting appliances handed down by Antiquity: pulleys, pulley blocks, and squirrel cage wheels (and one may think here of the famous cranes of Bruges and Luneburg). Similarly, the use of the worm or endless screw should also be pointed out, examples of which have already been provided by Hero and Pliny, particularly for screw presses. The Middle Ages extended its use to the jack, a perfect illustration of which is provided by Villard de Honnecourt in the first half of the 13th century.

We know little about the lathes used by the Ancients. If they possessed the brace, it is inconceivable that they did not use the the crank and wheel system, like those used in simple lathes, with a continuous drive, a good example of which is the hand-driven spinning wheel. By using a large wheel, a fly wheel could be formed, which if necessary left the driving hand free, at least periodically, for other uses.

Most medieval lathes were drawback lathes wood, and were therefore reciprocating. A few rare pictorial representations of these remain, and two types of lathes may be distinguished among them:

A) The single treadle lathe was based on a system of belts wound around the lathe's axle and linked both to the treadle and a spring (either in a wooden branch or pole). Operating alternately, the treadle and the spring turned the lathe in one direction and then the other (fig. 4, A and C).

B) The two treadle lathe was based on the same system, but the driving belt was activated alternately by one treadle and then the other. One of the rare illustrations of this can be seen in a stained-glass window in Chartres cathedral (13th century; fig. 4, B).

A typology of the uses of the water mill still remains to be done. We do possess, however, a few important elements. Direct transmissions with the possibility of reducing and changing planes were achieved by using cogwheels and lanterns with staver. Antiquity knew of these mechanisms and had even begun to construct the theory concerning them. On the other hand, the water mill could activate balanced apparatus and spring-operated systems, owing to the development of the cam. We may therefore classify the principal uses of the water mill into the following main categories:

A) Direct use of the circular movement with simple changes in planes, intensity, and speed: these are generally millstones for crushing or grinding. Our present knowledge enables us to draw the the following list:

1) the corn or wheat mill known from Antiquity;

2) the oil mill: quoted in the Forez in the 13th century;

3) the mustard mill: mentioned in the 13th century in the Forez;

4) the beer mill: vague texts in the 9th century in Picardy, and more certain texts from the middle of the 11th century;

5) the sugar mill: archeological examples from the 11th century in Morocco;

6) mills for grinding; the end of the 12th century in Normandy and in Beauvaisis;

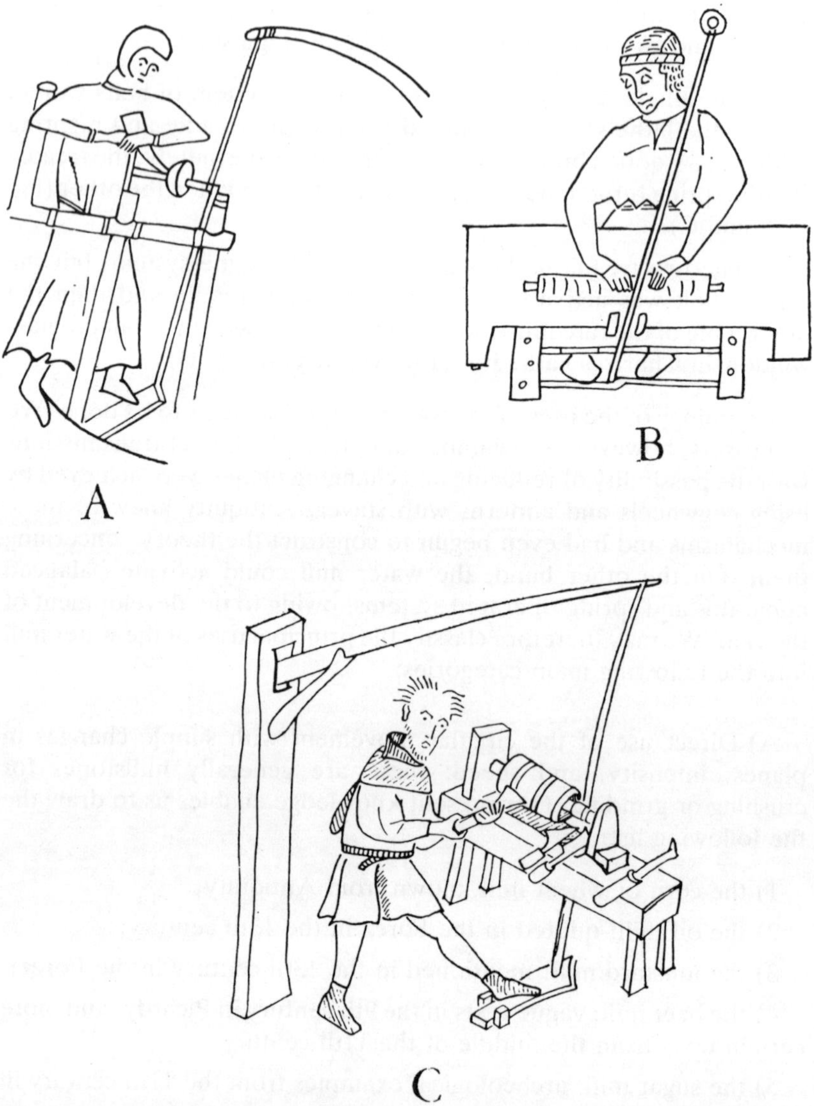

Figure 4. — Lathes. A, pole lathe (from "La Bible *Moralisèe*, early 14th century); B, lathe with a pulley and two treadles (from a stained-glass window from Chartres, 13th century); C, pole lathe (from the *Mendel Bruderbuch*, circa 1404).

7) mills for turning: these are only cited in the 14th century in the Dauphiné in France, but they could have been in existence before this date;

8) mills for raising water: these are attested to by Vitruvius, in the Near East in the 9th century, in Spain from the 10th century onwards (fig. 5, top), and from the 11th century further north where they were used less;

9) some mills using hammers which could have originally operated with millstones, as has been substantiated by certain texts: dyestuff mills from the 13th century in Namur; examples of tanning mills from the 11th century.

B) The use of the cam for balanced or spring loaded apparatus:

1) iron mills: in Sweden from the end of the 12th century;

2) copper mills: operating on the same system as the preceding, but we do not possess any ancient example of them;

3) fulling mills: from the end of the 12th century in Normandy, Piedmont, and in the Milanese;

4) paper mills: in the middle of the 13th century in Spain;

5) hemp mills: in the 13th century;

6) saw mills: the only example of a spring mechanism is illustrated by Villard de Honnecourt, circa 1250 (fig. 5, bottom).

Obviously our knowledge needs to be more detailed and international research should be undertaken to this effect, but this simple list demonstrates the way in which the water mill constituted, in many areas, a real industrial revolution, to use a rather ambiguous term but one which has been confirmed through use. Although the water mill was known before the 1st century AD, and although various mechanisms were known to the Ancients – then again they had not been applied to hydraulic energy – one notes a totally new and remarkable situation from the 11th and 12th centuries. It was only from this time onwards that techniques for using hydraulic energy really came into their own.

There was, however, a new type of mechanism in existence in the Middle Ages: war engines. Some of these, such as rams or revolving towers, were very old. But it is true that the Middle Ages had

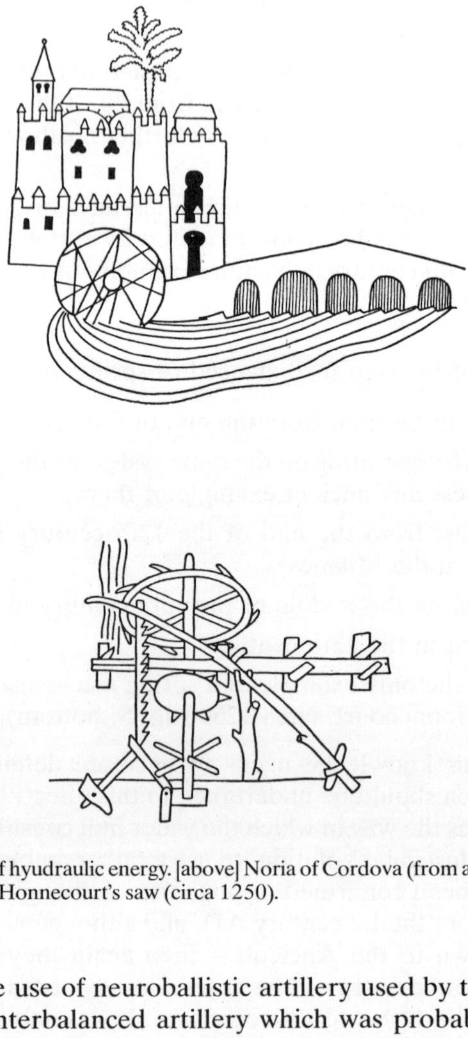

Figure 5. — Uses of hyudraulic energy. [above] Noria of Cordova (from a 14th century seal); [below] Villard de Honnecourt's saw (circa 1250).

abandoned the use of neuroballistic artillery used by the Ancients in favour of counterbalanced artillery which was probably more powerful. Trebuchets used the ballistic properties of the sling, by employing a suitable mechanism, with greatly increased dimensions. This in fact consisted of a very large pole which pivoted between two treadles and powered a huge sling and counterweight (fig. 6) The mangonel was a variant of the trebuchet. A straight long shot and a falling shot could therefore be obtained by using these two systems. Without doubt some essential rules had been established for firing a shot, taking into account the weight of the ball, the distance to be covered, the length of the pole, and the angle at which the ball should be released, all of which were certainly calculated in an approximate and empirical way.

In such a way, an arsenal of "engines" was built up in the West during the Middle Ages. Some new machines were created which had not existed in Antiquity, but more particularly they continued to extensively use mechanisms or machines which had already been developed, the employment of which had remained very limited. The water mill and its various applications constituted the bulk of this mechanization.

It is often more difficult to comprehend the extent of the problems involved in transport. Before even broaching questions which have formed the subject of well-known books - the fame of which has undoubtedly distorted the perspective of this topic – it would be useful to clarify certain points concerning harnessing:

A) This new equipment was designed for the horse, the harnessing of cattle not having been modified. All the research undertaken shows that on equal terms the horse is tougher and faster than the ox. Yet one wishes to be able to break down the uses made of one or the other animal.

B) A second important point arises. Where transport is concerned, it would be indispensable to know – albeit in simple figures – on one hand, the volume transported, and on the other, the method of transport employed. The following essential facts have emerged from certain recent works: (1) in the first place, the volume of exchanges was relatively poor, as can be clearly seen by the number of tollgates; this is also demonstrated by the obvious diversity of a region's productions, even if they were limited, which was designed precisely to avoid the difficulties presented by transport; (2) in the second place, the modern horse harness could have found its perfect application insofar as large carts could have enabled the transportation of a particular volume or weight: we know that the absence of a pivoted front axle system made a four-wheeled cart virtually impracticable. Moreover, this would have required a developed and well-maintained road network; (3) this second observation means that animals had to be used for transporting

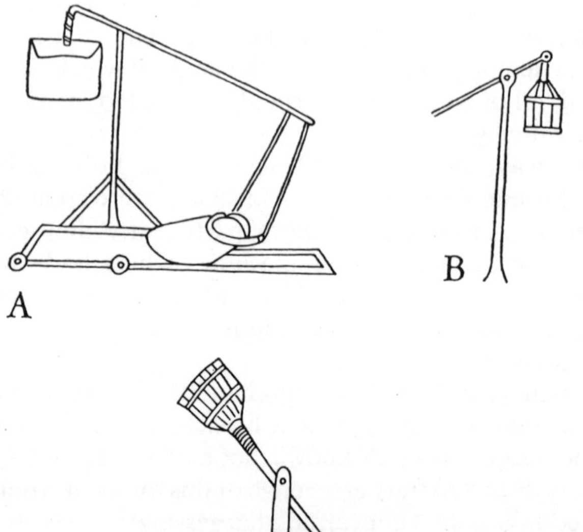

Figure 6. — Assay balances. A and B, from a Parisian manuscript from the early 14th century; C, from a 14th century manuscript from Cambridge; D, assay balance mounted on a boat, from the *Roman de toute Chevalerie*, Paris and Oxford, 14th century.

most goods (here the iron horseshoe could intervene) or were carried by waterways.

Thus, although the iron horseshoe was a valuable technique, modern harnessing, on the other hand, was certainly of little interest, apart from transportation over short distances, and was restricted to agricultural uses which we will examine later.

We have a few pointers on the subject of the iron horseshoe which may give a fairly accurate idea of the invention and its significance. From Antiquity onwards, it was well-known that the horse's hoof was relatively fragile. But apart from this, the initial idea of driving nails into a horse's hoof could not have been conceived without the knowledge of the composition of the various tissues of the hoof and their function. The complete absence of any mention of this in ancient texts proves conclusively that classical Antiquity never knew of the horseshoe. Horseshoes may have been at the most just protective devices, designed to prevent lesions of the hoof, in particular laminitis which was undoubtedly fairly frequent at this period. Anything said on the subject of horseshoes regarding the Barbarians is purely hypothetical.

Concrete facts to substantiate our claims are provided by reliable texts and objects. For the former, we actually have specific references: the Byzantine *Tactica* of Leo VI (886-912), and the *Miracula sancti Oudalrici* by Gerhard (973). References become much more numerous from the end of the 11th century. As for objects, the Crécy museum and the blacksmith museum at Saumur, in France, present examples which could date back to the 9th century: these are considered to be the oldest known horseshoes in existence.

Studies to date – and some worthwhile syntheses have appeared on this subject – enable us to suggest that the iron horseshoe could have been developed at the same time for the draught-horses in Byzantium and the West. Initially, small shoes seem to have been used: this is illustrated on a tapestry belonging to Queen Matilda. They are similar to Chinese horseshoes, although the latter were smaller, more delicate and fixed with violin peg nails. Shoes found in Norman tombs of the 9th and 10th centuries, measured three to six centimetres in length and were made of a metal plaque two centimetres thick. A point in the middle formed the calk, fitted with two awls which were driven into the surface of the hoof. Subsequently the horseshoe clearly evolved, as it increased in size and thickness. There are still copies of shoes for sick horses which are preserved on the doorway of Chablis church: it may be that it was not the crusaders who nailed their horseshoes to the church doors, but more probably people who were grateful to some saint for having cured their animal.

When the book by Richard Lefèbvre des Noëttes first appeared, there was much discussion surrounding the origin and the date of the modern horse harness. We know that this technique was based mainly on the shoulder collar, the breast harness and the reins. Improvement had certainly been made on the Antique form of harnessing, which restricted the animal's breathing, although one geographer has suggested that in Antiquity the horse was trained accordingly.

Many quarrels surround the origin of these devices. Lefèbvre des Noëttes was content to point out their appearance in the 11th century, based solely on iconographic sources, and their rapid development in the 12th century. More recently, André-Georges Haudricourt has observed modern horse harnesses on Han bas-reliefs of the 2nd century AD. Using a linguistic approach, he demonstrates that Europe knew of the modern form of harnessing in the 6th century. Again according to this authority, the shoulder collar only arrived in the West in the 8th or 9th century. And still more recently, in the light of archaeological discoveries, Lynn White asked if modern harnessing had not slowly evolved from Antique harnessing – an idea which rules out any contribution from the East. The three ideas are interesting, and future research should be directed around these.

Here again, whatever the case might have been, we should emphasize the late appearance of many precise texts and images. It is true, however, that the modern horse harness seems to have been extensively used during the 12th century. Above all it was used for pulling farm tools, ploughs, harrows and hay carts, rather than for actual transport. In such a way, the horse, shod and suitably harnessed, was considered superior to the ox. English agronomists of the 13th century pointed out that a horse cost more to keep than an ox, but unfortunately we lack any evidence to judge whether this was so or not. Even if it had only limited use, the modern harness constituted real technical progress, although less so than the iron horseshoe which saw more widespread use.

Although we do not yet possess a complete history of the medieval ship, our information is at least on the way to becoming more accurate and more comprehensive. In many cases, material data is insufficient, apart from some spectacular finds, such as the Bremen ship. It is worth pointing out that the inventory of pictorial representations of the ship is today almost complete (fig. 7).

Firstly, Nordic and Mediterranean vessels should be distinguished, the techniques of which remained very different from each other before they became combined, both in hulls types and rigging systems.

We know something of the beginnings of Nordic ships, which have been better studied thanks to the abundance of images and astonishing excavations. A stone engraving, found on the island of Gotland and dating from the 5th century, provides an image which almost matches the two boats found at Kvalsund in Norway, which date back to the 7th century. The largest of these boats was 18 metres long and 3.5 metres wide. The planking of the keelson had already become a true keel. The hull was U-shaped in the centre and thus ensured some stability without the use of ballast. There seems to have been a mast, although no mast appears in the engraving or was found in the excavation. Indeed, engravings show that Nordic boats were powered by sail from the 6th century onwards.

The engraved stones of Gotland from the 7th and the 8th centuries, appear to illustrate two types of ship. Some with sterns which formed a sharp angle with the keel, and others which were curved upwards. One might suggest that Viking merchant vessels required sail power, even more so than war ships, because equipment was limited, and were fitted with a bobstay piece. Sails were often diagonally rigged, so it was necessary to reinforce them. The hulls could not be decked, and the mechandise was stacked in the middle.

The Sutton Hoo warship (England) dates from the 7th century. Different planking lengths were used, denoting a greater skill in boat building. The ship is 24 metres in length and 4.2 metres wide. The hold of 1.35 metres resembles that of a flat ship so it was easy to run aground. Thirty six frames reinforced the hull, and there were nineteen oars.

Two major discoveries were the Gokstad and Oseberg, found in Norway. The oldest ship, the Oseberg, is generally thought to date from the 9th century. It measures 21.4 metres in length and 5.1 metres in width. It was built in oak, fitted with twelve planks on either side, and has a deep keel which is slightly arched in the middle. The ends of the frames are pegged on the ninth and tenth planks, the tenth plank

Figure 7. — The evolution of the ship. A, viking ship, reef sails (La Rochelle seal, end of the 12th century); B, viking ship with fore and after castles (Dunwich seal, 1199); C, viking ship (Bayeux tapestry, 11th century); D, viking ship with fore and after castles (Hythe seal, end of the 12th century or beginning of the 13th century); E, advanced viking ship with bow and straight stern (Stralsund seal, circa 1250); F, viking ship with fore and after castles (Pevensey seal, 13th century); G, viking ship (Calais seal, 1228); H, primitive viking ship (Gravelines seal, 1328) but certainly much earlier than the date of the seal; I, viking ship (Calais seal, 1270, from an older model); J, earliest representation of the Kog (Ipswich seal, circa 1,200) K, Kog (Wismar seal, circa 1250); L, Kog (Elbing seal, circa 1242); M, Kog (Kiel seal, 13th century); N, Kog (commentary on the *Apocalypse* by Alexandre le Petit, from a Breslau manuscript, circa 1242); O, ship with stern-post rudder (from the baptismal font at Winchester, 1180). Note that all the representations of the kog show a single stern tiller.

being thicker and L-shaped. They are fixed to the other planks according to the traditional system and lie freely on the keel. This boat had fifteen pairs of oars and carried a mast and sail. The mast was placed in the front half of the ship, on a crone which rested on the crossbeam in two sockets. The deck was on intervals between the beams and raised at the stern and the bow to make a platform. The Gokstad ship, which was built in the 10th century, was much larger and better suited to sailing: it was made of oak wood and was 23.33 metres long and 5.25 metres wide. The planking consisted of sixteen strakes fixed to seventeen ribs. The eight lower strakes were still fixed to the ribs with lashings which passed through lugs or cleats left in each plank, (lashings were made of fine spruce roots). The other strakes were fixed together with iron nails. Round headed rivets took care of the holding. There were nineteen pairs of oars.

All these ships were clinker-built: the planking was made of slim planks, each overlapping the one below, and caulked with thick cords and animal hair. It was this technique which meant the planks could be so thin. The ribs were made of curved wood, and these were also very light. There was only one mast, which could be detached. Sails seem to have been first used in the 6th or 7th century; they were definitely in use by the 8th century. The mast was taken down if the oars were being used, and the rudder was a lateral type. These ships did not have actual decks, but simply crossbeams resting on props which were fixed to the ribbing. Built at the end of the 10th century, Olaf Tryggvason's *Long Serpent* was 42 metres long, and therefore of the same dimensions as the galleys of the 17th and 18th centuries.

Long ships were derived from the Gokstad type and they can be seen on the engraved stones from Gotland and the Bayeux tapestry. They are also depicted in almost exactly the same way on the seals of La Rochelle and Lübeck. The ships depicted on the Bayeux tapestry are already slightly different: one has oars only in reserve, for example. Their shape is more rounded, less refined, with more bearing, however. The flatter filling floors accentuated the freight character of these ships, they also made grounding easier, and provided stability when the ship was on the beach. These ships could therefore be decked. Although they were less easily manoeuvrable, they were more useful. They were probably the successors to the *knorr*, a merchant ship which gradually became more widespread in the 7th and 8th centuries.

By the end of the 11th century, the various types we have just described were rapidly evolving, and the differences between them were becoming more marked.

Both ends of the ship were raised to create the fore and aft castles, which eventually went beyond the width of the hull. It is possible that at first these castles were simply added to the original structure – the seal of Dunwich (1199) indicates this, for example. On the seals of Hythe (late 12th century), Pevensey (early 13th century) and Sandwich (1238), the castles appear to be light platforms supported by a structure of lancets and semi-circular arches. The seal of Dover (1284) marks the progress ot the permanent castle, which was now incorporated as part of the hull itself.

Sails also changed, and this brought about changes in the types of mast as well. Sometimes the sail was furled on the upper yardarm, in the Roman fashion (as on the seals of Dunwich, Dover, Hythe and Winchelsea): the topmen were therefore obliged to sit astride the yardarm, and this is evident in most pictorial representations. The reefing points which enabled just a part of the sail to be used, appears to have been a Scandinavian invention. These are first mentioned by Wace, a Jerseyman brought up at Caen (1120-1183) They are either situated fairly low down, as on the seal of La Rochells – and this would indicate that the yardarm could be lowered – or fairly high up, as in a Parisian manuscript from the 13th century. It seems that they were usually placed lower down, however (as indicated by the seals of Hastings, Bergen and Dublin). A spar is evident on the front of a ship depicted on the seal of Sandwich, and this may have been a forerunner of the bowsprit. The latter is featured on a miniature made for Eleanor of Castille in 1279. At first it was probably only used to provide a fixing, or more likely, a return point for a pulley, on the bowline of the single sail. The bowsprit rapidly came into general use, and was to have a significant effect on the development of sails: it can be seen on the seal of Elbing (1242), Wismar (1250), and Kiel (13th century).

There have been many attempts to solve the problem concerning the type of rudders used, but these do not appear to have been successful. Technically speaking, this was a relatively simple matter. The two lateral rudders were replaced by the stern-post rudder, like the one still used today. Technicians are still not agreed about the actual efficiency of the new technique, especially since the developments made in the sails were the principal step forward to greater manoeuvrability, just as was navigation with oars. Then there is the question of dating and origins. The earliest document comes from the baptismal fonts of Winchester Cathedral, carved in Belgium around 1180. Some historians believe that the rudder portrayed there is a lateral one, and that the efficient instrument in fact is situated on the stern. A 13th century engraving on the roughcast wall of a church in Fide, Gotland, from the 13th century, depicts what is indisputably a stern rudder. The seal of Elbing, the oldest issue of which dates back to 1242, and a Spanish miniature of 1350 are both further proof of the fact that the stern post rudder was probably developed in the northern Seas, between the 12th and the 13th century. It was not in widespread use however, and the lateral rudder figures in many more representations until well into the 13th century. On the other hand, a mid-13th century Persian miniature actually depicts a single stern rudder. It is difficult, therefore, to reach any certain conclusion.

It is reasonable to suppose that henceforth all these changes gave rise to a type of cargo ship in widespread use by the end of the 13th century, and known as the Hanse's *cog*. We know for certain that this ship was derived from Nordic merchant ships, although there is less certainty about the actual nature of the ship, or its origins. The seal of Elbing (1242) depicts a ship with an upright stem of an indeterminate shape. The freeboard is raised, and the planking is clinker-built. Also noteworthy are the presence of both the bowsprit and the sternpost rudder. The seals of Wismar and Harderwijk and other towns controlled by the Hanseatic League, also depict the same kind of ship. The seal of Elbing of 1350 provides us with a clearer picture of this vessel. The prow and stern were rectilinear up to the keel, there was a long draught, and a long lateral plan. All these suppositions have recently been confirmed by the discovery of a ship of this type in the Weser sandbanks at Bremen. This ship incorporated a number of technical innovations: could it not have been this which made the Hanse's fortune?

It is much more difficult to study Mediterranean vessels, but two Greek manuscripts from the end of the 9th century provide us with a starting point. Small crafts resembling merchant ships were used, and these incorporated one indisputable innovation: the triangular lateen sail, of unknown origin. Mosaics in Saint Mark in Venice also provide evidence of another innovation: the splitting up of sails, and the use of several masts. Some of the ships have three masts and double steerage.

In 1268, Saint Louis ordered vessels from Venice and Genoa for his crusade. The Venetian ships were 25.8 metres high in full sail, the keel was 17.4 metres in length, 6.45 metres wide and 6.60 metres deep. They had two complete decks and two short rear decks, with provision for cabins. The Genoese ships were shorter, and this seems to have been more characteristic of the ships at the time. Figurative representations often depict ships with two masts, the foremast inclined towards the prow. There is a castle on the curved deck, as well as a forecastle on the stern. The rigging consisted of lateen sails.

We know very little about the medieval warships of this period. There were galleys with two wings which were raised in a loop-like fashion at the bow. The ram at the stern was placed well above the level of the sea. The trireme had been replaced by the dromon and selandra.

The main significance of this period is the interchange between the two techniques. The Mediterranean merchant ship arrived in Bruges at the very end of the 13th century, and in 1304, the Bayonne pirates entered the Mediterranean in cogs. From the beginning of the 14th century, the square sail and the stern post rudder were used in the Mediterranean.

The evolution of ships was slow, but there was more continuity with older techniques. These developments were quite significant, especially in the field of merchant shipping, which from the 13th century were to provide ships much more suited to the transport of goods.

Much progress has been made over the last thirty years in the history of iron and steel making, thanks to excavations undertaken in many countries. However, uncertainties still abound: the development of the Roman world is no doubt still better understood than that of the medieval age. The following interpretation is by no means generally accepted, since it is based on the still tentative conclusions drawn from archaelogical excavations and texts.

It is very difficult to situate the High Middle Ages (the period between the large-scale invasions and the year 1000) in the evolution of metallurgical techniques, but there are two essential observations to be made.

Some historians quite logically believe that the Barbarian invasions provoked a decline in techniques. It is certainly true that the Barbarians, accustomed to their own techniques, neglected those of the countries they invaded. According to Tylecote, Saxon techniques began at the point where the people of the first iron age left off. Excavations in Ireland have revealed the remains of low hearths of the shaft type in nearly all of the workshops between the 6th and 13th century. It was the same in Scotland. Some of these techniques even survived until the end of the Middle Ages.

In excavations in Zelechovicia, in Czechoslovakia, the remains of an 8th and 9th century workshop were found with twenty four kilns, A collecting channel was used to collect the scoria. The kilns were bored with loess and the interior was coated with a triple layer of bottom stone. The temperature at the end of the tuyere was about 1400 degrees. Provision was thus made for an artificial air flow. The Zerotin furnaces were built in the 10th century: they look like a dish-shaped hollow, coated with red clay. In Radetice, a forge a 13th century forge had shaft furnaces with an interior diameter of 30 cm and an exterior diameter of 130 cm. It was about 1 to 1.3 metres high. The walls were made of rock stone and there was an opening for the outlet of scoria. The air flow was artificial.

Excavations in England have uncovered several workshops. One in West Runton near Cromer dates back to the 13th or 14th century. The hearth was a tubular or tank type with a device for the outflow of the scoria. At High Bishopley in Durham, a 13th century workshop had two high furnaces, and a kidney-shaped crucible of fired clay: this most likely had an arched hearth with forced draught (fig. 8).

Beyond these traditional structures, there was a two-pronged evolution. The first led towards the Catalan furnace, one of which was found in Baysdale in North Yorkshire. Built in the 13th century, this forge has four kilns all of which are of the catalan type, made of clay and stone with hand bellows and a device for draining the scoria.

The second, more interesting course of evolution gave rise to the "sledge furnace." The forge at Radetice mentioned earlier, was no doubt a milestone in this evolution. The increased size of the furnace for greater productivity meant it had to be built above ground, with strong masonry and a hole near the bottom for extracting the lumps. This kiln became a permanent fixture. The apparatus discovered at Lantherthal in Germany and described by J.W. Gilles is from the 11th century. It is made of siliceous alluvium covered with a kind of basket made of boards, then fired. It was a conical-shaped well, therefore. The hearth was covered with sandstone taken from the ruins of Roman buildings. There were two blow holes (fig. 9).

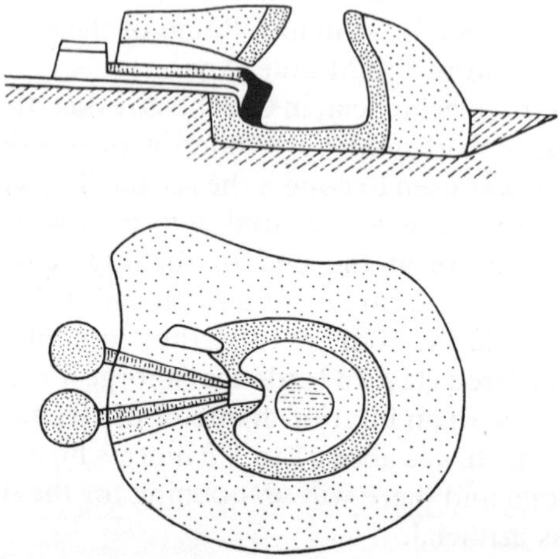

Figure 8. — Early type of blast furnace. High Bishopley (Durham) 13th century.

The apparatus gradually became larger and larger, and was eventually built entirely of stone. Unfortunately, we have no sources, written, pictorial or archaeological which would enable us to reconstruct the slow course of evolution that very probably ended in the 12th century. The development of the organisation of production, which went from the collective to the individual stage, and the many concessions granted for the development of the iron and steel industry which began in all western countries around 1140, are evidence of this technical evolution.

The last element is the appearance of the water-powered hammer which we noted above. This apparatus' productivity rate was much higher than forging by hand, and it provides ample proof of the increase in production. This new instrument appears to have been first used during the second half of the 12th century. It seems, therefore, that there was a transformation during this period, from the primitive forms of production to the beginnings of an industrial type of production. This was a major transformation of techniques and similarly, of production and the use of metal.

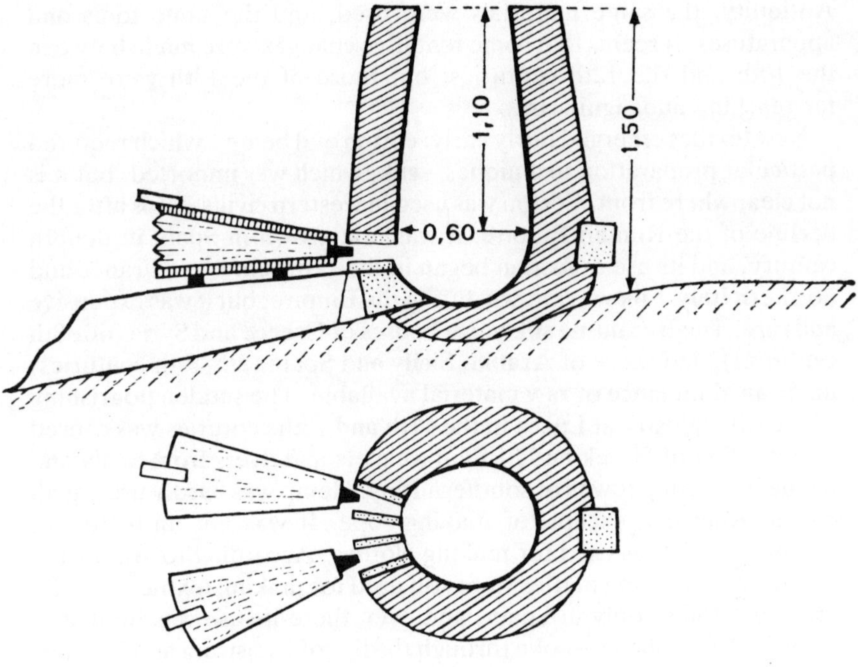

Figure 9. — Furnace at Lantherthal (Sarre) c. 1000-1100.

So much research has been done on the textile industry that it has been assumed that everything is known about the medieval techniques employed. Some elements are still obscure and not properly understood, however, but the recent work by W. Endrei has gone some way to clearing up a few of these points. The techniques employed in the High Middle Ages were virtually the same as those in Antiquity, the same materials were used, and the same tools and apparatuses. It seems that some tentative changes were made between the 10th and the 12th centuries, but those of the 13th were more far-reaching and significant.

New textiles emerged fairly early: cotton and hemp – which required particular preparation techniques – silk, which was imported, but it is not clear where from. Cotton was used in western civilisations after the decline of the Roman Empire: it was first grown in Spain in the 8th century, and its use as fustian began in the 12th century in France and northern Italy. Silk was known during the Empire, but it was expensive and rare. The Byzantine silkworm houses of Greece and Syria (6th-8th centuries), and those of Arabia, Sicily and Spain (8th-10th centuries) made an abundance of raw material available. The sudden flourishing of the silk industry at Lucca in the 11th and 12th centuries was caused by the influx of Greek and Sicilian weavers and dyers from Sicily and the neighbouring towns of southern Italy. Hemp was hardly used at all by the Ancients, except for making rope. It was not until the 8th century that it was used for making clothes. According to Albert the Great, even stinging nettles were still used for making textiles. Alfalfa grass was found only in Spain, and even there its use was limited.

Since Aristophanes spoke through the lips of Lysistrata and Catallus described the spinning Fates, there was no mention of any change in the techniques of spinning wool until the 13th century. The wool was still washed in urine and rinsed in water. Hemp and linen were retted in stagnant water. The woollen strands were separated by beating the wool with sticks or a whip made of ropes. Sartching was used with flax: excavations have revealed sartching troughs that probably come from the 10th century. We have no evidence of the brake being used at this time.

Carding and combing were already separate techniques by 1300, and from 1200 onwards the elements of the techniques used today were already evident. Carding and hackling were done with a fulling bur. Combing made carding largely superfluous, and it even dispensed with the need to brake flax. It had long been used with flax, for we have examples of Neolithic handle cards. W. Endrei believes that it was first used for wool in the 9th century, and its use became more widespread in the 11th and 12th centuries. The flax card depicted in the Chartres cathedral is convincing proof that the foot card was in common use in the 12th century.

The classic spindle was the instrument most often used in spinning, but it was not the only one. But we do not know enough about the processes used in the 12th century to develop this point. The distaff certainly existed before this time: there are many traces of it in Gdansk which date back to the year 1000. The first representation of a distaff figure is to be found on a manuscript of Rabanus Maurus, which dates back to c. 1023. A hook bobbin with an upper ring was used for the winding on.

Research into the products of spinning during the Middle Ages has been fragmentary and partial: it needs to be systematic and methodical. The warp and weft of most of the woollen serges produced during the Middle Ages were of uneven twists, until the advantages of having equal torsions were discovered by the Flemish draperers. By 1000, 70 % of materials were made with equal twists (these were known as S twist yarns), although draperers in Gdansk did not change to this until the 11th and 12th century. In comparison to Antiquity, the materials produced in the Middle Ages appear to be much less delicate. Before 1200 silk of a much finer quality was being produced. At the same time, the linen and hemp industries were in decline.

As for weaving looms, the dating is very uncertain, but it does seem that three important developments took place during the Middle Ages: the treadle loom was introduced, and after changes were made to cotton thread, the shuttle and the bobbin were also incorporated (fig. 10). The horizontal cotton weaving loom with treadles seems to have come from the Balkans: in the 12th century it was commonly used in Slav territory and in Scandinavia. The fustian weaving loom appeared in christian Catalonia from the 12th century – fustian is a heavy, tight cloth, the three or four treadles required a heavy sley and a solid structure.

These few developments transformed the textile industry by 1300. It would not otherwise be possible to understand the extraordinary expansion of the Flemish industry, or that of northern Europe in general. The use of different tensions for the weft and the warp, the

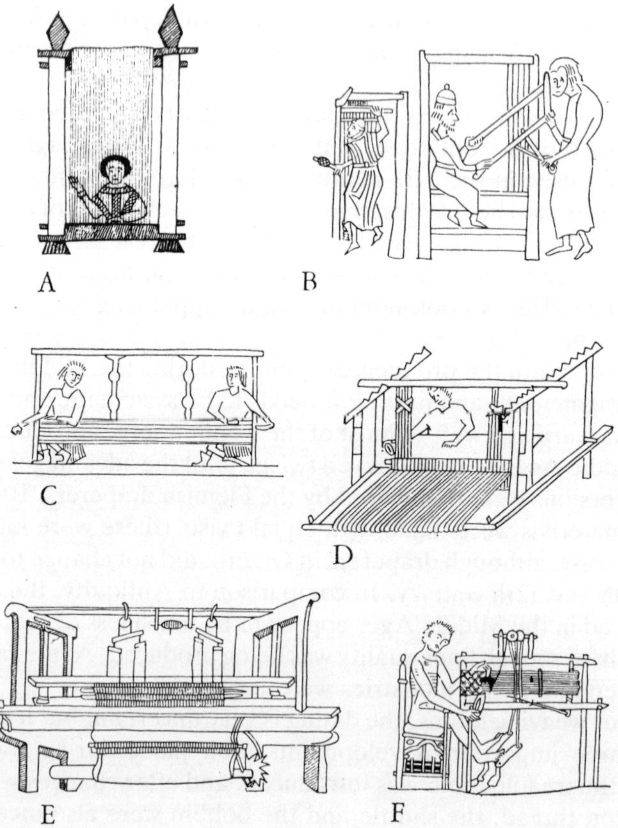

Figure 10. — The evolution of the weaving loom. A, vertical loom (from an 11th century manuscript); B, vertical loom (from a viennese manuscript from the 13th century); C, wide horizontal loom for two workers; D, horizontal look for one worker (Chartres cathedral, 13th century); E, large horizontal loom with four treadles designed for two workers (*Livre des métiers*, Ypres, 14th century); F, horizontal loom with two treadles for one worker (from a 13th century Cambridge manuscript).

emergence of the treadle loom, and the adoption of new cloths constitued a group of quite significant innovations.

Major changes were also to take place in the 13th century, so much so that one is tempted to suggest that as in England in the 18th century, the textile industry played a vital dynamic role in technical progress.

It was the introduction of carding with metal points that completely transformed spinning operations. This operation was called *carda*, the weaver's card. This technique was discovered either in France or Flanders. It was first depicted not in the Luttrell *Psalter* (1340), but in the marvellous statue – still intact – of the royal porch at Chartres, which would date it from the beginning of the 13th century. The qualitative advantage of carding with metal points resides in the regular formation of the web. At this time, the warp and weft of quality cloth were made of combed wool. The making of the weft by carding brought about a qualitative change. In response to economic pressures, this process was eventually used for the warp as well (1377 in France). The card was fixed between both the operator's knees and was pulled with both hands. The use of the brake in the transformation of hemp and flax was also extremely important. Everything points to the fact that this too became widespread in the 13th century, even though it only features in written sources from the 14th century.

The widespread diffusion of the spinning wheel has quite rightly been hailed as a revolutionary innovation. Some historians believe that it evolved in Asian countries between 500 and 750 BC, and that Moslem conquests brought it into Europe during the 8th and 9th centuries. It has also been said that the wheel was probably brought into France by Spanish fustian makers during the 12th century. The problems of using a spinning wheel, particularly knotting, for example, were not immediately apparent. But one factor was soon obvious: the spinning wheel was capable of a high productivity rate, so that spinners were tempted into going for quantity rather than quality, with a subsequent decline in the latter. Legislation became necessary, and several interdictions were passed: in 1224, restrictions were imposed in Venice and northern France, in 1268 in Paris (but only for cotton), in 1288 in Abbeville, 1292 in Sienna, 1298 in Speyer and Valencia (only for the warp), in 1305 in Douai, and 1308 in Champagne. Those countries which had a cloth but no cotton industry ignored the interdictions. There is no doubt that the unwillingness of

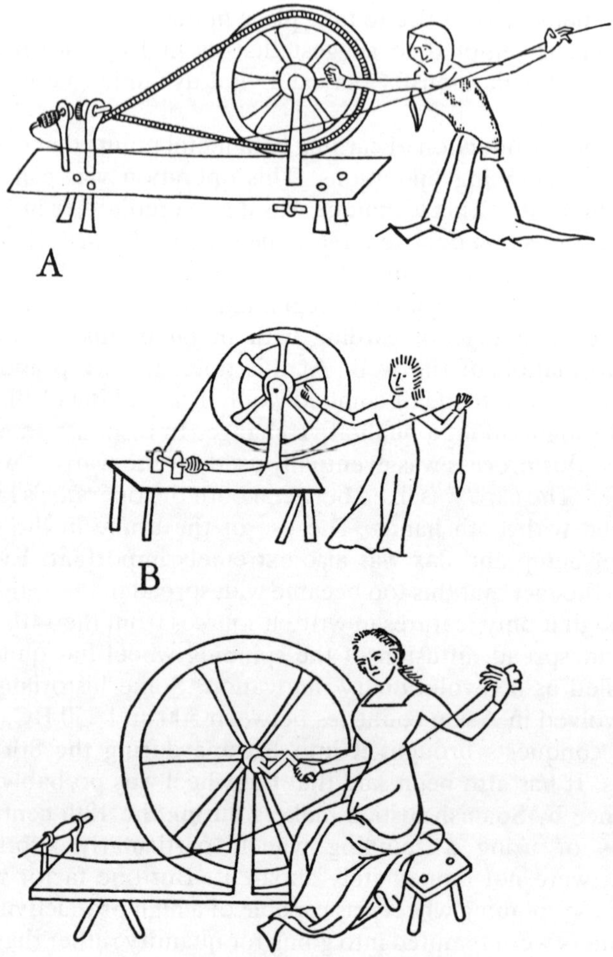

Figure 11. — Spinning wheels. A and B, spinning wheels with the spinner standing (A, from the Luttrell Psalter, circa 1338 and B, from a 16th century Parisian manuscript); C, spinning wheel with the spinner sitting (Lyons' manuscript, 14th century).

the drapers to use the spinning wheel stemmed from its original function and design as an instrument in the cotton industry.

The spinning wheel was mounted on a stool, like the treadle loom. The size of the wheel was considerably increased as was the distance between the hub and the pin. It should be noted that the wheel was turned by hand, and at this time it had no pedals, which were to be introduced later. This meant that the spinner had to work standing up and not sitting, as they did in India, even though some pictures do show spinners sitting down (fig. 11). It is possible that these modifications were designed to make working with wool with long fibres easier, as it is less resistant, but more flexible. In fact, the greater the length of the thread to be the more a strong rotation could make it straight. The increasing size of the wheel meant that it did not have to be constantly powered: it could run of its own momentum.

At the beginning of the 13th century, a primitive type of throwing silk thread was already practised. In 1221, Jean de Garlande's dictionary, and Etienne Boileau's *Livres des métiers* both list several types of winding instruments and the history of the evolution of these machines in Boulogne (between 1270 and 1280) is probably fairly representative. From the beginning of the 14th century, several documents in Lucca allude to the complex apparatuses then in use.

Other instruments were perfected or appeared at this time, such as the reel, first used in the silk industry. The spooling spinning wheel became widespread, and was first depicted on a Chartres stained-glass window. The toothed warping frame replaced the wall warp frame, along with the bobbin reel spool (depicted on a stained-glass window in Chartres and on a fresco in the Kunkelhaus of Cologne around 1300). It is possible that the toothed warping frame also came from the silk industry: it made the warping more regular, and increased the length of the part which was warped.

We noted above that the use of the treadle weaving loom was adopted before the advent of the 13th century. It is likely that the need in Flanders for an improved apparatus, capable of producing cloth of a superior quality prompted its use from an early date. Flemish cloth produced between 1000 and 1200 was the most expensive, but by far the best. Spinning techniques hardly changed, but the spinning wheel did lower the quality of the thread. Finishing techniques were much less advanced than those used in Florence, which gave to that city its wealth; wool was imported from Spain or England. The only field in which the Flemish could claim superiority therefore, was in weaving. The two-man weaving loom was a characteristic result of the technical revolution of the 13th century. In comparison to the narrow treadle loom, it was capable of a much greater productivity. From the 13th century, the capacity of the latter was almost as great as that of the flying shuttle loom. Ypres' *Keurebook* (Book of Trades), written in 1320, contains a perfect example of this. Around 1180, Alexander Neckam gave a dscribtion of the one weaver cloth treadle loom. The two-man loom was probably invented around the beginning of the 13th century in Flanders. It was the very first machine specifically designed for mass production.

There was a great deal of progress made in the textile industry, then, and there were several stages to this course of development. Some of these changes began to occur around the 8th century, which matured during the 1100's. In the first half of the 13th century, a second, perhaps more important technical revolution began to take place with the introduction of the spinning wheel, whose capacity was at least double that of the spindle. Other factors should be taken into account: the raw materials, their preparation, the equipment of the spinning wheel, and greater spinning skills. Furthermore, the spinning wheel was put into widespread use, and this was also an important factor. The increased productivity of the two-man loom was probably not quite double that of the older looms, but nevertheless the increase must have been considerable. If one adds to this the new products and new raw materials, one can expect that there was a very significant rise in textile production.

These were the most important medieval innovations; the others which we have not discussed are relatively marginal changes. The latter were merely substitutions of older techniques in that there was a greater diversity of raw materials than in Antiquity, and these were either cheaper, or of a better quality. An example of this is the glass workers' replacement of potassium by soda, or the use of a transparent coating, mistakenly known as varnish or glaze, in pottery. The same is true of architectural techniques, with the use of wood and half-timbering for houses, and the adoption of stone for use in more important buildings. The abundance of works on architecture enables us to paint a brief picture of the evolution of these techniques. At the very most it is worth noting – insofar as there was a revolution in technical thought – the solutions required by the great technical projects of this period. The wooden ceilings of the great Roman basilicas were replaced by the roman vault, then the gothic vault, for example. This transformation created many problems of balance, shoring and settling unknown during the Roman period, and it brought about major difficulties with stone-cutting. The history of these techniques is well known, all failures having been carefully noted in the finest detail, providing us with a wealth of information which we would look for in vain in more contemporary work on techniques. There are several striking similarities between the two periods which present a fruitful topic of research. Of course techniques do not suddenly emerge, they evolve from existing ones, and several historians have been able to trace the birth of the rib feature and the first rib-vault, particularly in Morienval. Everyone insists quite rightly however, on the importance of the Saint Denis basilica, which was begun by Suger in 1137 and which launched for this new type of architecture. The latter underwent its greatest development during the 13th century.

The use of hydraulic energy on a large scale, the practice of iron shoeing as well as modern harnessing, textile developments, the transformations in iron and steel making, and finally the appearance of new types of ships: this entire ensemble of techniques gave the western Medieval technical system a very different physionomy to the one inherited from the Ancients. An innovative attitude, a type of architecture and way of life which bore absolutely no relation to the past, and finally a more precise notion of technical progress seem to have been caused by the establishment of a new technical system. Bacon's letter, written in the beginning of the 15th century, and Guido da Vigevano's extraordinary project of using prefabricated material for distant campaigns, are striking examples of this.

The upheaval did not occur in all areas. Some techniques evolved more slowly, but none remained unchanged.

THE FIELD OF TRADITION

Insofar as the techniques of Antiquity were able to satisfy the needs of a more developed economy, it was not necessary to change them. At the very most, one might hope that adaptations would be made to the different natural environment – slight modifications which would then be integrated into a technology which had undergone a more profound transformation. Of all techniques, those of agriculture are the least susceptible to change. The reasons for this technical stability are evident: they are the techniques of scattered, less well-educated populations. It is true that an increasing population required more produce, but up to a certain point, more extensive cultivation – along with the great clearing projects which began around the mid-12th century – was able to solve this problem. The progress made in iron and steel making had increased the number of tools available to the peasant classes. This was perhaps the main secret of medieval agriculture, more so than the triennial rotation of crops which left a good third of cultivated land unused. Such practises would have been impracticable in an area with limited land and poorer soils. We have already discussed the other factors contributing to Medieval agriculture: the more abundant water sources, for example, and the thicker forests and richer grassland.

Nearly all the plants grown in the Middle Ages were already cultivated in Antiquity. Only a few new plants were introduced, and some slow improvements made in existing species. These were not enough to create any significant changes. The emergence of buckwheat is still a mystery: coming from Eastern Europe, this cereal was first cultivated in the West only at the end of the 13th century and by the middle of the 14th century it was still fairly rare. Hops are mentioned in a diploma of Pépin the Short in 768. Some historians believe that it was probably introduced into the West with the invasions between the 3rd and 7th centuries. For a long time it remained a secondary cereal used mainly in beer. Sorghum is mentioned in sources from the 12th and 13th centuries in northern Italy and southern France. Nearly all the other plants are first cited only at the very end of this period, such as rice in Italy and the south of the hungarian plain, and sugar cane on the shores of the Mediterranean. Edible plants were the same; there were about ten of these – starch was mainly provided from those gathered wild. At the very most can the Middle Ages be credited with having introduced chives, which feature in Charlemagne's *Capitulare de villis*, chervil, watercress, and horse radish. As a whole therefore, these innovations were of marginal importance.

The efforts to improve existing species were also limited and slow. The contribution of the Arab gardens in Cordoba was the object of a recent study: it seems the main area to benefit was that of medecinal plants. The authors of the treatises cited above, did insist on the importance of choosing the right seeds and on selecting them in order to obtain the best possible plants – but the Latin authors were already persuaded of this. Here again medieval agriculture was characterised by stability.

Triennial rotation was certainly an improvement on biennial rotation which had been practised in Antiquity, and which remained in practice in south-western Europe. The origins of triennial rotation, the time when it was first used, and the various stages of its expansion are still vague. Some believe this was a very old barbarian technique. We know that this process was based essentially on the alternation of a winter cereal with a spring one; during the third year the land was left fallow. There is evidence that this was practised in Germany from the 8th century. It was still rare in the Paris area at the time of the Saint-Germain-des-Prés' *Polyptyque*. England began using it only in the 12th century. Its progress was slow and was plagued by much resistance, especially at the time of the large clearing projects when the extension of cultivated land made it less necessary. Its adoption led to some fairly important changes. Leaving the land fallow provided conditions for animal husbandry, but no more so than forests where several types of animal pastured. Oats were an excellent spring cereal and were consequently extensively cultivated: the use of the grain – which had been limited in Antiquity – contributed to the upsurge in horse-rearing, which was much more practicable now that new harnessing techniques had been introduced. The widespread practice of triennial rotation which reached its height in the 14th century, also contributed to the expansion of wheat cultivation and the subsequent decline of millet.

All the techniques of drainage, and improving and enriching the soils were already practised in Antiquity: the Middle Ages brought with it no significant new elements.

It was the same with cultivation techniques, Most of the techniques and instruments used in medieval times had been used in Antiquity. Much emphasis has been placed on the evolution of ploughing instruments and especially the plough in the Middle Ages, i.e. the simultaneous use of the coulter, asymetric ploughshare and mouldboard (fig. 12). This problem was taken up again recently, even though in fact, it is a fairly straightforward line of research. We saw that the swing plough

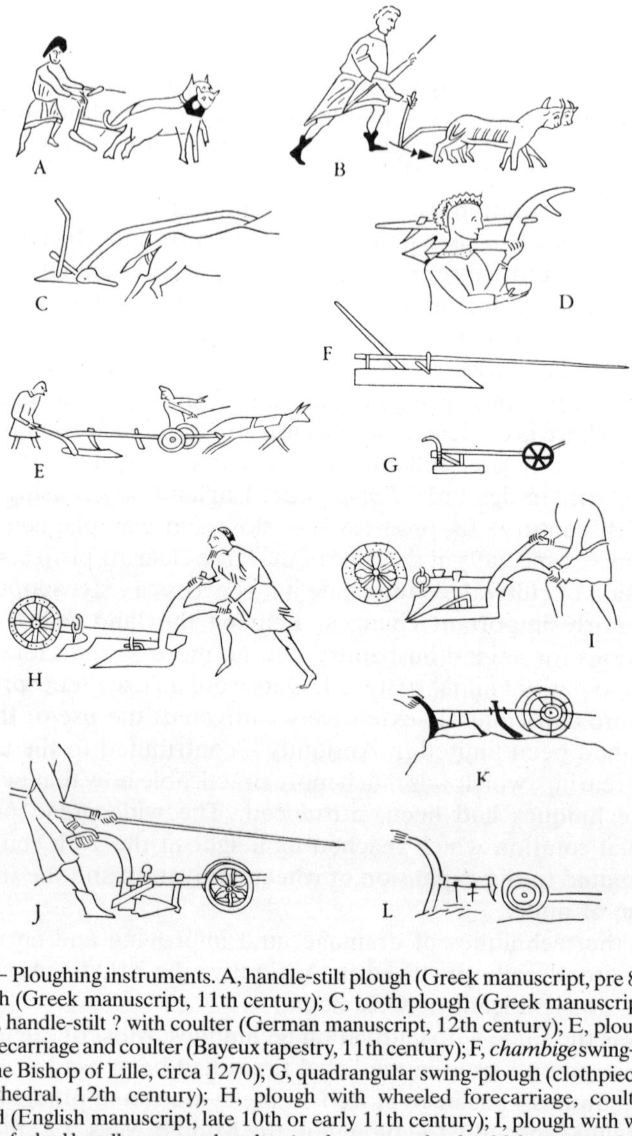

Figure 12. — Ploughing instruments. A, handle-stilt plough (Greek manuscript, pre 830); B, tooth plough (Greek manuscript, 11th century); C, tooth plough (Greek manuscript, 12th century); D, handle-stilt ? with coulter (German manuscript, 12th century); E, plough with wheeled forecarriage and coulter (Bayeux tapestry, 11th century); F, *chambige* swing-plough (terrier of the Bishop of Lille, circa 1270); G, quadrangular swing-plough (clothpiece in the Gerona cathedral, 12th century); H, plough with wheeled forecarriage, coulter and mouldboard (English manuscript, late 10th or early 11th century); I, plough with wheeled forecarriage, forked handle, prop and coulter (anglo-saxon calendar, 10th century); J, same as the preceding one (North of France manuscript, 12th century); K, plough with wheeled forecarriage and only one stilt — the coulter and the mouldboard can be discerned (English manuscript, 10th century); L, same plough as the preceding one (English manuscript, 10th century).

– symetric ploughing – was most widely used in Antiquity, particularly the tooth plough. There is evidence of the wheeled forecarriage, coupled with the swing plough in Rhaetia in north Italy, dating back to the 1st century BC. The Barbarians had no other instruments: no doubt they used the *chambige* swing plough in the North, and the handle-stilt plough in central Europe. The transformation from one-way ploughing to bed forming, which quite possibly began in the late Roman period, meant that the plough had to be inclined, and necessarily preceded by a coulter. The latter was a separate part at first, but it was eventually incorporated into the actual ploughing apparatus. The forecarriage helped this transformation.

One-way ploughing was practised in northern Europe from the late Middle Ages but the southern regions remained faithful to symetric ploughing, mainly because of the density of the soils. The one-way ploughing of heavy soils gradually led to changes in the type of instrument used. Some excavations in Czechoslovakia have brought to light asymetric ploughshares from the 8th and 9th centuries, even though symetric ploughshares were still being used. Graphical representations after the 10th century depict quite different instruments: a wheeled forecarriage plough, equipped with one or two handles and often with a mouldboard and a coulter. It seems that until the 12th century, the instrument used in the extreme north-west of Europe, therefore, was quite unique, with several regional variations. From 1200 onwards, the history of the plough is much more complex, while that of the swing plough seems to be relatively stable. The main changes in the plough occured during the second half of the 13th century. The birth of the turnwest plough is evidenced in drawings by the presence of the osier tie, designed to wedge in the coulter. Witnessed in Belgium at the end of the 13th century, its development was slow. The same was true of the control apparatus and the joints of the front and back wheels, which enabled the ploughshare's angle to be altered, and thereby the depth at which the land was ploughed. On the other hand, the instrument developed in England was the "crooked plough," with no forecarriage. Equally, Czechoslovakian excavations have revealed that the ploughshares used at the end of the 13th and beginning of the 14th century were getting increasingly bigger.

We have very little information about other techniques. Treading corn was abandoned, along with the roller – these were replaced by the flail. The reasons for this had more to do with climate than actual techniques. The wheat had to be threshed in a barn because the weather often prevented them from using a threshing floor.

Triennial rotation and the emergence of the medieval plough constitute, then, the two main innovations of the Middle Ages. But in fact these merely represent adaptations to different natural conditions. This is all the more obvious with regard to the cultivation of plants which hardly changed at all – viticulture is the most striking example of this.

Ultimately, we are obliged to measure the amount of progress made – if there was any actual progress at all. It seems at first sight that the increase in agricultural production was due to having richer soils, and by extending the land cultivated rather than by new techniques. Unfortunately, we do not have precise figures for all the different sectors of this subject, so that we cannot consider it in terms of global production and yields and establish the actual changes compared to Mediterranean agriculture after Antiquity. In a 13th century English manual, the yields are given as follows: rye, 7 for 1; barley, 8 for 1, peas, 6 for 1; wheat, 5 for 1; oats, 4 for 1. Thierry d'Hirçon's estate in Gosnay, Artois (France), produced a yield of wheat at 12.9 for 1. The results were probably better for animal husbandry, which was greatly improved by a larger and more well-balanced diet. Certain practises may have begun in the Middle Ages, one of which was transhumance, but this may already have been practised during the early Roman Empire. There is formal evidence from Bonnevaux Abbey of major migrations from the beginning of the 12th century (1122). There were still several difficulties with this field of agriculture, however: there was no permanent stabling available, and feeding was very difficult during winter.

Mining had been practised by the Ancients on a large scale, and their techniques remained virtually unchanged throughout the Middle Ages, despite the innovatory methods of organising new mines adopted in this period. At any event, it was at the end of the 12th century and even more so at the beginning of the 13th that there was an even greater upsurge in mining activity: mining charters – all of which are fairly similar – and the evidence of some excavations have provided us with some interesting information on this subject.

The texts, and the technical language of mines seem to indicate a marked German influence of which there are several examples. Apart from this, the documents available show that there was no significant technical progress made during this period.

The development of mines varied naturally according to the nature of the deposits and products mined. There were few gallery mines. Iron was generally obtained with open cast mining, but there were gallery mines established in Champagne and the Dauphiné during the 12th century, and later in the Pyrénées. The techniques used in areas with deposits of rock salt were quite unique.

It was invariably necessary to search underground for exploitable veins, and this meant building shafts. These shafts were usually vertical, but some were oblique. They had to provide easy passage for both the workers and the products (mineral and sterile). During this time, the shafts were of modest dimensions, measuring 1 to 1.2 metres in diameter. They were built in large numbers, probably to avoid long galleries and because of the extreme fragmentation of these developments. The interior was usually rectangular, making the timbering easier, or trapezoidal in cases where there was strong exterior pressure. There was usually a small hut at ground level which protected the entrance and the winch. The depth of these shafts varied considerably: in the copper mine at Massa in Italy, the access shafts varied from fifty to a hundred metres, so boring these shafts was a long process. The workers descended by means of long ladders fixed to the walls, and there were joists placed at intervals to enable them to rest.

The shafts opened out onto the galleries and the latter sometimes opened out onto the hillsides (Massa in Italy, Vicdessos in the French Pyrénées). There were several types of gallery, passages or rolling galleries, galleries for ventilation or the expulsion of water, and drift tunnels. The former were of various sizes, usually twice as high as they were wide. In Massa, they were 1.6 to 1.7 metres wide and 1.8 metres high. The walls were usually straight and well erected. At the Salins salt mines in the French Jura, the Amont shafts, built between the 10th and 13th centuries, had a straight barrel vault 10.30 metres high. The shafts at nearby Grès, were built with groin vaults, and were 11 metres high, 53 metres in breadth and 16 metres wide. The development galleries were less well-thought out, and the others were considerably smaller.

The layout of the galleries and their gradients admit to a fairly advanced knowledge of surveying, but we know nothing more than this. Timbering was certainly used as a support structure: traces of it have been found in the Massa mine. Mining regulations prove however that one gallery would often run into another, and they frequently collapsed. This was the inevitable result of having several more or less individual developments in the same field of deposits. There were accidents and the lives of saints and tales of miracles provide many examples of these.

The development itself was very primitive. The intersecting galleries and cross headings in the Massa mines were often very long, indicating that the miners of this time (mining here was stopped in 1350) felt very sure about the extent of metal bearing deposits both horizontally and vertically. Rock picks, pickaxes and levers, sledgehammers and wedges constituted the main equipment used by the miners. Many of the picks which have been found in Massa have a well-steeled and carefully designed head. Fire was often used to break up the rocks. The veins were developed in straight, inverted terraces in larger development chambers. Pillers were erected in these chambers, and the abandoned galleries were filled in with waste.

The lighting was very rudimentary. Iron and terra cotta lamps used with oil have been found in Massa, and in Vicdessos, the workers held a lamp in their mouths which was made of a small horn containing oil and a wick.

The ventilation galleries were used quite strictly for this purpose, and no-one was allowed to use them as a passage. Anything that needed transporting was usually carried on the mens' backs, either with baskets, as at Vicdessos, or with sacks made of buffalo skin, as in Massa. Hand-operated winches and ropes were used to pull loads up the shafts. Very little is known about how they dealt with the problem of water, but it is likely that they got rid of it by taking it up in goatskins through the shafts. This problem was more easily solved of course when the galleries opened out onto the hillside.

At Salins in France, rock salt was not brought out of the mines untreated. In fact, the water was used to impregnate the salt, and it was this water which was then taken out in vessels supported by enormous balanced poles. The salt water was then placed in heaters.

In fact, there was even less progress made in mining techniques than there were in agricultural ones. Every single one of the techniques that we have just discussed was known in Antiquity. It is even possible that in certain sectors, the techniques used in the earlier period were more advanced: this was true regarding the problem of expelling the water, for example. Mine drainage and bringing the materials out of the mine were difficult problems, and it was probably these which blocked development and forced mining techniques to remain on a fairly crude level.

As for prime movers, the equipment used was largely of Roman type, with few changes. Miniatures of the Middle Ages depict hoisting apparatuses which are the same as those found on roman bas-reliefs: simple winches, pulleys and squirrel cage cranes. In fact the cranes of Bruges or Lunebourg are exactly the same as those depicted on a roman bas-relief in the Vatican Museum.

The only innovation was the screw-jack, of which Villard de Honnecourt has given us a perfect representation: a wooden screw, ending in a capstan and fixed at the bottom, turns by means of pivots arranged in a stringer and capital. Two inclined uprights maintain the ensemble of horizontal pieces.

The same is true of presses: the machines used in Antiquity remained virtually unchanged. The winch or screw systems were used most often, and the frequent representations of wine presses in medieval iconography are evidence of this. Some of these presses are still in use today, the basic mechanism having undergone no change. Some texts mention other machines, but the actual type of machine and its structure is not always clear: a machine for correcting deformed structures, a machine for launching ships...

Metallurgy apart, fire using techniques evolved only very slowly. The furnaces and fuels – the basic elements – were still exactly the same. Only the choice of raw materials or the processes used could have provided possibilities for minor innovations. There is very little written informations on medieval pottery, but recent discoveries should mean that we shall soon be provided with some interesting studies on the subject. Matt pottery, made with a mixture of clay and blended sands was still used, while glazed pottery, which had been so popular in Antiquity had almost completely disappeared. The main problem was making the pottery waterproof. During the Carolingian period, a transparent lead coating was first used, commonly (and inaccurately) called varnish or glaze. It allowed the decoration of the pottery to remain visible. The varnished pottery was found in the Middle Ages, and was used for everyday objects (tiling, vases, crockery).

Glass working was rare in the late Middle Ages, or at least there are very few mentions of it in written sources. For the 11th century the monk Theophilus bears witness that medieval glassworkers were skilled in this field, however. The same raw materials were used: vegetable ashes (particularly beech and fern) and sand. They were able to make only small pieces, which explains the importance of stained-glass windows. Not much is known about how they obtained the colours, but it is possible that at the end of the 12th century, they stopped using potassium (obtained by burning vegetables) and used soda instead. They also used large amounts of alumina and iron oxide. According to Theophilus, the crucibles contained about 65 kg. of glass and produced small rotating blowpipes. The inadequate refining gave rise to bubbles and streaks in the glass, whilst the rudimentary stretching left lumps. In fact, the more widespread use of glass in the Middle Ages was not due so much to technical evolution as to a change in natural conditions. In the northern regions of Europe, blockages had to be more airtight – and glass made this possible.

Sea salt was very popular in Antiquity, produced by simply letting sea water evaporate in the sun. The same techniques were used during the Middle Ages, even in regions where there was less sun, like the Baltic, for example. Rock salt was also used, and there were salt mines in eastern France, in Germany in the region of Lüneburg, in Poland at Wieliczka, and in Transylvania. These were probably developed in the 12th century, and it might have been earlier in the case of Salins. We have already stated that the salt was generally dissolved in water. Once it had been brought up from the mines, this salt water was led via wooden channels to vats where by this time it was already more concentrated. It was then heated. The heaters were made of iron sheeting assembled with rivets (the latter were used from around the 11th century). The furnaces were embedded in the ground, and the operation took between 12 and 18 hours. The amount of heat required depended of course on the weather. The walls of the heaters were whitewashed in order to avoid melting the salt. These techniques were actually unknown in Antiquity, and must have evolved during the 10th century in western Europe.

Distillation was an ancient technique, and we saw that it was often practised by the members of the Alexandria School. The Arabs perfected the art and used it to make perfume, and it is through them that the techniques became known to the medieval West. The heart-shaped Alexandrian condenser was abandoned at this time – although in some areas it was still used right up until the Renaissance – and was replaced by the modern still, equipped with a spiraled, coiled or sprung tubular outlet, contained in a vessel in which cold water was circulated. The distillers of Salerno were thus able to make alcohol, one of the main bases of chemical techniques. This technique was further improved with the use of dehydrants such as potassium carbonate.

The chemical techniques of the Middle Ages, which were the subject of a valuable study by Berthelot, were very rudimentary. No rational techniques were possible as long as they were unable to determine the exact properties of substances. This meant that there were very few base products. Around the middle of the 12th century, they succeeded in producing nitric acid by distilling saltpetre, alum and vitriol. Nitric acid was the only acid to be regularly used at this time.

The most frequently used chemical products were colourants. These, along with drugs, constituted the only activities of something which could not yet be considered as a chemical industry. Mineral colourants were used to colour glass, such as cobalt oxide for blue, copper for red and green, and manganese for purple and violet. The textile industry of the Middle Ages used the same colourants as those used in Antiquity. Blue was obtained from the hard stone azur, but most of the other colours were obtained from plants. Woad produced indigo, madder supplied the red tones, and pink was obtained from braziletto. Yellow was obtained from weld and black from nut gall. They were not particularly skilled at dyeing, and the colours tended to be rather uneven. Alum was used to fix the dye so that the cloth could be dressed.

The main contribution of the Middle Ages to the field of chemical products was the discovery of gunpowder. Unfortunately, written sources and pictures are too vague to enable us to establish exactly how they discovered explosives. It might also have been imported from the Far-east rather than being of local origin. Albert the Great used saltpetre only for making nitric acid, and despite various peremptory claims, it seems that Bacon did not know about gunpowder. It is quite likely that gunpowder was discovered accidentally during the various processes used to purify saltpetre.

It is a great shame that we have no catalogue of the tools used in the Medieval period. As we said earlier, a systematic study of certain iconographical themes would make such catalogues possible. The value of such a catalogue is demonstrated by figure 13, which shows some of the tools used in building. If we had such a catalogue and were able to compare it with similar catalogues for Antiquity – of which

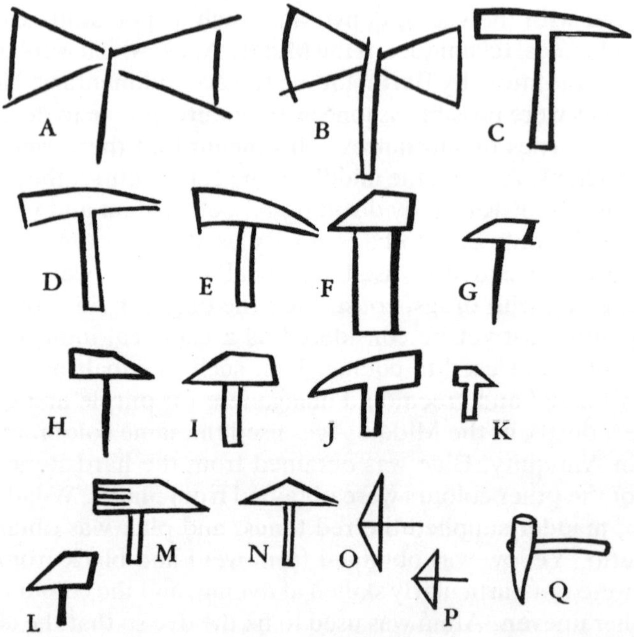

Figure 13. — Hammers for cutting stone. A and B, Schottenkirch of Regensburg (11th and 12th century); C to E, Maulbronn abbey (12th and 13th century); F to I, Freiburg cathedral (13th century); J, Strasburg cathedral (13th century); K, Regensburg cathedral (13th and 14th century); L, Nuremberg Sankt Sebald (13th and 14th centuries); M, Strasburg cathedral (1250-1275); N, Regensburg cathedral (13th and 14th centuries); O, choir of Diest Saint Sulpice (14th century); P, western gate of Diest Saint Sulpice (14th century); Q, Strasburg cathedral (1250-1275). All these drawings are jobbing-workmen's marks. (From Van Tyghem)

there are some excellent ones compiled by Blümner – we would probably find that they were very similar. It is thought today that the plane was used in the earlier period, but some tools were most likely developed only later, such as the bit-brace, for example. In former times only the bow drill and rope drill were known. Ancient representations of forging tools (from Greek pottery and Roman or Gallo-Roman stelas) are further evidence of this continuity, and the same resemblances can be found even in the plates of the *Encylopédie*. It was said recently that "All identifiable Roman tools subsist in virtually unchanged forms." Unfortunately, the lack of more precise documents prevents us from examining this subject on a deeper level.

Despite this general view of continuity, some historians believe that tools did change from the 13th century onwards, and that these changes became more marked until the 16th century. Which means that the end of this period represents the beginning of a new period of evolution in this field of techniques. It is difficult to be more precise about this, however. The large felling axe which was often used as a military weapon, had a head symmetrical and axial to the eye, which made it a very solid instrument. Was this the tool used in the great clearing projects? It is impossible to say. On the other hand, the two-headed axe, which was widely used during the Gallo-Roman period, disappeared around the 12th century. Roman saws remained unchanged and were used until well into the Middle Ages. It was after the 13th century – and some of the products made after this period are evidence of this – that a large number of other types of planes were added to the traditional planes with a rectilinear cutting edge. These planes were thinner, more narrow, and had a profiled cutting edge; they depended largely on the different types of moulding. Furnitures, doors and interior frameworks in dwellings were made in greater relief during this period, and this had required particular types of tools. Equally, different types of anvils evolved: the classic two-horned anvil became most widespread in smelting and the blacksmith's trade. These changes are evidence not so much that the present group of tools in use was being modified, but that it was being added to and completed.

If one considers the primary importance of agricultural techniques and other related techniques using tools in the Middle Ages, it is easy to appreciate the powerful effect of a slowly evolving tradition.

A global assessment is necessary, however, to get the right balance and establish just what was new about this technical system. The greater use of energy resources meant that there was a greater supply of raw materials needed in certain industries – take for example the tanning mills or woad mills which quite literally prevented "bottlenecks." It also meant that production processes, which had been perfected "upstream," could be fulfilled. In the latter case, there is no better example than the water-powered tilt hammer or the grindstone used in iron and steel making, which increased the production of metal, or the fulling mill which was able to meet the requirements of a new technical transformation in the making of textiles. The use of hydraulic energy in the form of corn mills, oil crushers, mustard mills and finally beer mills also helped the food industries to cope with the rise in population.

Elsewhere, innovation meant the more complete development of natural resources which had not figured in the technical system of Antiquity. The saw mill was useful in dealing with the expansive forests, and the richer soil of these regions was more fully exploited by the adoption of triennial crop rotation, the heavy plough with mouldboard, and asymmetric ploughing. In a sense, the technical progress of the Middle Ages consisted precisely in the ability to exploit natural resources which were different in both quality and quantity. On a parallel with this, the changes in transport techniques meant that this natural wealth was more widely enjoyed. Again, it was natural conditions which prompted the development of animal husbandry, in qualitative and quantitative terms.

Climatic conditions also had a role to play in some major transformations, and the rain and temperature were major factors in these changes. The construction and arrangement of houses, the shape of roofs, and the structure of frameworks were quite different to those used in Rome or Athens.

These were exterior factors which seem to suggest that the technical system of the Middle Ages was necessarily going to be different. But there were interior factors which lead to the same conclusion. An equilibrium is necessary within a technical system to maintain the interaction between techniques. Bottlenecks must be prevented, to avoid new techniques losing their viability at certain stages of production. It was impossible to increase iron and steel production without balancing up the different stages of manufacture. Furthermore, this upsurge in iron and steel making would probably not have occured without the increased demand for these metals, thanks to the development of iron implements, iron shoeing and weaponry.

It seems that there was a period of research which began – in certain fields – between the 6th and 8th centuries, while in other areas progress had been much slower, and continuous since the early Empire. But these elements of progress became effective only when an overall balance was obtained, and when the new technical system was almost completely formed. This technical progress finally "exploded" around the middle of the 12th century. It did not matter that certain techniques retained their traditional characteristics as long as they did not halt the general progress of the system.

The "development" of the system meant that it could progress as a whole, or that individual imbalances could be rectified – the textile industry has numerous examples of the latter. The progress made in weaving and dressing cloths, for example, made certain innovations vital, such as the new finishing techniques, and the invention and the adoption of the spinning wheel.

Ultimately – but this remains to be verified – it is possible that the major crises at the end of the 14th century arose as a result of latent tensions in the technical system. This imbalance was either between advanced techniques, or between advanced techniques and traditional ones (historians have used agriculture as an example of this, but it is also evidenced in mining, when profitable mines had been exhausted, and deeper veins could not be exploited until the technical means were found, this giving rise to a sudden lapse in production). Of course one should also take into account the intervention of other structures, and the fact that technical and social structures are very closely linked. When discord shows itself in one, it sends shockwaves throughout all the others, and there are several examples of such crises.

CLASSICAL SYSTEMS

About twenty years ago a colloquium took place in Paris which questioned the originality of the Renaissance. Amid all the problems raised concerning this subject, that of techniques was touched upon but hardly solved: allusions were made to printing, navigation and maritime construction, artillery, and fortification, but no conclusions were reached about techniques as a whole. It is inconceivable however, during a time when so many "inventions" in the wider sense occured, that there was no revival in techniques, when revivals were apparent in all other areas of interest, such as science and literature, the arts and politics. Research since the colloquium has not demonstrated the birth of a new technical system, but it has enabled us to build a more extensive schema, applying the same methods used to examine other technical systems. It would be helpful first to establish the general historical context, not in order to show the causal links, but to emphasise the degree of interdependency between all aspects of history.

The end of the Middle Ages – or the period which we call the Middle Ages – is marked by the major crisis of the first half of the 14th century, and by the prolonged depression which followed it. The movement of recovery was not uniform in time or space, or with regard to the different areas of human activity. It would be futile to claim that the course of history in the 15th century was a rigorously logical one.

The lack of adequate research and documentation makes the history of demography a difficult subject, and recent works are evidence of this. The losses suffered varied, and population growth and decline varied considerably from one area to another, but it is generally accepted that the period from 1380-1500 was one of a very slow, gradual recovery. This recovery began at different times in different regions: it is clearly marked in Spain and Italy from the beginning of the 15th century, but in Provence, Languedoc, Germany and the Low Countries it did not emerge until the last quarter of the same century. It does seem, however, that by 1500 conditions were on the whole more favourable.

After the loss suffered in some areas – England's population fell from 3.7 to 2.2 million – it would seem that there was a new take-off – which came early or late, and developed at an irregular pace – so that by half way through or towards the end of the 15th century, the figures were the same as those from the late 1200's. In countries which had been less hard-hit by epidemics and war, the movement was one of stabilisation. Slightly "smothered" movements would go some way towards explaining the development of mechanisation in an economy whose recovery was more clearly marked. This mechanisation was meant to attenuate the effects of an insufficient workforce, and the growing needs of a population which in fact was increasing only gradually, but whose demands might well have been much greater. There is still much to be discovered about this aspect which is a vital factor in explaining the technical progress which occured after the epidemics and troubles.

The economic renaissance which followed epidemics and disturbances is better understood, no doubt because it has been better studied. However, there are still some aspects which remain to be satisfactorily explained. Moreover, the phenomenon of the great discoveries was incorporated in a movement which was already well underway. The economy began to recover in Italy and Spain before the mid-14th century, and during the latter half of the century in the north. It is possible that the major discoveries of this time were in fact a direct result of this recovery: might they not also be the result of the emergence of a new technical system?

It is worth noting two phenomena related to this problem. The first one is a need for money, which was to take Europeans to Africa and then on to America. Also relevant here is the opening or reopening of numerous mines of precious metals all over Europe, especially in the central regions. It was during the 15th century that mines in Hungary and Yugoslavia were redeveloped – and this was only possible because of technical advances. Similarly, the mining discoveries in southern Saxony and northern Bohemia are generally supposed to have occured around 1422, in other words towards the end of the 15th century's first quarter. The economic recovery brought with it a monetary restoration and stabilisation. This movement began in Florence in 1464, in England in 1470, Spain in 1471, Venice in 1472, and France in 1475. These dates should be noted carefully; it will become apparent that they are symptomatic. The other major transformation was the birth of large scale capitalism, quite different to that of the large Italian banks of the former period, who became victims of the crisis in the mid-14th century. The Medici bank, whose role in the blossoming of the Renaissance was so important, was similar in type to the latter, however most of the newly-established concerns were run by merchants or bankers, but also industrialists. The bank established by Jacques Coeur in the latter half of the 15th century belonged to this category, as did that of the Fuggers in Germany, or the Hochstetter in England. As farmers, manufacturers and mine owners with abundant funds at their disposal, they assumed this role quite naturally, and were able also to encourage technical progress. It would be interesting to add to the studies on their contribution to the intellectual aspects of the Renaissance, a study of their contribution to the technical world.

These two economic phenomena must have had a direct influence on technical evolution in that they made investment possible and encouraged innovation. Capital was made available to the most highly skilled technicians, and money was invested in new processes and new, original machinery. It was these factors which have led economists to consider this period as the beginning of a time of economic growth.

A parallel development was the change and evolution of ideas and an attitude of mind which gave techniques a new dimension. The sheer extent of these changes illustrates that this was a veritable transformation, apparent in several different spheres.

It is widely accepted by historians that the economic and political upheavals of the 14th and 15th century greatly weakened the feudal regime. The first change was apparent in social structures. As feudalism came to be no more than a system of land-ownership, the non-landed classes acquired more freedom and mobility. The growth of towns during this period is striking evidence of this, and the latter development was in turn encouraged by the policies of the sovereigns, whose power was gradually being restored. The second or third generation of fortified enclosures also provides evidence of increasing urbanisation. The commercial life of the towns was revived and began to develop, and it was these towns which provided the setting for technical innovation.

These changes were symptomatic of a larger movement – the gradual formation of modern "States." Centralisation and the desire for power – the main characteristics of these new states – led to economic intervention, and a greater attention to both economic and technical affairs, in terms of both commercial and military interests. These new political frameworks began to take shape during the second half of the 15th century, and within these frameworks a new type of politics was born, that of mercantilism. Mercantile politics was to favour the introduction of new techniques, and that meant both inventions and new techniques acquired from neighbouring states. The initiatives of individuals or groups, and encouragement from governments further enhanced the possibilities of technical progress. The role of individuals was mainly one of actual innovation, while the function of groups and governments was mainly to encourage the more widespread use of new processes or techniques from abroad.

It would be useful to list the policies of states in the technical domain. Governments had always played an active role in the field of armaments, but their intervention greatly increased at this time and became more systematic. Louis XI of France hired Spanish crossbowmen, and inventors of machines which could dig ditches and trenches. All over Europe, princes were paying for the services of founders – the creators of modern artillery – and Italian architects who brought with them new means of fortification, as well as new ideas on urbanism and civil construction. One only has to consider the names of the great engineers employed by Francis I: Marini, Bellarmati, Castriotto, for example. Gerolamo Marini worked on the frontiers of Champagne (La Fère, Laon, Soissons, Epernay, Vitry-le-François, Château-Thierry, Troyes and Joinville); Bellarmati created Langes, Dijon, Besançon, Nuits, Vesoul and also Le Havre. Italian engineers could also be found in Spandau, Düsseldorf, and even in England.

Ivan III of Russia is a good example of this new generation of sovereigns whose aim was to "modernise" their country. He employed an Italian coiner, Gian Battista della Volpe, and the great architect and founder Fioravanti. In 1488, a mission was sent to Italy to recruit architects, silver and goldsmiths, founders and armourers. Three more missions were sent in 1493, 1499, 1527. Miners were recruited from Hungary in 1484 and Saxony in 1488, and printers were brought in from Denmark in 1524.

In 1450, Charles VII engaged a German, Claus Smerment to run the newly-acquired mines established by Jacques Coeur. In 1452, Henry VI of England recruited miners from Bohemia and Hungary. Louis XI recruited German printers and Italian silk manufacturers. The list of such examples is endless; again, it would be very useful to make a catalogue of such exchanges, to note the specialised fields of each country, and to examine the role of political motive or royal alliances in the making of such choices.

Beyond this, individual initiative and unofficial movements had a role to play in the expansion of new techniques. We still need to examine the movements of technicians, official or unofficial, their successes and failures, and the assimilation of these men, their techniques and their language. We have only very few examples of this type of study. It is true that erudite scholars have always been interested in this field of interest, and there has been some research into the displacements of printers and their setting up of new workshops. But there were many displacements like these, more modest perhaps, but which still warrant closer examination. Take for example the master salt-merchant from the Franche-Comté who travelled across Europe learning about foreign techniques of salt-mining, or the smiths from Germany and Lorraine who travelled to Berry in 1440. Language is extremely useful as an indication of such movements: a striking example of this is the number of French words in the English vocabulary of iron and steel making.

It would be unjust not to mention the Italian princes of the 15th century in this discussion. As sovereigns of small states, they had no real power, but they were the most remarkable craftsmen of a new spirit – the spirit of the Renaissance. These dynasties, which were founded on military strength or commercial power, (and it is possible that this might be a significant factor) tried to systematise and coordinate different areas of knowledge, just as the Alexandrians had done. These princes actually realised the importance of technical problems, and it is thanks to them that the material world was finally acknowledged as an integral part of global knowledge. They gathered around them not only artists and scientists, but also expert technicians. This was to become the hallmark of this period.

Francisco Sforza, the duke of Milan (1450-1466) called for the development of the water systems of the Po valley, and the digging of canals and diversions. He employed Fioravanti and Bartolomeo della Valle to complete the cathedral, and build the castle which still exists today. His successor was to enlist the help of the Filarete brothers, Bramante, and Leonardo da Vinci. He protected the Cardans and consulted Francesco di Giorgio Martini. Their library, whose catalogues still exist, provides evidence of their endless curiosity and concern to promote humanism, science and technical efficiency. The Malatesta from Rimini were not quite as cultivated: Valturio put their military knowledge into concrete form, (fig. 1) and Alberti built their mausoleum. It does seem however, that the Malatesta considered many military problems and their contribution to this field, especially concerning the techniques of war, was quite considerable. Urbino probably made the limelight with the Montefeltro family, who gathered a large collection of specialists in their court, just as they collected all the most interesting works for their library: this is strongly reminiscent of Alexandria in the 3rd century BC. Francesco di Giorgio was among the engineers, one of the artists was Piero della Francesca, and Luca Pacioli was among the scientists, to name the most important. Their library was perhaps the setting where the new civilization first emerged in a concrete form. Should the Medicis have a place in this discussion? As traders rather than industrialists or military men, their interests were not naturally directed towards techniques, in the strict sense of the term – their energy was invested primarily in the arts and sciences. Their entourage has also been studied less than those of other princes, and further study of this subject would no doubt correct this first hasty impression. These courts shared one single characteristic: they all manifested the same new spirit of the Renaissance. They may not have created this spirit, but it was they who were the first to perceive and benefit from it.

All this helps to explain the increasing importance of techniques in the early Renaissance, but here it would be helpful if we made some more precise points. Technicians on building sites or in workshops were confronted daily with substances, tools, apparatuses, and with social and economic problems, so that for them any new invention or innovation was never considered as an isolated factor. Other technicians had a more scientific attitude towards such matters. They reflected on technical problems, tried to solve them rationally, and wherever

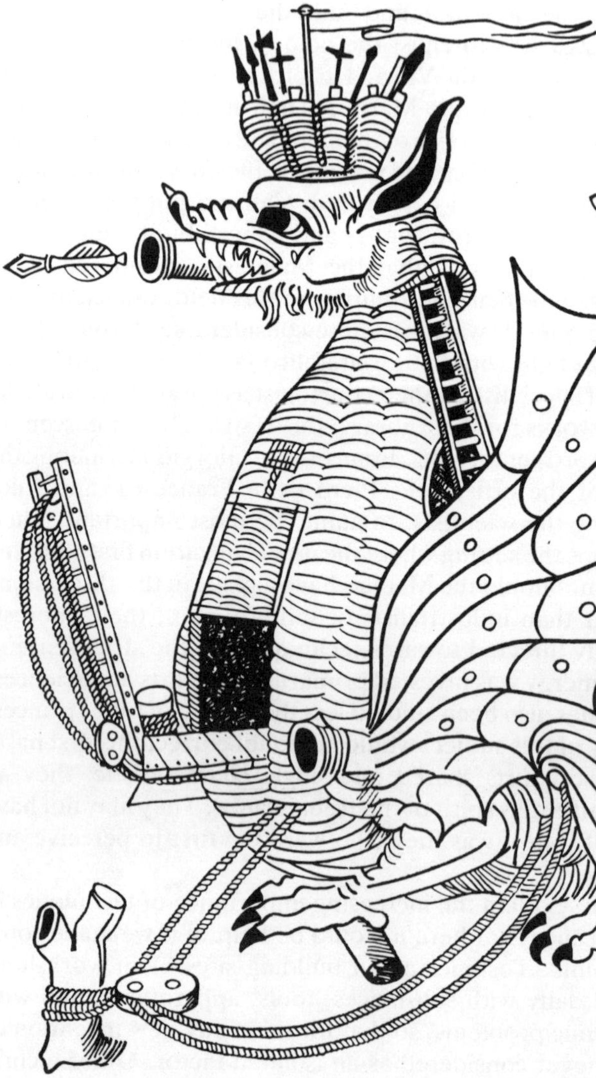

Figure 1. — Fantastic war machine (Valturio).

possible, they attempted to link technical knowledge with other disciplines. As we mentioned above, the technical mind underwent quite a significant transformation during this period.

The rise of the bourgeoisie from the rubble of feudalism, and the princes' desire for power encouraged a change from merely speculative research, to one which could be applied wherever possible. "The realist and bourgeois order" noted a German historian, "replaced the universal hierarchy, the abstract belief in a superior unity was abandoned and the understanding of existence became more direct and realist. The mind became less involved with the spirit, and more involved with the exterior world." The Belgian historian Fierens made the same points with regard to the history of art: "they broke loose from medieval spirituality, and became impassioned by the real properties of objects, the nature of things and the spectacle of the universe." This interpretation made in the context of art history could apply equally to that of the history of science, and with perhaps even more relevance. Realism, utilitarianism, empiricism, and eventually the growing appreciation of experimentation and mathematics were the moods of the age.

The teaching of mathematics gained ground gradually, first of all in Oxford, then Paris, then in a number of other universities. Here again, speculative mathematics was less favoured than applied mathematics. It was from these first elements that a pure, systematic mathematical science was born, and it is hardly surprising that some of the first true mathematicians were also technicians: the number of accountants and abacists are proof of this. Luca Pacioli lived in the remarkable centre of Urbino, and it was here that he conducted his research into the problems of accounting. Then there was the great Dutch engineer, Stevin. Mathematics and techniques were now inextricably linked; this was already apparent to Leonardo da Vinci: "Mechanics is the paradise of mathematics, because it is through the former that the latter is fully realised." And he added, "nothing is certain which cannot be explained by the mathematical sciences, or by those based on mathematics." Descartes used the same formulas, recognising that the principle function of mathematics concerned its application to the mechanical arts. Since Antiquity, geometry had been the science of land surveyors, architects and carpenters; arithmetics was the indispensable base of commerce. There is no need to give further examples of this.

The same was true of physics and techniques. The partial, isolated problems of physics posed by physicists at this time were a direct parallel of those encountered by technicians. They would avidly read all the ancient treatises: readers will no doubt recall the anguished search for works by Archimedes, for example. This is not the place to examine the emergence of modern physics, but it is worth noting the emergence of ballistics from the work of artillery men, and the contribution of hydraulic and general engineers to a science which had still yet to be systematised. From Francesco di Giorgio to Galileo, it was the technicians who were the most reflective. The creation of this new science meant inevitably that the technical mind and the principles involved were modified. We shall come back to this later.

The meeting (this term is a neutral one and avoids the fruitless disputes about which discipline is more worthy than the other) of a fragmented science and a group of disparate technical elements is again similar to the characteristics of the Alexandria School. Art, for example, is associated with perspective, with anatomy, the landscape, and with casting. According to Francastel:

> (...) the replacement of a universe of essences by an experimental universe was fundamental. The idea that the world was simply an attribute of God was completely abandoned. This conception of nature, of man as an actor on the theatre of the world, was accompanied by this extraordinary exploration of the universe which constituted the great contribution of those of the Renaissance. Consequently, they had a particular conception of invention. Invention implied the idea of a universal logic, and this explains the fact that one of the major objectives of inventors was not simply to enhance their technical knowledge of the disparate elements, but also to establish organised schemas and models of the universe. Their main effort thus consisted of selecting and segregating the possibilities of action and the theoretical possibilities of intellection offered by techniques.

This gave rise to perpetual confrontations between the various disciplines and activities, and furthermore, it encouraged, alongside the traditional channels of knowledge of the craftsmen and other practicioners, the emergence of what could now be called technology. Progressive stages were now evident: from the single operation, the process, the tool and henceforth the machine, – all of which were disparate, perfected elements – to the group, and the organised ensemble, to the tables formulated in answer to those drawn up by scientists, and finally to the explanations and the "reasons" which supplied the general framework within which all the various particulars were incorporated. This progression of stages is again reminiscent of the Alexandria School.

We know very little about those labour workers who probably contributed, in small but significant ways to technical progress, from partial, isolated improvements, to definitive transformations and from the correction of imbalances to the creation of an entirely new system. There is more information about those who collected particular cases and images and who from this "theatre of machines" managed to construct a more elaborate body of technical knowledge. The following discussion represents an attempt to trace the evolution of their thought.

Around the turn of the 14th century, the first anthologies were produced in Germany, devoted entirely to machines used in war. They are probably based on Konrad Kyeser's work, who was almost certainly a soldier. His *Bellifortis* is dedicated to the emperor Ruprecht of Palatinate (1400-1410). It was composed of ten books each one devoted to a separate subject: chariots; siege engines; hydraulic machines; elevating machines; firearms; defensive arms; wonderful secrets; ammunition; fireworks for war and for festivities; tools and work instruments. This may bring to mind previous works already mentioned, and indeed these were the very same themes dealt with in Byzantine encylopaedias, and even Hellenic works; Bacon also reflected upon these subjects, and they form the content of Leonardo da Vinci's letter to Prince Sforza. So this work was one of a long tradition. The same subjects are dealt with again and again: cannons, the first portable arms, assault chariots armed with artillery, mobile bridges, paddle boats, the archimedean screw, mills, drilling machines, and underwater divers. In fact, Keyser's work was similar in type to that of Guido da Vegevano, and it was to be followed by many others even into the 17th century, all coming under the category of "theatres of machines" (fig. 2). The so-called manuscript of the Hussite War, written around 1430, revealed the same preoccupations, but was slightly different. It was not an organised work; but consisted of a jumbled collection of information written down by a technician, or a civil or military engineer, who seemed to be trying to collect as many original ideas – old or new – as he could to enrich his technical knowledge. Most of the information concerned military machinery, but this modest work did deal with some civil machinery as well, such as mills, and machines for boring wooden pipes (fig. 3), which may well

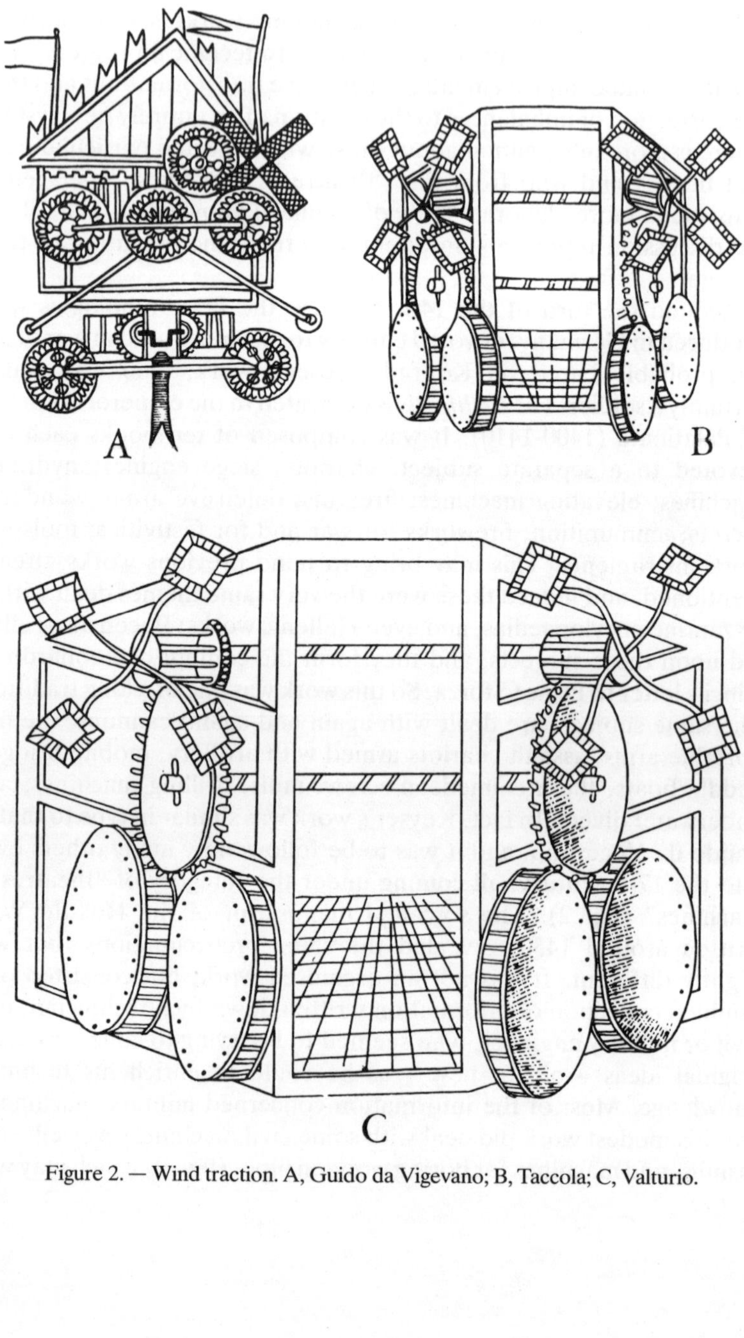

Figure 2. — Wind traction. A, Guido da Vigevano; B, Taccola; C, Valturio.

have been used in Nuremburg, and finally a machine for polishing gems, which was used in Venice (figs. 4 and 5).

This movement was more profound and widespread in Italy. Brunelleschi was a contemporary of Kyeser and he was the very epitome of the technical artist of the Renaissance. He began as a goldsmith and sculptor, then he became an architect. He also invented optical apparatuses, and built machines. Unfortunately, he has not left us any written works. He was the first of this generation of Italian engineers, and his training, his preferences and interests were markedly different to those of his German contempories.

Fontana was a doctor of the Most Serene Republic (Venice). His main interests were the natural sciences and physics, and he wrote some interesting works on these subjects. He was the author of the *Bellicorum instrumentorum liber*. The texts are punctuated with some brief remarks in cryptographical writing, and in fact this work is also a "theatre of machines." His interests were far from being exclusively military, and he dealt with hydraulics, fountains, water conveyance and particularly automatons (fig. 6).

Marianus Jacobus Taccola represents the height of this first generation. He was a great authority in his time and became known as the "sienese Archimedes." Like Kyeser, he was admired and imitated by many, and like his German predecessor he wrote a well-composed treatise *De machinis libri X* – a true "theatre of machines." He is perhaps the last representative of the Middle Ages. His preoccupations were the same: machines of war, siege apparatuses, and the use of hydraulic and wind energy. He was not an innovator, but he knew all there was to know about the techniques of his time. He was what we could now define as "technically erudite" (fig. 7).

Roberto Valturio merely compiled the military technical ideas of Sigismund Malatesta. But of course, the originality of a work is only relative. The main interest of this work *De re militari* resides in the fact that it is the first of such treatises to deal with the ensemble of a technique. It is not simply a collection of different elements; an effort has been made at systematisation. The large work by Leo Battista Alberti – philosopher, scientist, architect, and engineer – was written in the same spirit but concerned quite a different field. His *De re aedificatoria*, the first major treatise on architecture written during this period, was completed in 1450. Like his predecessor Vitruvius, he went beyond the narrow confines of architecture and dealt in precise

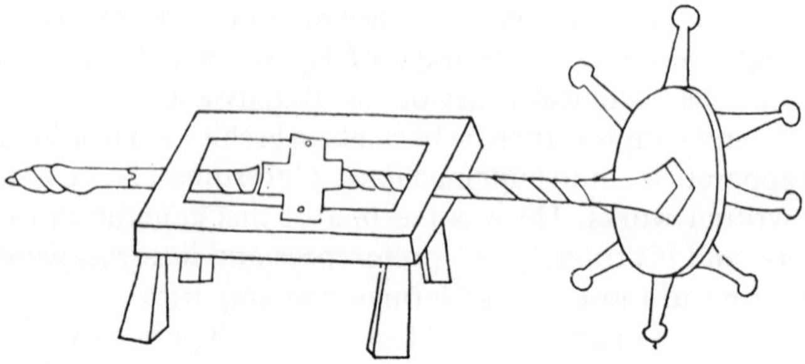

Figure 3. — Machine for boring wooden pipes (Taccola).

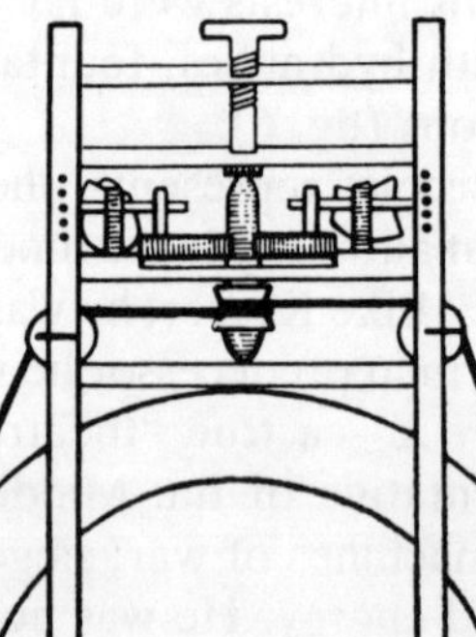

Figure 4. — Machine for polishing gems (c. 1475).

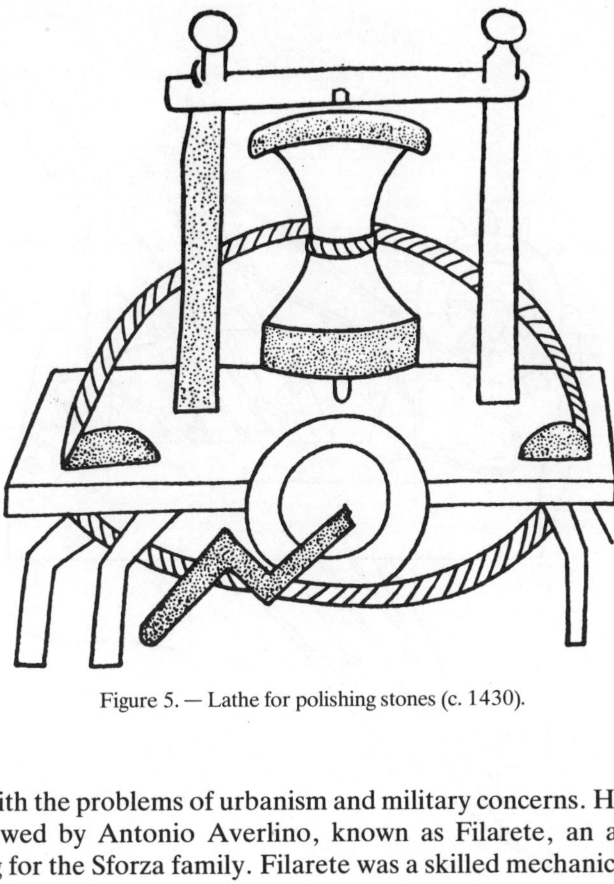

Figure 5. — Lathe for polishing stones (c. 1430).

terms with the problems of urbanism and military concerns. He was to be followed by Antonio Averlino, known as Filarete, an architect working for the Sforza family. Filarete was a skilled mechanician who had the same breadth of mind as Alberti, linking architecture with several other fields of activity. The two Fioravanti brothers did not write about their work, but they were successful architects and skilled mechanicians. Ridolfo Fioravanti, the younger, helped transport the Italian Renaissance to Moscow. Lorenzo Ghiberti has left us a valuable notebook full of precious indications about architecture, metallurgy, artillery and casting.

The age of the great technical treatises began with Valturio, Alberti and Filarete, but there are many others dealing with different topics. Some of these were written before this period, and they have already

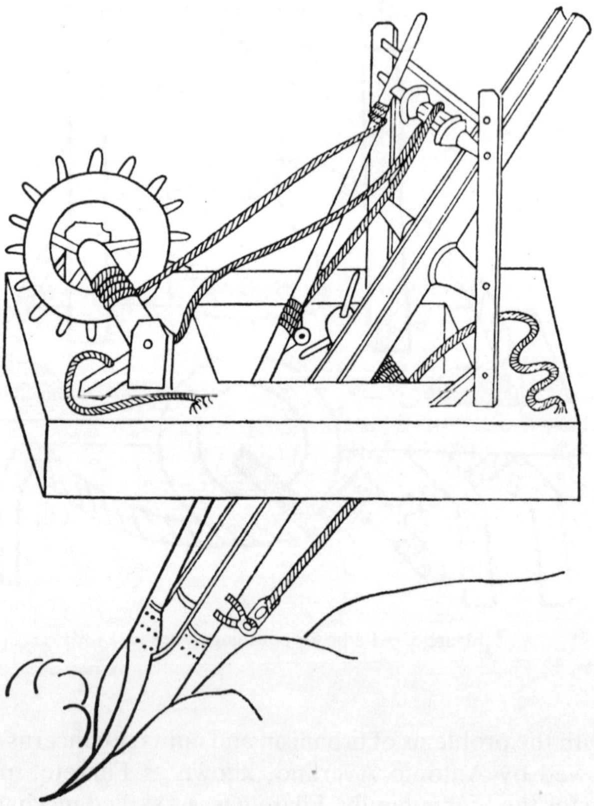

Figure 6. — Machine for removing rocks from the sea-bed (Fontana).

been mentioned, such as the large treatise on agriculture written towards the end of the Middle Ages and attributed to Pietro de Crescenzi. In the 15th century, there were treatises on astrolabes, and "books on canonry," which are still lying untouched on our library shelves, and which have yet to be satisfactorily researched. All these works represent an attempt to deal with the ensemble of a given technique.

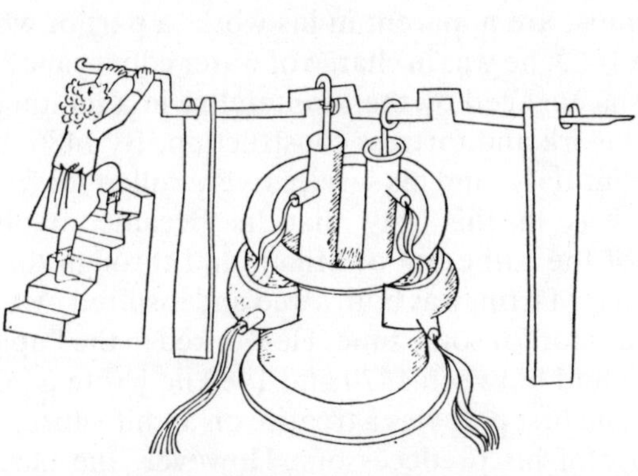

Figure 7. — Piston pump (Taccola).

One can now outline the profile of these engineers of the Renaissance: generally beginning as an artist, but an artist preoccupied with reality and the solving of practical problems. This concerned all aspects of art – painting was based on perspective, for example. Gold and silver working entailed the use of metals, and sculpting the use of various materials. Usually he was well practised in all the arts, or at least the most important. He was also an architect, and this of course made him a true technician, daily encountering such problems as stone-cutting, making buildings stable, and the problem of lifting apparatuses. As a sculptor, he also became skilled in casting when sculpture began to be cast in moulds. This inevitably led to learning about other types of casting, especially the casting of cannons. The combination of military architecture and casting cannons of course made him a soldier, and some added to this vast wealth of knowledge a preoccupation with hydraulics. The Renaissance engineer was curious about everything in a world whose evolutionary pace was accelerating so rapidly: he practised philosophy, he handled the first notions of a new science which he had helped construct with his own material knowledge. Between 1450 and 1475, it was his knowledge and outlook on the world which was to turn the face of the world towards modernity.

Francesco di Giorgio Martini is a perfect example of such a man; he was one of the great men of the following generation. He was born in Siena, where, if he did not know Taccola, he certainly read his works. He began as a skilled painter and sculptor, and some of his great skills, if not his genius, are apparent in his work, a part of which still exists today. From 1469, he was in charge of water conveyance to Siena, then in 1477 he was engaged by the Montefeltre in a military capacity, for architectural work and fortress construction. By 1486, he was renowned throughout Italy, and his services were called for from all over the country. It was in this way that he became involved with the completion of the cathedral of Milan, and through this he met Leonardo da Vinci. Martini was both awed and inspired by Leonardo, who became his mentor for some time. He worked in the Papal states and as far as Naples and between 1470 and 1480 he wrote a large treatise in three parts. The first part was a treatise on architecture, very similar in spirit to those of his predecessors. However, the part dealing with fortifications was much more advanced and in fact is the first real treatise on fortification. The third part deals with mechanics, and this, again is similar to those written previously, to the extent that it is another "theatre of machines" (fig. 8). This work was translated, copied and was very widely read. Other contemporary engineers had very similar careers." Take for example the Sangallo brothers, who are perhaps the most well-known.

Technical curiosity was a widespread movement, then, and it stimulated the research of both ancient texts, such as those by Archimedes and Vitruvius, as well as contemporary works. There were "editions" of Kyeser and the writers of the following generations to be found in all european libraries. Matthias Corvinus, the King of Hungary, the Montefeltre, and even the popes collected all the technical works available to line the bookshelves of their enormous libraries. Once more, this is reminiscent of the Alexandria School and its Library. And indeed, now that it had become possible, many of the treatises were then actually printed: this is indicative not only of the great importance attached to these works but also of the numbers of people interested in reading them. Some of the treatises we have talked about were the first to be printed: Valturio's work was published in 1472, and Alberti's in 1485. And to complete this vast store of knowledge, older texts dealing with traditional techniques and still felt to be useful were also printed: Pietro de Crescenzi's work was published in 1471, Pliny in 1469, Vegetius was published in 1471, and this work went through eight editions before the end of the century. The Latin agronomist works were published in 1472, Frontinus in 1480, and Vitruvius in 1487.

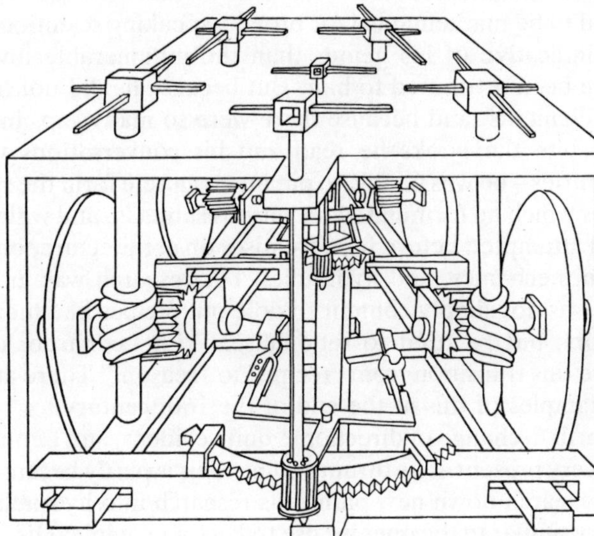

Figure 8. — Francesco di Giorgio's automobile vehicle.

These men were part of a new spirit, they had at their disposal a wealth of literature, their knowledge was more advanced. Their hopes were high, and the lure of success, glory and power all contributed to give techniques a place of the highest importance.

Leonardo da Vinci can only be properly understood if he is placed in this framework. He had the same qualities and faults as his predecessors and contemporaries. He had the same burning desire for knowledge, and perhaps because he was less erudite than the others, he felt the same boundless need to study every single work available, and to acquire its wisdom. He also had a passionate longing to work as a technician, perhaps, again, because he was less in demand than someone like Francesco di Giorgio, for example. Gradually, he became more withdrawn but his inspiration never ceased and if anything his ideas became more audacious as he became aware that they would probably never be realised. He became more reflective, and thus helped pave the way for the new explanatory type of technology. He sought the "reasons" behind things, and attempted to draw general principles, the base and formulation of which he understood to be mathematical, or broadly speaking scientific. This is far more indicative of his genius than the innumerable inventions which have been attributed to him. But because he did not have the necessary elements, and because there were so many gaps in knowledge – despite the books he read and his conversations with his contemporaries – he was unable to put into concrete form the systematic treatises which he formulated on various subjects, and which many others had attempted before him: treatises on architecture, on fortifications, on mechanics, and hydraulics. His research was more profound, he was not simply content to dole out recipes like most of his predecessors, but he tried to rationalise. With Leonardo, one can really sense this transition from "recipe" to "reason." There are some striking examples of this at the end of the 16th century.

Henceforth, a change of direction is quite evident, and Leonardo da Vinci was very present at its turning point. This is partly because he led technical research down new paths: his research into hydraulics used models very similar to the ones we use today for example. His research into gearing, and his appraisal of the workers' postures represented something quite new, and almost put technology in advance of science during this period. On the other hand, Leonardo met with several obstacles, and the sheer extent of his vision was one of them. It seems technology had become so wide that specialisation had become necessary once more. This was all the more important now that new methods had been developed, and these were able to bridge the gaps between the various techniques.

Consequently, it would be useful to continue the line of research which we have just outlined. A study of the technical mind of the 16th century would be invaluable. Some interesting work has been done on this subject, but it is not sufficient. Further study would explore the falling into place of the new technical system, the correction of some errors and the necessary adaptations made. A brief framework of perspectives for this is provided below.

The rather hazy pattern of trends at the beginning of this period was finally brought into sharp relief in the 16th century. The population increase, for example, became more marked and more general, and economic recovery was more apparent. It is worth noting, however, that when the large influx of precious metals and the first signs of inflation began in the mid-16th century, most technical transformations were already taking place. It is thus vital that we have precise chronologies for this period. One would have thought that inflation would have provoked people into finding ways of producing goods at a lower cost, but this does not appear to have been the case. There were many inventions in the former period, indeed, the latter is renowned for these inventions, but there were no technical mutations. Inflation caused an increase in production, but no significant changes in the methods used. A new technical system was expanding which had not yet been fully integrated in the western world.

The most difficult task was that undertaken by Leonardo da Vinci. The rationalisation of techniques without the necessary scientific tools was impossible. Every so often there is evidence of techniques being dealt with in terms of numbers, rather than empirically, but the transition was obviously difficult, and the method was still not scientific. This would explain the proliferation of tables. A theory of ships, for example would have been impossible at this time; it only became possible in the 18th century, when first developed by Euler. However, carpenters did use tables for the layout and the shape of ships, with varying degrees of success. This is clearly illustrated by the papers of Matthew Baker, a famous English shipbuilder working in the mid-16th century.

This represents a change in technical thought. In some rare cases, there is evidence of controlled experiments being conducted, rather than experiments controlled and initiated by the investigator – scientific experiments to replace the experience which the technician undergoes at work. At the very end of the 15th century, for example, on a beach near Naples, the artillery men of Charles VIII fired cannonballs (over a course which was marked out) through pieces of canvas placed at intervals; this was in order to establish the nature of the trajectories. There were several other experiments conducted to try and tabulate the ranges of fire. The work of Kricka, a Czechoslovakian caster and well-borer reveals similar attempts to avoid failures which were difficult to explain.

The "theatre of machines," so popular in the 15th century, emerged once more in the 16th. Much of the information was borrowed from the earlier works, and at the end of the 16th century, some designs were still being borrowed directly from the work of Francesco di Giorgio. Two famous writers produced works of this genre. Jacques Besson, who called himself an engineeer and mathematician, was professor of mathematics before he sought exile in Geneva. In 1569, he published his *Théatre des instruments et machines*, written in latin, but translated into French, German and Spanish. The Italian, Ramelli, was a military engineer in Charles V's armies before becoming advisor to Henri d'Anjou, king of Poland, the future Henri III. His *Livre des diverses et artificieuses machines* appeared in Italian and French in 1588. These two works are an anthology of various fields of mechanics, and often many solutions are provided for the same problem. As François Russo has noted, some aspects of these works were serious, and were designed to help experts, but for the most part they were an odd mixture of erudition and mechanical imagination which was applied largely for amusement and diversion.

Perhaps more interesting is the proliferation of long treatises dealing with a much broader range of techniques. It is through these treatises that one can perceive the emergence of a new technical system and its developments. They are also indicative of those sectors which have remained traditional and therefore fairly stable. Agriculture is foremost among the latter, indicated by the number of ancient agronomists' works published – thirty-one editions were printed between 1470 and 1550. Pietro dei Crecenzi was widely published and translated. But these treatises were dealing with ancient techniques which were largely invalid and they were also mainly limited to Mediterranean agriculture. It is not surprising, therefore, that the new treatises published in the 16th century reveal that techniques had made little progress since the Middle Ages: take the treatise of Fitzherbert, from England, for example, which was published in 1523 and went through eight editions before 1550, and the *Libro de agricultura* by Alonso Herrera from Spain (1539), and *la Coltivatione* by Luigi Alamanni from Italy (1546). The first editions of *La Maison rustique* which had a long-lasting success, were published just after 1530. The *Théatre d'agriculture* by Olivier de Serres does not seem to have progressed at all from the works of his predecessors. The most original work remains that of Bernard Palissy.

By contrast, the treatises on mining and metallurgy reflect considerable progress. They range from the short *Bergbuchlein* (published in 1505, with five more editions by 1540) to the long, "modern" treatise by Georg Agricola, *De re metallica* (repr. in Basel, 1969), which deals with both mining and metallurgy, and finally *Pyrotechnics* by the Italian Biringuccio who dealt exclusively with metallurgy. The treatises on chemistry were similarly successful. The *Liber de arte distillandi* by Brunschwygk, was published in Strasbourg in 1500, and this was widely emulated. A treatise by an anonymous author on dyeing, *Mariegola dell'arte di tintori* appeared in 1510. The first great treatise on dyeing was written by Rosetti and published in 1548 in Venice. There were many treatises on architecture, all of which provide evidence of the new style. The same is true of fortifications: the first main work was written by Albrecht Dürer in 1527, and this inspired some worthy imitations, by Battisto della Valle, for example, whose work went through ten editions from 1524 to 1558, and the work by Erard of Bar-le-Duc published at the end of the century. We mentioned above just how useful a full and detailed inventory of such publications would be to historians of this period and subject.

These works are interesting in that they caused some changes in the processes used at that time, and they are a testament to the technical transformation which was taking place. The publication of new editions and their continuing importance until well into the 17th century reveals not only that contemporaries were conscious of this mutation but that they believed no other transformations would take place. They sought less to innovate than to perfect the current system and extend it to its limits. Their only apparent objective was to expand the use of mechanisation, and to apply it to all fields of activity.

THE CLASSICAL SYSTEM

It is probably the concept of mechanisation which best characterises the technical system of the Renaissance. We may smile when we consider the mechanisation of this period in comparison to our own, but considerable progress had been made since medieval times. The only obstacles to mechanisation during this period were the materials available: gearings and transmissions were difficult because all machines were made of wood. The impossibility of constructing correct assemblies meant that after a while the machines became very rickety – this limited their power and reduced the speed at which they worked.

The discovery of the crank and connecting rod system, which converted rotary into reciprocal motion was at the basis of modern mechanisation. This transformation of movement was unknown in Antiquity and the Middle Ages, at least there is no mention of it in any of our written or pictorial sources. And if they had known of this concept, the mechanisation achieved during these two periods would have been considerably more advanced. This mechanism had several disadvantages. Mobile assemblies were difficult to construct from a strictly material point of view, and friction absorbed much of the machines energy. There was another difficulty which technicians and scientists did not fail to notice: the passage between the two dead points at each end of the diameter of the connecting rod's extension.

The first relatively crude example of this mechanism is found in Kyeser's manuscript: it is used on a grindstone, powered by hand; the crank being an extension of the arm. The development of this mechanism can be traced by examining the engineers' notes which we cited above. As they developed this system, they also tried to solve its drawbacks. The flywheel was incorporated to sustain movement and nullify the effect of the dead centres. Francesco di Giorgio came very close to conceiving of the ball and chain governor; the control of the apparatus was still being done by hand.

This apparatus was not actually used for quite some time. In all the earlier representations, the crank and connecting rod system was used only in hand-mills. It was eventually used in some pedal machines such as the spinning wheel, which according to all figurative representations was powered by hand, and from thence to all machines powered by pedals, like the grinding mill, for example. Later, the system was adapted for larger machines – particularly those powered by mill wheels whose motion was circular and which therefore required the conversion of motion – such as hydraulic saws and suction pumps. Sixteenth century works reveal that the expansion of this new mechanism was progressive but slow, on account of the difficulties in assembly.

We believe that the crank and connecting rod is one of the most important inventions of this period because all of modern mechanisation was to emerge from this system. The first pictorially represented pedal spinning wheel, which was in fact the first pedal lathe, is featured on a manuscript from 1470. Some drawings of hydraulic saws by Francesco di Giorgio appeared at about the same time. But it was not until the mid-16th century that Agricola depicted the first suction pumps powered by a water wheel with the crank-connecting rod system. But difficulties with the construction and running of these machines reduced the machine's capacity. The expansion of this mechanism was slow, but it gave the machine pride of place in technical progress.

The development of mechanisation prompted the need for more energy. The main resources of energy were the simple, relatively feeble water wheel, and wind energy which was hardly used. No other mechanical form of energy was conceivable at this time, and the claims about attempts to create a steam engine should not be taken too seriously. Leonardo's research and Branca's attempts to construct a steam turbine in the 17th century were merely expressing what had been known about the power of steam since Hero's aeolipile. Branca's machine did actually work on a small scale. It worked in the same way as roasting jacks, which were powered by hot air. The apparatus devised by Leonardo da Vinci which incorporated a cylinder and piston, could not work as long as the effects of condensation were still not properly understood. The concept of the void was denied at the time, and that of atmospheric pressure was unknown. No new "converters" were created, therefore.

The only way of increasing the amount of energy available, therefore, was by perfecting existing machines. It is in this light that Leonardo's much more interesting research into mill wheels should be appreciated, together with his considerations on the most efficient shape for the vanes and the angle at which they hit the water. This type of research reflects the beginning of a technology of the mill wheel – the first of its kind. It seems however, that the course set by the great Florentine was not pursued by any of his contemporaries or immediate sucessors.

There has been a lot of discussion about horizontal mill wheels, so much so that there is really no need to go into it any further at this point. It would simply be worth noting here that the passage from the wheel with vertical axles to the water turbine was a long one: these two machines were very different. There were some similarities – in both machines the wheel is placed under the water in order to prevent rotting and the effects of frost. The difference lies in the fact that the water acts on both sides of the wheel, but since the hydraulic turbine is enclosed and isolated from the water, and if the flow is correctly channelled, the water acts on only one side of the turbine. Moreover, it was necessary to include a suitable device which would put the vanes under maximum pressure from the water.

The first figurative representations of mills with horizontal wheels are to be found in the so-called manuscript of the Hussite War (c. 1430). This was simply a horizontal wheel, it was not a true hydraulic turbine. It was later declared that this type of mill was introduced during the 16th century just about everywhere, from Bavaria to the Pyrénées. Francesco di Giorgio went slightly further than this: his drawings provide a link with the origins of the hydraulic turbine. Almost all the elements were there: the horizontal wheel, the inclined blades, and the tube with appropriate fitting through which the water was conveyed to the wheel (fig. 9). It seems as if the wheel is isolated from the current, but we will never be entirely sure. This technique was not widely used: it seems that its first application was the installation of the Bazacle mills near Toulouse (which are beautifully illustrated in the *Encylopéedie*) in the 18th century.

It seems that the creative imagination was more effective in dealing with the windmill. The main problem was to cater for the changes in wind direction so that the blades would move continuously. The only solution for this in the Middle Ages had been to build wooden mills on enormous tripods. The stone mill first appeared in the 15th century, it was much more solid, and the roof turned on rollers. Of course, this had meant that the interior mechanisms had to change. The new type of windmill appeared around the beginning of the 15th century, although we cannot be sure about this. The tripod mill alone features in manuscripts from the end of the 14th century to the mid 15th century. Now that the main body of the mill was actually fixed and stationary, it could be used for purposes other than simply grinding wheat. And in the early to mid-15th century, windmills were used in Holland to power archimedean screws which were used to drain low-level land. Towards the end of this century they were used for sawing; and this again was only made possible with the crank and connnecting rod system.

The great expansion of the windmill came in Holland during the second half of the 16th century. It was only in the 17th century that it became more widely used in other countries. But in Spain in the mid-16th century, Ortiz was already advocating the extensive use of windmills in regions with few or unsuitable rivers. Readers are no doubt familiar with Don Quixote's struggle against windmills, which were seen as innovations running against the traditional order of things.

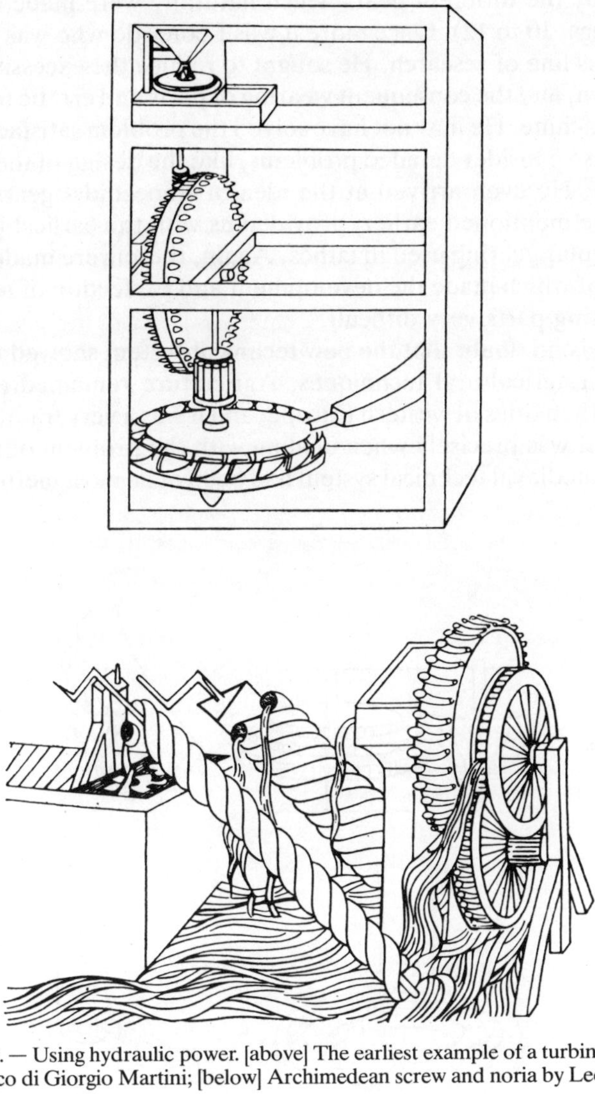

Figure 9. — Using hydraulic power. [above] The earliest example of a turbine by Francesco di Giorgio Martini; [below] Archimedean screw and noria by Leonardo da Vinci.

A parallel development was the attempt to reduce the absorption of energy by the different gears which naturally were made entirely of wood (figs. 10 to 12). Once more it was Leonardo who was the first to begin this line of research. He sought to reduce the excessive amount of friction, and the consequent wearing of parts and erratic functioning of the machine. He may not have solved the problem satisfactorily, but he did also consider detailed problems, like the design of the teeth, for example. He even arrived at the idea of trapezoidal gears. Besson, whom we mentioned earlier, provided us with the earliest pictures of the irregular gearing used in lathes. Again, these were made of wood, the use of which made the development and perfection of movement-transferring parts very difficult.

There is no doubt that the new technical system showed itself most clearly in agricultural techniques. Agriculture remained one of the essential activities of western Europe, and it was a very traditional one. Indeed, it was precisely when dealing with the problem of this sector that the medieval technical system had been at its most ineffective. We

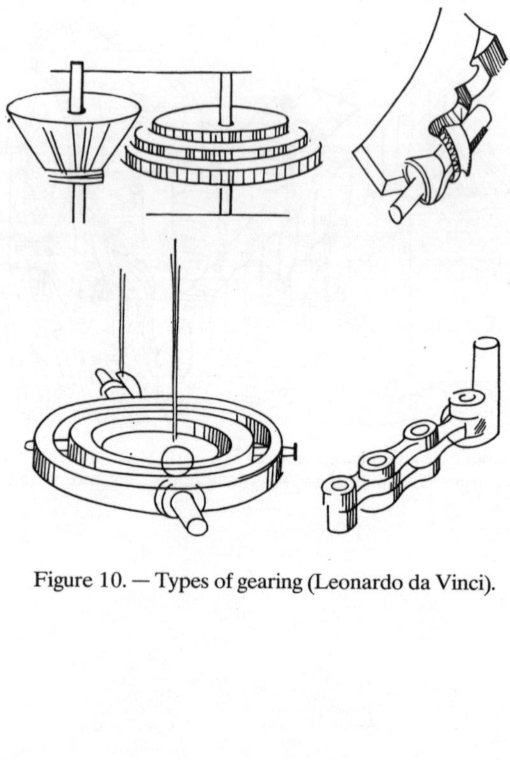

Figure 10. — Types of gearing (Leonardo da Vinci).

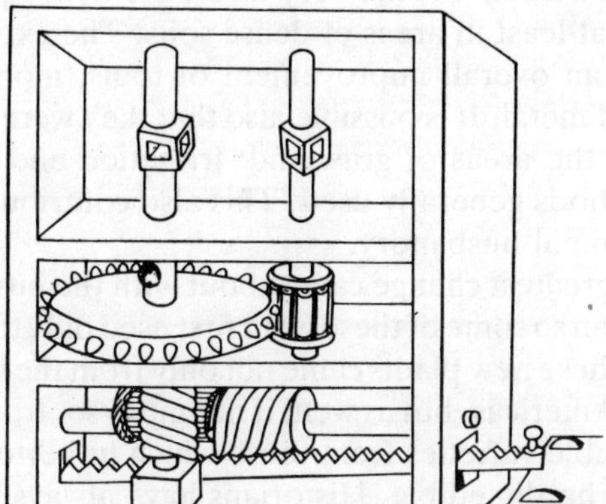

Figure 11. — Using the endless screw (Francesco di Giorgio).

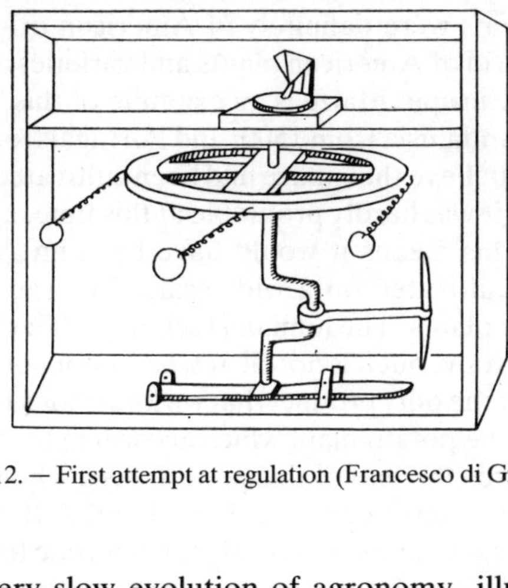

Figure 12. — First attempt at regulation (Francesco di Giorgio).

also noted the very slow evolution of agronomy, illustrated by the continued reliance upon medieval or even ancient works. We are still dealing with recipes rather than reasons.

The techniques of actual cultivation and crop rotation systems hardly changed at all. At the very most, the heavy plough was more widely used, at least in areas of dense soils. The expansion of metallurgy caused an overall improvement of tools, more of which were being made of metal. It is possible also that they were more efficient in looking after the areas of grassland: irrigation and dressing of land were the methods generally used. This also contributed to the development of animal husbandry.

In fact the greatest change came about with the introduction of new plants, but even so some of these were first used only towards the end of this period. These new plants came not only from the newly discovered world of the Americas, but as well, if not more so, from the creation of edible, cultivable varieties from those which hitherto had only grown wild, or were hardly edible. Historians have always been seduced by the contribution of the New World to such developments, and consequently not enough emphasis has been placed on the botanic revolutions taking place in Italian nurseries, for example, which in some cases acted as transit centres. The growth of botanic science was certainly important, but it occured after these discoveries, in the mid-16th century. In some cases, the American contribution has long been the centre of heated discussions, and the issues are far from being closed today. Some plants were definitely of American origin, while others seem to be a hybrid of American plants and varieties already known in pre-Columbian Europe. Maize is an example of this, and is known to have existed in Andalusia from 1500, and Portugal from 1515 onwards. Some historians believe that maize had been cultivated in Europe since before 1492, but it was hardly profitable at this time, and probably also hardly edible. This means it would have been an American variety which was first cultivated on a wide scale. The same doubts persist concerning other plants. The humanist scholar, Valeriano planted haricots in Belluno, Italy, but it is not clear whether or not these came from America. In fact, the only product that we know for sure came from the New World was the potato plant which according to many sources was introduced towards the end of this period. As a whole, then, it seems that the American contribution was fairly limited. It did not match the number of important plants sent by western Europe to the New World, which were to become a future source of great wealth to this area – such as coffee, sugar cane and certain cereals which had been unknown there before the arrival of Christopher Columbus.

Italy's achievements in developing and expanding the numbers of edible plants was probably a more important factor. Very little is known about this subject, which warrants extensive research. As far as we can gather from the information that we do have, it seems that in Italy two main channels converged. The first of these was the contribution from arab countries and from the East in general. Plants which had been grown or developed in these areas were imported into Italy either from the south, or from Venice, the two regions in contact with the eastern basin of the Mediterranean. A few examples of what would be a very interesting inventory will suffice here to demonstrate the importance of these plants. The artichoke – the thistle head – was gradually developed by the Arabs before it was introduced to the table. It was first introduced into Naples in 1466 from whence it passed into Florence in 1466, and Venice in 1480. It crossed the French border during the beginning of the following century. These dates are taken from well-known documents, but they are not necessarily completely accurate. And the same goes for the melon, which is supposed to have been known in Avignon by the end of the 14th century. Then there is the cauliflower, which came from the East in the 16th century.

Other plants came from different channels, such as Portugal, for example, and this was no doubt due to their importance as a maritime nation and especially their activities off the African coast: they probably introduced cloves and definitely cinnamon, which Vasco de Gama brought back from the Moluccas in 1498.

The other main channel was actually Italian. Many plants were improved or domesticated in the Italian botanic and kitchen gardens, and this aspect requires much more rigourous research. Our information on this subject is particularly scanty, so we can only talk in general terms. It seems that the cos lettuce and other lettuces already extensively cultivated at the beginning of the 15th century, the marrow, pumpkins and aubergines were first grown in these gardens. Other plants were developed – the carrot was one of these, which previously had been so woody as to be practically inedible, and the beetroot was developed from changes and improvements made to beet. The same progress is evident with regards to fruit. During the latter half of the 14th century, strawberries which had previously been gathered from woods, were grown in gardens for the first time. Raspberries and currants were also now grown in gardens.

Other plants were also far more widely cultivated in the course of the following centuries, but we know very little about them. At the end of the 15th century, some species of plants were cultivated which were to assure the later development of artificial grassland. Alfalfa had been known in Ancient Greece, and in the Middle Ages it returned to Venice then up to Lombardy in the 15th century. It finally made its way into France during the early 1500's. In a sense, this was a rediscovery. The same was true of sainfoin, and red clover was introduced slightly later.

Other developments reached their limits at this time: buckwheat was now known in all of western Europe, and hops continued to advance towards the West, gradually replacing barley. Older plants made way for these new ones: sorghum, panic grass, and what was left of millet disappeared in the wake of maize. Our task would be much easier if we had tables and maps to hand but these have yet to be compiled. They would enable us to measure precisely the extent of these botanical changes, which appear on the surface to have been very important indeed.

Furthermore, the significance of these changes is probably greater than they appear superficially. Food now consisted of a much greater variety of vegetables, salad vegetables and fruits, which enhanced the dietetic balance of the peoples of these regions. Unfortunately, there are some aspects about which we know very little, such as animal rearing.

The Renaissance does not seem to have brought with it many changes in the forests, and at the very most they may have introduced a hardier poplar. The main concern of the centralised states was to organise a more efficient use of forests – insofar as wood was still the primary resource. The first major reforms were passed in France at the end of the 14th century. In this same country, the organisation and administration of forests (i.e., their surveillance) were dealt with in the great royal ordinances of 1520 and 1544.

Nor were there many changes in fishing. Large shoals of cod had been discovered during the voyages made to America, which were apparently fished by the Basques – among others – from the middle of the 15th century. Guillaume Beukels is reputed to have invented the process of curing and barrelling of herring around 1447. The fish were prepared, salted and put in barrels on board the ships before they reached the ports. This greatly increased the ultimate yield. But this was also to introduce important changes in the activites of coastal peoples, for this work had been their livelihood.

Animal husbandry did not develop a great deal, although some botanical inventions probably improved it slightly. Jean de Brie's *Treatise on shepherding* was written in 1379 and printed before 1500. Nothing is known about the evolution of domestic breeds, but two new animals were introduced into the farm yard, the guinea fowl, introduced in the 16th century from Guinea, and the turkey appeared in the 15th century. Large amounts of game were served on the tables of the well-to-do, brought in by hunting. Other sectors of the population ate little meat, and usually this was pork. Again, our sources for this subject are very thin on the ground.

One cannot say that just because the Renaissance brought with it significant changes in certain areas of human activity, the same was necessarily true of agriculture. Even though current agricultural techniques had been in use for centuries, they were still capable of being developed. And according to Oliver de Serres, changes did take place, especially in the types of tools used. More progress was made in botany with the introduction of stocktaking, and more important perhaps, with the establishment of more botanical gardens: the one in Padua was created in 1533, that of Sologne in 1567, and Leyden in 1577. These botanic gardens evolved considerably; this is illustrated by the change in character of French botanical gardens, for example. In 1540, In Touvoie, near Le Mans, the explorer Belon built a garden for the bishop René du Bellay. Guillaume Rondelet established a garden near Montpellier, and the apothecaries of Paris established the *jardin des simples* in the *rue de l'Arbalète*. These gardens were designed primarily for curiosity's sake, or for the growing of medicinal plants. The first large botanic garden meant for wider purposes was built in Montpellier in 1593, in response to an edict of Henri IV. In Paris, the *Jardin des Plantes* was founded in 1626 and completed in 1635. Its original function was to provide medicinal plants, but it eventually became a garden for experimentation and demonstration. In fact, the main laboratory proved to be that of the common vegetable garden, which first emerged in the 15th century; previously vegetables had been grown in fields. Vegetables were cultivated in kitchen gardens with more attention and care, the work was more intensive, and effectively these almost became laboratories of plant research.

One curious fact is that the contemporaries of this period were actually conscious of the stabilisation of farming techniques in all its aspects. The sacrosanct virtues of experience still reigned, and innovations were incorporated only with difficulty. It would have been possible to give up digging the vineyards by hand and use light ploughs instead, but this technique was first used in the Bordelais only in the 15th century. It is also possible, therefore, that even the diversification of instruments and tools took place only very slowly. A systematic list of the figurative representations of tools would be very helpful on this subject. At any rate the general impression at the moment is that although instruments were developed, particularly by a more widespread use of metal, there was no actual increase in the number of different types of tool used.

Written sources bear this interpretation out quite clearly. At the end of the 16th century, Olivier de Serres - who has been mistakenly called an innovator – recommended the continued use of existing tools: "it is best that the ploughshare remain unchanged, so as not to incur the losses which always result from a transformation." This phrase is typically representative of the prevailing attitudes. On the other hand, Bernard Palissy noted that while the military engineers might have succeeded in improving all techniques in this field, from fortifications to armaments, it was apparent that farming techniques and implements held no potential for man's ingenuity, and so were looked on with disdain. The sometimes arbitrary transposition and simplification of traditional techniques meant that the practitioners of such techniques and those who advocated the use of them sometimes lapsed into error. Olivier de Serres advocated, as did Charles Estienne in *La Maison rustique*, the general use of the plough, and rejected the idea of the swing plough, thus failing to take into account the different nature of the terrain, the density of plant layers and topography. This attitude was simplistic, and was not without some major problems.

Here again, progress was made without the help of the theoreticians. The only notable development in ploughing implements was the introduction of the stock which was probably first used by artillery men. The first, remarkable example of this appears in a manuscript of the Duke of Berry (France) in 1404. The two-wheeled forecarriage has a support made of a forked stick, which can be adjusted by means of a dowel (peg), and used in five different positions, and on which is placed the leg of the plough. The closer the dowel to the coulter, the less deep the ploughing. A manuscript by Gilles de Rome dating back to the beginning of the 15th century, depicts another type of adjustment device.

As far as the other operations involved in cultivation are concerned, such as the use of the roller, or of the harrow with wooden or iron teeth Olivier de Serres and Estienne simply reiterated what had been known for centuries.

Insofar as some historians believe that the medieval farming system had arrived at its limits and that the recovery only began with the onset of a new technical system, it seems that farming techniques during the early 15th century hardly changed. Some variety had been introduced into agricultural production, but did this mean that there had been no quantitative increase? This, was still not really necessary at this point because the rise in population in fact had served only to compensate for the severe demographic losses of the 1300's, and this was so until the end of the 15th century. On the other hand, it would be worthwhile studying 16th century agriculture in that there was a significant rise in the number of people during that period, and this coupled with the increase in individual consumption meant that farming techniques were forced to change. Historians have shown little interest in this aspect of history, probably because of the lack of sources, but a study of this question would lead us to a whole new avenue of discovery, from which the history of techniques has everything to gain.

There was a profound transformation in the use made of the underground. There has always been a tendency to represent Agricola's famous treatise as a work dealing with medieval techniques, but on the contrary, his work epitomises the characteristics of the Renaissance. We mentioned above that many mines had gradually fallen into disuse by the end of the Middle Ages, but work was resumed in many of these towards the mid-15th century. Crisis and recovery were generated by many factors, one of which was the influence of techniques. Once the superficial veins had been fully exploited, for example, the development of deeper veins posed many difficult technical problems such as the extracting of waste and minerals and water drainage, to name the most important. Because of the lack of instruments or of the necessary machines, this work was either impossible, very slow, or very expensive.

Since this mutation also depended on economic factors, its evolution occured in stages rather than all at once. The retable in Annaberg, a town in Saxony which was situated right in the middle of the mining district discovered around 1425, and which was painted by Hans Hesse a century afterwards in 1520-21, reveals that the techniques used were still relatively crude. Galleries were built on a wooden base, winches were powered by hand, and ordinary pick-axes were used for extracting the minerals. The same is true of the manuscript of Sainte-Marie-aux-Mines, painted by Heinrich Gross around 1530: man-powered winches were used to raise the minerals, but the mining wagon, which rolled on wooden rails was also used. Agricola's work represents a very sudden transformation: the apparatuses are much more advanced.

Just as botany served agriculture, so geology, or at least the first elements of geology, were to prove indispensable for the research and development of mineral veins. Chance was certainly a factor in this as well: one only has to consider the number of discoveries attributed to miraculous intervention. The divining stick was henceforth no longer used – if it ever had been in the first place. Both the *Bergbuchlein*, the first editions of which date back to the beginning of the 16th century, and Agricola's work reveal the slow gradual evolution of geology. It was still incomplete of course, but they were already able to use it for prospecting.

All the manuscripts mentioned above, from those of Annaberg and Sainte-Marie-aux-Mines to the Kuttenberg manuscript (or Kutna Hora, in Bohemia – the manuscript is now in Vienna), that is, all those prior to Agricola reveal evidence of very traditional techniques being used, to bore wells, for timbering, for the layout of the galleries, and for the ventilation wells. It is traditionally thought that powder was used for the first time in 1527 in Chemnitz. It is also possible that they used compasses.

The technical revolution which is apparent in Agricola's work rests essentially on a more advanced mechanisation which began around the turn of the 15th century. This is also revealed in the evidence found in the Polish salt mines of Wielicka, which underwent a marked development in the first half of the 16th century. Mechanisation was developed at certain key stages of the mining process.

The extraction of water, which may have been the root cause for the eventual disuse of many mines, was achieved using machines which may have been devised before this time, such as the chain and bucket systems: there are examples of this in Agricola and in the Salins tapestry which appeared slightly earlier. The mechanism was powered by a horsewhim or other types of treadmills. But Agricola provides evidence of another machine – the suction pump powered by mill wheels and incorporating the crank and connecting rod system. At deeper levels, they had to use several pumps one above the other, linked by return systems to a single energy source. As long as the piping was correct – and Agricola was very interested in this aspect – the means of water drainage were much more effective than the ones used in the past, although since this machinery was made of wood, it was still relatively inefficient.

There were developments in the type of lifting machines used as well. The Kuttenberg manuscript, for example, shows a treadmill being used instead of hand winches. Agricola provides evidence of a somewhat extraordinary water machine (fig. 13). It is made of two wheels fixed together, whose blades are inclined at opposite angles. The water used to power it is supplied from a reservoir with sluice-gates. It was designed so that the enormous winch could be turned in one direction or the other, simply by opening one or other of the gates.

The three-wheeled wagons running on wooden tracks and known as "dogs" were now used for all transport within mines. Agricola has provided us with some very precise drawings of these machines. It is also possible that the wheelbarrow was more widely used.

It was this more advanced mechanisation and the effective use of hydraulic energy which created the upsurge in the mining industry. As we said above, the picture painted by Agricola should be qualified: the developments were only relatively advanced, and even these developments were not immediately used everywhere. It was only in 1531, for example, that the suction pump was used in Liège. Apart from the problem of water, other machines were only necessary if the techniques developed as well. There was no point in having more powerful lifting machines if the techniques used to extract the products were still those of the Middle Ages. Again, this needs closer study – if powder was used during the 16th century, then there was some justification in using lifting machines. In other words, the idea and application of such new techniques presupposes analogous changes in parallel techniques. It seems that most of these new techniques were first introduced by the miners of Bohemia, Saxony and even Liège. It is for this reason that they were so much in demand in western Europe, from France or England, and even from Russia.

Figure 13. — Elevatory machine with reversible motion (Agricola, 1556).

There is another element of prime importance. We are now dealing with large scale exploitations like that of Jacques Coeur, and no longer with disparate individual enterprises. This meant that the organisation was more rationalised, and that much more powerful means could be employed. The organisation of mining production was an important factor in the development of techniques, and this was evidenced by the manuscript of Sainte-Marie-aux-Mines, and later by Agricola. We noted earlier the importance of a particular type of capitalism regarding the passage to a new technical system, and the mining industry provides possibly the most striking example of this.

The developments in metallurgy are perhaps most representative of the new technical system. Here again, research so far has been inadequate. The dating is uncertain, and there are still many problems to be solved. We have nonetheless attempted to establish a schema for the emergence and development of new techniques, even if some of its elements are mere hypothesis which still need to be verified.

By the end of the preceding period, iron was probably being produced in blast furnaces which already constituted an advance on the old open hearths. Nevertheless, research and excavations in England, Bohemia and Hungary have shown that the old style furnaces were still used for some time in certain areas. The blast furnace was a large permanent apparatus. It could only be made larger if the air-supply was increased, and there were two major problems in this respect: the problem of the air-flow, and of transforming the operation so as to produce pig iron as opposed to ordinary iron. We shall not dwell too long on these unverifiable points.

There are hardly any written sources before the 16th century and the figurative representations are virtually non-existent or ambiguous. The dating of machines which we do actually still possess is also very arbitrary. There was no way of obtaining more air other than by using hydraulic energy. A text dating from 1323, from the Brie region provides us with the first evidence of this technique being used, and there are several more examples of this from engineers' notes dating from the mid-15th century (fig. 14). But the introduction of hydraulic air-pumping does not necessarily entail the production of pig-iron at the end of the operation.

It is very likely that the combination of new elements, the hydraulic air-pumping and the larger dimensions of the furnaces led to the accidental production of pig-iron, which was then probably put on the

Figure 14. — Water-driven bellows (Taccola).

scrap heap. If the production of pig-iron was deliberate, that meant that they knew how to refine the latter in order to produce pure iron. The first method was relatively easy, but the second was not without its difficulties. But in each case, some of the inspiration may have come from related techniques: we have already examined the importance of these technological transfers. Both bronze-smelting which was widely practised at this time, and copper-refining may have been very revealing in this respect.

It seems that the diffusion of the blast furnace, the new, essential element of the so-called "indirect" process can be situated in the last quarter of the 15th century without too many problems, and in fact, this does actually concord with other technical transformations. Smelting iron was probably first produced deliberately in Liège. The blast furnace was probably introduced to the rest of France from here, along with the "wallonne" refining method, and by the end of the 15th century it was in widespread use in Lorraine, Normandy, Champagne, and Nivernais. There is evidence of its being used in Sussex, from 1496 (fig. 15). During the 16th century, it continued to spread throughout Europe to the rest of France and Western Germany (Agricola had not heard of it), then into southern and eastern Europe during the second half of the century. In some areas, such as southern France and Spain, for example, it was not adopted until the modern era. It is difficult to understand why, but they continued to use the direct process known as the Catalan method until this time.

The adoption of the blast furnace completely changed iron and steel production. There was now a new product involved, cast iron, which was to force other changes. The production of cast iron balls was to

Figure 15. — A blast furnace (16th century English drawing).

cause a revolution in artillery, and similarly, cast iron was soon being used to make the cannons. Many other objects were made of this metal, from firebacks to cooking pots. We need precise inventories listing cast-iron objects to appreciate the full value of this innovation. With larger apparatuses, production could be significantly increased. And finally, the iron produced by refining was much softer than that made by the older method, and therefore easier to work. Here again, a list of the objects made of this new type of iron in the 16th century would be very useful in measuring the increased use of this metal. Research has hardly begun on this topic.

The increase in production and the use of materials which were easier to work risked causing a bottleneck during the stage at which the metal was transformed. The main instrument used in forging was the hydraulic hammer which had been incorporated during the preceding period. But both its structure and its function were now ineffective. Therefore is it not astonishing that we witness, in this domain and during the epoch which interests us here, considerable development in the mechanisation of iron and steel making methods. In fact we know very little about the evolution of the hydraulic hammer. It would appear that the Middle Ages only knew the so-called terminal hammer, the smallest hammer with the least output, at least for certain types of work. It is possible that during the Renaissance a heavier tool was developed, the so-called lateral hammer where the attack of the cam was operated between the pivotal axle and the head of the hammer, producing less stress on the shaft and subsequently enabling the weight of the hammer to be increased. It was henceforth possible to forge much larger pieces (fig. 16).

On the other hand, a series of instruments was to come into play, the development, dates, and place of origin of which we know very little. They were powered by hydraulic energy, which here again seemed to be the mainstay of advanced mechanisation. Let us try to review these tools, indicating all that we do know about them.

The rolling mill, composed of two cylinders turning in opposite directions to each other, is one of the instruments fundamental to modern iron production. The hot bar of iron is flattened and tightened by being continually passed between the two cylinders which moved progressively closer together. The operation could be continued for some time in order to produce sheet metal, which was also a large consumer product. This instrument raises two important difficulties:

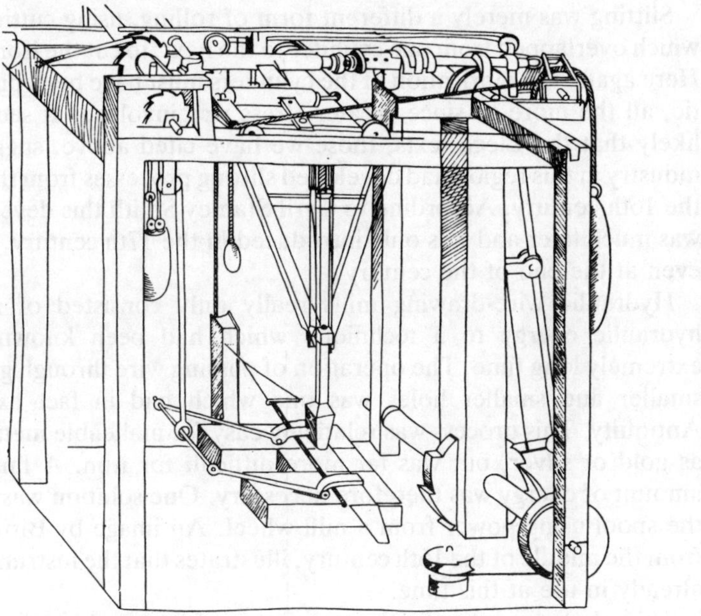

Figure 16. — Hammer for forging large objects (Leonardo da Vinci).

firstly, the way in which the cylinders were made, and secondly, the method of tightening. The existence of cast iron which could be moulded, facilitated the fabrication of the rollers, even though the cast iron may have been particularly brittle at this stage. Similarly, the use of cast iron made possible the creation of gears which transmitted the movement of the hydraulic wheel to the apparatus: it would have been very difficult to use wood in this process. Tightening could be done with the aid of nuts. It is difficult to know where and when the rolling mill was first developed. Many authorities believe that the first example of a rolling mill was produced in 1615 by the great engineer, Salomon de Caus. But it seems that this invention is probably much older than this, dating back to the last quarter of the 16th century, as Liège texts would seem to prove that the rolling mill was already in use in this region and at this date. The history of the rolling mill still remains to be completed, as we should trace the development of this mill from its origin to present day.

Slitting was merely a different form of rolling, using cutting edges which overlapped from one cylinder to the next, to cut the iron in bars. Here again, the fabrication of the cylinders must have been problematic, all the more so since greater stress was involved. It seems very likely that the Liège texts, those we have cited above, suggest that industry in this region had developed slitting processes from the end of the 16th century. According to Cyril Stanley Smith this development was much later and was only introduced in the 17th century, perhaps even at the end of the century.

Hydraulic wire-drawing mills really only consisted of applying hydraulic energy to a technique which had been known for an extremely long time. The operation of passing wire through gradually smaller and smaller holes was one which had in fact existed in Antiquity. This process was relatively easy for malleable metals, such as gold or silver, but was far more difficult for iron. A far greater amount of energy was therefore necessary. One solution was to drive the spool using power from a mill wheel. An image by Biringuccio, from the middle of the 16th century, illustrates that the instrument was already in use at this time.

This whole evolution in the techniques of ironsmelting thus contributed a more abundant output for a much greater range of products. The use of metal – and the new quality of irons should also be taken into consideration – could thus be increased in considerable proportions. We could even say that the metal age really began during this epoch.

Although our perception of the problems presented by the iron and steel industry is fairly keen, our knowledge of the development of non-ferrous metals is rather more vague, despite long passages which appear on this subject in treatises by Agricola and Biringuccio. Of course, techniques involved in non-ferrous metalworking obviously profited to a certain extent from the new techniques and machines used in ironsmelting. The discovery of America prompted some attempts to find precious metals. It was in 1451, however, according to a tradition which remains difficult to verify, that a certain Johansen Funcken developed a method of separating silver from lead and copper. The amalgam for the production of gold dates from the middle of the 16th century. Far more local metals were used, as the process of separating

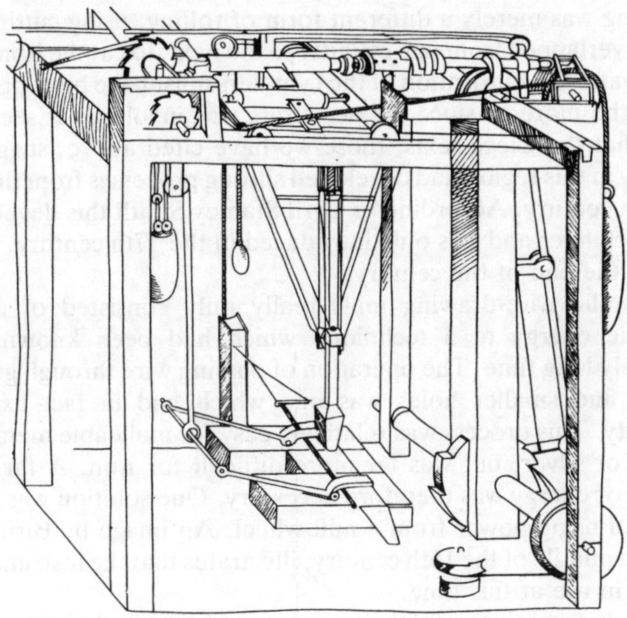

Figure 16. — Hammer for forging large objects (Leonardo da Vinci).

firstly, the way in which the cylinders were made, and secondly, the method of tightening. The existence of cast iron which could be moulded, facilitated the fabrication of the rollers, even though the cast iron may have been particularly brittle at this stage. Similarly, the use of cast iron made possible the creation of gears which transmitted the movement of the hydraulic wheel to the apparatus: it would have been very difficult to use wood in this process. Tightening could be done with the aid of nuts. It is difficult to know where and when the rolling mill was first developed. Many authorities believe that the first example of a rolling mill was produced in 1615 by the great engineer, Salomon de Caus. But it seems that this invention is probably much older than this, dating back to the last quarter of the 16th century, as Liège texts would seem to prove that the rolling mill was already in use in this region and at this date. The history of the rolling mill still remains to be completed, as we should trace the development of this mill from its origin to present day.

Slitting was merely a different form of rolling, using cutting edges which overlapped from one cylinder to the next, to cut the iron in bars. Here again, the fabrication of the cylinders must have been problematic, all the more so since greater stress was involved. It seems very likely that the Liège texts, those we have cited above, suggest that industry in this region had developed slitting processes from the end of the 16th century. According to Cyril Stanley Smith this development was much later and was only introduced in the 17th century, perhaps even at the end of the century.

Hydraulic wire-drawing mills really only consisted of applying hydraulic energy to a technique which had been known for an extremely long time. The operation of passing wire through gradually smaller and smaller holes was one which had in fact existed in Antiquity. This process was relatively easy for malleable metals, such as gold or silver, but was far more difficult for iron. A far greater amount of energy was therefore necessary. One solution was to drive the spool using power from a mill wheel. An image by Biringuccio, from the middle of the 16th century, illustrates that the instrument was already in use at this time.

This whole evolution in the techniques of ironsmelting thus contributed a more abundant output for a much greater range of products. The use of metal – and the new quality of irons should also be taken into consideration – could thus be increased in considerable proportions. We could even say that the metal age really began during this epoch.

Although our perception of the problems presented by the iron and steel industry is fairly keen, our knowledge of the development of non-ferrous metals is rather more vague, despite long passages which appear on this subject in treatises by Agricola and Biringuccio. Of course, techniques involved in non-ferrous metalworking obviously profited to a certain extent from the new techniques and machines used in ironsmelting. The discovery of America prompted some attempts to find precious metals. It was in 1451, however, according to a tradition which remains difficult to verify, that a certain Johansen Funcken developed a method of separating silver from lead and copper. The amalgam for the production of gold dates from the middle of the 16th century. Far more local metals were used, as the process of separating metals remained problematic, and a lot of silver must have been left in the argentiferous lead extracted from some mines.

Techniques employed in the production of copper appear to have been modified between 1450 and 1550, as copper – up until then used particularly for making bells and copperware – was suddenly in great demand for casting bronze cannons. The Germans called the factories where these new methods were in practice *Saigerhütte*, although we do not know exactly how the new techniques differed from the old ones. This explains the extent of cupriferous exploitations undertaken by Jacques Coeur and why many new mines were set up in Bohemia and Saxony. The use of hydraulic energy was generalized to power bellows and hammers. Shaft furnaces and open-hearth furnaces were employed for both lead and copper production.

The bronze used in cannons was different from that used in bells. The latter contained 23 to 26% of tin, while bronze used in artillery contained only 8 to 12%. The important factor in all these bronze-using techniques, from the traditional bell to the production of cannons and statues, was the introduction of the mould and the material from which this was made. Many works illustrate the way in which this was done. It is quite possible that the great demand for cast bronze pieces enabled these methods to be perfected.

A last development in the art of metallurgy may be attributed to our period. This does not concern an alloy but a combination of two metals, which was in turn to give rise to the new product of tin- plate – the use of which was limited in the 15th and 16th centuries but which was to enjoy a considerable application in the future. Mechanical problems were not involved here, although the fabrication of large sizes of sheet metal which were relatively regular in thickness could still present difficulties. What was most important was how a layer of tin could be stably fixed to cover this sheet metal. Products used to this effect remained secret for a long time. It is likely that empiricism was king at the beginning, and that all sorts of materials were used to fix the tin layer.

Our documentation contains virtually nothing on the origins of this process. It seems from later documents, in particular those which reflect all the efforts made by some nations to import this method of fabrication, that tin-plate was first developed in Germany. Tradition again says that the birth of tin-plate took place in Nuremberg in the second half of the 15th century.

Other fire-based techniques were also noticeably modified, but we must ascertain whether these were favoured by a change in the type of fuel used. Indeed, increased production could have resulted in the destruction of the forest, especially as wood was needed in greater quantities to provide more heating for a larger population, and also to reply to urban expansion and to the development of the navy which imposed major discoveries. The only other fuel was coal, the use of which had been known for a very long time, but which one was reluctant to employ because of the fear that it would introduce impurities in the materials with which it came in contact. Indeed, unrefined coal was often excluded from some operations because of its impurities, particularly in the production of sulphur: such was the case in the blast furnace although many attempts had been made to resolve this from the second half of the 16th century. Gradually, however, coal was introduced into a number of processes. Some authorities have suggested that coal was used in the forge in the Belgian village of Marchienne at the beginning of the 16th century. This date is undoubtedly suspect, but nevertheless it seems that fears about using coal slowly began to disappear: mining in the Liège region was already underway in the 16th century, and at, the end of this century and the beginning of the 17th century we witness considerable expansion in the coal industry in England. Such movements would not have taken place if the demand for coal had not risen.

The glass-making industry also saw some changes, a few of which we will examine, but without an overall and exact view. The increase in the number of stained-glass windows is surely an indication of this. More archaeological evidence enables us to put forward a few hypotheses on this subject. Modification was not so much in the construction of the furnaces, and a London manuscript from the end of the 15th century provides a perfect example of this, but the greatest change lay in the composition of the material which was modified approximately between 1450 and 1550. Soda gradually replaced potash, and we know that this gave a more even, white and fusible glass, which was both easier to work and to make smoother and purer. We should also note one factor here which has already been pointed out on the subject of metallurgy: obtaining a product which was easier to work and superior in quality, which in turn constitutes real technical progress – outside the question of the volume of the production and the cost of fabrication of which we know very little. Henceforth, glass could resolve better problems previously presented, seal larger surfaces and preserve increased transparency, although it was by no means perfect as yet.

A new product appeared in the second half of the 15th century, whose reputation was to grow rapidly throughout the whole of Western Europe and which still retains its lustre today – that is Venetian or crystalline glass. Only an as yet still unverifiable tradition, provides us with our first landmark: its invention or its introduction might be attributed to Beroverio, a reputable glassworker from Murano, in 1463. Other glassworkers, particularly Norman artists from Languedoc, claimed authorship for this. Venetain glass is a silico-alcaline glass (silicate of potassium hydroxide and lime), which was later to be replaced by a lead based form of glass (silicate of potassium hydroxide and lead) which resulted in the creation of a true crystal. Whatever the case, high quality products, especially sodas obtained by the combustion of certain plants, arrived in Venice.

There were also differences in the fabrication of coloured glass. If we compare the stained-glass windows of the Middle Ages with those of the 15th and 16th centuries, we may appreciate these differences easily, not only in style but also in the techniques applied. The fabrication of glass by doubling meant that a greater range of colours could be obtained. Some colours could be employed by mechanical means, and similarly different layers of colours could be added.

The widespread production of smooth or flat glass clearly demonstrates that aside from an increase in the number of glass-making concerns, some technical progress must have been made. Once again, we should note that research must have been undertaken with this aim in mind. The development of the mirror, particularly the Venetian mirror, illustrates that techniques in making smooth glass must have been improved and perfected. Air was blown in a cylinder to make crystalline glass. Here again, this invention has been traditionally attributed to the Del Gallo brothers, who mentioned this technique in a document of 1503.

Of all the fire-using techniques, pottery, we have noticed, remained the most traditional, to such an extent that it is frequently impossible to distinguish chronologically between some potteries made from the Gallo-Roman period to those of the dawn of the 19th century. Only the preparation of pastes and the techniques of varnishing might have evolved. These transformations may again be observed in Italy, for potters had progressed there through the stages of ceramics which were standard, unfinished, shiny or glazed, to earthenware which was to be developed to the product we know today. In fact, the main change lay in the coating which covered the clay. Although the paste used remained the same as before, i.e. clay, sand and chalky marl, the coating henceforth consisted of stanniferous enamel, the decoration having been placed on crude enamel, whereupon the colouring oxides mixed with the enamel during the firing process. This technique of producing earthenware may have been known to the Arabs, but we do know that this process took off in our regions from Faenza. Bernard Palissy introduced these techniques in France, using a lead-based enamel.

We again come up against the same problems of material life, of which we know so little. It is true that earthenware, owing to its solidity, impermeability and durability, was clearly superior to the pottery used up until then which was brittle, porous and difficult to clean.

Techniques involved in mechanical operations had undoubtedly progressed considerably more, thanks to the improvement of tools and to the development of the different types of mechanisms which we have discussed, the two factors remaining fundamentally linked together.

We have already stated several times that the history of the tool remains to be completed, and we are convinced of this yet again. By accumulating images we may establish some idea of the tools which existed at different epochs. Our methodogical introduction has demonstrated both the riches and the limits of such studies. The history of the tool contains several essential aspects: the material or materials, and the form and types. What we have already said of iron is still valid on the subject of materials. The blast furnace and refining process produced a metal which was both easier to work and to make into steel, insofar as or perhaps despite the use of this method of refining, the obtained iron was less pure than that produced by direct methods. We should also note that steelmaking seems to have been improved during the 16th century, since we have the tools from this period.

We have already observed that forms were considerably modified as of the 13th century. This evolution had in fact been extremely gradual in the beginning and probably accelerated from the 15th century onwards. During this century and the two which followed it, the forms and proportions of the tools were definitively established. The tool not only took its definitive form at this stage, although many had remained the same since Roman times, but also became more diversified as demands increased, for the number of operations increased with development in techniques, simultaneously requiring this differentiation in tools. The introduction of new or modified tools may also be noticed, although unfortunately the absence of precise inventories limits us to citing only a few examples. During the 15th century fretsaws, jigsaws, and hacksaws were introduced. Planes and rabbet planes, as well as concave and raised moulders, established their definitive qualities during the 15th and 16th centuries. Elsewhere cutting edges were improved between the 14th and the 16th centuries. At the end of the 14th century, the bit became detachable in boring or drilling tools, the breast drill appeared, and pliers were perfected during the 15th century. Note also the creation of better precision tools, such as the calliper rule which was developed in the 16th century. The whole method of working wood had been widely transformed by the modifications made in tools.

Apart from this, and perhaps because tools were improved, insofar as new mechanisms gave machines more flexibility, the union between these two elements became greater and more profound. To illustrate this statement one need only think of the movement contributed by the crank and connnecting rod to hydraulic saws: the machine therefore increased in power and speed (fig. 17). From the 15th century onwards the use of hand, horse-driven or hydraulically-powered tools increased. An example of this is provided by machines for reaming cannons, some extremely interesting drawings of which date from the 15th century. Reaming was done vertically, as is illustrated in Breughel's painting *The Fire*. The work of Biringuccio, on the other hand, shows that reaming was horizontal, the tool being operated by a squirrel cage. The tool itself had also evolved: here steel blades were used, mounted on wood and supported by a bronze ring, then the whole head of the tool was made of steel. In the manuscript on the crafts of Nuremberg, dating from the end of the 14th century, files were still cut by hand, using a criss-cross pattern of grooves. Leonardo da Vinci had already envisaged the use of machines for sharpening files and we know of all the improvements made during the 15th century to the form of this tool and to rasps. The manuscript from the Hussite war, around 1430, contains a drawing of a machine which appeared to be driven by a mill-wheel, used for the horizontal drilling of wood pipes. A vertical machine of this type also appears in Leonardo da Vinci's work. In the same manuscript of 1430, the first machine for polishing gems may already be noted. This machine must have been quickly developed, for a manuscript from the end of the 15th century provides an almost definitive image of it, showing the disk and stones fitted into small adjustable vices. A still unverifiable legend attributed to the German, Louis Berken, in 1476, the idea of polishing diamonds with their own dust. And finally, there were glass or mirror polishers, which may be seen in Leonardo da Vinci's sketches.

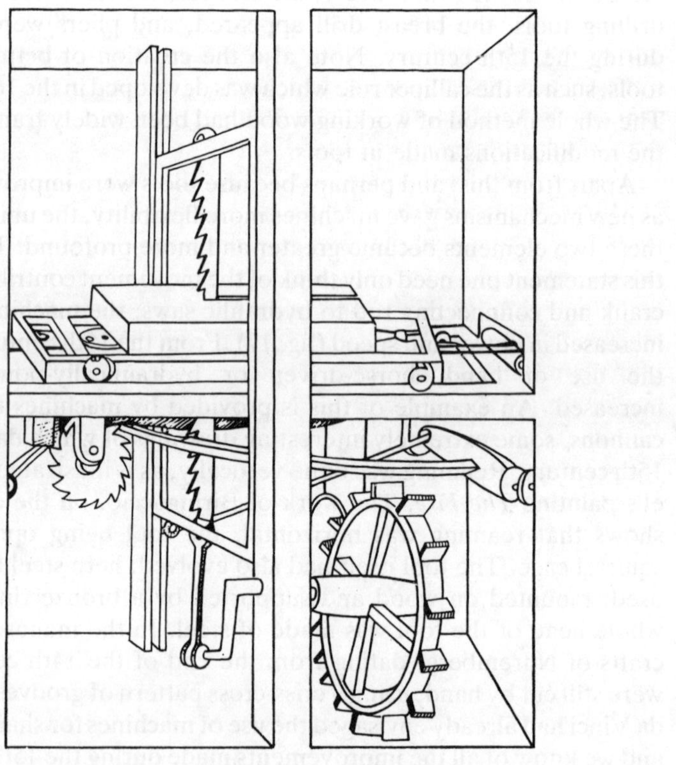

Figure 17. — Hydraulic saw (Francesco di Giorgio).

The system of the crank and connecting rod similarly made possible the multiplication and diversification of lathes, which were certainly already in use, but for which only reciprocating movements were employed. In 1470, treadle spinning wheels, which left the worker's hand free, were introduced. All the treadle grindstones also required a reduced work force. In the second half of the 16th century, Besson introduced the first irregular lathes, after Leonardo da Vinci's drawings which the artist had probably abandoned. Another problem arose regarding this which was to receive different solutions: the object either had to be introduced to the mobile tool, or turned on an immobile tool. The problem therefore lay in transporting the tool or the object. In the case of almost all the machines for working metal, it was the tool which was transported. In the case of wood, however, the object was mobile (fig. 18 and 19). Leonardo da Vinci supplied the famous drawing of a machine for cutting screws, probably made out of wood. Here, the chuck was the mobile piece, sliding on two endless screws. The object to be fashioned was placed in the middle and turned through the medium of a crank which also engaged the screw of the chuck. Accordingly, the tool itself was fixed on its chuck; thus two parts were mobile: the chuck and the part to be worked.

Although these machines were well designed and thought out, they did not yet function perfectly. Made of wood with rough joints, movements could only be jerky and they were probably very difficult to manoeuvre. Indeed, they must have been a far cry from the precision of the drawings. The brewers' museum of Antwerp and the Munich museum have preserved lathes from the 16th century and we may measure the gap which exists between the actual machines and the engravings in our treatises. Only machines made of metal, or at least partially of metal, were wholly suitable. Such examples may be found in the balance coining-press, invented by Cellini around 1530, and the printing press.

Here again, we require catalogues of these machines, images, and descriptions (notarized acts in particular) in order to ascertain the importance which they commanded from the second half of the 15th century. Such catalogues would enable us to fully comprehend the extraordinary development which these machines underwent during the 15th and 16th centuries (fig. 20). This was one of the domains where technical change implemented by the Renaissance was undoubtedly the most remarkable in extent and depth.

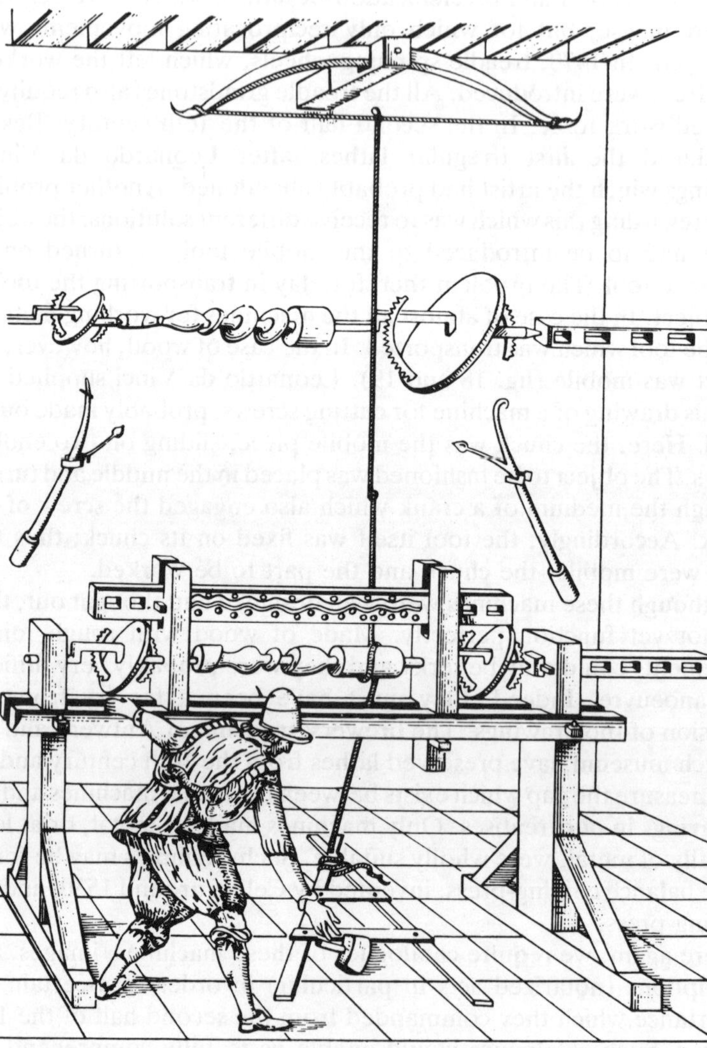

Figure 18. — J. Besson's lathe (1578).

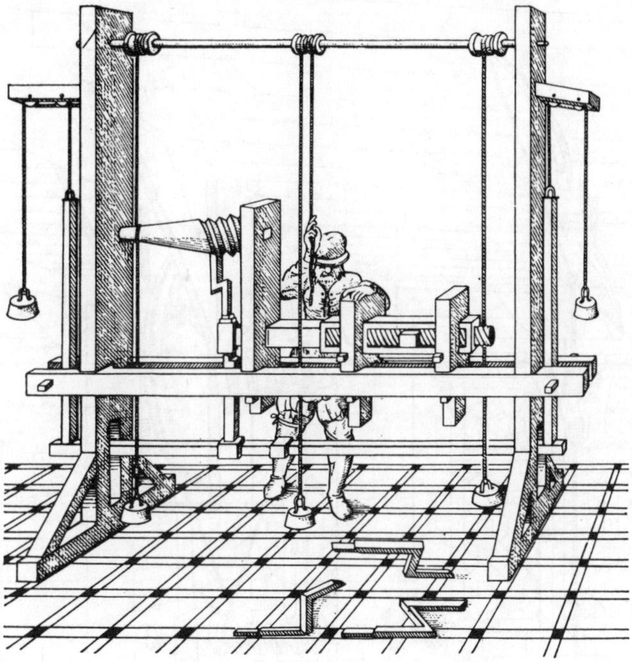

Figure 19. — J. Besson's lathe (1578).

Progress in the field of chemical treatments could only result from experimentation: only the creation of modern chemistry, in the second half of the 18th century, was capable of radically transforming the techniques involved. The introduction of the lathe in dyeing kettles, at the end of the 16th century, was a mechanical rather than a chemical contribution. In fact, the only domain in which some notable change took place was in that of explosives. The developments of the artillery were due as much to the improvement of explosives as to the weapons themselves. Gunpowder was made up of saltpetre, sulphur, and charcoal dust. A long series of manuscripts, mainly German in origin, provide us with the various compositions of the mixture from the 15th century onwards:

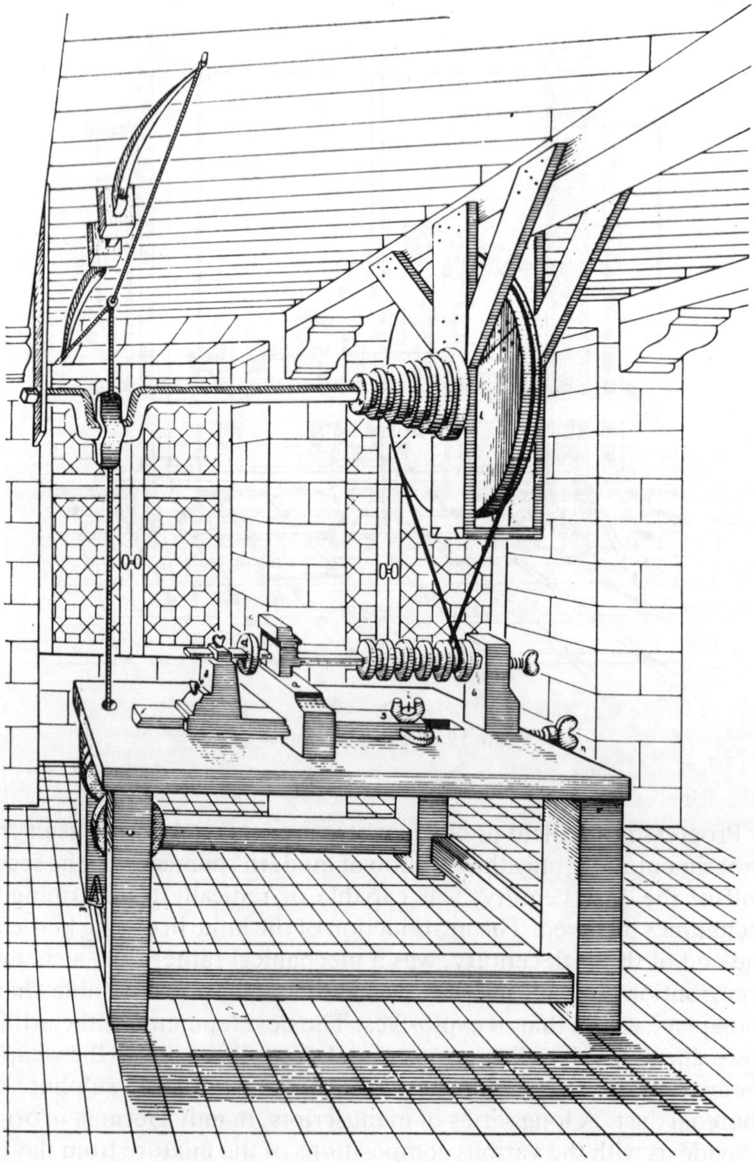

Figure 20. — Variable speed lathe (Cherubin, 1671).

	SALTPETRE	SULPHUR	CHARCOAL
1380	1	1	1
1420	3	2	2
1480	8	3	3

In France, this primitive gunpowder consisted of 75% of saltpetre and the rest was half sulphur and half charcoal dust: this mixture burnt but did not explode. From the 15th century onwards hydraulic mallets were used for grinding the mixture, which were similar in principle to the mallets used in paper mills. Virtually no noticeable progress was made in distilling.

The improvement of equipment enabled assembly techniques to increase in scope and scale. From the end of the 14th century, undoubtedly owing to the great crisis of the middle of the same century, less costly techniques, using products and machines generally prohibited by earlier regulations, were now looked to: such as the use of lower quality wools – particularly lamb's wool or slipe – the oiling of wools, then carding, the use of the wheel for spinning it, and multi-heddle looms. In the domain which interests us here, the most notable progress was made in the expansion of carding and the introduction of the treadle wheel and flyer-wheels. On the subject of silk making, the expansion of hydraulic spinning mills (fig. 21) should be cited as well as the use of the loom, apparently designed by Jean le Calabrais which was particularly widespread in the 15th century. Leonardo da Vinci was extremely interested in the complicated projects of designing thread- and rope-making machines, shearing machines, and in the curious idea of an automatic weaving loom (fig. 22 to 24).

It seems, however, that the weaving loom was only slowly improved, perhaps insofar as some of its parts were henceforth made in metal (fig. 25). The most spectacular invention was without doubt the machine for knitting stockings, invented by the English engineer, Lee, at the end of the 16th century. The actual method of knitting presented many questions which probably remained unanswered for some time. Weaving thread appears to have been developed in the 15th century, and one quickly changed from knitting by hand to using a machine of already advanced automatism, which Diderot was to praise even in the

Figure 21. — Machine for unwinding cocoons. (Florence, 1487).

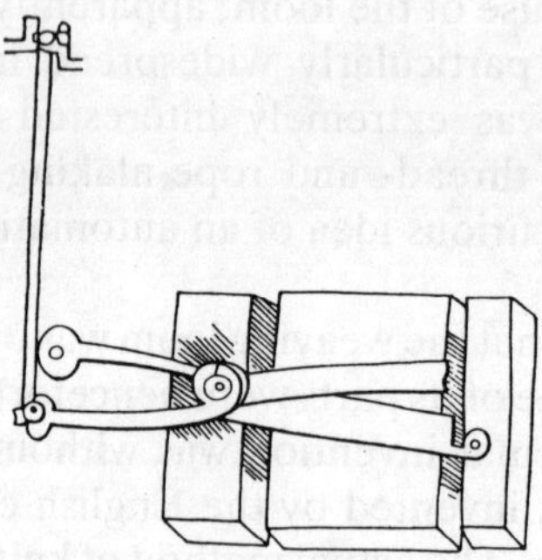

Figure 22. — Cloth shearing machine. (Leonardo da Vinci)

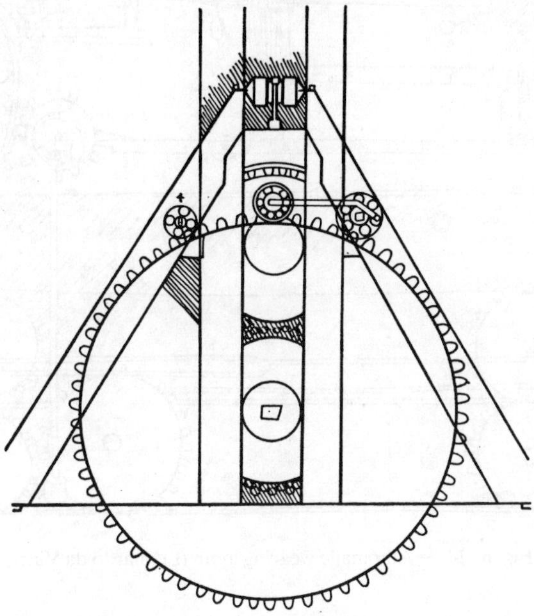

Figure 23. — Cloth carding machine. (Leonardo da Vinci)

middle of the 18th century. We know of the misfortunes suffered by the inventor who was rejected, indeed even threatened, because his invention overturned production methods and terrified the working population. As we have already noted, the printers' strikes in the middle of the 16th century were the first reactions demonstrated by workers faced with the domination of the machine, and this was to be one more proof of the importance of mechanisation during this period.

Similar stability may be noted in wooden constructions. From the 15th century, grooves and tongues were used to join boards and planks, made possible by the improvement of the plane. Although carpentry techniques had remained virtually unchanged, furniture, on the other hand, could evolve considerably, due to the greater ease in woodworking. Yet again, we require detailed research on the typology of furniture, the growth of which corresponds to the change in interiors

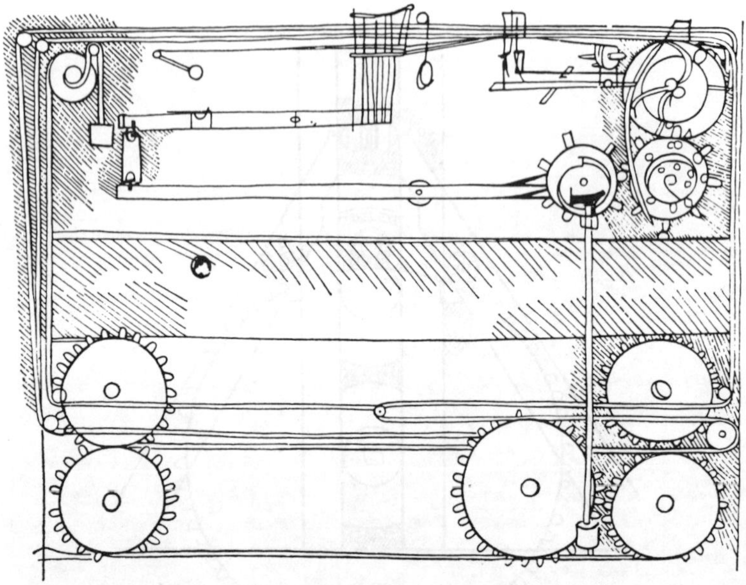

Figure 24. — Automatic weaving loom (Leonardo da Vinci).

and appreciably different life styles: furniture indisputably changed in structure, which naturally implies that there was parallel change in the structures of life styles and behaviour. The chest, or coffer, tended to be replaced by the cupboard or wardrobe in two parts, which we begin to see around the middle of the 16th century. The table had fixed legs and was more monumental in appearance, and the shape of the armchair was simplified to become a carver. The four-poster bed virtually became the symbol of the Renaissance.

In the domain of carpentry, there was little radical change in construction methods but greater adaption to new conditions. This was a result of the increase in the demand for wood – and we will return to this later – albeit for ship building requirements. One was therefore required to limit oneself to smaller sizes of wood for construction. The new method relied upon splitting posts which henceforth dictated the height of floors or levels: horizontal pieces were superimposed on vertical ones. Springers were assembled on posts and a string-piece was placed on the whole construction separating the levels. The impost

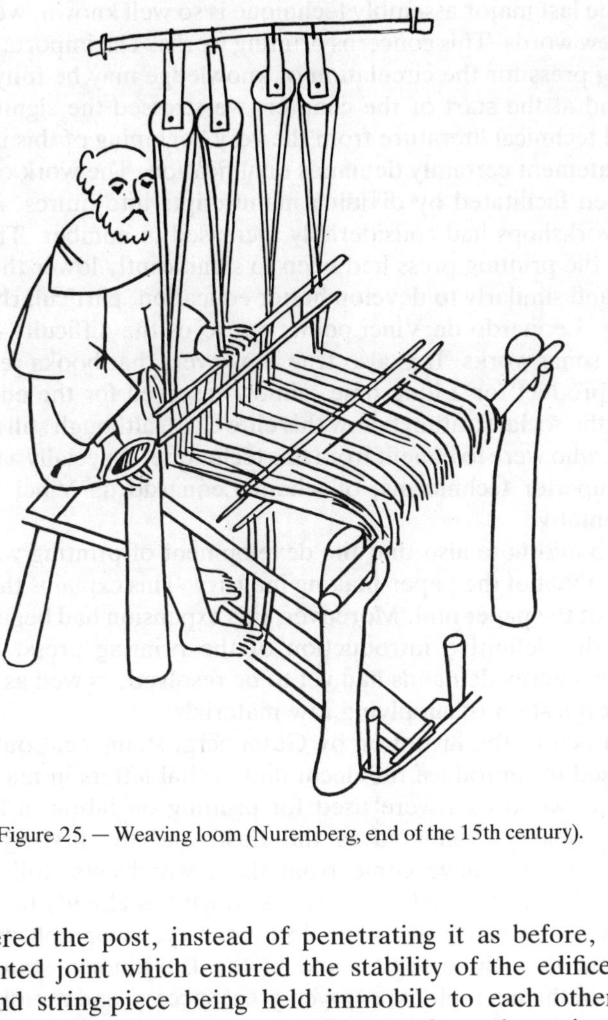

Figure 25. — Weaving loom (Nuremberg, end of the 15th century).

thus covered the post, instead of penetrating it as before, with a half-mounted joint which ensured the stability of the edifice – the impost and string-piece being held immobile to each other by a dovetail joint. The upper post was mounted on the string-piece, whereupon the angle was stiffened by a triangularly placed element (*pigeâtre*). These techniques were introduced at the end of the 15th century and at the beginning of the 16th century, as has been excellently demonstrated in Rouen. With these techniques, corbelled constructions could more easily be built and their use in urban buildings increased, many examples of which remain today.

As the last major assembly technique is so well known, we will only add a few words. This concerns printing works. The importance of the printing press for the circulation of knowledge may be fully appreciated and at the start of the chapter, we stressed the significance of printed technical literature from the very beginning of this invention. This statement certainly demands qualification. The work of copyists had been facilitated by dividing manuscripts into quires, and these same workshops had considerably increased in number. The advantage of the printing press had been to significantly lower the price of works and similarly to develop better education, particularly through reading. Leonardo da Vinci points out to us the difficulty he had in finding some works. It is also true, however, that books remained a luxury product for a long time, almost reserved for the elite: it was mainly the social stratum below this elite class, although still the upper classes, who were to benefit from this technique, especially a few of the more superior technicians, of whom Leonardo da Vinci is a good representative.

We should note also that the development of printing was closely linked to that of the paper-making industry. This explains the parallel growth of the paper mill. Moreover, this expansion had begun slightly before the definitive introduction of the printing press. Problems posed by energy demands had yet to be resolved, as well as the more difficult question of supplying raw materials.

Even before the invention by Gutenberg, stamps cut out in relief were used to reproduce the decorated capital letters in manuscripts. Similarly, wood cuts were used for printing on fabric, a technique which probably originated in the East, and the first xylographic impressions may have come from these wood cuts, following the process of technological transference which has already been discussed. From the first half of the 15th century, xylograph workshops existed in the Rhine region and in the Burgundian states. These images, in the main pious, enjoyed great success, and rapidly became associated with short legends. But the birth of the printing press and typography required separate letters, and the assembly and fabrication of these letters was in metal. Printing even brought to xylography thick ink and the press.

Printing, then, required movable metal type or characters of a certain life span, thick ink, and the press. The technique of engraving stamps had already been used by goldsmiths. The stamp was engraved in relief, whereupon it was punched in a less resistant metal mould (tin or lead) which provided the mould for the final character. Sand moulds may possibly have been used in the early days, made possible by the techniques of casting and medal making, these sand moulds being made with wooden matrices.

It is possible that research in this direction was undertaken in Avignon on the one hand, by a goldsmith from Prague, and in Mainz on the other, by the Gutenbergs, who were both goldsmiths and money makers. Other transferences of technology may be found at this point. In 1457, the *Mainz Psalter* was published, the first printed work of its kind and which was very high in quality. The take-off of new techniques was extremely rapid. Henri-Jean Martin has estimated that at the end of the 15th century, thirty five thousand editions had been published, representing, at the lowest estimate, fifteen to twenty million copies. 77% of these were in Latin, 45% were religious works, and two hundred and thirty six towns participated in this production.

The fabrication of characters was quickly perfected, giving birth to specialist workshops. The first moulds to be used were made of brass or bronze, and matrices were made of lead. Schoffer would have used the first steel moulds and copper patterns. Type characters probably took longer to develop a definitive form. In 1570, Aldo Maunzio changed the character in each volume. A tin based alloy was certainly used, avoiding a high proportion of lead which damaged the pattern, and adding a small quantity of antimony. But these early characters did not have any grooves to hold them in their correct position.

We know very little about the first printing presses. They may have consisted of a simple wooden press which came from the wine-producing industry. The German engineer, Husz, still used a huge screw press in Lyons, in around 1500. More elaborated forms were arrived at by a process of experimentation.

Major progress seems to have already been made around the middle of the 16th century. A sort of standardization of character types was introduced after the general adoption of Roman. Standardization of the composition, page and size was established when large workshops replaced the hundreds of small cottage printing works. The arrangement of the letters in the type-case, and the position of the latter in relation to the compositor was accompanied by modification in the presses. While the platen, placed at the bottom of the screw, was still limited in size, the table had become movable and slid on two rails, and could carry a much larger forme. This facilitated inking, and it was also easier to introduce the second part of the forme between the two movements of the bar.

Military techniques involve a much larger technical complex, involving a great many tributary techniques, a few of which have already been discussed. The development of artillery is a fact which dominates the whole history of military techniques. Although progress in this domain had been relatively slow between the middle of the 14th and the middle of the 15th century, development was to be considerable and very rapid after 1450.

Artillery was the first domain in which fire-arms were used. Early cannon were formed of iron bars shaped into cylinders and hooped. The powder chamber was screwed into the actual cannon. The whole was then transported on carriages or carts. This necessitated supplying ammunition, which consisted of stone cannonballs, peculiar to each type of cannon. In the second half of the 15th century, the introduction of cannon made out of one piece, due to the improvement of casting techniques and to the development of cast iron, which were more regular and used cast iron cannonballs, enabled artillery to fulfil its potential. Cast iron cannons were introduced in the first half of the 16th century and were henceforth to be the traditional armament of warships. Here again, the standardization of products was a great aid. The perfection in boring or drilling could reduce calibres and thus ensure a regular supply of ammunition. The invention, between 1550 and 1561, of iron or steel sights meant that these objects could last longer.

Henceforth artillery was permanent, while wooden machines of the past had usually been built in situ. Gun carriages without wheels were replaced by those with wheels, soon followed by independent front axle units, which permitted easier manoeuvring. Once a few more elementary laws of ballistics were understood – the study of which was begun at the beginning of the 16th century by the Italian, Tartaglia - the cannon was adjusted: the carriage was thus divided into two superimposed parts, the upper part could describe the arc of a circle and set a determined position due to the addition of a rack: thus the back sight adjuster was born (fig. 26). From this point on, artillery was far more widely used. The cannon of Charles the Bold, taken without great harm by the Swiss at Morat, in 1476, may be said to represent the first real modern weapons. Military historians believe that one may

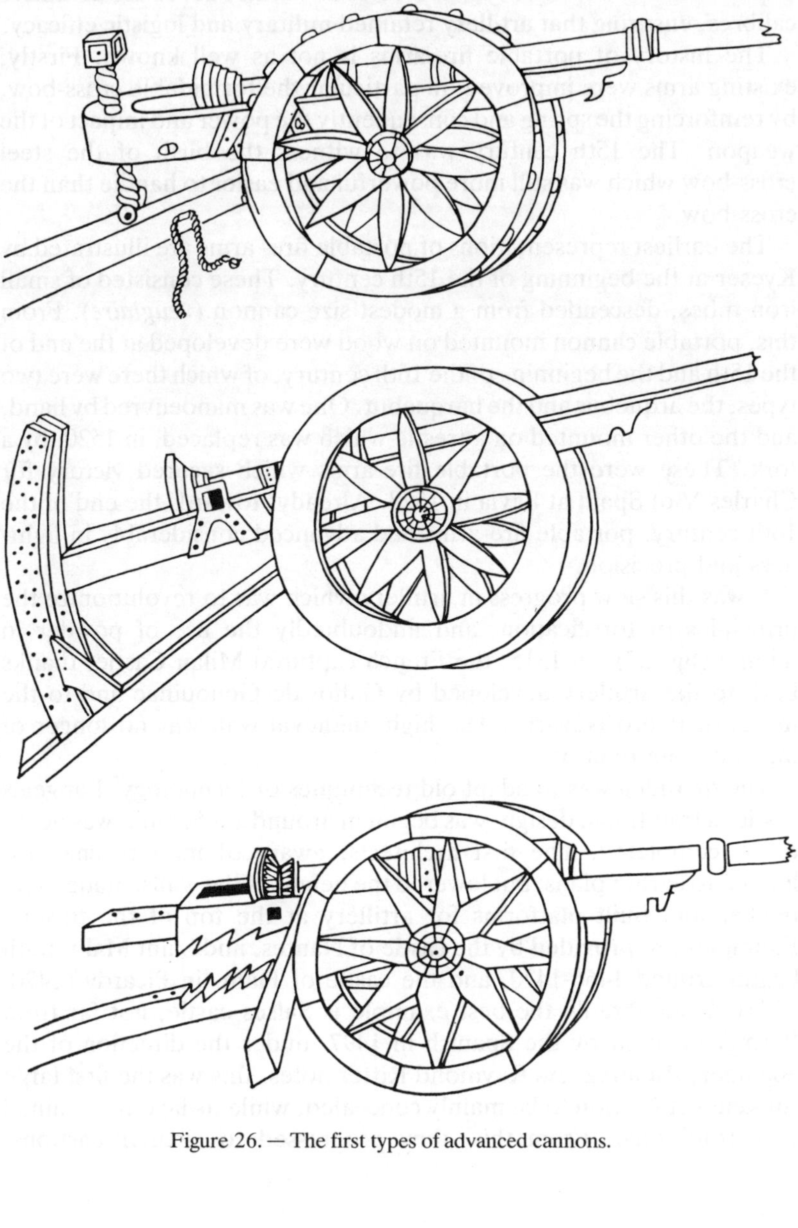

Figure 26. — The first types of advanced cannons.

speak of real artillery manoeuvres taking place at Ravenna, in 1512. At Melegnano, in 1515, the role of the artillery was already very important. The French royal decree of 1544 introduced six definitive calibres, ensuring that artillery retained military and logistic efficacy.

The history of portable fire-arms is not as well known. Firstly, existing arms were improved, in particular the formidable cross-bow, by reinforcing the spring and consequently the power and impact of the weapon. The 15th century was to witness the birth of the steel cross-bow which was still more powerful and easier to handle than the cross-bow.

The earliest representations of portable fire- arms are illustrated by Kyeser at the beginning of the 15th century. These consisted of small iron tubes, descended from a modest size cannon (*veuglaire*). From this, portable cannon mounted on wood were developed at the end of the 15th and the beginning of the 16th century, of which there were two types, the arquebus and the harquebut. One was manoeuvred by hand, and the other mounted on a trestle which was replaced, in 1520, by a fork. These were the portable fire-arms which secured victory for Charles V of Spain at Pavia in 1523. Already, towards the end of the 16th century, portable fire-arms had advanced considerably in lightness and precision.

It was this slow progress in artillery which was to revolutionize the principles of fortification, and undoubtedly the use of powder in mining (fig. 27). In 1515, the French captured Milan Castle, thanks both to the artillery developed by Galiot de Genouillac and to the mines of Pedro Navarro. The high medieval wall was no longer of interest from then on.

The first idea was to adapt old techniques or technology. Langeais castle, a traditional design, was begun in around 1465, but it was never to be completed. The first architects, aware of military changes, looked to earlier plans, but lowered the height of the walls, made them thicker and built platforms for artillery at the top of the towers. Examples are provided by the castle of Nantes, and Saint Malo, both begun around 1466-1470, and the castle of Ham, in Picardy (1470-1475). Undoubtedly the best example is Salses castle, not far from Perpignan, built by the Spanish in 1497, under the direction of the engineer, Ramirez. As Raymond Ritter notes, this was the first large sunken fortification to be mainly concealed, while its layout remained very traditional. From this, they progressed to circular bastions,

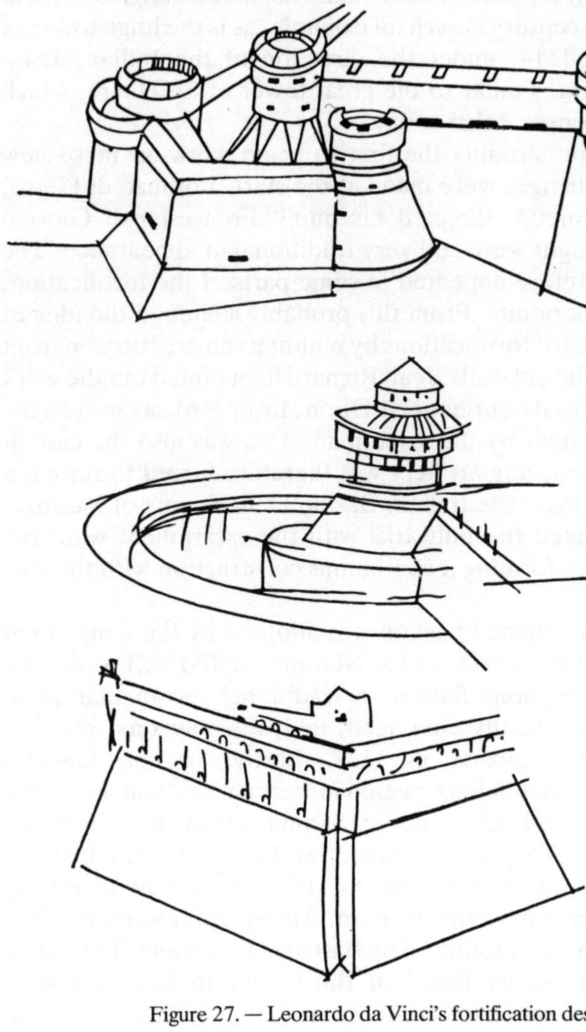

Figure 27. — Leonardo da Vinci's fortification designs.

descendants of the sunken medieval tower, with paved platforms for the artillery, but also with two vaulted block-houses with embrasures for law-built batteries: the bastion of Saint Nicolas de Langres, built at the end of the 15th century is such an example, as is the huge tower of Toulon, begun in 1514, under the direction of the Italian, Jean-Antoine Della Porta, similar to the great tower of Le Havre, which unfortunately no longer exists today.

The Italians were certainly the first to begin work on these new forms. Tentative changes were made at the start. Lorenzo di Pietro, known as Il Vecchietta, directed his pupil, Francesco di Giorgio Martini, whose designs were still very traditional in appearance. The *tenaille* system therefore appeared in some parts of the fortification, particularly in weak points. From this probably stemmed the idea of reinforcing old types of fortifications by building constructions in front of them to protect the old walls. Jean Richard has pointed out the work achieved by François de Surienne at Dijon, from 1461, as well as the castle of Fougères built by the same man. This was also the case at Rhodes in 1496. Besieging artillery was therefore forced to retreat a long way back. In 1521, the Italian, Basilio Della Scuola of Vicenza, still at Rhodes, linked the bulwarks with the escarpment wall, the bulwarks henceforth forming a continuous construction with the curtain.

The ace of spades shaped bastion was adopted by the Sangallo at Civita Castellana (1494-1497), and at Nettuno (1501-1502) where ace of spades shaped bastions flanked a traditional rectangular plan. Antonio da Sangallo finally envisaged, in 1515, polygonal plans at Civitavecchia, then Ancona in 1527. Lastly, orillion bastions appeared. The della Maddelena bastion in Verona was built according to this plan, in 1527 by Michele Leoni. Sanmichele made it his pride. These designs were adopted in France, at Troyes in 1524-1529, at Saint-Paul-de-Vence, and at Navarrenx (1543-1569). One might say that modern fortification was finally born. Albert Dürer was one of the first theoreticians in 1527, followed by Battista Della Valle. The whole technique was codified by Erard of Bar-le-Duc in *la Fortification réduite en art et démontrée* (1594). In this he defined the methods that provide shelter using features of the terrain, instituted inclined planes designed to avoid the surprise of downward views and succeeded in concealing the flanks of the bastion from the enemy by building carefully placed curtains. He also invented the cavalier and standardized the thickness of the ramparts. We are approaching the era of Vauban.

The greatest work of the Renaissance was certainly in the conquest of geographical boundaries. The considerable growth of trade and traffic and simultaneously the organisation of space, were mainly the preoccupations of technicians and after them various authorities.

The first problem concerns transport. The centralization of kingdoms and the extension of the known world required modified means of transport in order to reply to the growth of trade and the length of the journeys which had to be undertaken. Although harnessing had been modified – at least only a few details remained to be solved – the vehicle on the other hand, which was in fact hardly used and only then over short distances, was badly adapted to fulfill increased demand. Several innovations were to enable an evolution in the vehicle. The first, as yet not fully studied, involves the pivoted front axle unit, which considerably facilitated manoeuvring the four-wheeled vehicle. The first example of this is illustrated on Francesco da Carrara's seal, dating from the very end of the 14th century, although the image is somewhat clumsy and difficult to interpret. A more detailed image may be found, however, on a manuscript dating from around 1470. This shows two carts, both with shafts. The cart at the back is fixed by a shaft to a pivot placed on the cart in front. But this vehicle could easily topple over due to the lack of balance, and could only take very wide turns, not sharp corners. It was to be some years before the new system was established. We know that Ravaillac was able to commit his infamous crime because Henry IV's coach did not have a pivoted front axle unit and had difficulty in turning the corner of the *rue de la Ferronnerie*. At the end of the 16th century, however, the pivoted front axle unit was already in everyday use.

The other problem was that of suspension. This not only concerned transporting people, but also all the jolts and bumps suffered by the vehicle as it went along the road. Firstly, the vehicle's chassis was suspended by chains or straps, as can be seen on a drawing of 1405. Some authorities have been able to identify "swaying carriages" in this drawing which were attested from 1398 onwards. It seems that more recently, even older representations of suspension have been found: in a South German manuscript, dating from 1330-1350 (a miniature representing the flight from Egypt), and in an English manuscript from the second quarter of the 16th century. The most significant improvement made in the middle of the 16th century, would therefore have been to attach chains or straps, not to the framework, but to the large springs resting on it. A German manuscript of 1568 shows a very clear representation of this.

Other inventions or discoveries remain to be discussed. Attempts were made to improve wheels, and during the 16th century two particular techniques came into being. The wooden wheel was protected by nailing plates onto it: examples of these date from the Middle Ages, and the Ancients already used bronze plates. Emergency hoops which were placed on vehicle wheels were a later development made in 16th century. The second improvement was what has been termed dishing wheels, in other words inclining the spokes of the wheel on the hub at the angle. Use of this system meant that the great stress exerted on the wheels could be avoided. This technique was undoubtedly quickly employed for carriages used for carrying artillery.

Land transport and roads are closely linked. Truth to tell, the techniques used in road networks could hardly be modified except on the level of organisation: infrastructure in this domain is absolutely essential, as is so today. In France, to take but one example which has been the object of extensive research, the general road network began to become a governmental consideration from the time of Louis XI, with paving, the supervision of particular encroachment, and even the creation of new routes. It was at the express command of the king that the Mount Viso tunnel was built, designed to provide a communication route between the Dauphiné and the marquisate of Saluces. The tunnel cut through the mountain side at an altitude of more than two thousand metres, was two metres high and two and a half metres wide. An error in layout resulted in a slight reorientation. This work, executed from 1478 to 1480, was the major event of the century: it was the first great tunnel of modern time. Bridge building techniques which increased at the end of the 14th century should also be studied, but unfortunately no work of any scale has as yet been devoted to this subject.

River navigation was still a convenient and extensive means of transport during the period in question. Here, technical transformations in navigation played a smaller part in realizing significant progress than the organisation of river and waterways. Renaissance engineers quickly perceived that river trade, which was extremely important from the beginning, could be improved by undertaking large scale work. Hazards of the terrain and the vastness of the work to be done often halted even the most daring projects. Levellings, embankments and supplying water, all presented difficulties which were not easy to overcome. In cases of different levels, the problem of closing locks remained without any practical solution. It seemed that the most noticeable progress involved, on the one hand, the digging of canals (and we know that Leonardo da Vinci was particularly interested in this), and on the other, the question of locks. Leonardo da Vinci worked on the Milan canals under the direction of his elders, who must have used techniques about which we have little information. We know almost nothing about the techniques involved in banks and making them watertight. And as for locks, historians have tried to shed some light on this problem. Locks had already been used in building the canal linking Bruges to the sea, constructed in the 13th century. Drawings by Renaissance engineers show that there were in fact two solutions: the inclined plane upon which small craft landed, and actual lock gates. There would also have been mobile gates on the canal created by Jean, Duke of Berry in 1394, which went from Niort to the Atlantic. At the same date, similar locks may have been used on the Juine, in the earldom of Etampes (France). Improvements would have taken place during the 15th century, such as flaps on the lock gates, and henceforth, the sluice gate and the lock were to establish their definitive appearance. Filippo Visconti had these built in 1440, and the Venetians installed them on the Piovego in 1481. Alberti wrote a perfect description of this construction. The opening and closing systems, which had formerly been the principal difficulty, now enabled this expansion.

Henceforth, and because of the existence of centralized and relatively wealthy states, it was possible to create a canal network where natural river ways were unusable or inadequate. The flow and course of natural waterways could even be regularized. From 1468, the regularization of the Loire, begun some time earlier, was taken up again, and the work continued down a whole series of tributaries which were to make the Loire a main communication route in an extensive territory: the Auron river for Bourges, the Brenne river to carry wood to Tours, the Clain river to link Poitiers to this network, then the Loire, Maine and Sarthe rivers. The same was also achieved in Lombardy. During the 14th century, the diversion of the Ticino was continued right up to Milan, and the Po was partially regularized. Marble for the cathedral could be transported, owing to the use of locks built in 1395. The diversion of the Martesana was begun in 1457, initially to supply water to Milan, before being used for navigation: Leonardo da Vinci worked on this under the direction of Bertola da Novate. At the beginning of the 14th century the Stecknitz river was regularized up to Lübeck, where a canal linked it to the Elbe, for the first time crossing, in 1391-1398, the dividing line between the two basins.

Major finds have supplied evidence on the progress made in maritime navigation. This progress extended in several directions: firstly, of course, the ship, then navigation techniques and finally the building of ports. Although many works exist on this subject, the history of the ship has yet to be completed. Evolution was certainly slow and progressive, without any sudden changes. The Nordic vessel was fairly widespread, but it in turn was to benefit from certain developments in Mediterranean sailing ships. During the 15th century, cargo ships differed little from the vessels of the 13th century. The fore and aft castles were definitively incorporated into the hull which remained very broad and flat to facilitate running aground. In order to maintain the solidity of these hulls, exterior reinforcements were used which reduced sailing. At the beginning of the 16th century, the hulls tended to become lighter by dispensing with the carrack bow and lowering the prow to the level of the bridge, while the aft castles increased in height with progressive displacements. The caravel was introduced after a long evolution, finally taking its definitive form at the end of the 15th century; while a more refined hull was developed which facilitated sailing and was more resistant to drift. The essential factor seems to lie in the mixture of Nordic and Latin sails. The ship was henceforth to possess three masts. The foremast carried a square sail, and the mainmast formerly the old large mast, also carried a square sail. The mizzen-mast, placed on the aft castle, had a lateen sail. Only the main mast and its sail had any importance, as the others were to act as a counterbalance, to maintain the ship's course, and were to evolve separately themselves. Between 1450 and 1550, this rigging was to be divided and the mizzen-mast was coupled with a counter-mizzen-mast, also carrying a lateen sail. The bowsprit was to be fixed in front, underwhich the spritsail was hung; this was to facilitate both the manoeuvre of tacking and to maintain holding a steady course. Throughout the whole of the 16th century these ships were to be continually improved upon and were to increase in a variety of types;

the galleon, which established routes to America, appeared in the second half of the 16th century.

The art of sailing was to be fairly radically tranformed even more than the ship itself. Discoveries made at the end of the 15th century and the beginning of the 16th obviously modified cartography, but it was not until Gerhardus Mercator that flat maps were transformed into the proper maps we know today, the first of which dates from around 1569. Henceforth, it was possible to trace a route on a map. Throughout the 15th century trigonometry tables were perfected. As Guy Beaujouan notes, they enabled calculations of distances in terms of a boat being taken off course by the wind, but which could keep to its route by following a chosen rhomb. The differences of latitudes between ports could also be calculated. Astronomical observations on the land were made for some time, while navigation was essentially coastal. Before 1480, there was probably no astronomical navigation. The nautical astrolabe appeared for the first time on a map dating from 1529, now preserved in the Vatican. The impossibility of establishing a point of latitude on a map did not require extreme precision, moreover. The first declination tables of the sun were printed in Venice in 1483. Until the middle of the 16th century, navigation remained a matter of guess-work. It was still difficult to ascertain an exact route and speed. The use of the compass, known in the preceding epoch, the hour-glass, and then the clock – albeit difficult to use at sea – and the invention of the log, mentioned for the first time in 1577, was gradually to solve the problem of loxodromics and to ensure more precise navigation.

The multitude of ships, and the increase in their tonnages presented the problem of harbours. Obviously this problem was even greater in tidal seas. Consequently, the method of landing by running aground continued to be employed. The only real large port of northern Europe was Bruges, but the canal which linked it to the sea was straight, fairly shallow and silted up slowly but surely. Dredging, the creation of flush ponds, and even of a new canal did not help, and from the end of the 16th century, Antwerp had finally taken its place. In Antwerp most of the Channel ports, Harfleur, Granville, and Port-en-Bessin, it was necessary to harness rivers and create these tidal basins which were to increase in number throughout the 15th century. For a long time, ports were cleared with spades and barrows, before the improvement of dredges facilitated this work. These flush reservoirs enabled the creation of the port of Le Havre at the beginning of the 16th century. Similarly, it seems that the construction of dikes and jetties was mainly perfected in the north, and likewise in the south where a more rocky coast without tides made the creation and development of ports difficult. Large scale work was undertaken in this direction in Naples in 1447, in Palermo in 1541, and in Livorno towards the end of the century. This was also the same at Genoa where the jetty was extended and given a lighthouse. All the techniques involved in harbours and ports need to be studied in greater depth as they were an essential element in the development of maritime navigation: research does not really appear to have been undertaken in this domain, the importance of which is obvious for the period between the end of the 15th century and throughout the 16th century.

The organisation of space was extended to other techniques, which were also made easier by some evolution in techniques themselves, as well as by political centralization. Draining had been put into practice for some time, as had been the case in the Poitou marshes and on the shores of the North Sea, to take but two examples. In the 15th century, these works were systematized and large overall programmes were established. Work was restarted in the Poitou marshes between 1438 and 1443, and the general plan was launched in 1526. Between 1440 and 1460, large scale work was undertaken on the Venetian coast, doubling the outlets of the great rivers which flowed to the coast. Between 1530 and 1540, a network perpendicular to the lines of the slope were established. After the formation of the Zuyderzee, around 1300, and the great floods of 1421 which broke the dike built in the 14th century – perhaps due to the sluice gates of the river Maas – a systematic attempt was yet again undertaken in this direction, facilitated by the use of windmills. From 1435 onwards, polders were created with a control and a series of superimposed dikes. The great dikes of the Island of Walcheren, dating from the end of the 15th century, were nearly four kilometres long, one hundred metres thick, and towered four metres and eighty centimetres above the high seas; they were nevertheless broken through in 1530, but rebuilt in later years. The Friesland huge dikes in Frise were constructed around 1570 by the Spaniard, Caspare de Robles. Protected from the sea and the great rivers, the lower terrains could thus be drained. The Dutch created a vast administration which was directed by a team of remarkable men, such as André Verligh, from 1552, and then Simon Stevin.

Past masters in this art, the Dutch were summoned almost everywhere: in France for the Vernier marshes on the lower Seine, and in the mouth of the Vistula, between Elblag and Danzig, between 1528 and 1562.

At the same time, extensive irrigation work was undertaken in dry countries. An example, in a field which has been little studied, is that of the great dam of Almansa in Southern Spain, which undoubtedly dates from the middle of the 16th century: it is 20.69 metres in height and 89 metres in length (fig. 28). The Tibi dam, also in the south of Spain, would have been built in the third quarter of the 16th century by the architect of Escorial, Herrera.

All these great works – ports, drainage systems, irrigation networks – certainly presuppose that the methods used were perfected (fig. 29 to 31). Unfortunately we know very little about these. Although Leonardo da Vinci left us many drawings of equipment, particularly a large canal excavator, several lifting machines, and machines for driving in posts, and although other drawings illustrate dredgers, we are still not sure how embankments were made. It seems certain that a developed form of mechanization was introduced, especially in the first half of the 16th century. Engineers who thought of these machines or executed them may be cited almost everywhere. One looked to inventors. From 1413, the people of Marseilles engaged the services of a Genoese engineer who invented a new machine for cleaning out the ports.

Lastly, urban techniques, or town planning was one of the major concerns of Renaissance men. Although some of them have been studied in detail, there were others of which we know nothing. Certainly the birth of theoretic urban planning is proved conclusively to have existed: like Leonardo da Vinci, Alberti, Francesco di Giorgio, and Filarete prolifically wrote and drew plans of regular towns, orthagonal towns, with new elements such as the public square, the first examples of which are provided by the cities of Corsignano and Vigevano around 1460, including regulations on the length and width of roads, the height of houses, and the cleaning of public edifices. Some new towns made their appearance where town planners adapted regular plans to terrains which were not. We know of the development of Rome between 1464 and 1540, the creation of Corte Maggiore between 1470 and 1481, of Le Havre by Bellarmati in 1541, and of the new town of Nancy, established by Citoni in 1596.

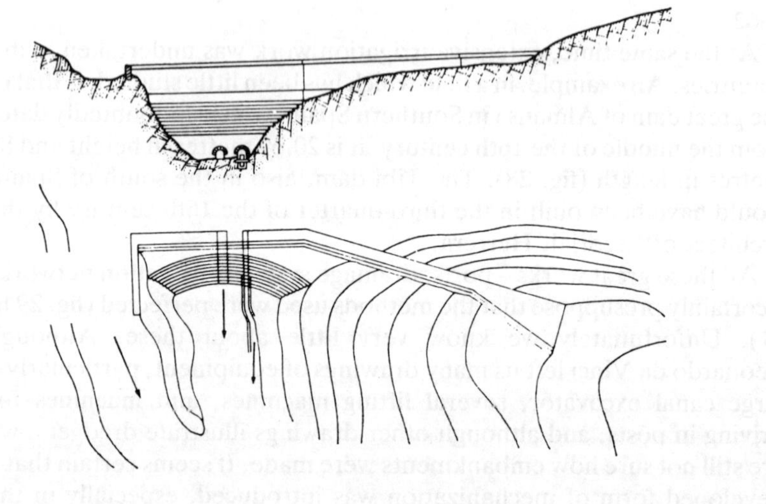

Figure 28. — Cross-section and elevation of the Almansa dam (circa 1550-1560).

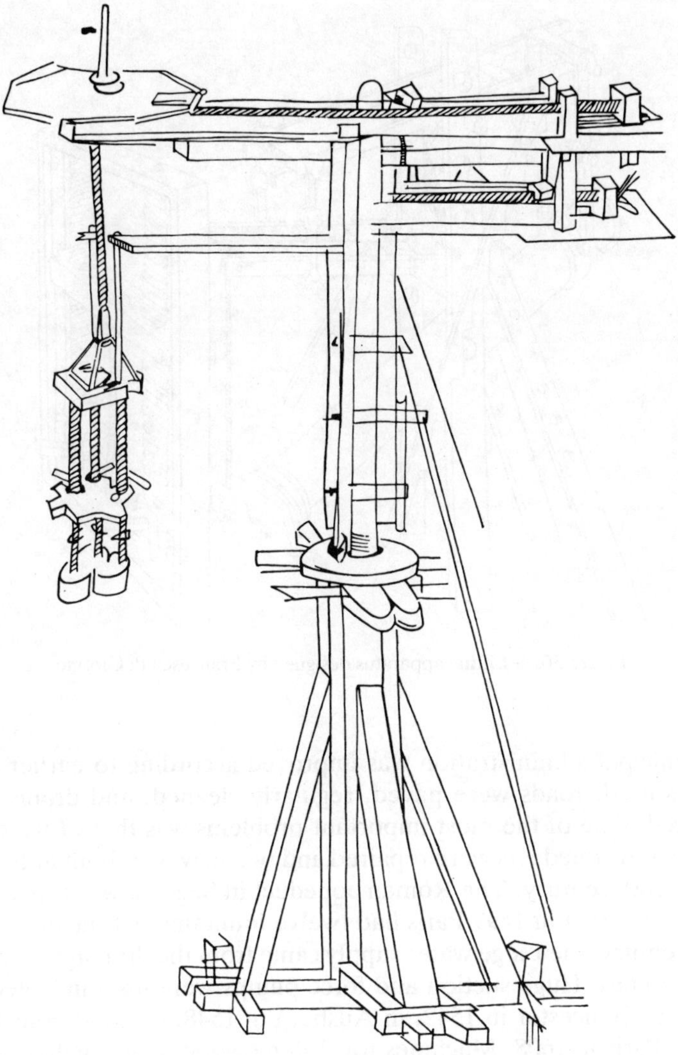

Figure 29. — Crane designed by Leonardo da Vinci.

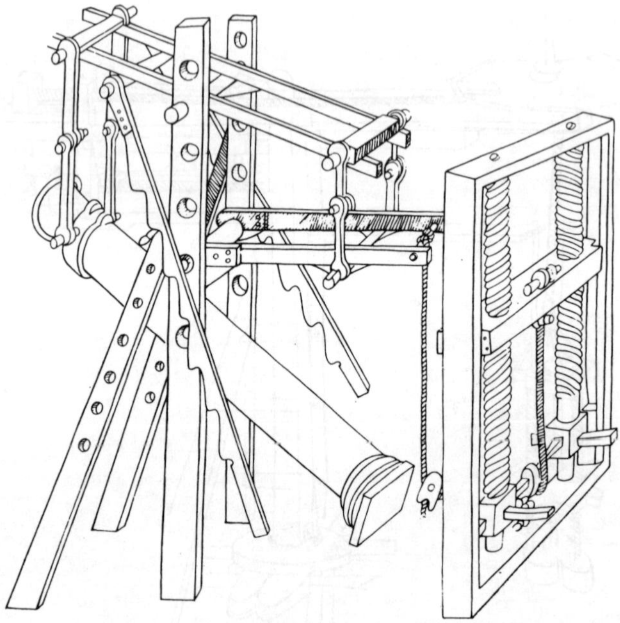

Figure 30. — Lifting apparatus designed by Francesco di Giorgio.

Municipal administration was improved according to earlier techniques used: roads were paved, regularly cleaned, and drains were serviced. One of the most important problems was that of water. In Rome, old aqueducts were repaired and two new ones built at the end of the 16th century. The Roman aqueduct in Segovia was repaired in 1481. Note that in 1499 Paris had twelve fountains. At the end of the 15th century, the Liège water supply came from the drainage channels in the mines. Huge suction and force pumps were used in Toledo in 1526, in Gloucester in 1542, in Ausburg in 1548, in London in 1582, and in Paris in 1608. Machines for drilling wood pipes facilitated the distribution of water.

We will now return to the original subject of discussion. The Renaissance was a time of profound technical change: major discoveries and economic progress were positive proof of this. If we look at the situation in the second half of the 16th century, we will see that there was considerable change in a great number of more diverse techniques. In many such instances, this involved not only the continuation and development of earlier techniques, but also the introduction of completely new features. Highly developed mechanisation introduced with the discovery of the crank and connecting rod system similarly invaded large scale works, such as mining and public works, as well as industrial productions. Many procedures were transformed in metallurgy in general, with the development of casting iron using the indirect method, and with the amalgam for gold. The "industrial revolution" was also very profound in the domain of transport. And a new technical system was born, insofar as all the new inventions were complementary to each other. The appearance of the world had really changed.

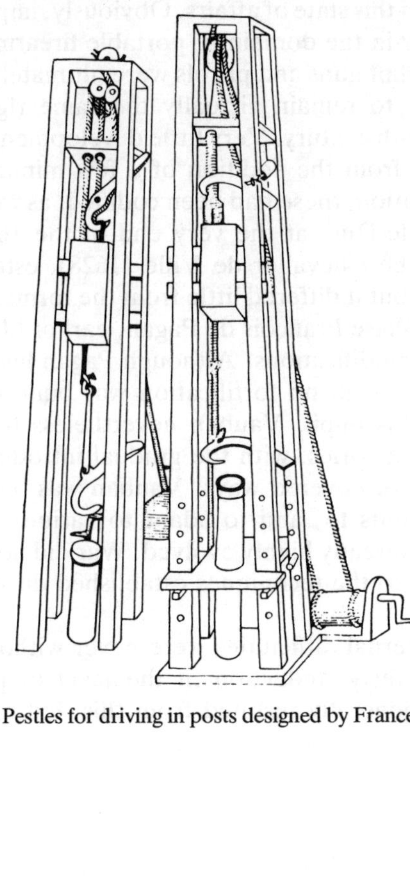

Figure 31. — Pestles for driving in posts designed by Francesco di Giorgio.

DEVELOPMENT AND LIMITATIONS

It was upon this classical system that the whole of the 17th century was to be founded, and a good half of the 18th century. Terms, such as the halt of technical progress, and the blockage of technical thought, have been used in order to grasp a situation which is difficult to comprehend, and the causes of which are even more difficult to distinguish.

Some important facts may certainly be highlighted, and have been so, which partly explain the depression which followed the extensive growth in the 16th century. Among other problems were political difficulties which weighed very heavily. There were various internal wars, especially in the second half of the 16th century, and then long wars which were to devastate much of Europe in the first half of the 17th century. Again one would have thought that military techniques would benefit from this state of affairs. Obviously, improvements were made, particularly in the domain of portable firearms. It was in fact around 1630 that flint guns and pistols were ultimately developed and these forms were to remain virtually the same right up until the beginning of the 19th century. Very little development was to be made in artillery, apart from the addition of a few minor details. As for systems of fortification, these had been codified, as we have just seen, by Erard de Bar-le-Duc, at the very end of the 16th century. *Les Fortifications* by the Chevalier de Ville (1628), established a more elaborate system, but it differed little from the former. The *Traité des fortifications*, by Blaise François de Pagan, earl of Merveilles (1645), introduced a few modifications. Although Pagan had never directed any work, and although no fortification was built according to the "Pagan system," his pupil, Vauban nevertheless based his science upon his master's theories, with the gradual introduction of greater defence and exterior covered ways. Vauban was only to contribute limited improvements to, and to adapt to various circumstances a science which was already highly evolved. War did not contribute any profound changes to the techniques established during the 15th and 16th centuries.

Political and internal difficulties were never without demographic consequences. Poverty, reduction in the level of production, and epidemics which generally resulted from this, led to sudden demographic losses. To such an extent that one recent author has termed this period, as have many others, the "terrible 17th century." The plague again became endemic in Western Europe between 1620 and 1640. And once again, the problem of the relationship between technical progress and the demographic level presents itself. It would certainly seem, however, that in the circumstances decrease in demand and long stagnation of production enabled some sort of classical technical system to persist, even without significant developments, and to survive without major drawbacks, without resulting in economic and social imbalances and tensions.

We lack a comprehensive history of technical ideas, and this is all the more regrettable for the 17th century as this period could certainly offer fundamental elements to such a study. We have considered the privileged position enjoyed by technology in the second half of the 15th century. But this technology had remained largely empirical in nature, rewarding those who did successfully practice techniques with a considerable audience. The men working at the beginning of the Renaissance do not appear to have had any great successors in the 16th century – apart from exceptions, of course. Mankind was preoccupied with religious matters, or in spirituality in the widest sense of the term, and were somewhat detached from material concerns which seemed to have reached a peak. Men were now more used to the introduction and circulation of new techniques, which no longer excited the same astonishment or admiration. Rabelais' pupil, interested in material problems, visited workshops and craftsmen. Montaigne's pupil was completely different. On the whole, the 16th century represents a void of interest in techniques and in the goverment of a material world.

But from the end of the 16th century and until the end of the 17th century, technology began to take off in another direction, despite all the troubles and poverty of the period. It did not, however, inspire the sort of enthusiasm which was prevalent during the time of Leonardo da Vinci: proof of this fact is exemplified in the drop in the number of technical treatises printed in the 17th century. Henceforth, the technical spirit was to change completely. We do not intend to write a history of the development of scientific thought as that may easily be found elsewhere. We should nevertheless clarify that it was precisely at the end of the 16th century and throughout the 17th century that the modern scientific spirit was to be born. Although some sciences, such as chemistry, were to remain the same – and yet still to make wonderful discoveries in their fields – physics, mathematics, and astronomy underwent profound change.

It seems therefore that the relationship between science and technology was turned about to some degree. At the end of the 16th century, some scientists – we should not generalize here – still based their theories on techniques. The names of Galileo, Stevin – and we could cite other men who were no less effective – remain to prove this. But their science very rapidly rose to a superior level. Unlike their 15th century predecessors, these men knew how to draw scientific facts from their observations on techniques, and to integrate these elements into a more organised, logical and more universal science. Was it not Galileo, fountain-maker to the princes of Florence, who reasoned out and proved the theory of atmospheric pressure and the existence of a vacuum? And did not Stevin make conclusive progress in mathematics while pertinently dealing with problems of draining land in Holland?

This reversal was to quickly come about at the beginning of thc 17th century. Henceforth, science was to be more autonomous in nature, initiating its own developments which were to lead, only in a limited measure it is true, to technical action. The explanation for this technical activity must lie either in the method or the machine. Combinations still existed, however: Roberval was certainly an engineer, but Desargues literally created the science of cutting stone. Huygens invented the regulating balance or pendulum in clocks and then the spiral hair spring. Formerly the technican had instructed the scientist, be it the same person or not, of the nature of the problems to be solved. During the 17th century, the scientist taught the technician, meeting less and less in the same person, of the theory behind his technique.

This change is important, although it entails drawbacks which should be stressed. It is certainly interesting to study these reasons and they may serve to lead either to the implementation of improvements, or to the choice between the various possible courses of action. On the other hand, and we have noted this at the beginning of this work, in most cases, science was still at a level where it could not initiate real technical progress. Science attempted to explain but could not support innovation. Furthermore, science sterilized technical progress, insofar as technology seems to become dependent in some way upon science, and where science could not cross certain technical barriers. Naturally one meets the as yet insurmountable problem of making the transition from scientific imagination to technical imagination. We will try to illustrate this with some examples.

Firstly, an important fact must be emphasized. We have already stated that technical progress was naturaly dependent upon the notion of the nature of techniques on the one hand, and on certain economic conceptions on the other. Merchants based their politics on a static or unchanging state of affairs. To put it another way, throughout the world, all quantities (population, production, and commerce) were fixed. A nation could only become richer if it took from others. But as everything is stationary, techniques must also remain the same. From this results the notion that no technical progress is possible, but only the simple improvement of the existing techniques may be effected. So the effort did not consist of initiating other techniques, but of imposing those which did exist on other countries and not at home. The policy taken by Colbert, and by a great many princes, consisted of importing manufacturing processes of products which formerly had obliged non-producing states to export large quantities of gold.

We have tangible evidence of this desire for perfection. For reasons of military power, Colbert wished France to possess the best ships. Carpenters from various arsenals were therefore required to make written records of their science, and to do a sort of project on ship building. Some of these compendiums have been preserved and merit detailed studied. A "commission" was then charged to choose all the best elements from these projects so as to construct *the* definitive ship, *the* ideal ship. The latter was actually built and, not surprisingly, was not very successful.

Knowing the "reasons" behind a technique was certainly one way of perfecting it. Similarly, it is not astonishing that considerable attempts were made, not so much to rationalize techniques, but to establish their scientific foundations. At the end of the 16th century and the beginning of the 17th, Galileo and Stevin were already on the way to doing this. To give another example, did not Wren, Huygens and Wallis study the hammer in the middle of the 18th century? Similarly, the effects of projectiles and pile drivers were studied. It would be interesting to know more about the beginnings of a technology which Renaissance men had only begun to glimpse, and which could only take shape once science had already been formed. "Science in the middle of the 17th century was sure of itself," writes Claire Salomon-Bayet, "believing that the domains of the life sciences and the science of machines could be opened to geometric jurisdiction (...) Mathematical "science" is mature whereas experimental "technology" is occasionally practical, always curious, and almost never illustrative in nature."

Inasmuch as technology "aims, according to Guillaume and Sebestik, for the constitution of a discourse on technical operations like a scientific type of discourse," Salomon-Bayet has demonstrated that it has progressed through three stages. In 1675, Buot envisaged this discourse as a catalogue; from 1693 to 1695, Filleau des Billettes considered the conditions of the possibility of such a discourse which he wished to be both logical and methodical; in 1699, the reform of the *Académie des sciences* marked the failure of this attempt. Indeed, in 1675, the king of France charged the *Académie des sciences* with the task of "examining the methods of writing a treatise on mechanics, with an exact description of all the machines which were currently used in industry in France and throughout Europe." The work of developing this was given to Buot who envisaged a two-part treatise on mechanics: one part was to be theoretical, setting out principles, and the second part was to explain everything which concerned the practice and application of the theory of machines. This idea does not appear to have progressed much further, and in any case Buot died in 1677. In 1692, the Abbé Bignon took up the project. He established a group comprising three specialists named by the king: Father Sébastien, Jaugenon and des Billettes, as well as an engraver, Simonneau, a "consultant," Anisson, director of the royal printing works, and a technical adviser, Granjon, engraver of moulds and patterns. For the first time on the 16th of January, 1693, Bignon gathered together what might have become an Academy of arts. "A very precise treatise on the arts which was almost to be a type of Encyclopedia." It was these men who began a description of the arts, designing patterns for the plates. They began by describing the printing press while numerous plates were engraved. The reform of 1699 brought into the *Académie des sciences* some of the members of this abortive *Académie des arts*. The rules prescribed by the *Académie* revised the examination of machines, which was to become the initial section of a school, and the continuation of the work directed by Bignon. We know that it was only in 1762 that the first volume was published, after the scandal concerning the plates of Diderot's *Encyclopédie*.

Description was certainly a beginning but demonstration could also have been employed, in the same way that a physicist or a doctor operates by demonstration in order for his science to be understood. From the first half of the 17th century, the idea of an exhibition of machines was implemented. Arthur Birembault has soundly assessed an exhibition of machines which was held at Paris, in the *rue de la Harpe*, in 1683. It was an exhibition of twenty-one scale models which had been executed from some of the drawings from the "theatre of machines," by Besson, Böckler, Ramelli, Salomon de Caus, and Strada. The other models represented the inventions of two inventors.

The explanation of Machines and moving Forces must be considered to be a serious, important and very useful subject to the Public, for by theory and practice everyone may quickly acquire a level of perfection in his craft or profession. It is an illustrative means which instructs by a single glance, and which is determined by concrete and effective experiment.

Demonstration on the one hand, and experimentation on models on the other, were therefore the means by which one hoped to achieve perfection. This whole programme was to be continued up until the *Conservatoire des arts et métiers* was established, born from Vaucanson's model collection, and to the research undertaken by Watt on the scale model.

Let us recapitulate all that: scientific description, demonstration and experimentation were the real means by which technology aspired. And like the Encyclopedists of the 18th century, the only difference being that they merely justified their ultimate purpose, we wonder what the value of technical literature was towards the end of the 17th century, as the latter did not have the same lustre which it had known from the end of the 15th to the end of the 16th century.

Technical literature of the 17th century often resumed works which were more than a century old, such as the various editions of *La Maison rustique* or out-dated formulae such as the "theatres of machines." Among the latter, we may cite those of Della Porta (1601), and of Branca (1629) in Italy, the works of the Germans, Zeisingk (1612), Schott (1657), Strada (1629) and Böckler (1601), and the Frenchman, Salomon de Caus (1615). Moreover, other works may be found up to the first half of the 18th century: Leupold (1724), Weidlerus (1728), Konig (1752) or Polhem (1729). In more specialized sectors, aside from works on architecture, there are no more of these treatises which had been so numerous and diverse in the 16th century. One may cite a Spanish tract on metallurgy by Alvarez Alonso Barba (1640). And even if valuable works had been produced at the end of the century, they virtually all disappeared in the wave of mediocre works which were easy to read and often well illustrated. Such were the works of Barnaby Googe (*Four Books of Husbandry*, 1577) who had already realized the revolutionary role of weeding arable land and artificial meadows. The 17th century thus appeared to lack a certain technical vigour. This may explain why evolutions were so slow, or at least the lack of of any significant change.

A historian of agriculture recently characterized the 17th century as having "a deficiency of innovatory spirit." Certain trends were carried on: buckwheat gradually spread throughout the areas where millet had been grown, hops spread from the continent to Britain where it enjoyed a remarkable circulation. It would seem that it was Flanders and Holland who had, at this epoch, begun the evolution which was to create the wealth of the 18th century, by advocating the practice of weeding crops in the fields and introducing fodder plants, particularly clover and lucerne. Ship building presents an almost identical picture. The shape and conception of the ship had slowly evolved until the most perfect results, such as the "Sovereign of Sea" by Pett (1637) and the French "Couronne" (1638) were established examples. But at the same time, research was continued in order to build crafts on a more scientific basis. Thus we witness the *Hydrographie* by Father Fournier (1643), although a point was reached where the titles of such works were more ambitious than the content, but this nevertheless expresses the hope for a rational technology. In Britain, Anthony Deane produced a *Doctrine of Naval Architecture* (1670), in France, Father Paul Hoste wrote a *Théorie de la construction des vaisseaux* (1697), where, for the first time, the centre of gravity is located above the centre of hydrostatic pressure. The dry dock was this age's only novelty, which facilitated repairs considerably.

This is also true of the techniques used in the exploitation of resources. Mining techniques had been hardly modified at all. By comparing Agricola writing in the middle of the 16th century and le Morand writing in the middle of the 18th, one is aware that methods and means had remained stationary. At the very most we should point out the mechanical devices used to bring up products of either open cast or deep mines developed by the Swede, Polhem. Barely any modifications were made in the iron and steel industry apart from improvements made to some machines, such as splitting machines. As for the rest, the technology described by Diderot in the *Encyclopédie* remained exactly the same as when it had been established at the end of the 15th century.

Technical initiative was not completely lacking, however. In many domains there were partial inventions which made possible the development of old techniques which had not progressed at all. A few examples of these will suffice here. It seems that chemical techniques were noticeably perfected from the middle of the 17th century, although no radical change took place. A historian of chemistry has recently demonstrated that there was no actual modification in the methods used, but arrangements were better and conceived in a more rational way: as was the case for alums and aquafortis, and mineral salts. A number of mineral compounds used for dyeing were introduced: borax, for example, or tin chloride designed to make a brilliant scarlet mordant from cochineal. The perfecting of candle making led to better lighting and made possible public lighting at the end of the 17th century: we know that in France, in 1697, public lighting was recommended for every town of a certain size.

Some techniques even demonstrate remarkable vitality. The start of mechanisation was a result of many small, often repeated operations. Such was the case for ribbon weaving and stocking knitting: one stocking required 150,000 to 157,000 stitches. The bar loom, invented by Willem Diericksoon Van Sonnevelt, from Hondschoote, then a great textile centre, was apparently established in 1604, enabling several ribbons to be woven simultaneously on only one loom and operated by one worker. Up to twenty-four ribbons could be produced with these synchronized movements. We have already discussed William Lee's loom for knitting stockings (1589), which was improved, by Hindret's addition of fairly complicated transmissions, in France in the middle of the 17th century. Cloth-crimping machines, rollers for glossing sayettes, and gig mills henceforth produced continuous operations, although movements in wool weaving machines were still not synchronized. In cloth shearing, stopping and starting procedures were made easier: this machine was to lead to some problems in Leyden, in 1680. The best idea was the crimping machine, set up in Troyes in 1678.

It was also at this date that the automatic weaving loom was introduced, designed by M. de Gennes, a French naval officer, an account of which appeared in the *Journal des Savants*. It was undoubtedly inspired by the ribbon loom, except that there was no shuttle. Whatever, the progress made is measurable: indeed we begin with a series of old projects, many of which are illustrated in Leonardo da Vinci's notebooks, continue with the first sketches of the beginning of the 17th century, as is demonstrated by Zonca, and then finally to the achievements of the middle of the century. Thus, technological lines are perfectedly continued.

What had formerly been potentiality in techniques in the domain of turning at the end of the 16th century, was to become reality during the 17th century. A clear evolution may be noted starting with Besson, in 1578 when the first mandrels and the first back rests were introduced on a lathe, to *L'Art de tourner* by Father Plumier (1693). The spherical glass making machine, where the glass was stuck on the nose of the mandrel was produced by Father Chérubin in 1671, as was the first sketch of a face lathe. Besson and Ramelli had developed drawings for conical gears by the end of the 16th century. In 1650-1660, Gérard Desargues envisaged the epicycloid line of the gear cogs. Shortly afterwards, Philip de la Hire was to write a *Traité des épicycloïdes et leur usage en mécanique* (1694). In the following year the same author was to publish a *Traité de mécanique* which to some degree reunited the preoccupations of the powers that be and the desires of the French *Académie des sciences*.

The strength of centralization, and better organisation of public services enabled "national and regional development" perhaps more than technical inventions, to an extent which had been formerly unknown. We have cited the introduction of dry docks and the massive work on ports which the increase of ship building required. We know that it was during the 17th century that the idea of linking these great river basins was put into practice: from the beginning of the 17th century, the famous Briare canal linked the Loire and the Seine basins. In 1626, a plan was initiated to link the Rhine and the Meuse. Need one emphasize the canal in Southern France, which links two seas with twenty-six locks from the Atlantic to Toulouse and with seventy-four locks and a tunnel from the Mediterranean coast?

The process was similar for water conveyances. Although large suction and force pumps were still in use, albeit only usable for a limited flow, improvement was made in the method of canalization. The first cast iron pipes seem to have been made for the waters at Versailles, and in the elevator machine at Marly, so cast iron pipes favourably replaced wood or pottery pipes. The difficult problem of relaying pipes over a great distance, that is to say transporting energy over some distance, still had to be practically solved. At the end of the 16th century, the Spaniard, Turriano, had designed mechanisms based on articulated parallelograms, a system which the Swede, Polhem, was to use at the beginning of the 18th century. Salomon de Caus partially saw the solution to this but could not develop any practical application. It should be noted that the surveyor's level with telescopic sights by Picard (1675) considerably facilitated canal construction. But we know how they wavered between the various solutions for the waters at Versailles, at the expense of quite useless work.

It was possibly in the domain of clock-making that absolutely fundamental inventions were made. Huygens is credited with the development of the balance, using the principle of small isochronous oscillations (1657), then the regulating spiral spring (1675) which increased precision fivefold.

In fact, the 17th century lived well on the new techniques and ideas launched during the Renaissance. They sought "perfections," developments, installations, and concrete achievement of all that had been previously envisaged, rather than new ideas. Let us go even further: the *Encyclopédie* by Diderot and d'Alembert more or less represents the techniques of the classical epoch (in particular the steam machine which is almost hidden in the article on "Fire"). Upon reading it, one may gauge the limits reached by the technical system of this epoch. It also demonstrates how little the crank and connecting rod system was used in various machines, the difficulties of energy production, and the extent to which the use of the small tool was henceforth fixed. "The artist's hand is essential," declares d'Alembert's preamble. Because the research carried out by the Encyclopedists was done in "workshops," we have a reflection of a system which was traditional rather than the current system of the time, and the *Encyclopédie* therefore represents the preceding system. The notion of technical progress, and the idea of a new technical system were perfect strangers to this publication which was so rich in information in other respects.

Classical technology is the technology of a static population. It is also the technology of a strict world, half closed in on itself. Demographic pressure after the first third of the 18th century, and outside expansion – no longer in the form of exploitation but in the direction of the conquest of new markets – were to produce tensions which generated progress.

THE INDUSTRIAL REVOLUTION

The term "industrial revolution" is now an accepted part of historical vocabulary. However this does not mean that it is no longer the subject of disputes or that it no longer needs justification. But if we examine this factor in the same perspective as the rest of the book, the justification is self-evident. A new technical system was created, which was one of the fundamental elements of a new economic recovery. This is evident in Great Britain from the end of the 18th century, and in other European countries from 1825-30, to 1850-60. It seems, therefore, that this framework does actually coincide with reality. Some historians tend to confuse industrial with technical revolution, and since they consider that technical progress was a gradual continuous process, they deny that there was any industrial revolution. This is probably just a matter of semantics. If the history of techniques is merely a list of inventions, then of course it will appear continuous. If however, – and this is what we have attempted to do – one considers the progress made in the 18th century it is evident that all the main manufacturing processes, and not such-and-such process or machine, represent the end result of new techniques applied during the 1780's. Furthermore, it was around this time that a balance was established between the interrelated techniques. So it was only after this period, after 1780, that is, that the new techniques reached their full potential: and the same phenomenon is evident in other periods. An industrial revolution must be dealt with in global terms, not in terms of isolated disparate elements.

This discussion will not be concerned with the details of new techniques and their development, even though such research is fascinating and extremely informative. We shall be concentrating more on the conditions surrounding the changes, and on their global aspects and consequences. And since this is in a historical context, it would be useful first to establish a chronological framework.

The long economic recession which historians generally believe to have begun about half way through the 17th century, continued for almost a century, and it was only between 1730 and 1750 that conditions began to look more favourable. There is little need to dwell here on two other important phenomena: the steady decrease in the influx of precious metals, and the gradual lowering of prices.

It is difficult to be precise about the period at which this recovery began, but this is hardly surprising, since it relates to different sectors, different regions and different countries. At any rate, the upturn only became really evident after 1780, and in some places not until after 1800.

Similarly, it is possible to identify several different attitudes which appear to be quite important on a technical level, in that they form the very basis of a politics of technology. One of these attitudes is best described as "stationary," and Colbert's policies are a perfect example of this. A stationary state necessarily implies a stable level of quantities, in terms of both population and production and this is on a global scale. Realistic policies can succeed, therefore, only in modifying the apportioning of these quantities, and that implies that no further technical progress is possible. Efficient policies consist in perfecting what already exists, developing techniques to their full potential, and importing advanced techniques from abroad. We saw that with a few rare exceptions, i.e., those who were preparing for the ensuing transformation, this was the politics of monopoly.

The lowering of prices created difficulties for the entrepreneur of various kinds, but alongside this, the cost of research was also decreasing.

The latter attitude was reinforced by two important factors which seemed to have played a major role, and which therefore deserve more precise investigation.

The first of these is the exhaustion of some raw materials which were at the basis of preceding technical systems. One example of this is the shortage of wood which affected many countries, in particular Great Britain.

The second factor is the change in colonial policies. It was no longer a matter of the economically most advanced countries using the raw materials of other parts of the world, but of their attempting to create markets in these areas for their products. This expansion was double-edged: not only did it concern the discoveries made in preceding periods, but also the creation of new colonial empires. The consequences of the treaty of Utrecht in 1713 was of primary importance in this respect.

All this led to adaptations in techniques, which in turn led to the formation of a new technical system. The interlocking of techniques and the establishment of a satisfactory equilibrium required the perfection of certain techniques, inventions and innovations, and the integration of older inventions which had been unable to find a place in preceding systems. We made this point earlier: a technical system is not simply the result of inventions made during the particualar period when it is developing, but also of anterior elements.

CONCORDANCES

We need to consider the circumstances surrounding these developments, just as we did for other epochs. This seems more effective and significant than trying to establish hypothetical lines of causality, especially since these causal relationships are usually reciprocal. We are not claiming circumstances are all-important by giving them priority; they are simply providing a framework for a clearer discussion.

Several different conclusions have been drawn from research into demographic movements. The population in England rose considerably during the second half of the 18th century to reach 5.5 million inhabitants by 1700. This figure was six and a half million by 1750, and nine million by 1800.

As for France, opinion is more divided, but it is generally thought that there was a sharp increase in population during the second half of the 18th century. Some historians believe that this rise simply served to fill the gaps created by the severe losses sustained in the 17th century, and that during the last decades of the Ancien Règime, the population figures stagnated or actually declined in some areas. It is generally thought that the populations of Spain and Germany rose during this period, but more precise studies are needed to confirm or deny this very general view.

The causes of these demographic movements still need satisfactory assessment. One important reason however, seems to be a marked decrease in mortality. This is an important point, particularly with regard to the evolution of techniques.

But any study of a population cannot confine itself to quantitative research. And from our point of view particularly, the distribution of population is very important. The proportion of able-bodied men, for example, is very important – and the decrease in mortality is particularly interesting in this sense. The size of the professional class is also significant. There is a view prevalent in history today that rural workers in England were forced by enclosures to leave the countryside and enter the towns. The large available work-force provided the conditions for a growth in industry. It seems, however, that the reality was not quite so simple. However, it is no less true that if techniques are not sufficiently developed, a large workforce is needed to compensate for the lack of technology at certain stages of production. Even if the depopulation of the countryside was not as systematic as it has been made out to be, nonetheless, the influx into the towns was sufficient to meet the needs of industry. Aside from this, technical progress is certainly sustained by demographic rises, but too sharp an increase can have adverse effects.

As for economic developments, there is little need to discuss them here: there are already plenty of excellent studies on the subject.

Recent studies have revealed that the main thrust of economic recovery took place between 1730 and 1760; this was the very period when there was a renewed impetus to technical progress, indeed, the latter began just after this period. One point is clear, therefore. The economic recovery preceded the establishment of the new technical system. The research of Dean and Cole has shown that there was a marked change in economic development from about 1730, at a time when none of the great inventions had been put into productive use. This leads us to believe that before the main movement of technical progress began, there was an intermediate stage of imperceptible progress coinciding with the increase in population and other phenomena. It was only after this stage that both the technical changes and the main demographic increase actually took place. And was this not so of the 15th century before the enormous expansion of the 16th century? Research in this field would be absolutely invaluable to the history of techniques.

As always, this general view needs to be qualified. Growth needs to be measured in quantitative terms, but the qualitative changes also need to be examined. Production during the first decades of economic growth was very different to what it had been at the beginning of this period, and this was partly due to technical progress.

Technical progress is always conditional upon investment, which means that there had necessarily been an accumulation of capital. Some countries had the benefit of a stable currency which had been established at the beginning of the century: this was particuarly true of Great Britain and to a lesser degree, France. The extent of this increased wealth is not clear. Global enrichment was necessary, but so too was the accumulation of capital in the hands of potential investors. Here again, the mechanisms are difficult to define: some believe that the capital accumulated in commercial sectors, and this was probably true of Great Britain, while others believe that the landed sectors were more significant – and this in turn was probably true of France. However, this problem is a marginal one, but it is worth noting because it may help explain the differences in technical activity between the two countries.

But even this is not sufficient to explain such a complex movement. Investment is partly controlled by mechanisms of a legal nature. The new technical system necessarily changed the dimensions of enterprises, and this in turn generated changes in the latters' legal organisation, which after all represents a field of techniques. Less bound by rigid structures, Great Britian was able to adapt fairly rapidly. But France was able to change only by bending the law (this was the illegal beginning of the public or limited liability company, in the modern sense of the term, which was originally called the "false" partnership) or by modifying it (the Miromesnil project at the end of the 18th century)

We shall not dwell any more on this exterior aspect of technical progress. In order to make clear certain aspects, however, some related problems should be considered.

It is difficult to appreciate the man of eighteenth century Europe in his relation to techniques. It is no less difficult to do this in a few words.

It is perhaps easiest to establish a brief schema which contrasts two opposing attitudes. On the one hand there is Diderot's "Encyclopédie" which holds an important place in technical preoccupations. On the other hand there is the English classical school, of which Adam Smith was the prime exponent. He had only a mild interest in techniques and their effect on the economy: at the very most he was responsible – apart from the famous concept of the division of labour – for the idea that the "secrets of production" could affect prices, independently of supply and demand. On the one hand there were the good mercantilists, the governments whose policies were actively in favour of technical progress, while on the other, there was the general public, who were more or less indifferent to the transformations taking place.

In other words, it seems that technical progress was largely the concern of small groups of people who, according to the different countries, came from different environments. In Great Britain, despite all that was written on this subject, technical progress was mainly the work of men who were already engaged in production, or at any rate working fairly closely with it. By far the majority of inventions were due to entrepreneurs or foremen, or to machine builders or carpenters. There is no better example of this than the Darby family in the field of iron and steel production, or the association between Boulton, a famous entrepreneur, and an instrument maker employed by the University of Glasgow called Watt. Another example is the Wilkinson family, this time in the field of machine tools. Exactly the same kind of examples can be found in the textile industry. True, there is a priest among the list of these inventors, but he came from an industrial region where he could not possibly have been unaware of certain technical difficulties. The main characteristic of this phase of British technical history was the individual nature of these efforts. Boulton and Wilkinson represent perhaps the only examples of technical research being undertaken by large enterprises. The coherence of advanced techniques depends largely on a country's economy.

In France, on the other hand, the work was done on a collective level, and was much more systematic. No doubt this was partly due to the lack of dynamism on the part of entrepreneurs, who had to be spurred on by government policy. The best example of this is a text by Turgot written in 1773, which passed almost unnoticed:

> After removing all taxes on the production, transport, selling and consumption of foodstuffs, there can only be one thing left to do to help trade. That is by training, and by the encouragement of scientific and artistic research which may help further development, and especially by helping to extend the knowledge of processes which have been kept secret through plain avarice.

Turgot's method was to use colbertist policies: this consisted essentially of sending missions abroad to learn about advanced techniques. Various missions were organised to Britain, and also to Germany and the northern countries.

This collective role of the government was expressed as Turgot hoped it would be, in the provisions made for technical training. Their primary function was to establish a state body of technical personnel, but this was in fact largely superseded by the state itself, which played a large part in the promotion of technical progress. A centralised state needed a trained technical personnel, and this gave rise to the famous higher training schools of today, such as the *Ecole des Ponts et chaussées*, which emerged from a small office of designers, and all the military schools, (engineering, artillery, and shipbuilding). At the same time, a body of manufacturing inspectors was established to enforce very strict industrial controls. In fact, all these institutions had been planned during the time of Colbert. These executives were soon closely involved in industry, through their attention to products supplied to the state, and they paid increasing attention to techniques in particular. There is no better example than the creation of the *Ecole des mines* at the end of the 18th century, which was designed not only to establish a body of inspectors, but also to improve the technical knowledge of those developing the mines.

Technically trained personnel and state inspectors naturally enhanced technical progress. And considering the vaguely defined position of the civil servants at this time, it is easy to understand the close interaction between representatives of the state and those of the private sector which is especially apparent in the latter half of the 18th century. A considerable number of the inspectors came from this private sector. After they had been employed on a mission abroad they joined the public sector on a permanent basis. In an industrial country which was hardly inclined towards technical development, state interference was a vital factor of progress. Other industrialised countries were in a similar position, such as Germany, for example, where a mining school was set up in Freiberg at the beginning of the 18th century. Similar schools were also set up in Sweden and Austria.

These differences were also apparent in other areas, one of which was technical literature. French industrial controls were just as much technical as they were economic or administrative. It was only natural, therefore, that they should seek to codify the use of techniques. This gave rise to the publication of treatises which promoted the correct rules; we shall see later how this idea was systematised. Britain was more pragmatic – the government simply trusted in the technical minds and imagination of British practitioners. Only one technical treatise was produced in the 18th century, an important one nevertheless, written by Jethro Tull and entitled *Horse-hoeing Husbandry*, the first edition of which was published in 1731. Even though the author was interested in certain new crops, such as beetroot, sainfoin and alfalfa, "he does not" according to A.-J. Bourde, "consider any way in which they could be included in a type of agriculture with tranformed structures of cultivation." He considers them only as tributary crops, outside of the traditional patterns of farming, but protected against common grazing land. It is true that Britain had a long tradition of such treatises. But is it not surprising that she failed to produce during this time a single treatise on naval construction?

France, on the other hand, was steeped in technical literature which, furthermore, was of a somewhat erratic quality. In effect, it represents the course begun by Colbert and culminating in the industrial and technical controls to which they were so partial. It was not a matter of progressing, but simply of perfecting that which already existed, and establishing the rules. Such was the nature of the technical part of Diderot's *Encylopédie*. The *prospectus* itself actually recognised the impossiblity of producing a systematic technical literature, or of a technology which did not simply consist of descriptions:

> Furthermore, the workers are the artists, and it is not by consulting books that people learn their trade. In this work, craftsmen will encounter perspectives which may be new to them, and observations which may otherwise only have become apparent after many years of work. The studious reader is offered that which they would have learned from observing a craftsman at work in order to satisfy their curiosity; we offer craftsmen, on the other hand, the chance to learn from philosophers, and thereby hopefully to advance further towards perfection.

Which brings us back to descriptions once more. It would be interesting actually to establish just how much of the *Encyclopédie* was original, and how much was borrowed from the technical literature which it so scornfully frowned upon, and that is not to mention the pillaging of the plates from the *Académie des Sciences*. Diderot's greatest merit is to have established a table of different types of knowledge which includes techniques. This was also attempted by the *Académie des sciences* with the *Description des arts et métiers*, begun at the end of the 17th century but never completed. The publication of the *Encylopédie* provided the necessary impetus. The first volume, *L'Art du charbonnier* (the charcoals burner's art) appeared in 1761. But these works had been in preparation for a long time, and some of them were quite obviously out of date in technical terms – the same was true of the *Encylopédie*, because much of that was copied from older works. Between 1761 and 1788, seventy-eight volumes were published, dealing with techniques of all kinds. Only a part of them dealt with industrial techniques, or with the ensemble of the techniques of an industry. Many of them examined artisan techniques, and finally a smaller portion dealt with single techniques on their own. The first volumes were quite traditional, while the latter reveal a greater sensitivity on the part of the authors to technical progress. The last major collection was begun at the very end of the 18th century, and was finally completed in the early 1800's. This was the *Encylopédie méthodique* of which two parts are of most interest to us: the *Arts et métiers mécaniques* (1783-1791) and *Manufactures, arts et métiers* (1784-1790). There is also some valuable information to be found in the scientific volumes, however. Although this new publication was meant to pay hommage to Diderot, it was also a critique of the *Encylopédie*. The new *Encylopédie* was a compilation and revised collection of previous works – furthermore they were also able to include new elements which had emerged as a result of the technical progress made in Britain.

If the historian of techniques studied nothing else but these anthologies, they would be justified in thinking that there was a fixed level of techniques, set, precisely, at what it was before the great technical transformation of the 18th century. However, there were many publications besides these. One of the most prolific and competent writers with a truly encyclopedic mind was Duhamel du Monceau, who was appointed *inspecteur général de la marine* (general inspector of the navy). Apart from the seventeen volumes that he wrote for the *Académie des Sciences*' collection, he published some of the greatest works on French agronomy. His *Traité de la culture des terres* (1756) was adapted from Jethro Tull, and the *Eléments d'agriculture* (1762) was the result of his own experience. He converted his land in Denainvilliers in the Ĝ*f*atinais, into a veritable laboratory for farming experiments. His work with shipping had made him an expert on wood, and this is reflected in his *Traité des arbres* written in 1755. He also wrote *Eléments de l'architecture navale* (1752), which was a basic text used in the *Ecole des constructeurs de navires* (school of naval construction), founded in 1765. These few indications are quite sufficient to convey the prime importance of these works in French technology.

The abundance of technical literature is equally apparent if one reverses the approach and deals only with one particular technique. There were twelve treatises on naval construction written in the 17th century, and sixteen in the following one, many of which were important works. The first main work on this subject was written by the Dutch Witsen, in 1671. The first theory of ships was expounded by Bernoulli in 1714, and this was followed in 1749 by Euler's work. Many of these works represent a two-fold consideration, the technicians' practical experience, and their knowledge of theory. Such is the nature of works by Bouguer (1727 and 1746), Duhamel (1752), Chapman from Sweden (1768), and Vial de Clerbois from France (1776).

We would need an entire chapter to discuss adequately the technical literature of the 18th century. There are famous works dealing with all types of industry, both traditional and modern: those of Swedenborg, Bouchu and Courtivon for metallurgy, Morand for mining, Berthoud for clock-making, Monge for cannon-construction, and Belidor for "engineering science" and hydraulic architecture. It should be recognised that the contribution of France to technical literature was an important one, no doubt for the reasons we discussed above. The Swedish and the Germans were also important in this respect, expecially for their work on mining and metallurgy. The Dutch wrote some important works on naval construction. Again, we can only repeat that analytical inventories would be of great use to historians of techniques: some of them are being compiled at the moment.

We have already emphasised the close links between science and techniques. A science which is in the process of being created, or that has just undergone some transformation needs technical support, either to provide problems to be solved, or to permit experimentation. It is no less evident that at a certain stage of development, a technique needs the support of scientific notions which can solve problems and allow techniques to be considered on a general level and not just in terms of particular cases. Studies by Maurice Daumas have shown how much science has benefited from techniques which have provided the former with every type of equipment. On the other hand, it would have been impossible to conceive of the steam engine without some scientific knowledge of atmospheric pressure. Again, it is not a matter of deciding which of the disciplines is more important than the other, but simply of establishing the reciprocal relationships between them, and showing how each can favour or hinder the development of the other.

Of course as far as our particular point of view is concerned, it is true that in many cases, scientific knowledge is not necessary. During this period, for example, it was not required in the textile industry. In other cases, science could have helped techniques but the latter was able to advance anyway without the support of the former. Metallurgy was also able to develop before there was proper scientific understanding about the different *Etats de fer* (states of iron), to use the title of the definitive pamphlet published by Berthollet, Monge and Vandermonde in 1788. However, the phlogiston theory had considerably delayed this moment and had sent Réaumur, for example, completely up the wrong paths. The steam engine was developed at least half a century before the theory was expounded.

Most links between science and techniques are situated somewhere in the middle of these two extremes. It is obvious, for example, that the chemical industry could not develop until after the new discoveries and techniques were made available, until people were familiar with the work of Lavoisier, Priestley and Cavendish. This was an indispensable condition. Watt reaped great benefit from the scientific milieu at Glasgow University and from the works of his professors, especially Black, and it was this that enabled him to complete the development of his steam engine. In the first case, science was an indispensable base, while in the second, it was simply a support. Another example of the latter case, are Bernoulli and Euler's works on ships which may have helped develop ship construction in a certain way, but certainly did not completely transform it. Euler's important work on reaction wheels did not bear fruit until the early 19th century.

There are various different points of contact between technique and science, then. There is often chronological concordance between the two courses of development, and also concordance in terms of activity and current areas of interest. Once more, we are bound to note both the serious lack of research on this subject, and a lack of serious research: studies of this period have always separated the two disciplines, to the extent that they are now extremely difficult to link.

We have therefore been forced to build a new model, which will certainly not be the last. The best approach would be to define the common thresholds of science and techniques, and also the break points where science and techniques become autonomous, and where there is no possibility of reciprocity.

The creation of a chemical industry during the last part of the 18th century was a direct result of the emergence of modern chemistry: the production of artificial soda and the use of chlorine to bleach materials would have been inconceivable before this time. But apart from this, modern chemistry has had little effect on other techniques. The role of carbon was clearly established in iron chemistry a year before the French Revolution, but this discovey had no effect whatsoever on either traditional or modern techniques: all descriptions of iron production in technical literature make it quite apparent that those actually involved in iron production went entirely by how things looked, and nothing else. Réaumur's research was different in this respect. however. Réaumur was actually able to control some aspects of the treating of iron and steel, (the running of blast furnaces and quenching, for example) by examining and analysing these "appearances," such as the breaking of a test-tube, for example. But it was years – not until the beginning of the end of the 19th century – before this research was to yield effective results for iron and steel production. This industry was finally delivered of one stranglehold when the phlogistic theory was finally abandoned.

The developments made in physics were useful only insofar as they enabled phenomena to be quantified, which in turn helped provide solutions to certain problems. A few works written in the first half of the 18th century, like those of Bernoulli and Euler, for example, made some important developments possible. Black's discoveries about latent heat put Watt on a fertile path. But all these relationships were limited. In most cases, numeric controls established empirically were sufficient for technics, and the building of boat hulls is a good example of this. Duhamel did not trust the work of theorists and had little confidence in the application of mathematics. Black, who was a member of the Royal Society, attempted to apply certain calculations to the dimensions of steam engine cylinders and their effect: these attempts did not provide any new effective methods, and Clapeyron's charts were not assembled until 1834. Agriculture reaped few benefits from the progress of botany and plant physiology.

So in the vast majority of cases, techniques remained an empirical activity, with no permananent or significant links with science. Techniques certainly helped the progress of science by extending its experimental field, but it did not actually dictate what course these experiments should take. There were still two types of logic: scientific logic, and the logic of an accumulation of results of experience, rather than experiment. There is no better example of this than ship building. Colbert, for example, attempted to take the best elements of all ships in order to build the ideal vessel. Some of these notes still exist today, and they deserve careful research. In the 18th century, both the British and the French had different conceptions of the design of the hull, for example, but both studied the other's ideas carefully, and took the best elements from each. Technical progress was the fruit of observation, experiment and imagination – not the result of constructive reasoning, except in rare cases. It was for this reason that the emergence of the principal inventions was a slow process. The technique which emerged in Britain at the end of the 18th century, had been in embryo stage for over a century. It finally evolved as the result of productive failures and sucessful imitations. A study of technical failures would be extremely instructive.

There was a vague sense towards the end of the 18th century, especially in France, that science and techniques were very closely linked, and that only their reconciliation could bring about the establishment of a true form of technology, which was not mere discourse or description, or based on exterior appearance or empirical tables. The creation of the *Ecole Polytechnique*, which was to provide scientific training for civil servants whose occupations would be extremely diverse, is a striking concrete result of this awareness.

What we need most of all is an inventory of all the points of contact between science and techniques, but no-one has yet even begun such a project. There were some technicians during this period which could almost be called scientists, like Duhamel du Monceau, Belidor or Grignon. And there were also scientists who were close to being technicians – the most famous of these was Bernoulli and Euler. A study of the different works of these men would provide valuable indications about the links between science and technique. It is possible that from the middle of the 18th century, scientists were able to provide technicians with refined models which potential users only managed to avoid.

GLOBAL PROGRESS

The ever-increasing interdependence of techniques, coupled with technical progress and the impetus created by economic growth has meant that the effects of successive technical revolutions have become more and more profound and far-reaching. There have been partial changes, affecting exclusively a number of important sectors whilst other sectors have developed and still maintained crude levels of development, without creating too many awkward distortions. The more technical advance that is made, the greater the hold of each individual technical development. Henceforth, technical changes were to become more and more global.

It is true that mechanisation began before the 18th century: we have already seen its remarkable development during the Renaissance period and the strong desire to develop mechanisation as much as possible. However, mechanisation assumes another dimension during the period with which we are presently concerned. This is no doubt due to several factors, which should be outlined.

Firstly, there was the question of materials. However well they were made, machines made out of wood could never be effective. The play of the assembly's fixings, the friction and the subsequent wear of the moving parts were major problems which were difficult to overcome. These ramshackle machines were erratic and did not work well. They deteriorated rapidly and since they were made of wood, they were incapable of performing heavy duty work. Furthermore, the machine parts frequently broke and there were continual stoppages.

These problems are sufficient to explain the small amount of progress made in the field of mechanisation, considering its great potential resulting from the refinement of new mechanisms discovered during the Renaissance period. A brief survey of the illustrations in Diderot's *Encylopedia* will confirm this point.

The idea of constructing machines using metal could have come quite naturally and without a great deal of effort to the engineers of the classical period. However, there were several factors making this impossible. The first one is concerned with the nature of the metal itself. For a long time, the iron produced was of a poor, and sometimes extremely bad quality. It was irregular, brittle, difficult to weld and difficult to work. The progress made in ferrous metallurgy during the

18th century brought about the production of a metal which was both more resistant and easier to work, as well as being available in greater supply. Secondly, the implements used to work metal were incapable of creating even those assemblies which could be made using wood.

The working of metal, using manual tools, was, moreover, a longer and more costly process than that of working wood. The production of cast steel, which began in the middle of the 18th century – which we will discuss later – and subsequently, the production of machine-tools for the working of metal, in particular Wilkinson's reaming machine (1772), threading machines (for screws and nuts), and planing machines, finally paved the way for machines made of metal. (figs. 1 and 2).

The development of mechanisation was limited, in the second place, by the fact that although there was an abundance of energy available overall – we are thinking here particularly of hydraulic energy – it was broken up into small units of production with only a minor capacity; it was dependent upon natural conditions, and was necessarily localised. Apart from a few exceptions, the only example of which is the Bazacle mills near Toulouse which were run by a series of very primitive turbines, the units of production rarely surpassed ten horsepower. Furthermore, the irregular flow and freezing of rivers provoked prolonged interruptions. Factories had to be installed beside the water sources, and this entailed either high transport costs or the creation of large hydraulic plants.

The steam-powered machine, which has become the symbol of the new technical system of the 18th century, freed the production of energy from many of its restrictions. There was now a possibility, albeit a progressive one, of obtaining larger quantities of energy, which could be produced anywhere. Similarly, however, this led to the choice of other sites or to the development of new means of transport in order to diminish the cost of the fuel, and at the same time it created the conditions for the concentration of manufacturing.

Finally, it is possible to point out at this stage a change in fuel types, and this concerned a whole series of operations (the production of mechanical, thermic and chemical energy). In certain regions, this was an urgent requirement, due to the notable disappearance of plant fuels. Thus, the basic trilogy of the new technical system, within which there were many interactions, finally came into being: metal, coal, machines powered by steam. The history of the steam engine is now

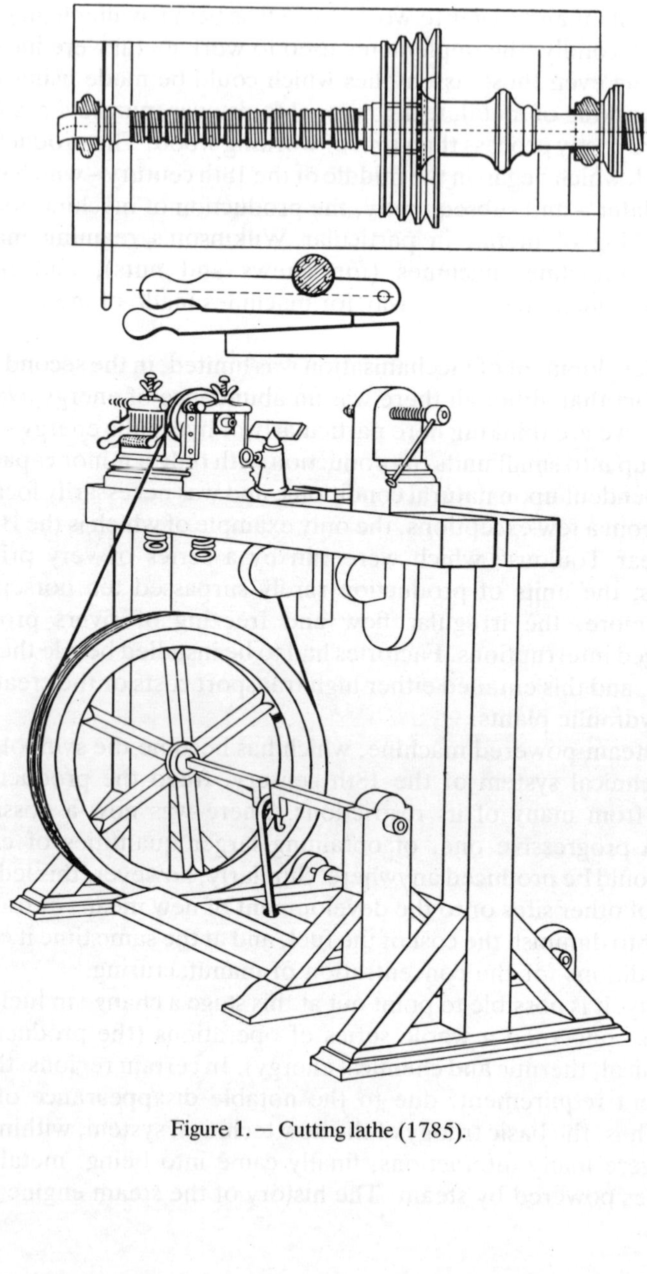

Figure 1. — Cutting lathe (1785).

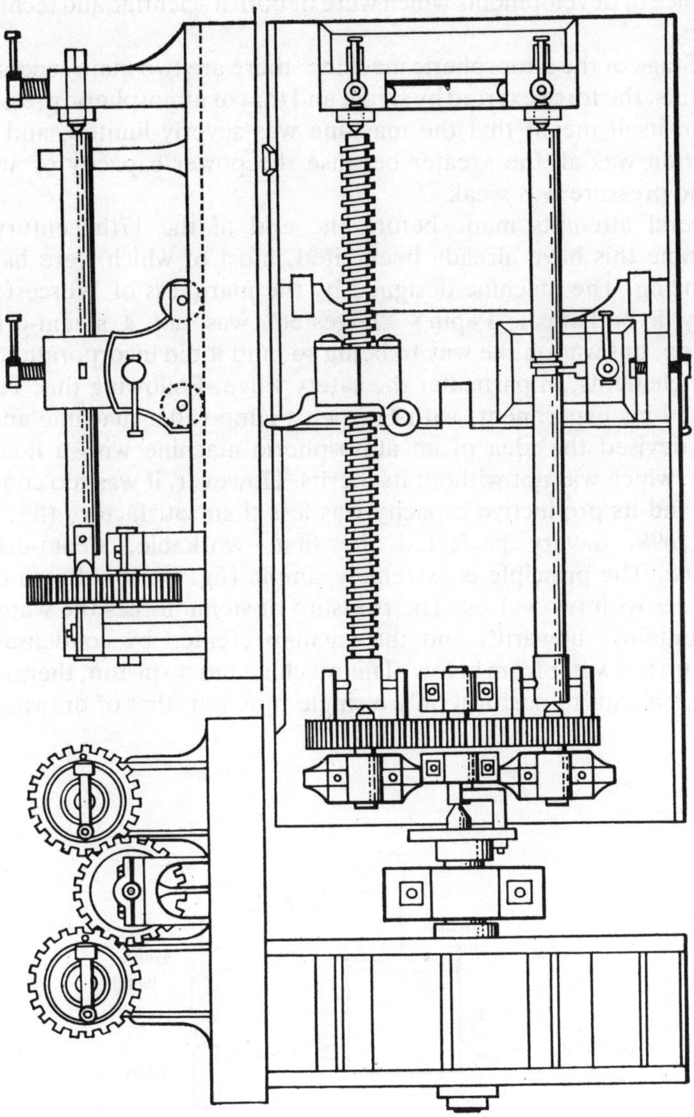

Figure 2. — Wilkinson's cutting lathe (1798).

well-known. There were several stages to this process, due to a sequence of developments which were of both a scientific and technical nature.

I – Stage of the atmospheric machine: there are two main successive elements: the force exerted by steam and that of atmospheric pressure. This in itself meant that the machine was severly limited, and this limitation was all the greater because the power capacity of atmospheric pressure was weak.

Several attempts made before the end of the 17th century to overcome this have already been cited, most of which were hardly convincing. The machine designed by the marquess of Worcester is merely hypothetical. Papin's "Digester" was not a steam-driven machine, but was on the way to being so, and it did incorporate some useful elements, in particular the safety valve. Following this, Papin resumed his experiments with Huygen's gunpowder-machine and in 1707, devised the idea of an atmospheric machine with a floating piston, which was not without its merits. However, it was too complicated and its productive capacity was less than satisfactory (fig. 3).

In 1698, Savery perfected the first, workable, steam-driven machine. The principle is extremely simple (fig. 4). It is based on a container with two valves. The pressure of steam forces the water in this container upwards, and the vacuum created by condensation draws up the water from below. This machine has no piston, therefore, and it is meant to perform only a single function, that of drawing up

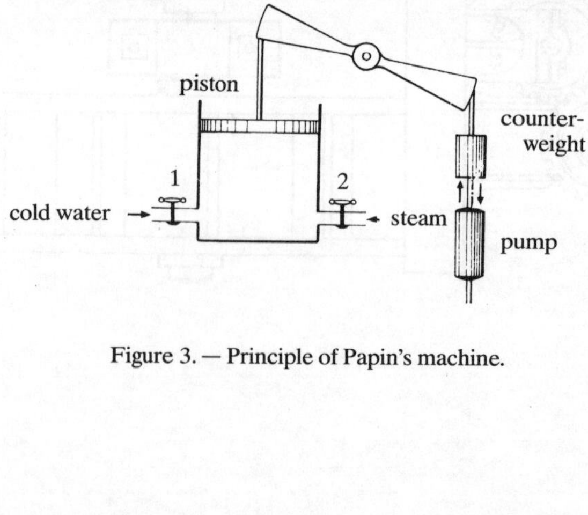

Figure 3. — Principle of Papin's machine.

water from the mines (fig. 5). No doubt, it was with this machine in mind that Papin, in 1707, reviewed his ideas and re-introduced the piston. This machine was capable of up to two-thirds horse-power, and apart from the idea of placing hot irons in the cylinder to heat up the water, Papin had some interesting, but as yet inapplicable notions, such as the superheating of steam, and the bell-jar, the idea of which he borrowed from Mautsch of Nuremberg. In fact, the latter served only to draw up the water and to let it flow down again onto a hydraulic wheel. The first, practical, steam-driven engine was designed by Newcomen in 1712. Similarly, the basic principle was very simple (fig. 6). The pressure of steam forces a piston upwards, then the vacuum created by condensation enables atmospheric pressure to drive it back down again. The cylinder is, therefore. open at the top. Newcomen still had the same idea in mind: the pumping of water forced out again by steam. This was logical, since it was necessary to maximise the use of the more powerful form of energy. A whole balancing system served to counteract the superfluous weight, notably the rods, the piston and the entire apparatus of the pump.

II – Following stages considered by Watt: when Watt took up this problem, the steam-driven machine had already undergone consider-

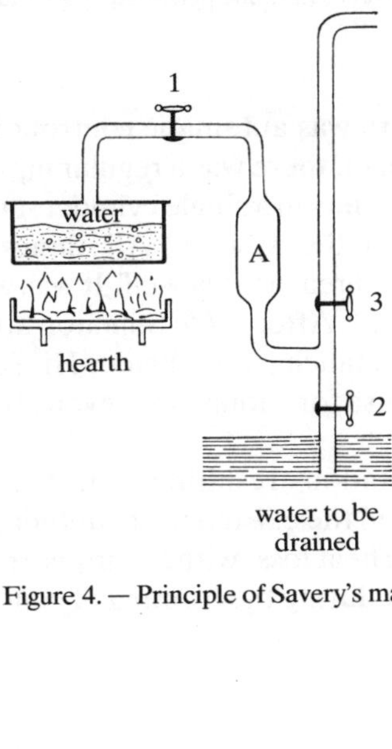

Figure 4. — Principle of Savery's machine.

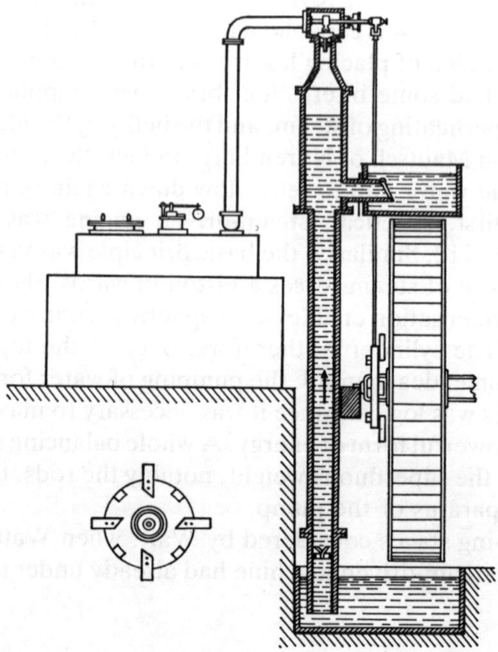

Figure 5. — Savery's machine perfected by Kier and Wrigley.

able development: there was automatic control of the circuit's valves by using the balance beam; there was a regular input of steam provided by the use of multiple boilers in a single cylinder, and Carneau's boilers were incorporated into the system. All things considered, Newcomen's machine worked reasonably well. It was widely used, particularly in mines in Cornwall. After 1765, Smeaton attempted to dispense with the cumbersome balancing crossbeam. His success was such that his design remained in use for a long time, even after the discoveries of Watt.

Apart from the limitations resulting from the use of atmospheric pressure, Watt perceived the existence of another problem, which was that of the considerable heat loss, without any corresponding mechanical effect. Heat was lost during the forming of condensation; and there

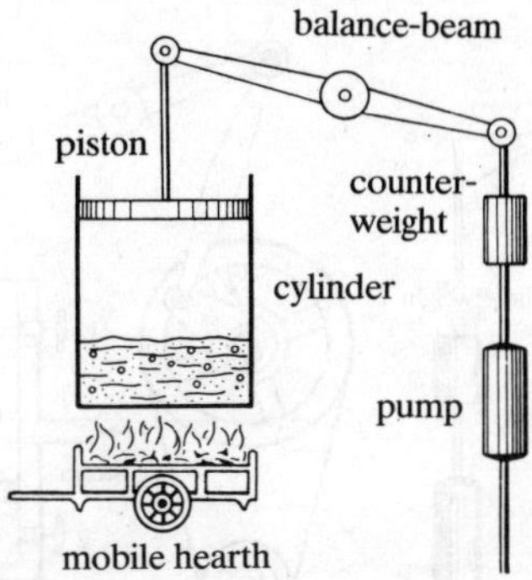

Figure 6. — Principle of Newcomen's machine.

was a loss of steam when it was introduced into a cooled cylinder. The solutions to these problems were the following: the process should take place in a closed, not an open, cylinder, and condensation should take place outside the cylinder, not within. Between 1765 and 1769, Watt perfected several vital elements: (*a*) a separate condenser was incorporated, which meant that the cylinder would be kept at a constant heat; (*b*) the heat within the condenser could be maintained by using an interior lining and steam circulated between this lining and the cylinder; (*c*) water was ejected by means of a condenser pump; (*d*) atmospheric pressure was no longer used, it was replaced by the action of steam on both sides of the piston (fig. 7).

The capacity of the new machine was infinitely higher, which considerably lowered fuel consumption and therefore energy costs. Furthermore, the construction of the machine had vastly improved after 1775 with Wilkinson's invention of the reaming machine for the production of cylinders. This apparatus was costly however, and was only profitable for use with larger machines.

The steam engine acquired its definitive aspects between 1780 and 1787: (*a*) the beam was made redundant by the use of two cylinders and the double effect of steam and its release on two sides of the piston (fig. 7); (*b*) the invention of the articulated parallelogram assumed the beam's function of assuring the smooth play between the valves (fig. 8); (*c*) the ball governor regulated the running of the machine (fig. 9).

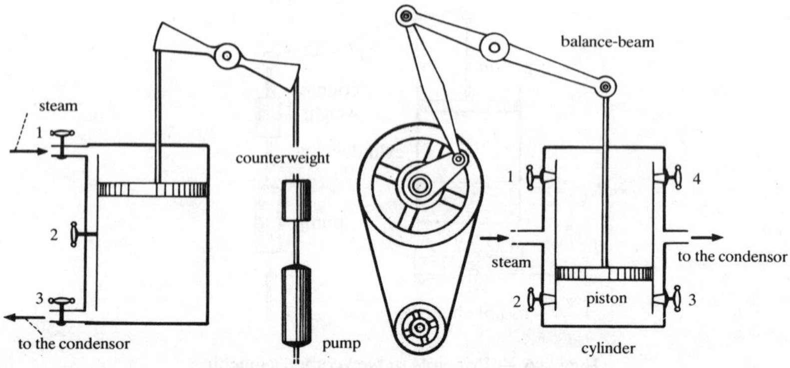

Figure 7. — Watt's improvements made to Newcomen's machine. (From Furia and Serre, *Techniques et sociétiés,* A. Colin, Paris, 1970)

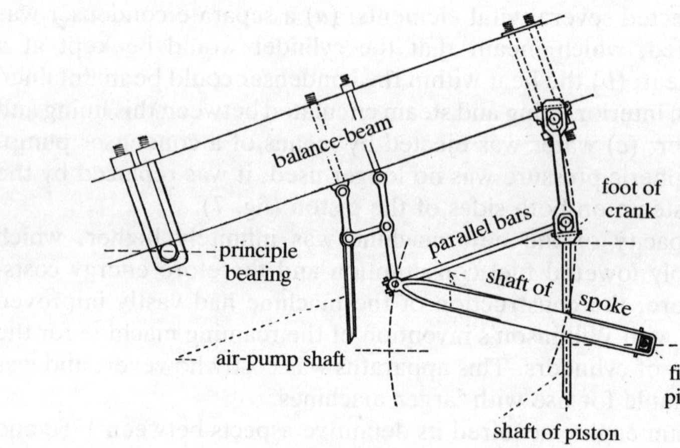

Figure 8. — The articulated parallelogram.

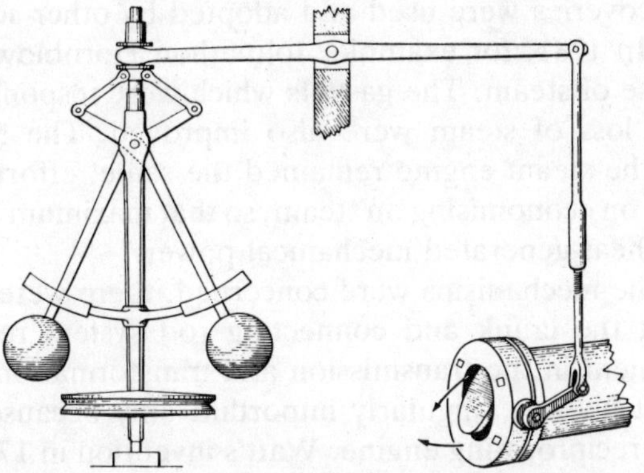

Figure 9. — Ball governor.

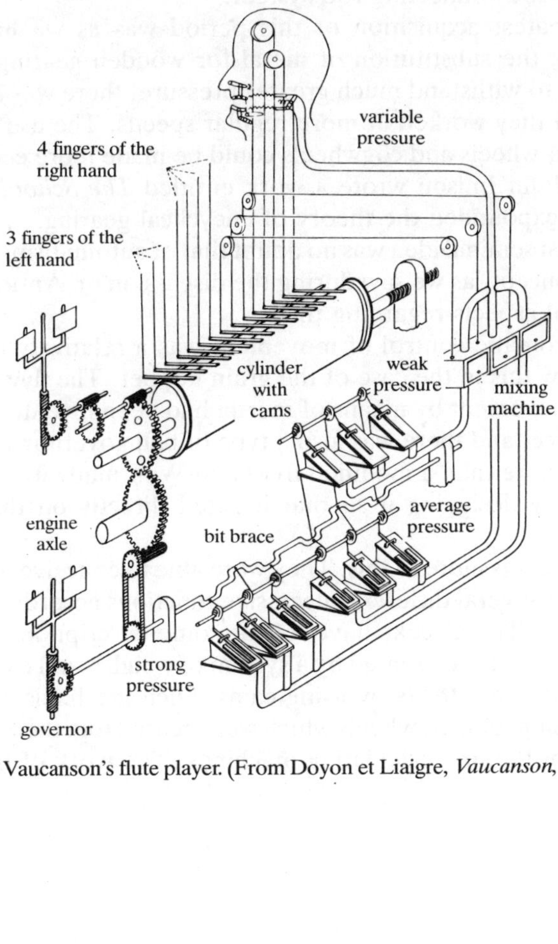

Figure 10. — Vaucanson's flute player. (From Doyon et Liaigre, *Vaucanson*, Paris, 1966)

Watt's discoveries were used and adopted by other scientists and technicians. In 1781, for example, Johnathon Hornblower used the double release of steam. The gaskets which were responsible for the considerable loss of steam were also improved. The fundamental principle of the steam engine remained the same; efforts were now concentrated on economising on steam, so that maximum use could be made of the heat-generated mechanical power.

As far as the mechanisms were concerned, there were no marked changes, and the crank and connecting rod system remained the essential element in the transmission and transformation of primary movement. It became singularly important only because the steam engine was a reciprocating engine. Watt's invention in 1781 of epicycoidal gearing, otherwise known as the planetary system, was in fact an aberration made necessary only by the excessive patenting in 1779 of the crank and connecting rod system.

The greatest acquisition of this period was as we have already suggested, the substitution of metal for wooden gearing. Machines were able to withstand much greater pressure, there was less using of parts, and they worked at more regular speeds. The use of cast iron meant that wheels and cogwheels could be made more economically. In 1787, John Imison wrote a work entitled *The School of Arts*, in which he expounded the theory of encyclical gearing.

The most seminal idea was no doubt that of automatism. This was an ancient concept, as we saw during the discussion on Antiquity. There are several aspects regarding this.

The automatic control of movement was a relatively recent idea, and we saw this in the case of the grain hopper. The flywheel, which controls movement by means of inertia had been invented during the Renaissance, and there was also a type of ball governor which in fact was little more than a modified flywheel. Watt made it into a perfect governor by designing it so that it acted directly on the intake of steam.

But there was more to this. They were able to conceive of a machine performing several different successive functions according to a given programme. The Greeks have left us some descriptions of automatons: movement was created by a system of weights and ropes, and the various different effects by using cams. Such mechanisms were also possible using toothed wheels which were geared together: this was the principle of the first calculating machines. The work of the German

Schickhardt (1592-1635), Pascal's machine (1642-1645), and findings of Leibniz (1673-1710) were milestones in this course of history.

The mechanical clock was also based on analogous principles. The only difficulty came with controlling the movement, but the adoption of the spiral spring in the second half of the 17th century almost solved this completely. The main issue was deciding which metal to use for these springs: it was with this object in mind that Huntsman created the method of forging steel in a crucible.

Mechanisms used in musical instruments and automatons both brought into play diverse types of other mechanisms, within which the use of cams was paricularly important (fig. 10) – Vaucanson from France and the Droz brothers from Switzerland were particularly important in this respect.

The use of cams meant that programming was possible, and this development was further advanced when the punch card was invented. making programming possible – the latter was indispensable if automatism was to develop. This was devised between 1728 and 1734 when Falcon, after Bouchon, used perforated cards to replace the draw-boys in weaving looms for tailored cloths. imitating perhaps the stitched cards used in the cylinders of hydraulic organs.

The oldest known automatic machine for industrial use was Jesse Ramsden's circular dividing engine devised in England in 1773. "The circle is turned through a determined arc by depressing the treadle, then the scriber traces the division on the plate."

All these factors – more powerful energy, more control, better methods of construction, and more diverse mechanisms - indicated the emergence of machine-tools. This had taken some time to get started, but the development was well under way by the end of the 18th century.

In 1751, Vaucanson built the first metal slide-lathe; in 1775, Wilkinson produced the first reaming wheel – and the importance of the latter has already been indicated with regard to the use of cylinders in steam engines. In 1795, Senot from France produced the first threading lathe, which similarly was made entirely of metal. This was before the lathes devised by Maudslay (1797) and Wilkinson (1798). In 1761, the planing machine was devised by Focq.

Henceforth, industrial machinery was to come into widespread use in all industries. On the 17 March 1799, Japy, a watch maker applied for a patent for ten machines, in a quite remarkable effort to mechan-

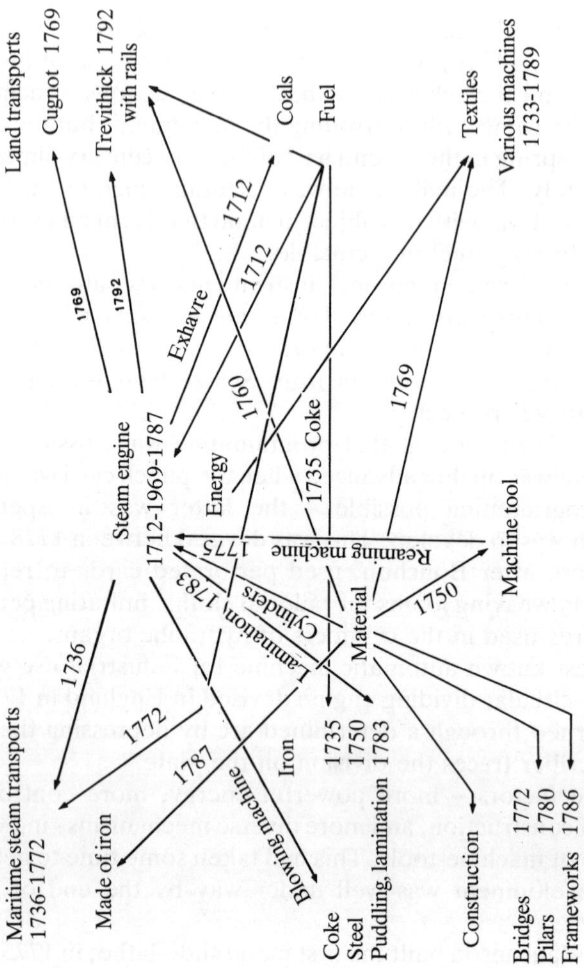

Figure 11. The technical system of the 18th century.

ise his industry. Also relevant here are the Hollanders used in making paper which in 1798 led to the invention of the continuous paper-making machine by Robert, from France.

Also noteworthy in terms of this rather summary chronology is the mechanisation of all types of transport. After attempts had been made by many others to achieve the same object, Hulls, an American, succeeded in powering a ship with steam for the first time. This was in 1736. After Auxiron, Jouffroy d'Abbans from France also conducted some successful experiments on the Saône in 1782; he was emulated by several others experimenters, such as Fitch, Rumsey and Fulton. The first steam-powered land vehicle was devised by Cugnot (1770), and it incorporated an atmospheric machine. This was followed by a similar machine invented by Trevithivk (1796-1797). The first locomotives were built shortly after this, as we shall see below.

The trilogy of iron, coal and steam was by now clearly the basis of the new technical system which lasted throughout almost the entire 19th century. Two basic elements of the preceding system were thereby abandoned – these were wood and water. This transformation was reinforced by a whole series of radical changes in many techniques: the dates concerned are again in perfect concordance with the general view that the basis of this transformation was completed during the decade of 1780 to 1790.

The structures of development techniques were changed quite considerably during the 18th century, but to varying degrees. The transformations in agriculture and the iron and steel industry were marked, but that of the mining industry was apparent only indirectly, outside of the mining industry.

"British" agriculture was not a sudden development, it evolved gradually and was only fully incorporated in its entirety in the mid-18th century. Its emergence was linked to general changes, the most important of which was no doubt the considerable expansion of animal husbandry. Another important change in agrarian structures was the use of enclosures, which allowed for the birth of new techniques of cultivation. As a contemporary historian put it: "land had become the property of a few, and it was this small sector which developed the land, thereby creating veritable farming monopolies. The association between the lord and the large farmer in the 'landlord tenant system' became a major characteristic of the British countryside, and one of the elements of technical progress." Three new developments allowed the incorporation of this new type of agiculture to take place:

A) The first of these was the emergence and expansion of root crops which cleaned and enriched the earth. Maize was unable to grow in the North for climatic reasons, but the potato was a hardy crop and was suited to colder regions. The main expansion of potato-growing took place during the 18th century, when the plant was improved and finally became an acceptable form of alimentation.

B) The turnip represented a veritable revolution. It probably came from Belgium, but neither Googe (1577) nor Weston (1645) who both advocated growing it in fields and using it in artificial fodder had any success. Turnips were not used for ordinary food, but only for animals who used to pasture on fallow land at this time. Turnips were widely grown in Norfolk and Sussex to feed sheep. J. Tull advocated that turnip seeds should be sown in rows and only half-ploughed.

C) The idea of artificial grassland had been known for a long time (clover, sainfoin, and lucerne), but it had hardly been used. In fact, the obsession with wheat meant that fields were very rarely given over to plants considered as more or less useless. The combination of these various factors clashed with the earlier system of cultivation and with its surrounding social and agrarian elements (particularly the practices of fallowing land and using common grazing land). The first step was to grow root crops instead of fallowing land, which was now done on only half the scale in the some regions. The great agronomist Tull, passed by this essential transformation, and it was only in the the mid-18th century that the new type of triennial rotation between artificial grassland root crops and cereals was first practised. The main idea of this was to provide richer food for cattle, especially in winter.

This fundamental revolution was accompanied by more detailed changes: land which had not been cultivated because it was unsuitable for the earlier types of crops grown was now cleared and the soils were improved by drainage, marling, and liming. On the other hand, there was little change in agricultural implements. At the very end of this period, the first threshing machines appeared which were used for shelling, winnowing and sifting the grain. Their use was still limited, however.

Animal breeding, or at least one form of breeding underwent profound change within the agricultural system, but here again, the evolution was slow and gradual. It was only gradually that they began to import the best breeds, and practise cross-breeding by selection – in a far more rigourous way than in botanic science. The creation of new ovine and bovine breeds, not to mention the birth of a new thoroughbred English horse born from an Arab one and brought into England at the end of the 17th century, was simply a matter of time.

These technical innovations in agriculture, however vague, were extremely important. They led not only to a more effective use of land, the development of animal rearing and breeding, and consequently to the possibility of improving their diet, but also to the discarding of social structures and habits which had played such a major role in the civilizations of the Ancien Regime.

In contrast, the development of mines seemed to have conserved the technical systems which emerged during the Renaissance: the basis of these techniques was the same as it had been in Agricola's day, i.e. in the mid-16th century. There was some progress, but this affected only two areas.

The first was water drainage, and it was this which prompted the emergence of the steam-powered machine. The latter was first developed on a wide scale in this industry, and this in itself vastly improved mining development. The mining industry was particularly important in the expansion of the steam engine, because the latter was less profitable than other forms of energy, but it was the only one that was technically feasible for use in mines.

The second aspect of progress was also a marginal one, and concerned the extraction of deposits. Wagons running on rails were used to transport coal either within the galleries or to the larger transport network beyond the mines – such as the canals and ports – and this practice represented a considerable economy. The first actual rails were made of wood, so that originally they were used only inside the mines. The drawings of the mine at Sainte-Marie-aux-Mines depict wooden planks which were used to prevent ruts. The first true streamlined rails with an adaptable wheel were first used in 1630. When metallurgical techniques were finally capable of producing a more durable metal, cast iron rails were adopted. Jars indicates that they were being used in Britain by 1765. They were also being used in Le Creusot by 1783. There is a map of Newcastle in 1788 which clearly demonstrates the advantages of using such a technique for transporting the goods to nearby rivers. This technique was also less costly. The impact of this new method is a well-known phenomenon and does not need discussion here.

On the other hand, the techniques and tools of the actual mining processes underwent no significant change. The main difference was the gradual adoption of powder, which was being used in Chemnitz by 1625.

Of all those industries concerned with transforming matter, the iron and steel industry probably underwent the most complete and total evolution. There were three basic elements to the equilibrium of classical iron and steel production: minerals and wood were used for fuel, and the motor force was water. The new form of this industry was based on minerals and coal. A detailed discussion of this is unnecessary; there has already been a lot of serious research into this subject. We shall simply include a short summary of the evolution of this important aspect.

It was the shortage of wood which precipitated the need for a new fuel, and this could only be coal. The latter cannot be used directly because it contains sulphur which makes iron brittle, however if it is burned in isolation it produces coke which is perfectly suitable (there are still some types of coal from which coke cannot be obtained, or only with difficulty). The solution to the refinement of coal – or "desulphering" as it was known in the 17th century – was first sought in the 16th century. It is possible that coke was already being used by the end of the 18th century to dry malt and in the copper industry, however we do not know enough about this to be sure. One interesting point however, is that Abraham Darby had worked successfully in the malt and copper industries before he became interested in iron and steel production. And it was he who in 1709 was the first to use coke in a blast furnace to produce cast iron. This new process was put to industrial use only later, around 1735-40. Most other areas were slow to adopt it after this time. Coke was more resistant to compression than charcoal, and furthermore, it could be used in much larger furnaces, thereby increasing production. Finally, the progress of the mining industry and transport reduced costs. Cast iron now became a cheaper and more useful product than many other materials (figs. 12 and 13).

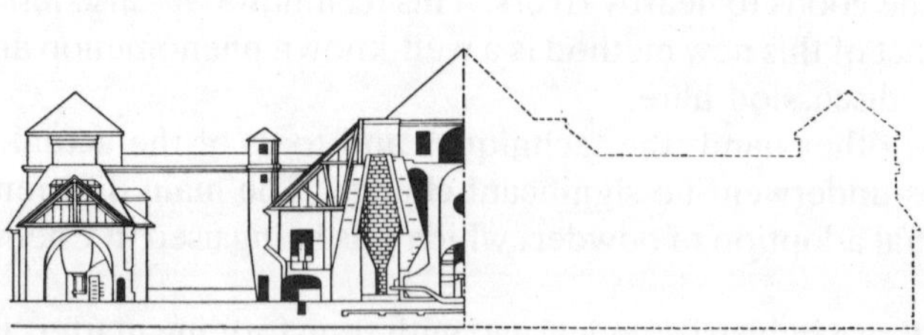

Figure 12. — A large factory of the 18th century: Le Creusot (1785). The blower.

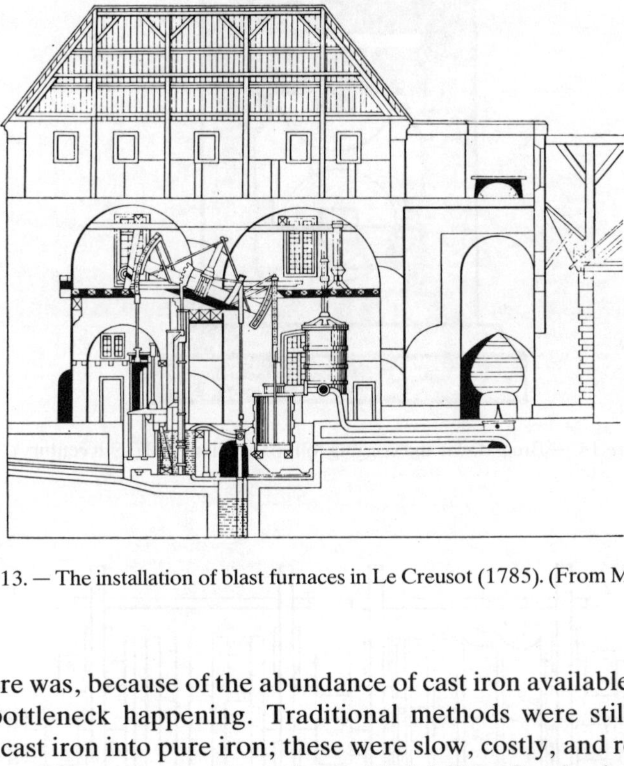

Figure 13. — The installation of blast furnaces in Le Creusot (1785). (From M. Daumas)

There was, because of the abundance of cast iron available, the risk of a bottleneck happening. Traditional methods were still used to refine cast iron into pure iron; these were slow, costly, and required a large workforce. It was Cort who in 1783-1784, after having tried in vain, was lucky enough to discover the process of puddling. This consisted of mixing the melted cast iron on a hearth in a refractory furnace with the decarbonising oxygen of the furnace's atmosphere. In this way the metal had no contact with the solid part of the fuel without having to use air-pumping apparatuses. Henceforth it was possible to process with coke all the cast iron produced.

Now that iron could be puddled after coke smelting, it was also necessary to find another instrument to forge it. The hydraulic hammer currently in use was unable to produce a sufficiently high yield. Once more it was Cort who solved this problem by putting the puddled iron balls through the cylinders of a fluted rolling mill. He patented this idea in 1783 (figs. 14 to 16).

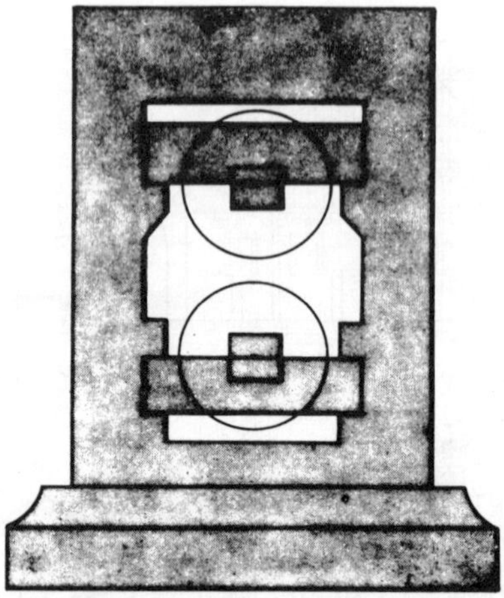

Figure 14. — Breakdown and cogging tolls of a rolling mill (19th century).

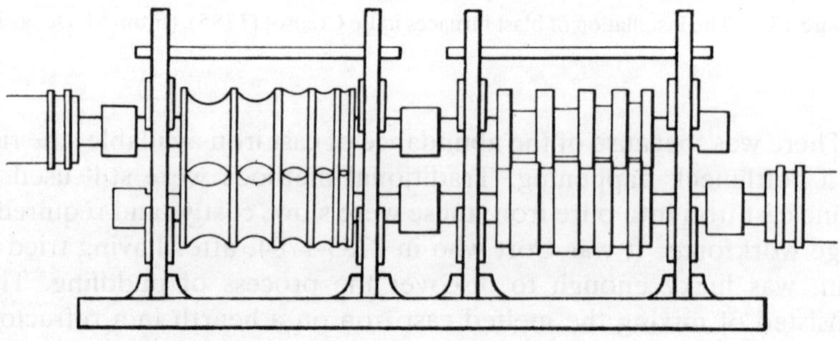

Figure 15. — Breakdown rolls [left] and cogging rolls [right]; early 19th century.

We should also note that the adoption of the steam-powered machine for use in furnaces and rolling mills, finally dispensed with the need for hydraulic energy.

The last invention in this group was destined for a less general use. It was in order to produce steel springs for watches that Huntsman, from England, developed the production of crucible steel, an old technique

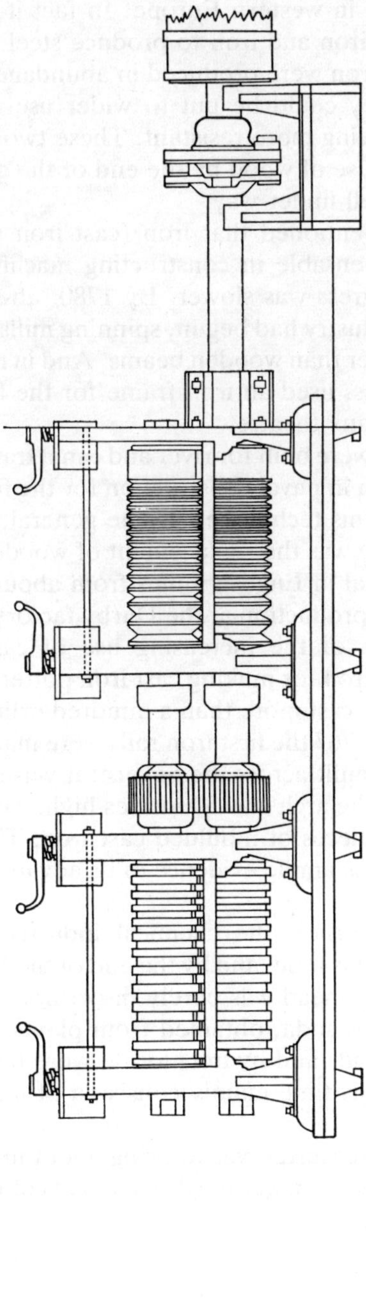

Figure 16. — Finishing equipment at the Rybrick forges (c. 1830). (From Daumas)

which had been known for centuries in some civilisations, but was hardly practised at all in western Europe. In fact it was simply the process of fusing cast iron and iron to produce steel.

Both iron and cast iron were produced in abundance by the end of the 18th century. They could be put to wider use, and they were cheaper than wood, being more resistant. These two metals had not yet fully replaced the use of wood by the end of the century, but the transformation was well under way.

We have already mentioned that iron (cast iron most probably) rapidly became indispensable in constructing machines, but in the building trade its progress was slower. By 1780, after the mechanisation of the textile industry had begun, spinning mills were equipped with iron columns rather than wooden beams. And in the same year, a French architect, Louis, used an iron frame for the first time in the construction of Bordeaux theatre.

In 1787, iron barges were built for river and canal traffic. This period also saw the use of iron in naval construction for the first time. It was still sometime before this technique became general, however.

Then of course there was the replacement of wooden for iron rails for the transport of coal in English mines from about 1765.

The changes in iron production at the Darby factory in Coalbrookdale provide evidence of the increasing use of cast iron. Darby conceived of smelting iron for making cast iron pottery, and by 1758, the factory had already cast more than a hundred cylinders for steam powered machines. In 1767 the first iron rails were made. In 1779, the first metal bridge was built across the Severn: it was 90 metres long, 7.30 metres wide and the arch was 30 metres high. The entire bridge was constructed with pieces of moulded cast iron. This increasingly diverse use of cast iron is ample evidence of its advantages over other metals.

As we mentioned earlier, the chemical industry was only just beginning to evolve at this time. and by the end of the 18th century the brilliant future that lay ahead was barely discernible.

The base product was soda, obtained from plants. Sulphuric acid (then called vitriolic acid) and hydrochloride were hardly used. The chemical industry at this time mainly consisted of making soap and vitriol.

A whole series of discoveries was to bring about major changes in this industry. The first of these was the discovery of chlorine by Scheele in 1774. Eleven years later in 1785, Berthollet established all the different properties of chlorine, and found that it was extremely useful for dyeing cloth – a process which, up until then had been extremely slow and laborious. From 1777, the Javel factory near Paris began the large scale production of chlorine, which itself required large quantities of hydrochloric acid.

The second problem was how to produce soda in greater quantities. For some time, attempts were made to produce artificial soda, but then in 1776, the *Académie des Sciences* in Paris offered a prize to the person who could find a process for extracting soda from calcium chloride. This process was finally discovered by Leblanc in 1790, but there were still some difficulties. Sulphuric acid was costly, and there was also a problem with by-products, especially hydrochloric acid. This was to be solved with the discoveries of chlorine and lime sulphide.

Sulphuric acid was produced from iron sulphate or by burning sulphur in a bell jar. In 1736, Ward attempted the same process using carboys, but the yield was no greater. In 1746, Roebuck developed the concept of lead chambers while trying to find a way of using a metal on which acid was ineffective. This process was introduced into France in 1774, by Holker the younger.

These completely new techniques had been practically applied then, by the eve of the French Revolution, but they were not so advanced that they could support a very extensive chemical industry.

The techniques used in the textile industry are a familiar subject to historians, largely because of the great deal of progress made in this field of industry. We shall therefore limit our discussion once more to simply providing a schematic framework.

The reasons for the renewed impetus of the iron and steel industry (the lack of fuel) are easy enough to understand, as is the development of the steam powered machine and the chemical industry. It is not so easy to explain the reasons behind the technical transformations of the textile industry. There had been no significant increase in the population figures, nor had there been any marked change in social or industrial sectors in the first half of the 18th century which would help explain the radical changes taking place. The problem remains to be answered: in this case technology does not appear to have been an induced phenomenon. Once the first discoveries had been made, the later changes occured quite naturally. The imbalances between the various stages of the manufacturing process provided the impetus for complementary inventions; from this point onwards, there was a kind of linking process which restored the broken equilibrium. A more or less definitive balance was finally achieved towards the end of the century.

It is therefore quite valid to present the main inventions in the following fashion:

I – John Kay's flying shuttle (1735) made it possible to weave larger fabrics. It meant that only one worker as opposed to two were required to work the loom, and this represented a considerable economy. The weaving process was made much faster by this invention, and the widespread application of this process caused an increase in the demand for thread.

II – The developments in spinning processes were made in successive stages, at relatively short intervals of a few years.

A) The date of the first invention, by Wyatt and Lewis Paul (around 1733), indicates that it was separate from John Kay's. Besides, this machine was not put to widespread use (fig. 17).

B) The next two great inventions occurred in quick succession: the spinning jenny was invented in 1765 by Hargreaves, and Arkwright invented the water frame in 1767 (fig. 18). Mantoux noted that the first represented a transition from manual labour to mechanisation. It was in effect a multiple and partially mechanised spinning wheel – a simple and fairly uncostly machine. The water frame, which was conceived at around the same time, produced a much stronger and more resistant cotton, solving the problem of mixed fabrics.

C) Crompton's mule jenny (1777) was a hybrid machine, developed on the principles of the two preceding inventions. It produced a strong, fine thread, and furthermore, it was completely mechanised. Around 1783, more sophisticated machines were being built with cog wheels and metal cylinders: by 1790, they were able to support up to four hundred spindles.

III – Once more, the process had become imbalanced – spinning had developed more rapidly than weaving. Mechanical weaving had long been a major preoccupation of inventors. The first machine of this kind, devised by Cartwright in 1785 was difficult to work but successive improvements solved most of the problems. And once this was

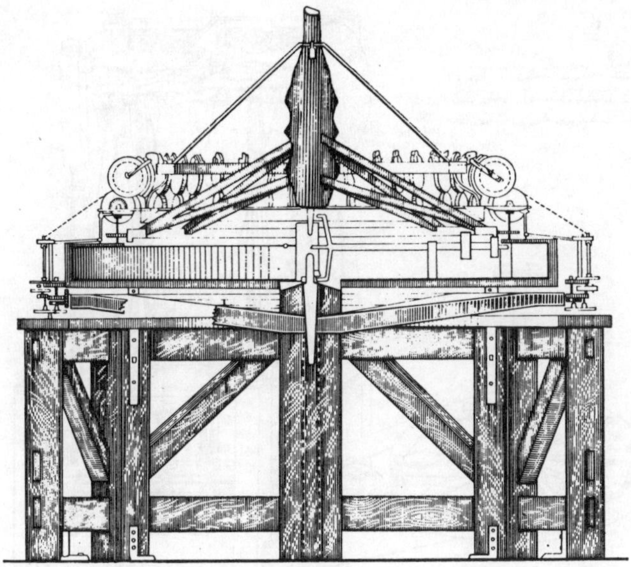

Figure 17. — Threading machine by Paul and Wyatt.

achieved, an equilibrium was attained once more (fig. 19). From 1789, the steam-powered machine was used during all stages of textile manufacture.

Such was the technical revolution of the textile industry. These machines were first conceived for use with cotton, but they could easily be adapted for use with wool. Other inventions completed the arsenal of textile machines: the carding machine was devised by Bourn in 1748, and fitted with two spools by Arkwright in 1775. Despite the fact that a machine was devised for combing in 1792, this process was much more difficult to develop. We have already mentioned the machines using punch cards invented by Bouchon and Falcon for silk working, and these were developed by Vaucanson in 1775 (fig. 20). Vaucanson also devised a machine for making taffetas by 1744. It was only in the following century that all these machines were incorporated into a single one (fig. 21). It was Bell, from Scotland, who invented the first process of mechanical textile-printing (1783).

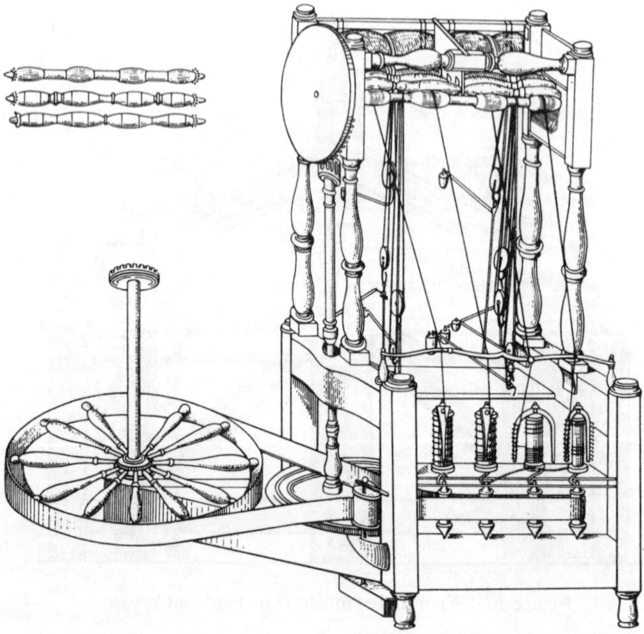

Figure 18. — Threading machine by Arkwright.

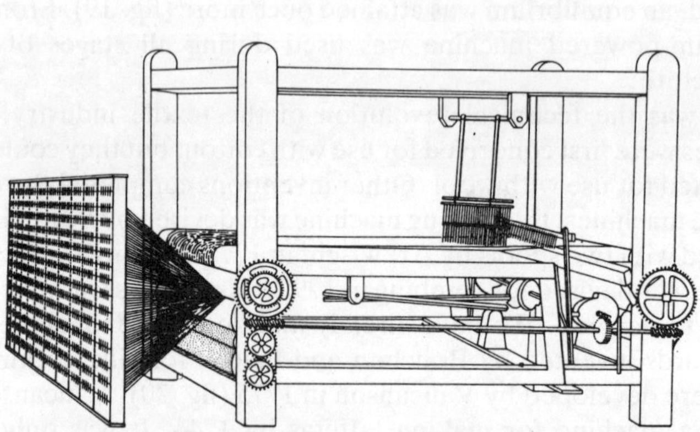

Figure 19. — Threading machine by Cartwright.

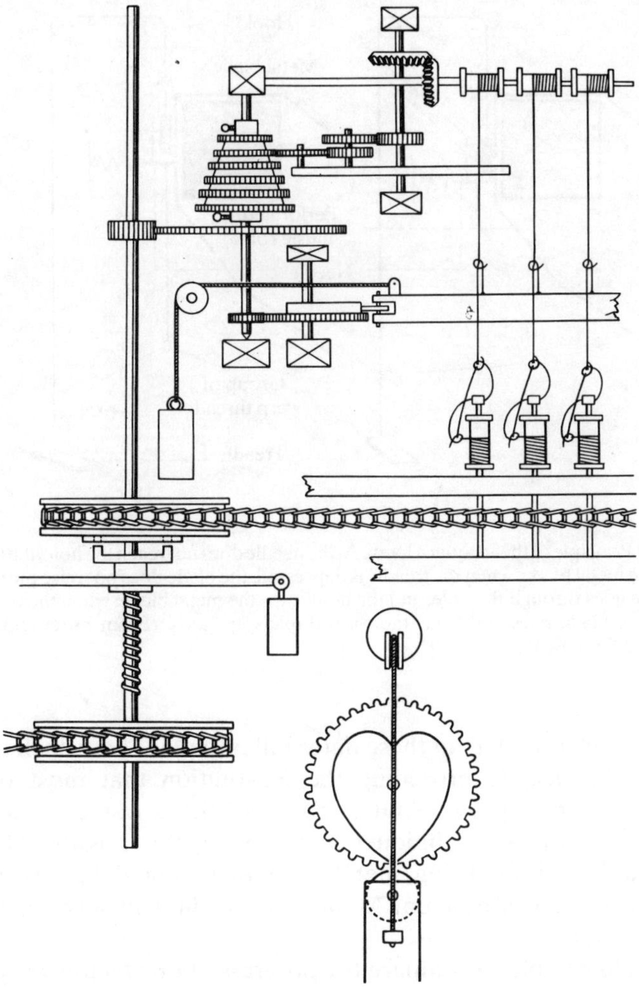

Figure 20. — Diagrams representing the command devices of silk mills by Vaucanson.

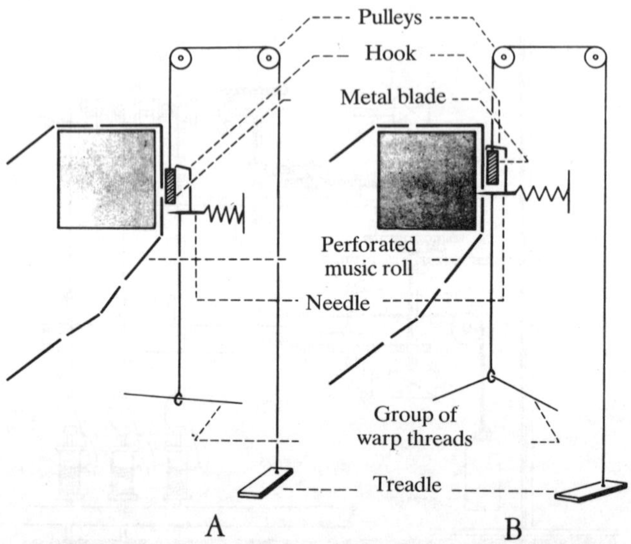

Figure 21.—Principle of the Jacquard loom. A, the needle does not reach the hole, the hook is not above the metal blade: when the treadle is depressed, the blade does not raise the thread; B, the needle goes through the hole, and the hook caps the metal blade: when the treadle is depressed, the blade rises and leads the warp threads upwards. (From Furia and Serre, *Techniques et Sociétés*, Paris, 1970)

It appears clearly that all these dates fall into place with each other. It was in the decade preceding the Revolution that most of the developments took place. And as we saw, there was considerable interaction of diverse techniques which made this possible: bigger looms could be built using metal, and these could be driven by machines powered with steam. At this point we find ourselves in a new technical system.

We should be able to compare the progress of one technical system with the preceding one, but unfortunately this is made difficult by the lack of precise figures. Furthermore, there were usually benefits to be found further afield, beyond the simple increase in production. There is no better example of this than the steam powered machine. It was more powerful, certainly, but there were other advantages besides.

The energy supplied by such machines was constant, it was not subject to natural conditions, nor was it confined to certain geographical areas. It is not easy to integrate these notions into a purely mathematical model.

Of course, we do have some numerical data, but are these valid in terms of production as a whole? And insofar as progress was slow and gradual, there must have been occasions when changes were simply matters of substitution. If, for example, a small number of modern factories replace a large number of traditional ones, it is quite likely that there will be no marked change in the figures of global production. The new technical system modified structures – it did not change production as a whole. Now that we have stated our reservations, we can provide a few significant examples. After it had been in use for some time, the coke-burning blast furnace was capable of up to three times the yield of the wood-burning blast furnace. The capacity of the puddling furnace was fifteen times that of the old refining processes. In 1787, the Crawshay, from Cyforthfa, had difficulty in producing 500 tons of iron bars – by 1812, this output had increased to 10,000 tons. The figures are even more striking in the textile industry, where progress had been even more marked. One small spinning jenny did the work of six to eight workers using traditional instruments. The growth in productivity was even more significant once steam-powered machines were used, which meant that machines could be built to unlimited sizes.

The period from 1785 to 1790 was quite clearly one of industrial revolution (fig. 22). From the technical point of view, the innovations had been developed to their fullest potential, and each technique had acquired an internal equilibrium. It was at this time that the first links between the various techniques became apparent. The steam-powered machine was in widespread use in the mining industry, and in the iron and steel and textile industries. Iron was put to widespread use, for making rails and various types of machines, such as those used in the textile and construction industries. These radical changes only became really apparent some time after they had actually begun. As is to be expected, this transformation was initially more indicative than real. It would be useful to study the transformations of British industry and particularly its structures, in order to appreciate the rapidity with which the new technical system imposed itself. This transformation was particularly marked in certain areas, such as the iron and steel

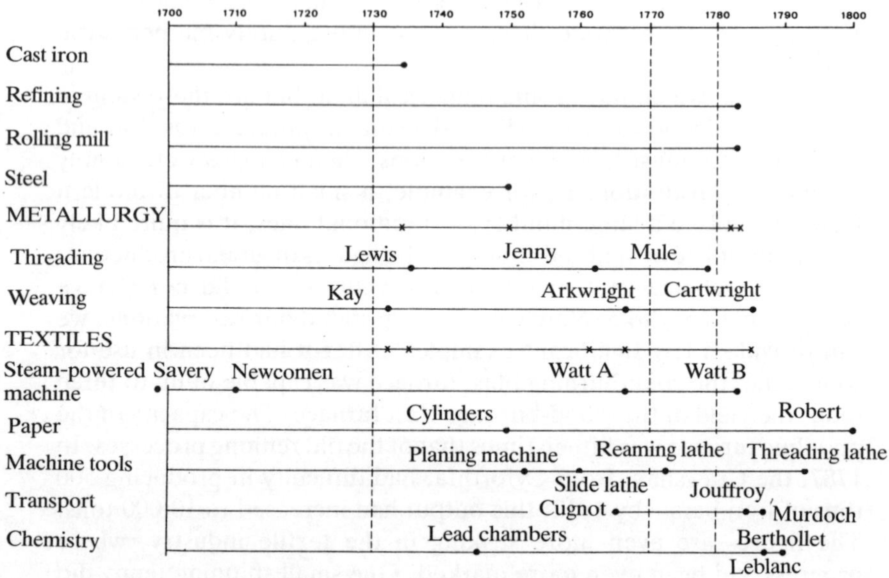

Figure 22. — Chronological scheme of technical inventions in the 18th century.

industry for example. It is true that between 1790 and 1815, other factors, of a political and economical order, were equally important. Most of the major imbalances had been corrected. However there were still several problems to be resolved.

DEVELOPMENT

The new technical system was in operation by the end of the 18th century, but had not reached its definitive state. It was still capable of development and further progress in some areas. The progress made was gradual and steady – there were no sudden transformations. This mutation continued well into the mid-19th century, albeit at a slower pace, now that most of the major elements were already present. It was not until after the Napoleonic Wars that the movement was well under way, however. Besides which it was necessary to complete the assimilation of the first set of innovations.

One of the main consequences of the industrial revolution of the 18th century was the recognition of technical and economic problems. This was a new phenomenon, and was clearly evident in certain areas. M. Courthéoux noted that the notion of real prices and its use in measuring technical progress was a familiar one to the first exponents of the classical school. Adam Smith opened *The Wealth of Nations* with a "book" relating to the "causes which perfected the productive faculties of labour." In his work *Principles of Population*, Malthus wrote at length on the "inventions which reduced the work load of the work force, and which were considered as a stimulant to the constant increase in wealth". But as we have already stated, these statements represent only a passing interest in technical problems: the latter would never be at the centre of a branch of thought whose central preoccupation is the mechanisms of market forces.

From the beginning of the 19th century, there was an ever-growing interest in machines. Jean-Baptiste Say, in his *Traité* and *Leçon* went from *Essais dans les Arts* (Essays on the arts) to *Service des machines* (Servicing machines), *Emploi des moteurs aveugles*, (The use of blind engines) and the *Etalonnage dans les manufactures*, (Standardisation of Manufactured Products). Smith was concerned with the conditions of an increasing productivity (accumulation of capital, acquisition of skills, use of machines, improvement of the land, division of labour), while Ricardo insisted on improvement of the land, as did Stuart Mill. But it was Say who attributed the notion of productivity with its technical characteristics. Curiously enough, remarks Courthéoux, it was his works, rather than those of the classical school, which were imbued with the spirit of the industrial revolution. "To obtain a greater productivity with the same amount of labour – that is the main objective of industry." From this stemmed his idea of technical progress contributing to the lowering of production costs.

The question of unemployment as a consequence of technology was considered at a very early date. One of the first people to deal with this idea, Sismondi, did so with vehemence, writing that the invention of any machine which assumed the task of a single worker amounted to a catastrophe. Bastiat opposed this, maintaining that the rather less obvious advantages – the saving of money, for example – outweighed the importance of the more obvious ones, such as removed or displaced labour. Say's position on this was somewhere between these two interpretations. He claimed that the appearance of a new machine was favourable to the working class, but it sometimes had "the rather serious consequence of changing the nature of working class occupations." Ultimately, however, although the introduction of expeditious machines did sometimes make some jobs redundant, "they do not reduce the chances of the working class to earn a living." Ricardo noted the changes in employment patterns during times of technical progress, and notably the movements from the primary sector to the secondary or tertiary sectors. He also tried to relate innovations vis-à-vis "net profit" and "gross profit": "Some innovations increase the net profit (i.e., the real profit), as perceived by the capitalists, even though they reduce the gross profit (i.e., the sum total of the products) "

At any rate, all the theoreticians were surprised by the fall in prices resulting from technical progress. This fall in prices was an indispensable element of the development of the consumer market, and the expansion of new machines and processes. General welfare was improved as a result, and the increased demand created a price war which inevitably brought about the diffusion of new techniques. Say was very interested in this particular aspect of technical progress, but the same was not true of Ricardo and the English classical economists who estimated that the developments were unable to check the price rise which followed the decrease in yields. The consumer may possibly be able to benefit from technical progress, but Ricardo and his compatriots believed that technical progress also caused a decrease in revenue and the erosion of capital. The latter observation clearly highlights the problems concerned with the limits of technical progress.

So it was the British industrial revolution which led economists to attempt to integrate technical issues into a global theory of economic movements and balances. There were theorists of other disciplines who noted the "immense progress" made in Britain, and who also began to tackle the problem from their own particular point of view.

The development of technology, which was still in embryonic form during the 18th century, and the relationship between science and techniques changed a great deal in the first part of the 19th century. The birth of a scientific technique necessitated the forging of links with other purely scientific disciplines – but the connections were still tenuous. The progress of the chemical industry, for example, is closely linked to that of science, but in the field of mechanics, technical progress often precedes scientific explanation. As we stated earlier, Carnot developed the theory of the heat engine only in 1824. Also noteworthy, however, are the efforts made at systematisation which inevitably brought about the use of scientific data. At the end of the 18th century, Beckmann, in spite of his promise to treat the subject scientifically, contented himself with simple empirical descriptions of several different trades which were linked neither to each other, or to scientific principles. In 1806, with his *Entwurf der allgemeinen Technologie*, he abandoned the descriptive divisions so characteristic of all the treatises and encylopedias of this period, and actually pursued a single principle theme, which enabled him to regroup all the operations used in different trades, and put them into separate classes. As stated by another author:

> It is the general view of technology which is changing: the use of technology is no longer confined to informing administrators and functionaries – non-specialists – about the operations of a trade by using simplified schemas. Technology is now capable of contributing to the development of the actual trades themselves: the listing and comparison of different methods used to achieve the same object made possible the transference of operations from one trade to another. The importance of this concept of technical transference cannot be over-emphasised: it makes it possible to go beyond the limits of a given trade, – in which classical technology was confined – and to establish links with other, very different occupations.

Hachette's *Traité élémentaire des machines* (1811), led directly to the *Cours de mécanique appliquée aux machines* (1836) by Poncelet. Mathematical calculation and notions of physics found practical applications and all the more so because the works of Monge had provided representations which made a certain amount of rationalisation possible. Thus the perfect image and global thinking substituted themselves to the plates featured in the *Encyclopédie*, and in the *Descriptions* of the *Académie des Sciences*.

But contrary to the introduction of the *Encyclopédie*, the artist's hand no longer does everything, and so it was possible to develop education at every level. One was more preoccupied with attaining the highest levels possible, the lower level naturally being linked to general primary education which was difficult to establish. The French example, which was more systematic, was copied widely.

The *Ecole Polytechnique* was the result of an idea of providing precisely scientific information fundamental to crafts and professions which were later taught in schools of instruction. The *Ecole Polytechnique* had been created by the *Convention*, at the same time moreover as the *Conservatoire des arts et métiers* which itself remained based upon the old principles of description and demonstration. Did not Vaucanson's collection of models, begun at the end of the 18th century, follow on from the plates in the *Encyclopédie*? Did not Diderot and d'Alembert prove that they too could make small scale models? And did not Watt himself work on and from models? This encompasses the whole difference: the model or the drawing, and the *Ecole Polytechnique* or the *Conservatoire*. The former was finding its boundaries with the major state corps, but its pupils very rapidly began to slip away towards the private sector. The need was felt, however, for the creation of an industrial school and it was in 1829 that the *Ecole centrale des arts et manufactures* appeared.

Such examples involve a number of other countries. Great Britain retained its empiricism, and Germany set up more polytechniques: at Karlsruhe (1825), Münich (1829), Dresden (1829), Stuttgart (1839) and in Hanova (1831). Austria created polytechniques in Prague (1806), and Vienna (1811). Almost everywhere, the image became reality.

At the end of the 18th century, La Rochefoucauld-Liancourt had created the first school for arts and crafts. This idea was taken up under the Empire at Compiègne, later at Châlons, to form what Napoleon had called non-commissioned officers of industry, in other words foremen. The Restoration was to increase the number of schools for arts and crafts which were in fact to become engineering schools, owing to the dearth of management for an industry which was definitively "taking-off." Louis-Philippe was to create one last school in Aix-en-Provence. Lastly, there were schools, originally on a lower level, which were restricted to the mining industry: the schools of Douai, Alès, or Saint Etienne really trained foremen. In this domain, other countries did not seem to adopt the French system, the higher levels of which they were content to copy.

Finally, education was introduced for the working masses. As technology was growing ever more complex, so the requirement for minimum training became essential. Almost all over the world the old practice of apprenticeship undoubtedly continued. But this did not in fact provide the "worker's hand," to use d'Alembert's expression. Henceforth, notions of calculation, the ability to know how to "read" a drawing, and certain mental processes were required. The problem was situated on two levels. It was necessary to found what today we call primary education. It is significant that we may observe from a survey done in 1834, that French textile manufacturers pointed out that one of the superior qualities of British industry was precisely the existence of a working class which possessed the foundations of elementary education. In France, the Guizot law of 1833 endeavoured – as yet imperfectly but it was to become more effective – to establish the foundations for primary education. Elsewhere this was not attempted until the second half of the century. Apart from this, it was necessary to replace the practice of apprenticeship which had never or no longer existed in more industrialized sectors. Typical in this respect were private associations in France and Germany designed to provide the working classes with the basic elements of calculation and drawing which would be useful to them. These associations enjoyed a marked success, indeed so much so in France that on the eve of the 1848 Revolution these "evening classes" had enrolled over 100,000 students. Like the famous school of La Martinière, in Lyons, these were to become one of the first professional establishments.

Training and information are closely linked. Therefore, it is not astonishing that technical literature saw considerable development at the beginning of the 19th century, and especially in countries which had yet to learn their apprenticeship from the English industrial revolution. Need we emphasize that we still lack the bare minimum of knowledge indispensable to assess this phenomenon? Hence, we may only make a few general statements on this subject.

Obviously, the use of earlier knowledge or experience was continued or resumed. Such was the case for the *Encyclopédie méthodique*, begun at the very end of the 18th century, the last volumes of which were published in 1834. Thus, because of the absence of more recent works, 18th century treatises in particular were frequently republished. The treatise on ship building by the Swede, Chapman, translated by Vial de Clerbois, was still being republished in Paris in 1839. Moreover, we should compare these old and more modern editions in order to observe the differences between them.

Many industries, however, provided material for more recent works, especially industries which had been more or less revolutionized by British inventions. We need only look at the techniques of metallurgy for proof of this: old works were discontinued and replaced by the works of the Frenchman, Hassenfratz (*la Siderotechnie* ..., 1812), and of Karsten (*System der Metallurgie* ..., Berlin, 1831), before the appearance of the major work written by the Englishman, Percy, which dates after 1850. It would be useful to trace this production, industry by industry, in order to analyse both the exact technical content and the general spirit of the age.

The taste for encyclopedias, even on the subject of techniques, had not died. In the first half of the 19th century, we still see a large number of dictionaries, some of which are very general, and others which are given over to technical data, such as the *Dictionnaire technologique, ou nouveau dictionnaire universel des arts et métiers*, published by "a gathering of scientists and artists" in Paris in 1835. Bessemer's only education was gained through reading one of these dictionaries, published in London in 1831. Again these require detailed research so that we may establish the exact nature of their content.

Lastly, one more factor remains to be considered, the importance of which has continued to grow – technical journals which appeared at the same period and which were a major contribution to the diffusion of established progress and to the preparation of future progress. The French Revolution gave birth to a number of these, some of which were even official publications: *Journal des Sciences, Arts et Métiers* (1792), the official *Journal des Arts et Manufactures* (1795-1797), or the remarkable journal published by a state organization, the *Journal des Mines*, the publication of which began in 1794. Some publications produced by certain associations mindful of technical progress should also be cited, such as the *Bulletin de la Société d'Encouragement pour l'Industrie nationale* (from 1801). To a certain extent, this movement, which was unquestionably born in France although its origins may be traced back to the 18th century, was imitated abroad. Among these important and regular publications was the *Polytechnisches Journal*, which first appeared in Berlin in 1820, and the *Mechanics Magazine* published in London in 1823. It would be too tedious to make a list here of all the titles of these publications. They served many roles, the most important of which was to provide information on techniques and processes which had been developed. A recent study on French ironmasters has revealed the enthusiasm with which the *Journal des Mines* was, received which enabled unfruitful attempts and many unnecessary expenses to be avoided. Aside from this, the comparison of industrial experiments, frequent in these periodicals and easier than in well written treatises, led to both indispensable adaptations and progressive developments, indeed even to the revival of practised techniques. This circulation of technical knowledge was to become the source of future progress.

The two actions, the formation of frameworks and the diffusion of technical knowledge, remain closely linked in this evolution, giving rise to a new type of man which the 18th century could only imagine. Whatever systems were adopted in different countries, the "engineer" was born out of this progress, not only from the progress completed in the last years of the Ancien Régime, but from all the developments following it. He is no longer tied to the state, as had been the case in the 18th century, but he knows all that is necessary for the direction of this new institution, be it a factory or a large mining business. He must also understand equally well the fabrication procedures of his specialist product and the workings of machines or the construction of the necessary buildings. Although he is not polyvalent, as some authorities would have us believe, he is at least multivalent. It would be extremely interesting to know exactly how these men and their careers were moulded. French schools have been thoroughly studied. In Britain, some businesses were breeding grounds for engineers, and herein lies the strength of the country: the firm which produced the gifted engineer, Maudslay, the "Maudslay Nursery," as one contemporary author aptly calls it, represents this type of British factory where technical progress was achieved. Invention and innovation were at the heart of business in Britain, while this was not always the case on the Continent where attempts and achievements took longer to establish.

Technological developments required other considerations which only saw the light of day in the first half of the 19th century. Let us look again at our simple outline. The new technical system had virtually been formed at the end of the 18th century. It was only applicable, however, in modified economic structures. Requirements for this are complementary to each other: mass production was henceforth possible in a great many industries, but mass production must take place at the heart of more and more powerful, but less and less numerous centres of production. We will return to this factor later. This concentration of production centres must also be matched by a parallel concentration of capital. Thus, on the one hand, the small family business disappears, while on the other hand, more and more abundant capital is assembled. Business techniques as such, and financial techniques must consequently be adapted to these new circumstances. The engineer, businessman, and entrepreneur similarly change, as does the legal framework within which his action develops. Changes also affect the structures of the circulation of money. The promulgation, particularly on the Continent, of the 1807 French code of commerce, should be pointed out, as this created the sort of industrial or commercial societies whose capacity was to advise – and here we should also stress the rapid and enormous expansion of transferable assets – and which stabilized social capital without hindering the mobility of fortunes. We should, however, clarify that the late development of some structures, such as banking structures for example, did influence the problems of adopting a new technical system, for the state was no longer an effective support for new businesses which it had been during the 18th century. The state could therefore only deal with infrastructures indispensable to economic development.

Concordances of a demographic nature are less clear cut. Although the populations of more advanced countries continued to grow, birth rates were to reach their peak around the years 1820-1830, whereupon they began to decrease. If these elements are carefully examined, we realize, particularly in Britain, that industries compensated for this less dynamic demography by considerably increasing exports. This consumption certainly continued to grow in industrialized countries, all the more so as populations aged, but the development of exterior outlets is another dimension altogether.

Economic growth has now been the subject of so many works that we need hardly labour the point. The intervals which may be observed, however, are most significant. Let us take the example of the production of cast iron (fig. 23). The following figures will illustrate the considerable differences that exist: between 1800 and 1850, the production of cast iron in millions of tonnes gives the following results: Britain, from 600 to 2249; France, from 200 to 406; Germany, from 70 to 144; Belgium, from 50 to 144. To put it another way, Britain almost quadrupled its production, Belgium tripled it, and the other two countries merely doubled theirs. But are gross figures indicative of the situation? Do they have any significance whatsoever? Was this cast iron produced from coke or wood? And above all, for our purposes, what is the significance of the new techniques? It is evident that if

Britain enjoyed this considerable increase in production, it must have been due to the application of new techniques. The situation is completely different in other countries and old traditional industry had to make an additional effort. As far as we can judge for the latter and for this industry, the initial increase was not necessarily the result of adopting the new technical system. The natural conditions and the natural adaptation of these new techniques was to take some time. In other words, growth, at least at the beginning, was the product of enormous tension within the old technical system. The few studies we have on the subject of cost prices prove this conclusively. The Continent had adopted puddling more easily than the production of cast iron using coke (fig.24).

We should therefore study, in what we call development, two series of distinct actions. Some were adaptation to production conditions, other than those which existed in countries which initiated new techniques. Within this arises an important problem to which we will return later. Secondly, we must also consider the repercussions of partial discoveries and the necessity of re-establishing wholesale balance. This will be our first consideration.

It was in the first half of the 19th century that the techniques of the industrial revolution of the 18th century were improved and took their potential form. Inevitably, improvements were most often implemented after the Napoleonic wars. Economically, this fact is easy to understand, as the return of peace, the end of the Continental system as well as the absorption of earlier innovations were necessary. This therefore explains, to a large extent, at least why the completion of technical innovation only came about between 1815 and the revolution of 1848, whereas technical research per se remained a continuing process.

We should note that the development of new techniques was henceforth not reserved to Britain alone. Now aware of technical problems and of industrial power and its strength, other countries acquired new techniques and more willingly participated in progress. As we have already pointed out, a certain enthusiasm was born and its inevitable counterparts were the movements of romanticism and love of nature. With the return of peace, every enlightened man became aware of the importance of Britain and the lead which this country had over the others. Here too, the reaction was very sudden and was manifested by the introduction of all sorts of customs' barriers to

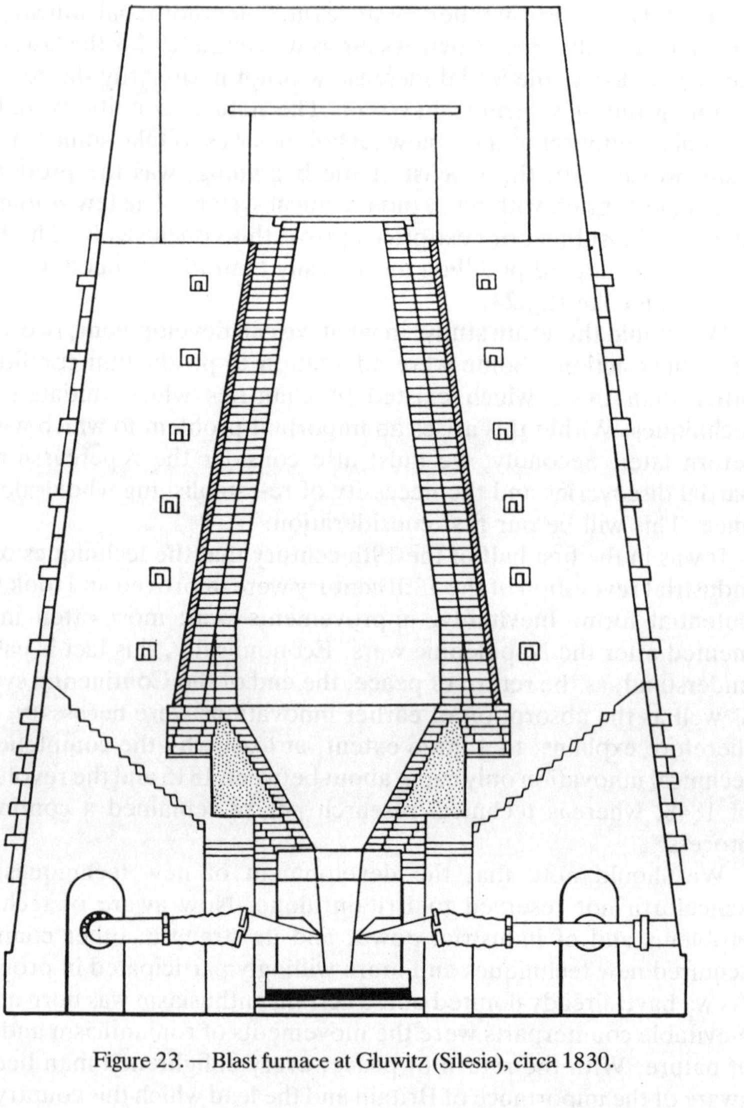

Figure 23. — Blast furnace at Gluwitz (Silesia), circa 1830.

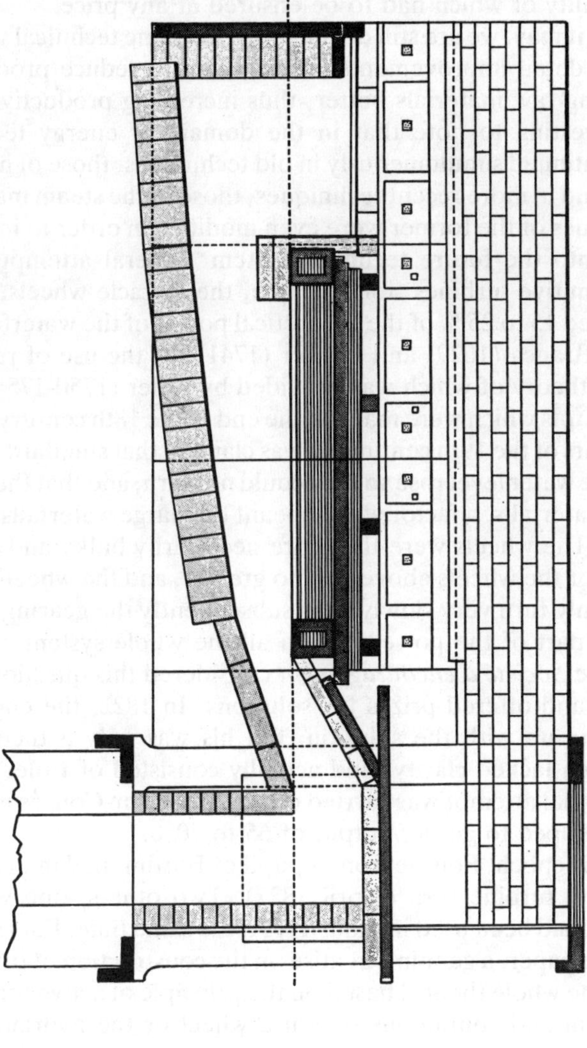

Figure 24. — Cross-section of Blatna's puddling furnace.

protect oneself from foreign competition, from a country which had already realized industrial change. In an economy based on profit, any technical transformation could only succeed with massive investments, the profitability of which had to be ensured at any price.

Be that as it may, we are still dealing with the same technical system. Essentially, detail improvements are required to reduce production costs by using raw materials better, thus increasing productivity.

It is interesting to note that in the domain of energy technical progress continued simultaneously in old techniques, those of hydraulic energy, and in more recent techniques, those of the steam machine. The techniques of the former were even modified in order to incorporate them into the future technical system. Several attempts were made in primitive turbines: for example, the Bazacle wheels, which only produced 15 to 25% of the theoretical power of the waterfall, the projects of Branca (1629) and Barker (1741) for the use of reaction wheels, the theory of which was provided by Euler (1750-1754), and several attempts which were made at the end of the 18th century and in the early years of the 19th century. It was claimed that standard wheels flooded if the water level rose so they could not turn, and that they used only small waterfalls, a factor which meant that large waterfalls had to be divided. The wheels were therefore necessarily bulky and heavy, the output for the wheels above was no greater, and the wheels at the side could only turn very slowly, and subsequently the gearing absorbed a large part of the power. All in all the whole system was not efficient. The *Société d'encouragement* considered this question in its programme and offered prizes for solutions. In 1822, the engineer, Burdin, came up with the solution, but his was only a theoretical concept which lacked clarity, and actually consisted of Euler's 1754 machine. A trial attempt was carried out at Ardes-sur-Couzes in 1828, where they hoped to get an output of 65 to 70%.

During this epoch, Fourneyron, a pupil of Burdin, had managed to make his first turbine turn (April 1827). Two others, one with 50 horsepower, had been used in industry. At the same time, Fourneyron had written a paper, a genuine treatise on the construction of turbines, containing the whole theory, based on the principle of active forces, of the "universal and continuous pressure wheel or the hydraulic turbine." Fourneyron had managed to distribute water without any pressure drop, thus retaining the potential energy to within a few centimetres of the tail race; the water was directed at a suitable angle into the receptor where it was stripped of all its energy in a very short time and the length of the distance was reduced to a minimum, reducing the speed absolutely necessary for its evacuation at the exit. The new turbine could thus turn in a flooded environment, unaffected by the rise in tail water and frost. This turbine, set up at Pont-sur-l'Ognon (Doubs), powered a rolling mill. It provided 6 horsepower under a waterfall of 1.40 metres. The relationship of theoretical effect to produced effect reached 80% in the first experiments, and 87% during later experiments. The two other turbines were used in the bellows in the blast furnaces at Dampierre and with 50 horsepower in the ironworks at Fraisans. Over 80 horsepower was the power achieved in a textile factory at Inval, and this rose to 220 horsepower in a spinning mill in Augsburg. In 1835, the facilities of two waterfalls measuring 108 metres and 114 metres in height were studied for the spinning mills situated at Saint Blaise, in the Black Forest: turbines were installed in 1837 and 1838.

This invention was significant insofar as it could compete with steam engines in regions which were situated too far from coalfields. In any case, it prolonged the use of water power, and in some countries, such as the United States of America, it enabled industrialization without having to recourse to coal.

The steam engine also saw some important developments which improved its working or output. Boilers were operated at high pressure in order to use the whole reduction in pressure, and the system of steam distribution or piston-lagging was improved. High pressure was adopted by Trevithick (1797), and then by Evans (1804). Arthur Woolf finally completed the high pressure double expansion engine, or the "compound" engine (1811) which enabled a fuel saving of around 50 %. It was not until 1830 that Seguin envisaged the tubular boiler with a large heating surface. In 1797, Edmund Cartwright first used a non frictious metal for the pistons. In 1816, Barton designed segments pressed against the surface by small springs. Distribution by four-way taps, and then by sidevalves (Murdock, 1799) simplified the distribution instruments. In 1836, Farcot was to patent the first distribution system which practically realized variable reduction in pressure by the regulator. It was the British engineer, Maudslay, who in 1807 managed to abolish the bulky balance wheel, the rod of the piston being guided vertically on two sliding channels, thanks to a wheel which was placed on the end. In 1815, Manby obtained the first patent for a machine with an oscillating cylinder. There are many such examples of patents which tried to find an improvement for the steam engine. Neither the principle, however, nor the basic details were changed, but it remained fundamentally the same as Watt's original engine: attempts had only succeeded in removing imperfections which its first realizers had already noted. Better adapted materials and some ingenious ideas had done the rest.

Progress was slow and partial in the domain of exploitation techniques: they remained on the same lines as those realized in the 18th century, without revolutionizing whatever improvements may have been made.

As for agriculture, it was mainly the tools themselves which benefited from the progress realized in neighbouring techniques. The use of iron rather than wood instruments, and the mecanistion of certain operations were the major new ideas. Mathieu de Dombasle at his Roville farm, and Bella at the school of Grignon, were concerned with perfecting ploughs and diversifying their types according to the tasks to be undertaken and the nature of the land to be worked. These tools were already made entirely of iron, In 1825, however, Fondeur finalized the swivel plough, which had a wooden beam and triangular ploughshares with wooden helicoidal mould boards, an ancestor of the machines used all over the world today. In 1837, the American, John Deere, introduced the first steel ploughshare.

Mechanisation appeared in the second half of the 18th century. This merely entailed the improvement of machines and making worthwhile practical tools. Around 1800, James Smith constructed an articulated ploughshare seeder which was called the "king of seeders." Throughout the first half of the 19th century, there were many types of seeders. Robert Meares and Taylor had inaugurated reseach into the modern harvester. The first major harvesters which worked suitably were those introduced by Mac Cormick (1831-1833) and Hussey (1833). In 1834, Harris Morre introduced the first combined harvester, the threshing-machine itself dating from the attempts made by Meikle in 1775. Similarly, the first plough powered by a steam engine and a cable was put to the test in 1834.

Progress was even more gradual and imperceptible in crop growing and soil improvement. One may note, however, considerable efforts made to protect arable lands but no worthwile results had been obtained as yet. Despite this, we should point out the success of the attempts made by Raclet, in 1828, in his battle against pyralis of the vine, by scalding the wood before the vegetation started to grow.

The success of some breeds ensured the development of animal domestication and breeding. These breeds were mostly created by the British. The Dishley breed, imported into France in 1833, was crossed with the merinos in 1837. From the famous Hubback bull, the Colling brothers had launched the Durham bovine breed in 1785, also imported throughout Europe (e.g. in 1825 in the French province of Nivernais). Pig breeding was similar: the Yorkshire breed arrived in France from 1819. Between 1815 and 1830, the French introduced two major breeds of draught horse, one from the Boulogne region for heavy draughts, and the Perch for lighter draughts.

All these efforts served to stabilize and perfect the agronomic system developed in the 18th century. Linked to agrarian and social changes, this agronomic system proved to be perfectly suitable for rich regions in temperate zones.

Progress was extremely restricted in mining techniques, and to some extent this progress was a lateral one. Humphrey Davy's invention, in 1816, of the metal cloth lamp – the safety lamp for firedamp – visible improvement of ventilation techniques, and the appearance in 1833 of the first hoisting cage, developed by T.Y. Hall are virtually the only developments which may be cited. In the middle of the century, metal linings were also introduced. The mining industry had to be content with profiting as best it could from neighbouring techniques.

More significant technological modifications were made in the iron and steel industry. They all had the same aims: to reduce the consumption of raw materials, particularly in the blast furnace, to increase productivity in other operations such as puddling or lamination, and to create new tools for use in ironworks.

The blast furnace undoubtedly retained the same structure which it had originally been given in the 15th century. It was, however, to receive two lateral developments which resulted in a considerable reduction of production costs. The first invention was that of introducing hot air into the furnace, developed by the Englishman, Nelson, in 1828. From 1830, air was blown into the furnace, raising the temperature to 315æoæ C. Naturally this involved taking several precautions, particularly in cooling the twyers. This procedure could perceptibly reduce fuel consumption. In the ironworks on the Clyde, coal consumption was reduced from eleven tonnes in 1811 to eight tonnes in 1828, and from four in 1831 to two and a half in 1832. The apparatus used was quickly developed for general employment. In 1833, the Frenchman, Cabrol, also succeeded in saving the same quantity of fuel with slightly different apparatuses. The second invention, apparently French in origin, was no less important. The recovery of throat gases provided a source of thermic energy which had previously been overlooked. This was used both to reheat the air blown into the furnaces and for steam engines: here again was another important fuel saving innovation.

At the same time, the apparatus itself, the blast furnace, attracted the attention of technicians. The form of the blast furnace had long been a matter of discussion from the 18th century, but no worthwhile conclusions had been arrived at yet. In any case everybody agreed that the shape of the furnace was important for the regular supply of loads and, subsequently, for the smooth running of the furnace. In a pamphlet, published in 1839, a Staffordshire ironmaster, John Gibbons, recognized the importance of the modern profile of the blast furnace: he believed that the gradient of the bosh could be reduced and that the bosh should be longer with a steeper incline.

The puddling furnace envisaged by Henry Cort was not yet a perfect instrument. In 1816-1818, the English engineer, Samuel Rogers, suggested that the hearth of the furnace should be lined, and a few years later the inner walls also received the same treatment. The drawbacks of the primitive furnace, which had necessitated prerefining, were therefore abolished. Weekly production in the furnaces increased from eight to twenty tonnes. In 1825-1832, Joseph Hall devised hot puddling, once again in order to reduce fuel consumption.

The rolling mill was gradually improved, but unfortunately its history remains vague. On the other hand, the power hammer was introduced in the first half of the 19th century in 1839-1841, the paternity of which is disputed between the Frenchman, Eugène Bourdon, and the Englishman, John Nasmyth (fig. 25). Other engineers had already considered this invention: among others, Watt and the French builder, Cavé. The essential purpose of the new tool was to enable large objects in the ironworks to be forged, particularly the steam-ship propeller shafts.

If we add to this list the progress of some complementary techniques, such as the fabrication of coke, and ore roasting, we notice that the iron and steel industry had too encouraged the almost complete

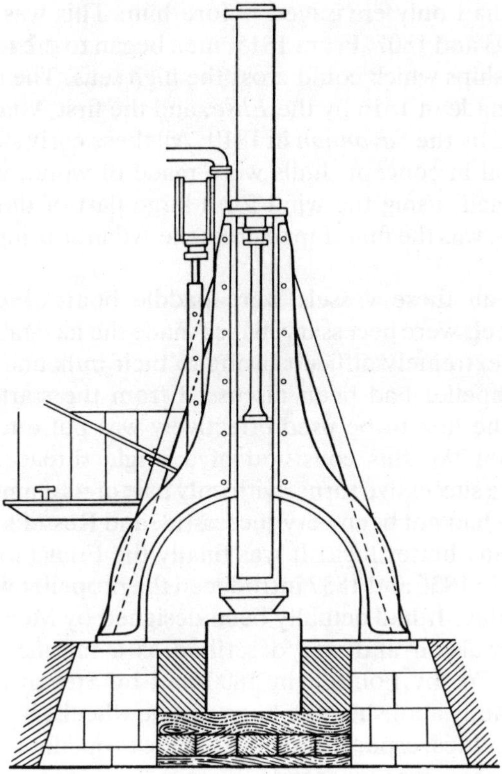

Figure 25. — The first power hammer by Bourdon.

and total development of its own techniques, following the lead exemplified by the British in the 18th century. From this resulted a very clear reduction in production costs and a growth in productivity. Only then iron- and steel-using techniques could in turn expand and enjoy every possible development.

Consequences of that situtation were undoubtedly most remarkable in the transport industry. Mere indications of this at the end of the 18th century were to become profound change in the first half of the 19th century.

The first steam-powered ships were river vessels. Indeed, Jouffroy d'Abbans was the first to launch the steam-driven boat in 1783, an idea which others had only envisaged before him. This was also true of Foulton in 1803 and 1807. From 1815, men began to dream of building steam-driven ships which could cross the high seas. The first Channel crossing was made in 1816 by the *Elise*, and the first Atlantic crossing was completed by the *Savannah* in 1819. All these early ships were still very traditional in concept: hulls were made of wood, and they still depended on sail, using the wind for a large part of the journey. In 1837, the *Sirius* was the first ship to cross the Atlantic using nothing but steam power.

Originally, all these vessels were paddle boats. We know that enormous wheels were necessary, which made the navigability of these constructions extremely difficult owing to their bulk and weight. The use of the propeller had been discussed from the start of the 18th century, but the first to be used effectively was put into practice by John Fitch in 1796: this consisted of a single thread screw which described three successive turns, with only part of its diameter submerged. Attempts had not been very successful, and Ressel's spiral (1812) did not have any better luck. It was finally the Frenchman, Frédéric Sauvage, who in 1836 and 1837 introduced the propeller with blades as we know it today. It had actually been designed by Meusnier in 1784 for aerial propulsion and was described as a marine propeller by Bushnell in 1776, by Foulton in 1800 and by Steven in 1804. The propeller certainly abolished the huge paddle wheels, but also presented other difficulties, especially in making the propeller shaft for which the power hammer was invented.

The last development, linked in part to the preceding one, was the iron hull. Here again, the first iron ships were river craft: the first appears to have been launched in 1777, on the river Foss, in Yorkshire. Barges were constructed later, part of the frame of which was still made of wood. The first steam-driven boat was the *Aaron Manby*, which took the name of its creator (1822). It was sent to France in kit form, where it was constructed for use on the Paris to Le Havre route. The second vessel, the *Commerce de Paris* (1823), built by the same man and for the same journey, was also supplied in kit form. In 1818, five iron steamers were in use on the Seine river. As Charles Dollfus notes, wooden boats could only carry out twenty journeys per year between Rouen and Le Havre, transporting a total of 3300 tonnes, whereas iron boats could make thirty-six to forty such trips, carrying 5000 tonnes. Britain and the United States were to quickly follow suit.

Maritime navigation was also to use iron. The first ship made completely of iron to sail the open seas was probably the *Iron Sides*, launched in 1838, and she successfully crossed the Atlantic. In 1839, the *Queen of the East*, a vessel weighing 2618 tonnes with an engine of 600 horsepower, established a service between London and Calcutta. The results of all these combined efforts was the "Great Britain," a propeller and steam-powered vessel, made wholly of iron, built by Brunnel in 1844. It still had sails, however. The sheet metal plating was twelve to fifteen millimetres thick, and the engine had a capacity of 1500 horsepower. It could travel at a speed of eleven knotts, and cross the Atlantic in fourteen days instead of the twenty-eight days taken by the *Savannah*. The hull of the *Great Britain* still survives today, evidence of the quality of the metal. The majority of these early propelled boats were constructed of iron, which is indicative itself of the twofold progress made.

The appearance of the first railways, and the perfect application of steam traction to land vehicles is so well known that we need hardly dwell on this for long. After the early attempts made by Trevithick in 1801, Stephenson built the first practical steam-powered locomotive in 1814, the final model of which was the *Rocket* of 1829. Shortly afterwards, the tubular boiler by Seguin contributed an important development. The earliest bogie locomotive was constructed in 1833. At the end of the period, the *North Star* haulled trains at a speed of 64 kilometres per hour.

The development of the railway industry presented many and varied problems: financial (the organisation of a large amount of capital), legal (the compulsary puchase of land), institutional (the establishing of monopolies) and more directly technical difficulties. Other than complementary techniques (and we are thinking here of the techniques involved in public works which had to be visibly improved for the *ex nihilo* creation of new tracks), the creation of railways necessitated extensive modifications in the iron and steel industry. The iron rail was infinitely superior in quality to the cast iron rail which was too brittle. It was therefore necessary to finalize lamination methods, with fluted cylinders, directly providing the large quantity of rails required with better adapted forms. Improvements were made both in the loads which had to be supported and in the speed of the trains. The track shape corresponded to that of the wheels of the wagons and the engines. From his first locomotive of 1814, Stephenson had devised flanged wheels, while some tracks, on the other hand, were built with flanged rails. Stephenson's idea prevailed. The problem of wheel resistance rapidly arose and it was only at the end of the period that the wheels were shrouded in order to avoid huge expenses, and a little later shrouding without welding was developed. The wagon brakes presented yet another problem.

Perhaps the importance of the notion of a technical system may be better judged at this point. More diverse techniques using extremely varied material or materials came into play in the railway. It was therefore necessary on the one hand, that all the different techniques were at the same level of development, and on the other, that continued improvements should visibly reduce costs. Similarly, one may easily imagine the pressure which the railway could exert on techniques "upstream," in particular on iron and steel making and mechanical techniques.

The widespread use of metal, and the considerable increase in demand for this material was to compel parallel changes in the domain of machine tools. Certainly, the main machine tools had been in use before 1800 (lathe tools or planing machines, drills, reamers, taps and milling cutters). Even so, certain important achievements had been introduced, although this was just a modest beginning. Machine builders endeavoured,more than ever, on the one hand, to ensure the solidity of their equipment by generally using metal, cast iron or iron, and on the other, to transfer the skill from the worker to the machine. The British were real pioneers in this domain: Henry Maudslay, Roberts and Sharpe, Nasmyth, Whitworth and Fox were to give the machine tool its real modern concept. We should also mention Calla, Decoster and Cavé, the Frenchmen who also worked towards this aim.

Carriage lathes were, moreover, well known. These were to be replaced by parallel lathes (Roberts in 1817 for the lathe used for planing), the feed-rod (Fox), and the separation of the chuck and the sliding poppet. When progress in automatic functioning was achieved and adopted in the screw-cutting slide lathe by Whitworth (1835), the worker became a mere supervisor (fig. 26). In 1839, Bodmer patented the vertical turning lathe (the circular planing machine). Whitworth's lathe for wooden screws (1835) exploited the mechanisms which resulted in greater automation: the feed-rod, the clamping by pliers controlled by a conical socket, and the camshaft. Vertical boring machines were also designed (1840). Planing machines by Fox (1814),

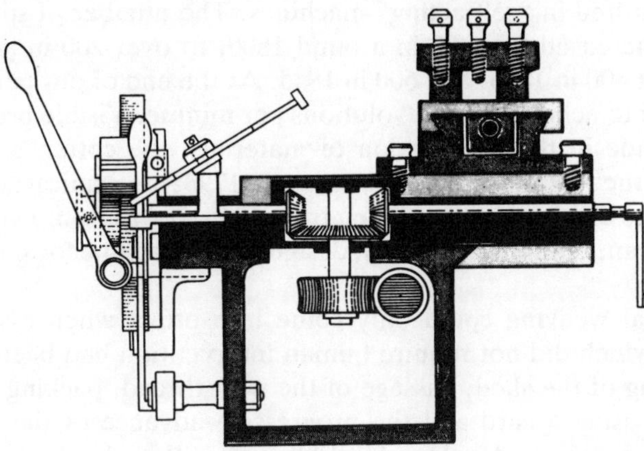

Figure 26. — Whitworth's lathe (1835). (From R.S. Woodbury, *Studies in the History of Machine tools*, Cambridge, M.I.T., 1972)

Roberts (1817), Clement (1820) and Whitworth (1835) mark the stages in this rapid progress.

We should also make a systematic study of the evolution of the machine tool. Some research has been done or is in progress, but no well-expounded work has yet been attempted with regard to this. Two points, however, must be emphasized. The first concerns the actual structure of the machine: improvements in gears, perfection in the method of producing the gears (machines for making gears were developed by Pihet in 1827, Whitworth in 1835, and Decoster in 1843), and the improvement of kinetic transmissions (here we should recall Galle's patent in 1829 for his articulated chain). The second point, which has already been stressed, is the pressure exerted by other techniques. The development of the railways and ship building should also be cited: wheel-making lathes, riveting machines and machines for curving metal were developed to fulfil the requirements of these industries.

Progress in the textile industry had been extensive in the 18th century but much remained to be done. The 18th century had developed not perfect but at least highly advanced machines for spinning. The use of metal, the perfection of mechanisms and locking systems, the increase in the number of bobbins, and the greater perfection of products resulted in "self acting" machines. The number of spindles per frame increased from 80 in around 1800, to over 200 in around 1825, almost 400 in 1835, and 600 in 1845. At the end of our period it was possible to achieve 6000 revolutions per minute. Visible progress was also made in the preparation of materials: the cotton gin was invented by the American, Elias Whitney in 1793; mechanical gearing for long fibres by Mac Carthy, dating from 1845. It was from 1810 that the Frenchman, Philippe de Girard, designed the machine for spinning flax.

Mechanical weaving could only come into effect when essential operations which did not require human intervention had been realized: opening of the shed, passage of the weft thread, packing down the threads using a card and the progressive advance of the warp. Cartwright's machine, developed in 1785, was still made of wood and as yet far from perfect. The use of metal and the perfection of mechanisms was the work of the first half of the 19th century: Horrocks, Roberts and Hattersley in Britain, Heilmann, Koechlin and Magnan in France, and Schonherr in Germany were the principle authors of this work. Control mechanisms were gradually introduced and established. In 1796, Miller had invented the catch which blocked the shutter if the shuttle broke down in the middle of a row. Heavy, broad looms appeared between 1823 and 1845. But evolution was still slow due to all sorts of resistances to progress, which we will discuss later. In the silk making industry, Jacquard synthesized the looms of Falcon and Vaucanson: he introduced the mechanical operation of a series of cardboard rectangles which had been devised by Falcon, almost a century before. Dressings also received important improvements. Mechanical fulling dates from 1838. Shearing machines were first built by Samuel G. Dorr, of Albany, in 1792. R. Dorr used helicoidal blades in 1807, and in 1819, Ellis developed the mechanical shears.

This constant progress virtually led textile techniques to their temporary completion. The great boost which the British had given this industry in the 18th century therefore continued over almost a century.

The chemical industry could only guess at what its future might hold at the end of the 18th century. But during the first half of the 19th century this industry was to be really established. The industrial development of the manufacture of artifical soda took place between 1810 and 1820. At this point sulphuric acid was to find a veritable opening: lead chambers considerably increased in size. The combustion of sulphur needed to be improved, however, by using a carriage which could easily be reloaded. Sulphuric anhydride was absorbed by water placed in layers on the ground. This water was then trickled down the walls of the chamber, and lastly, between 1820 and 1825, the water was pulverized by a slope situated above. Successive chambers were developed, and in order to reduce cost prices, attempts were made to improve the copper pyrites: the inventions of the French chemist, Perret, rapidly spread throughout the chemical industry.

Following this, Leblanc's process was also perfected by using a dual reverberatory furnace: in one, the chloride became sulphate and in the other, it became sulphate carbonate. The arrangement of the kilns and methods of soda leaching at the exit were gradually modified. Similarly, when the production of soda visibly increased, it was necessary to collect the hydrauchloric gases. In 1830, the Englishman, William Gossage, tried to pass this gas through a coke-filled tower in which water streamed. The solution provided thus enabled the preparation of chlorine which the textile industry required in ever greater quantities.

At the very end of the 18th century, the process of distilling coal in order to obtain an inflammable gas was developed. Between 1790 and 1800, William Murdock in England and Philippe Lebon in France succeeded in collecting and burning the distillation gases, giving a flame for lighting. From 1802, this gas provided the necessary light for the Soho workshops of Boulton and Watt, and from 1805, lit a spinning mill in Manchester. The first development company was created in 1812. Gradual improvements were necessary in order to arrive, around the years 1820-1830, at the more or less definitive apparatus required for distillation and storage. They quickly learnt to use the by-products of the distillation process, in particular the various forms of tar.

These are the specific circumstances which led to the appearance of new techniques in the domain of foodstuffs. Both war and the Continental system had dried up the sources of colonial sugar. At the end of the 18th century, the German, Achard, had described the process of extracting sugar from beet. After unsuccessful attempts to manufacture grape sugar, the Frenchman, Derosne studied the method of extracting sugar from beet, from 1806 onwards and, in 1812, the first sugar refinery at Passy was built. From this was born the crushing machine and the hydraulic press which Périer introduced in France. Purification was first done using sulphuric acid, and later lime, until 1811 when Pierre Figuier discovered the absorbing and decolorizing properties of bone black. Evaporation rapidly required the mechanical processes of stiring and pumping. Throughout the first half of the 19th century, modifications were to constantly improve the production of sugar beet.

Long voyages as well as the resupplying of great armies in the field presented the problem of resupplying provisions. This resulted in the formation of the canning industry which presented a dual problem: the preparation of the contents on the one hand, and the discovery of a container which was less fragile than glass on the other. We know that it was the jam maker, Nicolas Appert, who, under the Empire, used pressure-sealing to destroy fermentation germs. Around 1812, the English employed soldered tin cans which resolved the second problem. Tinned food was born, but only for a limited number of foodstuffs.

Total coherence of the new technical system was therefore established, which, around 1850, seemed to have developed as far as it was able. The analysis of the evolution of many other techniques, which would require many more pages, confirms this conclusively. This new system of techniques was based on three essential elements:

A) The generalized use of metal enabled evolution in a great number of techniques: a real era of mechanisation could be developed, and transport techniques were completely revolutionized. Iron began to invade construction work: after the structure of the Bordeaux theatre, the *Pont des Arts*, and the dome of the corn exchange in Paris, a new era was inaugurated, further testified by the Polonceau farm. Henceforth, metal was the universal material, dethroning wood once and for all.

B) The steam engine liberated energy production from the constraints of nature and also became a universal energy itself due to the advance of mechanisation. We have already discussed the benefits which were now available with the mobile engine. Despite some drawbacks, those of weight and bulk, it was successfully applied to both industry and transports.

C) Lastly, coal provided the link between the two preceding elements. It was adapted for use in the iron and steel industry and in all the operations involved in thermic energy, as well as being the ideal fuel for the steam engine.

If we add to this the benefits contributed by the chemical industry which offered a multitude of new products, we may estimate the extent to which the balance of the world had changed. Even within more traditional sectors, such as agriculture, techniques underwent transformations which raised them to the level of all other techniques.

CONSEQUENCES OF THE TECHNICAL REVOLUTION

Apart from a few limited sectors, the technical history of this technical revolution is today complete, but perhaps less profoundly in its developments in the first half of the 19th century where there are still many obscure areas. But other aspects appear to have been neglected to a greater or lesser degree, the emphasis having been placed mainly on the links between science and techniques, and on the obvious relationship between technical change and the beginnings of prolonged economic growth. In many domains, however, research remains very vague.

The reader must forgive a little disorder in this account: studies on this subject are on the whole still too inadequate to permit any strict synthesis. In many cases there are only impressions which are more or less justified, but there are too many gaps to enable a perfect account.

The consequences of technical progress are of two orders: quantitative and qualitative. The latter has undoubtedly been emphasized, although the former is equally important. Difference of quality presents itself under two points of view. Firstly, the product is not the same in any of its properties as the product produced using old techniques. The respective value of iron smelting with wood and iron smelting with coal has been the subject of long tracts. At the time of the 1829 survey on iron types in France, the Englishman, Wilson, an associate of Manby, both of whom had become directors of Le

Creusot, held the common opinion that iron produced using wood was better. And at the time of the 1860 survey on the commercial treaty with England, the great industrialist, Talabot, still claimed this to be so. This was one of the "brakes" to the expansion of certain new techniques.

Aside from this, new techniques could still provide products which the old methods of fabrication could not. The idea of making iron rails using classical iron and steel making techniques was unthinkable. New techniques therefore offered to the public a wide range of products which had until then remained unknown, be they the process of adding to old techniques, such as finer numbers in spinning, or enabling small or great substitutions. Unfortunately, we are in an area where research is very poor.

The increase in quantities, on the other hand, is better felt if not better known. Among the most important consequences of technical progress was the growth of productivity and the consequences arising from this, the most significant of which is the reduction in production costs, and therefore prices in certain conditions. Calculations, moreover incomplete ones, have been made for textile techniques. While the "water frame" and the "jenny" may have been inferior in output at the beginning compared to the spinning wheel, the possibility of operating a machine, on the other hand, with twenty-five spindles and only one spinner, increased productivity eight or tenfold (figs. 27 and 28). The improvement of machines and their perfection was to increase these figures one hundredfold. According to the figures we have for other industries, while progress was not as spectacular, it was at least considerable. Mechanical carding represented five times the productivity of carding by hand. Cartwright's carding machine increased the productivity of one worker from 5.33 to 133.33 pounds per day.

According to the report by Héron de Villefosse on the French iron and steel industry, a blast furnace using wood produced on average 416 tonnes of cast iron per year, and a blast furnace using coke produced 1325 tonnes, with a far lower fuel consumption which was to be further reduced still the adoption of hot air and the recovery of throat gases. We have observed that output with puddling was fifteen times higher than that produced by old refining processes.

Increase in productivity, and the frequent use of raw materials obtained at a better price or consumed in smaller quantities, was to

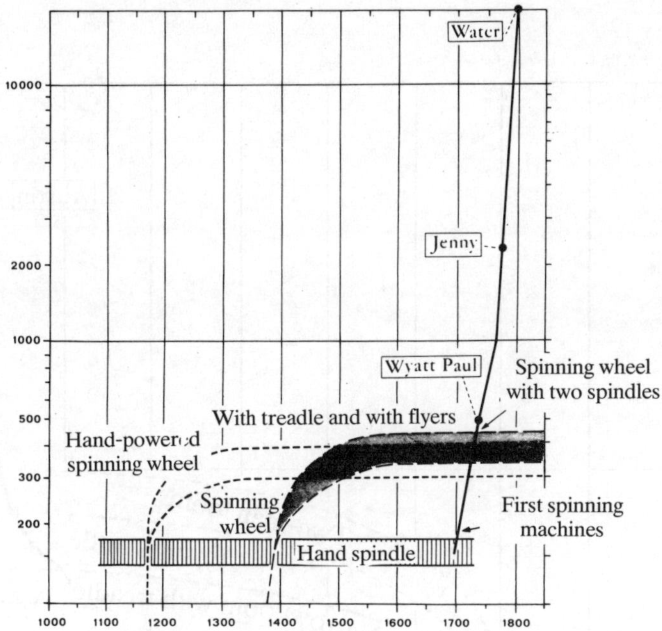

Figure 27. — Change in spinning yields. Production expressed in hour-metres per person. (From Endrei, *L'Evolution des techniques du filage et du tissage*, Paris, 1968)

contribute to the reduction of production costs. The price of thread number 100 was still 36 shillings in 1786, falling to 10 shillings in 1795, and to 6.9 in 1807. In France, the thread number 3 which was worth 12.60 francs per kilo in 1816 fell to 6.40 in 1824, 5.60 in 1834, and 3.60 in 1844. A survey made in 1834 gives the following figures for France:

Finished products	1816	1834
Calicos....................	2.60	0.70
Percalines.................	4.60	0.65
Dimity.....................	2.60	0.75
Muslins....................	2.60	0.45

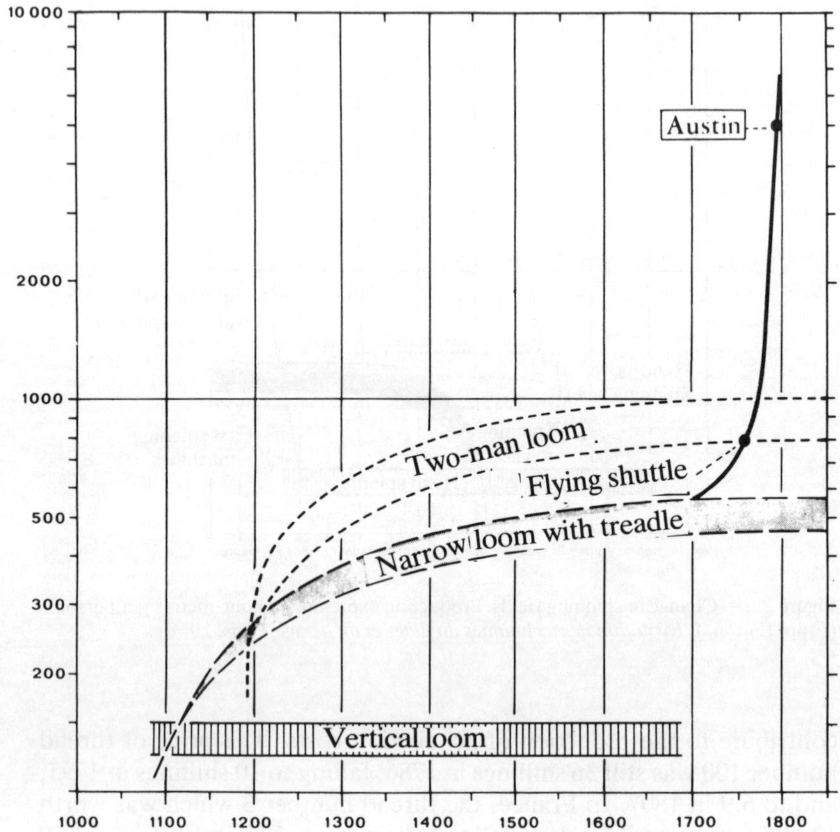

Figure 28. — Change in wearing yields. Production expressed in metres of weft inserted per hour per person. (From Endrei)

Confronted to underconsumption, which was widespread at the beginning, these new prices ensured extensive outlets, more important perhaps than those resulting from a population growth or the expansion of outside trade. Whatever the cause, all the sales added up.

Particular circumstances may arise which may ensure the manufacturer and the producer very low cost prices compared to those where old techniques were used, but may also limit the fall in market prices. This might have been the situation described by Turgot, likewise by Ricardo, of an entrepreneur finding himself with the monopoly on a new technique. This could also be true of a country sheltered from foreign competition, in fact British competition, and aided by protective, indeed prohibitionist rates. These rates could only be calculated by comparative study of the cost price of old techniques: that is to say those who initiated innovations therefore had the opportunity – by aligning their selling prices with those of traditional entrepreneurs – of receiving considerable profit margins, and sources of self-financing which often necessitated a very strict capital market. In 1828, iron produced in the Champagne region was worth 280 francs per tonne, puddled and laminated iron was 200 francs, and British iron from Cardiff was worth 175 francs. Note further more that for limited markets such as that of rails (in France in 1842 there were only fourteen possible suppliers), customs' protection enabled the formation of policies specifically intended to maintain high prices. It is also true that throughout the first part of the 19th century, there was a constant and notable fall in prices in the French iron and steel industry. The price of a quintal, in francs was as follows: 1820: 46.0; 1830: 42.5; 1840: 34.7; 1850: 24.0.

In some industries, the growth of production and demand was to prompt consequences on the technical level itself. At this time when a much larger quantity of products was put on the market, it was necessary to standardize and normalize them. Here, technology itself required this process, and fabrication methods were only conceived with this aim in mind. It would have been unthinakble that different types of rails be used on the same line. The links between networks necessarily demanded a uniform type and a uniform gauge of track, which Britain had been quick to realize. The production of the rolling mill ensured the implementation of this uniformization at the very beginning.

Here, this uniformization was a necessity in itself, but it was a major contribution in facilitating the use of products and subsequently the sale of products. As André Garanger has noted, the consumption of nuts and bolts in the early days of mechanical construction was so low that each mechanic had to make whatever he needed. Maudslay was the first to research and establish standardization of threads between 1800 and 1810. In 1841, the "Withvoth thread" was adopted, in other words true standardization was imposed. Meanwhile in the United States, Elias Whitney studied the problem of mass production, having received an order for several tens of thousands of guns, and to solve the problem he constructed the first milling machine in 1818. At the same time, Blanchard invented a copying lathe to produce a large quantity of rifle butts. Normalization was thus imposed upon certain mass produced goods: but what was quite natural for nuts and bolts or the various parts of a gun was still not so for large machines. Furthermore, the comparatively frequent change in types, as was the case in the construction of locomotives, did not permit standardization.

What was born out of necessity in some cases, could become a new method of production in others, where standardization and normalization were not imposed. Glove and shoe making could only be individual work, as each product was destined for one particular person. Around 1835, the Grenoble glovemaker, Jouvin, and the Châteaurenaud shoemaker, Pinet, had the idea of reducing production to a particular number of patterns and sizes established after a survey: hence the birth of sizing. From this point on it was possible to devise mass fabrication and to realize the machines necessary for this (the first were machines for cutting several layers of material for gloves). From individual production, it was henceforth possible to develop mass production. At the very end of our period, the founder of the *Belle Jardinière*, a department store in Paris, tried the experiment on clothes and subsequently introduced what is now called the clothing industry. This was aided by the invention of the sewing machine in 1845 by the Frenchman, Thimonnier.

The increase in tool productivity and procedures, and the cost of necessary new plants inevitably led to important modifications in production structures: the large scale business was to replace small individual producers and family businesses. Although the notion of industrial concentration was not new in itself – and we have already seen that Colbert had established this with his privileged manufacturers who brought into play techniques which had formerly not been applied within the country – it had not yet been put into practice.

Henceforth, the manufacture of rails or the use of large textile machines could no more be the work of small enterprises.

This concentration had begun on a modest scale at the end of the 18th century, even in France where the technical revolution only infiltrated very gradually. In an account of 1785, out of 87 factories belonging to the cotton industry, 69 manufactured less than ten thousand products per year, but 11 produced ten to twenty thousand, 6 produced twenty to thirty thousand and only one produced over thirty thousand objects. At Abbeville during the same period, Van Robais employed more than one thousand, two hundred factory workers and ten thousand people working at home. The huge silk making operations in the provinces of the South-East, the paper mills in Essonnes made possible by the use of Dutch cylinders, and the Le Creusot furnaces, signal the transition from one type of business to another, conceived on a completely different scale. Saint Gobain had already announced semi-monopolies of businesses in the mirror making industry.

This movement was far more rapid in England, and from the first decades of the 19th century concentration was already considerable in the iron and steel, coal or textile industries, although in the latter this was less so in weaving than in spinning. There are many such individual examples, but no study of the whole has as yet been undertaken. The large enterprises of Boulton and Wilkinson may be cited, however, dating from before the French revolution, which were both concentrated and integrated, even becoming, as has already been pointed out, initiators of technical progress.

Some examples of concentration in France, where the process has been more thoroughly studied, prove the impact of technical progress on the evolution of structures. Although coal production was less technically advanced than other industries, it was, however, very rapidly concentrated. Around 1840, the ten leading businesses produced over half the total value of French coal production, exactly 55.5%. With fifteen businesses in operation, this percentage increased to 62.86%. The leading company, Anzin, alone produced 23.24%. In 1851, the two leading companies were responsible for almost half the production (48.8%), and the ten larger companies accumulated about 82% of the total production. In most instances the technical pumping problems were the cause of these successive concentrations.

Figures are less clear for the iron and steel industry, doubtless owing to the lack of sufficient information. In 1810, the ten major companies would have represented 43.6% of the turnover of the French iron industry, but this result should be considered with caution. According to more reliable data, in 1828 the ten leading companies represented 22.2% of the cast iron production, and 24.35% of the iron production. In approximately 1840, this percentage was to fall to 13.44% for production value. This phenomenon is more easy to explain than one would think. Customs' protection, poor geographical distribution of raw materials, and the lack of developed transports to some extent constituted a setback to technical innovation. The overall statistics for production demonstrates that faced with almost complete stagnation of iron produced using wood and the noticeable growth of iron produced using coal, cast iron continued to be produced for the most part using wood. The growth of certain large modern or semi-modern firms clearly illustrates the direction in which the balance swung (growth rate during the period 1830 and 1848):

Global production	60%
Wendel	350
Le Creusot	200
Fourchambault	178.3

We do not possess similar tables for other industries, especially for the textile industry. We may suggest, and we will return to the particular case of France later, that although large scale enterprise had not yet succeeded with new techniques, it was nevertheless on the way: it was only when all the techniques were at the same stage of development, particularly for those of transports, that the triumph of industrial concentration was to become evident.

It is remarkable that the appearance of these large production centres aroused some interest in the organisation of labour, which was to be termed industrial economy for a long time. Certainly, Adam Smith had praised the division of labour, but this was merely a purple passage and the manufacture of pins was probably more symbolic than real. It was in London in 1832 that Charles Babbage published the first work on the economics of machines and factories (*On the Economy of Machinery and Manufactures*). It was undoubtedly still ahead of its time, as despite some success this work still did not have a long lasting effect.

There is another significant aspect of the technical revolution, that of the transfers which are implemented in different sectors of economic life. Although the transfers implemented in the world of labour astounded both contemporaries and historians, it hardly referred to transfers put into effect in capital and incomes. Obviously, industries existed which previously operated if not without any capital at all, then at least with a limited amount of capital. Such was the case in the textile industry: virtually no machines were used in the spinning mills, and they were still of little value in the weaving industry. The introduction of better looms, and the necessity for large buildings to house a more concentrated industry, naturally presented the problem of capital. Again the amount of capital required was not enormous and could be put together by an individual, indeed by a family. Throughout the first part of the 19th century, the textile business was still largely a family business; companies with capital were very rare. Families of industrialists existed both in France and England: originally either craftsmen or even farmers, they gradually became distributors of raw materials and collectors of finished products, even renting the equipment required for production. These men were also to become the textile industrialists in the early years of the 19th century. By accumulating capital they could participate in the innovation of techniques. In France and many other areas, the secularization of the property of the clergy enabled industrialists to acquire large buildings (particularly abbies) very cheaply (which may be recalled in the case of Ourscamp, Royaumont and Senones: Cistercian abbies which had advanced hydraulic systems were perfectly suited for these uses).

The problem was rather different in the iron and steel industry. This transformation was progressive in England, so that old ironmasters could adapt their trade to technical change. On the Continent, however, this transformation was more sudden and more global because it took place much later. It was therefore necessary to rally together considerable amounts of capital. A large number of enterprises were subsequently formed in France (Société de Saint-Etienne, Alais, Decazeville); but the disposition of capital and the organisation of financial markets hardly facilitated these operations. We will see later that the distribution of raw materials, which had largely favoured technical change in England, were on the contrary in

France a considerable restraint on progress. Guy Thuillier has emphasized another difficulty in the province of Nivernais (but this may also be encountered elsewhere): most ironmasters were tenants and the relatively short length of leases was an obstacle to technological investments. The rigour of the capital structure naturally curbed any transfer.

We require more detailed studies in order to measure the strength of the links which existed between technical transformation and the development of capitalism, of the techniques of capitalism: the chemical industry, now in the process of being studied, provides a good example of an entirely new industry. Several years ago, a symposium organized on the beginnings of steam sailing demonstrated the existence of identical facts. The development of various techniques was to aggravate the problem to some extent: no single investment but repeated investments were made in order to adapt quickly to technical change. Here again, we suffer from the absence of precise studies. In 1834, a Parisian millowner, Griolet, observed this phenomenon: "In industry, when there is no progress there is regression." In the same year, Vayson, a carpet maker in Abbeville, had already changed his equipment twice since the beginning of the Restoration. Others, however, testified of the limits with which they collided. This was so in the case of the great industrialist from Lille, Mimerel: "We hesitate to replace the old machines with new ones, we repair our old equipment." Only customs' protection, enabling prices to be aligned on the highest rates imposed in a country, allowed repeated investments by self-financing. It is easy to comprehend the motives behind these protectionist politics, declared by some authorities to be provisionary, which from 1821 spread throughout the whole of Western Europe: they precisely ensured technological change by directing capital for investment towards new techniques at a time when Britain, which had mainly taken the lead, had almost exhausted her first industrial transformation.

Propensity to innovate and invest are naturally bound together. Even during this period, men were perfectly aware of this. One of the most clear-sighted men in this respect was the Parisian banker, Laffitte. From 1824-1825, he firmly upheld the two fundamental mechanisms by which he intended to achieve capital transfer. The first was reduction in the rate of public loans: by acting thus, loans were directed towards a more private rather than industrial province and more mobile capital was released for investments in new economic sectors. The second was the creation of the *Société commanditaire de l'industrie* (Limited company for industry), destined to favour the indispensable accumulation of capital. According to Laffitte this rate conversion was "to reduce interest in our provinces, contribute capital within them and with capital aid the provinces could then develop means of communication, enable the formation of industrial establishments and revive work." And he added significantly: "This penchant for work is naturally accompanied by another constant factor, that of innovating and improving, a no less favourable tendancy than the former towards industrial progress." One could hardly make it clearer. The object of the *Société commanditaire* was to "contribute and participate in the success of any enterprise, any invention and any improvement concerning agriculture, industry and trade." it was later specified that enterprises which solicited its support were to exploit either a new branch of industry or an "industry improved by new methods." In the first half of the 19th century, no other position as logical or any other programme as coherent as that of Laffitte's existed.

The problem of transfering revenue has not been the subject of any research, either as a whole or in part. Nevertheless, this question would seem to be significant. We will return later to the subject of moving manpower, which really is the transfer of salaries. Owing to the absence of any work on this subject, we will look at just two examples.

The textile industry was mainly a cottage industry. We have just said that at Abbeville, Van Robais employed 1200 workers in his factories and 10,000 employees working at home: most textile industries were based on the same system; at the end of the 18th century or at the beginning of the 19th, Ternaux or Richard-Lenoir found themselves in exactly the same situation. The mechanisation of textile operations and the concentration of labour removed a source of revenue from the country which, although was only a supplement to other resources, was nevertheless a means of income. This was also true of the concentration of silk making. While the 18th century had searched every possible avenue to provide rural communities with subsidiary income (as was the case in France for the expansion of the spinning wheel), technical progress tended to upset a balance which had been established with some difficulty. According to the proportion which this supplementary revenue represented in global resources, men could either follow or not follow machines.

The second example is provided by the 1829 survey on the iron and steel industry in France. A simple problem was presented and a more wide reaching and more complex question was evoked. The simple problem was quite clearly the loss which proprietors of woods and forests experienced or were to experience through the expansion of coal production, the iron and steel industry being their only outlet. And already the total loss was calculated at around thirty million francs. The second argument was more specious. By maintaining high prices, customs' protection imposed considerable charges upon consumers. Freedom of trade would have given a country much cheaper iron supplies. Moreover, the capital invested in all these modern enterprises, estimated to be between 70 and 94 million francs during this period, would have been better used in improving agriculture. It was even claimed that the French iron industry led to the increase of woods and forests to the detriment of arable lands.

The acquisition of new techniques by non innovatory countries was the subject of a very recent synposium. Presented in this way, the question may reveal two principal aspects. Firstly, one general fact may be observed: the introduction of new techniques, whether they imported ideas or not. We have already recognized the changes in structure which could encourage the adoption of a new technique. In the period which concerns us here, we have seen that the most important phenomenon was the concentration of industry which was henceforth to become the necessary corollary. We have already examined the financial consequences of this, and there are others which more particularly involve man.

The first and the best known factor is what is termed technological unemployment. From the beginning of the 19th century, the question arose as to whether technical progress actually suppressed labour or only induced transfers. Although today we are generally persuaded that the second proposition is the most likely, the first has often violently manifested itself from the 16th century onwards. Stories of inventors being threatened and machines being smashed have punctuated the history of technical progress and this behaviour may be found throughout every industrial sector: these facts have been highlighted often enough, so we need not debate this point any longer.

The transfer of the working class and the formation of these masses have not been studied in as much depth. Some research has demonstrated the difficulties which the French, Belgium or Austrian iron and steel industry experienced in assembling the manpower needed in large modern factories, and the fears generated by this concentration of workers, who were part of the "dangerous classes." Although it is wirth being underlined here, this problem is not part of our investigation.

The transmission of a technical system from a pioneering country to other countries is somewhat different, and we need only look at Britain in the 18th century for proof of this. The first objective was to acquire an understanding or knowledge of the new techniques. Certainly, many techniques were circulated throughout Europe which illustrates the technological advance enjoyed by England after 1815. In order to preserve this lead, the British tried to check the export of machines: they were quickly convinced of the fact that it was difficult, if not impossible, to block the spread of technical progress. A famous Parliamentary Report of 1825 provides plenty of information on this topic. Equally well known are the beginnings of what could be called industrial espionage, about which there are almost legendary accounts: the Frenchman, de Wendel, became a worker in a metallurgic factory in Wales, and Scrive, of Lille, smuggled out the plans of English textile machines by sewing them into his jacket; there are many other such examples. The distribution of technical journals quickly solved this problem.

Difficulties could vary in nature. While some industries merely depended upon knowledge of new procedures or equipment, there were others in which natural manufaturing conditions played an important role. A good example of this is provided by an account written by the French ironmaster, Rambourg, in 1815:

> Those who so easily propose the substitution of mineral coal with vegetable coal do not seem to comprehend that this entails changing virtually everything in furnaces, refineries, mechanics and workshops, and that one must be near the coalmines supplying the suitable type of coal, have the mineral within range of the fuel, and able to employ workers capable of carrying out this work. We would only be deluding ourselves by claiming that most of our ironworks rival the price of English iron. Favoured by nature for fuel and minerals which may be found in the same pit, they convert that into coke and, by using steam and rotary engines, they give their cylinders enormous power with the capacity to draw this former product into bars. Numerous channels are then used to receive these irons and to facilitate their transport to the sea. All these advantages, which we are far from possessing, are huge sources of economy for the English.

This explains why innovation was often partial in this sector of French industry. Although large factories, Saint Etienne, Alais or Decazeville, adopted global innovation, coke furnaces, puddling workshops and rolling mills, within which there were still serious problems to be solved, the French were content to build ironworks to British design, supplied with iron produced by smelting with wood: such were de Wendel's factories at Châtillon and Fourchambault. Of the thirty-one "English" ironworks which existed in France in 1828, only two belonged to factories which had adopted all the new technical ideas. In many cases the new transport techniques were necessary in order to reunite the two essential elements used in making iron with coke – coal and minerals -, at a suitable price (that is to say prices which were lower than those produced in the old metallurgy process which used wood). In the Austrian-Hungarian empire, the first blast furnace was only built at Vitkovice in 1836, when the railway supplied Styrian minerals to the coal basin at Ostrava.

Originally, it was also indispensable that the purchasers of new techniques could employ machines at the very moment when a national industry of mechanical construction took over. National production of textile machines seems to have been rapidly established: plans and designs rather than machines themselves were sought in Britain. In other industries, on the other hand, only the importation of British machines enabled the adoption of new processes, the construction of heavy mechanical equipment being far more difficult to put into practice due to the limited market. England therefore supplied France with equipment for ironworks, railway material and sea going machines up until 1835 and even later in other countries.

The last point involves men. From the 18th century onwards, entrepreneurs and workers, many of whom were British, contributed recently effected techniques to the Continent. After John Kay, the inventor of the flying shuttle summoned by the French government, Holker who became *inspecteur des manufactures*, and Wilkinson who participated in the founding of Le Creusot, other Britons went to France after 1815: Jackson contributed the techniques of cast steel to Saint Etienne, and Manby and Wislon created an "English" ironworks ard a workshop for mechanical constructions at Charenton and also took over the Le Creusot factory for several years. In Belgium, another Briton, Cockerill, founded the first modern iron factories and equipped the textile industry of Belgium and the Rhineland, indeed even Poland and Russia. British builders constructed railway lines almost everywhere. All this has been amply demonstrated.

Apart from entrepreneurs, workers also travelled abroad to set up machines and operate them until the local workforce could take over. Many British workers were transplanted abroad, particularly in heavy industries, metallurgy and mechanical construction. In 1828, in the factory of Fourchambault in the province of Nivernais, the substitution of French personel for British was virtually completed: only ten puddlers and ten laminators remained of the latter. "The French employees having less skill and less dexterity have cost us greater fuel consumption and an even greater waste of cast iron beyond that which we allowed them." At Charenton, where a factory had been founded by the British, all the workers were originally this nationality. At the time of the survey, half the puddlers were French, but almost all the laminators were still British "because the work required a dexterity which the French had great difficulty in achieving." The major industrialist, Boigues, summed up the opinion of the French ironmasters thus: "Workers from England brought up from childhood in the ironworks had acquired, without trying, a skill which was of use to them. When the children of French labourers had the strength to work they proved themselves to be as good as the English." The Lillonais mill owner, Mimerel, also considered that English workers had more experience and were industrially better educated. "They produced more and better quality goods for the same price."

This situation was typical of most European countries who wished to introduce British techniques into their industry. When these advanced techniques were introduced into the forges of Vitkovice, in Moravia, we see that foreign workers represented around 20 % of the workforce. Shift workers already appear to have been used, and there was a large proportion of highly qualified Prussians and Poles among these foreigners. As for the British, there were extremely highly qualified specialists, particularly the puddlers and laminators, who were instructed to train local workers in the new techniques.

No one isolated problem of techniques exists, indeed it cannot, owing to its very nature. The adoption of a new technical system, in whichever country, requires an overall framework of adapted structures in every domain.

It is therefore certain that an industrial revolution took place in the true sense of the term, insofar as every structure became coherently integrated. It is difficult, however, to specifiy which sectors played motivating roles as events were so interwoven.

Historians will probably continue to dispute and interminably debate this subject. Similarly, some industries only enjoyed partial progress, at least in the first phase of development, and countries also existed where production conditions, especially material conditions, did not produce an industrial revolution similar to the one in Britain which became the example for other countries. The Czech, Purs, has demonstrated the expansion of new techniques in Bohemia, favoured by the Continental System which provided this region with all the Central European markets. In 1799 there were 40,283 spinners, 12,000 in 1815, and 5391 in 1825 when production was considerably increased: mechanisation of spinning was finalized from the 1820's. After 1830, mechanisation was established throughout the textile industry. Railway building after 1835 involved the iron industry and mechanical construction (the Englishmen, Evans and Lee undertook this operation in Prague). The slow development and gradual use of the steam engine has been cited against this thesis: in fact some countries employed large hydraulic resources. In 1848, in the district of Liberec, energy supplies represented 1895 horsepowers in terms of hydraulic energy, and 91 horsepowers in terms of steam engines. But if Britain had enjoyed the availability of these hydraulic resources would she have developed the steam engine to such an extent? In North America, the same processes were met in the first stage of industrialization. As one authority has recently suggested, "Waterfalls were the primary source of industrial energy in the United States up until the third quarter of the 19th century." Major businesses, Lowell and Manchester, were founded upon hydraulic energy. We have said that the turbine had raised hydraulic power to the same level as steam energy. In order for an industrial revolution to come about, it was indispensable that all the elements which had contributed to the wealth of England be reunited.

The adoption of techniques involved in modern production, industrial concentration, and the appearance of a certain form of capitalism are conclusive proof by which the adoption of a new technical system may be defined.

Bertrand Gille (Translated by P. Southgate and T. Williamson)

THE MODERN TECHNICAL SYSTEM

The history of how the modern technical system came into being – that same system which is in the process of disappearing today – has not yet been adequately studied, apart from some instances, at least not in its entirety, and still less in its interpretation. And even in detail, there are a great many points which have yet to be properly clarified. Historians of economics seem to have been blinded by the industrial revolution in 18th century England, and have neglected to evaluate the remarkable technical change which took place in the second half of the 19th century. It may be that the relatively recent notion of economic growth has played an important role in this respect. This long and continuous period, which has been considered as one of the motivating factors of the English industrial revolution, has been highlighted. But the very continuity of this movement defined in this way conceals the possibility of another transformation as marked as the first. Most works, even the most recent, have consequently insisted on a parallel continuation of technical progress without distinguishing any precise phases.

We will try to break at this point with this oversimplified pattern and replace it with another, based precisely on the existence of another industrial revolution which could have created, in the second half of the 19th century, a technical system which was very different from that of the end of the 18th century or of the first half of the 19th. Although the techniques in use on the eve of the First World War have not been the subject of serious study, one may observe that they were completely different from those which were generalized around 1850;

even though certain old sectors, so to speak, were carried over into the new and more advanced technical system. The aeroplane, oil, thermal turbines, steel, organic chemicals, and electric power, to cite but a few examples, were manifestations of a system of techniques which was completely new by comparison with the system of 1850 with its reciprocating steam engine, iron and coal, etc. This view of the relative decline of certain economic powers, such as England, may be explained in terms of the country's devotion to techniques of the old system, while the rapid expansion of other economies, such as those of the United States of America and Japan, provide a striking image, the development of which coincides exactly with the appearance and the immediate adoption of a new technical system.

Let us now list these developments, using approximate dates. The new "industrial revolution" took place in two distinct phases. The first phase may be placed roughly between 1855 and 1870. In order for a technical system to be definitively established, a period of adaptation is necessary which is both of a technical and economic nature, as indeed we have already noted for the 18th century. It is therefore logical to assume that a second stage was necessary, and this can be situated approximately between 1880 and 1900. This was finally followed by all the developments and improvements of which the modern system was capable, possibly stimulated by the demands of the First World War.

It is agreed that economic growth or development is necessarily aided by technical progress. In other words, we should seek to find out if this continuity of growth, unanimously and statistically observed and evident everywhere, was compatible with the permanence of a technical system, even when it had reached the final stage of its development. Would it not be better to say that at a particular stage, accounting for a certain number of circumstances which have not yet been adequately emphasized, economic growth would not have been possible without some change in the technical system? The very notion of a technical system implies, to a certain extent, that a global transformation would have had to be general and systematic, rather than piecemeal progress in one or more series of independent inventions. This explanation, arbitary though it might appear, nevertheless merits greater consideration, and it is in this way that we intend to approach the problem.

ENVIRONMENTS AND CONDITIONS OF AN INDUSTRIAL REVOLUTION

As we have seen repeatedly, any technical mutation depends upon both internal and external causes. And it is from these causes that stem the necessity and the possibility of mutation. Apart from these causes, technical change takes place within a framework and a context, as indeed we have already observed, where cause and effect are far less evident than convergence and concomitance. It is these circumstances that we would first like to try to apprehend.

We have already stressed for the 18th century, the possible importance of demographic expansion in the introduction of a new technical system. But the second half of the 18th century undoubtedly saw no such expansion in industrialized countries, at least in those where the new technical system was likely to arise. The worldwide population growth of these privileged countries was infinitely less than it had been between 1750 and 1850. But we will not dwell on facts which are relatively well known. The birth rate dropped constantly, only compensated by the decline in the mortality rate, which resulted in the aging of the population. A superficial analysis of this would hazard a complete separation of the population movement and technical mutations.

We have already corrected this first assessment of the effects of demographic expansion for the 18th century. Obviously the increase in the number of people affects the volume of demand. In societies which were less advanced in the domain of consumption, as was so in the case of Western Europe at the end of the 18th century, the population growth was effectively something important, indeed of paramount importance. But in more advanced countries, such as those of Western Europe in the second half of the 19th century and all the more so in our day and age, the number of consumers certainly counted less than the volume of consumption. The constantly growing wealth of these countries and the aging of their populations contributed considerably to the increase in the level of consumption. This may be seen in the greatly increased consumption of industrial products by comparison with food. In France, the annual consumption of wheat flour per head was 1.76 hundredweight in 1841-1850, increasing to 2.45 hundredweight in 1891-1900. The following figures are for cotton: 1.7 kilogrammes in 1850, against 4.1 in 1900. Although these figures apply to a slightly larger population, they are nevertheless still highly significant. Thus despite rapidly reduced demographic expansion, the pressure of demand remained very high. One need hardly multiply these examples as the figures for other countries in the same stage of economic expansion are analagous.

There is another phenomenon which is equally important. Increase of demand was not only felt on the local level, inside industrialized countries, but there was considerable expansion of foreign markets. The decision to open the Suez canal dates from 1854, and it was completed in 1869. The penetration of Eastern countries, such as China (1860) or Japan (1858-1868), and the beginnings of great colonial expansion in Africa, both by France and England, also provided new outlets which required equipment and consummer goods. And here, on the contrary, it was the amount which compensated for the poor level of individual consumption. We encounter yet again influences which may contribute to demographic evolution taken in a different sense from the one usually considered.

Economic conditions are no less important factors. It is mundane to state that technical progress, or a new technical system must be compatible with particular economic institutions. There are two essential elements for this economic organisation: the accumulation of capital and the growth of major enterprises. The former is indispensable, and the second may be a consequence rather than a pre-existing necessity.

Although self-financing had mainly been and was to remain an effective way of accumulating capital, it soon became apparent that sources of capital external to the enterprise were indispensable, particularly as technical change stimulated new industries and therefore new businesses, such as the electrical and automobile industry. We should immediately note that this accumulation of capital required not only movements of capital within countries, but also the international transfer of capital that is so important today.

Several phenomena may be observed, the chronological concordance of which was almost perfect. Between 1854 and 1860, restrictions on interest rates had been abolished throughout most of the Western world, although England had attempted this from 1834. This liberalization was to greatly facilitate the movement of capital. And between

1856 and 1867, governments generally lifted restrictions on limited liability corporations, giving a freer hand to the directors of enterprise, while at the same time the public grew accustomed to the new forms of investment. The major international expositions, the first of which was in 1851, definitively launched the great industrial and technical era, even fostering a taste for technology. At the same period we may notice this change of attitude in the face of action or obligation, and also in face of state loans. The major projects of the railway companies between the years 1845 and 1847 had been an indication of this, but due to the interest shown, the action of the railway could almost be compared to the loans invested in it. Henceforth, began the great adventure in finance and industry. In 1850, the Paris stock exchange listed 42 stocks; by 1869, the number had already grown to 182 by 1869.

Equally significant is the appearance of well developed banking systems. Here again the dates of this development match those of the new technical system. On the Continent, the founding of the French *Crédit Mobilier* in 1852, marked the birth of the commercial banking network: especially devoted to financing industrial innovation, it was infinitely superior to similar establishments of the preceding period. The deposit bank was first created in France in 1859, with the *Crédit Industriel et Commercial*, quickly followed by the *Crédit Lyonnais* (1863) and by the *Société Générale* (1864). This example was quickly followed by the whole of Europe, and there was a veritable banking explosion as Germany, Austria, Scandinavia and other countries followed suit. In the second phase after 1870, the proliferation of counters, made possible by branch banks, contributed to the systematic draining of small deposits to create capital for investment. Once again we will not dwell on a phenomenon which has already been emphasized several times.

The concentration of enterprises is another aspect of this structural revolution in the economy. Truth to tell, it appeared both as a condition and as a consequence of technical progress. It was a precondition in that only large concentrated enterprises could initiate major technical innovations; it was a consequence in that the new techniques were inconceivable without the concentration of production. Today we are fairly accustomed to this evolution. The availability of more abundant capital naturally facilitated this increase in the size of business, remembering that in France this concentration of business was considered by the Péreire brothers to be one of the essential factors in the policies of Crédit Mobilier. In the same country, the first great wave of concentration was set in motion in 1854-55, by the formation of the first great railway networks from 1852.

Changes were only possible if intellectual and cultural values and attitudes encouraged them. After 1850, romanticism was stifled, the love of nature exhausted itself and was replaced by other values of the age. It was undoubtedly from this time on that Europe felt the industrial spirit that had permeated England in the first half of the 19th century. Refusal to accept industrialism was manifested almost until the end of the reign of Louis-Philippe in both France and Belgium, and fears of industrialism disappeared in their turn. Industrial power was no longer only power but it was now the source of happiness as well as profit. After 1850, a new generation came to economic power, which had first demonstrated its power in the economy during the reign of Louis-Philippe, and even practised a literary, political and economic sort of romanticism. This generation took over many industrial enterprises from individualistic founders, supporters of these small isolated and scattered units, who had only been belatedly aware of the first industrial revolution and had even feared it. The family firm gave way to a specialized body of management. Henceforth, they were to leave the ways of their ancestors, concentrate their resources, break out of the old confines, transform manufacturing processes as well as the method of labour itself, open enterprise to the outside, and, in short, revolutionize "industrial organization." It would be interesting to study the employers class from this angle but this research remains to be done. Once again this is a matter of cause and effect, and one may observe that social and mental attitudes and values persisted far longer in sectors where technical transformation was more gradual and less radical. The textile industry is probably the best example of this.

The new men had also been trained in a different way. Apart from the *Ecole polytechnique* in Paris and its administrating schools, new engineering schools were founded throughout Europe between 1830 and 1848. It was therefore after this date that the first graduates arrived at commanding positions in industry. Both the subject matter and the spirit of instruction had changed.

Around 1834, a French industrialist attributed the economic success of England partly to the general standard of education of the working classes. The development of primary education (remembering that the Guizot law was created in 1833), and the creation and development of adult education enabled the introduction of technical knowledge, which was not yet a closed world, to more and more people on all levels of society. This was one of the motivating factors of technical progress. It became relatively easy, even for a self-taught person to improve his education to the standard of a useful and working knowledge.

Yet another fact merits more detailed research. This concerns the extraordinary diffusion of scientific knowledge and its technical application. Reviews and encyclopedias proliferated, and scientific and technical journals became abundant, but a critical list and an analysis of their content is required. In other words, it was possible or had become possible to acquire the knowledge necessary for certain inventions even without a university education. Bessemer learned much from these encyclopedias. James B. Francis, who improved Fourneyron's water turbine, was an educated and learned man. He gave an account of his experiments published in *Lowell Hydraulic Experiments* in 1855. And Pelton, who took the turbine beyond the achievements of Francis, was only a village carpenter, working on water mills, but he had bought and studied a copy of Francis' book which gave him an elementary training (figs. 1 and 2). Gramme is yet another example; he had learnt his skills from Ruhmkorff, since the oral transmission of knowledge was still practised; he may have attended Becquerel's lectures at the *Conservatoire des arts et métiers*, and he had read Ganot's manual with the aid of a dictionary. So although Gramme was a mechanic, he also possessed more than a little useful scientific knowledge. The diffusion of a certain level of scientific and technical knowledge to a broader public already provided with primary education was of fundamental importance to technical invention.

A particular place should be given to the domain which is situated half way between science and techniques: that of measurements. Measuring instruments, created partly with a scientific aim in mind, were quickly found to be indispensable for techniques. Most of them were created before our period, but in refinements and new methods obtained by experimentation, measurement was all important. It has been suggested that American hydraulics could not have developed without Prony's brake, and in the the mechanical industries, there was the invention in 1849 (at least this is the date of the patent), of the gauge equipped with "screw and circular vernier" by Palmer. Although some detailed studies have been done, research still remains

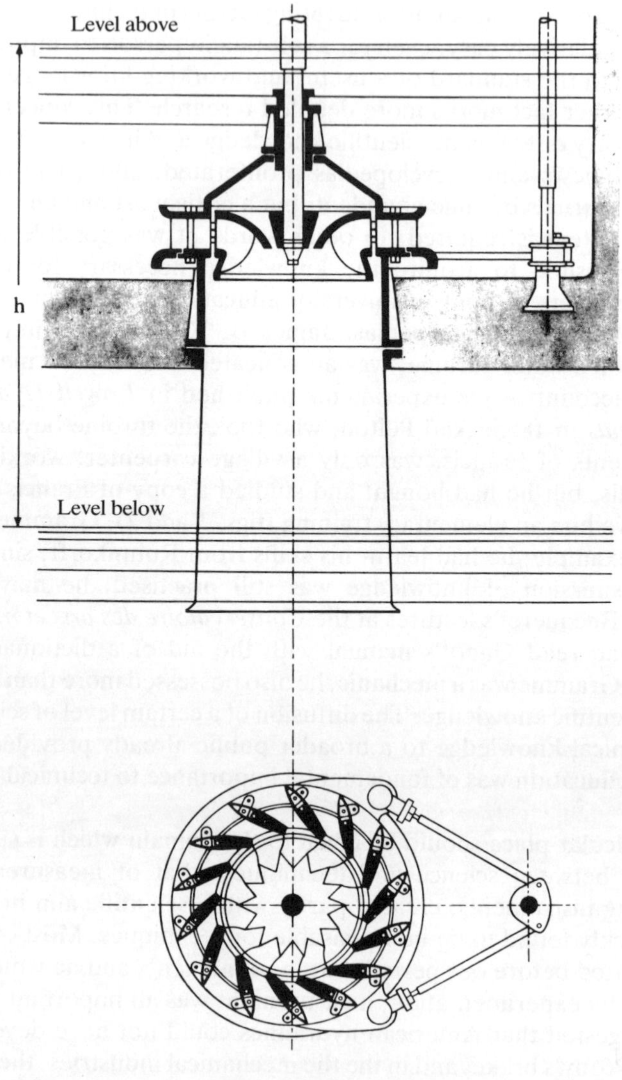

Figure 1. — The Francis turbine. (From Lehr, *Les Moteurs*, Paris, 1970)

equilibrium being applied to the reduction of ores in the blast furnace and to the phenomena of deoxygenation of steel: the law of the displacement of equilibirum in 1884), which permitted slag masses and their composition to be taken into account in the problems of development in the blast furnaces and steelmaking; and finally, the preparation of alumina for polishing metallographical samples which was followed in 1900 by the construction of a microscope with inverted lens.

Meanwhile, in 1885-1886, Floris Osmond created two methods of investigation. The first was microscopic metallography, concluding the research undertaken by Réaumur. The images he obtained made it possible to determine the correct treatment of steels, notably specialized steels. He used the second method to determine the critical points of steels as well as to characterize them. The theory of tempering had been provided by Gore in 1869, and were refined by Barrett in 1873. Osmond, together with Werth, completed the theory in 1885. In 1894, he defined the peculiarities of ferro-nickels. In 1895, he determined the qualities of martensite, used for quenching and austenite, an element of super-quenching. So gradually a scientific technology of ferrous metallurgy and its alloys was formed.

Let us examine this history of scientific metallurgy to its conclusion as it is a valuable example and underwent a fundamental evolution. As the technology of iron and steelmaking became increasingly scientific, so invention could no longer be accidental, but on the contrary, invention was now sought in order to bring about a specific result. The entrepreneur thus anticipated future development and now had only one way in which to implement his plans: to set up a laboratory and direct technical research according to a scientific rationale. When in 1867, Boussingault, a chemist known for his work on metals, arrived at the Holtzer metallurgical plant near Saint-Etienne, he completely changed the previous research methods. Helped by the engineer, Brustlein, Boussingault had by 1877 developed ferro-chromes. By mixing them with oxygen-free charges of the crucible, chromed steel was obtained. Other firms followed suit. By 1888, similar research was being carried out at Imphy, south of Nevers, especially on nickel steel. On the request of the International bureau of weights and measures, the correct proportions were calculated which were to lead in 1897 to the production of new steels, invar and elinvar, whose coefficients of elasticity did not vary with temperature. Once again we require the exact and detailed history of these factory laboratories, their results, their methods and their policies of research, both from a scientific and technical point of view, in order to appreciate precisely the way in which science and techniques are related to each other. And this applies to every kind of industry involved. These laboratories designed for technological research were no longer to be the fruit of individual initiative, but were soon to be organised on a collective level, as in 1911 a research laboratory was linked to the chair of the textile industry of the *Conservatoire des arts et métiers*.

As one advances, so technology becomes an extension of science. The development of the steam turbine and its successive improvements were the result of many applications of the theory of thermodynamics and fluid mechanics. It would certainly be quite simple to determine precisely those zones of scientific restriction that exist within different techniques, but as yet no one has attempted to trace these limitations.

It is therefore in this important context, where every chronological division may be found, that technical change could be brought about. At this point of our anaylsis we should emphasize the exact convergence between techniques and economy.

We have already begun to see such a convergence: at this level, continual growth naturally requires a new technical system. For technical systems have their limits, and if the pressure of economic requirements continues to grow, technical mutation becomes imperative. Inside the technical system in use in the middle of the 19th century, we may observe a certain number of tensions resulting in several ways.

An increase in demand must result in an increase of production. The first problem is therefore a problem of quantities. Greater quantities may be obtained without a change of techniques simply by increasing the size of individ-ual enterprises or by increasing their number. But this is not always possible, if one considers production factors, owing to difficulties of securing raw materials, assembling labour, and a greater amount of capital investment. And any given technique imposes limits of marginal utility on these various factors of production. A few figures will prove this statement. For example, the production of pig iron in the most advanced countries of Europe increased considerably between 1850-1854 and 1910-1913, as is shown in table I, in figures of thousands of metric tonnes.

TABLE I

	FRANCE	GERMANY	ENGLAND
1850-1854....	561	245	2716
1910-1913....	4664	14,836	9727

Without a general technical mutation in the whole complex represented by pig iron production, in other words in the machinery itself, in the mining of ore and coal and coking, etc., no such increase in production would have been possible. It would have been necessary to increase the number of blast furnaces in France ninefold, and sixty times as many would have been required in Germany. These figures also reveal the difference in demand between the three countries.

The problem of cost demands no less important research. To produce a constant quantity at a lower cost was a temporary solution, as sooner or later demand would have increased to absorb the full capacity of the existing plant. As everybody knows, the ideal is to produce more at lower costs. Here again the technical system must eventually reach its limits and these same limits come into play when more is produced at constant costs. It would therefore seem logical to assume that two fundamental principles are valid. Logically, then, economic growth could not have continued without fundamental technical mutation or progress, otherwise the limits of the existing technical system would have brought it to a halt. On the other hand, the very notion of a technical system implies an overall mutation to assure the necessary equilibria at each level of the system.

The historian should therefore look first to the tensions which were manifested towards the middle of the 19th century, or which would have been manifested had there been no technical progress. He should then explain how other factors responded to this, either by improvement of old techniques where possible, or by the adoption of new methods. Finally, the existence of the new technical system should be demonstrated. Here again this is merely wishful thinking as no research to date is adequate enough for us to make definite conclusions. We are therefore compelled to limit the discussion to a few examples.

Consider the production of energy in around 1850. Hydraulic energy had made considerable progress since Fourneyron's development of the water turbine and since the improvements made to it by the Americans, Boyden and Francis, on the eve of the period in question. Efficiency had been maximized, and the power could be increased only within certain limits. Moreover, hydraulic energy required its users to be located close to an adequate water supply, and without heavy investments and planning, they were at the mercy of natural hazards. The first of Boyden's turbines (1844) produced 75 horsepower with a trial efficiency of 78%; the second (1846) developed 190 horsepower with a maxiumum efficiency of 88%. More progress was possible, particularly in horsepower. Even before 1850, 500 and even 700 horsepower had been reached. Better results were obtained between 1850 and 1890, but only using low falls and slow wheels owing to the fact that the materials could not bear any greater stress. Manufacturing techniques were improved, enabling the cost of turbines to be cut considerably. From more than 200 dollars per installed horsepower in 1850, the Boyden turbines were available at 70 dollars per installed horsepower in 1861, and then 5-8 dollars twenty-five or thirty years later. About 1890, a power of 1000 hp was attained, presenting the final limit, but this had enabled both the power to be increased and the cost to be reduced.

Consider the steam engine, the old alternating engine which was the glory of the preceding industrial revolution. Obviously, it could never escape the limits it was born with, from both a purely technical and an economic viewpoint. Either considerable time was lost starting it up after each shutdown, or else it had to be kept running continually. Its register, or range of power, was severely limited for various reasons: much like all other alternating machines, it could only be used for large power demands that did not exceed 5000 horsepower. Maintenance of this engine, especially lubrication, was costly and continuous. Stoking the boiler and removing cinders required a large workforce. To these drawbacks should be added the costs of construction, installation, and housing, as well as the difficulty of adjusting its running to variations in demand for power, and in some cases excessive congestion (transports). Lastly, there was the problem of poor efficiency of 6 to 10%. These limitations meant that the alternating steam engine could only be used in certain well-determined sectors: industries with heavy concentrations of mechanised equipment, and heavy land transport for which securing fuel and water presented few problems. The steam engine was difficult to use in land transport where slight power was needed. In maritime transport, its space and weight requirements imposed strict limits on the speed and size of ships. Lastly, the alternating steam engine proved itself to be poorly suited to driving electric generators.

Nevertheless the classic steam engine did undergo some important improvements, but these simply pushed back the limits rather than removing them completely. The change to high pressure and the use of expansion in the compound engine had been the first such developments in 1830-1834. At the beginning of our period, after improvements in accessories (the boiler to take but one example), engineers concentrated on regenerated steam first considered by Seguin at the Academie des sciences on the 3rd January, 1855, and in the attempts made by Siemens. This brought about a fuel saving of two thirds. Then came superheating, a process in which a portion of the steam produced was taken out and carried to a very high temperature and pressure. When this steam was returned to the boiler it turned the drops of water that had escaped the first heating into new steam. The 1855 exposition demonstrated its effects. But note here again the necessity of parallel developments in complementary techniques, particularly in the metals necessary for very high pressure, which did not increase the weight, bulk or cost of the machine. Efficiencies and, simultaneously, horsepower produced, were therefore able to increase considerably, (in terms of kilogrammes of coal burnt per horsepower per hour):

The Newcomen Engine	8.40
The Watt Engine	2.90
The Trevithick (high pressure)	1.50
The compound engine	0.90
With superheating	0.325

But we should repeat that such developments made in a certain technique could not have taken place without the appropriate changes in neighbouring techniques and the aquisition of the scientific knowledge necessary. Induced progress is more important than autonomous progress. Indeed, it could be argued that the progress of the steam engine was induced by advances in complementary techniques. Neglect, as we have already noticed, of such interdependencies among technical sectors has long served to distort the history of techniques. But in any case the limits of the steam engine could not be overcome:

ever greater horsepower, even lower costs, and more diversity and flexibility of operation were indispensable for continued economic growth (fig. 4 to 6).

Let us take another, equally striking example of the limits met by techniques. In 1850, the railway was already in operation. Investment in the new means of transport had been very large, and so long as costs remained fairly low, it was to contribute a major factor in future economic growth. François Caron has aptly defined the situation:

> For the engineer charged with managing the great transport service of the railway, technical progress provides the only form of escape from the contradictions in the centre of which he finds himself. Administrators and supervision demand that he ensures an increasing traffic and better services which cost less while maintaining profits. The only way of achieving this is to reduce cost prices. There were three problems: firstly, the problem of wages, particularly after 1860; secondly, the unlimited expansion of the railway network necessitated employing methods of exploitation for reduced traffic which were in proportion to the length of the network rather to the traffic itself. And thirdly, there was the risk of a glut or obstruction on the main lines, on which rising traffic could only be handled by the immediate accumulation of expensive operating methods. These were the three reasons which compelled the smallholder to determine a reduction in the workforce, fuel, and installations. But, in order for innovation to be adopted, annual expenses brought about by loans could no longer be higher than the economy of the business which it made possible.

In the early years of the railroad operation, the replacement of worn out iron rails was a financial abyss. Nor was the cost of broken axles and wheels negligible. Failing some sort of technical solution, the growth of transport would have been abruptly halted. The advent of cheap steel in large quantities was to solve this problem. Meanwhile, considerable progress was made in tractive power which resulted in the perfection of the machines. A Crampton engine of 1850 weighed 125 kilos per horsepower; a 1894 locomotive weighed 72, and in 1913 only 50 kilos per horsepower.

Maritime transport presented a similar case except that the limit of its development would be more quickly reached. Any increase in speed required a greater increase in power, to take but one criterion, and the power required had to increase by approximately the cube of the speed. Thus for faster and bigger ships, engines needed to be vastly more powerful without becoming too heavy and bulky. Around 1900, the engine of a big liner comprised of alternating steam engines and cylindrical boilers capable of holding a large volume of water, which were consequently extremely heavy, not to mention the coal and water

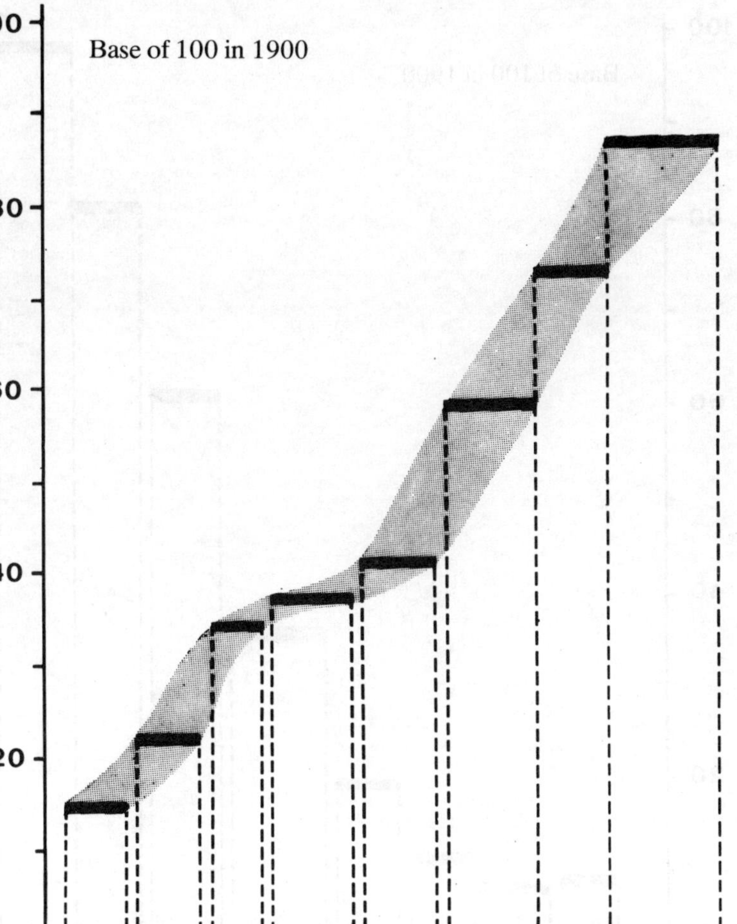

Figure 4. — Growth in productivity in the mining industry, with two phases of quickening pace: 1825-1850 and 1870 to the end of the 19th century. (From C. Fohlen, *Qu'est-ce que la révolution industrielle*, Paris, 1971)

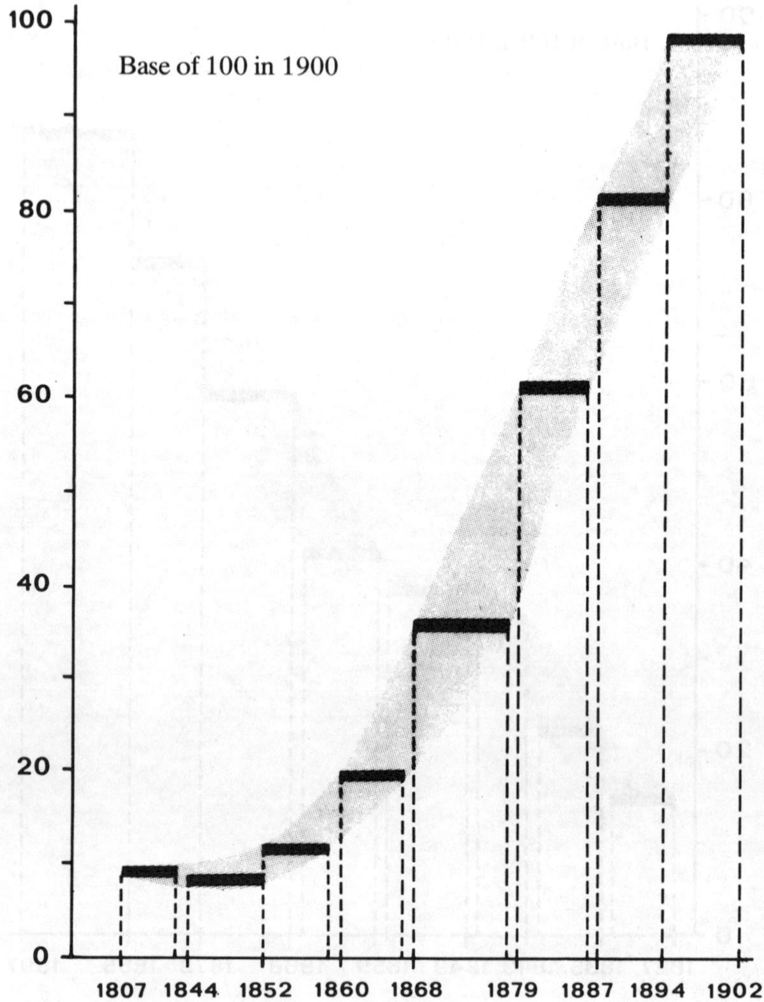

Figure 5. — Advance in productivity in pig iron making in Germany. An acceleration is visible after 1850. (From C. Fohlen)

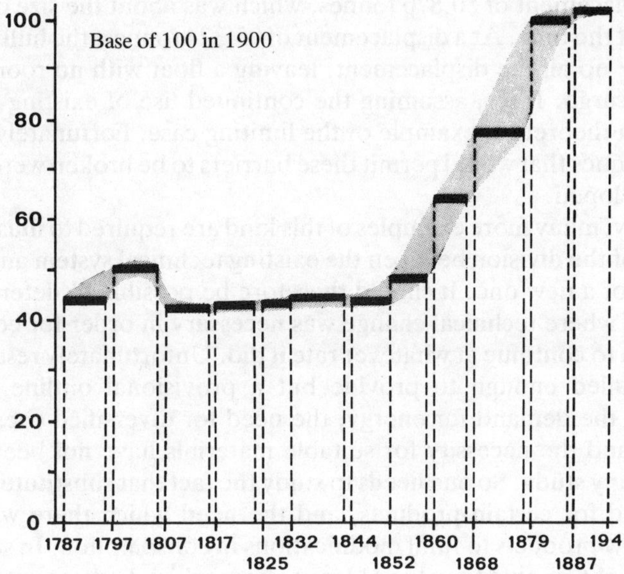

Figure 6. — Stagnation in coal mining productivity in Germany before the sudden increase from the mid 19th century. (From C. Fohlen)

supply the ship required. The total weight per installed horsepower was approximately 150 kilos, and the machinery of the engine took up between a third and a half of the ship's usable cubage. It is not surprising in these conditions that between 1907 and 1914 the French naval engineer, Bertin, following actually the principles set forth by Jacques-Auguste Normand in 1897, demonstrated mathematically the limitations of these displacements. According to Bertin, the stress bearing upon the hull of a large ship would require it to be made of increasing thick metal as displacement was increased. The fraction of the weight of the metal hull of a liner in relation to the total weight was expressed by the simple formula: $C = 0.2166 + 0.01176\ P^{1/3}$. He calculated that the optimum displacement, giving the best proportion of the available fraction of displacement (engine and load), that is where the engine occupied the smallest fraction of the hull, was in a ship of 29,588 tonnes, and that the maximum load would be reached with a displacement of 50,876 tonnes, which was about the size of the big liners at the time. At a displacement of 95,324 tonnes, the hull itself would take up all the displacement, leaving a float with no room for engine or cargo. Here, assuming the continued use of existing techniques, is a theoretical example of the limiting case. Fortunately, the new techniques that would permit these barriers to be broken were just being developed.

Certainly, many more examples of this kind are required to illustrate the limits of the division between the existing technical system and the beginning of a new one. It should therefore be possible to determine the sectors where technical change was necessary in order for economic growth to continue at whatever rate it did. Unfortunately research is not detailed enough to provide but a provisional outline. The analysis of the demand for energy, the need for diversified chemical products, and the necessity for suitable materials have not been the subject of any study. So one needs to study the fact that substitutes had to be found for certain products, and the need which there was to develop new products to fulfil modifications in consumption. In such a way, we might be able to value old inventions which had not yet been put to any use, and explain the attempt made to find new discoveries.

A series of imbalances, a series of tensions caused by economic growth itself, whatever rate it took, was thus leading towards new techniques. But due to the necessary links between diverse techniques, mutation was general. The automobile is not just a horseless carriage with a motor, with all its diverse components made of appropriate materials, such as the rubber tyre. It is also the network of fuel stations, and the road system designed for the automobile. Electrification was not merely electricity, but it was also used for hydraulic turbines operated by high falls and great speed, or Diesel engines. Economic progress takes place through technical progress, and technical progress could take place only through the supersession of one technical system by another.

This kind of analysis should also show all the pressures that economic and technical progress exercised upon the social system. The call for ever increasing numbers of workers led to the decline of the rural population, and the exodus from the countryside was on a larger scale after 1850. Compensating institutions were sought in the face of economic expansion and technical change which modified the relationship of strength. Here again the dates match almost exactly; In France the law against unions was repealed in 1864 and full labour legislation was passed in 1884. We find the same stages once again. Coherence between systems cannot fail to exist.

THE GREAT MUTATIONS

A description of all the elements which the new technical system enforced between 1850 and 1900 in two major stages as well as the continuation of its development, perhaps owing to the First World War as well as to the Great Depression beginning in 1929, would fill volumes. We must therefore limit ourselves once more to indicating the principal lines of development and to the main outlines of events in the most important sectors. Among these sectors we should emphasize the problems of energy, as well as those of natural resources, the modifications made in materials, and lastly, the main instruments of industrial production.

We have already pointed out the constraints of the preceding system on the subject of energy production, and we have also discussed how the evolution of other techniques, particularly those of materials, had enabled energy convertors handed down from the old technical system to be best used.

The classic alternating steam engine was therefore able to continue its career. After some early attempts in superheating shortly after 1850, it was mainly in its fabrication, in the use of specialized steels that the steam engine could provide greater sources of power and higher efficiency. Everybody knows, however, that few further developments could be made in this direction. By contrast, improvement of the hydraulic turbine was to be continuous and significant. Progress was at first slow after the modifications made by Boyden and Francis shortly before 1850. As Louis Hunter observes: "These turbines were different in the originality of their shape, the simplicity and solidity of their construction, their large scale manufacture at reduced costs and to standard sizes, and their output which was soon to exceed that of the wheels carefully designed and constructed by Boyden and Francis." Engineers concentrated their efforts on turbines using falls of 15 to 20 feet, developing moderate horsepower of 35 to 75 hp. They had to be cheap, reliable, and efficient. In the wheel elements, the precision of the sizes, manufacturing and hand fitting was reduced to a minimum. As Thurston has observed, American turbines were usually cast with their buckets in place and could be put into operation as soon as they left the factory.

But these turbines still suffered certain limitations when in use: low speed and efficiency when the gates were partly opened, and a tendency to become clogged by debris because of the small size and opening of the buckets. Low speed goes hand in hand with large volumes and weights compared with the energy produced and with a proportionally high cost of installation. These limitations were finally eliminated by the successful development of the inward flow turbine: speed was increased by deepening and then enlarging the buckets and extending them towards the centre of the wheel, but this left little room in the centre to evacuate water. So the buckets had to be orientated towards the bottom to guide the water: the blades were curved towards the bottom and the outside in such a way that the water passing through the wheel ran out in a gentle and continuous spiral. Moreover, reducing the number of buckets increased power and avoided obstruction by debris. Thus in half a century, from 1840 to 1890 the turbine designers obtained the same power in half the space and for one fifth of the cost.

Final improvements sought to save water by increasing efficiency at half gate. A Swain turbine with an efficiency of 75% at full gate would drop to 66% at three quarters and to 50% at half gate. By 1879-1880, engineers had virtually eliminated efficiency loss at half gate. In the mines of California, the most improved tangential wheel was Pelton's creation; it was developed so much in size and power, attaining efficiencies of 80% and 1000 horsepower, that it could replace steam engines, saving at least two thirds of the energy cost. But here again, it was not possible to go any further than this.

As materials with better physical properties, steel and then specialized steel, became available, it was possible to use the hydraulic turbine in different conditions. As we have already seen, the use of high falls with great power and volume had been envisaged from 1848, but cast iron wheels limited both the pressure and speed of rotation. The development of wheels and conduits made of steel meant that high falls could be used. The first attempts seem to have been made in

France to drive heavy papermaking machinery which required a considerable amount of power. Matussière in Domène (1870), Frédet in Brignoud (1871), and Horteur in Saint-Rémy-de-Maurienne (1878) were forerunners in this field. In 1882, Bergès succeeded in harnessing a fall of 500 metres. The Pelton turbine appeared just in time to be used as an energy convertor, and was to be used to great advantage in the production of electricity, as we shall see later.

The new energy convertors were in fact directly descended from two main earlier machines: the old alternating piston and cylinder steam engine and the turbine. This is a perfect example of technological continuity. Lenoir used the basic mechanisms of the steam engine in his internal combustion engine, and Diesel's great idea was to invent a high efficiency steam engine. Parsons and Laval substituted steam for water to turn their turbines. The reason that these ideas took so long to implement and were often spread over more than one century was due to the fact that essential elements were not discovered until the period in question now, that is to say after 1850.

The Abbé de Hautefeuille and then Huygens outlined the first ideas of an engine powered by internal combustion before the actual development of the steam engine. But until the end of the 18th century, with the work of the Englishman, Barber, inventors hesitated between explosion and combustion, between direct use of this combustion or indirect use through the intermediary of hot air. Moreover, they worked towards the use of the mechanical power of hot air: the work of the Englishman, Stirling, in 1826, was taken up and completed by the Swede, Ericsson. But this only involved a single acting machine driven by hot air and therefore entailed serious drawbacks. The firebox was dangerous in a small machine intended for use in the home, the coal it used as fuel was cumbersome, and the cylinder buckled under the heat of internal combustion. In 1824, the Englishman, Brown, used a mixture of air and gas after the many suggestions put forward at the beginning of the 19th century, by Lebon among others.

The real engine was the result of a convergence of several partial inventions shaped by clearly defined design criteria. In fact the main problem which had to be solved was that of a small engine to drive machinery in the home, and for particular mechanisms which had recently been developed, such as the sewing machine. It needed to operate at low power, take up little space as well as use less fuel and avoid fire hazards. The most practical source of energy was illuminating gas, which had come into extensive use since 1815. Here was a widely distributed fuel, one which required no storage. The machine itself was quite simply put together from the components of the classic alternating steam engine. Air and gas had to be mixed: Barber had sketched the idea of the carburettor at the end of the 18th century. In 1843, Father Eugenio Barsanti thought of using the Voltaic pile to ignite the combustible mixture with a spark. In 1852, this same Barsanti together with the engineer, Félix Mateucci, studied the problem of the proper proportions of the combustible and also discovered the ground-return. Henceforth anything was possible. The Belgian, Lenoir, developed the motor in 1859, and patented it in 1860. The patent was for a "motor worked by hot, expanding air created by the combustion of gas ignited by electricity and capable of replacing steam as a driving force." The combination was extremely simple. "I use a part of the main pieces of a horizontal steam engine, that is, a fly wheel, a crank-shaft, a connecting rod, a slide bar, a cylinder and two cams to activate the slide valves." The spark plug was made of two platinum wires insulated with porcelain and connected to a Ruhmkorff coil used to change the tension of the current supplied by the batteries. One slide valve was used to let the mixture of air and gas into the cylinder, and the other to exhaust fumes after combustion. The mixture was 95% air and 5% gas. The cylinder was cooled by water circulating between it and a cast iron jacket.

It is curious to note the remarks made by Lenoir's contemporaries. Lenoir had at first contemplated inventing an electric motor. "But," wrote Figuier in 1870, "it did not take long to recognize that this was hardly possible because of the insignificant mechanical power produced by electromagnatism." Lenoir's gas engine was by no means perfect. The slide valves, particularly when they overheated, were frequently problematic. Lastly, as Figuier remarks, "this engine was not economic." Lenoir had originally estimated that one cubic metre of gas was needed to produce one horsepower. At a cost of thirty centimes, this meant a cost of three francs per ten hour day, while even a mediocre steam engine required five or six kilogrammes of coal per hour and per horsepower. Lenoir's design certainly saved in bulk, avoiding the cumbersome boiler, fuel and smoke. In fact it consumed eight times more gas than expected. The estimated heat power of that gas being put at 5000 calories, the mechanical equivalent of heat being put to 635 calories per horsepower per hour, its efficiency was barely

3.52% to 4.65%, this for the two types which were tried. This engine could only be used for very small powers, of one-half to two horsepower.

But Lenoir's internal combustion engine was open to improvement. Firstly, it required a good theory to substantiate it, and this was provided by the Frenchman, Beau de Rochas, who in his patent of January 16, 1862, defined the theory of the four-stroke engine, made possible by the work on thermodynamics developed since Carnot (1842), by Joule, the principle of equivalence formulated by Mayer (1842), and the studies of William Thomson, and Clausius. The notion of preliminary compression of the combustible mixture existed in embryo in the patent of another Frenchman, Degrand, (June, 1858), and secondly in the Italian, De Cristoforis' suggestion of compressing the air before it was mixed with the gas (1859). But it was Beau de Rochas who proved, shortly after Lenoir had built his engine, how the use of a four-stoke piston could achieve better efficiency: suction, compression, explosion and expansion, with the evacuation of the burnt gases.

Research was also being carried out in other areas. In 1862, the German, Nicolas-August Otto, built a four-cylinder motor with floating pistons behind the driving pistons, hoping that the floating pistons would complete the evacuation of the burnt gases. He had noticed that if the cylinder was poorly filled, the ensuing explosion was not sufficient enough to drive the piston through a full stroke. Indeed, it was pulled back in the opposite direction by the depression left when the burnt gas was exhausted. In other words, the mechanical effect was not completely used, and the atmospheric pressure was coming into play even before the piston reached the end of its stroke. So Otto modified the gas-air mixture, filled the cylinder completely with the mixture during almost all of the stroke, and let the crank-shaft turn in the opposite direction in order to compress the mixture, setting the ignition to take place at the point of maximum compression. To his great surprise, the flywheel completed several rapid revolutions at the same time as developing much more power. On his own he was discovering the virtues of prior compression. This is why he conceived a four-cylinder motor with cylinders opposite to each other, two by two, and then decided upon four cylinders, in order to have a uniform torque. The motor operated in January, 1862, but sudden shocks caused during ignition led to its failure. Nevertheless, Otto's intuition was correct, but his theoretical knowledge was poor. The French engineer, Pierre Hugon, whose research began in 1853 and was completed in 1858, used the two-stroke cylinder, without compression, and developed motors using direct and indirect action. His patent of 1860 demonstrates in fact a motor driven by atmospheric pressure, the piston operating in a water flooded area which enabled regular strokes. In March, 1865, Hugon adopted the two-stroke motor, with a vertical cylinder for water injection and supply bellows.

Between 1864 and 1867, Otto secured the help of the engineer, Eugen Langen, and developed a motor with a vertical cylinder where the expansion could continue until the pressure it was in was less than the atmospheric pressure. This long expansion improved the efficiency of the the two-stroke engine – a remarkable achievement at this epoch. Its fuel consumption was reduced to two thirds of that of Lenoir's engine.

Thus by 1870, three internal combustion motors were in existence: the Lenoir, the Hugon and the Otto-Langen, as well as Beau de Rochas' clear definition of the four-stroke cycle. Fuel efficiency had been improved but was still poor. These small engines were used for low power requirements despite the fact that they were extremely noisy when in operation. Otto's last engine had an efficiency of 13.7% at 106 revolutions per minute, and 8.25% for 75 revolutions. Fuel consumption was still large so this form of energy was not cheap but these motors were nevertheless the most practical in certain conditions. Virtually no progress was made between 1867 and 1876. But old ideas were to be resumed, for in 1859 De Cristoforis suggested the use of petroleum instead of illuminating gas. In 1872-1873, a Münich watchmaker, Christian Reithmann, built the first real four-stroke motor, and in America between 1872-1876, Brayton devised a carburator using petroleum: a sponge soaked with petroleum through which air passed driving fine droplets onto a metal screen behind which combustion was continuous without explosion. This was therefore the first internal combustion engine using petroleum.

In 1876, Otto, aided by Daimler and Maybach, adopted the four-stroke engine, and the definitive engine, at least in concept and main design, appeared in 1877. It was still not perfect and minor improvements followed. Daimler perfected the carburettor, Forest devised magneto ignition in 1885, and in 1890, water cooling was adopted. Around 1892, the four-stroke motor was complete, having also benefited from progress made in metallurgy and the change in fuel which gave it greater mobility. Henceforth, its purpose was to change completely. Originally conceived as the engine to power small industry in the home, it was now to become the essential instrument in a new means of transport which was to enjoy great success.

Rudolph Diesel's main idea was to build a motor that would come as close as possible to the Carnot cycle. In 1883, he began working on steam engines using ammonia in place of steam. The theory was dreadful, but he gained technical experience in attempting to put this idea into practice. In 1890, he departed in another much more fruitful direction:

> How the fundamental idea came to me of replacing ammonia with a true gas, i.e. highly compressed hot air, of slowly injecting fine particles of fuel into it and letting it expand during the combustion of the particles, so that the greatest possible amount of the heat produced was turned into mechanical work – that I do not know.

In any case he arrived at several conclusions: (*A*) superheated steam or gases are required to reduce heat losses through cylinder walls; (*B*) high pressure is necessary to achieve the great temperature drop indispensable to the high efficiency envisioned in Carnot's theory; (*C*) combustion must take place in the cylinder itself in order to avoid heat loss in transmission; (*D*) the temperature of combustion and ignition must be reached by compression alone.

From this followed a perfectly logical cycle, comprising: (*A*) a first phase of compression kept isothermic (20°C.) by injecting water; (*B*) a second phase of adiabatic compression raising the temperature above the flammable point: 250 atmospheres and 800°C. Then came the ignition of the gradually introduced fuel; (*C*) a first phase of expansion with fuel being injected and isothermic combustion: to 90 atmospheres with the temperature at 800°C; (*D*) a second phase of adiabatic expansion. Then the return to the original 20° in less than one atmosphere of pressure.

Such a machine was extremely difficult to realize, as it entailed very high maximal pressures against a very low average pressure. Diesel therefore changed the cycle by eliminating the first phase of compression, which made it possible for ignition to take place at an appreciably lower pressure. He also had to retreat from the extremes of theoretical compression and expansion and settle for nonisothermal initial combustion and incomplete expansion. With the support of a large mechanical firm and of Krupp, he was able to build his first motor where no cooling of the cylinder was provided with a view to using mechanic fuel injection. Attempts to do so took place in August 1893. While the principles were sound, the engine worked poorly. Water cooling and pneumatic injection were therefore adopted, and the engine built in 1895 worked much better; its 20 horsepower successor of 1897 was fully developed, achieving an efficiency of about 26% But he had to abandon all hopes of achieving greater efficiency by implementing the theoretical Carnot cycle.

The steam turbine was based on an idea which had already been tried, the hydraulic turbine, and combined it with another old idea, that of energy produced by steam in certain conditions. The development of this new technique took place in several stages, as the theory was perfected (fig. 7). In 1884, Charles Parsons was the first to build a steam turbine which was multicellular and developed 10 horsepower at 18,000 revolutions per minute. Here again it had been necessary to wait for the development of adequate materials. In 1890, the Swede, Laval, independently of Parsons, built a steam turbine consisting of a single disk turning at 30,000 revolutions per minute. In order to develop the idea, he had to overcome many difficulties, such as the design and machining of precision gearing, and a novel rotor shaft which was thin enough to be flexible (fig. 8). All that, together with considerable stress and high speeds, required machine tools and materials that had recently appeared in other sectors of the modern technical system. The superiority of the turbine over other machines was its rotary movement which enabled the construction of machines with almost an unlimited range of horsepower. Power for power, it compared with the classic alternating steam engine, and moreover, at any horsepower, it was lighter and required much less space. And this made it particularly advantageous as a marine engine. Maintenance costs were low; its lubrication required only a third the grease consumed by the steam engine. The only disadvantage at the beginning was a considerably higher consumption of steam.

Further development was therefore aimed at affirming the advantages of the steam turbine and eliminating any drawbacks. In 1896, Curtis increased efficiency by building a vertical axis turbine that used not only the pressure of steam, but also its expansion. Rateau had begun to work on the problem of ventilation in the mines, and therefore turned his attention to rotative mechanics. In 1892, he

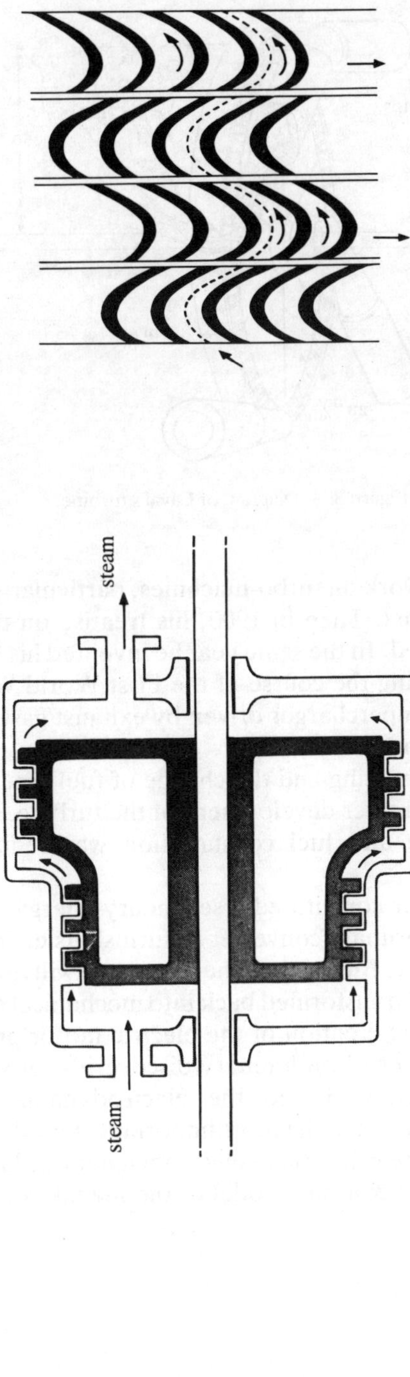

Figure 7. — General diagram [left] of Parson's turbine and diagram of the blades of the turbine [right].

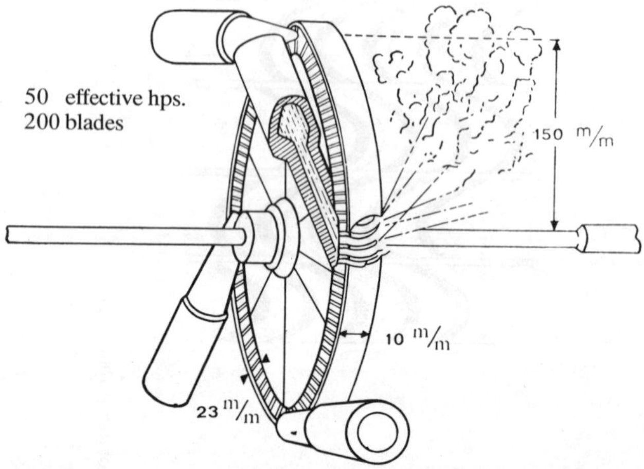

Figure 8. — Diagram of Laval's turbine.

published the first work on turbo-machines, particularly in connection with mine ventilators. Then in 1900, his treatise on turbo-machines proper was published. In the same year he invented his first multicellular turbine and during the course of the First World War he built his remarkable turbo-supercharger driven by exhaust gases, thus liberating the turbine from steam.

The use of superheating and the change of fuel types, from coal to oil, was to lead to greater development of the turbine. Improvements were rapidly made and fuel consumption was reduced as far as possible.

The electric motor constituted a secondary energy converter. It is dependent upon a primary converter to furnish its energy in the form of mechanical energy. But on the other hand, we will see that this form of electricity may be transformed back into mechanical energy without an intermediary. The creation of the electric motor presupposes the existence of a number of preliminary discoveries, of which the electromagnetism of Galvani and the electrodynamics of Oersted, Ampère and Faraday were the most important. Faraday was the first, in 1828, to explain the principle of electromagnetic induction. In 1832, Pixii created the first working model of the magneto-electric generator, inducing the current in the two coils of a stationary electromagnet by rotating the field magnet beneath it and then rectifying the current with a simple commutator. In 1834, Moritz Hermann von Jacobi, inventor of galvanoplating, published a paper on the application of electromagnetism for powering machines. Then in 1843, Stöhrer, and Woolrich in 1844, designed a large steam-driven electromagnetic machine, finally built in 1867. In 1849, Clark set up one of the first electromagnetic machines, quickly followed by Little, in 1852, and the Dane, Sören Hjorth, in 1854. Meanwhile, in 1851, Wilhelm Josef Sinsteden built the first monophase alternator with an exciter magneto. We may therefore observe that progress was slow in domains where both scientific discovery and technical knowledge played their parts.

Antoine Pacinotti was looking for a practical way of measuring a current. He finally had the idea of creating an electrodynamic device connecting a stationary circuit and a mobile circuit, both traversed by the current to be measured. The device was practically impossible to build, but the inventor understood that some slight changes would turn it into a direct current electromagnetic machine. He wound wire in a spiral on a revolving ring, and on each side of this ring armature he placed two sliding contacts; then he placed two magnetized steel bars with their poles near the ring. On January 10, 1859, he attached a galvanometer to the sliding contacts and turned the ring between the opposing poles of the magnets. The needle of the galvanometer jumped to the end of the scale. This was the first time that a continuous current had been induced by a non-unipolar machine. An improved model was built in 1860: the permanent magnets were replaced by a single electromagnet provided with two arched polar pieces, each one of which curved around, on opposite sides, a little more than one third of the circumference of the ring. The continuous spiral had been replaced by sixteen coils mounted in series and equidistant from each other, with projecting pieces of iron between them to reduce the distance between the ring and the polar pieces. Electrical contact with the coils was effected by a cylindrical commutator, centred upon the axle of the ring and carrying sixteen metal segments, each communicating with the sixteen pieces of copper wire which linked up the sixteen coils in series. Lastly, the sliding contacts were replaced by metal rollers. But Pacinotti never succeeded in building the full size machine.

Work by Werner Siemens contributed to the decisive stage. Working at the same time as Pacinotti, Siemens had been trying to finalize the development of the electromagnetic machine since 1857. In 1865, he devised the dynamo using the principle of the electromagnet with an auxilary generator to excite the magnet. Together with Charles Wheatston, Siemens succeeded in producing some currents of strong intensity.

But it was the Belgian, Gramme, who finally succeeded in building the first practical dynamo by putting together elements that had already been used several times by his predecessors. By doing so, he ousted the battery, the storage battery developed by Planté in 1859, and the alternator (fig. 9). After his first patent of February, 1867, for several improvements to alternating current machines, he took out the definitive patent in November, 1869, "for different improvements to magneto-electric machines," and this was completed by two certificates, one of which, the certificate of April, 1870, contained the theory of the machine. The first major generator was built in 1875, with a small axial dynamo exciting the magnets. At the International Exposition at Vienna in 1873, the Frenchman, Hippolyte Fontaine,

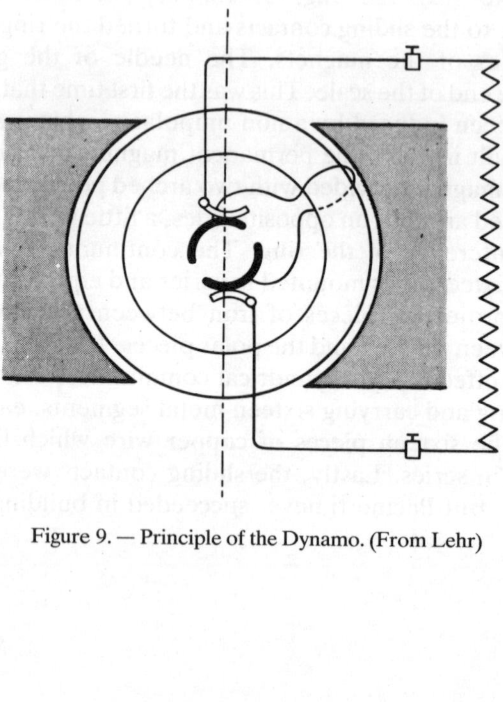

Figure 9. — Principle of the Dynamo. (From Lehr)

inadvertently connected up a dynamo with another machine running 250 metres away. If this was indeed the way in which the reversibility of Gramme's machine was discovered, he would also have found that electricity could be transmitted over some distance. Marcel Deprez's successful demonstrations of transmitting electricity between Creil and Paris in 1883, and Gaulard's invention of the transformer in 1884 gave electricity its real dimension.

But the electrification of the modern system – and we will see this in a number of domains – remained dependent upon energy converters. The hydraulic turbines powered by high waterfalls, the steam turbine – and Parsons' first turbine was precisely intended to turn an electric engine – and Diesel's engine were all developed just in time to be used to produce electricity. In fact, Parsons presented his first steam engine in 1884, the same year that Gaulard built the first hydroelectric generating plant driven by the waterfall at Bellegarde in the French Alps.

The problem of energy was therefore completely transformed. The old prime energy converters which were limited, inflexible, and stationary, were replaced by superior converters with almost limitless ranges of power, some of them mobile as well, enabling the transport of energy over long distances, thus fulfilling one of the most fervent dreams of mankind.

The second great advance of this industrial revolution seems to have been made in the field of materials, for it was materials that largely conditioned many other technical innovations of the period, and conversely new materials naturally depended upon advances made in the energy and chemical sectors. This involves a whole history of events, and the reader will not be surprised to read that once again, despite the existence of a particularly rich area of documentation, much remains to be done. There are, however, some exemplary cases, such as steel and man-made textiles, but there are many others which are equally not well documented and we are therefore obliged to make rather vague statements.

We have already said that iron imposed strict limits in certain technical fields. Although in other cases, such as the framing of bridges and buildings, which were relatively few in number, iron would increase in use for some time to come. Large scale production of rolled iron, particularly of "T" and "U" beams, supplied directly from the rolling mills and no longer laboriously forged, had begun about

1845-1848. Thus, the iron frame, which had until then been almost an object of curiosity, became a standard building technique. The first train sheds of the preceding period, the development of which goes hand in hand with the railway, were henceforth used in more diverse buildings: from the roof of the *Bibliothèque nationale* constructed by Labrouste to Baltard's famous umbrellas of iron and glass over Les Halles, to name but two in Paris. And if one considers the late second half of the century, one rediscovers the dominance of iron, at the Eiffel Tower, completed in 1889.

The pressure of certain requirements was, however, to direct the iron industry along other roads and towards a new material, even in circumstances where iron could have sufficed. One should point out from the first the fact that the appearance of a metal which was already known, but was produced in cheap large quantities, naturally entailed a number of modifications in the equipment used in this industry, both "upstream" and "downstream" of the actual production of steel itself. Steel was indeed the solution to many essential requirements and was therefore to be produced in ever greater quantities. "Upstream" lay the problem of pig iron production, which as we shall see presented special difficulties. The blast furnace was the same as before, but output had to be increased: its mechanisation, especially of the charging operation, was to reach its peak by using electrification. The operation was improved, with the aid of technological and scientific research, so as to obtain the products necessary for the fabrication of steel. Around 1875, at the Terrenoire works, Pourcel adapted the blast furnace to produce the ferromanganese indispensable for this process: he eliminated the corrosive action of the manganese on the crucible's siliceous linings, by lining it with carbon bricks, and also found that the furnace had to be heated to a very high temperature and with a slag which showed a high index of basicity.

The most spectacular innovations led to the mass production of steel. During the time of the Crimean War, Henry Bessemer, a self-taught and prolific inventor, became interested in the problem of artillery. At first he intended to create a aerodynamically shaped shell, which he patented in 1854. Rebuffed by the British War Office, Bessemer persuaded French authorities to carry out trials at Vincennes where it was quickly apparent that only steel cannons were strong enough to fire this new shell. The problem was now how to produce this steel in large quantities and at a cheap rate. Bessemer quickly set to work and began by trying to produce more steel with the processes already in use. His first patent of January 1, 1855, dealt with nothing more than the very old process of cofusing pig iron and wrought iron, to produce a median, steel. The second patent of June, 1855, also kept to well beaten paths. A third series of experiments took him closer to his goal. Bessemer injected air and water into molten raw iron. This was in fact the old process of refining pig iron already resumed by James Nasmyth in a patent of May, 1854. The effect of injecting air or steam was to agitate the molten raw iron and cause chemical reactions with the carbon and sulphur in the iron owing to the introduction of oxygen and hydrogen. But the injections also caused cooling and excessive oxydation, thus involving a fire risk. The water steam was first used to remove the sulphur, following Nasmyth's quite erroneous notion of the role of hydrogen. The converter first appeared in the patent of December 5, 1855 (fig. 10). Although the problem of chemical reactions was important and could only be resolved by successive empirical attempts, another essential problem existed, that of the molten iron's container, which affected everything else. Following the 18th century example of Huntsman's crucible, in which malleable iron was obtained by the injection of air, Bessemer devised a crucible of refractory material housed in a wrought iron casing. On the advice of his friends, Bessemer demonstrated his converter before the British Association for the Advancement of Science at Cheltenham on August 11, 1856. The response was enthusiastic and the London *Times* covered the whole exhibition on August 14. A picture of the converter was reproduced in the *Illustrated London News* three days later. Immediately, the large ironmaking firm of Dowlais, envisaged an annual production of 70,000 tonnes of steel. Bessemer quickly sold licences, taking care to deal only with large firms which were capable of perfecting an invention which was still far from its definitive development. Then a series of tremendous setbacks ensued.

The reasons for these setbacks should be sought. Indeed they reflect the limitations of the inventor. Bessemer's knowledge only came from reading popular technical encyclopedias. Every time he had a problem he immediately consulted all the patents which had anything to do with his invention. Although he managed to build his converter, he could not do much beyond this. The converter presented many problems, the most fundamental of which was the nature of the lining used in the retort. Bessemer therefore sought the advice of two prominent met-

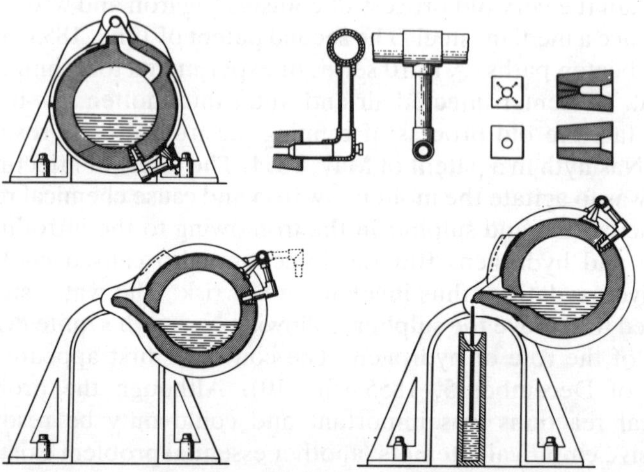

Figure 10. — The Bessemer converter from the 1855 patent.

allurgists, Riley and Percy who advised that the raw iron should be completely free of phosphorous. But neither the use of pure hydrogen or boric acid, nor the use of metallic oxides and many other remedies were successful. The Swedish metallurgist, Jöran Frederik Göransson, who had obtained a licence as early as 1857, increased the volume of air and injected it at the bottom of the converter rather than from the side. A successful trial took place in July, 1858. It was Robert Forester Mushet, a professional metallurgist of renown, who perceived that an excess of oxygen was causing the molten iron to be too completely decarburized. He therefore added a mixture of wrought iron, manganese, and carbon, and obtained a good quality steel in September,1856, and then again in 1857, by stopping the decarburation earlier. The addition of manganese also facilitated the operation. Muchet took out patents in 1856 and 1857, but neglected to renew them in 1859, and Bessemer took out similar patents in 1861. But by that time no backer was willing to bear the expense for the development of the definitive version, so in 1859 Bessemer was forced to set up his own pilot steel plant at Sheffield. At the same time, Jackson, whose father had introduced crucible steel in France, bought a licence and set up another pilot plant at Saint-Seurin, near Bordeaux. There, the converter was given its definitive form and mode of operation, for which the additional patents were taken out. Note that Bessemer meanwhile had an American emulator, William Kelly, who may have discovered the same process between 1847 and 1851 but failed to bring it to the industrial stage.

The Bessemer process suffered from two major disadvantages. With a siliceous, that is, acid lining, neither phosphorous nor sulphur could be eliminated, which considerably limited the range of useable ores. A large part of iron ore deposits contained various quantities of phosphorus. Moreover, as it received no thermal heating whatsoever from the outside, the converter could neither remelt or reuse the scrap iron.

May we digress for a moment to discuss another subject. Fuel, firebox, and heat were the essential elements in metallurgical techniques. All the techniques used could only reach a certain temperature, which presented serious limitations upon the methods employed so that steel could not be made out of scrap iron or steel as they both required a high temperature for fusion. Wilhelm Siemens took up an idea which had already been patented in 1816 by the Scott, Robert Stirling. Friederich Siemens, brother of Wilhelm, took out a patent on December 2, 1856 on the regeneration of heat. They suggested capturing the hot gases as they travelled up the furnace chimney and using them to preheat the air going into the combustion chamber. Thus part of the lost heat would be used to raise the combustion temperature sufficiently without much increase in operating costs. The "regeneration" chambers were lined with bricks and the captured hot gases were directed through chambers, and directed back and forth among them by a system of vanes. The inversion of the vanes enabled fresh air to be drawn into this exchanger and heated before being blown into the combustion chamber. Wilhelm Siemens saw the possibility of using these heat "regenerators" to attain temperatures high enough to melt metals which were difficult to fuse, and took out a patent for this in May, 1857. But trials conducted at Sheffield with coke gases failed owing to the bad quality of the refractory material. So the apparatus was only used in glass and porcelain industries. In January, 1861, the Siemens brothers patented a gas-producer with inclined walls which could produce gas from bituminous coal. An incomplete combustion with little air yielded an oxide of carbon which was combustible at 500°C and was a very versatile fuel; it could even be preheated to 800° or 1000°C. Siemens' regenerator apparatus was quickly put to use to reheat the air blown into the furnaces by using the throat gases. Cowper built the first apparatus at the factories at Clarence in 1860. But due to the inability to decoke the stacks, the process was abandoned. It was only in 1869 that Whitwell and Cowper succeeded in developing better purification and designed ways of keeping the apparatus from clogging so that the regenerators could be used again.

In 1861, two French metallurgists, the Martins father and son, gave up their own research on gas-generator and the regeneration of heat. They then took out a licence from the Siemens and the Siemens company sent them one of their own engineers, Nehse, to convert the form and material of their reverberatory furnaces, using siliceous material. On April 23, 1863, the younger Martin succeeded in remelting a piece of crucible steel on the hearth, without using a crucible. Meanwhile, in February, 1863, following the work of Louis Le Châtelier, the Fourchambault Company took out an analogous patent on the methods of making steel from mixtures that later became today's two major processes, called the scrap iron and ore processes. Nehse once again travelled from Germany to develop the processes, but he failed because of the bad quality of the refractory material. It was Martin who persevered and succeeded on April 8, 1864, in melting a mixture of pig iron and wrought iron, conforming entirely to the method Réaumur had advocated a century before. His experiment resulted in several patents, from August, 1864, to March, 1865. Bessemer had never experienced any temperature problems as the combustion of carbon in contact with air was sufficient to raise the temperature of the smelting bath. In contrast to this, Martin required a high temperature to use the principle of cofusion. Thus Réaumur's idea and the Siemens' apparatus formed the basis for producing Martin's steel.

Bessemer steel and Martin steel both had their advantages. The first method was quick and relatively cheap. The second was slower but sure, allowing time to make adjustments and control the quality of the steel during the operation. Both processes were limited, however. The linings used in both were siliceous and hence acidic, which meant that only very pure ores could be used, with no trace of phosphorus, the elimination of which presupposes the use of a basic slag rich in lime. This had been perfectly demonstrated by Grüner in 1869. Research in this area was undertaken immediately. From this date, Müller advocated the use of magnesian refractory material. In 1872, Snelus came to the same conclusion. But trials carried out at the Krupp works in 1877 were a failure. Two cousins, Sidney Gilchrist Thomas and Percy Gilchrist, presented their discovery on May 28, 1878, at the London Iron and Steel Institute amid shouts of laughter. They used dolomite, a mixture of lime and magnesium. The first trial took place in the Bolckow and Vaughan factories and the first basic casting took place on April 4, 1879, in a Bessemer converter with the new lining. In France, Pourcel at Terrenoire and Walrand at Le Creusot did the same with the Martin process. The Thomas process was first used in the De Wendel plants at Hayange, a region rich in phosphorous ores, and his method was definitively perfected in 1880-1881. From 1874, the steel rail took the lead from the iron rail which had virtually disappeared by 1885. Sheet steel supplanted sheet iron from 1891. But only in 1900 did commercial steels surpass the tonnage of commercial irons which in 1913 still represented a third of the total production.

But the process went further. Once steel had been developed, it had to be endowed with various properties for particular functions. These properties were sought for in one well determined area, that of alloys, which had been made possible by the discovery of numerous metals and the development of the industrial processes to produce them. The fabrication of armour-plating may have been the basis for these efforts. Boussingault, working for Holtzer near Saint-Etienne, began chemical research and gave an engineer, Aimé Brustlein, the task of working out the industrial applications using the crucible furnaces in the plant. Brustlein began by making a ferro-chrome by using a charge of chromite and charcoal with tar as an agglomerant. The steel obtained titrated at 10 to 84% chrome and 2.5 to 11% carbon. By incorporating them into the charges of the crucibles, the first chrome-steel was produced in 1878. These steels proved to be self-hardening. The age of special steels had thus begun.

From 1883, Marbeau began to develop nickel steels, and research continued in this direction. Le Creusot took out a patent on the production of ferro-chromes in October, 1888. At the same time, the Englishman, Harvey, was developing nickel steel with only 2 to 3% nickel content, destined for armour-plate production. By 1888, Robert Hadfield of Sheffield was working on the systematization of special steels. He finalized the process of making maganese steel, and introduced silicon steel. Thus science could resume the rather disorganised relay in industrial research. The laboratory at Imphy, in the Nièvre region, made indilatable steels in 1897, then alloys with particular physical properties (elinvar), a steel with 34 to 36% nickel and 12% chrome. From 1891, alloys of steel-nickel-chrome were available. Towards 1900, it was the turn of rapid cutting steels, developed by White and Frederick Winslow Taylor, with 18 to 20% tungsten and 5 to 7% chrome. Henceforth, steelmaking was no longer subject to trial and error; the industry now possessed the scientific knowledge necessary to carry out systematic research.

The introduction of steel as the basic material of the iron and steel industry, and alloy steels with special properties were henceforth materials vital to the modern technical system. Machines conceived of but previously impossible to build could now be produced: electrification and the diverse new energy converters, the new means of transportation, and machine tools immediately took advantage of the new discoveries which also revolutionized other techniques, particularly military techniques.

The domain of chemistry and chemicals, which also played an important part in the materials necessary to other industries, was to be transformed no less radically. Until 1850, the processes discovered shortly at the end of the 18th century and in the early years of the 19th century to make hydrochloric acid and sulphuric acid, the two essential products of the industry, remained almost the same. Soda, chlorine, and sulphuric acid were the fundamental trilogy. Despite some improvements in the equipment and some variations in chemical processes, the chemical industry had not changed much. Solvay's discovery in 1866 of the fabrication of ammonia-soda was of more economic than technical interest because it produced only soda and not hydrochloric acid, which was an important product. As for sulphuric acid, the combining of the Guy-Lassac tower (used industrially at Chauny in 1842) and Glover's tower of 1859 allowed certain by-products to be reused, especially nitrous products.

The chemical "revolution" in fact resulted from several important inventions which we will briefly list:

I – The chemistry of dyes was completely transformed by William Henry Perkin's 1856 discovery of aniline which was used to produce mauve dye. Gradually all the other dye colours were to be produced in the same way: red in 1859 and black in 1863.

II – This was to lead to a chemistry of synthetics that would first replace vegetable dyes and other natural products and then very quickly produce a range of products unavailable in nature except in limited amounts. Synthetic quinine was followed in 1879 by Baeyer's synthetic indigo, and in 1875, vanilline opened the way to synthetic perfumes.

III – It was only a step away from the development of the first artificial materials, that is to say, materials unknown in nature before, which could become the basis, and the fundamental basis at that, for other techniques, or better substitutes for natural products owing to economic rather than purely technical advantages (greater physical or chemical qualities and greater diversity). In 1868, Hyatt developed the process of producing celluloid, the first artificial material. In 1897, galalithe replaced animal horn, and bakelite was developed in 1909 to replace natural resins. We must also cite the discovery of artificial silk in 1884, by the Count of Chardonnet, and dynamite in 1867, by the Swede, Nobel. These were only a few of the new products.

We should point out that there is no good history of the chemical industry. Such a history would have to begin by listing the products obtained and the processes by which they were produced. This would reveal their technological lines of descent and clarify the inevitable interdependencies which were intensified by the reversibility of most of these production methods. Such a history would analyse the reciprocal effects between these new chemical and other technical sectors, and the industries that employed them. This would establish the necessary links between the technical and economic needs of the user industries and the chemical industry's efforts to improve its old processes, create substitutes for existing materials, and invent new ones. We should emphasize that we are a long way off from such a study. And what of the race for greater effectiveness, such as the advance from dynamite (1867) to melinite (1884) and then to tolite (T.N.T., 1910), or, in a broader context, the race between complementary techniques, such as the endless battle between armour-plating and firepower; similarly, the birth of cement, then reinforced concrete used in contruction work, superphosphates for agriculture, or all the successes of photography after Eastman (1886).

In the domain of the production of materials, we will witness another "marriage," that of electricity with a certain number of fabrications, owing to the former's thermal and chemical properties.

The first electric furnace was patented by William Siemens in 1879, and it was quickly discovered that electricity had certain properties which could decompose particular bodies and deposit the different elements on one terminal or the other. In 1800, Carlisle and Nicholson had succeeded in decomposing water by electrolysis, but only as a laboratory experiment. Industrial application naturally had to await the advent of cheap electricity in large quantities and the availability of the materials necessary for such fabrications. Meanwhile, research continued throughout the first two thirds of the 19th century, especially on anodes and cathodes.

To return to a history, the beginnings of which we have outlined, it was between 1880 and 1885 that electrolysis was applied industrially to produce a soda derivative. The first work to manufacture aluminum by electrolysis, patented by both Hall and Héroult, dates from 1886. A few years later in 1892, Moissan used a variant of Héroult's apparatus to reduce metal oxides with carbon, and to produce carbides or ferro-alloys. But Moissan only used electricity as a source of heat. At the same time, Chaplet, in conjunction with the company *La Néo-Métallurgie*, endeavoured to industrialize Moissan's furnace and his scientific work. A patent resulted from this work in 1895, concerning a process to alloy refractory metals with more easily melted metals. One of the latter was melted first, whereupon the refractory metal was introduced in the oxide state, mixed with the amount of coal necessary to reduce it. Paul Héroult resumed his research and adapted his furnace for making iron alloys and then steel. In 1898, he patented the process for making chrome irons with a high chrome content, by treating any chromite in the electric furnace, keeping the current on "after the reduction and melting are complete, until the total or partial volatilization of the iron. As the boiling point of chrome is higher, the product gains in chrome while losing in iron;" in 1900, the patent for the production of steel was issued.

All electric arc furnaces comprise of a tank with a refractory lining; the electric current is brought in by a carbon electrode through the top. In the systems designed by Héroult and Keller, the current is provided by other vertical carbon electrodes, suspended in such a way that the electrode tips nearly touched the bath of molten metal. Flashing between one electrode to the other, the electric arc passed through the bath of molten metal. In Chaplet's design the current returned through the bath, then through an "ad hoc" canal to a steel electrode on the outside of the tank. In this case the arc flashed between the positive electrode and the bath. Paul Girod's mechanism, published in 1898, was similar to the preceding one, in that the floor of the tank was a conductor in the circuit of the returning current. But it was Héroult's design that stood the test of time.

Ernesto Stassano, an artillery engineer in charge of the steel founderies at the arms factory at Terni, north of Rome, tackled a different problem. He tried to use electricity to make steel directly from the ore. In his patent of March, 1898, he adopted Moissan's arc-furnace, using the thermal energy of the arc to effect the physico-chemical transformations involved in ore-based production. The patent specified the preparation of the charge and the form and details of building the furnace. It was to some extent, the combination of an electric furnace and a blast-furnace, controlled to obtain a quality product. He succeeded in obtaining a product with only a few traces of sulphur. But for economic and practical reasons, like those designed by the Swede, Kjellin, in 1900, these furnaces were quickly forgotten.

Electricity was therefore an important contribution in the development of certain materials, apart from its role in the transmission of mechanical energy. Until the beginning of the 20th century, it was above all thanks to electrolysis, that electricity could contribute to this major work, the creation of new and economically viable materials.

In any case, these new techniques depended upon improved exploitation of natural resources, that is, precisely the field where technical progress had always been more or less slow. Although progress may have been gradual and hardly noticeable, technical changes were nevertheless extremely striking. And if our manuals, or "histories" barely mention them, it is probably due to the fact that this progress affected the technical surroundings of these sectors more than the central techniques of natural resource exploitation. These surrounding techniques were in turn heavily dependent upon other tributary techniques.

Agriculture is certainly one of the best examples of this phenomenon, and it is particularly regrettable that we have no good history of agriculture in the 19th and early 20th centuries, as it would surely demonstrate progress taking place during this period, and the predominant influence of other techniques in almost every field. This would show, if necessary, how the notion of a technical revolution is closely linked to the research undertaken: everybody is aware of the "agrarian revolution" of the 18th century, as recent works have defined it, but there is a complete lack of information for later periods. In agriculture itself, there are many long lasting problems, with slow and gradual solutions: such is the case of the species cultivated, linked now to botany and to agricultural experiment stations, be they official or not. Today, the increasing number of hybrids allows much more land to be cultivated, with much higher yields. But during the period of the modern technical system, the real advances were undoubtedly of another sort, three of which should be emphasized.

The first was soil regeneration owing to the use of fertilizer. This made it possible not only to maintain yields but also to cultivate newly developed high yield plants. Of course the use of fertilizers had been known for a long time, as we have already noted, but their effects were still limited and only natural fertilizers had been used. After 1850, guano from South America was used on a growing scale. Phosphates were also used on an increasing scale, coming from Wissant after 1857 and Tunisia after 1873. But chemistry soon came to the aid of agriculture by furnishing it with man-made fertilizers: hence the development of superphosphates after 1870.

The second advance was also linked to scientific progress. It involved the protection of plants. Traditional methods were gradually superseded by scientifically developed techniques. Small wonder then that Pasteur was interested in this. First, it was necessary to diagnose the ills to be treated and then seek the appropriate treatment. After 1850, oidium was treated with sulphur, and after 1869, milldew was combatted with sulphate. The battle against parasites was far more difficult. We know that when phylloxera made its appearance, threatening to destroy the vineyards of Europe, all sorts of extreme cures were tried, even submersion. But in the end the vineyards were to be saved only by grafting on American vine stocks.

One of the most resounding manifestations of agricultural progress was the development of machinery. Some farm machinery had certainly appeared before 1850, but its real development, dependent upon mechanical improvements and special energy converters, actually came about in the last half of the 19th century. Considerable progress was made in the machines themselves, developed for more and more comprehensive functions. The development was as follows: the reaper of 1835 became the reaper-baler of 1852, which became the harvester-thresher of 1855. In 1858, 73,000 reapers did 70% of the harvesting west of the Alleghanies in the United States. Seeders became effective, and fertilizer spreaders were introduced. Even before 1870, the main agricultural machinery was in use, and it was to be developed and improved as it came to be powered not by animals but by engines. Steam engines, first used in 1849-1850, were widespread by 1870; petrol-driven tractors first appeared in the United States in the 1890's. It was this group of innovations that overcame the scarcity and high cost of labour which was to enable such vast expanses of land to be cultivated in America, where production could still be increased by bringing new land under cultivation. Although the fundamental structures of agriculture could hardly be changed, at least its complementary structures benefited in this period from numerous useful inventions.

Similar changes may be observed in the mining industry. Here again neighbouring techniques contributed essential elements in technical mutation. Firstly, we should point out the use of new materials. Thus wooden structures were replaced with cast iron, wrought iron, and then steel structures. This evolution began shortly before 1850. In 1838, Triger suggested lining shafts with tubulated cast iron joined with lead instead of the old caulked wood lining. The development of iron framing allowed wood to be replaced in all its traditional mining uses. In 1899-1900, at the end of our period, Portier made the shaft lining water-tight with injections of cement. Progress was also made in the sinking of shafts in soft or watery earth. In 1840, the same Triger perfected a method of digging shafts with the aid of compressed air. In order to sink a vertical mineshaft so that the lining could be inserted at the same time, Triger instructed the workers to dig in a compartment at the bottom of the pit, where compressed air, maintained at a bearable pressure of three kilogrammes per square centimetre, overcame the pressure of the water, forcing it out through a pipe. This compartment was separated from the open air by a double closing which acted as a lock-chamber.

Much more important were the advances made in the actual mine-working itself and in the mechanisation of a whole series of accessory operations. And these advances were quite naturally the result of the progress of techniques in neighbouring techniques. The compressed air percussion drill was invented by Sommeiller in December 1858, for drilling the Mont-Cenis tunnel. It was quickly adopted by the mining industry, in replacement for the old pick and crow bar. The subsequent increase in productivity was enormous, but at the same time it became necessary to mechanise the operations following this process. Electric motors were used for transport within the mines, and for manoeuvring the cages. The same was true of the operations of washing, grading and crushing. Furthermore, the problems of water drainage and ventilation also benefited from the progress made in motors. It was at the end of the 19th century and at the beginning of the 20th that the first Rateau turbines were used for ventilation purposes.

We need only mention briefly the new source of energy, oil. Oil "sources" had existed for a long time, but the real beginning of this industry came with the drilling of the first oil wells by "Colonel" Drake, in Titusville, U.S.A., in 1859. There were two main branches of techniques in this industry. The first of these was concerned with extracting the oil, and progress in this field was extremely rapid between 1860 and 1900. Suitable materials were soon found for the drills and pipes. We shall consider the problem of transport further on. The second branch of techniques was concerned with refining the petrol. Originally, the petrol was used for lighting, and the manufacturing of paraffin lamps meant that little or no use was made of the by-products. The development of the internal combustion engine brought about the use of petrol to its fullest potential. At this point the oil refineries had to perfect the processes used, and vastly increase their output. Furthermore, this had to be done fairly rapidly.

There were profound changes in transport techniques, and this was in fact vital because as production was developing, there was more and more traffic to be catered for. Happily, this is one area where there has actually been a lot of research, even if there are still some sectors to be investigated.

The railway was one of the areas of techniques which maintained from the preceding period a slow but steady progress. We have already examined some aspects of this progress, and established that this particular form of transport might have reached its limits if a new technical revolution had not introduced some critical elements in its development. The use of new materials, particularly steel for the rails, and for the wheel bands which were made without welding, and also for the boilers used in locomotives, was a great economy for the industry, and this also encouraged the development of the actual trains and the speed at which they ran. Other forms of progress, such as technical transfers, supplemented this first group of innovations.

Most of the research was concerned with traction. Giffard's injector, which provided an automatic supply of steam to boilers (1858), the locomotive with coupled axles (1864) and Anatole Mallet's compound machine (1878) all paved the way for the large modern machines whcih were approaching their limits of development. The "Pacific" locomotive built in 1907, used superheated steam, weighed 100 tons, and was capable of producing 3500 horsepower. But the speed and weight of trains meant that further safety measures were required. In 1869, an American, George Westinghouse, developed the compressed air brake. The automatic engaging of shuntings and signals was devised by a french engineer, Viguier and industrialised in 1880 by Saxby and Farmer from England. After 1885, electric signalling was introduced, and this meant that much higher speeds were now possible. Thus, despite appearances, the railways at the end of the 19th century were in fact considerably different to those of the mid-century. But as far as steam traction is concerned, there was no possible further development.

The modern ship began to evolve during the mid-19th century. At this time, ships were made of iron and powered by steam. The large paddle wheels had been abandoned in favour of screw propellors. Not all ships were the same, however; there were three main types. There was the warship, the transatlantic liner, which was first built around this time, and finally, freight ships. Speed was the essence of the first two, and maximum carriage was the priority of the third. Of course, the speed and power of a machine are interlinked, and it is well known that at this time ships were usually powered by a steam-powered reciprocating engine, heavy and cumbersome, especially if one includes the freight. Since most maritime transport did not require great speed initially, it was natural that the older types of sailing ships continued to be used with fairly weak auxiliary engines. In contrast, part of the available carriage space had been sacrificed in liners and warships, to obtain higher speeds. At any rate, there were still several problems to be overcome. Unfortunately, the history of shipping has usually been concentrated on the more unusual vessels, so that the overall problem is difficult to assess.

Progress was naturally concentrated on the more exceptional types of ship. After a certain time, the differences between liners and immigrant ships became more marked. There were problems in two main areas, on the one hand with speed and tonnage, and on the other with the power and bulk of the engine apparatus. Inevitably, efforts were concentrated on solving the latter, whilst at the same time trying to reduce as much as possible the weight of the hull. Only the improvements made to the classic steam engine and the emergence of new materials were capable of solving the problems created by the development of maritime trade – the ship's power could now be increased without any corresponding increase in the size or the weight of the machine. It was also possible to increase the ship's tonnage, with lighter, but more resistant hulls. The progress from 2500 ton ships in the mid-19th century to 20,000 ton vessels at the beginning of the 20th century, was slow but steady. The same was true of warships, and all the more so after the technique of armour-plating was developed after 1854. Sometimes, engineers were too quick to introduce new elements, and this usually resulted in failure. Take the case of Brunel's "Great Eastern" for example, which was launched in 1857. It had a capacity of 19,000 tons and was 211 metres long. It could carry up to 4000 passengers, and 6000 tons of freight. One cannot say that he was ahead of his time; it is more likely that there was not a large enough market as yet for this type of ship. We should also point out that it was still powered by enormous paddle wheels measuring 17 metres in diameter, and each one weighing 185 tons. They were unable to provide it with a powerful enough motor, and they had to compensate for this by using large sails. The limits of existing techniques had been reached.

The adoption of the steam turbine constituted a major step forward. The "Lusitania" and the "Mauritania," built in 1907, were the first large liners to be equipped with these. This meant that suddenly these vessels were capable of 68,000 horsepower, and a displacement of almost 40,000 tons. Sails continued to be used for a long time on slow freight ships, however. In 1850, forty sails were used for every three steam engines, in 1870, sixteen sails were used for every three engines, and by 1885, this ratio was one to one. It is possible to consider the revolution in maritime transport as complete by this date.

The combustion engine gave rise to other means of transport. People had dreamed of creating means of road traction, and despite all the difficulties involved, Cugnot had devised the first steam-powered motor vehicle (1769-1771). There were many attempts throughout the 19th century, which culminated in the various vehicles devised by the Bollées, built in Le Mans, and which were derived precisely from the research on the steam engine. The first of these was the *Obéissante*, built in 1872, which was a twelve-seater weighing 5 tons and capable of reaching 40 km/h. Then there was the *Mancelle*, (1878) designed to carry 16 passengers and capable of going at 42 kmh. This vehicle was already much more "modern" in appearance. The *Rapide*, built in 1881, reached speeds of up to 60 kmh. Obviously, other avenues had to be explored.

Lenoir had already considered using his engine in road vehicles. But it was not until engines were improved, or petrol was first used, along with lighter materials, or until certain other tributary elements developed, that more practical, manageable vehicles could be produced. The first three-wheeled vehicle was created by Benz in 1886, then Daimler attempted to devise a workable model with Maybach in 1889 and Panhard from France in 1891 (fig. 11). Peugeot also had a role to play in these developments (1887-1890) which ultimately succeeded. Motor races, the first of which was from Paris to Rouen in 1894 and the motor shows which began in 1898, both contributed to the expansion of these techniques. As the reader is no doubt aware, it would be very useful to write a comprehensive history of the automobile, with a thorough technical analysis of all the constitutive elements of this new technical ensemble, from the motor and transmission, to the bodywork, and from the materials used to more detailed aspects like the discovery of direct drive by Renault (9th February 1899). The Mercedes of 1901 had the framework of the chassis stamped out, 4 cylinders developing 35 hp, gear change and a cellular radiator. The automobile industry has a large archive collection, which makes it almost easier to study in fact than more recent industries. The abundance of written sources helps to compensate for the lack of large motor museums, and we are thus able to obtain information about virtually all the models which have ever been made. There are other factors worthy of study, such as the economic and social context, the imperatives and needs, and all the by-products of the motor car, from the motorcycle to the lorry. Also significant are the technical requirements which have conditioned to a considerable degree the nature of certain projects.

The history of aviation is more well-known, probably because it is more spectacular. Ever since Montgolfier devised the first aerostat, people had been trying to conquer the skies. The balloon and airship were interesting side developments, but they remained outside of the

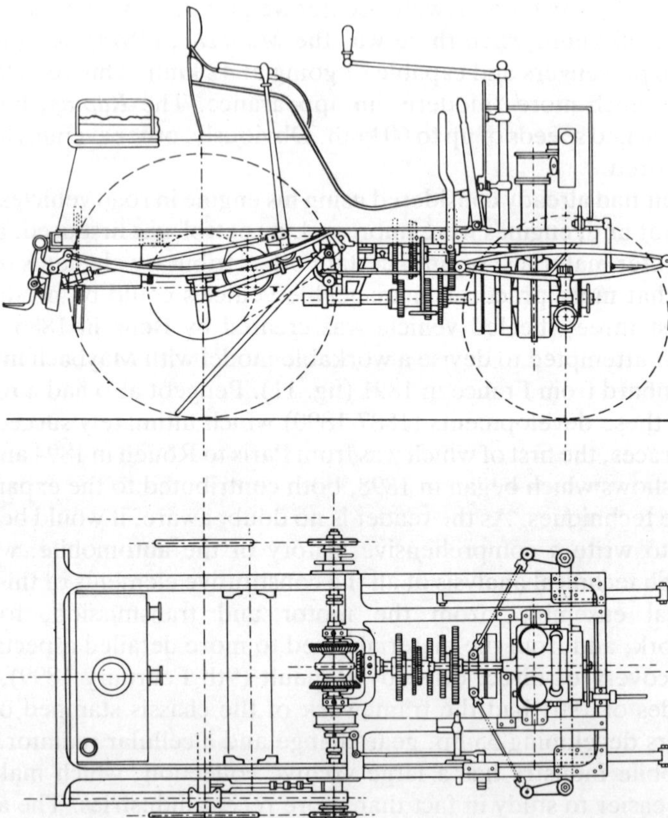

Figure 11. — Global mechanical organisation of a Panhard and Levassor dog-cart (1891-1894). The engine is a V-type two-cylinder 4 h.p. engine (2 cylinders 80/120).

mainstream aviation industry. The first airships were launched in 1851 and 1852 by Arnauld and Giffard, then by Le Bris in 1857, and Penaud and Gauchot in 1876. It was Clement Ader who devised the first motor-powered aircraft, in 1890. The *Eole* as it was known, was a 20 hp machine, weighing 15 kg in horseweight, and it actually succeeded in taking off. In 1897, the *Avion* beat this by travelling over 300 metres.

This apparatus was shaped like a bat, it was 16 metres wide, and had two propellers each driven by a small steam engine. Exactly the same processes were used in the automobile – the only difference was in the problems they had to solve. In fact, Ader's experiments made some elements much clearer: firstly, the engine should be as light as possible (even though it was a steam engine, it weighed only 3 kg per hp.) Then there was the frame and the stability to consider, and finally, the science of piloting. It was also necessary to abandon the idea sustained by the myth of Icarus, and perpetuated by Leonardo Da Vinci, that anything heavier than air should be built like a bird. Mouillard in 1881 and Marey in 1889 were still convinced of the truth of this theory. The experiments conducted by Lilienthal which ended so tragically in 1896, led to its refutation and this greatly benefited the work of the American Chanute.

The two brothers Wilfred and Orville Wright adopted a scientific approach to all aspects of the problem, and this was to prove successful. They began their research in 1889, resuming the studies conducted by Chanute. They adopted the biplane and modified the rudder apparatus. Finally, they incorporated a key element, the internal combustion engine. In 1903, they succeeded in taking off, but they only publicised this fact in 1908, by which time, several others had succeeded in doing the same thing, people such as Ferber, Esnault-Pelterie, Voisin, and Bréguet. Levavasseur devised the V-type engine, and the *Gnôme* motor was also developed. This was a rotary motor which proved to be very successful.

The historians of aviation have contented themselves with simply naming names, dates and performances. This is insufficient – what we need is the same catalogue of successive, global developments which finally led to the first successful flying machine. We need to examine the kind of motor used, the quality of the materials used, the frame and the rudders, the control systems and the methods of flying. The characteristics of the early flying industry cannot be over-emphasised. It was a sport, planes were made in a makeshift fashion and there was a sense that anyone could have a go. It was this, not economic considerations, which was the principal dynamic of this industry.

The methods of transport and communication had thus been completely transformed, and they were to benefit from new resources. Some were exclusively technical, while others were effective in more indirect, complementary ways. These should be considered more carefully.

The new complementary techniques available stemmed largely from the new methods of traction, i.e. the engine. There are two main principles to this, and in chronological order these are electricity and the Diesel motor. Since the discovery of the reversibility of the Gramme machine, it was possible to conceive of means of transport powered by electric motors. Deprez' work on the transmission of electric energy showed that it would be possible to convey electricity by aeriel lines. Siemens conceived of the first electric locomotive in 1879, but originally it was used only for urban transport, largely because they were unable to consider electricity being conveyed over long distances. Tramways, above ground, were used for the first time in Berlin in 1879, and the underground trains ran for the first time in London in 1887. It was only some years later that the process was applied to railways, that is, over larger distances. In 1907, the Austrian line of St. Pölten to Mariazell was one of the first to be electrified.

The second means of traction was the Diesel motor, whose range of power was such that it was suitable for a number of uses. In 1903, the French constructor Sautter-Harle used diesel motors in barges, thus radically tranforming the conditions of navigation on rivers and canals. The logical progression of this was the use of Diesel motors in maritime navigation. It was also the first step in another major branch of navigation, that of submarines. At the end of the 19th century, the *Gymnote* and the *Gustave Zédé* worked with the aid of electric motors linked to accumulator batteries. These were extremely heavy, and relatively weak for their size. The transformation to autonomous submarines, with the Laubeuf's *Narval* and the American *Holland*, was effected by using a Diesel motor, the function of which was essentially to recharge the accumulators between two dives.

The changes in methods of transport did not simply change the conditions of transport, but they also extended the network of trade. This gave rise to new problems, concerning with transport itself, with stocking the merchandise and with safety measures. One of the first problems solved was the storing of grain, which of course was very important for reducing the problems caused by bad harvests. Originally, the grain was simply piled up which resulted in fermentation, making it unfit for consumption. Between 1850 and 1860, the first modern silos were devised, which kept the grain constantly moving so that it was unable to ferment. The use of electricity greatly helped this new process. The second main problem to be solved was the transport and conservation of perishable foodstuffs. The creation of large types of refrigerators using ammoniac or compression, was to provide an easy solution to this. In 1876, Charles Tellier devised the first ship with refrigeration. the *Frigorifique*. He worked not only on refrigerating transport, but also on all types of cold systems. Again, electricity was to play a major role in this field.

The second problem was transporting the new type of fuel, petrol. Petrol was first transported in crates, or "barrels," but for land transport at least, these were soon replaced by tankers. The main concern regarding maritime transport was fire, and in particular, the obstruction of the Suez Canal if this problem should arise. The first crossing of the canal in 1892 by the *Murex*, which transported the petrol in bulk, inaugurated the era of oil tankers. On land, pipelines soon replaced other forms of transport.

Transport is necessarily linked to space, such that the use of the latter needs to be planned and developed if transport is to be possible. As soon as the railways were built, supply routes had to be established. The supply of coal was not too much of a problem, but water supply was more difficult. They went so far as to consider digging channels between the rails, and having a shovel which could be lowered to force water up to the tenders. Even more difficult was the problem of the ports of call for steam ships. The invention of tinned food had given sailing ships a considerable degree of independence, but steam ships needed to be supplied regularly with more fuel and fresh water. The political geography of the world is a strong reflection of this. The Diesel motor did away with the problem of water, but the problem of fuel remained. This was even more acute given the distribution of depots, which posed many political problems some of which are still important factors today. Cars, of course, also needed to be refuelled with petrol, and furthermore, suitable routes needed to be built to accomodate them. This is a prime example of the interaction of techniques, and is worth noting here – a technical problem can never be an isolated one. The notion of a technical system does actually reflect a fundamental reality. Any line of research which is too isolated does not allow for the technical requirements of inventions, and technical changes. The history of technical progress will be fully developed as a discipline only when historians are made properly aware of this interaction. The reader should realise that we have schematised this discussion deliberately: we shall return to this point later on.

Labour processes were to benefit from the progress made in the manufacture of certain materials – particularly the production of special types of steel – and of new energy convertors. Again we are restricted to illustrating this with only a few succinct examples. The history of implements used in forging still remains to be written. Around 1880, the large steam power-hammers had reached their technical limit. These enormous hammers, the first of which had been installed in Le Creusot in France and in Terni in Italy, weighed one hundred tons. It was around this time that the first large presses were used, which were to reach up to 20,000 tons – this would have been unthinkable with the old machines. Once more, then, it had proved possible to go beyond certain limits. At the same time, electricity was to lead to the creation of reversible rolling mills. This dispensed with the need for delicate manoeuvres, which had limited the size of the pieces it was possible to work. It would be helpful to have an analysis of the effect of these changes on output, something which would not be so difficult, given the wealth of statistics to hand. Between 1850 and 1900, blast furnaces increased in capacity from 150 to 750 cubic metres. The increased speed and power of all equipment used in mechanical processes, caused output to be raised tenfold. In other words, some techniques may appear to have remained stable, in that the principles on which they were based did not change, but changes in neighbouring techniques considerably modified the conditions of production. The refining of waste gases using the Cowper apparatus, the production of coke, which developed between 1850 and 1860, and the emergence of reversible rolling mills are very important factors, which usually go unnoticed.

This is not such a problem with the study of machine-tools, where the relationships of reaction and interaction are multiple and much more obvious. A machine has the external supports of materials or energy, but as they become more sophisticated machines are actually capable of providing themselves with the required pieces. We have already noted several times that this phenomenon occurs fairly frequently in the history of techniques. In 1873, Spencer invented the revolving lathe, a semi-automatic machine, introducing several economies in the assembly and dismantling of tools, to work the pieces. More sophisticated forms of milling machines and grinding machines were created and put into widespread use. Because of its large output, the milling machine was more widely used in mass production than any other machine tool. And at the Philapdelphis Exhibition of 1876, it was demonstrated that the grinding machine had obtained more precision than ever before. With greater power and an increased productivity available, tolerances became more narrowly defined. Rathenau used to claim that up until about 1880, work done with hand tools could still compete with that done using machine tools, but after this, his claim was no longer true (fig. 12). Semi-automatic machines were replaced by machines which were fully automatic. The high speed steel used to replace carbon steel invented by Brustlein, Taylor and White, increased efficiency by one hundred per cent: carbon steel permitted a cutting speed (the relative speed of the tool on the surface being worked) of ten metres per minute. By 1912, this figure had risen to between thirty and fifty metres a minute. The history of machine-tools is marked by more and more radical advances: Browne and Sharpe's universal milling machine (1862); Hartness' semi-automatic lathe with revolving turret (1862); Spencer's automatic lathe; Gleason's machine for cutting bevel gears (1874); Brown and Sharpe's grinding machines (1876); Fellow's machine for cutting spur gears (1890); Kreutzberger's machine for sharpening milling cutters (1874); and finally Reinecker's machine for backing off tools (1882).

We need an inventory of all these tools, the context in which they were used, and the effects they had on other techniques. From this point, the machine tool was an indispensable part of production.

A global inventory would require not just a single chapter, but several volumes. We had only one choice, and that was to use examples which perform two functions. They not only demonstrate the existence of a new technical system, they also help explain the meaning and the significance of technical changes. We would like to emphasise here the importance of certain phenomena.

The first of these is the incredible diversity of the new materials available. We have already cited some of the latter, but added to these were alloys, which were a fundamental part of the new means of transport. These alloys were usually metals which combined lightness with other properties. One of these was iron brass, which was created by Dick in 1883 and used to make ship propellers. Aluminium bronzes were perfected by Sainte-Claire Deville and Debray in 1855: this metal was an alloy of 90% bronze and 10% aluminium, and had a high chemical resistance. In 1908, in a factory at Düren, Willm discovered duralumin, which was finally perfected by 1910. This metal was a

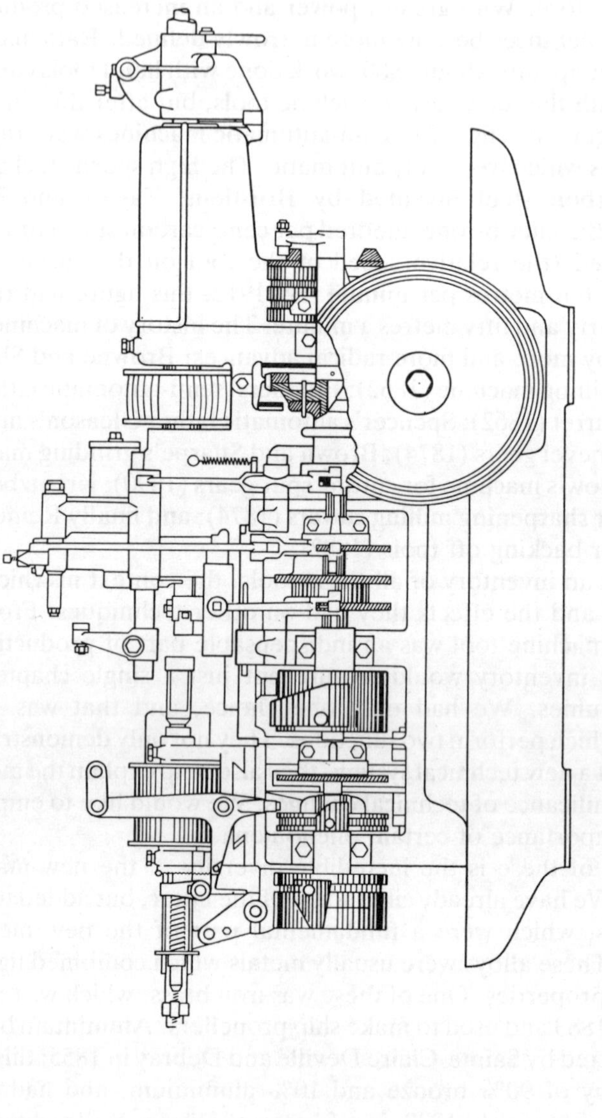

Figure 12.—Automatic lathe controlled by cams.

mixture of 93% aluminium, 5% copper, 1% manganese, and 0.5% magnesium. If nickel was added, this metal could be drawn. It was eventually used in the transport of electricity. Perhaps it would also be relevant to mention here the addition of new properties in quite a different area, that of food. Take for example the creation of an "artificial butter" or margerine, by Mège-Mouriès in 1869. Calcium carbide, which was discovered by Moissan was immediately used for several purposes. The discovery of acetylene in 1892, was to revolutionise the use of blowtorches.

A second aspect of this revolution was the increased output and therefore speed of production. Transport had to adapt to these new speeds, and so did the various means of communication. Morse conceived of the electric telegraph in 1843, but it was in 1851 that Brett succeeded in sending the first telegram across the Channel. Telegrams were first sent across the Atlantic in 1858. Branly and Marconi succeeded in making the first wireless transmissions in 1897. In 1866, Marinoni assembled the first rotary presses for continuous printing, just at the very time when paper pulp was being made from wood, and no longer from rags. The next advance after this was the creation of linotype in 1886, and then monotype in 1900. The first attempts to construct writing machines go right back to 1714, but it was only after Wheatstone and Remington's work, from 1851 to 1860 and from 1876 to 1878 respectively, that these machines were actually created. Ideally, any work on this subject would be divided into three different stages. The first would consist of a monograph – and some excellent ones have already been written, particularly in the last few years on machine-tools. Secondly, all the applications of any given invention should be demonstrated. Take the widespread application of electricity for example, or the incandescent lamp (1879) or Alexander Graham Bell's telephone (1876). The final part would be the other end of the spectrum, determining all the necessary conditions for the emergence of a new technique. Divergent and convergent studies could provide an effective framework for studying the problem of the technical system.

Complementary research would be concerned with certain aspects of the installation of the new technical system. We still do not properly understand the dynamics or the resistance of these structures, and this problem has rarely been dealt with. The discrepancy begins when an invention becomes more and more a scientific construction, such that the entrepreneur is no longer able to function alone – at this point, any transformations are made by teams of managers, rather than by the heads of firms. But the latter tend to think in terms of investment, of materials, and of the problems of obsolescence, for example.

Bessemer's discovery aroused even less enthusiasm because the research and development had been a long, drawn-out process. There were also problems with materials – very pure iron was needed. It is not surprising, therefore, that there was a great deal of reluctance about the whole project. It is true that the Bessemer convertor was quickly adopted in certain cases, in the Marine steelworks at Assailly in 1862, and in the ironworks at Terrenoire. In 1863, the directors of the Firminy company acknowledged the usefulness of the new process, but judged it to be unsuitable for use in their factories. The directors of the Alais factory even went to England, where they noted the limitations of the process regarding the different types of cast iron. "It seemed to us that this process only worked with certain types of cast iron. There is currently a major experiment being conducted into this process, and it includes all the elements of a normal large scale production of steel." It seems that most people preferred to wait for the results of the experiments conducted by others, and furthermore, to wait for the price of the licence to go down. In Allevard, they spoke of "a new, rather strange process, six or seven years old, with apparently no future, and which has been rejected both in theory and in practice." So everyone waited to see what everyone else would do. "Manufacturers seem reluctant to adopt it, and those who have given it a try are not fully satisfied. It seems that this type of steel is only suitable for limited purposes, – for rails and axles. The Terrenoire, La Marine, and Fourchambault companies have all bought licences, and they seem to have faith in this new method of production. We shall certainly examine this method, and we will also visit England in order to get a clearer perspective, but so far there is little evidence to persuade us that we too should adopt this new process." Schneider, in Le Creusot, waited for the results before he too embarked upon this new adventure in 1864. The steel works at Saint Etienne admitted that they did not have the necessary capital to make the investment.

In fact, it seems that the adoption of the new process was the work of a monopoly. One of the only ores providing suitable cast iron was found in the Algerian mines at Mokta. Now the latter were organised by a group which were part of the *Société générale de France*, and furthermore, this group was also comprised of the Marine, Fourchambault and Le Creusot ironworks. In 1869, Firminy, who was not yet linked to this group, and Alais (1867) adopted the open hearth steel plant. The directors of Mokta had wanted to profit from this situation and create concentrations and regroupments. It was thus an entire complex of particular situations which demanded the adoption of the new process.

The same was true of the Diesel engine. Several industrialists were approached from 1894, before the engine was finally developed in 1897. While Krupp and the *Machinen Fabrik* in Augsburg in Germany, and Sulzer, a construction engineer from Winterthur in Switzerland quickly applied for licences, thus financing the last stages of research, the main French building companies, in particular Cail and the *Société alsacienne de construction française* still had their reservations. "We have adopted the policy" wrote the managing director of the Say refinery, "of concerning ourselves only with those aspects which concern our particular industry." Schneider's response was similar. "We have examined your report with interest, but we believe that the complexity of our current business will not enable us to satisfactorily follow up your communication."

A category of authors have found their pleasure in highlighting a certain type of intellectual myopia – yet, wasn't it Pechiney himself who was uncertain about the future of aluminium? And what about the banks which were often called in to put up the capital needed for investment? Some banks established services for the funding of industrial and technical research: the *Société générale* established such a service in 1867, and the Rothschilds in 1870. There is another interesting turning point apparent here. It was not only techniques at stake in these situations, but mental attitudes and prejudices, which were far more difficult to change. There were concerns of all kinds, financial, industrial, and sometimes even technical, and of course that good old trust in tried and tested methods. It was audacity, the possibility of making a profit from a monopoly, even if it was only temporary, and a fondness for risk which proved to be the impetus for technical progress. The evidence seems to point to the fact that it was around the period between 1870 and 1880 that a completely new conception of the material world evolved. This deserves far more attention from historians than they have given it so far.

Amongst all the other problems caused by technical progress and the necessary adaptations to be made within technical systems and other, non-technical systems, one problem in particular is paramount. The passage from manual labour with tools to machine labour brought with it changes in labour capacity, and consequently, changes in the organisation of labour in the workshops and factories. After 1865, there was a massive wave of immigration into the United States. These immigrants were mainly unskilled (they were often unable to read or write in their native language) and as such they were unable to supply the needs of American industry. "The lack of skilled workers," noted Charlotte Erickson, "the difficulties of recruiting workers from abroad and training them and the ill will of the unions, means that American management has been forced into a policy of mechanisation, so that they do not have to meet the demands of skilled workers from Europe." The machine tool would no doubt have been introduced into America eventually, but progress would certainly have been slower if it were not for these problems. This explains, then, the intensity of the Americans' efforts to create and perfect machine-tools.

All that remains to be discussed now is organisation. Most people are familiar with Frederick Winslow Taylor, worker, foreman, then director of a turning shop in an iron and steel factory. Taylor's main preoccupation was with machine output. First of all he explored the technical possibilities: the composition of steel used for tools, the cutting speed, the tools' angles, the depth of the cut, the length, etc, and he was to solve many of these important problems. He realised that besides the use of the machine, there were other variables which were just as important. No matter how perfect the machine, it was only was good as the worker that was running it. This research, begun in the 1880's, – and Taylor was not the only one to be concentrating on these elements – did not become a cohesive body of doctrine until the early 20th century. The starting point was the timing of all the various operations, and Frank Gilbreth added to this the study of movements. It was necessary to understand that the machine was not simply an isolated unit: it soon proved necessary to extend this research to an examination of the workshop as a whole, and then to the entire business. They also established subsidiary bodies to study general methods and planning, so that the global output could be maximised. C. Barth lent his mathematical knowledge, and H. Gantt, his methods of graphical representation. S. Thompson contributed to the improvements made in the timing of operations. "As well as these pioneers" writes H. Pastermadjian, "other names deserve a mention, such as Person, who elucidated the actual principles of organisation, and C.B. Thompson and the great industrialist H. Dennison who both advocated their application.

We have been forced to schematise, but it is no less true that a new technical system was set up during the second half of the 19th century. Table II illustrates the rise in productivity per worker (base 100 in 1900).

TABLE II

Periods	Cast iron (Germany)	Oil (Germany)	Mines and factories (U.S.A.)
1833-43	9	46	22
1844-52	8	45	35
1853-59	11	49	37
1860-67	19	65	41
1868-78	35	78	58
1879-86	60	100	72
1887-94	80	101	86
1895-02	97		

These figures illustrate the technical progress made during this period. They also explain to a large extent the reasons why growth was maintained.

DEVELOPMENTS

On the eve of the First World War, the new technical system had become reality. Two observations can be made from this fact. The first is related to its worldwide expansion. Not surprisingly, in many cases, even the most useful inventions had not replaced old techniques or methods. The statistics to substantiate this would certainly be easy to find, and would prove that in some sectors the "industrial revolution" was already well under way. By 1914, Ford had already manufactured

240,000 vehicles, a considerable number which cannot be ignored. By 1913, 1388 millions tons of coal had been mined, and approximately 45 million tons of oil had been extracted. The production of electricity was still relatively small. Even though the large liners had begun to use steam turbines between 1907 and 1914, almost all merchant vessels were still driven by reciprocating systems. One must also consider the number of telephones, houses with electric lifts, and petrol tractors in use. It would be most welcome to have an assessment of the widespread application of the modern technical system for the year 1914.

The second point is equally important. The way in which the new technical system was formed in 1914 had the potential for often considerable development in almost every sphere. There is little need to emphasise this fact as concrete examples immediately come to mind: the automobile, the aircraft, wireless telegraphy which was still at this point only telegraphy, and electric and hydraulic installations, were all open to considerable improvement and advance before fulfilling their potential.

Many of the advances made in technology were unquestionably the result of the war effort between 1914 and 1918. Te pressure exerted by the requirements of the military, a systematic expansion of technological research which was encouraged and supported by governments, visibly accelerated the improvement of the latest modifications and the earliest refinements of the system. Mass production, the assembly line, the search for superior materials in both armaments and transport, presented problems many of which were not resolved; in most cases these problems were to pave the way for fecund developments from which the post-war period was to profit.

Although statistics for the end of the 19th century are available, the period between August 1914 to the beginning of the Second World War is far less clear. Both historians and econominsts gave up in the face of increasingly more complicated problems. Technicians appear to be more interested in remote periods of history than in recently outdated technical systems. Consequenly the nature of this work must remain more and more allusive, and the account must be restricted to a few particularly significant examples, althought these may in some respects distort the argument.

On the subject of energy, the discussion still concerns the same type of energy converter: steam-driven turbines (progressively replacing the old reciprocating engines which were consequently disappearing almost completely) (fig. 13); hydraulic turbines, and petrol and diesel internal combustion engines. Each of these became more powerful and increased in output capacity by almost imperceptible refinements which, principally relied upon support systems, by the use of ever superior materials and by a more efficient manufacturing of the different components. The change occurs in the elements around these engine systems, resulting from the use of new combinations; as for example, the coupling of the diesel and electric engine, which meant that an inflexible energy could be converted into a more adaptable form and thus respond better to the variations of demand. This also increasingly involved the substitution of hydrocarbons for solid fuels. Henceforth, however, such developments were more laborious, halted by necessary adaptations and by all the difficulties created in the process. Perhaps the most considerable efforst were made in the development of the aircraft engine, from which stemmed the most important results. The paths for the development of both the power and the weight of the engine were very limited. In 1920, the power to weight ratio was already less than one kilo per horse power. The problem of pressure took longer to resolve for when over an altitude of 5000 metres, the power would fall by half. The discovery of the altimetric corrector, superfueling and supercharging enabled a breakthrough in previously very restricted boundaries. If the questions

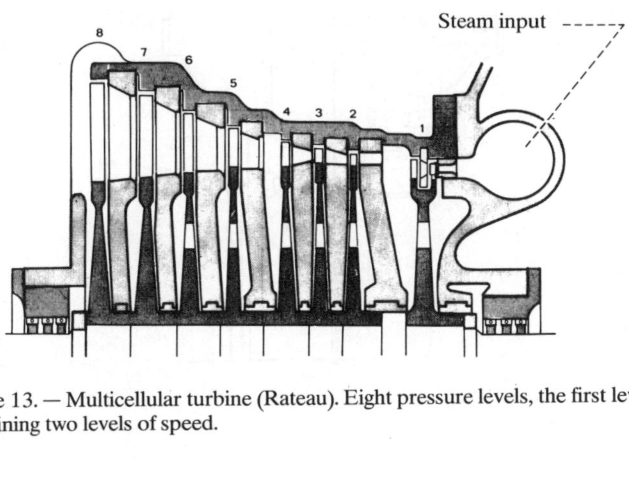

Figure 13. — Multicellular turbine (Rateau). Eight pressure levels, the first level containing two levels of speed.

of efficiency in fixed installations was the only one to arise, mobile cnverters on the other hand, essentially those used in transport systems, could only advance either by resorting to very different techniques, as regards the developments in materials and complementary systems, or by the reorganisation of installation, which included the instruction of system operators, a repair network, a fuel distribution network, etc.

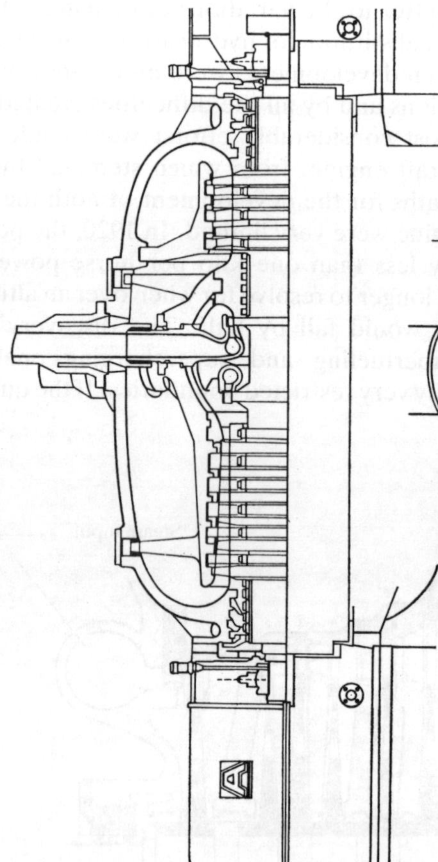

Figure 14. — Cross-section of the HP body of the Alsthon turbine of 600 MW. (From Lefort, *les Turbomachines*, Paris, 1969)

Without doubt, the best illustration of the method we have suggested is provided by the techniques of transport. In this area in particular transport technology brought into play an extensive environment. It appears that from the end of the First World War, transport was in a sense conditioned by two important factors; on the one hand, cost, and on the other, speed. Cost is certainly an extremely complex notion, both technically and economically; technically insofar as the productivity of materials aims for maximum output (it would be fascinating to study the strength of the materials used, the efficiency of the machines, and many other such problems); economically, insofar as everything must come within a cost system which is not entirely dependent on the technical one. It is necessary to cite only one example to illustrate this last point. The production of coal in France after 1918, shows the distortions which are contingent upon a large number of factors. An evenning out of the price of the various mineral fields was then established in order to make life more convenient for certain of them. It is the classic example of economic imbalance created by technical development, and by particular conditions. In fact the main reason for this policy results from the lack of workforce mobility. Similarly, to use a present day example, oil entails many political problems now that it has become a major source of energy. Once again, consistencies must not only exist within the technical system itself but also within the whole system network.

We have already discussed the extent to which the railway was able to expand rapidly due to secondary systems. Apart from these first innovations, which have already been mentioned, one must also add the track circuit, developed in 1879, and the block system in 1892. Henceforth the organisation of traffic was to reach its perfection: the arrival of centralised controls, and "dispatching" in 1930-1933 completed the evolution, so to speak. The second problem, other than the actual technology of transport itself, is the light weight of the material: the development of specialised steel, and the advance in metallurgy led to appreciable weight saving advantages. This consequently enabled the development of the first metallic cars. At the same time the growth of the power of the locomotive influences load and speed. In 1900, on the Paris-Bordeaux line, and engine of 1000 horsepower pulled a load of 255 tons at a speed of 90 km/h. In 1925, an engine of 2000 horsepower pulled a load of 450 tons at 95 km/h, and in 1935, an engine of 4000 horsepower pulled 760 tons at 130 km/h. There was definite progress made in the engine itself, but also in secondary systems, for instance that of automatic lubricating, to take but one example.

The most important development was that made in the transformation of the locomotive. Electric traction was generalised particularly after the First World War. But the first electric diesel engines also appeared at this point. They certainly did not show an immediate technical improvement, but they did bring about savings on investment, which must be taken into consideration.

The most important development was that made in the transformation of the locomotive. Electric traction was generalised particulary after the First World War. But the first electric diesel engines also appeared at this point. They certainly did not show an immediate technical improvement, but they did bring about a saving of investment which must be taken into consideration.

Although the diesel engine represents a weight of 4 kg horseweight, and although the electric diesel engine was able to exceed this by 20 to 30 kg, as well as this, the starting and stopping modes were very much facilitated by the flexibility of electric energy. We also remember the first "michelines" [named after the Michelin tyres they used] or "railcars" produced in 1931, although this is still a lateral innovation. The grip of the rubber tyre is almost three times that of the adhesions of steel to steel. In order to finish what is essentially an inventory, we should examine the development of aerodynamics, which is relevant to a good many other forms of transport: it was possible to achieve speeds of up to 100 km/h, and to reduce air resistance to 60%. This brings us back to the essential problem: the limits of the railways are the result of factors which are not necessarily of a purely technical nature.

The development of maritime transport was slightly different, even though the nature of the results was the same – both speed and tonnage were increased. However, like the railways, the performance of ships meant that the volume of traffic was fairly low, and of a particular nature. We have tried to schematise these developments and their particular aspects in the schematic table III. Insofar as we are dealing with performance – which we shall consider later – obviously techniques are of primary importance. According to this recent text:

The development of engines from 1920 to 1930 was considerable. Oil-firing was adopted and the use of geared turbines meant that engines turned faster and were thus more productive, while the propellor turned more slowly. Water heaters were made lighter, and this meant a higher pressure could be obtained (materials): the original heaters produced steam at 15 kg in saturated vapour, and they had a coal consumption of 640 to 700 grammes hp per hour. The gross weight of the engine was thus reduced from 125 to 75 kg horseweight. The fuel consumption fell from 700 g of coal to 465 g of oil in direct turbines, then to 450 g in the first geared turbines, with saturated steam. The pressure of steam went from 15 to 30 kg, and when superheating of 350 to 400°C was introduced, fuel consumption fell to 325 g hp per hour. In brief, the weight of the propulsive apparatus had been reduced by half since 1900. Added to this was the reduction of the amount of fuel required, which represented a drop of over 50% for an Atlantic crossing.

Special types of steel were used to make hulls lighter, but they were still as solid. And even if Bertin's theory about the maximum size of ships was still true, nonetheless, these boundaries had been widened.

TABLE III

Year	Ship	Tonnage	Speed in knots	Machine (type)	Power (hp)
1858	Great Eastern	28,800	13.4	2 reciprocating machines and wheels	
1863	Scotia	3,871	15.3	*ditto*	
1893	Campania	19,000	22	reciprocating, propellor	
1914	Aquitania	51,700	24	turbines	60,000
1935	Normandie	68,960	28.64	turbines, reciprocators	160,000
1935	Queen Mary	4,000	29.68	*ditto*	200,000

The development of warships was very similar. The problem of materials was more important, perhaps, because of the need for armour-plating. Steel had already reduced the weight of ships by 50 to 55% in comparison to wooden ones, and by 30 to 45% in comparison to ships made of iron. Specially made steel and welding reduced this weight even further. By contrast, merchant ships hardly changed. Their tonnage was modest, and the old alternating machines were still used with coal. Speed was not so necessary for these ships, and Diesel proved to be perfectly suitable for this kind of transport. The *Palmyre*, which is representative of this type of ship had a tonnage of 22,000 tons dead weight, and 30,000 tons load displacements.

The car had begun to conquer the world, now that it was no longer the work of an artisan but the product of an assembly line. The war had been a major factor in this. The French army, for example, had 6000 vehicles in 1914, and by 1918, it had 92,000. Ford had an annual production of 240,000 cars in 1914, and this increased to 950,000 in 1919. Once again, materials were an important factor, as much for the engines as for the bodywork and the chassis. The French *Omega six* was the first car to be made almost completely with special steel. Sheet metal was first used for the bodywork in America in 1920. However, the "lateral" progress was probably more important. The electric starter was invented by Bossu in France in 1905, and developed by the American, Bendix in 1912. After 1923, four-wheel braking was finally adopted, and the addition of a servo-assisted brake increased braking power without having to increase the pressure on the pedal. Low pressure tyres, larger than the older type, were now used in all cars from this period. The first "balloon" tyres caused severe vibration on the front axles (which came to be known as the "shimmy" after a fashionable dance of the time). As a historian of cars wrote recently, "they had to completely revise the types of suspension, shock absorbers and brake systems; they soon discovered that a car's road handling depended on all these factors, which all interlinked." Tyres could cover distances of 1500 km in 1906, 4000 km in 1914, and over 8000 km in 1925. Several changes completed the definitive look of the car by 1934: independant suspension on the four wheels, along with suspended engines and gearboxes which had been highly improved from 1925, pressure lubrication throughout the chassis, were widely used, and finally in 1924 cellulose varnish replaced oil paint. The greatest revolution was in 1934 when the postioning of the major organs of the car was called into question and completely rearranged, so that cars now could have front wheel drive. Another important fact is noteworthy, – despite the big increases in prices at this time, the price of cars in 1925 rose by only 1.5% to 2% at the most.

There is no doubt that the war had made aeroplanes a familiar sight in the world. Between 1914 and 1918, work in this field had been characterised by haste and improvisation. But the requirements of war meant that only planes of a particular type had been built. It was only when the war was over and so many crafts were available, that they were first used for civil purposes. The first air services were operated between Paris and London, for eleven passengers, and Paris and Brussels, for five passengers. These journeys were made with the

Goliath F 60, but this was simply a military plane which had been adapted, whereas quite different planes were required. As in the case of cars, competition encouraged increases in both speed and distance. In 1920, the Spad 20 A attained a speed of 265 kmh. At the end of the same year, the Nieuport-Delage 309 kmh, which was thought then to be the fastest speed possible. In 1924, a speed of 448 kmh had been reached, and this remained unbeaten for ten years. In 1927, the first crossings were made over the north and south Atlantic.

We saw that the power and weight of engines were developed, but the very conception of these apparatuses also changed. The monoplane with internal armature was widely used and the retractible was adopted. The airframes became more streamlined and comfortable. After 1930 light alloys and special steels were used, dispensing with need for shrouding. The modern plane had arrived.

The plane was an ideal way of transporting mail, and it was soon used for carrying passengers on a wide scale. A German company was formed to link Berlin, Hamburg and Leipzig in 1919. The first British company was founded in August of the same year. A few statistics are sufficient to show the development of this new form of transport – the size of the air network in 1919 was 5150 km, in 1929 it was 202,455 km, and by 1937, this had more than doubled to 536,717 km.

There had been a veritable revolution in transport from the end of the 19th century, then, but the development and consolidation of these changes was a long and difficult process.

Other techniques had developed more slowly. We are familiar with the general outlines of the evolution of these techniques, but the various stages still require more precise investigation, especially in terms of chronology. The lack of precise data in this sense means that we are unable to isolate the break points as we did for the development of transport. We can only repeat here that what is needed above all is a distinction between technical research, which is becoming more and more continuous, and industrial techniques. If such a history was written, this would no doubt substantiate our view that technical systems are part of industrial reality, that these systems have limits, and that the passage from one system to another can only take place insofar as there has been prior research in one system which has – still establishing balances between all the different areas of progress – led into the following one. Research was continuous between 1920 and 1939, of that there is no doubt: penicillin and nylon date from this period, but they could not be incorporated into the current technical system. There was even organised research on a large scale: take for example the *Office central d'étude de matériel de chemin de fer*, (Central Office for the research on materials used in railways), the research department of the main French railway network.

The iron and steel industry was able to make considerable progress in this way, without modifying the structures of its apparatuses and processes. Between 1900 and 1950, the diameter of crucibles in blast furnaces more than doubled in size, raising the daily production rate to 1000 tons - the equivalent to the annual production of a blast furnace in the 18th century. The output of Thomas convertors was raised from 30 to 60 tons. The first open hearth convertors had an output of 15 to 20 tons, by 1950, this figure had gone up to 550 tons. By 1930, coals which were unsuitable for use in the iron and steel industry could be coked. Rounds bars could be laminated by 1900, at first only with up to diametres of 150, and at an hourly rate of 8 to 10 tons. By 1950, diametres of 550 were possible, at an hourly rate of 80 to 100 tons. Continuous laminating of sheet metal was introduced in Teplitz, in 1892. In 1916, in the United States, an iron strip 600 mm wide could be laminated in a continuous rolling mill, whose shaping groove consisted of a universal mill. From the 100 tons of the power hammer, they had progressed to the 20,000 tons of forging presses.

Many of the processes in mining and agriculture were mechanised. The transport of energy was generalised by the use of electricity, and the latter was used more and more extensively for its heat as well as for its chemical properties. We could only make an accurate assessment of the development of certain techniques if we had maps of the various transport networks of different periods (electricity, telephone, pipelines etc). The first projects to build tidal power stations date from 1918: research at the Arcachon basin (1921), at La Rochelle (1919), and on the Severn (1918). Once again and as we shall see in the following chapter, productivity is perhaps the best measure of technical progress in this field.

In the 18th century, the labour of one rural worker provided for 2.5 persons. This figure had gone up to 3.7 persons by 1925, to 4.3 in 1929, and to 5.1 in 1939. Progress was moderate until the eve of the Second World War, therefore. Table IV illustrates the rise in productivity in the United States, taking a base of 100 in 1929:

TABLE IV

Year	Rate of production	Production	
		Per worker	Per hour
1869	7.3	34.1	27.3
1879	11.2	39	32.2
1889	21.6	49.4	42.0
1899	32.1	59.7	51.4
1909	49.6	66.1	59.3
1914	56.6	71.6	66.1
1919	63.4	63	61.5
1929	100	100	100
1939	103	80	128

Table taken from J. Fourastié, *Le Grand Espoir du XXe siècle*, Paris, 1972, p. 42)

We shall return to this table later because it warrants fairly detailed explanation. This somewhat abstract table is in a sense translated into concrete terms in Table V:

TABLE V

Years	Energy consumption (million of tonnes of coal equivalent)	Steel production (million of tonnes with iron included until 1990)
1860	200	7
1880	390	19
1900	765	41
1920	1450	70
1930	1730	110
1950	2520	160

These figures pose one final, important problem, that of the discrepancies between advanced and undeveloped countries. It is a commonplace to say that the economic development of various countries is inextricably linked to technical transfers: "international technological transfers are an essential factor of industrial development." We prefer the term "acquisition" of techniques by non-pioneering countries, reserving the term "transfer" for the passage of a technical process from one industry to another.

In fact, the problem of the acquisition of techniques by non-pioneering countries is one of the very cases where the notion of compatibility between systems is vital. In other words, a technique, no matter what type, presupposes the existence of a social and economic system which can accomodate it. The existence of a working class – one which has had some education, that is – the training of management, possibilities for investment, the need to supply certain consumer demands, all these are significant factors in the assimilation of a new technique. The period we are currently discussing is extremely informative on this subject, but there is still a great need for more research. The technological development of the United States needs further study as do those countries where technology has suddenly become an important factor,such as Russia, Italy, and particularly in this respect, Japan. The importance of capital from abroad – and the end of the 19th century was the most important period for foreign investment before the finanacial crises caused by the First World War – the "export" of factories, of technicians, the "imitation" of foreign practices, the missions abroad, and the system of licencing are all areas of research which have been left virtually untouched. And these must all be considered in relation to particularly delicate monetary problems: the industrialisation of Italy was certainly facilitated by its forced currency.

It is difficult to schematise the nature of the technical world between 1929 and the beginning of the Second World War. But we can be sure in saying that by 1939, only a few vestiges of the British technical revolution of the 18th century remained. Only the steam locomotive, which was giving way to electric or diesel traction, and the steam boat remained as a testament to the old technical system. The plane, the car, electricity and petrol, the telephone and wireless telegraphy, Linotype, steel, margerine, and artificial silk are typical representatives of the new technical system.

TOWARDS A CONTEMPORARY TECHNICAL SYSTEM

The historian is always slightly disarmed when required to assess his own age, for he is not distanced from the subject and often fears misapprehending the events, incorrectly analyzing the phenonema observed, and misjudging the dimensions of the world which surrounds him. So it is with some hesitation that this chapter has been written.

We are confronted by yet another difficulty, for documentation on the subject is now overabundant, so much so that technical problems play an important part in our civilization, our intellectual and social culture, and even in our daily lives. Documentation varies widely and is often factually conflicting, as is conclusively proved by all the literature which concerns nuclear centres. These contradictions also become all the more apparent in reality. Perhaps man has progressed too quickly in some sectors of the major movement which is currently taking place within the technical system, and may also have taken paths which appear to be closed to him today. We know how much debate surrounds the subjects of "Concorde," supertankers, and the nuclear engines used by the merchant navy.

We should first classify these questions, and more importantly present them in clear and precise terms, which is not as easy as it sounds. This will also explain why this final chapter is so long. It is a that chapter will soon become outdated, moreover so we hope that the reader will be indulgent.

We will continue to apply the outlines established at the beginning of this work, even though certain problems remain unsolved. In other words, we must first seek the reasons to explain why the preceding technical system was no longer viable, or at least why it appeared to be no longer effective in some conditions. Herein lies the most difficult aspect of research, one which has therefore been rarely attempted. We are subsequently forced to fall back upon suggestions and suppositions, but a considerable effort should be made in this direction in order to reach valid explanations.

We should then examine some of the major technical mutations existing within the technical ensemble, particularly those mutations which had brought about the evolution of other techniques. Attention should be drawn to some particular aspects of driving zones, as the innovations which are manifested represent these "centres of development" noted by economists. Obviously these must be listed and examined.

We will now turn our attention from major innovations to look at their industrial application, indicating all the transformations which resulted from this. Not only were products changed and new ones launched, but subsequently an evolution in the structures of production took place, which was no less important. Here again, a choice must be made, as volumes could be written on this, only some of which have already been published on subjects where the infomation has rapidly become outmoded.

The transition from invention to innovation is far more rapid today than it was during the last century and during the first half of the 20th century. The problems do not only involve technical development, but also extend to the financial and social levels. The same compatibility between different systems is at stake here. Although much research has been completed in this domain, a lot more still remains to be done.

The political class was not the only one aware of the problems presented by technology, as had long been the case. Entire populations were now variously concerned with technical change, as technical progress had numerous repercussions on daily life, the environment, and on economic or political evolution. Indeed these repercussions were so numerous that they are difficult to distinguish. This vast and important question inevitably leads to reactions being centred on two contradictory opinions, with fears on the one hand and hopes on the other. Analyses are once again numerous and diverse in their conclusions, and it is increasingly evident that progress does not only manifest itself as it advances.

THE "REASONS"

As we have just said, the reasons for a technical revolution are the most difficult to explain, particularly as technical systems are improved. An analysis must therefore be focused on several points:

I – Firstly, the extent of requirements and the possibility of supply must be studied, from the point of view of quantity, quality, and price, or at least short-term prices.

II – Secondly, saturated techniques should be determined in the same way, and the so-called bottlenecks which arose in this domain should be distinguished.

III – Supply and demand curves should be deduced from technical innovations which are naturally linked to technical progress and fundamental research.

In the words of Rostow:

> Viewed in this way, the volume of resources, including human talent, devoted to fundamental science and to invention in a given society at a particular period of time can be symbolized by quite conventional supply and demand curves. A demand curve would exhibit the expected yield to be derived from the application of additional resources to the pursuit of fundamental science, with the existing stock given. Since the results of pure science do not enter directly into the private economy (except in certain contemporary sophisticated industries with their own huge laboratories), achievement may reflect the premium in prestige and academic status a given society attaches to such achievement, or public subsidy. A supply curve would exhibit the volume of resources actually offered by a given society to the pursuit of fundamental science at a particular period of time in response to such incentives. A particular individual may or may not be responsive to the expected yield from the pursuit of fundamental science. Some men are driven to search for new knowledge by inner compulsions not closely related to external reward, but for substantial numbers of human beings one can expect that talent would be responsive in degree to the rewards, financial and otherwise, a society offers for scientific achievement. A similar pair of curves would exhibit the demand for inventions and the supply of talent and resources offered in response to the expected yields. Here we are closer to the marketplace. Therefore, for invention, one can probably presume somewhat greater elasticity of supply in relation to expected yields, and this is borne out for our period by the fluctuation of inventions with peace and war, prosperity and depression. But we are also dealing with men with an instinct to express a creative talent, and the shape pf the supply curve for inventors may also reflect non-material rewards.

This is a rather interesting model insofar as it incites reflexion, and insofar as we must go beyond it. Firstly, we should emphasize that the period to which Rostow refers is not the same which interests us here and that there are considerable differences between the two epochs. It is not easy to trace the curves of supply and demand as they are not concerned with simple quantities or prices, but with manufacturing methods or quality. So outside this useful if rather too sophisticatedly established pattern, we deem it necessary to give prominence to a certain number of major phenomena which have encouraged fundamental research, improvement, and innovation. In other words, the same reasons continue from one technical system to another, and the same coherences are established as quickly as possible within the new technical system.

Need we even mention the fact that authors do not agree on the way in which the new technical system was enforced. We must therefore be content with some vague suggestions, accompanied by many question marks.

There are really three major phenomena, the first of which is undoubtedly the great crisis in the Western economy during the period 1929-1932. Numerous diverse and conflicting reasons have been provided to explain this, and we do not intend to present yet another. Let us firstly look at the idea put forward by the Czech writer, Richta: "Several Marxist economists have suggested that the extent and depth of the crisis of the 1930's was linked to the prodromes to the scientific and technical revolution which caught the capital system of the epoch unprepared." Such a view obviously admits the concept of a "scientific and technical revolution," which some authorities may challenge in the same way that the notion of a late 18th century "industrial revolution" has also been challenged. It is vital if we are to accept this argument, not to deny that we are faced with a new technical system in the process of development today. One might then recognize as a correct interpretation that the great depression of the 1930's may have been due partly to the saturation of the preceding technical system. Certainly Galbraith is not far from this idea in his analysis of *The New Industrial State* and the crisis of 1929.

One may suggest that in the face of fierce demand dominating the world's economy from the end of the First World War to the Wall Street Crash in November, 1929, technology could not fulfil this very keen growth either quantity or price wise. The production of steel in America almost doubled, soaring from 31 million to over 57 million tonnes in 1929, and without any basic change being made in the techniques of the industry. Such a leap could not have been produced without some effort. And this fact is not only reserved to the iron and steel industry, as similar proportions may be observed within a great many industries. Some technical progress had been made between the period of 1914 to 1939, such as continuous hot laminating (1924), then cold laminating, and the Ugine-Perrin method of refining (1933), all of which were rapid and complete developments. Similarly, the capacity of the apparatus continued to grow, and around 1940 large blast furnaces could produce 1000 to 1200 tonnes of iron or steel per day, while the Martin furnaces reached a capacity of 30 tonnes and rose to 100 tonnes in the United States, the Thomas retorts produced 20 to 40 tonnes, and the electric furnaces produced 15 to 25 tonnes per day.

No one would deny that some technical progress was in preparation during the period between the two world wars. Although nylon, the jet engine, and penicillin were introduced between 1930 and 1940, their real entry into the economy dates from the war or the post-war period. It is not surprising, moreover, that the great depression led to research in some sectors before 1940. On the very eve of the war, a few of these inventions had already been industrialized.

The second phenomenon, the war itself, again lacks the benefit of study. We have an overall impression, indeed some facts, but no details or proof on this score. We require information on the state of technical research during the decade before the conflict, as well as a history of technical research during the war, in order to assess the effect which the war had upon particular areas of research. Obviously everyone knows of the development of rifles and jet engines. Wiener was directed to expound his research and succeeded in developing cybernetics, but much remains to be uncovered. We have already noticed in a preceding chapter that the First World War did not lead to great technical change, but this is not true of the Second World War. There is a notable difference but it is difficult to explain.

A explanatory summary may be useful though not absolutely valid. In 1914, the existing technical system was still capable of further development and it was probably in this direction that progress was made, but progress was not to be the same in 1940. As we have just seen, the great crisis of 1929 had already started a process of change within the technical system. The enormity of the conflict, which took place throughout most of the Northern hemisphere, required infinitely superior methods to those which were mobilized during the First

World War. Man had passed from an essentially European war to one which was truly worldwide, involving enormous distances and different climates. It is certainly useless to quote at length everything concerning the techniques of armament.

The third phenomenon is undoubtedly the enormous increase in demand. In the United States, the annual increase in production per head was as follows: 1.3% from 1913 to 1929; 1.6% from 1929 to 1950; 2% from 1950 to 1965; and 3.2% from 1965 to 1970.

The beginning and end figures of the whole period are 0.8 to 2.4% for Britain, 0.9 to 3.8% for France, and 0.5 to 5.9% for Germany. Steel production in the United States of America remained at 57 million tonnes in 1929, but in 1974, it was over 93 million tonnes. Figures are similar for other countries and other sectors, and one need only peruse all the works of retrospective statistics to see this. We have already observed several times, that the only possible reply to a massive increase in demand was to transfer from one technical system to another, and here the demand for invention comes into full play. As for supply, this first brought into play the whole stock of existing inventions, but unused for various reasons, and then required considerable investment both in basic and applied research.

Apart from this explosion in demand, competition between the two worldwide political systems was also taken into account. These systems demonstrated that they were both equally capable of resolving technical problems with which they were presented. Technical competition, as much on the industrial level itself as well as on the level of prestige, certainly contributed to the development of certain techniques, other than military techniques of course.

Thus a "causal network" was established, by which we may try to explain the technical mutation which we have witnessed: a crisis which emphasized the inadequacies of the preceding technical system, a worldwide conflict which required radical transformations, and lastly, an explosive demand which necessitated new production methods. But other phenomena intervened in the logical formation of a new technical system. Naturally, there were inevitable time-lags in the rate of these mutations' progress, some taking place more quickly than others, and to an even greater extent, certain mutations producing the disappearance of other techniques by a sort of ricochet effect. Let us look at an example of this which has recently been stressed.

Techniques used in chemistry were essentially based on coal. In effect, the chemical process of extraction, which today is not so important, was replaced by the chemical process of transformation, which introduced artifical silks and celluloid, then progressed to organic chemistry, linked to the process of synthesis, the development of which dates from the middle of the last century. This form of chemistry employed three main resources: (a) the carbonization of wood which gave acetic acid, methanol, acetone, and formol...; (b) the fermentation of alcohol which produced ethyl alcohol; and (c) the carbonization of coal which was the source of benzene, naphtalene, anthracene and more generally a large proportion of aromatics.

After 1918, the carbonization of coal opened a way to the development of aliphatic products using the process of obtaining acetylene from calcium carbide. "Coal is therefore at the root of the development of industrial chemistry which may be schematically characterized as that of acetylene and aromatics."

This technical ensemble was linked to other techniques, such as the fabrication of town gas from coke, and the production of metallurgic coke. Chemical resources no longer existed on the day when town gas had largely been replaced by natural gas, when metallurgic coke had been mainly replaced by heating oil and oxygen, when electricity was oil-based, and when the steam engine was virtually no longer in use. When this happened, the balance upon which the economy of the chemical industry was based was broken. And as we have already pointed out, at the same time the increase in the demand for basic products and major intermediaries was a completely new phenomenon. "The inadequacy between requirements and availability offered by coal gave rise to extensive mutation in the field of petrochemistry and the chemistry of natural gas." And in addition to this, both these fields – petrochemistry and the chemistry of natural gas – in turn opened up the way for the development of new products.

These re-establishments of previously lost balances and the search for a new coherence with other techniques are relatively numerous and it was through them that technical progress spread throughout the whole system. There were, on the other hand, sectors where progress was slow and where the number of obstacles increased, as for example in construction and building where dispite some progress there were still many delays.

These analyses are inadequate, but by presenting these problems in a very broad way we hope to open up the field for more research which will help us to comprehend better a recent past and perhaps foresee an immediate future.

GREAT TECHNICAL MUTATIONS

Everybody, or almost everybody, agrees that we have entered a new technical era, and some authorities talk of a second or third technical revolution. This revolution or new era may be broadly characterized by the mutations implemented in a certain number of technological sectors, that is nuclear energy, the electronic revolution, and new materials.

A detailed examination of the implications of these new techniques reveals that they did indeed revolutionize the material world, differing completely from the its earlier form before the great crisis of 1929.

A new technical system was indeed created where the most significant factors were already established, forming the coherence which is indispensable within any system. We will see that the coherence between various techniques in relation to each other was imposed in a very restricting way. This fact is all the more remarkable when the latest technology or advanced techniques are involved.

It is also true, and we have noted this fact in the case of each of the preceding technical systems, that delays and discrepancies persist, leading to inevitable imbalances and distortions which are translated equally well on the purely technical level in relation to other systems. What should be revealed here, is not the fact that leading sectors continued to progress, leaving other sectors way behind them, as has been done all too often in the subject of technological expectation, but more particularly whether the imbalances produced were finally adjusted and if these distortions were eliminated, as this is the most serious point of the evolution.

We will observe that technological research could henceforth only be undertaken by very rich countries, and therefore of a certain size. A "technological gulf" was consequently created between these and other countries which was fairly profound depending upon the degree of development in the latter, surmountable for industrialized countries, but often catastrophic for Third World countries.

Disequilibria have been known to be produced within the technical system of a developed country. Agriculture had followed technical progress, but was considerably slower in momentum. This is also true of building techniques, where building sites seemed to be far less advanced and consequently relied upon a greater workforce. Contrary to this, the rapid advance of technical progress in some sectors stimulated mutation in the structure of the profession of employees, indeed even in global social systems: such was the case, for example, in the development of automation, although its advance was far more gradual and less widespread than most have thought. And we need only consider the town congestion caused by cars, roads saturated as soon as they have been built, and traffic jams. There are certainly hundreds of examples.

Now that research has been undertaken in different countries, we are in a position to understand better the influence of technical progress, or in this case the transition from one technical system to another, on economic structures and systems. But here again, mutations are long and frequently incomplete, and any adaptation is subsequently difficult to put into effect.

Much more still remains to be discovered. Essaies by Lewis Mumford and the Belgian, Henri Janne, still only offer us descriptions, which despite their interest are not capable of providing us with elements for profound and systematic analysis. This analysis should obviously begin by studying the technical mutations in existence and then try to bring out in each case the points of interaction with other techniques. Only after this technical analysis may one consider the consequences of change within the global system.

Energy

The events concerning petroleum at the end of 1973 have made the public aware of the importance of energy problems and the difficulty experienced in trying to resolve them. As the reader may have noticed, we have used the plural for "problems" as these are numerous and very diverse in nature. The first and foremost is undoubtedly the problem of quantities which has captured the attention of the general public, and has presented itself in several forms. In October, 1973, several factors became apparent, such as the importance of oil costs, the selling price and all the repercussions entailed by these costs in every economy generally based on oil producing techniques. But at the same time, and in accordance with well known economic laws, the increase in the price of petroleum created competition between energy sources which had not been competitive before, such as coal and underwater deposits, insofar as new techniques added to a purely economic phenomenon. The second point is of a political nature: oil subjected a whole group of developed countries to the mercy of foreign powers and decisions: the way to national independence seemed to pass through energy autonomy. There were warning signs of this last aspect of the problem before the crisis came about. Some authorities have indeed shown that classic energy resources were likely to be exhausted, and that as a result of this the world was rapidly heading towards a more or less pronounced energy shortage.

The qualitative aspects of these energy problems have undoubtedly already been emphasized. There is not just one worldwide energy problem but several. Firstly, and this is true of all forms of energy, the availability of energy, and subsequently its transportation and storing are essential considerations. This is partly why oil became such a successful form of energy, as it was easy to store and to transport in pipe-lines. This is also why electricity, which might be called an intermediary energy, was successful, except of course that it is impossible to store. The ways in which energy is employed is also an important consideration. Energy may be used in a directly employed caloric form: such is the case in the fuel necessary in a blast furnace, and for every industry which consumes heat (tile-kilns and bakeries), and it is equally necessary for individual heating which represents a very large percentage of the overall energy consumption. Note that electricity was indirectly employed for the same purposes, apart from mechanical uses upon which we need hardly dwell, at least regarding their quantitative significance.

An energy source may be converted directly into mechanical power: such is the case in hydraulic energy, and had been for some time. Almost every other form of energy is necessarily derived from caloric energy. This fact emphasizes the essential role played by energy converters, each possessing its own special qualities and a particular way of resupplying an energy source. But they also have their own specific uses. Energy is required for industrial production and is necessary for transports, distinguishing in the latter the railway from the ship, the classic ship from that which demands major autonomy in distance and time, and rockets which require special fuel types as the classic combustion engine cannot be used in space where there is no air.

We will try to divide these energy problems into two main groups:

A) The problem of energy sources, that is to say the improvement of techniques used in classic sources of energy and, on the other hand, the new sources or forms of energy.

B) The problem of energy converters, in other words not only engines but also the techniques involved in the use of caloric or chemical energy.

Production of energy As we have seen, the first type of energy used was hydraulic energy, and its role became almost exclusively to produce electricity. Attempts were made primarily to improve tributary techniques, particularly those in the design and construction of dams and pressure pipe-lines, as well as in the development of energy converters, turbines and electric engines. True, there were some spectacular developments, but these techniques were really almost totally saturated and any further improvement was virtually impossible. The real problems lay in financing these ventures and the development of land space.

The use of tidal energy, which had been long known since the first tide mills cited in the 13th century, also received technicians' consideration. In 1737, the French engineer, Belidor, developed the first industrial project using this form of energy, and between the two world wars, British and American technicians began work in this direction. It was only after the Second World War that a project to power a factory using the tides of the Rance, in France, came into being. The techniques used here were similar to those employed for river dams and entailed the same problems.

Coal had become the main source of energy from the end of the 18th century, and it was in this domain that mining methods and the way in which coal was transported to the surface underwent the most profound change, from the 1930's and particularly after 1944. These techniques were naturally determined by the aspect, slope and potential of the coal beds. Mechanical methods were extensively developed for coal beds where the gradient was less than 45 degrees.

The pneumatic drill replaced the pick and henceforth gave way to cutting and planning machines. Two parallel galleries were dug through the coal bed, approximately 200 metres apart, giving access to the workforce, enabling coal to be brought out and making possible ventilation. The cutting machine worked gradually from one gallery to another, along the whole length of the coal face. This powerful machine dug a collection of picks into the coal and pulled out a layer of coal about 70 centimetres wide. The cutting machine was then placed approximately the same width away, and worked in the opposite direction until the site was exhausted. In order to avoid rock falls, the "walking support" was now used, and jacks were linked to the cutting machine. The planning machine, a smaller machine which was haulled the length of the coal face by an endless chain, was used for soft coal. Conveyor belts were now used to bring up the coal, replacing the old railways. All these new techniques enabled infinitely greater output than that recorded before 1940.

Here again we are probably confronting techniques which had reached their saturation point. The space required to store this fuel, the difficulties and the cost of its transportation, generally condemned it to be used in situ, either to be burnt and transformed into electricity, or to provide raw material for a chemical industry.

At the end of the 19th century, petroleum took the lead in the total energy consumption, and this percentage was to increase still in the coming years. Above all, techniques in prospecting and drilling were to advance, and were often completely replaced. Continuous seismic recording had been inaugurated in 1956, followed by gravimetric recording in 1957. Systematic magnetic, gravimetric and seismic surveys made possible known reserves to be increased considerably. Although drilling equipment remained the same, the quality of the materials used and the automation of mechanical plants led to greater efficiency in production.

The most significant new technique was the introduction of drilling and extracting deposits from below the seabed. The first tentative attempts had been made in 1880 on the Californian coast where a few derricks were set up on wooden piers in order to find underwater extensions of inland deposits. In 1930, permanent plants were established in lagoons in Venezuela and near the Louisiana coasts. In 1938, drilling went to a depth of five metres under the sea, and similar attempts were continued in 1939 on the shores of the Caspian Sea.

One major new innovation lay in exploiting deep sea deposits some distance from the coast. The first steel platform began work in 1947, 30 kilometres from the coast and 17 metres under water. The first oil-drilling ship was put into operation in 1951, and progress in this technique was rapid up until 1960, whereupon advance continued at lightning speed until 1970. In 1968, the American ship, *Glomar Challenger*, discovered traces of oil at a depth of 3580 metres in the Gulf of Mexico. The first auto-elevating platforms saw the light of day in 1953, followed by the semi-submarine platform in 1962. From 1971, all anchorage systems were dispensed with, and the floating supports were maintained above the drilling wells with the aid of transversal propellers controlled by a computer.

The most difficult problem was certainly in trying to support platforms in bad weather: waves in the Persian Gulf can be up to 7 metres in height. At Ekofisk (a diagram of which we have reproduced) the secular wave average might go up to 24 metres, 28 metres at Forlies, and 30 metres at Brent, all deposit areas in the North Sea. One of the platforms at the last deposit site mentioned, *Sea Quest*, set up at latitude 60 degrees north off the coast of Norway, is triangular in shape, dominated by a 45 metre derrick. Its legs rest on three laundry iron shaped floats, 30 metres in length, 15 metres wide, and submerged 20 metres underwater, while the platform towers 35 metres above the sea. Drilling must stop when the vertical movement of the platform reaches two metres. It has become possible to create true miniature floating towns, as has been done at Ekofisk (fig.1). New techniques, such as underwater concrete reservoirs, are being tested.

All these new techniques therefore enable the exploitation of deposits which had previously been inaccessible. But many of these techniques are not profitable, as the price of oil has increased out of all proportion. Today, we can drill to a depth of 800 metres (research carried out off the coast of Thailand), but we cannot exploit deposits over 150 or 170 metres below the sea. These off-shore deposits provided 9.8% of world oil production in 1960, 14.7% in 1970, 16 to 17% in 1975, and one was hoping that they would provide 25% in 1980. At the beginning of 1976, there were 75 platforms in operation, and 52 in construction.

In 1937, methane was discovered and exploited in the south-east of the Po Plain, in Emilia. In 1952, the production of natural gas in Italy represented a fifth of its coal consumption. A vast "gas pipeline"

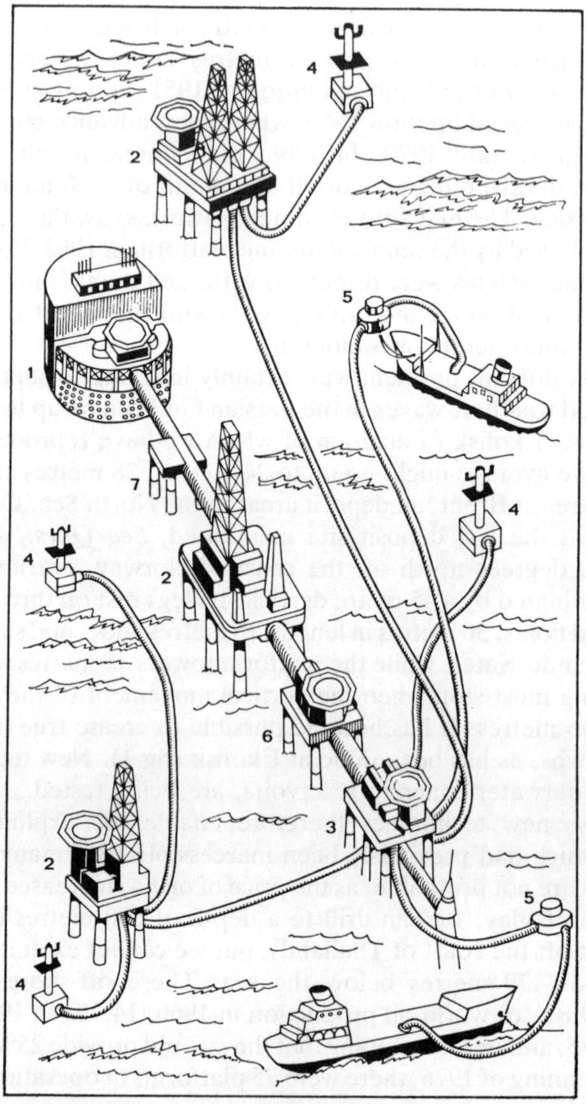

Figure 1. — The Ekofisk complex. 1, reserve supply of 160,000m^3; 2, 3 drilling platforms for new wells; 3, processing and accommodation platform; 4, wells currently being exploited; 5, loading bays; 6, living quarters; 7, pumping platform. (From a drawing by P. Ferran printed in *Le Monde*)

network was constructed for the distribution of this energy throughout the whole country; in 1971, 13 million cubic metres of natural gas had been distributed using a pipeline network of approximately 10,000 kilometres, and, at the same date, another 5000 kilometres of pipeline was in the process of being built or was in project. Discoveries of gas in Lacq in France, and Groningen in the Netherlands, Russia and Algeria, have provided Europe with a rich new source of energy. Although gas pipelines borrowed techniques used in distributing oil, they have on the other hand developed the techniques of liquefaction and the transport of methane by sea.

Natural gas, gas used in blast furnaces, and the gas used in coke-works, are now supplied by extensive networks of gas pipelines. In 1946, France counted some 546 old type gasworks in its towns. Only 14 of these were left in 1963, just one in 1968 (that at Belfort), and none remain today. Radical change in the techniques used therefore took place.

Of the other energy sources (developed either recently or a while ago), some have seen the improvement of the techniques used to exploit them, while others are still waiting for their techniques to be developed. The Frenchman, Georges Claude, suggested using the thermal energy of the sea, produced by the temperature difference between the surface of the sea and that several fathoms below the surface or on the sea bed. Attempts have been made in the Caribbean Sea where the temperature was 30°C on the surface and 7°C 500 metres below. This difference is relatively slight and subsequently the machine's output was mediocre, less than 10%. Furthermore, the operation was destroyed by a storm. Similarly, a project to construct a pilot factory with an approximate capacity of 7000 kW, at Abidjan (Ivory Coast), also had to be abandoned suddenly. At the time of writing about these techniques, there does not seem to have been a real source of power.

Men have long dreamed of using solar power. The main difficulty stems from the low energy flow. At sea level, its annual average is in the nature of 0.1 to 0.2 kW per square metre, which would require a collecting surface of two hectares, in order to supply a working plant of 1000 kW operating with an output of 30%. The furnace at Montlouis, in the French Pyrénèes, found that this energy could be concentrated on one point using mirrors. The harnessing of solar energy has been used, as we will see, in space engines. Silicon, solar cells which provide electricity on board satellites are not very efficient, around 11 to 12% output, and are extremely expensive; cadmium sulphide based cells have a slightly higher efficiency, around 15%, but they are very fragile. The improvement of solar pumps (air heated by the sun gives up its calories to another fluid, propane, for example, and the gas operates a motor), and the development of solar cells are undoubtedly avenues for future progress, particularly as the technique of using mirrors presupposes a limpid atmosphere.

Geothermal energy was first used by men cooking eggs on hot lava flows. During the Middle Ages, some houses in Chaudes-Aigues, in the province of Cantal (France), were heated with water which had a temperature of 81°C to 83°C degrees when it came out of the springs. These dry steam deposits are in fact extremely rare, indeed only five are known to exist in the world; one in Italy, two in the United States, and two in Japan. Hot water springs are a far more frequent occurrence and usually lie deeper underground. The deposit used in Melun (France), and which lies under the whole Paris basin, is at a depth of 1800 metres. It has been estimated that this deposit may be used for some thirty to forty years – the use of this source of water in fact reduces the temperature of the layer by 1°C every year – and it is possible to supply heat to 2000 houses by using 100 cubic metres per hour. Warm, dry rocks exist in regions with a high thermic pressure gradient, but their exploitation raises great difficulties and may lead to considerable seismic risks. Despite an extensive programme costing 600 million dollars, it is thought that the proportion of this source of energy would never contribute more than 4 % of the total energy production in the United States. Moreover, this energy cannot be stored or transported.

Bituminous schists have also been exploited for many years. Bringing up oil has always entailed prohibitive costs, and research has continued into finding means of improving the techniques involved. This sector will therefore probably remain a minor source of energy.

We will just mention in passing the burning of household waste used by some urban heating systems. Although there was concrete development of this technique from the end of the last war owing to the low cost price of the fuel, on the other hand the cost of installations and the loss of steam heat transported in the pipes are significant factors, and there does not seem to have been any room for technical progress.

All things considered, these marginal sources of energy could only have been of secondary importance, and were undoubtedly temporary ones for the most part. Even with the development of these secondary energy sources, they could only ensure, at the very most, the transition towards the definitive development of other techniques.

The greatest hope lay in the advance of nuclear energy. We have seen that a whole century of research had led to the discovery of atomic fission, and subsequently to the liberation of a considerable source of energy. In 1938, three German physicists, Hahn, Strassmann, and Lise Meitner discovered and explained a new phenomenon, this being uranium nucleus fission. Although the first nuclear bombes suddenly unleashed energy produced through fission, it was necessary to achieve chain reactions and to slow down the process in order to use this energy practically and peacefully. Several types of usable nuclear reactors were developed between 1944 and 1955. The French, who for military reasons required plutonium production, concentrated on a graphite-gas network: natural uranium reactors, slowed down with graphite and cooled with carbonic gas, were developed at the beginning of the 1950's. Successive failures in these nuclear centres led to this technique being abandoned in the middle of 1968. The Americans designed two channels, both burning enriched uranium, where water was used as a regulator and a peddling fluid, which enabled a chain reaction to be produced and evacuated the calories formed by fission of the fuel (fig.2). Pressurized water centres were also created (P.W.R. or P) where high temperature water is produced, but is prevented from boiling by being maintained under pressure (fig.3), as well as boiling water reactors (B.W.R. or B fig.4). A heat exchanger was necessary in the first type: primary water is raised to a high temperature through its passage in the core of the reactor, whereupon the heat is transferred to a secondary water supply which is transformed into steam for use in the turbines. Each of these channels has its advantages and disadvantages (the cost of installation, the risk of radio-active waste, etc.). Their unit power is wholly comparable: from 1000 to 1200 megawatt.

The increased use of nuclear energy presented several problems of a technical or economic nature. Firstly, one must have the necessary fuel resources, and here again the questions of energy independence come into play, as uranium deposits are not as well distributed as large petroleum deposits. Consequently enrichment factories had to be built: this in turn entailed both technical problems – there are several

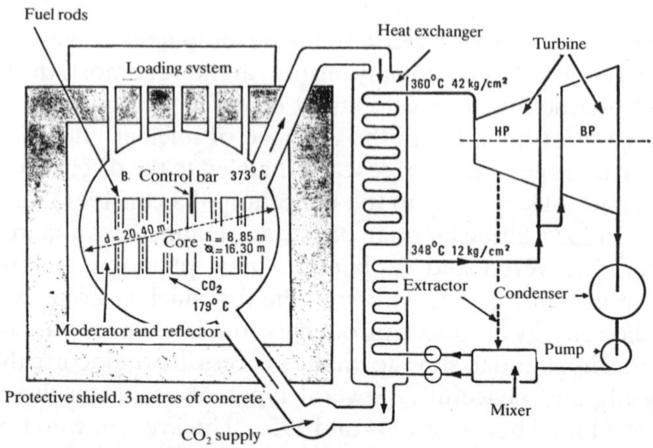

Figure 2. — Diagram of the nuclear power station at Hinkley Point, fueled with natural uranium. (From G. Parreins, *les Centrales nucléaires*, Paris, 1967)

methods of enriching uranium – and investment difficulties for countries with only mediocre economic power, particularly Europe where either French or English technology had been favoured. In the meantime, this option has exerted a quasi-domination in the United States.

The generalization of power stations was not any easy process to implement, as there were several internal problems. The productivity of nuclear power stations was less than that of thermal power stations. Nevertheless, the price per kWh of energy produced by nuclear power stations seems to have become very competitive with that produced by thermal power stations. In June, 1974, the cost price per nuclear kWh was estimated in France at 4.5 *centimes* against 8 to 8.5 *centimes* for one thermal kWh, but the cost of investment required for nuclear power stations was 50 to 60 % higher than that for thermal power stations. Financing a programme which involved the replacement of a large number of thermal power stations with nuclear power stations seemed exorbitant, and essential technical progress was therefore staggered over a number of years. These nuclear power stations must be established on very strict lines: a large quantity of water is required, indeed so much so that the possibility of building power stations in the sea has

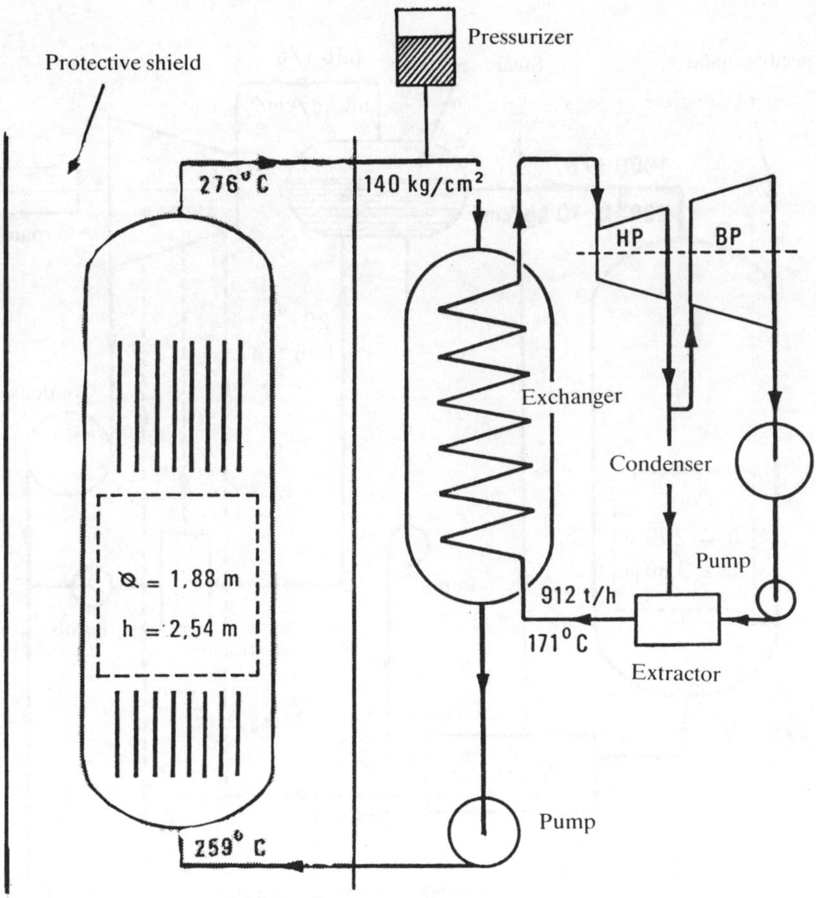

Figure 3. Diagram of the Yankee power station P.W.R. Protective shield: radially, outside the caisson, there are 1.20 metres of water and 3 metres of concrete. Dimensions of the caisson: diameter – 2.80 metres, height – 9.60 metres, thickness – 2.16 metres. The sphere of protection contains the reactor and the exchanger: it is 38 metres in diameter and weighs 800 tonnes. (From G. Parreins)

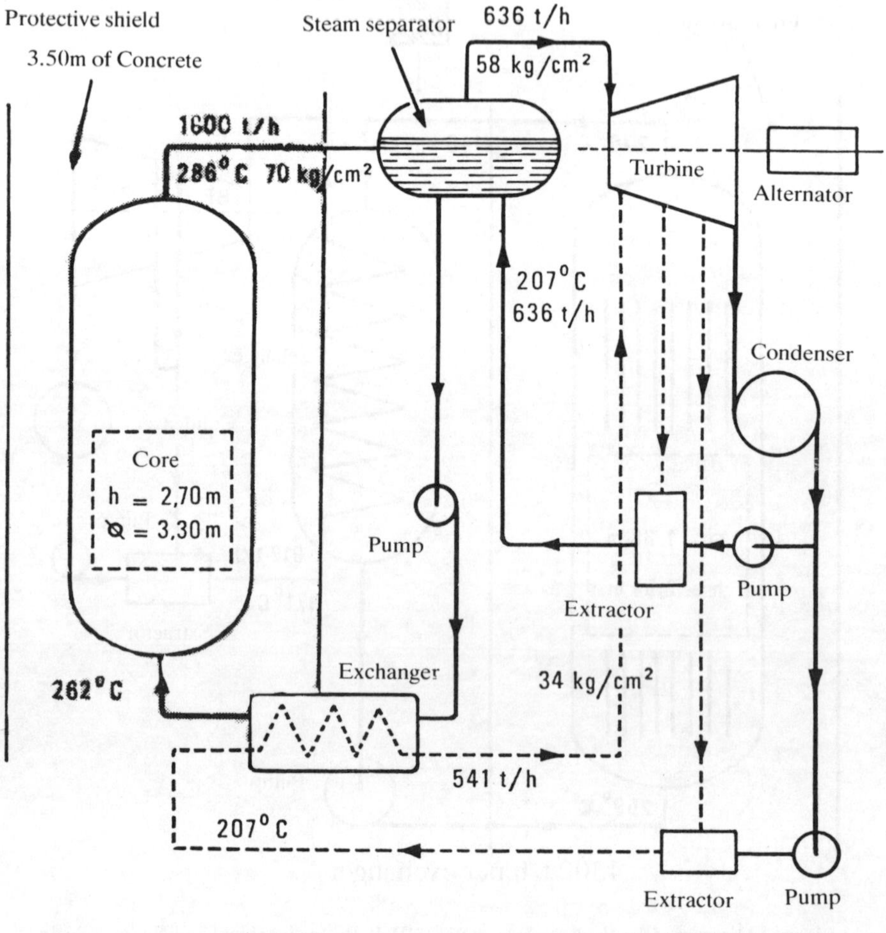

Figure 4. — Diagram of the Dresden power station B.W.R. Dimensions of the container: diameter — 3.7 metres; height — 12.2 metres; wall thickness — 1.4 metres. Dimensions of the protective sphere: diameter — 5.8 metres; location of its centre — 17 metres from the ground, wall thickness 35mm. (From G. Parreins)

been studied. Although it is not our aim here to discuss the "nuclear threat" – we will return to this subject later – some of the unknown effects of nuclear power have given rise to fears about pollution. Firstly, pollution from thermal power stations: two thirds of the energy produced goes in heat, for water and air. Thus a power station of 1000 megawatt requires a removal of 40 cubic metres per second and heats the air and water by 10°C. It has been calculated that at a higher low-water mark, the temperature of the Rhine was increased by 3°C, that of the Rhone would increase by 5°C, and that of the Loire would increase by 7°C. Chemical pollution is linked to noxious products being released into the atmosphere and river ways. Finally, there is a risk of radio-active pollution together with the liberation of radio-active elements, and that is without considering the question of waste. There are undoubtedly more surprises in store as yet.

But the technical evolution is by no means complete. The greatest change will probably be in the development of different fuel types. Studies have begun in developing what have been termed super-generators, where heat is produced by the atomic fission of plutonium and not uranium 235 as before. This would certainly constitute a major advantage, for as the plutonium burns so new plutonium is formed, so to some extent, the power station produces more fuel than it actually burns. The quantity of heat given off per unit of volume is far greater than that produced in standard reactors and the heart is recooled by liquid sodium. So far, only the U.S.S.R. has managed to produce kilowatts from a reactor designed in the image of the super-regenerator stations of the future. The Americans expressed the opinion that these new techniques were dangerous, as sodium is highly noxious. Similarly, the construction of the first power station of this type, at Oak-Ridge (Tennessee), was broken off in the middle of 1973, upon a judicial decision. In France, the super-regenerating power station, Phénix, at Marcoule, was coupled up to the national electric network in December, 1973. Indeed it was thought that the plutonium produced by power stations using enriched uranium could supply the new type of power station. But whatever the case, this was still only a demonstration station, working at reduced power (250 megawatt). The British were to quickly follow this example by building a similar power plant in Scotland.

Meanwhile, hydrogen-using techniques were gradually being developed. Since the development of the first thermonuclear bomb in 1953, physicists have dreamed of being able to control the reactions of atomic fusion. The idea of using hydrogen is not new, however, as town gas once contained approximately 50 % of hydrogen, and the first hydrogen powered engine dates from 1927. Hydrogen is the most common element in nature, and may be partly made from water. Its combustion in the air produces only water and very little nitrogen oxide, and like natural gas it is both transportable and easy to store: the United States have developed advanced techniques in hydrogen liquefaction and storage. At the moment, hydrogen is produced by the distillation of coal, although production is naturally limited, and by electrolysis of water which requires large amounts of electricity. The second process is therefore very expensive and only partialy efficient, giving around 60 %. The simplest method of producing hydrogen would be to split water molecules, but this has only taken place at 2500°C. Even the most advanced nuclear power stations with high temperature gas reactors (H.T.G.R.) can only obtain a temperature of 800°C. Thermo-chemical methods are currently being studied, but this technique is still far from being perfectly controlled.

Thus, research is still being carried out in ever more delicate areas. Israel filed a patent on laser enrichment of uranium in March, 1974, and the Japanese have allocated considerable funds to research the development of an iron and steel manufacturing system, replacing classic blast furnaces using coke from coal, with nuclear reactors. Moreover, one should note that evolutions have already taken place in this domain: it is hoped that steel may soon be produced directly from a blast furnace using fuel and oxygen. After long trial periods, the high temperature Dragon reactor in Britain, has succeeded in providing its cooling fluid, in this case helium, with temperatures between 1400°C to 1500°C. But change is still a long way from being wrought.

In fact, we have still not really entered the atomic age. Promising developments exist – as is so in the miniaturization of atomic reactors so they can be used in certain forms of transport – but there are many bridges to cross as yet.

Let us end by looking at one particular technique. Most of the engines in use today operate using air as a combustive, whereas space rockets cannot use air and must consequently carry a fuel supply with them. The new fuel therefore appears to be a source of thermochemical energy, the energy of which may result either from the action of a reducing element on an oxidizer, or from the catalytic decomposition of a body. Such an energy source is known as a propellant, made of two separate components. When two components are stocked in separate reservoirs (nitric acid and rocket fuel for example), they are known as liquid propellant. Solid propellants are kept in a single reservoir, as is the case with ethylene oxide and nitro-methane. Compound propellants also exist. Such discoveries made in the field of modern chemistry have enabled the development of the all important fuel required to send rockets to the moon.

All these modifications have naturally been recorded by statistics which reveal the changes implemented (in percentages of world energy consumed):

	1900	1938	1958	1966	1975
Coal	92	64.2	44.7	37.6	31.7
Lignite	3.9	6.7	7.6	5.8	31.7
Oil	3.5	18.6	28.7	33.6	44.7
Gas	1.2	5.2	12.7	17.1	20.9
Electricity	0.4	5.3	6.3	5.9	2.7
Millions of TEC	818	1904	4074	6309	7971

This provides some sort of numerical image of the technical revolution which is in process.

Energy converters In the domain of engines and energy converters, one becomes aware of the restrictions of technical imagination. This is obviously due to the limited number of possible combinations.

Of course, certain machines disappeared altogether: mill wheels are only used in some under developed countries, or are tourist attractions. This is also true of the reciprocating steam engine, the classic steam engine from the "first" industrial revolution. Here again, its preservation, particularly on some railways, is proof of a technological and economical halt. In advanced countries it has disappeared completely.

In a great many domains, especially in mobile machines and in those designed for use in industry, a three stage system is currently favoured, consisting of the production of energy, the transformation of this energy into a mechanical form, and then the latter's transformation into electricity.

Electricity is an ideal intermediary because of its flexibility, and its capacity to rapidly adapt to variations of demand, even though it reveals significant losses in the course of these tranformations and the transportation of energy. But the ease with which it may be transported and used, although there was no way of storing it, ensured the success of electric power, as storing could be done at the level of the primary source of energy. We will look at a few examples to illustrate this.

We have seen that electric traction was an old idea used in the railways, and it continued to develop as current supplies were generalized. In recent years, the use of high voltage currents (25,000 volts) has been developed, which reduces any energy loss from overhead cables and, at the same time, locomotives could run on lines of different voltages. The only drawbacks of this technique lie in the massive investments required to establish the infrastructure. Electric diesel traction has virtually replaced steam traction in advanced countries. The electric petrol-traction torque had been introduced around the 1880's, but had not been developed any further owing to the bad performance of petrol engines during this period. Some railcars, however, enjoyed a solid reputation around the year 1914; it was in Sweden in 1913, that the possibility of replacing the petrol engine with the diesel engine was considered. The diesel engine had still to be perfected before it could be used properly: it was subsequently made lighter, and the speed at which it operated was increased. The electric diesel engine appeared on main American lines in 1930, and in 1934, a light and powerful diesel engine was launched by General Motors. In fact, in many countries, it was the existence, after the Second World War, of collections of old fashioned locomotives that ensured the success of the electric diesel engine. We should add that this success was also due to easier fuel manipulation.

As for maritime transport, the diesel motor had mainly replaced the steam engine in merchant vessels which hardly required great speed, while faster ships preserved the turbine, having adopted the fuel used in steam production. Naturally, the transition was made from diesel to electric diesel engines which facilitated manoeuvring.

After the Second World War, the use of atomic energy to provide the necessary steam was envisaged. This technique presented many advantages in certain sectors, particularly in the military domain where it made possible the development of submarines with almost infinite range. Combined with atomic missiles, a virtually elusive force of perpetual movement was possible. As for the merchant navy, the process was undoubtedly less interesting. In fact, only four nuclear propelled ships were built. The first was the ice-breaker, the *Lenin*, which sailed in the Arctic from 1959. The second was the American cargo boat, the *Savannah*, which from 1962 to 1970 travelled more than 800,000 kilometres before she was laid up. The German ship, the *Otto-Hahn*, with a 25,000 tonne capacity, was launched in 1969, but her reactor was replaced with a better one in 1972 after she had logged over 350,000 kilometres. From 1967, the Japanese began building the *Mutsu*, but Japanese fishermen, aware of the atomic risk involved, immobilized the ship for nearly two years. When she entered circulation again, there were indeed radio-active leaks. As fuel was low in price at this point, these nuclear ships were hardly worthwhile. Nevertheless, research was restarted as soon as the cost of oil began to rise again. The Germans and the Japanese worked together on constructing containers, developing from 80 to 240,000 horsepower (against 10,000 produced by current nuclear ships). Here again, the techniques involved had not yet reached the point of perfection.

Like the classic steam engine, the petrol engine appeared to be a saturated technique. In the domain of the automobile, detailed though often minor improvements were made, or were made in annexed instruments, in particular the gearbox.

The only new technique to be introduced was the rotary engine, or the rotary piston engine as it is known today (fig. 5). The reciprocating machine was now defunct: the petrol engine had progressed beyond the evolution undergone by the steam engine. The piston, or rotor, with a triangular and curvilinear shape, turned in the trochaical cavity, or stator, on an eccentric shaft to which it transmitted the movement. Its edges were in contact with the inside walls of the stator, and by moving forwards and backwards from the walls of the trochaic, the sides of the piston delimited the volumes of the working chambers where the operations of the four stroke cycle were performed (admission, compression, explosion-release, and exhaust). The cycle is completed when the piston has gone only a third of its course, while the shaft completes its own cycle, owing to the ratio of the gears. Thus, a whole piston rotation corresponds to three cycles and three turns of the driving shaft. The rotors in the Comotor engine, are made of a light alloy, with an interior covering of nickel-silicon, with segmented iron rotor sides of steel and the edges of the segments in a calcined material. This less pollutant motor was more efficient than the classic engine owing to the elimination of the connecting rod and crank system. Although this was adopted in some machines, its definitive improvement was probably not completed.

In the domain of electric engines, the use of the classic engine was still very widespread, with the inevitable improvements in detail or the use of new material. There were some new innovations in this field, however, one being the linear engine. It was composed of a metallic structure supporting coils of electric wire which passed through notches. The metallic structure and the coils encased a central metal rail fixed on a concrete line. An electric current was sent to the coils, thus creating a magnetic field between the coils in the air-gap in the rail. This magnetic field induced electric currents in the rail which tended to oppose the currents which were produced. A force was therefore created which moved the rail motor and furnished the thrust. The engine was not totally perfected, as the metallic structures of the engine which supported the weight of the coils and received the stress of the thrust and the guides throughout the length of the rail needed to be made lighter. Problems also arose in the way in which the engine should be attached to the vehicle, and finally, the method by which the current was harnessed. The engine was supplied by a three-phase system which had to be furnished from the exterior by three trapezoidal shaped aluminium rails, one of the sides of which was covered with a fine stainless steel sheet. The major problem was ensuring that the engine and the harnessing shaft did not suffer too much from the jolts and shakes of the vehicle.

It was this engine which equiped Gometz's experimental hovertrain, and was then tried in the prototypes of the "Urba" vehicle, designed for town transport. The engine was to be mounted on the experimental air cushion vehicle devised by the American firm, Rohr, the "Tracked Air Cushion Vehicle" (T.A.C.V.), and could produce a speed of 250 kilometres per hour on 4 degree slopes and confronted to a wind speed of 80 kilometres per hour. Here again some disappointing results were had, but the Grenoble firm which had developed this technique had good reason to expect satisfactory results.

But it was undoubtedly in aircraft engines that the greatest progress was made. Until the eve of the Second World War, only the classic

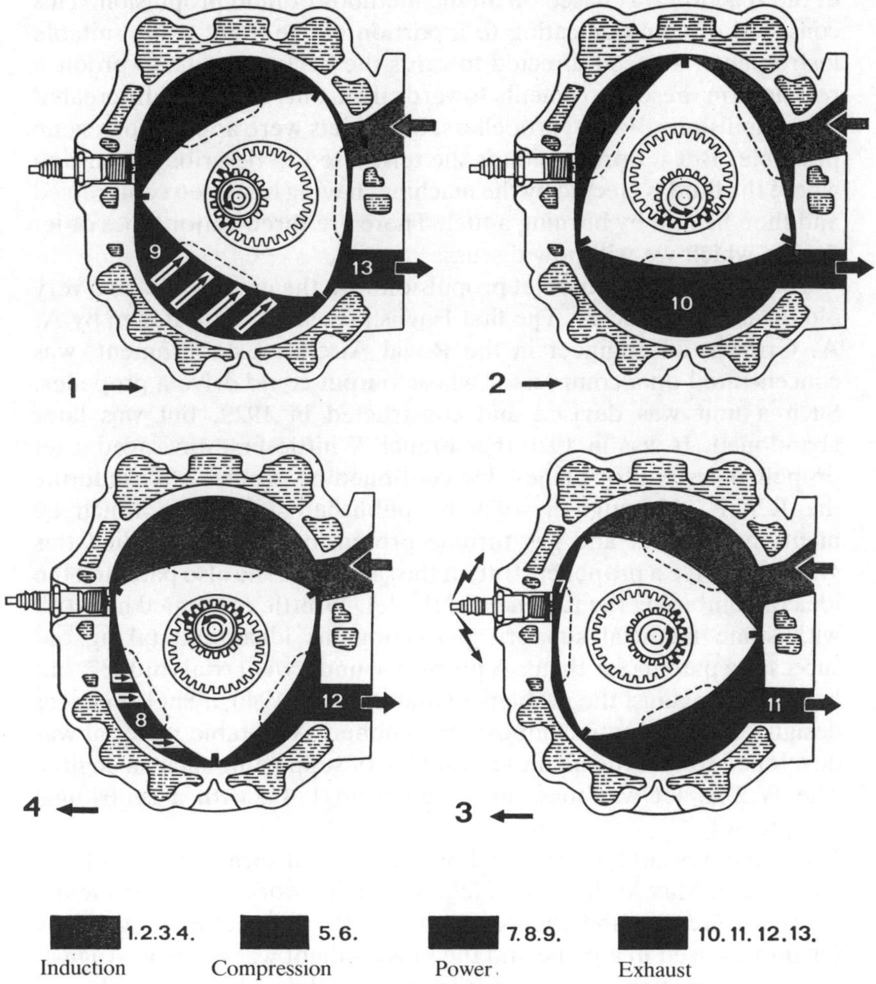

Figure 5. — Principle of the operation of the rotary engine. (From *Le Monde*)

piston engine had been used, and the range of the apparatus, its speed and its working altitude were measured in the same way. The principle of the reaction was based on all the methods of fluid propulsion. This consisted of communicating to a portion of the fluid, using suitable instruments, a speed directed towards the back and thus to produce reactions in these instruments towards the front, which finally created the propulsion. Blades, propellers, or gas jets were applications same principle. But jet propulsion is the term used to describe propulsion where the fluid is ejected by the machine, having first been compressed and then heated by burning a fuel. There are three major types of jet engine which we will now discuss.

The turbojet combines jet propulsion with the gas turbine, two very old ideas (fig. 6 and 7). The first English research, undertaken by A. A. Griffith, an engineer in the Royal Aircraft Establishment, was concentrated on a compressor whose output could drive a propeller. Such a unit was devised and constructed in 1929, but was later abandoned. It was in 1920 that Franck Whittle first envisaged a jet propelled system for planes. He continued his studies after entering the R.A.F. Although in 1929 he published a work in which he mentioned rocket and gas turbine propulsion, he still applied this system to drive a propeller. But, at the same time, he also patented the idea of combining the turbine and the jet. Whittle continued his work with some financial support, perfecting his ideas and taking out successive patents for them. A prototype underwent trials in 1937, but he came up against the problem of materials. In 1940, a Shell engineer designed an adequate combustion chamber. A suitable material was developed so that the plane was capable of supporting such an engine. The W.2.B. jet was mass produced from 1943, with a centrifugal compressor.

German research was carried out in a similar direction, with Hans von Ohain, Max Müller, and Helmut Schelp, working independently of each other around the years 1937-1939. A reactor designed by Ohain was used in a plane and the first jet flight was made in August, 1939. This reactor was very similar to Whittle's, although both men were unaware of the other's findings. They both came up against the same difficulties, those of combustion and the strength of the material. Rivalry was fierce between engine and aircraft constructors, and the official authorities were sceptical of abandoning the piston motor. Although mass production began in 1943, it was only in 1944 that large

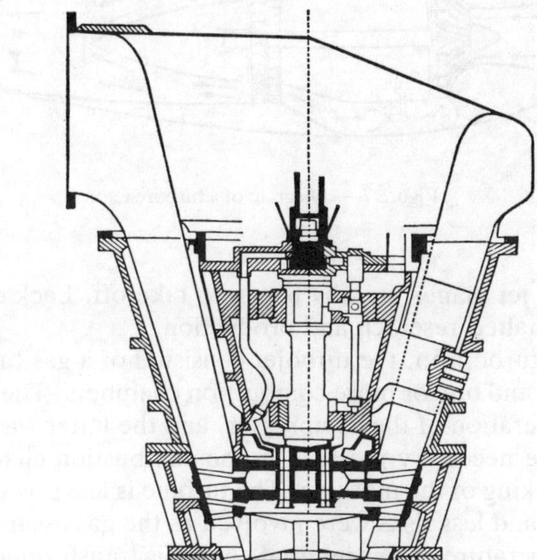

Figure 6. — Cross-section of a gas turbine. (From P. Lefort, *Les Turbomachines*, P.U.F., Paris, 1969)

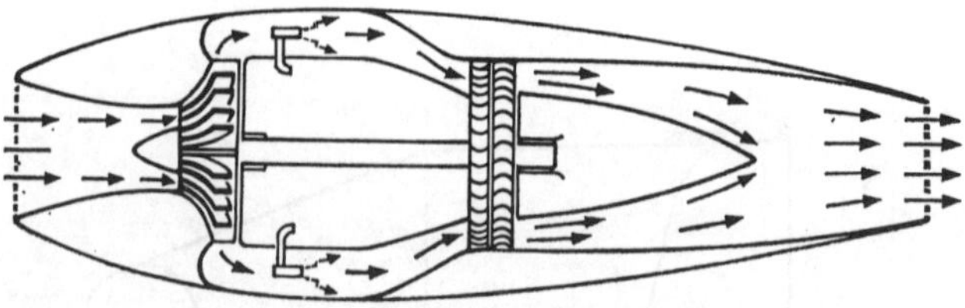

Figure 7. — Diagram of a turboreactor.

numbers of jet planes literally began to take-off. Lack of nickel had previously halted research and production.

Like the turboprop, the turbojet consisted of a gas turbine and its compressor and one or more combustion chambers. The turbine only used the operation of the compressor, and the latter was intended to establish the necessary pressure in the combustion chamber for the smooth working of the machine. The turbine is less powerful than the turboprop, and less risks were involved as the gas escaped at a much higher temperature. Their internal energy is transformed into kinetic energy in a propulsion nozzle where they reduced in pressure at the same time as they cooled, and it is their speed of ejection which generates the thrust which propels the apparatus. One major drawback lay in the fact that power of movement was insufficient during take- off: jet propulsion had poor efficiency at low speeds. It was therefore necessary to create a richer mixture without raising the temperature in the turbine which its blades could not withstand.

Displacement of a body or mass naturally produces air pressure proportional to the square of the speed. This pressure may be used to compress the air to be employed in the gas turbine. The use of such a source was not ignored in turbojets, which equally eased the compressor. Air intake was an important consequence of this. Although speed was sufficient, one could, at a pinch, bypass the compressor and the turbine which drove it. Subsequently the ramjet was developed, an extremely simple machine as it does not possess a single mobile part. Divergent air intake is achieved by a combustion chamber and a jet nozzle where the burnt gases reach their speed by being released. This apparatus was patented by the Frenchman, Lorin, and was also the subject of Guillaume's patent of 1920 for the turbojet. The first ramjet was built in 1944 for use in the German V1's: their speed was insufficient for the production of adequate compression, so the engine was operated by propulsion, using a phenomenon of resonance and flow of exhaust gas. The ramjet cannot operate under static conditions as without thrust there cannot be any speed.

Finally, the last type of engine was the rocket engine. As in the other engines, propulsion developed by this engine is created by the high speed ejection of a fluid towards the back. Air is no longer required: this fluid is entirely produced by chemical reaction inside the machine. We have mentioned above this new "fuel" type. The Germans were the first to study propulsion in fighter planes, assisted flight take-off, radio-controlled bombs, the propulsion of underwater torpedoes, and the V2 missile. The rocket engine is characterized, apart from its exceptional power-weight ratio, by a extremely high consumption of energy, in the sense that it does not draw from its oxidant, but it has to furnish it with the fuel. It is most often used in the propulsion of air to air, air to ground, or ground to ground flying missiles, with immediate impact, and for satellite launching, the first of which took place on October 4, 1957. In the last case, usually three engine sections are used, released successively so as not to overload the whole rocket. The first section or launching rocket corresponds to what we have just said. We will return to this on the subject of satellites.

Although new results have been obtained in engine performance, they have not changed totally from the engines of the old system. Moreover, it is all the more remarkable to state as we have here, that in most cases very old ideas become concrete reality. Tributary techniques therefore contribute considerably to the development and use of new machines, particularly in the creation of new materials without which nothing would have been possible. This is also true of the phenomena of combustion: for example, in the ramjet where the flame in the combustion chambers stabilizes with mixtures varying richness in proportion from 1 to 30, depending upon the flying requirements. Research should also be carried out on the fuel types used, but does an Institute of fuel combustion exist?

The New Materials

This is probably one of the lesser known aspects of the contemporary technical revolution, but it nevertheless one of the most essential. In many cases, material may provide an absolute restriction to progress being made. In other, new materials have succeeded in replacing traditional materials owing to their superior qualities.

On the eve of the Second World War, the history of materials is already an old one. Alloys, bronze, and brass date back to Antiquity. In the second half of the 19th century, considerable progress was made, with the development of specialized steel from 1867 onwards, and the search for all kinds of metal alloys, such as duralumin on the eve of World War I. Synthetic materials were also introduced, and we know the part they were to play: celluloid (1870), Galalith (1900), and Bakelite (1902). It has been suggested that half of the materials in use today were not known thirty years ago, and that almost all the materials which will be used at the end of our era have yet to be developed.

The actual transformation of ancient methods used in the production of classic materials is also important. For the production of pig iron, one tended to adopt a semi-acid or acid process. But the greatest innovation in this domain was the use of super-oxygenated air, combined with a simultaneous injection of steam, hydrocarbons, or with an increase of the temperature. The importance played by hydrogen was recognized, the use of which facilitated balancing thermal exchanges. The whole accessory apparatus, electric or pneumatic plugging machines, burners and Cowpers' hiving, automatic regulation of air temperature, and electrostatic purification, had all been improved as well as the controls. The blast furnace now became an enormous aparatus, which meant that fewer were required. We should also mention the experimental kiln developed at Liège, which operated with an injection of liquid fuel and an increase in the air temperature to 900°C. Fuel may be replaced by high methane content gas used as a reducer.

Similarly, direct ore reduction was also attempted, with the aim to liberate the process of coke production, but this did not eliminate the necessity of remelting the ferrous material on the furnace floor. The choice of reducing agent is essential. The Swedish engineer, Wiberg, employed warm gases from the carburettor which produces these gases from the mixture of carbon and steam. The Krupp-Renn process uses a mixture of low grade minerals and carbonated products in a revolving furnace which produces a pasty fusion, and from which a silicated slag containing globules of iron is obtained. The iron globules, extracted after cooling and crushing, are then introduced into the blast furnace. The Norwegian process operates by reduction, by simple contact with iron oxide and carbon, without fusion in the revolving furnace heated by gas, fuel, or coal. The final product, called sponge iron, is cooled, crushed, and compressed into billets, and then passes through the Martin furnace or the electric furnace.

As long as instruments were changed, so the methods of fabrication were greatly modified. The Ugine-Perrin method was definitively developed in 1934: this was a refining process involving contact of slag metal. Today, two liquid materials are prepared seperately, the slag is then poured into pockets and the metal is introduced in a fine trickle, thus there is close mixing and instantaneous reduction. Rich gases from coke furnaces, natural gas, or oil, are now used in classic furnaces, which enables the elimination of gas producers. Fuels no longer require preliminary reheating. The growing use of pure oxygen in steelworks remains the major post-war innovation. This enabled the acceleration of operations, growth in the hourly production rate, and visible reduction in fuel consumption. Research at the Ijmuiden centre into flames has led to much shorter, hotter and thereby more efficient flames. In the Thomas furnace, the blast of air enriched either with pure oxygen or a mixture of oxygen with a neutral gas, means that high temperatures can be avoided. The Austrian LD process consists of blowing the oxygen at high speed onto the surface of the bath. In the Swedish Kaldo process, the oxygen is blown into a cylindrical apparatus turning at high speed, by means of a nozzle tangential to the level of the bath. The length of this operation, which took six and a half hours in a low hearth furnace, has been reduced to an hour and a half.

In many areas, progress is still at the research stage. By using a blast furnace, it was hoped to achieve direct production of steel with enriched, already reduced ores. In steel production, many more processes were to use oxygen. In any case, it seems that the role of coal in iron and steel making was declining.

In principle, forging processes remained unchanged. Continous trains were becoming much more important in rolling, but continuous production itself was an old idea beginning with the coninuous production of paper first introduced by Robert in 1797, and which was eventually used for all types of processes, such as the continuous casting of sheets of glass. The first sheet metal continuous trains were tested at Teplitz in 1892, and considerable improvements were made to these in the United States by 1916. These were break-down and finishing mills and were used for military production. The progress made in automation has made them even more efficient: steel travels in ingots of 14 tons, and conveyor belts now run at a speed of 45 kmh., which may eventually rise to 100 kmh. Servo-systems provide a far more precise check on the quality of products than formerly.

The history of industrial alloys still remains to be written, and would be instructive for many reasons. The industrial alloy was born from the need for certain materials possessing specific qualities which could be used in certain techniques: these qualities were strength, resistance to corrosion, elasticity, creep resistance, etc. These materials were needed to make steam turbines, and are also required for turbojets or the shields protecting satellites when they re-enter the earth's atmosphere. We also noted above the emergence of special steels in the laboratory of the Jacob Holtzer firm at the very end of the Second Empire. Around 1902, ferro-nickels were produced which were chemically hypersensitive, and thus extremely useful. Research during the inter-war period concentrated mainly on thermal treatment and on the structural hardening of special steels. Steels with the alloy iron-aluminium-nickel constitued a veritable revolution in the production of magnetites. In 1938 the Inland Steel Company developed the difficult steel-lead alloy.

After the last war the lack of certain metals brought with it the first alloys with a low grade of rare metals. In France the first alloys were produced with less than 1.6% nickel, and this was supplemented by chromium, manganese and copper. Molybdenum was also becoming rare, so this too was reduced and chromium and vanadium added. In this way more and more binary or tertiary alloys were produced which brought out materials possessing particular properties from a mechanical, chemical or thermic point of view. It was this proliferation of steel-based materials which made certain elements of technical progress possible in the field of engines, telecommunications and time-keeping. There is no point in discussing all of these and their specific qualities in detail, but it is worth pointing out the imoprtance of the support that precision metallurgy provides to a number of technical sectors. It is also worth noting that the iron and steel sector was not the only one to contribute to this growth of new products, other types of metallurgy also contributed, with multiple alloys, to this effort. Again, we are forced to reiterate that we lack an adequate historical schema of this fairly continuous trend of progress, born of very conscious needs.

All aspects of metallurgy are directly linked to alloying, from auto-tempering to the forge, the indispensable complement to the fabrication of materials. In many cases it was also necessary to improve the measuring or examining apparatuses. The laboratory of an iron and steel enterprise was quickly transformed into a laboratory of fundamental science.

Who cannot be aware of the prodigious rise of plastic – or to use a more general term, synthetic materials (fig. 8)? At first, natural materials, such as cellulose or casein were used, which gave rise to the first new materials, celluloid and galalith. Bakelite, the last one of these new materials to be created, was in fact the first purely synthetic material. In 1907, Baekeland began with formol and phenol, and with an alkaline catalyst he obtained a condensation resin, with exactly the same properties as that of gumlacquer. This was to replace substances as diverse as wood and porcelain. It was at this point that the importance of these new materials were first realised, materials which were not derived as before from extracts of animal or vegetable products, or from natural minerals, but which could be artificially created by in the laboratory by shrewdly combining analytical and synthetic processes which the progress made in chemistry now made possible.

It is worth specifying a few points, here. Some research was done into the possibilities of replacing traditional materials difficult to obtain by these new, synthetic ones. This means that some of these materials are merely replacements: the Germans substituted rubber with synthetic rubber, for example. On the other hand, it also proved possible to create entirely new materials which not only replaced older ones, but could also be put to completely new uses. Bakelite may have replaced wood in some respects, but it was responsible for the emergence of the ink pen.

There are several types of elaboration of a plastic substance:

I – Cellulose, a macromolecular body is extracted from vegetable plants. After nitration, the cellulose is made plastic by another natural compound – this time with a relatively simple molecular formula – that of camphor. This is done in the presence of a solvent, alcohol, which is removed afterwards through evaporation.

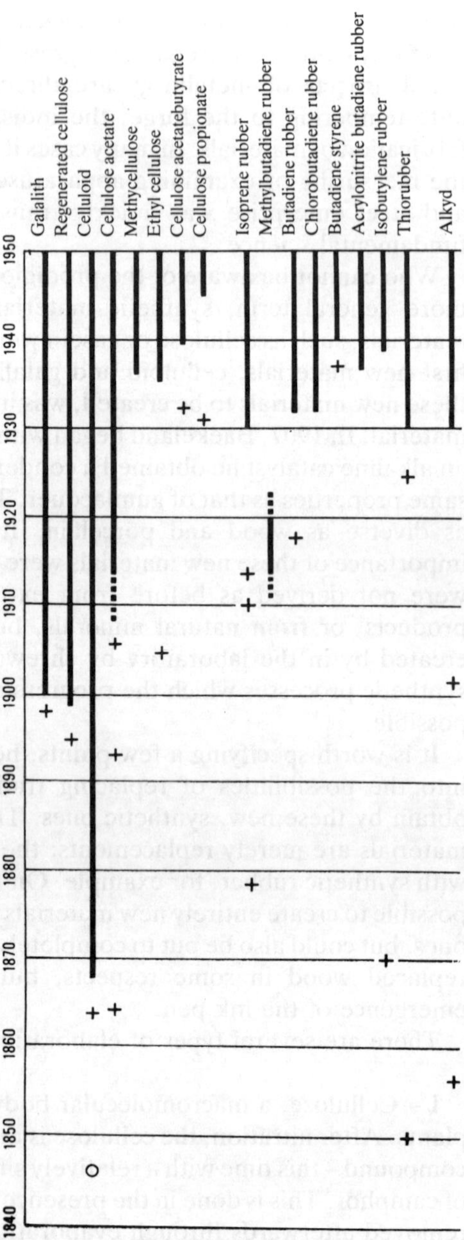
Galalith
Regenerated cellulose
Celluloid
Cellulose acetate
Methycellulose
Ethyl cellulose
Cellulose acetatebutyrate
Cellulose propionate
Isoprene rubber
Methylbutadiene rubber
Butadiene rubber
Chlorobutadiene rubber
Butadiene styrene
Acrylonitrite butadiene rubber
Isobutylene rubber
Thiorubber
Alkyd
1840
1850
1860
1870
1880
1890
1900
1910
1920
1930
1940
1950

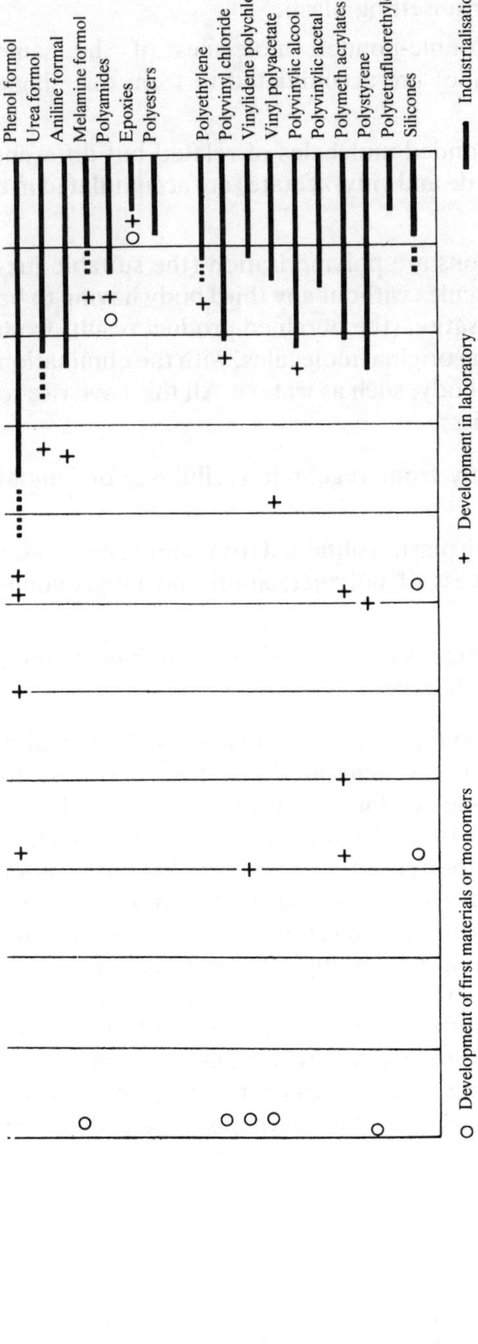

Figure 8.—The birth of plastics. (From Fournier, *L'Ere des matières plastiques*, Paris, 1955)

II – Phenol and Formol, two bodies of not very complex molecular characters, are isolated from natural substances (or prepared synthetically). They are then combined by an appropriate heat treatment to produce a thermosetting plastic.

III – Several simple double-bonded molecules of the same compound (polyvinyl chloride) are accumulated to form one single chain of molecules.

IV – Numerous double-bonded molecules of related but different compounds (polyvinyl chloride and vinyl acetate) are accumulated in a single macromolecule.

The two essential operations are polymerisation (the suitable juxtaposition of monomer molecules without any third body having to be eliminated) and polycondensation (the obtained product results from the generalised reaction of the original molecules, with the elimination of a third, simple molecular body, such as water). All this gave rise to three main families of plastics:

I – Plastics derived directly from vegetable (cellulose) or animal (casein) substances.

II – Synthetic thermosetting plastics obtained from simple molecular chemical types, by the process of polymerisation and polycondensation.

III – Synthetic thermoplastic resins obtained by the polymerisation of simple molecules of a single type.

Cellulose plastics developed rapidly. Celluloid itself had one major drawback – its inflammability. Regenerated cellulose gave rise to matter in the form of pellicules or films, like cellophane which was developed in France around 1900. Then came the cellulose acetate plastics. The first synthetic thermosetting resin was bakelite; other phenoplastics have been industrialised from around 1925. Aminoplastics are obtained by the reaction of polymerisation between formol and an organic amine body (urea for example): the Czech H. John discovered this in 1920, and the process was industrialised by the Austrian F. Pollak. The subsitution of glycerine for formol gave rise to glycerophtalic resins – these were eventually to be used extensively as resins. Finally, indocoumaronic resins – part of the group of synthetic thermoplastic resins – were used by the Germans during the First

World War. Vinyl resins are obtained by the polymerisation of slightly comised double-bonded molecules. The reaction of hydrochloric acid on acetylene gives vinyl chloride from which polyvinyl chloride can be produced through polymerisation. This substance is remarkable for its resistance to a large number of chemical agents. Styrene was prepared for the first time in 1831. But it was Staudinger in 1926 who showed that it was possible to obtain from polymerisation a whole range of products, from dimers to polymers consisting of enormous macro-molecules, which serve as excellent insulating material.

Before the end of the Second World War, then, there was already a large number of products available whose properties were both extensive and diverse. In many cases, the definitive developments took place during the 1930's, and the industrial production of many of these products had hardly begun. Throughout the entire the Second World War, all the belligerent industrialised countries paid close attention to these new materials, perfected them and discovered new essential products.

Polythenes or polyethylenes are obtained by the polymerisation of ethylene, at first under considerable pressure. The major British company Imperial Chemical Industries began research into this around 1928. Their research was supplemented by help from the Dutch scientist Michels who worked on hyperpressures in chemistry. Using apparatuses devised by Michels in 1931, an experiment was conducted in 1933 to effect the polymerisation of ethylene. Something caused an abrupt fall in pressure, and as a result a small amount of white powder was discovered which was endowed with remarkable properties: it could be stretched when cold, it could be moulded, be transformed into fibres or leaves, and it was resistant to a large number of chemical agents. In 1937, a continuous running apparatus was set to work. The superiority of polyethylene over gutta percha was soon recognised, particularly for insulating cables. Its qualities also proved to be one of the essential materials needed for research on radar. In 1950, the German Ziegler discovered new ways of preparing polyethylene in normal temperature and pressure conditions: the patents, taken out in 1954, indicate that polymerisation should take place with a catalyst which was aluminium. The chemists at Philips Petroleum used chromium. Production began in 1956 in Germany and the United States, and in 1960 in Britain. The mechanical properties of polyethylenes are sufficient, their electric properties are remarkable. And the same is true of their chemical properties. They can weld together, mix with other substances, and produce synthetic rubbers.

Silicones are synthetic resins identical to the ones we have just discussed, except that the carbon is replaced, atom for atom, either by silicon or by oxygen. A reaction between two compounds was effected in a laboratory in 1904, but it was only in 1939 that two major American companies, Corning Glass Works and the General Electric Company first became interested in the industrial use of these products, some of which were as transparent as glass, and which possessed important electrical properties. Corning Glass works began industrial production for military purposes in 1942, while GEC began production in 1946.

We saw earlier that the large firm Du Pont de Nemours decided to enter into fundamental research in 1927, and for this purpose they employed the services of a chemist, W.H. Carothers, who had already made a big reputation for himself in the field of polymerisation phenomena. He proposed studying the structure and synthesis of long-chain molecules and his first studies dealt with polyesters which he was the first to synthesise. In 1930, an observation took place on a polyester obtained by the action of suberic acid on ethylene glycol. It was noticed while trying to take a sample of polyester that this material could be stretched into long fibres, and what is more, once cooled, this fibre could be stretched even further to several times longer than its original length; furthermore, far from being harmful, this operation actually increased its solidity and elasticity. Its only drawback was that hot water made it pasty. So it was necessary to look down other avenues. The obstacles were many and Carothers decided to abandon the project. However he was persuaded to continue, and it was at this point that he turned his attention to polyamides, which have the same chemical arrangement as polyesters. The team succeeded in producing a polyamides known as "No 66," which gave an elastic resistant fibre capable of withstanding temperatures up to 260 degrees. The first threads, produced for an experimental batch of stockings, were prepared in April 1936 and July 1938. In December 1939, industrial production began. Nylon was born.

The Germans had done similar research. but they started later than the Du Pont scientists. Their work resulted in the development of a very similar fibre called perlon. In general, polyamides were used at first almost exclusively as textiles: they are now also used as plastics.

They are used notably in extrusion and grinding by injection or compression. Nylon is used today to make diverse utensils for the kitchen or toiletries. Rilsan, a polyamide produced in France, obtained from ricin and more resistant to water than nylon, is used extensively in packaging and conditioning.

Carother's results encouraged researchers to pursue further on the same line. It was as a result of this that J.R. Whinfield and J.T. Dickson, both scientists at the British Calico Pinters Association, discovered terylene. They succeeded, where their precursor had failed, in producing a fibre form polyesters. They discovered this at a high fusion point, and during development they discovered by chance that this fibre had other remarkable qualities. In 1941, this substance was finally developed in England. The first of the two required products, ethylene glycol, was already produced industrially, the second terephtalic acid was only a laboratory product. The solution to this problem was found after the war and industrial production began at the I.C.I. factories in Wilton in 1955.

Research continued in the field of polyesters, polycarbonates and polymaleates. These have undergone considerable development since 1956: their main use is in the production of laminated mouldings, which can be made either rigid and very hard, or in a supple form.

In 1938, American chemists demonstrated that di- or poly-isocyanates could react with poly-alcohols to give substances with plastic properties. Licences were taken out in Germany after 1939 in this field. This reaction led to the formation of polyurethanes with a linear structure, which could be used as thermoplastic resins.

The final series to be developed recently is that of fluoro-plastics. In monomers identical to vinyl chloride, but in which the hydrogen and chlorine is partially or totally replaced by fluoride, it is possible to produce polymers which can be used in technology, and which constitute the group fluoro-plastics. The most important of these is teflon, a plastic with truly exceptional qualities, resistant to all acids, and which conserves its solidity and form at extremely high temperatures. It has even been used in mechanical parts. It is still an expensive commodity, however.

Carother's earliest work had been concerned with polyesters. But given their low melting point, they were abandoned infavour of polyamides which led to the discovery of Nylon. The two English scientists, Whinfield and Dickson discovered a polyester with a high melting point. The first licence for the new textile fibre was granted in July 1941. Because of the war, however, this was not followed up. The technical aspects of the discovery were published in July 1946. It was then that two firms, one British, Imperial Chemical Industries, and the other American, Du Pont, applied for licences and devoted a considerable amount of capital to the industrial development of what was eventually called Dacron.

The number of synthetic materials is now considerable, and we could multiply the examples. We shall mention only two more cases, whose importance is considerable.

It was Julius A. Nieuwland, a priest of Belgian origin who emigrated to the United States, who pursued the slow course of developing synthetic rubber. He began his study into acetylene in 1920, and at the end of many years of hard work he obtained an acetylene polymer, called divinylacetylene, formed by the union of three molecules. For his part, in 1925 Dr. E. K. Bolton, working at the research centre at Du Pont, proposed exploring the synthesis of rubber from acetylene. Involved in an erroneous line of research, he was able to correct it when he came into contact with Nieuwland's work, and with Carother's help, he was able to produce neophrene, a synthetic rubber whose qualities were far superior to those of its natural counterpart.

A Swiss chemical company specialising in the manufacture of synthetic colourants, began research on the production of moth repellents and they were eventually successful in producing one of these. The chemist Paul Müller continued this line of research to look for a general insecticide. He managed to find such a substance in 1939, by using the action of monochlorobenzene on chloral in the presence of sulphuric acid. This discovery was blocked by the war, but in 1942 it became known to the English legation in Switzerland, whereupon it was immediately adopted for industrial use both in Britain and the United States. It was discovered later that it had been first produced in 1874 by the chemist Zeidler, who had not recognised all its qualities. It was this process which generated the manufacture of the product known as D.D.T.

We should be clear about all this however. These synthetic products are all made from natural ones. It is simply the origins of these industries which have been displaced. In former times, natural products such as wood, horn, wool and cotton, etc, were used directly, whereas today, most products are chemically produced, using a few major natural products, of which petrol – although not the only one – has become the most important. Salts and pyrites which were widely used when the chemical industry first began, are still relatively important.

The Electronic Revolution

It is widely thought that the computer is becoming one of the fundamental symbols of comtemporary techniques, and of the contemporary technical system. In fact, however, it is probably better to talk in terms of an electronic revolution, of which the computer is only one aspect. In many areas electronics is clearly one of the fundamental aspects of the new technical system.

The origins of electronics go back a long way. Ever since 1832, the great English physicist Faraday, had suspected the existence of electro-magnetic waves. In 1833, Faraday himself actually observed that the resistance of silver sulphate possessed a negative coefficient of temperature: he was thus on the way to discovering semi-conductors. Every history of science teaches us that Maxwell provided theoretical proof of the existence of these waves, while Rudolph Hertz, in 1887, actually demonstrated their existence experimentally. We are also told that Branly invented the practical device used to detect these waves which was known as the coheror, and he also conceived the antenna. In 1895, Alexander Popov created the first system of wireless telegraphy, of which the first installation was set up by Marconi in 1896. By 1901, he had succeeded in sending long-distance signals.

In 1879, Edison created the first electric lamp using a carbon filament. In 1883, when Edison was trying to correct certain defects with the lamp, he noticed that in certain vacuum conditions and under certain voltages, a bluish light appeared inside the lamp: he discovered that an unexplained current passed between the two wires which fed the filament. J.-J. Thomson showed that the Edison effect was due to the passage of electrons from the negative electrode to the positive electrode. The theory of thermionic transmission was established in 1903 by O.W. Richardson. Shortly afterwards, Ambrose Fleming invented the diode, the first device putting freed electrons to work: the era of electronics was born (fig. 9). Branly's coherer was thus replaced by a radio-wave receiver. In 1900, Lee De Forest begun his research into finding a more effective detector of radio-electric signals. He introduced a third electrode, shaped like a gridiron, between the cathode and the anode: the triode thus made its appearance in 1906 (fig. 10). It was thus possible to control the flux of electrons. The diode and the triode were both devised before the technique which was to make them both more effective, and which was evolved between 1910 and 1925. It had also proved necessary to perfect the method used to obtain a vacuum (the molecular vacuum pump devised by the German Gaede in 1910), and find a suitable material for cathodes (metallic

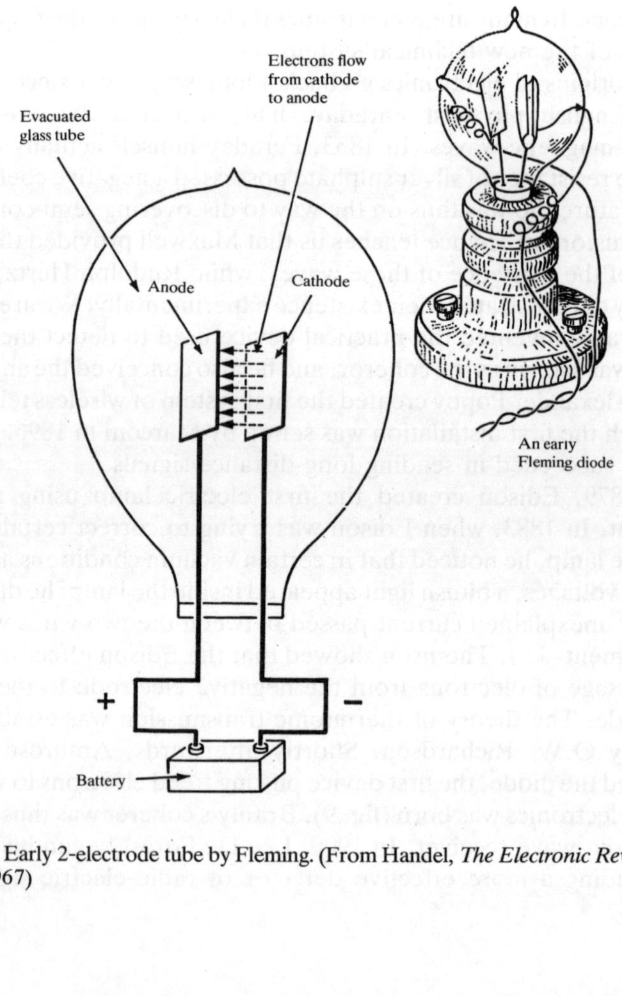

Figure 9. — Early 2-electrode tube by Fleming. (From Handel, *The Electronic Revolution*, London, 1967)

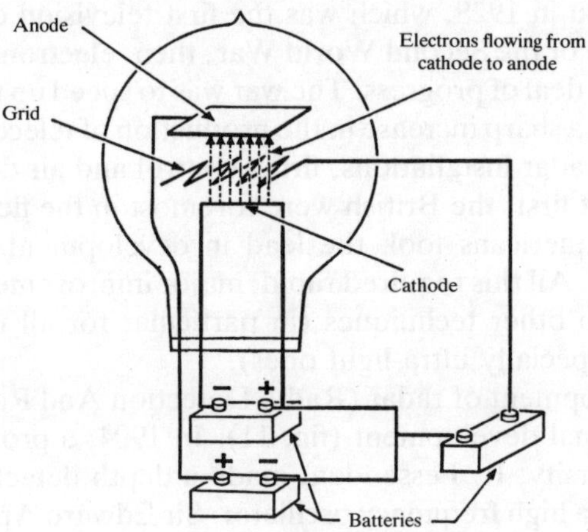

Figure 10. — The triode. (From Handel, *The Electronic Revolution*, London, 1967)

oxydes cathodes were introduced by Arnold in 1915). In 1927, A.C. tubes were devised which were connected, via a tranformer, to the main supply. After this, the tetrode (1928), then the penthode (1929) both enabled radio signals to be amplified even further.

In the same year that Thomson announced the discovered of the electron (1897), Karl Ferdinand Braun, of the University of Strasbourg, created the first cathode-ray oscilloscope. Since it is possible to convert almost any physical phenomenon into variations of current or voltage the latter in their turn serve to vary the deflecting fields of cathode-ray tubes.

The photo-electric effect – the discharging of electrons from atoms by means of light photons – was known before Planck and Einstein provided the theoretical background. Hertz had discovered that under the action of light, electrons are emitted by negatively charged alkaline metals, such as sodium or potassium. This gave rise to the the first photo-electric cell developed by Elster and Geitel in Germany in 1905. After various improvements made by 1920, these became, like cathode-ray tubes, a basic element of electronics. Zworykin, a Russian emigré living in the United States worked on amplifiers which could be used to reproduce film sound tracks, and on ways of converting images into electric currents. The latter research resulted in the iconoscope, first produced in 1928, which was the first television camera tube.

By the eve of the Second World War, then, electronics had already made a great deal of progress. The war was to speed up this process, by precipitating a sharp increase in the production of telecommunications equipment, radar installations, firing control and air defence apparatuses, etc. At first, the British were foremost in the field of research, whilst the Americans took the lead in development and industrial manufacture. All this required rapid, major improvements and always in relation to other techniques (in particular for all that concerned materials, especially ultra-light ones).

The development of radar (Radio Detection And Ranging) proved to be a seminal development (fig. 11). In 1904, a professor of Pittsburgh University, R. Fessenden, made a depth detector which used the pulses of a high frequency oscillator. Sir Edward Appleton and the teams of researchers at the Carnegie foundation used radio waves to determine the height of the different layers of the ionosphere. Waves were relayed to the object to be detected; the waves were reflected off this object back to the starting point. If the speed of the waves was known, it proved easy to calculate the distance of the object. There was a great deal of work to be done between the principle and its practical application, and Appleton's research was continued after 1930 under the direction of Robert Watson-Watt. They also had to miniaturise the apparatuses so that they could be used in planes and ships, and also to make sure that they were capable of producing powerful pulses of high frequency waves. In 1939, in Britain, J.T. Randall and H.A.H. Boot invented the magnetron, which was developed by the M.I.T.

All electric tubes use freely-moving electrons in as absolute a vacuum as possible. It had been easy to understand their action in magnetic and electric fields. For a long time, components of solid crystals had been in use, the main advantage of which was that they were better conductors of currents in one direction than in the other. This was true of the famous galena receiver, composed of a crystal on which rested a fine metal point. This was effectively a kind of diode which rectified the current. These crystal valves, using silicon crystals, were much more efficient rectifiers for high-frequency radar signals than any other thermionic valve which was available before the war. Even so, it was not clear how they worked. They were all built with

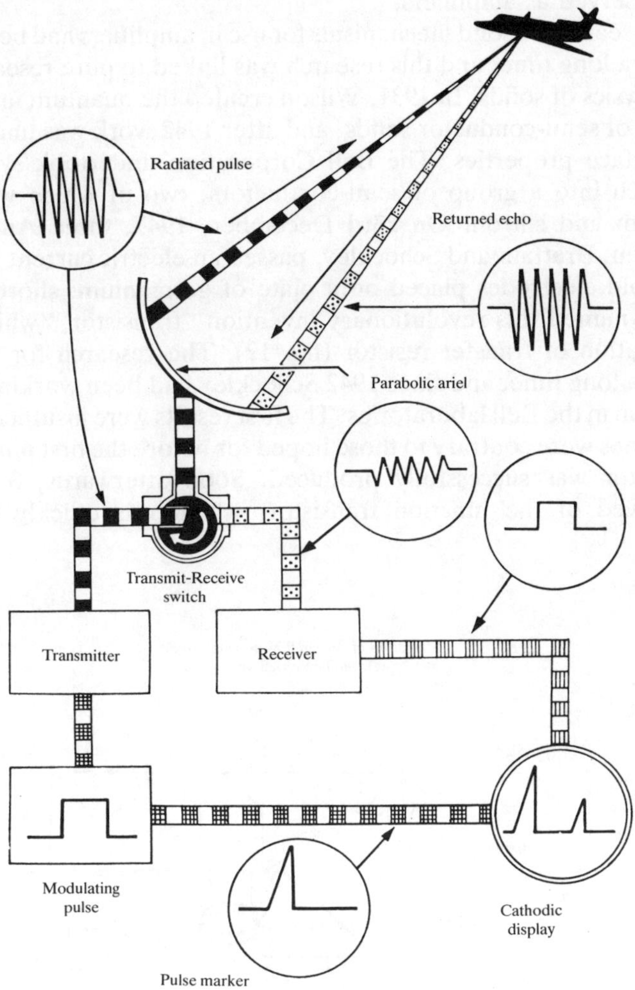

Figure 11. — Diagram of radar. (From S. Handel)

what we call today "semi-conductors," positioned half-way between the conductors and insulators. They were used only as rectifiers, and never served as amplifiers.

The search for solid mechanisms for use in amplifiers had been going on for a long time, and this research was linked to pure research into the physics of solids. In 1931, Wilson created the quantum mechanics model of semi-conductor solids, and after 1942 work was undertaken on surface properties. The Bell Corporation undertook systematic research into a group of semi-conductors, two of which were germanium and silicon. On 23rd December, 1947, three Americans, Bardeen, Brattain and Schockley, passed an electric current between two gold electrodes placed on a plate of germanium: shortly after, Pierce named this revolutionary invention "transistor," which is an abreviation of transfer resistor (fig. 12). The research for this had taken a long time, and since 1942 Schockley had been working on the problem in the Bell laboratories. The first results were insufficient, and later ones were contrary to those hoped for before the first pin-contact transistor was successfully produced. Soon afterwards, Schockley conceived of the junction transistor, which was quickly adopted

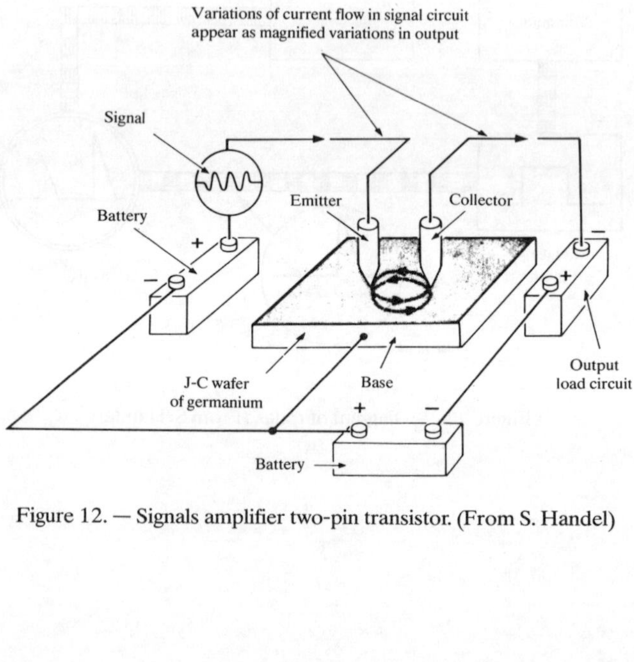

Figure 12. — Signals amplifier two-pin transistor. (From S. Handel)

universally. This represented the embryo of a major revolution which was to have consequences in many different areas, from transmission to atomic bombs, from computers to artificial satellites.

Now that currents could be controlled, the data had to be recorded. Magnetism solved this problem, in that certain materials, once magnetised, remain so. In an American journal of 1888, one Oberlin Smith described a system of magnetic recording which was finally realised, ten years later, by Valdemar Poulsen. Smith advocated the use of a rope, a thread, a ribbon or a chain made entirely or partially of tempered steel. He even suggested that these could be made of "silk or cotton whose fibres would be mixed with quenched steel filings."

The main emphasis in research, particulary that concerning computers, was on materials and systems which would increase memory capacity and the speed of access to these memories. The most simple method was to use bars which were magnetised in one direction: the use of the binary system meant that negative and positive poles could be used. Obviously, it was not possible to stock a large number of these bars no matter how small they were. Each grain of matter constituting these bars itself acts as a bar. It was therefore possible, – and Smith had advocated this – to pulverise these bars and coat cylinders with the resultant substance: hence the magnetic drum. The next step was to produce magnetic tape and magnetic discs. It was hoped that this would increase the memory capacity and that the unwinding of tapes or disc rotation would provide more rapid access.

Eventually, static magnetic memories were produced, and usually, these were circular rings of magnetisable material, placed concentrically to the magnetising current. These magnetic cores were comparable to the bars mentioned above, and as such were compatible with the binary system. A matrix of cores is constituted by a lattice of *n* parallel and equidistant wires, and *m* wires, parallel to each other, and perpendicular to the former. Today, central memories are composed of magnetic cores and ferrites, the latter no doubt to be replaced by other materials.

It is clear that research in all areas has been constant, and in some cases it was begun a considerable time ago. But we must still await the time when all the elements become complementary and coherent before a more or less global technical mutation can take place.

A NEW WORLD TODAY

Everyone is aware of the extraordinary transformation taking place in the world as a result of new techniques. The introduction of the latter "breaks the spatial equilibrium, modifies the framework of existence, which in geographical terminology constitutes both landscapes and lifestyles." It would take volumes – and there are already many of these – to analyse this evolution, the privileged and the deprived areas, the leading and the backward sectors, the imbalances and the compensations. We shall therefore limit this discussion to some of the major areas, which are quite clearly dominant in the installation of the contemporary technical system.

"Chemicalisation"

This term is not very elegant, but it does manage to express a reality which is apparent in all areas. From furniture to medecines, from clothes to cars or pleasure boats, everything, or nearly everything, comes from chemistry. The chemical industry sells 20% of its production to the consumer directly. The rest goes to other industries in the form of intermediary or subsistence goods. A base of 100 for the year 1962, rose to 311 by 1973 for the chemical industry, as against 197 for industry as a whole. And using the same base, we have the following figures: organic chemistry, 502; mineral chemistry, 198; parachemistry (pharmaceuticals, paint, varnish), 208.

A few base products were derived from some raw materials. Sequences of chemical reactions resulted in a formidable branching out (fig. 13). At the top of this tree are several dozen base products, and at the bottom several thousand products. "The sequence of reactions tranforming a higher product into a lower one constitutes a channel of production. The ensemble of these different channels forms a network which illustrates at once the complexity of chemical production and its logic, its flexibility and its rigidity." The capacity of chemistry to produce such diverse products has meant that its use is now all pervasive. The authors of the 6th French Plan were perfectly aware of this: "By the next generation, about 80% of the industries of developed countries will be chemically based."

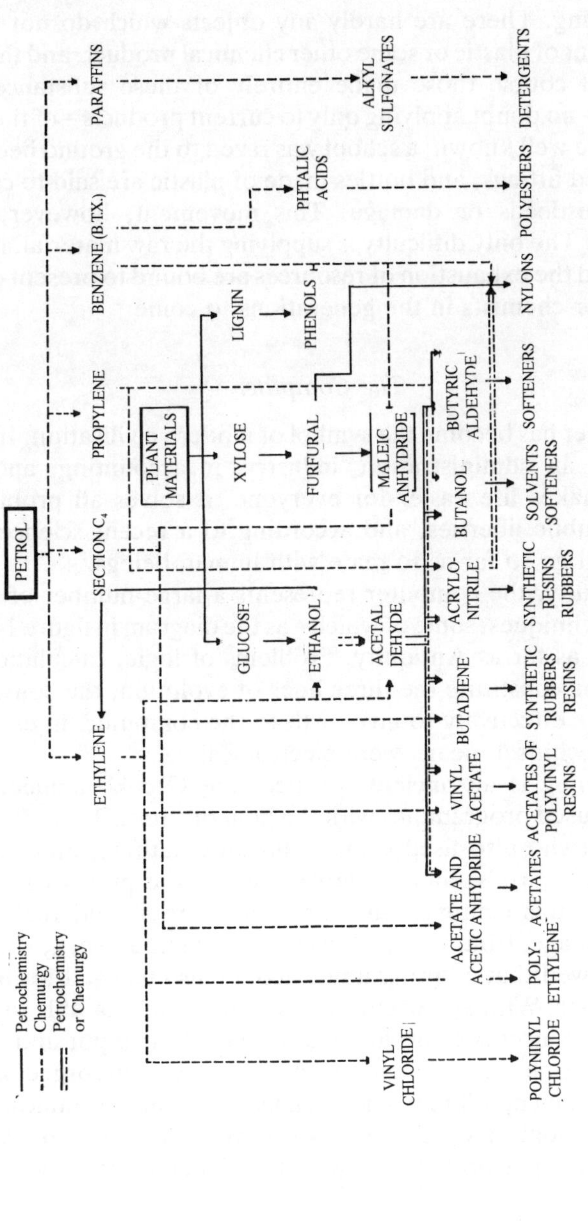

Figure 13. — Petrochemistry and chemurgy. (From *Le Monde*)

Products of the chemical industry are used in nearly all types of manufacturing. There are hardly any objects which do not contain some element of plastic or some other chemical product, and this is not including of course those made entirely of these substances. The limitations – no doubt applying only to current products – of these new materials are well known: a school was razed to the ground because of its plasticised fittings, and bottles made of plastic are said to cause all kinds of misdeeds or damage. This movement, however, seems irreversible. The only difficulty is supplying the raw material, i.e., oil: oil crises and the exhaustion of resources are bound to present difficult problems for chemists in the generations to come.

The Computer

The computer has become the symbol of modern civilisation. It is used everywhere, in administration, industry, in accounting, and space flights. It makes life easier for everyone, it solves all problems, it threatens public liberties, and according to a recent science-fiction novel, it will go so far as to mate with human beings.

In its totality, the computer represents a large number of diverse ideas and techniques, some of which – as the diagram in figure 14 shows – date back as far as Antiquity. Problems of logic, calculation, and programming constitute the three lines of evolution, the conjunction of which was effectively to give birth to the computer, in essence at least. The technical means were electronical.

Programming is an ancient practice: the Greek engineers from Alexandria used programmes with the help of cams. Basile Bouchon, Vaucanson (who also used cams in his automatons), and Jacquard used perforated rolls, and Hollerith created the punch card at the beginning of this century. But our admiration should really go to Charles Babbage, from England, who, in his famous work, *Analytical Engine*, showed that a "programme" could also be used for intellectual operations. With an extremely fine analysis, he was able to extract all the basic concepts of an automatic calculator: input and output, memory, programme and logic. A link had now been forged between something which up till then had been used only in mechanical organs and abstract logical operations. Moreover, he used an extended version of the perforated roll which had been developed by the French.

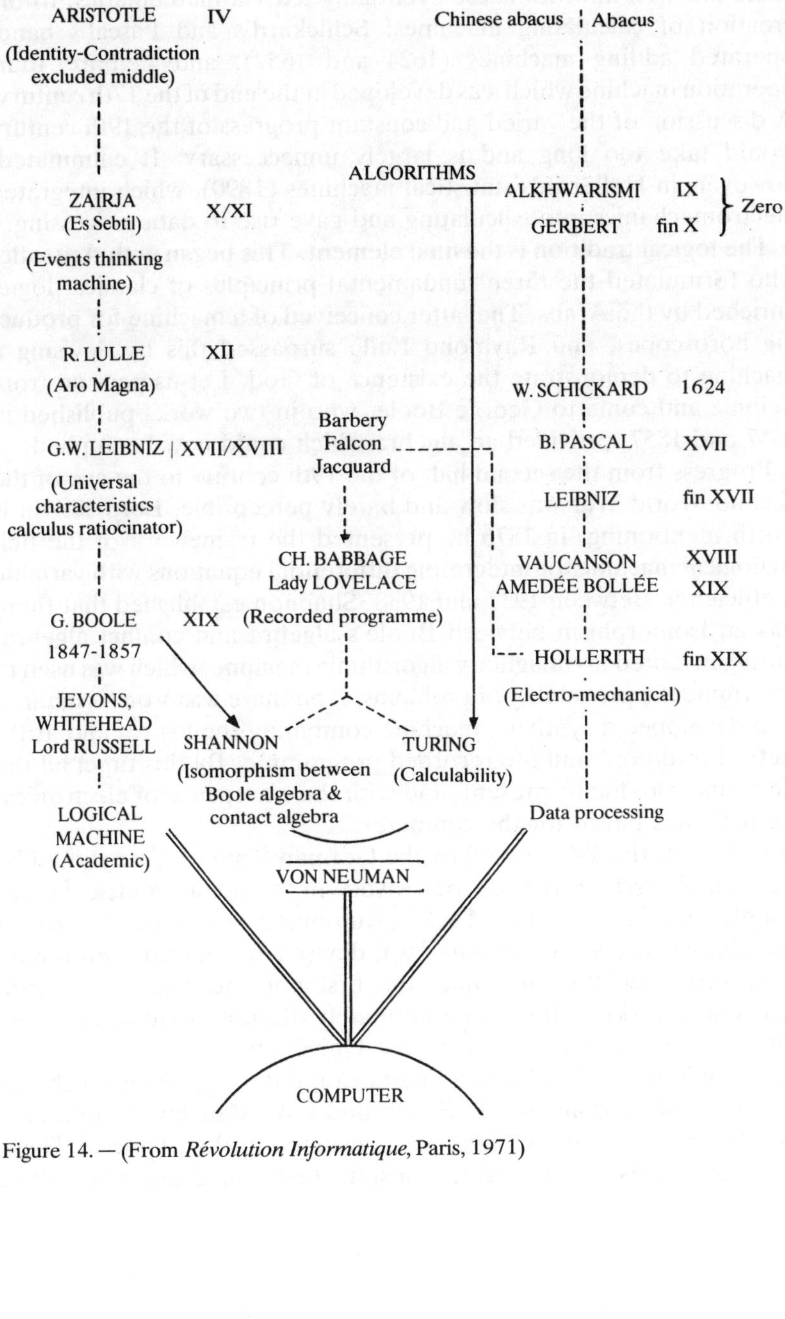

Figure 14. — (From *Révolution Informatique*, Paris, 1971)

Calculating methods also date back to many centuries ago. Abacuses are well-known. These eventually led via mathematics, to the creation of calculating machines: Schickard's and Pascal's hand-operated adding machines (1624 and 1652), and Leibniz' four-operation machine which was developed at the end of the 17th century. A discussion of the varied and constant progress of the 19th century would take too long and is largely unneccessary. It culminated, however, in Hollerith's statistical machines (1890), which integrated electromechanics into calculating and gave rise to data processing.

The logical tradition is the final element. This began with Aristotle, who formulated the three fundamental principles of classical logic, enriched by the Arabs. The latter conceived of a machine for producing horoscopes, and Raymond Lulle surpassed this by devising a machine to demonstrate the existence of God. Let us pass on from Leibniz and come to George Boole, who in two works published in 1847 and 1857, provided an algebra which could now be applied.

Progress from the second half of the 19th century to the eve of the Second World War was slow and barely perceptible. Lord Kelvin is worth mentioning: in 1876 he presented the framework of the first analogical machine for integrating differential equations with variable coefficients. Between 1937 and 1938, Shannon established that there was an isomorphism between Boole's algebra and contact algebra. Turing invented his imaginary algorithmic machine, which was used to determine the provability of problems. And there was Von Neumann, who developed a synthetic machine combining the logical and arithmetical tradition, and the recorded programme. By this time, all the elements were finally present, and with the emergence of electronics, the path was paved for the computer.

No doubt, the *Z4*, devised by the German Konrad Zuse should be mentioned first; this was destroyed in a bomb attack before completion. In 1944, the *ASCC* (Automatic Sequence Controlled Calculator), also known as *Mak I* was devised at Harvard. This was an enormous calculating machine, the first non-specialised, universal machine of its kind, but which ultimately did not prove to have any notable innovations. Pennsylvania University produced *ENIAC*, (Electronic Numerical Integrator and Computer), in 1946, which was the fruit of work done by J.P. Eckert and J.W. Mauchly. It contained 18,000 normal electronic valves and consumed 100 kw of power. These two apparatuses were veritable monsters, taking up the space of entire buildings, and consuming as much electricity as a small factory. They required enormous refrigeration installations, and the constant replacement of the valves meant that maintenance costs were inordinately high. These two apparatuses provided a testing ground. The choice had been between analogical computers, which tended to be imprecise, or numerical computers. They contained an input device, a memory, a calculating centre, and an output device. All of this based on the binary system, i.e., a device which could be in one of two states, 0 and 1, a switch, a relay and a semi-conducting diode. The memory recorded the data and the results, and on the other hand, the programmes. The memories were also magnetised in two states. Every soluble problem would be divided up into a series of elementary logical operations. These were controlled by the programme which was either on punch cards, or magnetic imprints.

Despite its limitations, *Mak 1* was extremely important regarding the development of this technique. From 1948 to 1960, Harvard and the M.I.T. developed a series of machines which were gradually perfected, particularly under the direction of Howard Aiken. *EDVAC* was created in 1948 by the same team which produced *ENIAC*. Then the National Standard Office in Washington developed *SEAC* in 1949. Both of these computers used cups filled with mercury for the memory, each one of which was able to store about 400 binary elements. In fact, *EDVAC* was the first true computer. All that was needed now was to simplify the machine and reduce its size. *EDSAC*, developed in Cambridge in 1949, contained only 3000 valves and worked six times faster. Besides this, it was the first computer with internal programming.

At this point, progress was still at the research stage, and principally academic research. It seemed, initially, that there was no commercial network which could be organised around these types of apparatus. In 1950, however, the ENIAC team founded a commercial company, UNIVAC, the only one until 1955, at least in the United States, to try and commercialise computers. We should note, however, that the first commercial computer was *LEO*, created from *EDSAC*, and acquired by Lyons of London in 1953. It was after this that IBM were tempted into this field and called upon Remington Rand to develop a marketable computer. It is curious to note that only the scientific and military openings were being considered, and that the problem of administration had been completely forgotton. These efforts resulted in the *701* of 1953, and the *650* of 1954, which proved to be unbelievably successful with over a thousand sold.

Research continued from 1951 to 1958. The British *ACE*, and the Harvard *Mak* series represented progressive developments. It was now that research began into creating other types of memories (1950-1954), notably electrostatic ones where the information was stored in the form of electric pulses on the surface of an insulator. The magnetic drum used in *ADEC* and *Mak 3* was the principal example of this.

The computer of this first "generation" was still no more than an enormous calculator. It worked in milliseconds, was clumsy and therefore very slow, it was virtually blind and dumb, and with a somewhat short memory. There was hardly any input of data, or output of results. But there was still a potential future for computers, however; latent developments were being made, and a market was gradually forming. *UNIVAC 1*, the first alphanumeric machine, was also the first to be industrially produced. As we mentioned earlier, Lyons opened up the commercial sector.

The second generation, which began around the years 1958 to 1960, represented a major transformation. The machines' characteristics had completely changed: memory ferrite magnetic cores (around 1955), transistor logic, and a new organisation of information exchange by semi-conducting elements mark this stage. Rapid progress was made regarding the components, and new assembly techniques led to an enormous reduction in size and weight. The IBM *709* memory had a capacity of more than a million binary elements. Two years later, the transistorised version, the *7090*, proved to be four times as fast. Bull's *Gamma 60*, – as happens so often in France – was a technical success, and a major commercial failure. There were also major technical failures. The *IBM 7030* was designed to perform a million logical operations per second, but it had to be abandoned, as well as Ferranti's project *Atlas* at Manchester University, which ran into serious difficulties. On the other hand, the simultaneity of operations was achieved around 1959-1960.

Computers were by now working in microseconds. Transistors and ferrite memories not only facilitated a reduction in size of the equipment, but they also increased the machine's speed and complexity. It was now possible to effect many thousands of operations per second. A parallel development was the increase in speed of the printers. The perforated reading tapes were developed, as were magnetic tape heads. Their capacity for memorising and re-reading large amounts of data was still relatively limited however.

The third generation of computers made its appearance around 1964, and developed rapidly from 1965. The micromodule was now part of this technology. The classic chip was composed of a ceramic wafer of 2 square centimetres, supporting miniaturised components. These wafers were layered on top of each other, and linked by vertical wires which served both as electrical connections and as a support. Densities to the order of six components per cubic centimetre were now possible. Components were no longer manufactured in isolation: they were produced en masse in the form of films on glass or ceramic, at the same time as the transistors and the connections which linked them. These "integrated circuits" were made using techniques such as diffusion, engraving, alloying and evaporation. "It was now possible," states the work from which we are borrowing these details, "to obtain densities to the order of sixty components per cubic centimetre, and with this a security of operation hitherto unobtainable with earlier methods." Microchips and integrated circuits are at the base of this third generation.

The structure of the apparatus had been developed and miniaturised; it was now necessary to transform the way it worked. The passage was made from the cabled computer, i.e., capable of performing a series of operations in a given order, to the cabled command: it seemed sufficient at this time to cable elementary functions. This meant working at the most elementary levels, which limited electronic circuits to a few different types. A parallel result of this was the introduction of microprogrammes, which reduced the bulk of central memories. On the one hand, the number of programmes to be run simultaneously had to be reduced, and on the other, the programmers were requested to reduce the size of these programmes as much as possible. A new type of "fixed storage" memory, meant that the programme instructions available to the user could be separated from the internal structure of the central unit. If this was a "read only" fixed storage memory, it was purely and simply a matter of changing the physical support – in general a board of electronic circuits of a few squared decimetres.

Multiprogramming denotes the capacity of a computer to contain different programmes in its memory at the same time, programmes of a type such that they do not all use the same input and output units simultaneously. Multiprogramming has become more extensive with the introduction of time sharing. This is a particular process for use with programmes which allocates to each an equal period of running time that is sufficiently brief to make it appear that they are all running "simultaneously." It was possible to interweave the running of programmes so that they ended at the same time: it seems in this way as if several computers are contained within a single main one. With multiprocessing, the separate programme are run rigourously at the same time, as if several computers are being run together.

The performance of the new apparatuses needed to be improved, and the memory is an essential, but onerous part of this. Added to this is the increase in size of the central memory caused by the incorporation of multiprogramming and time sharing. The processing unit is used for logical operations and calculations. It works in conjunction with the central memory containing the instructions and the data relative to the programme being run, and in contact with all the elements of the calculator, and in particular with the entry and exit peripherals and auxiliary memories. The latter are slower, but with a far greater capacity. The power of a computer depends therefore on the functioning characteristics of the various units and their ouput, but also indirectly on the capacity of the central memory. Until this time, there had been only one data address, this was the real address. IBM added to this by creating the virtual memory, so that there were now two addresses that could be used. "Dynamic address translation" (DAT), means that the passage from the address to the central memory can now be effected automatically. Today, large capacity auxiliary memories can be used.

Progress has also been made in two other areas thanks to these improvements: in remote data processing and in the networks of interconnected computers. A visualisation console or simply a telephone means that the computer can be set to work from a distance and the appliance used for this is called the terminal. A multi-processing computer with time sharing can be used by several people at a time, therefore, which considerably reduces the costs of computer installation. The vast majority of computations by computers is done with data recorded on files or in memories. If someone wants to make certain calculations, then they address themselves to the computer which possesses the right data. The process is exactly the same when a given programme is required. Files, memories and programmes are no longer assigned to a given computer, and interconnections enable several computers to use a group of files and programmes. This means that it will be possible to use not only one computer, but a whole group of computers, which could also be of different makes. The most difficult problem is translating from one language into another, but this is not insoluble. The first computers network was established in the United States in 1968.

The next arrivals were pocket calculators. These were made possible only by integrated circuits and elementary miniprogramming. It was possible to use betwen four (the four fundamental operations), and ten programmes, and as fallback, a small memory.

When IBM presented their new 360 line of computers on 7th april 1964, they opened up completely new channels. Before this, there was the maxi computer for universal use and more simple industrial computers with limited and fixed functions. After this, universal mini-computers were created: they were incorporated as components in industrial systems and could be given fixed allocations.

Of course, everyone knows [Transl. Note: written before 1977] that the evolution of computers is incomplete and that the laboratories are currently working on the fourth generation. The hallmarks of these future computers are said to be super integrated circuits with hundreds or even millions of transistors, large capacity optic memories, improved visualisation systems, and new programming methods. The future computer will be even smaller, faster and more effective. It will be easier to use and more varied in its functions, and cheaper. One can easily imagine the fourth generation performing super-multiprogramming by generalising the interconnections, with each one being considered as a peripheral of the others, or with all of them working together. Direct access to these fourth generation computers would be a matter of course.

Is it not also possible to foresee the limits of computers? The computer will never be able to do everything: it uses an algorithmic method to solve problems, and Markov has shown that there are some "classes" of problems which are insoluble in this way. And even if some problems are soluble by this method, no machine would ever be able to resolve them in practise. The problem of chess has been mentioned in this context. However, in August 1974, the first world championship of computer chess programmes took place in Stockholm. "A limited intelligence which means that they are not much higher than the level of a good amateur. But for all that, it is an intelligence nonetheless, this faculty of determining in a situation where the possibilities are almost endless, if not the best move, at least the most plausible, and of gradually outlining a winning strategy." Futile research no doubt, but programming research has been useful in other areas. In any case, the technical future of the computer is still open.

Automisation

This is a relatively old practice, and one which developed rapidly after the Second World War, to the point that people are still unsure as to which terms to use. "Automatism," "automation," and "automisation" are representative of a vocabulary which is not yet fully established. It has been said that automation is the integration of a number of automatisms. Others define automisation "as a means of replacing and expanding man's physical or intellectual labour with the labour of machines, by performing operations of analysis, organisation and control. One could say that automisation brings a new dimension to production. It is distinct from other forms of technical progress in that it is not actually a process of production, rather it improves and accelerates existing processes, thereby paving the way for new improvements."

A United Nations report distinguishes three periods in the evolution of automisation:

(The first, which began in the forties, was the) development of a pneumatic or electric switch in three positions (neutral, positive, negative), which was used to control various industrial processes; originally, they were of general use in automatic loop controls of temperature, pressure, levels, concentrations, and physical and chemical variables. At this stage, the different automatic loop controls worked independently according to a prearranged programme, so that several control operations and manual intervention were required for their coordination. In the second stage, automation was extended to these levels of control and coordination. This was achieved by developing control instruments and connecting them to detectors which were able to maintain constant checks on the working of the production operations. In other words, the different loop controls act on one another, and react automatically to unforeseen disruptions or changes in the production process so that the latter can work correctly according to the prescribed pattern. The first practical application of these controls in industrial processes and numerically controlled machine tools dates back to the period between 1955 and 1960. The third stage, which began during the sixties, represents the development of electronic computer control. The use of a computer enables the preestablished pattern of the production process to be changed according to management requirements and the production programme.

Automation has made considerable progress in industrial production. Of course continuous machines were first produced centuries ago: they were used in the production of paper by the end of the 17th century, and the blast furnace, in use since the 15th century, is also a continuous machine. The same is true for the production of ice, but this is a more recent development. Hot or cold continous rolling mills were also used in manufacturing, and the same generalisation and extension was apparent in the field of machine-tools. These were first made automatic after 1840. The apparition of the electric engine greatly eased the expansion of automation. Engines, by means of mechanical controls, gears, racks, shafts and other mechanisms were used to perform the necessary tasks. The work could be done faster, products could be standardised more easily, and it became possible to manufacture much larger pieces. It was on this basis of automatism that automation was founded.

Let us take a production process, of any kind. It begins with the raw materials, and ends with the final product. The function of automisation is to coordinate the various existing automatisms, and compensate for any problems of automatism which might occur. Machines are thus linked to one another as if part of a chain, and they are regularly supplied with raw or semi-worked materials. From the first, three different aspects were apparent: handling (movement of goods within a factory), manipulation (presenting the pieces to the machine), and transfers (the movement of pieces from one machine to another). Since the Second World War, most of the effort has been concentrated on these problems, thus consolidating those machines which were already automated. The most important elements of automatic handling installations are sensors for circulation, sensors for flow of materials, gates for tele-controlled shunting, carriage conveyors, continuous and discontinuous indicators of level to control the filling of funnels, balanced feeding of conveyors and balanced loading , measures for the regulated discharge of funnels, as well as apparatus for acoustic and optical signals. Generally, all these installations can be controlled from a central control post. It is worth remembering that it was D.S. Harder, one of the directors of Ford, who coined the term "automation", precisely with regard to automatic handling.

After this came the automisation of those elements which were already mechanised. Classical machine-tools were replaced by reproduction lathes or milling machines, which work by reproducing a model, and by grinding machines, of which the grinding face is controlled automatically by optically enlarged profiles. Today, these machines are being replaced by numerically controlled machine-tools. Since these machines can be automatically supplied, they can be set to work alongside each other, making continuous production possible: transfer machines made their appearance at the end of the last war, and were used principally in car production. The difficulties involved in mechanising certain assembly operations makes it impossible, at least in mechanical construction, to conceive of the complete automisation of this type of manufacture.

In the iron and steel industry, a great deal of progress has been made through the impetus of automisation (fig. 15). Minerals are prepared (agglomeration) with automatic weighing and loading machines, the roasting is controlled. Today, the automisation of cast iron production in blast furnaces is under way. Automatisms control the regenerators, the flow and the temperature of hot air, the temperature of the molten iron and they automatically distribute the wind in the pipes. They control the level of the bath with a continuous probe or the isotope method, and the gas heating. They also control the loading of coke, dolomite, minerals and flux, as well as the apportioning of the mixture and bell control. On a more experimental level, it has been found possible to control the process by computer which automatically adjusts the characteristics to their nominal value. In completely automated rolling mill trains, the rolling mills are controlled using punch cards or electronic memories. Again on the experimental level, steelworks have been established which are run from an incorporated calculation center. The decisive innovation of automisation was the introduction of the closed circuit.

The Newstead and Ormonde collieries in England were automised in 1964-65. This made possible the extraction of four tonnes of coal per minute, with the moving support. The Pittston Company had already spent 30 million dollars by 1958-60 in order to mechanise a single shaft in West Virginia. This increased output to 30,000 tonnes of coal per day. We should keep in mind, however, that these installations depended upon particular natural conditions.

The transfer chains used for cutting the multiple cylinder pieces and engine blocks were first put into operation in Detroit in 1948. They were capable of performing 550 operations in 15 minutes. This technique was eventually used for engine blocks, for the casting of the

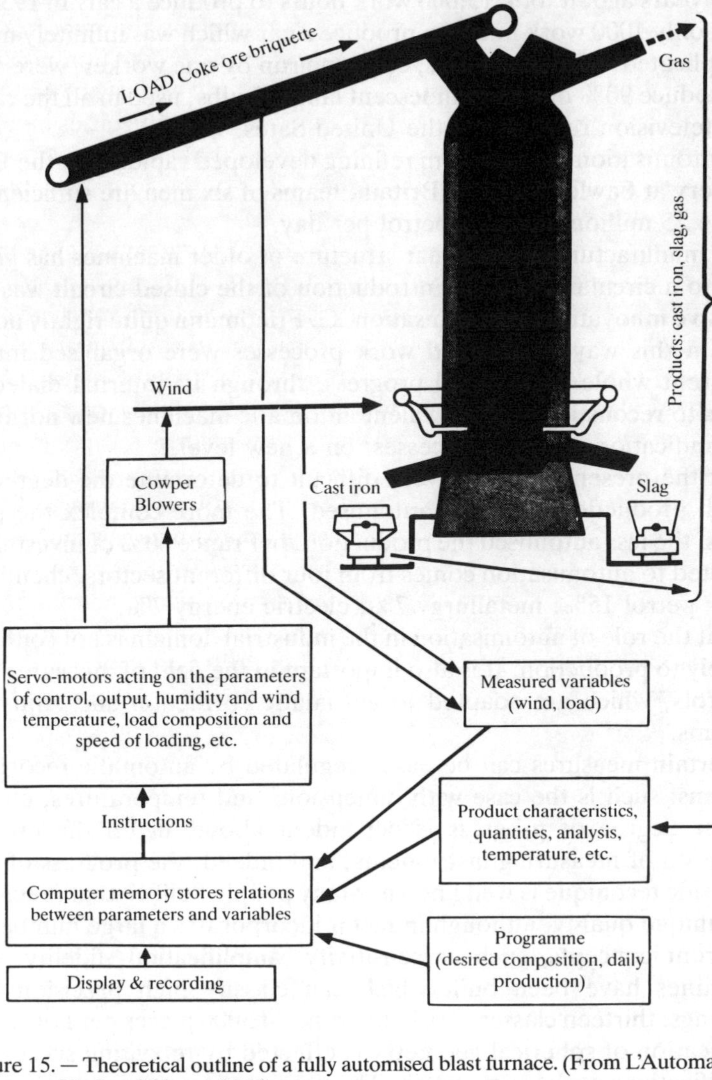

Figure 15. — Theoretical outline of a fully automised blast furnace. (From L'Automatisme *Science et Vie*, Special issue, 1964)

blocks, and the mounting of the engines. It used to take 400 people 40 minutes to create a engine block; today it takes 48 people 20 minutes. Fifty years ago, it took 15,000 work hours to produce a car. In 1957, it took only 1000 work hours to produce a car which was infinitely more complicated. Forty machines, each one run by one worker, were able to produce 90% of the incandescent lamps' bulbs, used in all the radio and television receivers of the United Sates.

Automisation in petroleum refining developed rapidly. At the Esso refinery at Fawley in Great Britain, teams of six men are sufficient to refine 25 millions litres of petrol per day.

In manufacturing, the linear structure of older machines has given way to a circular one. The introduction of the closed circuit was the decisive innovation of automisation. G. Friedmann quite rightly noted that in this way fragmented work processes were organised into a coherent whole. "Technical progress, through its internal dialectic, tends to reconstitute in polyvalent automatic machines new norms in the unification of work processes, on a new level."

At the present time it is very difficult to determine the degree to which production has been automised. The more complex the products, the less automised the production. In France 60% of investment devoted to automisation comes from four different sectors: chemistry 30%; petrol 15%; metallurgy 7%; electric energy 7%.

But the role of automisation in the industrial domain is not confined merely to production. It is also important in the field of measures and controls, which are adapted to automatic correction and adjusting systems.

Certain measures can be easily regulated by automatic recording systems: such is the case with dimensions and temperatures, etc. A higher degree of progress is dependent above all on the greater precision of measuring instruments, and indeed, the progress of this scientific technique is well known. Many people believe that precision is a unique quality, although in fact it incorporates a large number of different concepts, such as sensitivity, amplification, fidelity, etc. Machines have been built which can classify high precision ball bearings: thirteen classes of 0.25 microns, at 5000 pieces per hour. The verification of spherical geometry is effected by repeating successive classifications of the same batch. The progress in mass production has necessarily meant that the tolerances of dimensions and weight have become smaller and smaller. The tolerances in the machining of parts are between 10 and 20 microns. In this way, all the parts can be regulated automatically. In the most simple cases, the parts are sorted out and grouped into three categories: good, bad, to be adjusted. Continous rolling mills contain automatic sequence control which engage and extract the spools, as well as automatically adjusting them. With billet shears, the optimal length is determined by an internal electronic computer which reduces the proportion of off-cuts to a minimum, the latter being collected in a circular conveyor belt. In cold rolling mills containing three, four or five continuous stands, the thickness of the plate is now automatically controlled. One X-Ray gauge is situated after the first gate and controls the thickness of this cage by releasing the mechanism controlling the tightness of the screw. This rough-cutting action is completed by a second X-Ray gauge placed after the second gate. The degree of precision obtained is excellent: the desired thickness of the plate can be obtained to within a single micron.

The quality of products can also be controlled. Here again, the methods of observation have been vastly improved, and with the help of electronic circuits it is now easy to establish whether or not the qualities detected correspond to the production programme. These controls can sometimes be followed by automatic adjusting, and in this way automisation permits not only a growth in production, and a better use of the apparatuses of production, but also an improvement in the quality of the products.

One of the latest areas where automisation has been applied is that of transport. Again, the earliest attempts were made a relatively long time ago. Take for example, the automatic "block" for trains running on the same line, divided up into sections each of which is protected by a signal. Progress in this field was quite appreciable in the immediate post-war period. Points are now provided with certain automatic devices. Inasmuch as the points can now be controlled from a distance, it is now possible to automise the running of trains within a given sector according to pre-established programmes. In France, this method was first tried in the Dole-Vallorbe sector. It is now in general use in areas of heavy traffic, particularly in the suburbs of large towns.

Repeat signals were first adopted with the famous "crocodile," in France in 1872. This stage was superseded by the introduction of electric circuits and electronic transmissions, such that trains could now be run entirely in this way, as could the control centres on the lines.

It was now possible to conceive of trains being run without drivers entirely by remote control. The first attempts at this were made during the 1960's with electric locomotives, in well determined conditions. There are still some problems to be solved. After the automatic underground network in San Francisco was put into operation, an accident occured which forced the authorities to put a driver back on board. However, Lille's future underground is to be completly automised and there will be no drivers. This type of automisation in overground trains is only possible on lines with no possible obstacles, no level crossings, no possibility of game crossing the line, etc. There have been only a few attempts at this.

The automisation of engines used in air and sea transport has been made possible thanks to radar which can provide information on position, direction and speed. Automisation in ships is internal, and therefore only partial. Even so it drastically reduces the size of the workforce. It is mainly the bigger ships which are automised, particulary the medium and large oil tankers. Some instruments, especially the gyrocompass, enable the ship to be navigated by light and astral observations inside a closed space. It was in this way that an American submarine was able to cross the Artic underneath the icecap.

The honour of having devised the first automatic pilot worthy of the name goes to Elmer A. Sperry. In 1914, Sperry flew to Europe at the controls of a Curtiss hydroplane. The plane was equipped with an electric system of gyroscopic stabilisation. This ancester of automatic pilots stabilised the plane around the roll and pitch axes, and itself drove the ailerons and the elevator. By the eve of the Second World War, these now improved systems had been put into general use on airliners, at least in the United States. Hydro-pneumatics assured the tranmission of commands, but these were only developed after the advent of electronics. The first electronic autopilot was devised by Honeywell in 1941, and was used in several thousand bombers. In September 1947, the U.S. Air Force used a transport C-54 to go from Newfoundland to Great Britain, and the flight was made entirely on automatic pilot, including take-off and landing. The successive improvements made to the equipment has meant that they are now considerably lighter, and as such can be used in smaller planes.

Today, the use of the automatic pilot is standard practice, as is the automatic landing with no visiblity (fig. 16). Operational minimums, vertical and horizontal visibility are being reduced to ever-decreasing levels. The same is true for air traffic control, which today is almost completely automised. Control networks cover almost the entire earth, except in those regions with little traffic, or high mountainous areas where radar, radio and television are largely useless. The operating conditions, as with all techniques used in transport, depend essentially on the infrastructure.

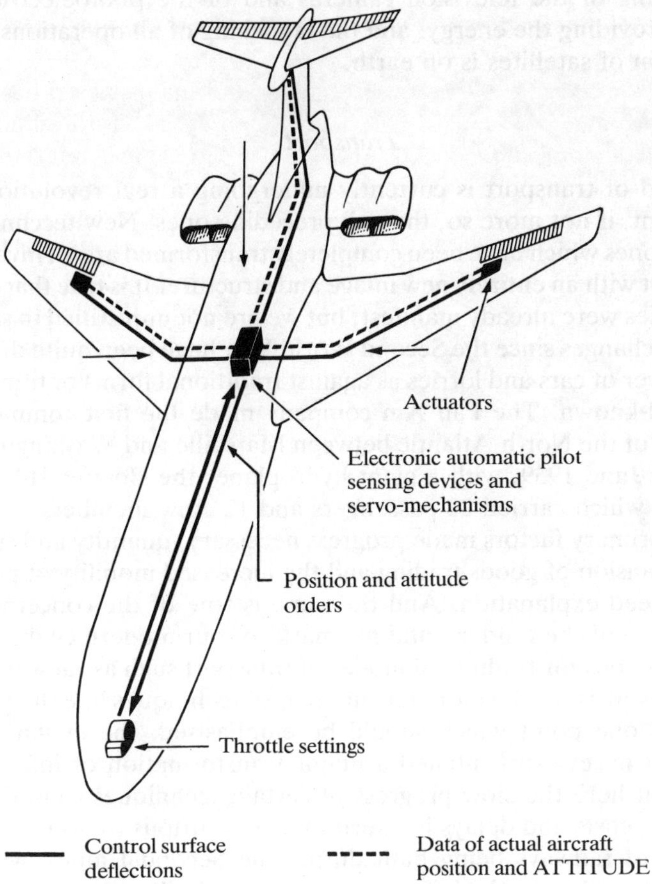

Figure 16. — The principle of automatic piloting. (From S. Handel, *The Electronic Revolution*, London, 1967)

Automisation is the key to space. Guidance, computers, altitude control, power sources, telecommunciations, instruments and structural sub-assemblies are all subject to unprecedented standards of extreme lightness, high efficiency, rigourous precision and reliability. These missions are conducted automatically from take-off, the pilots, if any, having the option of taking control. Of course, the large majority of these spacecrafts navigate in space without pilots. Everything within these crafts therefore, must be automised: the positioning of the television cameras and of the photoelectric cell plates providing the energy, and the launching of all operations. The real brain of satellites is on earth.

Transport

The field of transport is currently undergoing a real revolution, as important, if not more so, than all preceding ones. New tecchniques and old ones which have been completely transformed are all investing transport with an entirely new image and structure. It is true that some tendencies were already manifest, but we are not unjustified in saying that the changes since the Second World War have been quite drastic. The power of cars and lorries as against traditional forms of transport was well-known. The Pan Am company made the first commercial crossing of the North Atlantic between Marseille and Washington on the 28th June 1939, with a giant hydroplane, the Boeing 314 Dixie Clipper, which carried 22 passengers and 12 crew members.

Two primary factors made progress necessary: quantity and speed. The expansion of goods trading and the increased mobility of people hardly need explanation. And the same is true of the concern with speed, one of the fundamental hallmarks of our modern civilization. And that goes for traditional modes of transport such as railways and ships as well as for the more recent areas of techniques like the plane. There is one point which should be emphasised: the evolution of transport necessarily entailed a major transformation of infrastructures: but here the slow progress of certain techniques caused both mounting costs and delays in execution. It is curious to note that the number of railways being built during the Second Empire was far greater than the number of motorways being built today.

Ships have certainly changed a great deal, but it seems now that naval technology has reached its saturation point. In any case, the current tendencies are clearly marked. The regular passenger services have almost completely been replaced by the airlines. It is therefore the merchant ship which attracts the technician's attention, and this from several angles.

Firstly, the types of transporters have grown more and more specialised. Ore tankers and liquified gas tankers (fig. 17), and banana boats have been added to the older petrol tankers, etc. In 1954 in the United States, Malcolm Mac Lean ran articulated lorries between New York and Houston, crossing eight states whose road regulations were all different. In order to avoid this problem, he hit on the idea of carrying the trailers on adapted petrol tankers. The container ship was born. In the mid-sixties the system was adopted in Europe bringing with it important savings in handling, and thereby generating more sea-traffic. The time taken to load a ship weighing 10,000 tonnes fell from seven days to fifteen hours. From this point onwards, 80 to 90% of the goods transported between Europe and the United States were carried in containers, representing 61% of New York harbour's traffic, and 23% of Le Havre's. The first ships travelled at a speed of 20 knots and carried 700 to 1000 containers. The third generation travels at a speed of 26 knots and they carry 2800 containers.

It is through partial and numerous developments, and by the use of more suitable materials that it has been possible to supercede limits which were originally thought to be insurmountable. The reduction in weight and bulk of the unit of installed power, and the reduction of the weight of the hull meant that the size of ships became almost limitless: some oil tankers for example weigh more than 500,000 tonnes. It seems however that ships are no longer being built as large as they could be, partly because of the dangers of pollution in the event of the loss of a ship, and partly because of the blocking of certain passages due to the extension of territorial waters. The re-opening of the Suez Canal has generated a rise of medium-sized tankers, weighing only 150,000 tonnes.

Despite the fact that the protection required for a nuclear engine makes it bulky and heavy, it is nevertheless admirably suited for powering ships. It has one principle advantage, which is that the refuelling is much less frequent. The attraction of autonomy and an almost unlimited field of action tempted the naval services to adopt this form of power, particularly in nuclear submarines. Half-way through 1975, some two hundred and fifty nuclear-powered ships were

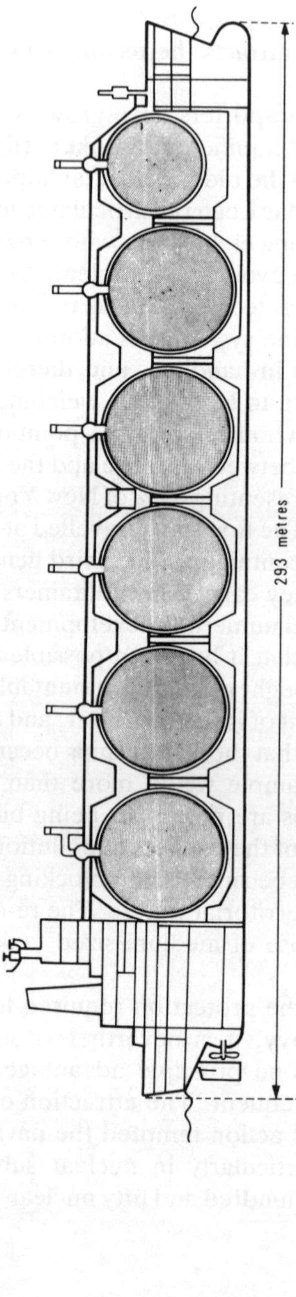

Figure 17. — The independent spherical tank system used on the liquified natural gas carrier of Gotaas-Larsen longitudinal section.

counted, (three of which were soviet ice-breakers). The problem is quite different for merchant ships: a nuclear reactor is only profitable for very fast or very large ships which consume large amounts of power. It was calculated in 1975 that a reactor was only economically competitive with ships capable of at least 45,000 horsepower, and in some cases this figure was as high as 60,000 to 90,000 depending on the price of petrol. This would be equivalent to petroleum tankers of 250,000 to 500,000 dead weight, travelling at speeds of 16 knots, or to container ships of at least 50,000 tonnes dead weight, travelling at a speed of 27 knots. The first nuclear powered ships, the American *Savannah* and the German *Otto Hahn* did not produce any significant results. And most people are familiar with the problems encountered by the Japanese. People's general dislike of using nuclear apparatuses, and the decreasing consumption of petrol have compounded this already difficult problem. The transformation in this domain can only be envisaged as a long term one.

It seems that in the last few years, the railways have been in decline, unable to compete with cars and planes. In almost all countries, passenger routes are being closed, and in the United States the number of towns no longer served by railways continues to grow. In France, from August 1934 to the 1st April, 1968, the length of railway lines closed to passenger traffic went up to 13,475 kilometres. The same is currently true for the traffic of merchandise, and in 1971 in France, more freight was carried by road than by rail. At this time the total bulk of traffic was shared between the different modes of transport according to the following proportions: roads 38.7%; rail 37%; pipeline 17.3%; waterways 6.9%. And of course more and more freight is being carried in planes.

The evolution of techniques still leaves the railways some hope, however. Some of these techniques have yet to fulfil their potential, like the hovertrain for example, which can travel at 400 kmh. but which has yet to find a market and an engine. Progress has been much more obvious in other areas. The electric Diesel train, developed in 1924, has definitively replaced the steam engine in the most advanced countries since the Second World War. Electrification is also making great headway, and with ever higher voltages. On the 25th April, 1967, the French national railway company, SNCF for the first time, ran a turbotrain which is a self-propelling element whose propulsion is assured by a gas turbine originally developed for the aeronautical industry. In this way, under independent traction a rapid and light vehicle was created which was capable of sustaining a much higher average speed thanks to its strong mechanical thrust. (a speed of 240 kmh. was obtained on the 21 February 1969; [Transl. Note: however, early in 1975, for economic reasons the SNCF decided to replace the gas turbine with a 6400 kW electrical group, which propels the present high speed train, the *T.G.V.*] fig. 18).

But over average distances, i.e., those distances for which the railways could reappropriate those passengers lost to road and sea traffic, these speeds have required new railway lines. There are several examples of this. The first express line was built between Florence and Rome, with impressive results. The most spectacular line was the one built in Japan over a distance of 1000 kilometres. Several constructive works were necessary, 55% of the line is a tunnel, there are no level crossings. At first, when the line was not fully completed, 30 trains ran on it each way. Today, there is a flotilla of 2000 trains, 129 of which run

Figure 18. — High speed train T.G.V. (From an SNCF document)

both ways each day, carrying 560,000 passengers daily. This line has obviously been very beneficial. The final example is that of the line between Paris and Lyon. The speed reached by the *T.G.V.* is 260 kmh. This advanced project has incurred the ecologists' wrath, however, (it means encroaching on conservation areas, opening up 110 cuttings, and more noise etc).

On the other hand, less soil is needed for this than for a motorway, and it would no doubt have been necessary to enlarge existing roadways. Furthermore, it dispenses with the need for new airports.

The rise of the aviation industry is a widespread concern. There have been major increases in speed and the capacity to carry passengers and freight. We shall limit our discussion of engines to a couple of points. As with the car, the classic engine with pistons was part of a saturated technique in that beyond a speed of 500 kmh. the classic engine and its propeller gradually become less and less efficient. The output of the propeller falls to 75% at 550 kmh., and to 42% at 1000 kmh. Air friction heats the frame and makes ventilation impossible. The highest possible speed with this engine was 750 kmh., achieved by a German in 1939. History has clearly shown that it was military requirements which led to the various solutions, and that even if the principles were understood, there were still refined developments to be made on the technical side. Aviation was given a new dimension with the turbopropeller and the turbojet from which the commercial industry was to reap many benefits. It was now possible to travel at faster speeds, go for longer distances without refuelling and carry heavier loads. As well as all these advantages, the costs were getting lower and lower.

Lindberg took thirty three hours, twenty-nine minutes and thirty seconds to travel alone across the North Atlantic. In 1946, it took an Air France DC-4 carrying 24 passengers twenty four hours and forty five minutes to make the same crossing, with two stops in Iceland and Newfoundland. In February 1960, the four-engined jet Boeing 707 carrying 144 passengers crossed the Atlantic in seven and a half hours. It took the Boeing 747 approximately the same time in 1970, but with 357 passengers. On the 26 September, 1973, *Concorde* linked Paris to Washington in three hours and thirty three minutes. These figures are sufficient to indicate the type of progress achieved.

Over long distances, the plane certainly comes out on top over other forms of transport. The number of passengers crossing the North

Atlantic by air in 1963 was 2,956,000; in 1969 it was 8,452,000; and by 1972 it had reached 13,040,000.

The proportion of those preferring to go by sea were as follows: 21.5% in 1963, 3.8% in 1969, and 0.8% in 1972.

The advantages of travelling by plane diminish as the distances get shorter. It is more expensive and airports are invariably more awkward to get to than railway stations. The new line between Paris and Lyon has shortened the distance time to two hours, which is on the way to competing with travelling by air. It is very difficult to establish the boundaries here, because assessing journeys simply in terms of time is not sufficient to assess the advantages and inconveniences. Going on a business trip is not the same thing as going away on holiday. Traffic jams on roads and long waits in airports (in France, a third of annual traffic takes place in three months, a third of weekly traffic takes place in three days) are persuading people who are tired of waiting to start using the railways again, especially now that these have been vastly improved (some trains now have sleepers and carry cars as well for example).

During the winter of 1948-49, about a hundred American DC-3's dispatched some 150,000 tonnes of freight to the isolated city of Berlin over 22,000 trips. This was a major first for air freight transport. From 1945 to 1967, the volume of air freight transport quadrupled. In 1966, the weight of goods transported over the North Atlantic by cargo planes exceeded the weight of passengers carried in airliners. And we should note that this freight consisted largely of small parcels. In the spring of 1972, Boeing 747 cargo planes were put into operation on the line between Frankfurt and New York. These planes carried at least 100 tonnes of freight, and they are also container planes. It is nonetheless true that most merchandise is still transported by sea, however. It seems that the plane still has several difficult obstacles to surmount with regard to this. We are still in a relatively new field, therefore.

We are going to finish with two completely antithetical examples. Urban transport is still one of the blockages of contemporary techniques. In truth, solutions to this problem are hard to find, and the usual method is to apply measures of constraint (restricted parking, traffic lanes for different vehicles, car parks outside towns, etc).

There has been shortage of research or ideas, however. Usually, these are concerned with automatic systems of urban transport, consisting of comfortable, fast vehicles, the size of which varied between that of an ordinary car and a bus, running on clean tracks and operating automatically with no drivers. Hundreds of such systems have been dreamed up since about 1969; a dozen or so have been developed and two are already in operation in the United States, not without growing pains however. The automatic underground system in San Francisco, the *BART* (Bay Area Rapid Transit), is around 130 kilometres long and passes through 34 stations. The system in use at Dallas-Fort Worth airport is based on the same techniques. On the other hand, the system in Morgantown (West Virginia) which was originally to link several university buildings, uses smaller vehicles. At first these carried four to six passengers, but the places now number from twelve to twenty. Plans are currently being made to demolish existing works on this project, however. In France, research is being conducted into a system of small automatic vehicles (*Aramis*) and the automatic metro in Lille. In fact, this research has brought about the modernisation and automisation of older forms of transport, in the same way as the hovertrain prompted the rise of the the turbotrain.

The second example is infinitely more spectacular: transport in space. From the launching of *Sputnik* the first inhabited satellite, in 1957, to man's landing on the moon in 1969 and the even more distant explorations of today, progress has been steady and rapid.

It is not difficult to see that such enterprises require the coherence of a large number of very advanced techniques: materials with very specific heat and mechanical qualities, the rockets and fuel, computers for calculating the trajectories, televisions for the transmission of visual data, automation techniques and an infinity of details, each one as important as the other. Liquid propellents are generally used for the launch stages and the correction of the flight path. Solid propellents are used during the stages requiring less power. At the moment, research is being conducted into nuclear-powered engines derived from the Nerva model experiments in the United States. The essential element is the launch thrust. The *Saturn 5* rocket used in the Apollo lunar mission had a thrust of 3400 tonnes, 140 times greater than the earlier german V2's." The integration of the various stages of a rocket takes place in a special compartment situated at the tip of the final section. The equipment concentrated in this compartment maintains the coherence of the ensemble from firing until the separation of the various stages and the successive firing of the latter. Fifty-five cameras are placed in a ring 90 cm high and 2.65 metres in diameter, weighing

2 tonnes. It is here that the guidance system operates, that the problems of navigation are resolved, that the current source is to be found, and all the equipment necessary for protection against the surrounding environment and against drops in temperature.

The live load is composed of the satellite or of a special harness which must be led to a precise point in space. The vessel *Apollo* consisted of, starting from the top, the command module, (the manned Apollo capsule), the service module (the propulsion system to assure the movement of the ensemble in space), and the lunar module. The whole of this load represented about 45 tonnes. Thus is it that man, although incapable of creating an adequate system of urban traffic, is able to navigate in space and arrive on the moon.

The Transmission of Thought

The effect of printing, telegraphy, and the telephone and radio on our lives is of course fundamental, and in the second half of the century, all these forms of communication underwent several transformations. Their importance in the world of techniques is widely acknowledged, probably because everyone is directly affected by them.

There are no doubt some sectors within which procedures may have been perfected, but they have kept their basic overall structure. This was so of printing, photography, the radio and telephone, and telegraphy. Their basic principles remained the same. Apart from a few major innovations which we shall discuss below, it seems that it was mainly existing techniques which were developed, although they did not reach saturation point. This phenomenon is also apparent in earlier periods. Coherence can be maintained between fundamental transformations and potential development.

Photography is a good example of this. Of course it is clear that the cumbersome, heavy concertina camera of 1920 was nothing like the slim compact camera of 1976. But apart from the reduction in size of both the camera and the film, there were many other changes: in winding on the film, in the automatic control of distance and aperture, etc. The only innovatory camera was the *polaroïd*, with instantaneous developing, based on a principle discovered by Edwin H. Land in 1909, which was developed in 1947 and finally perfected in 1961. There was steady progress made in the type of film and paper used, and colour film was developed in 1935. However it was only in 1942 that these colours could also be printed on paper. In a sense, apart from these major elements of progress, the principal transformation was the creation of the modern camera made of moulded plastic; sometimes even the lens was made of plastic as well. The latest research has been concentrated on long distance transmission photographs, by the transformation of variations in luminance into variations in electric current. By using satellites, photographs can now cross space with perfect results. This is a far cry from the first apparatus produced by Belin (1907).

Printing underwent much the same form of development. The principles remained the same, the machines were widely perfected. But here again, there have been innovations which have still not been fully exploited. One cannot over-estimate the importance of the photocopier in administration. Readers are no doubt familiar with the progress made in the use of offset for printing, and reprinting ancient works. Telecopying was first thought of very soon after telegraphy – in 1860 it was already possible to reproduce a signature or picture at a distance. Some people believe this is the only form of telegraphy possible for those civilizations which use ideograms, like the Chinese, for example. There are some telecopiers of a limited size in existence today. Of course they all require a transmitter, and a receiver but this process could be made more automatic. If the apparatus could be equipped with an answering machine linked to the telecopier receiver, this would dispense with the need for any person to do the work. This form of communication is still in its research and development stages.

The use of electronics has profoundly changed newspaper editing (fig. 19). The editor, seated in front of a visualisation console similar to a computer terminal, types the text out using the keyboard. One by one the letters appear on the screen. By moving the cursor around with special controls, they can modify the text until they are completely satisfied with it. When they have finished, the computer tells them the exact length of the text. If it is too long, the editor can remove an entire part of the text, or a sentence or a few words. The remainder of the text automatically fills up the gaps. After having keyed in typographical instructions, the editor can then justify the text. The computer indicates the exact size of the text, indexes it, and then stores it. If some new elements need to be added, and the making up has still not been done, the editor can bring the article up onto the screen and edit once more. The sub-editors responsible for the making up check the articles on file as they go along. If for any reason they wish to change either the typeface or the justification, all they need to do is change the information at the top of the text, and the computer will alter the latter accordingly. The make-up is also done by computer. It transmits the article electronically to a photocompositor which produces an offset plaque, for example, which is then ready for use on the rotary press. Absolutely no paper has been used until this stage, except of course at the proof stage. Progress is advancing extremely rapidly in this field.

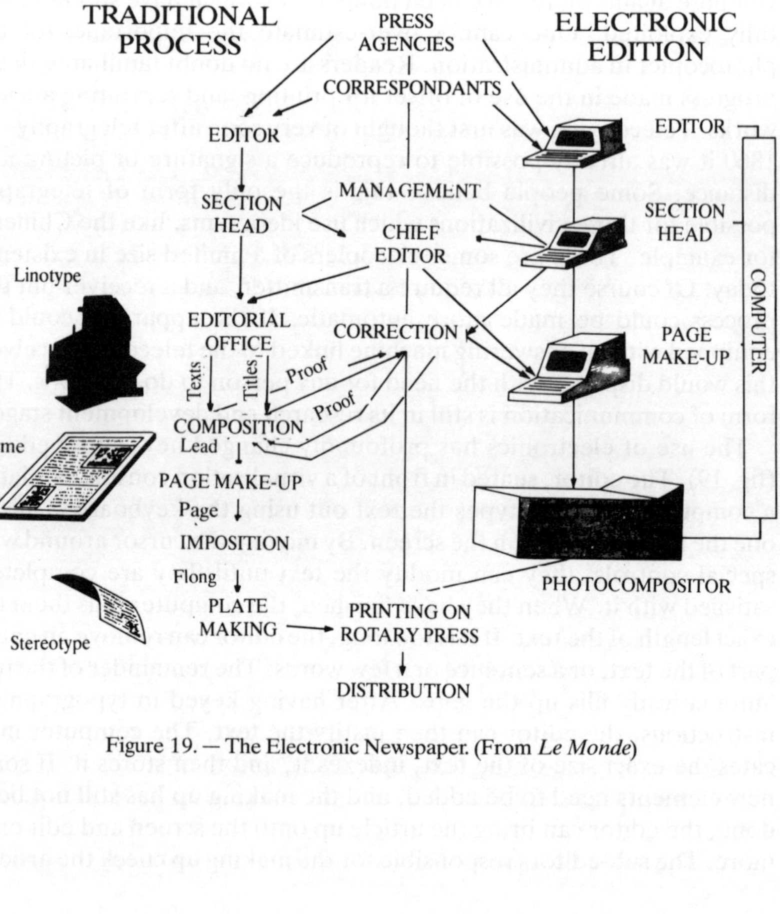

Figure 19. — The Electronic Newspaper. (From *Le Monde*)

Forty-five per cent of the sixteen thousand American dailies now use word processing for printing and editing, and this has meant that the content and style of newspapers has profoundly changed.

We shall not dwell at length on the telephone and telegraphy. Once again, the principles did not change, but the apparatuses and installations were profoundly tranformed. Automatic telephones and the development of automised exchanges represented considerable progress. The main development of the telephone was the change from electro-mechanical switchboards to electronical ones. In France, this movement began only in 1960. American companies had already adopted electronic switchboards based on the principle of spatial commutation. This was a classic mode of transmission, the variations of the human voice being conveyed by variations in the intensity of the current. France adopted temporal commutation whereby the variations of the voice are measured and coded in binary. Several conversations can therefore take place on the same pair of lines connected to an exchange. In the first case, the exchange connects two callers with mobile magnetic contacts. The latter are not used in the second type of transmission. In both cases, computers control the connections and they can also inform us of the state of the lines and junctions at all times, as well as all the signals coming into the exchange. The exchange has now become a complex data processing system, and the main difficulty today is to perfect the programming. Modern exchanges can cater for up to 50 to 65,000 subscribers. There is also another problem, that of the connection with the old network.

The number of radio cables over large distances, particularly overseas, had doubled by this time, and they were now relayed by means of telecommunications satellites. These are able to relay radio signals across the world, and are thus extremely useful for the transmission of both the spoken word and the printed image. High frequency hertzian signals have an optical range: this is therefore limited by the curvature of the earth to around 150 km. But since these signals can be retransmitted by satellites orbiting 50,000 km. above the earth, they can actually cover distances of up to 100,000 km. There are passive satellites, which simply retransmit signals, or satellites with a recording mechanism. The first telecommunication satellites were disappointing, from the *Score* (Signal Communication Orbit Relay Experiment), released in December 1958, to the *Echos*. The *Telstar* group were the first successful satellites to be used. *Telstar 1*, which was launched in

July 1962, was a bowl measuring 87 cm in diameter, and weighing 77 kilograms. It has 72 facets, 60 of which are equipped with 3600 photo-electric cells for constant recharging. Three other facets carry mirrors for optical location. It contains 1064 transistors, 1464 diodes and 2528 semi-conductors. Giant antennae receive and transmit signals. It orbits between 953 and 5637 km above the earth, and revolves on its own axis three times per second, It can be used 9 times a day for 40 minutes. *Telstar 2* was launched in May 1963, and works at between approximately 1000 and 10,000 km. The third generation of satellites, begun in 1965, was the *Intelsats* (INternational TELecommunications SATellites). The *Intelsats 2*, launched in 1966 and 1967 weighed 168 kg and had 240 telephonic channels. The *Intelsats 3* were launched in 1968 and 1970 – these weighed 287 kg, and had 1200 telephonic channels. The group of *Intelsats 4* was launched from 1971, and these weighed 1312 kg, and were equipped with 6000 telephonic channels. The entire telecommunications system had thus been completely transformed. Furthermore, one country managed to create a monopoly of this new system of communication. The French and the Germans may have succeeded in launching one small communications satellite, but it was the U.S.A. which put the operation into practise. We shall discuss this problem in Part 3 of this book.

There is still a great deal more to be gained from these techniques, none of which has yet fulfilled its potential. Current research is concentrating on direct broadcasting satellites, which will no longer require relaying, like the one located at Pleumeur-Bodou in France. Ordinary television sets could therefore receive programmes directly.

One could hardly deny that the television, along with the computer and atomic energy, is one of the symbols of the new technical system. Its history is already a fairly long one, but the definitive improvements and its sudden widespread expansion have in fact taken place only in the last thirty years. In 1877, Constantin Senlecq from France, elaborated a quite feasible method of television. The first experimental television programme was broadcast in Britain by the B.B.C. in 1929, thanks to Baird's pioneering system. The wave-length was 261 metres, and they produced twelve images per second. The invention of the iconoscope and the dissector tube was one more important step. In November 1936, the B.B.C. established the first official television studios, and a television receiver at one of the highest points in London. In the same year, the American company R.C.A., did exactly the same on the Empire State building. Most of the basic technical problems concerned with transmitting television pictures had been resolved by 1940. All that remained was to perfect these techniques. The war interrupted this research, but it did encourage intense research in the field of radar, for example, which indirectly, was to bring about major improvements.

By the end of the Second World War, then, television was not a new concept, it was simply being developed and expanded. It expanded at a much greater pace than cinema, with which technically, it is closely linked. Colour television was the next step, and one wonders how soon it will be before we have television in 3-d. Furthermore, we need hardly point out the primary importance of television, not only as an instrument of information and cultural diffusion, but in many other fields as well, such as transport or industry.

Landscapes and Lifestyles

The new technical revolution is radically transforming the physical world and all aspects of daily life, but no definitive statements can be made about these effects as yet, because this technical revolution is still not complete. Consequently, there is a lack of equilibrium both in the use of space, and in the organisation of daily life.

This profound transformation in the way of life is not always perceived to its full extent. No doubt, for example, it would be relatively easy to re-build a kitchen as it would have appeared, say, in 1929, with all the everyday household objects in use at that time. Such a collection would no doubt surprise many people, and not only those who were born after that year. It is likely that many people who do still remember this period would also be surprised by the objects they used to use and which they have probably completely forgotten. Some of these are no longer made, while others are rarely used: the pestle and mortar for making pureed potatoes, the antique coffee maker in two sections, the galvanised iron wash bowl, and the mushroom boiler for laundry. Then there are the old heavy sewing machines on cast iron supports and worked with treadles which made Singer so famous, or the magneto telephone for calling the switchboard, or of course the phonograph which also belongs in this category. Manufacturing catalogues, or those of major department stores are extremely useful for measuring the passage from one technical system to another in the context of everyday life.

After the Second World War, all these individual, everyday objects were replaced. Some objects or processes date back to the inter-war period, but nonetheless the post-war revolution was quite brutal – in the industrialised countries of course. Plastic, miniaturised electric motors, and even electronics all completely changed the nature of pre-war household equipment. Furthermore, the perfection of certain techniques and the use of new materials lowered prices and made these objects more readily available to the whole population. The more expansive use of heat sources, such as gas and electricity meant that hot water was now supplied straight to the sink. This meant that the old galvanised iron washbowl gave way to a totally new type of bowl made of plastic. The washing machine was introduced around this time, and in the fifties, the vegetable mill was devised. A further important development was the programming of washing machines and cookers. The number of different types of household appliances also increased.

The following is a list of household appliances currently in use. The information is taken from the catalogue of a large trading firm:

coffee grinder (some are adjustable);
electric percolators;
toaster (some automatic);
electric mixer;
electric whisk;
citrus fruit squeezer;
fruit juice or vegetable centrifuge;
multi-purpose mixers;
electric slicing machine;
electric knife;
peeler;
combined tin-opener and knife sharpener;
plate warmer with thermostat;
hot-plate;
table tidy;
ice cream maker;
kettle;
immersion heaters;
electric boiling plates;
meat grill;
sterilizer;
electric deep fryer.

The same was true of other types of objects: the portable transistor radio was developed, which replaced the cumbersome, stationary lamp radio. The biro was invented – which has not only destroyed the classical ink pen, but also the metal nib and lead pencil. The pocket calculator was also invented. The entire system of communications between people was profoundly changed.

It is noteworthy that the disparities between countries became much more marked at this time. J. Fourastié has demonstrated this with some telling statistics: in 1950, a small washing machine with no heating apparatus cost 35,000 francs in France, the equivalent of 500 hours of a cleaner's salary. In the United States it cost 50 dollars, or 50 hours of cleaner's salary, and in Egypt and India, the cost of such a machine was equal to 3 to 5000 hours of cleaner's salary. Since this time, in industrialised countries at least, prices have dropped considerably. The same type of washing machine in France in 1962 cost the equivalent of 230 hourly salaries. In 1950, a 200 litre capacity refrigerator cost the equivalent of 1000 hourly salaries, by 1962 this figure had dropped to 420.

As J. Baudrillard wrote, "in an advanced industrial society, manufacturing products are not only different to what they were in the past, they also fulfil another function." Social practises concerning these objects, as well as the mythologies based upon them have been transformed. Giedion has provided us with a brief history of furniture, not of its style, but of its main functions. We lack many other such histories which would be invaluable to help us assess this progress.

The world outside the home was changing in similar ways, if not to the same degree. Working in offices had changed drastically since the 1930's, for example, but public transport had retained most of its inconveniences. Having a television enabled people to stay in all night, and both the television and the radio dispensed with the need to buy newspapers, or even books. On the other hand, the car, of which the cost was getting lower and lower as technology progressed, meant that people could now escape from an urban nightmare to the tranquility of the countryside.

The evolution of landscape – or to put it more generally, the organisation of space – was even more profound. In effect, distances have become much shorter. A map of equidistances would be extremely useful in this sense. It can take up to three or four hours to cross Paris on a Friday afternoon, and the same time to go from Paris to New York. There are many, many more examples like this. It is getting more difficult to travel shorter distances, while the time taken over longer distances is constantly decreasing. The expansion of agricultural machinery is emptying the country of its inhabitants, and transforming property by the regrouping of lands. Industrial areas are not at all what they were fifty years ago. There have been changes in two major aspects. The first of these is the concentration of manufacturing areas. The optimum productivity of a factory in relation to the highest returns seems to be closely linked to technical conditions. Steel production in one factory before the war was between 500,000 and 1 million tonnes. In the early post-war period, this had risen to between 1 and 1.5 million. By 1960 this figure was 2 million, and according to most estimations, by 1968, 5.5 million. In 1970, production was at 8 million tonnes per annum, and this rose to 20 million in the next ten years. The Soviet Union, the United States and Japan all developed approximately in the same way (fig. 20). The same progress is apparent in chemical industry. The enormous size of factories stems from technical progress which was best served by integrations.

The second point is no less spectacular in its manifestations. Not only had the units of production been increased, but they had also been displaced. This stemmed as much from the location of sources of raw materials as from technical progress. This was especially so after coal was no longer used, the transportation of which had been very expensive. Factories were no longer automatically built near coal fields, but closer to areas where raw materials were cheaper and more accessible. Take the iron and steel industry, for example. After several experiments, some of which date back to the Second Empire, this industry has now found its definitive place by the sea. This process began after 1950. The chemical industries installed themselves close to sources of petrol, and many such factories are close to refineries. The transport of petrol by pipelines has made the choice of location a much wider one, however.

It is clear that many of the balances of the technical system have been broken, and that it is difficult to restore such balances. This concerns peoples, individual daily lives as well as the industrial world. Many people have dealt with these problems, and a few solutions have been found. There is still much to be done, however.

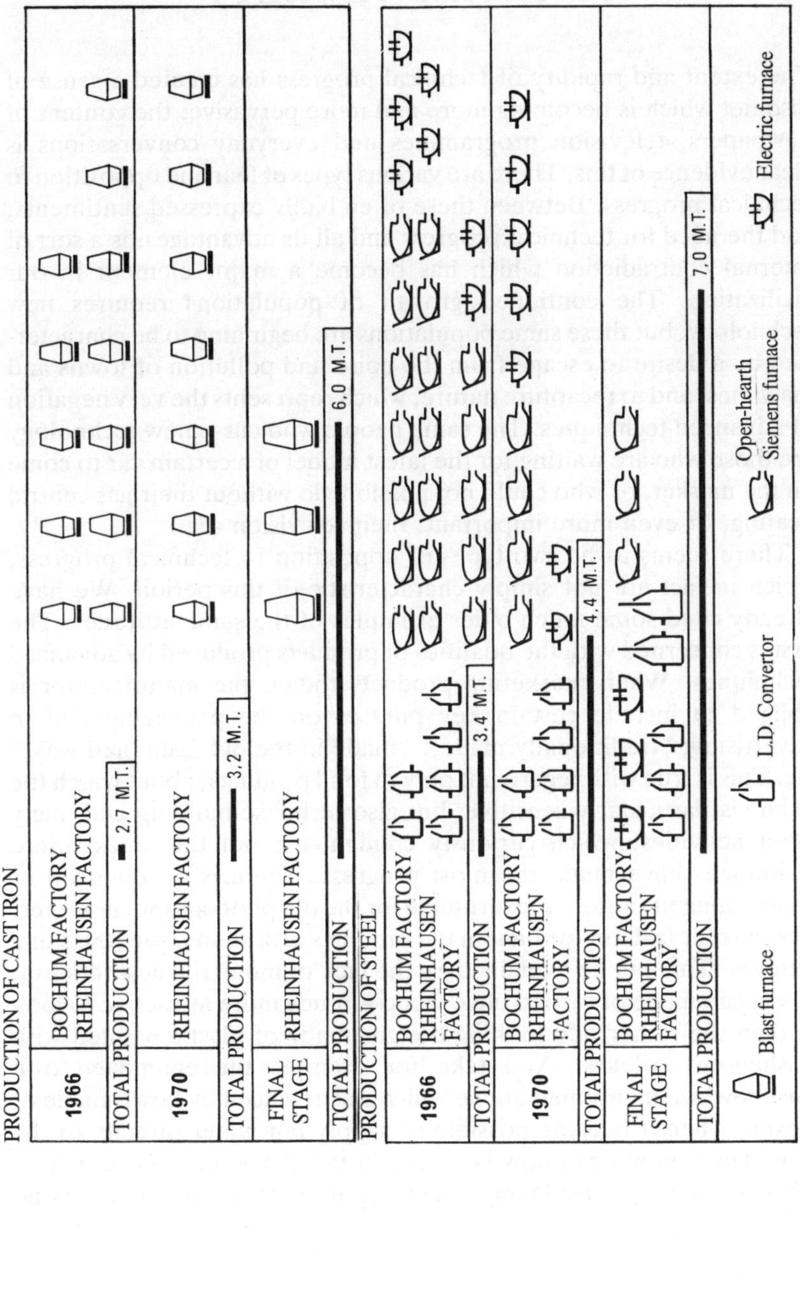

Figure 20. – Concentration of production (in millions of metric tonnes) in Krupp factories. (From a Krupp document)

FEAR AND ANXIETY

The extent and rapidity of technical progress has created a sense of disquiet which is becoming more and more pervasive: the content of newpapers, television programmes and everyday conversations is clear evidence of this. There are various types of fear and opposition to technical progress. Between these often badly expressed sentiments, and the need for technical progress and all its advantages, is a sort of internal contradiction which has become a major element in our civilization. The continued growth of populations requires new technology, but these same populations are beginning to be characterised by a desire to escape from the noise and pollution of towns and machines, and to recapture nature, which represents the very negation of advanced techniques. The same people who curse new technology are those who are waiting for the latest model of a certain car to come on the market, or who could not possibly do without their gas central heating, or even more important, their television set.

There seems to be two types of opposition to technical progress, which in fact are not simply characteristic of this period. We have already cited some much older examples of the same attitudes. The first is concerned with the qualities of products produced by advanced techniques. When marketing products today, the manufacturer is obliged to include certain key phrases on the packaging and in advertising: "traditionally made," "made in the old fashioned way," etc. This is true with regard not only to food products, about which the public is particularly sensitive, but also of house-building and many other activities, which curiously enough, are not the areas where techniques have made the most progress. There is no question of abandoning the automatic turntable for the old phonograph, however. The current fashion for handmade products also stems from the same attitude. This can be explained by the lack of industrial aestheticism: an eggcup made of turned olive wood is much more attractive to look at than the rather odd-looking designs made of plastic painted with washed-out colours. And take for example the enamelled iron washbowl we mentioned above which was replaced by bowls made of plastic, when it became possible to supply hot water directly to the sink. These bowls can now be found in the shops once more, with a label prominently displaying "made entirely by hand." There is no need to cite any more examples of this, they can be found on the labels of all sorts of products.

In a very general sense, this attitude is linked to another, which concerns the quality of life – to use a common expression – and the fact that it is seriously deteriorating, largely due to technical progress. The latter is responsible in that it causes the expansion of industrialisation and the proliferation of harmful products. In part, this probably represents our version of the myth of the golden age. This attitude goes beyond the mere factors of environmental nuisances, which include pollution, and insofar as it is not based on concrete grievances, it is somewhat vague and ill-defined. It is supposed that there is a fundamental contradiction between technical progress and peoples' happiness. In a sense, technical progress is condemned because it represents a distancing from nature, because the artifical is assuming the place of the natural, and this is all the more important because the differences between them are sometime far less clear. The opposition to technical progress is based on other factors as well. We have already stated that there is a close relationship between technical and social structures, such that technical progress inevitably brings with it the evolution of society. Some see these changes in society – or from a more global point of view perhaps, the changes in social values that are implied by technical progress – as being the result of a one-way cause and effect relationship between techniques and society. and they believe that technical progress can have only a detrimental effect.

Both these attitudes question the value of technical progress, and much more, they actually warn us of its dangers. There is an anti-scientific attitude which has been the subject of specific studies, and an anti-technical one. According to an enquiry held in 1972 by the *Délégation générale à la recherche scientifique et technique* in France, the latter is more prevalent than the former. In the scale of social utility, the scientific researcher has a privileged place just below that of the doctor (which reveals an unconcious egoism since the doctor is more directly useful to oneself), but above that of the engineer. This enquiry revealed a bipolarisation of attitudes towards science which is considered implicitly as a major source of technical progress.

The person in the street tends to associate science and techniques with political, social and economic power. This vision of things means that there is less trust in those involved in scientific research when the latter are identified with authority, and as a result they experience the same distancing and mistrust as those in political authority.

The same attitude is held by people who are highly trained, but in fields other than sciences, such as law or literature for example. Those who have had a scientific training, on the other hand, think the opposite: "science is a world quite distinct from politics."

Apart from these general fears, there are other, more precise concerns. They are partial in the sense that they do not condemn technical progress out of hand. We shall consider a few precise examples.

The fear of nuclear power is now a familiar one. It has been propagated by campaigns in the U.S.A. and Europe for many years now. Every argument has been made against the harmful effects of nuclear radiation from nuclear power plants. But this of course is not the only side of the question: everyone knows that the idea of nuclear war must be considered; it cannot be ignored. In the event of a leak in a nuclear plant, everthing within a ten kilometre radius would be affected, from new-born babies, to fish and vegetation: asparagus would turn red – the colour of danger. A survey conducted in the area surrounding the nuclear plant at Chinon in France, revealed evidence of wild speculation, obvious or badly concealed panic, unverifiable attitudes, and organised protestation. The authors of this report concluded that fears of the bomb had been directed towards nuclear plants, and they also noticed that protests against the nuclear bombs were less violent and more sporadic than those against the peaceful utilisation of nuclear power. Fear of radiation is just as great as that of nuclear explosions. Even the clearest explanations about the plants' safety are plagued by the aura of mystery which surrounds all nuclear plants' operations. The real problem is not these plants, but the bomb, and beyond this, the political power of citizens and peoples.

Other surveys revealed that this fear was not quite so general as some had made out. One questionnaire carried the following question, with the corresponding distribution of answers: If the construction of a nuclear power station for the production of electricity was being considered, and if scientific experts affirmed that there would be no danger, would you:

trust them completely?	19	58%
tend to trust them?	39	
tend not to trust them?	22	37%
not trust them at all?	15	
no opinion expressed		5%

What can we conclude from all this? The issue is a concrete one, nuclear power plants do exist. However, the mistrust and defiance, and the fear and anxiety are not so clear cut. They even become linked with other issues, which are quite different, at least superficially so in a technical sense. Thus, the fear and opposition to new techniques have been transformed into much vaguer notions of political power, and indeed, it is true that in some areas, techniques can give an enormous amount of power to the sectors which control it. Some countries are very much aware of this, and they have acted in a new spirit of freedom – which goes beyond traditional criteria – and created what are now called balancing powers. The role of these arbiters – which were first established in the northern countries – in our rigid technical administrations is becoming more and more important. Unfortunately, no-one has yet thought of writing a history of freedom, or of describing the evolution of this notion.

The solutions are not easy to find, or to apply. Take for example, the effect of television on a population where visual media have become more important than audio ones. In most cases, this instrument of unbounded power over peoples' minds is in the hands of governments. There are some cases, however, where the problem could possibly be solved. In the United States, television is not controlled by the government, but it does therefore depend on private interests, and inevitably it has fallen into the same mould as the press. Commercial viability is the most important factor, to the detriment of culture, or even an honest information service. This has reached the stage where people are now attempting to start a channel which is completely independent, subsidised only by large foundations. In Britain, there is a fairly satisfactory balance between the political power, which funds one of the two main television companies, and the company itself, which is conscious of its duty and the importance of public opinion. Here again, technical progress is clearly linked to structures which have still yet to be clearly defined.

"When industrialisation enabled every home to have a television," wrote a scientist recently, "one could see that society was no longer the same." And he added:

In spite of this disquieting power, television was installed in an experimental fashion, subject to tradition and circumstance (..), television became a new tool in the arsenal of the executive power. It was neccessary to define the objectives before creating the structures. People today watch more and more television, and the televised image has taken over from other cultural medium except in certain rare cases which we can link with the opposition to technical progress that we mentioned above. Some people believe that this "intellectual elite," which was able to propagate its thought by now outdated media, simply saw its monopoly pass into a new set of hands. And to its horror this new set of hands belonged to those in political power. Furthermore, people were trading in their freedom for comfort, solutions and certainty.

There is another field of techniques which generates the same type of fear and anxiety. A recent debate added fuel to these by bringing to light certain facts which had hitherto been unknown by the general public. According to Pierre Laroque:

In the past, machines have affected only the production, transmission and distribution of goods and services. It has not affected the exercising of political power or the freedom of individuals (even though this does require a certain amount of qualification). The emergence and development of computers is creating an entirely new situation, in that information can be collected and stored which concerns all aspects of public and private life, and collective and individual activities.

"Information, on this collective level, is an instrument of power." This is apparent at all levels: the power of a university, a firm, or a state, depends largely on the number of its computers, and their capacity. The embargo placed on the exportation of certain types of computers in the U.S.A. is clear evidence of this. A lot of capital is required to build as big an information system as possible, except for business computers. So this becomes a political problem. In order to prevent monopolies forming, some countries have nationalised the press agencies: the same question is being posed about public access to these information systems, and therefore to computers.

But there is more. "The computer enables information to be collected concerning all aspects of an individual's life, placing more and more restrictions every day on the autonomy and privacy of the latter." The computer, by means of this diverse collection of somewhat vague and unverifiable data, enables authorities to "synthesise social controls, and with all the infallibility which is attached to machines, it imposes a perpetual conformism." The system may enable one to cross-check, and thereby prevent the falsifying of information, but it encloses the individuals within their past, and within a past which is stereotyped by the language of programming. Ultimately, "the technical, economic and financial aspects of information systems is a legal and political problem, a problem of the exercising of power, and in particular, in the control of the latter." Once again, we find ourselves back with the problem of institutional structures whose evolution is necessarily linked to technical progress. The basis of most fears is wondering which side will win.

Of course the issue of individual freedom is paramount, but there are other questions of undeniable importance which are more closely connected to everyday material life. These are situated on various levels, sectional or general.

The replacement of people by machines is an obsessive fear which has existed before our time. The luddite movement which began in the first half of the nineteenth century in Britain, was a violent manifestation of such fears held by the working classes. In the second half of the century, it seemed that this fear had abated somewhat. The enormous increase in production in contrast to the lesser development of mechanisation, and the emergence of relatively efficient compensatory powers were probably the reasons for this. But during the inter-war period, and in particular after the Second World War, the acceleration not only of automatism, but of automation revived this opposition and disquiet. There were many ambiguities in this area of activity, and this might explain the sometimes contradictory positions assumed by successive groups. Automation unquestionably reduced the number of jobs in those areas where it was applied, or more accurately, perhaps, it transformed the distribution of tasks. After the Second World War, American unions were extremely concerned about the effects of automation, particularly in the automobile industry. The British unions soon followed suit in opposing such moves. However, they were soon appeased by the fact that the number of those made redundant was not as high as they had thought.

However, there was, and still is an awkward transition to overcome. It is true that the umemployment caused by automation is being compensated for by the expansion of the mechanical and electronic sectors. But any transfer of a workforce from one industry to another is an extremely delicate operation. It is difficult to turn a mechanician into an electronic engineer, especially beyond a certain age. In a study of labour at the Renault factories, it was shown that many of the trades no longer existed (there were far fewer joiners for example) but that others were emerging according to the development of new professional sectors.

Technical progress is completely transforming professional structures, and consequently, a major part of society. Some professions are becoming obsolete, while others are emerging, but as well as this, the different types of trade are becoming more and more simplified, and the hierarchy of these trades is becoming more marked. Intermediary trades are tending to disappear, such that we seem to be progressing towards a kind of professional polarisation: there is the unskilled worker on the one hand, and the technicians on the other, who have been properly trained and therefore know everything about the sector in which they work. There is thus a polarisation in that the former cannot progress beyond a certain level, while the latter is obliged to seek ever higher levels of achievement.

The concern about the constraints imposed on social structures by technical progress is understandable. Or course, one could say that these sorts of constraints have always existed during such periods of adaptation. But insofar as progress is constant if of variable intensity, the problem is no less worrying. This permanent questioning of social and heriarchical structures cannot provide even the minimum of security to societies where change is so difficult.

The most potent image used against technical progress is that of pollution. It is no longer simply the freedom of individuals and the organisation of societies which is affected by technical progress and industrial growth, but life itself and the whole material environment which supports this life. Of course, pollution has existed – under various names – for a very long time. During the First Empire, it was the object of special legislation concerned with establishments which were considered a potential hazard to public health. This legislation advocated holding a public enquiry into the setting up of industrial establishments which might have a harmful effect on public health, and especially on the people living nearest the installation. As technical progress advanced, and as new industries were set up and old ones transformed, the list of such establishments was regularly revised.

The extent of pollution is certainly due to technical progress, insofar as the latter creates new industries and encourages the development of others. Technical progress and industrial expansion are closely linked. Population growth is another factor, and the concentration of industrial activity, or in other words, the rise of urbanism. The expansion of the large urban agglomerations is causing more pollution, and the latter is also exacerbated by the products of certain industries, such as oil-fired central heating and cars and planes, to name only the petroleum products for example. Then of course there is the acute problem of waste (nuclear waste, red mud etc.) The report issued by the Club of Rome was quite justified in making the following statement: "pollution is a complex function of population, industrialisation and technological development."

In actual fact, we do not know enough about certain aspects of this subject. There is the problem of direct pollution, but what about indirect pollution? Then there is immediate and long term pollution, such as that in the air, for example. Some types of pollution disappear quickly, others remain for a long time. "Our ignorance about the limits of the earth's capacity to absorb pollution" affirmed the Club of Rome's report, "is sufficient reason for us to limit its sources."

The models used can be illustrated by a few examples. The first concerns DDT, whose harmful effects, as well as its short term efficiency, have long been recognised. Over 100,000 tons of DDT is used in crop spraying every year. Part of this evaporates, and is carried by winds to land much further away, or to sea. In the latter, DDT is absorbed by plancton, then fish, and then by humans. At each stage of its cycle, DDT can either be degraded and transformed into harmless substances, or it can become concentrated in the tissues of living organisms, each stage representing a certain delay. The graph in figure 21 was part of the Club of Rome report. It was based on the hypothesis that the consumption of DDT had reached its maximum level, and that it would decrease to reach level zero by the year 2000. According to this graph, it will take another twenty-five years before the DDT content in fish organisms is the same as it was in 1970.

Other products have similar cycles: such is the case with two metals which are being increasingly used in industry. These are mecury and lead.

There are two other graphs in the same report which are just as telling. The first represents a forecast of the amount of nuclear waste which will be produced, as a function of the capacity of nuclear power plants (fig. 22). It is clear that the two curves match. The second curve represents the oxygen content of the Baltic Sea (fig. 23). The Baltic is a more or less closed area; there is no great movement of the waters. Organic waste accumulates in this area, therefore, and it is likely that in certain depths there is no longer any oxygen at all.

There are many such examples of this; numerous works have been devoted to demonstrating the extent and the depth of that multi-faceted problem which can no longer be ignored. Anti-pollution

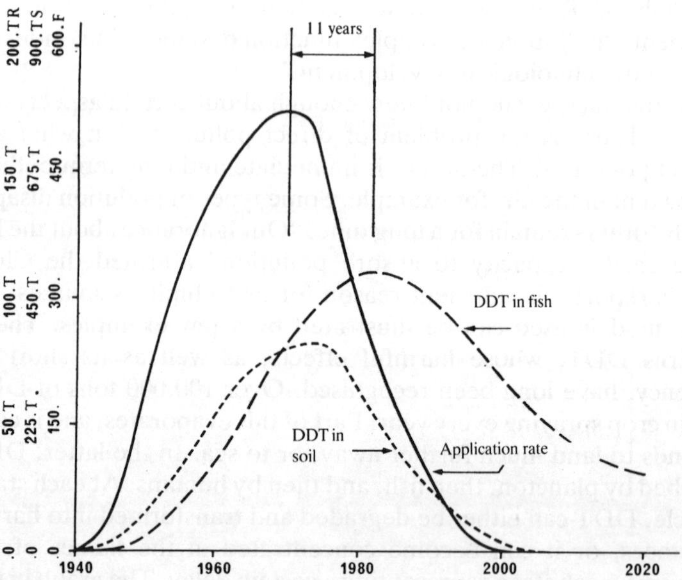

Figure 21. — DDT flows in the environment. (From Meadows et al, *Limits to Growth*, 1971)

bodies have been set up by governments and even by private companies. But simply to recognize the problem is not sufficient. A few of the solutions offered to solve the problem of pollution are extreme: some people believe that industry and technical progress should be halted to prevent further pollution, for example. Others believe that the quality of life should still be improved, but that pollution should be reduced. The former envisage a total impasse, while the latter feel that choices must be made.

Two systems have been devised to reduce pollution, which are not necessarily interchangeable: taxes or constraints. The difference is that constraints create what economists call annuities. Taxes go straight to governments, whereas constraints benefit the polluters and the land owners. The result is the same either way for the consumer, however.

All this is ultimately a matter of cost. The effects of anti-pollution measures are multiple, both directly and indirectly. They depend on

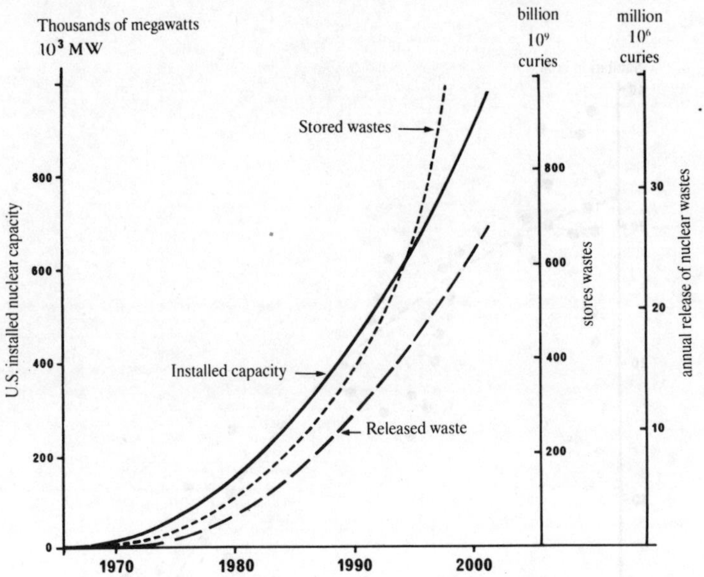

Figure 22. — Nuclear wastes. (From Meadows et al)

the flexibility of supply and demand, of changes in production, on consumption and market structures. Models have been created in an attempt to examine the effects of anti-pollution measures in very precise sectors in Italy, the United States and Japan. The following are some figures taken from these studies:

	Cost increase	Price increase
Chemicals	5.0%	5.0%
Sugar	3.8%	3.4%
Alcoholic Drinks	3.4%	3.0%
Plastic Materials	3.0%	3.0%
Petrol Derivatives	2.5%	2.5%
Pharmaceutical Products	2.0%	2.0%

The level of expenditure neccessary to deal with the scale of these problems is almost unthinkable. General Motors and Ford each spent 350 million dollars in 1972 on controlling the extent of car exhaust fumes. This resulted in an increase not only of car prices, but also of the

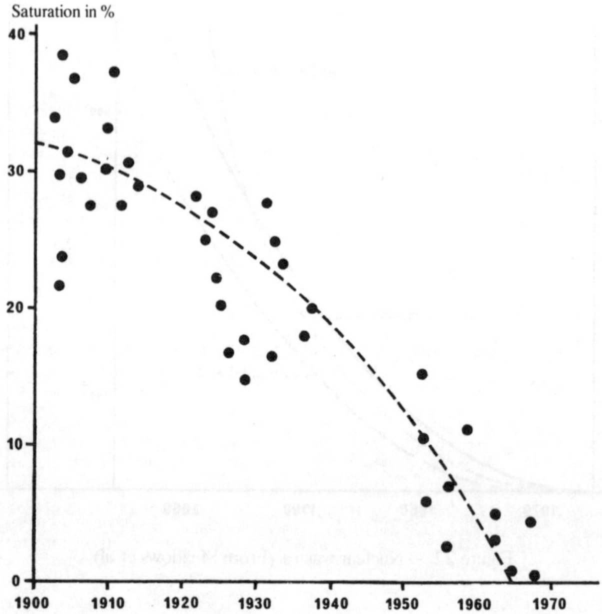

Figure 23. — Oxygen content of water in the Baltic Sea. (From Meadow et al)

cost of maintaining the cars and in the rate of fuel consumption. The price of smaller cars went up by 25%. We shall return to this problem later.

The red mud tipped into the sea by an Italian company also had to be dealt with. The anti-pollution measures taken in this case caused an 18% rise in the price of titanium.

In its report of September 1973, the American Council for the Environment estimated that 274 billion dollars were spent on controlling pollution between 1972 and 1981. The break-down of these expenses is as follows:

atmospheric pollution .... 105.6 billion
water pollution .... 121.3 billion
solid waste .... 41.8 billion

In 1968, the damage caused by atmospheric pollution alone cost 16.2 billion dollars to deal with. In 1977, the lack of established controls led to this figure being raised to 25 billion dollars. Some scientists even believe that the density of carbon dioxide layers resulting from the burning of hydrocarbons is gradually forming into a screen, which could eventually weaken the effect of the sun, and ultimately bring about considerable changes in the earth's climate.

Fears about pollution are so strong that some factories have had to close down. A paper mill near Gérardmer (Vosges, France) used 200 cubic metres of water for every tonne of paper produced. The used water contained cellulose fibres and starch, which destroys the flora in rivers and canals. It was estimated that the cost of refinement apparatuses would be 2.5 million francs. This was felt to be unprofitable, so the installation was abandoned. Some industrial projects have been rejected out of hand before they were even built. One remembers the "wandering" refinery in Brest (Britanny), the one in the Lyonnaise region, not to mention the nuclear power stations. Once industrial congestion and pollution are no longer tolerable, new areas of development will have to be found. "Come and pollute our country," became a catchphrase in Brazil, where it was clearly understood that France's iron and steel complex in Fos, on the Mediterranean shore, was the last one that would ever be allowed to be built in the country.

Technical progress is in part responsible for this poisoning of nature. It was the technical advances made in drilling and distillation which brought about the development of the petroleum industry, and technical advances which brought about the reduction in car prices, and consequently, the increase in the market, and the number of cars on the road. The same is true of all new materials, particularly plastic-based ones. The accumulation of waste is striking evidence of this. As the novelist Chabrol once remarked during a television programme, the proportion of waste was once very small, today it is considerable. In any case, the statistics show that the volume, if not the weight of urban waste is continuing to increase at an alarming rate.

There are no specialist studies of these different phenomena. We know very little about the effects of these various types of pollution on the environment, in what way the ecological system of the world might be changing as a result, or how these problems could be mitigated. We also need to know how this would effect other factors.

Pollution is not the only drawback of industrial growth. There is also the factor of environmental nuisances, or "nuisability." There are various types of nuisances, and some of them in fact are age-old problems. We shall consider only two examples which are characteristic of our modern civilization.

A recent medical congress dealt with the problem of "urban sickness." Technical progress, industrialisation and especially the increase in communications have brought about rapid urban growth. The problem is all the more pressing since actual urban techniques have not developed at the same pace. Atmospheric pollution, of course, is a major problem. Most factories were built near urban areas after the mid-19th century, then there is central heating, and vehicles, all of which are prime factors. The prevalence of carbon dioxide, nitrogen, sulphur dioxide, lead salts, hydrocarbons, ash, tar, and various gases have all resulted in above-average levels of morbidity and mortality. As well as this, it has been estimated that 14 million Americans are subjected to the noise of aeroplanes. 47 million planes passed over the United States in 1968, and 120 million in 1978. Traffic, public highways, transmissions, and the vibrations of buildings built on the cheap to make them more profitable, are all factors of stress in a town-dweller's life. They are all the more dangerous for being constant, and because the city dweller's fragile nervous system makes them all the more vulnerable. It is not even safe to listen to jazz: the 125 decibels of a jazz orchestra is enough to burst an ear drum.

The symptoms of depression, anxiety, neurosis and madness are the concrete manifestations of life in our giant cities today. Promiscuity, noise, competition, crowds, drugs, alcohol, suicide: these are the most frequent hallmarks of urban life and modern technology. Research has confirmed this popular view, films have depicted it on screen, and books have denounced it. And since technical progress is considered as a prime factor of urbanism, it is also invoked as the essential cause of this malaise, being a central part of these nuisances: traffic congestion, noise, telephones, etc. etc.

The second example of such environmental nuisances is the car, which technical progress has made available to us all. At the end of 1971, there was one car in France for every 3.3 persons: nearly all those that professed a dislike to cars therefore probably had one. A newspaper article recently highlighted the attacks made against cars.

Autophobes are legion. And noisy. In the Holy of Holies, a car showroom in New York, leaflets are distributed bearing the provocative title "the car is anticity." The mayor of New York, John Lindsay, joins a thousand cyclists on Fifth Avenue to protest against pollution; their example is followed in Paris, where on the 22nd April, 10,000 young cyclists – without the mayor, for they don't have one – demonstrate "contre la bagnole qui pue, qui tue, qui encombre, qui est moche, qui abrutit, (against the car which stinks, kills, congests, deadens the mind and spoils the view.)." (*Charlie Hebdo*); on the 11th May, 15,000 people demonstrated on the lawns at Rungsträdgarden, Stockholm, shouting, "We need trees, not cars, to survive!" In London, on the 3rd July, the Council suggested that people using cars in the city between 7 am and 6 pm should pay for the privilege; in Tokyo, on the 6th August, traffic between certain hours was forbidden in areas where schoolchildren had been affected by exhaust fumes; in Florence on the 8th september, the socialist mayor forbade the use of motor engines in the large park at Cascine, and replaced these with municipal bicycles. Etc.

All that happened in only one year, 1973.

Autophiles are keeping a low profile. They are accused by all the intelligentsia of civilized countries of perpetuating these evils. They are being forced to apply draconian measures to prevent pollution and accidents as soon as possible. As the enemies of their workers in the factories, the car princes have finally succumbed to the guilty delights of sinistrosis. In the summer of 1970, Henry Ford II, the grandson of the man who invented the car, announced: "The golden age is over." On 24 February, 1972, Umberto Agnelli, the manager [sic] of Fiat, declared soberly to the Rotary Club of Turin: "Our real mistake was not looking ahead." Of course, not all the great car manufacturers agree with this fatalism. Some, such as Fiat, fight it with audacious advertising: "Now that the car creates more problems than it solves, where do we go from here?" And Volvo, the Swedish firm, observes: "For many people, the car has become synonymous with envrionmental nuisances."

We could quite easily repeat the whole of the discussion on pollution in relation to the car. However, for the sake of brevity, we shall consider only certain aspects. In the first place, its manufacture:

The manufacturing of the piston alone requires both welding and machining operations. The hourly rhythm of welding in 1920 was 40; today it is 450. The number of machines needed to produce 2000 of these has been reduced from 16 to 3, and the number of workers from 34 to 2. The number of machining operations has fallen from 14 to 5. The machining of a piston used to take 33 minutes; it now takes one and a half. The number of machines required has been reduced from 47 to 5. Even more important is the fact that it used to take 94 skilled workers to do this work; it now requires only 20 unskilled workers and 3 setters. Originally, the skilled workers used lathes and drills. Today, special machines for each operation devised by technicians and built by skilled workers are used by the unskilled workers, whilst the work is overseen by adjustors. All the components required by the production planning are manufactured in this way (...) Thus the limited number of skilled personnel still available in an economy do not produce the car anymore, but make the machines used by unskilled labour to produce the necessary parts. Unskilled workers can be trained in a few hours or a few days. It takes three to ten years to train a skilled worker.

In less than a decade, then, the social structures, the professional training, and the hierarchy of values in this sector of industry were completely tranformed. And these changes were not confined to the car industry alone. In 1950, it took three hours to make a shoe, for example, it now takes only 32 minutes. We could cite endless examples of this, but the car industry is the most important manifestation of what many believe to be detrimental changes.

Furthermore, the subsequent lowering of prices had a considerable impact on the family budget. On average, 27% of a family's budget in France is spent on the car. The burden of such an expenditure is enormous, which many people believe will only get worse.

The car is also a major factor in pollution, especially in the towns where it is thought that the vast majority of the population will be living in fifteen years time. Three toxic gases are emitted by cars: carbon dioxide, hydrocarbons, and nitrogeneous oxide, and that is not including solid particles, notably lead. It was estimated that in the U.S.A. in 1969, cars were responsible for 45% of hydrocarbon waste, 65% of carbon dioxide, and 37% of nitrigeneous oxide. Car pollution is worse when the vehicles are moving slowly, which is obviously the case in towns, and the above report stated that 80 to 95% of the carbon dioxide in towns was produced by this form of transport. The pollution is greater from multi-cylinder and older cars.

The car represents an environmental nuisance in another more concrete, and more dramatic sense. In towns, there is an increasing proliferation of cars which are often empty, and congestion in these areas is getting worse. We are all familiar with the problems of traffic jams – which make pollution even worse – the impatience and irascibility of car drivers, and pavements being taken over; they hardly need more discussion here. One very striking factor is the number of accidents occuring each year. As an indication, there are 200,000 cases of litigation per year in France as a result of car accidents, and there are not enough tribunals to deal with all the cases. This is not the most serious problem, obviously. Again, there is hardly any need to mention the wholesale slaughter caused by cars. If we are not aware of it at first hand, then there are plenty of statistics to remind us of the problem. Every slogan, every picture has been used: "All the children born today will be involved in some way in a car accident; 2% will be killed, 25% will be slightly injured, and 15% will be injured seriously. Can we call ourselves a 'society,' where one person out of two will be either killed or injured in a vehicle accident?"

This of course is a choice. Apart from other car nuisances, there are also the changes in personality and the effects on the human body.

Perhaps it is worth noting at this point that the car is a necessary evil. There can be no question of getting rid of cars: technical progress is irreversible and the car is too important a factor in our society and economy. In France, for example, it is estimated by some that the car represents 7% of all salaries. Others believe this figure is even higher, as much as 15%, if one includes all those who work in the car industry. In 1960, the exportation of cars represented 16.58% of the total goods exported from France, in Germany the percentage was 12.06%, and in Britain it was 10.81%. Accidents are profitable for the car sector: there are 40,000 car dealers and repairers providing work for 245,000 people. Cars and tractors have caused the disappearance of other trades, as well. How many blacksmiths were there once? And how many are there today?

It is not difficult to list the advantages of cars, for getting to work easily, for drives into the country, and for avoiding the inadaquecies of public transport. The car does also have some psychological value. It is a mass phenomenon: 20% of french households had a car in 1953, 58.3% had one in 1972. A short while ago, a regional french newspaper reported that young people held the driving licence to be "a passport into society", more so than the *baccalauréat*. In an article judiciously entitled "Moteur et bonheur" (Engine and happiness), P. Viansson-Ponté quite rightly asked the question: "what can justify the damage to the environment and the slaughter?"

The car is often the substitute for a cramped, and unhealthy dwelling. It is an annex and an extension of this dwelling, which extends beyond the road, the workshop or the office. It recreates the privacy of a family. It restores the role of the father, it reinforces the faith in his competence and skill. It enables him to display his *savoir-faire*, and this can even be shared with the mother or the elder son, if they are allowed to take the wheel. The car represents escape, a change of scenery, adventure, freedom, and the dominical revenge against the monotony and colourlessness of daily life (...) Traffic jams give the illusion of equality.

Apart from the desire for independence, the car also represents a thirst for power, and the concretisation of social rank.

Car manufacturers are well aware of the different aspects of the situation. The Paris car exhibition of 1972 projected a rosier future for the car. One reviewer called it "a showroom of purity." The car is not responsible for urban congestion (already pointed out by Boileau in the 18th century), it is simply the towns which need adapting. And as far as pollution is concerned, cars are beaten by heating. The car alleviates the suffering of man, enables them to escape and return to nature.

Measures to solve the problems created by cars had to be taken by public authorities. They were less radical than those proposed by car supporters, however. The two points of view held by these two camps were diametrically opposed. Car manufacturers and their supporters felt that related techniques should be made coherent within the techniques of the car industry. Neither the towns or the roads are suited to cars: that is where we should concentrate our efforts. The building of motorways has gone someway to solving the problem, but solutions within the towns are not so easy to find. In some fifteen years time, 85% of French people will be living in towns. Existing towns will be extended, and new ones will be designed which will be able to accomodate such large numbers of cars. As the president of Fiat declared in an interview, "It is the old towns which will be condemned. They will be preserved as historical centres and museums. People will live in new towns, and the problems of traffic will be resolved from the very beginning."

Some groups believe that it would be better to act in a more direct way. For example, public transport should be improved, which would go a long way to reducing the number of cars on the road. Of course, there are many technical problems concerning public forms of transport which have yet to be resolved. Such a transformation would also entail many changes in peoples' routine, and not only the routine of those who would use this transport (like shopkeepers for example). Statistics show that such a transformation would have to take place over several years. The following is a list of statistics for Paris in 1972, with a base of 100 for 1962:

bus	75
train	115
metro	120
car	150

In the same report it was noted that current institutions were ill-equipped to deal with such problems.

Of course, small cars are becoming much more popular, and recent research has concentrated on designing models from an aesthetic as well as from an economic viewpoint. But there are still several problems, some of which are psychological. The Fiat 500 was a useful car for a country with a relatively low standard of living. Eighty percent of those bought in France were to be used as a second car. The same was true of the Austin *mini*, despite its being considerably more expensive.

Steps are also taken to reduce the harmful effect of cars on the environment. The first problem to be considered was the pollution resulting from the burning of hydrocarbons. The United States has certainly taken some strict measures. The Federal Agency for the protection of the environment ruled in 1975 that the pollution from cars had to be reduced by 90%. Research was thus undertaken by both the petroleum industry and the car manufacturers. The former are dealing with the type of fuel rather than the engine; they are attempting to homogenise fuel mixtures and make them less flammable. The Japanese (Honda) are concentrating on engines with stratified charge which is leading to the construction of entirely different engines. From the construction point of view, one needs to think in terms of a series of devices to be adapted to existing engines. But it would be difficult to perfect the various parts of engines: they have reached their saturation point. The cost of this research is enormous, and this will no doubt effect the prices of cars later on. It is thought that this rise could be as much as 15 to 20%, with an even greater proportion for small-engined cars.

Attempts have also been made to increase the safety of cars, and a large programme was launched to this effect in the U.S.A. called E.S.V. (Experimental Safety Vehicles). But the schedules of conditions in almost all countries have been very ambitious, and they entail vast sums of expenditure. Also, the norms were set on the basis of cars which were two or three times larger and more powerful than the existing American car. Again, the solutions adopted were little suited to medium or small sized cars. As R. Guillain wrote in *Le Monde*:

The original American concepts did not include the essential notion of "aggressiveness": to ensure safety by increasing the size and rigidity of the vehicle means putting veritable tanks on the road, which are safe only for those inside – to everyone else they are extremely dangerous. The notion of compatibility should also be considered: cars are of varied size and strength, and the shock should be taken by both cars in the event of an accident. Furthermore, such measures should allow for the passengers, and not just the cars. There is only one conclusion: efforts should be made to strengthen smaller cars rather than the larger ones. The American system of simply strengthening the front of every car is not sufficient. The E.S.V. programme did not take into account what happens in a reinforced vehicle after an accident. The air-bag method was expected to solve this problem, but the results of tests were disappointing and deceptive. The cost-benefit ratio was insufficiently taken into account, which means that prices could increase by as much as 40 to 50%.

What else could be done? The only possibility was to fall back on older solutions. The Federal Agency for the protection of the environment advocated tax measures to encourage the public to buy smaller cars. This was as much to reduce fuel consumption, a pressing problem in times of energy shortages, as to reduce pollution and the number of accidents. The other measure is now generally accepted: speed restrictions. In most countries, there are restrictions on both minor and major roads. This represents the opposite of what the car manufacturers have been striving towards ever since this industry began.

Some people doubt the effectiveness of such solutions. We have already noted the price increases caused by such conditions. Safety measures imposed increases of 28% for average-sized cars, and 40% for smaller ones. At any rate, the latter, which in fact represent the most satisfactory solution to the problems of urban traffic and pollution are tending to disappear. In France, the proportion of small cars on the road was 27%, in 1972, this had gone down to 23%, and it is forecast that by 1980, this figure will have shrunk to 17%. Statistics have shown that eight many times as people are killed in accidents on motorways as in urban zones, and twice as many as on ordinary roads. Electric cars are expensive, and the rotary piston can only bring marginal improvements. So does this leave us in a vicious circle? In 1955, there were 3,300,000 vehicles in France, in 1965, there were 9 million, there were 12,500,000 in 1970. In 1975, there were 16 million, and in 1980 there were 19,280,000. It is forecast that there will be 24,340,000 by 1990, and 29,100,000 by the year 2000.

Some figures are even more illuminating. In a period of five years, between 1962 and 1967, car production increased by a third. From 1967 to 1972, it rose by two-thirds. If this rythmn is maintained, by the end of the century, the rate of car production in France will be 40 million cars and lorries per annum. In the course of the last ten years, the number of vehicles on the road has doubled. If this trend continues, there will be 120 million lorries and cars on the road by the end of the century. From 1971 to 1973, the number of new road projects more than tripled, and similarly, if this rate of growth is sustained, then by the year 2000, there will be projects for motorways being set up for a total distance of 4 million kilometres. These figures show that this type of growth will not always be exponential.

It is possible, therefore, to define the waste created by cars in terms of pollution, of the dead and the wounded, and also in terms of occupied soil. To carry the same amount of traffic, the railways take up 215 square kilometres, while roads take up 3400 square kilometres, or sixteen times as much land. And this is not to mention the land taken up for motorways and car parks, etc. Or the absurd amount of energy consumed by cars in relation both to the amount of energy resources as a whole, and also in relation to other forms of transport.

The conclusion to such a state of affairs is given in the following paragraph:

> The dangers of such an unlimited expansion urgently require counter-action: stabilisation. The first phase of this process would concern only the production of cars and lorries, and this would still enable the number of cars on the road in France to increase by 900,000 per annum. In the second phase, the number of cars on the road would actually be stabilised, but such a policy does not require either authoritarian measures or constraints with regard to consumers. All that is necessary is to reduce the demand for cars by making them more costly, and less neccessary.

So it is not merely a policy concerned with production and markets, but one concerned with development, and the competition of railways and public transport etc., which will lead to success.

If one considers that the car problem is only one of the many caused by technical progress, even outside the transport sector which is such a sensitive subject, then one soon realises the extent of the problems to be resolved. The solutions can only be found by systematic, global research, and this will not be an easy task. If this is not done, however, then we run the risk – as the journalist author of the "Automobile, croissance zéro" paper (Automobile, zero growth) affirmed – of being governed only by catastrophe.

The latest fears concerning technical progress and production in terms of the world as a whole have been concretised in the recently published [at the time of writing this book] famous Meadows report, which is the fruit of research conducted at M.I.T. at the request of the Club of Rome. This report stirred a certain amount of controversy which has still not been resolved. Its conclusion is centred on the "collective death" of the world at the beginning of the 21st century.

The research was based on the exponential character of growth. “Any consideration of problems concerned with human activities inevitably meets with phenomena of an exponential nature.” The M.I.T. team chose five major parameters: population, food production, industrialisation, the use of non-renewable resources and pollution.

I – The demographic considerations were based mainly on the schema provided by Malthus. The growth rate of the world population in the 17th century was 0.3%, which means that the population was doubling every 250 years. In 1970, this figure had risen to 2.1%, or a doubling of the population every 32 years. In 1970, the world population was estimated to be 3.6 billion. So by 2000, this will have risen to over 7 billion.

II – Taking a base of 100 for 1963, the figures for industrialisation are as follows: 30 in 1930-40, 50 in 1950, 70 in 1958, and 140 in 1968; the rate of production per head is growing at a slower pace. A comparison between the two rates of growth: 7% for industry, and 2% for population is fairly hopeful, and in fact the major part of economic growth does not concern industrialised countries.

III – In 1970, there was an average of 0.4 hectares of arable land allowed to each individual, or in other words, a total of 13 billion hectares of land was cultivated. These figures are sufficient to prove this is not enough. In fact, 0.9 hectares per individual is necessary. It is therefore possible to calculate the required land surface for a growing population (fig. 24).

Until the beginning of the 20th century, the theoretical ceiling of cultivable land surface was 32 million square kilometres. This ceiling has lowered because of urbanisation, industrial installations, and its associated infrastructure. There is a threat of food shortages before the year 2000 therefore, and the same is true of the supply of soft water.

IV – Non-renewable resources, principally mining resources, are diminishing, while the demand for such resources is still growing. Inevitably, then, there will be shortages. It is possible that there will be no silver, tin or uranium by 2000. The current known reserves of chromium are estimated at 775 million tonnes. The annual extraction of chromium is 1.85 million tonnes per annum, which means that in 420 years time the reserves could be exhausted. But the annual consump-

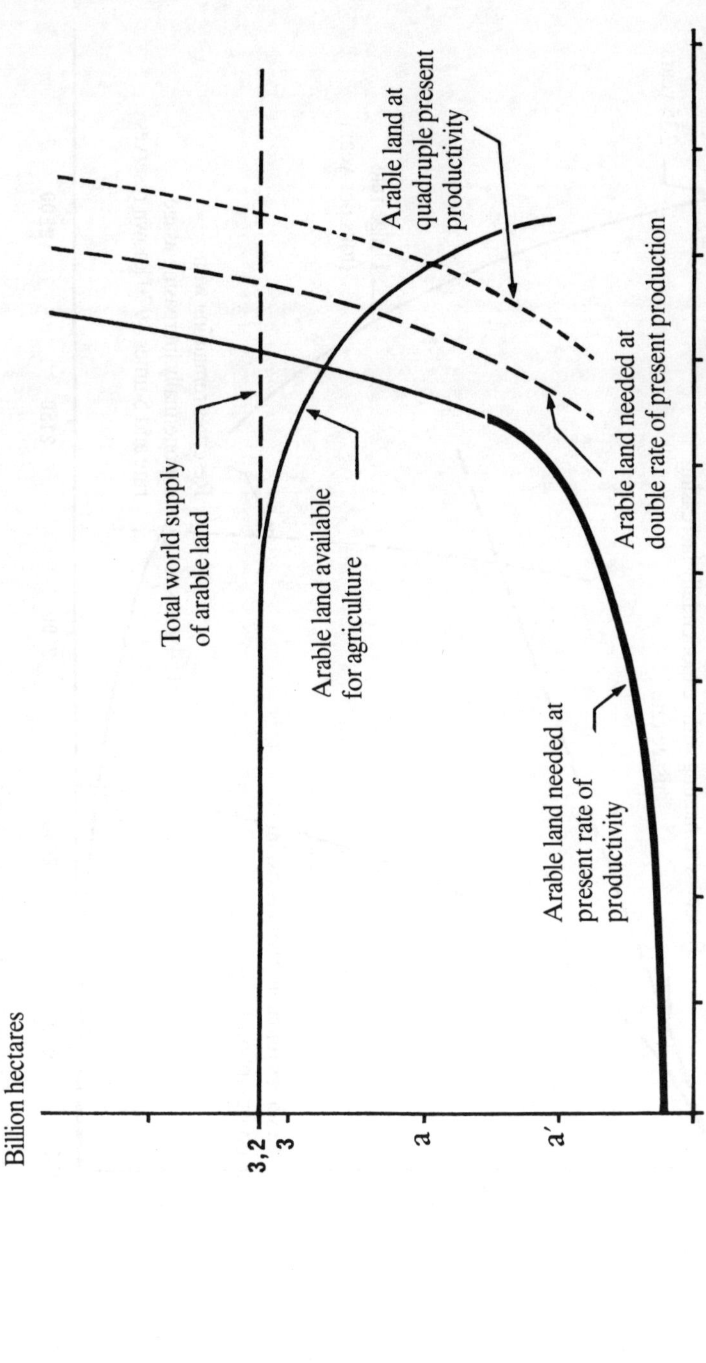

Figure 24. — Arable land. (From Meadows et al *Limits to Growth*, 1971)

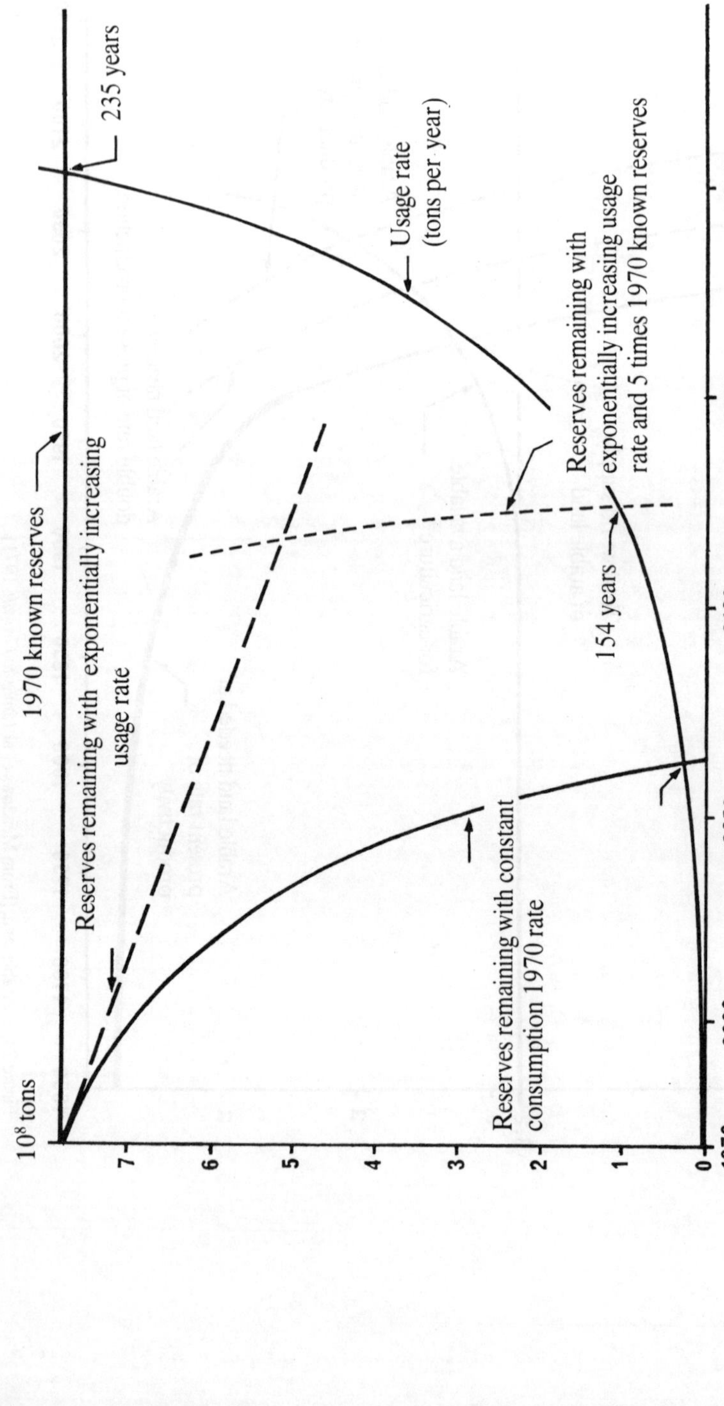

Figure 25. — Chromium reserves. (From Meadows et al)

tion of chromium is increasing by 2.6% per year, which means that these reserves are likely to be exhausted in only 95 years (fig. 25). The same calculations have been done for petrol, and the most optimistic estimates claim that the reserves of hydrocarbons will last for only another twenty years.

V – And finally there is pollution, the growth of which is also exponential. Carbonic acid gas from fossil fuels, heat energy and radio-active waste, which represent only three of the disruptive elements that men have introduced into their environment are still increasing, and this increase is exponential. The same is true of other types of industrial waste, such as pesticides, for example.

We live in a world of limited possiblilities. If all the factors are considered on a global scale, the conclusion is that growth cannot continue beyond the year 2100 (fig. 26). We cannot avoid catastrophe. And what is more, technical progress can do nothing to remedy this.

Technology is not included in this model claims the report issued by the Club of Rome, as an independent variable. It is multiform in the sense that it can issue from a particular sector, or it can actually be an influence on it. Any element emerging as a result of technology (the contraceptive pill, high yield seeds, television, drilling platforms) has a distinct role in a given sector, and a precise part in the way a system works.

In fact, all the experts at M.I.T. are persuaded that technological progress can only postpone the final result.

Each hypothesis concerning almost unlimited energy resources (fig. 27), pollution control, and an increased agricultural yield and birth control fails to present a more optimistic outlook. The model continues to show the same results, it is simply a matter of time (fig. 28).

"The fundamental workings of the world ecosystem is defined by the exponential growth of population and investment, followed by a collapse." The authors of this report believe they have "demonstrated with their global model, that any attempts at solving with technology the problems of the exhaustion of resources, pollution or food shortages are bound to fail in terms of the essence of the problem: exponential growth within a closed, complex system."

According to the M.I.T. model, a stabilisation would require draconian measures. The reduction in the birth rate, and production investment for consumer goods is seen to be "against nature" and "beyond

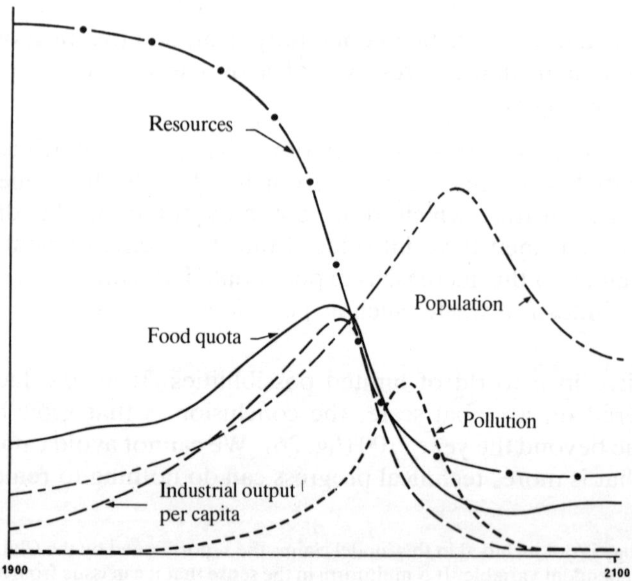

Figure 26. — Hypothesis on the basis of maintaining current trends. (From Meadows et al)

understanding" because its effectiveness has been put to the test, and has never been seriously envisaged.

"Everything points to the fact that of the three possible hypotheses: unlimited growth, the voluntary limitation of growth, or the natural limitation of growth, only the last two are plausible." Gradually, the definition of a state of equilibrium is reached, characterised by a constant global population and capital, the forces which tend to diminish or increase the latter being carefully balanced.

"Nevertheless, technical progress would be necessary and welcome in a stabilised world." Of course, this would have to be carefully controlled, so that progress took place in certain precise fields, and not in an arbitrary fashion, which could destroy the equilibrium. In fact, in the list of examples provided, it is merely a matter of progress bringing about indispensable changes to re-establish the equilibrium. There is no longer the growth of technical progress, just the control of detri-

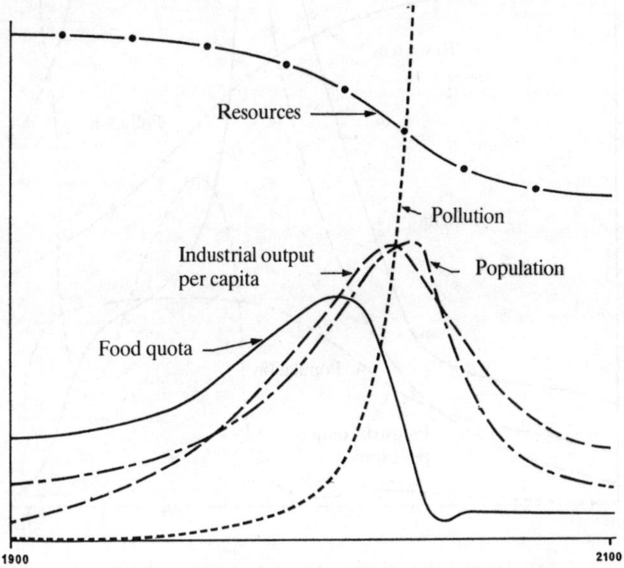

Figure 27. — World model with natural (unlimited) resources (recyclable). (From Meadows et al)

mental tendencies: (*a*) new methods of collecting waste to reduce pollution and facilitate recycling; (*b*) more effective recycling methods to reduce the consumption of basic raw materials; (*c*) a different conception of manufactured products, making them more durable and easier to repair, in order to reduce the rate at which capital depreciates; (*d*) the domestication of solar energy, the least pollutive of all energy sources; (*e*) the more efficient means of neutralising harmful pesticides founded on a better understanding of ecological intercalation; (*f*) progress in medicine to reduce mortality; (*g*) effective contraceptive methods to bring the birth rate in line with the rate of mortality.

It is clear that this technical progress is very limited and rigorously directed.

There is thus widespread anxiety throughout the civilised world about the coming of the year 2000. And for some, this anxiety is very real. The vice-commissioner from the Holland at the E.E.C. Commis-

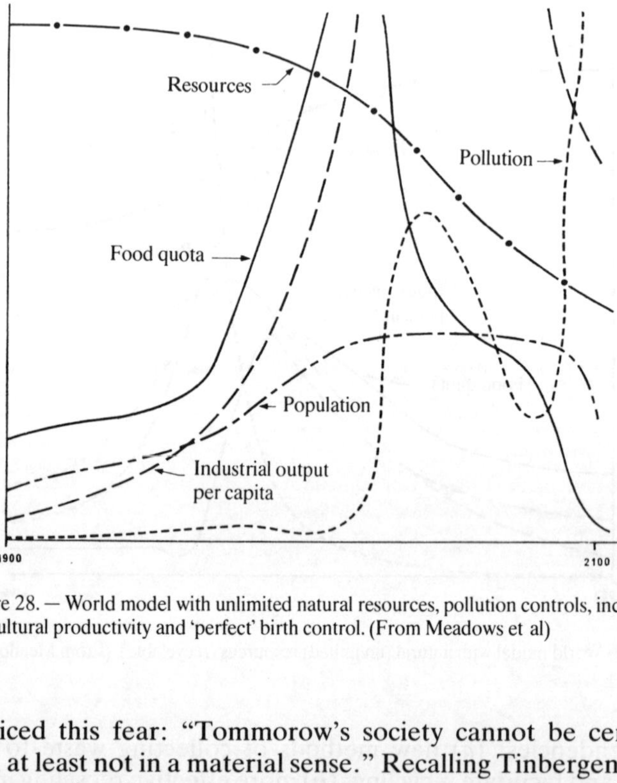

Figure 28. — World model with unlimited natural resources, pollution controls, increased agricultural productivity and 'perfect' birth control. (From Meadows et al)

sion voiced this fear: "Tommorow's society cannot be centred on growth, at least not in a material sense." Recalling Tinbergen's notion of "gross national happiness", he proposed replacing the "gross national product" with "gross national utility." The primary objective of this was to safeguard ecological equilibrium and reserve for future generations sufficient energy resources. This picture of impending catastrophe clearly reflects the pessimistic attitudes to technical progress. The only thing lacking was a scientific formulation. This was provided by the Rome Club and M.I.T.

HOPE

In the face of doubt and anxiety, there is necessarily hope. Some of this hope of course is naive and simplistic. There is a belief in an indefinite technical progress, which in fact is a relatively recent phenomenon – with all due respect to Jules Verne – and furthermore, a belief that technical progress will bring with it human progress. This is simply faith, a kind of revealed truth which is not based on any concrete fact. Besides these beliefs, there are other, more well-founded attitudes. We have already said that technological forecasting has undergone a certain impetus and that some of its claims have already been substantiated. It is possible that the problem has not been considered carefully enough, and the famous report of the Club of Rome, although partisan, has certainly stimulated reflection on this subject.

We have already raised some ambiguous points which we would do well to point out once more. The most important of these concerns the notion of technical progress itself. Is it true that any concrete invention, useful or not, profitable or not, or dangerous in some way (such as those which cause pollution for example) and which brings some positive element to a technical process, constitutes an element of technical progress? It does if it is a link in a chain of discoveries, but not otherwise.

Some recent examples which have been the subject of a great deal of discussion, reveal the limits of technical progress as such. *Concorde* is unquestionably an example of technical success: it has the advantage for commercial aviation of greatly reducing the length of flights. The flight from Paris to New York has been cut by three and a half hours thanks to this plane. But who wins? Those on business can go to more meetings: which means that they will live at a faster pace than ever. But is this a good thing? For someone involved in research, or going on holiday, this gain represents nothing at all. A reduction from 5 days to 8 hours was very significant, but is the same true of taking off another three and a half hours? The cost of this marginal advantage should also be considered. This apparatus is only marketable if it pays off the research and development costs and if companies are allowed to buy it on hire purchase with a governement guarantee. On the other hand, railway lines judged to be non-profitable are being phased out almost everywhere. The slow decline of the railways is a general process, and has reached large proportions in the United States. Those who oppose such a decline usually invoke the notion of public service, which should be assured "at any price." The authorities responsible have replied with the notion of "public interest," arguing that a national community should not have to bear the expensive brunt of such high costs which favour only a peripheral group – the public interest being identified here with the national one. In spite of their similarity, the two cases are in fact diametrically opposed: in the first case, technical progress has taken place, while in the second, there is a lack of technical progress which might recreate the profitability of a means of transport (it is also possible that the latter case is the effect of a reduced demand due to demographic changes brought about by changes in economic activity). But as far as the general interest is concerned (one might almost, like Sicco Mansholt, call it national utility) the two cases are very similar, since the questions concern only restricted groups. Unless of course, one considers performance and prestige as elements of national utility.

These precisions reveal two fundamentally different types of attitude among the optimists. Sometimes, the technician seeks only one element of what we have termed narrow technical progress. In a technological lineage which has not yet become saturated, possible future plans of action can be foreseen. Research is concentrated on certain areas, and science and technology are such today that technical developments appear to be possible. It was thus for the first satellites and so on until the first walk on the moon. Besides this, there are some people who have a wider outlook, whose optimism is more reasoned, and who take into account factors which are not necessarily technical. It is the latter that we shall consider, but while still bearing in mind the forecasts of the former.

The first basis for hope is having succeeded in dispelling most fears. But it is not sufficient to calm anxiety with words, and this is not a responsible attitude. "With the help of wild extrapolations and fabricated statistics," wrote a reader of a national newspaper, "the economists and engineers, the magi of our time, foresee all sorts of evils: take for example the ramblings issuing from the M.I.T. echoed recently on television by M. Mansholt."

A more interesting aspect of this opposition was the criticsm levelled at the method employed by the M.I.T. researchers. The discussion began around their statistical analysis (and cases have been cited, where factors intervened in the rate of mortality which were quantified in a particular way). Furthermore, it was noted that there was some confusion between the notions of simultaneity and causality. "For example, does the birth rate really depend only on income, industrialisation and hygiene?" The central focus of the critique was the choice of variables, however. Some factors would not have been relevant a few years ago, such as pollution, for example (fig. 29). Is it not then reasonable to suggest that in a few years time other variables will have to be taken into account?

Current methods of quantification lead to a mechanistic, deterministic vision of the world's future. The birth of ideologies and new attitudes cannot be foreseen, but they should be allowed for. The latter would create new curves within the model, and therefore a different vision of the world.

It is clear of course that the model is partial and not global, as some would have us believe. A global model cannot take into account local variables. It is also immutable, and cannot be adapted either in terms of techniques or otherwise. We can draw some conclusions from this.

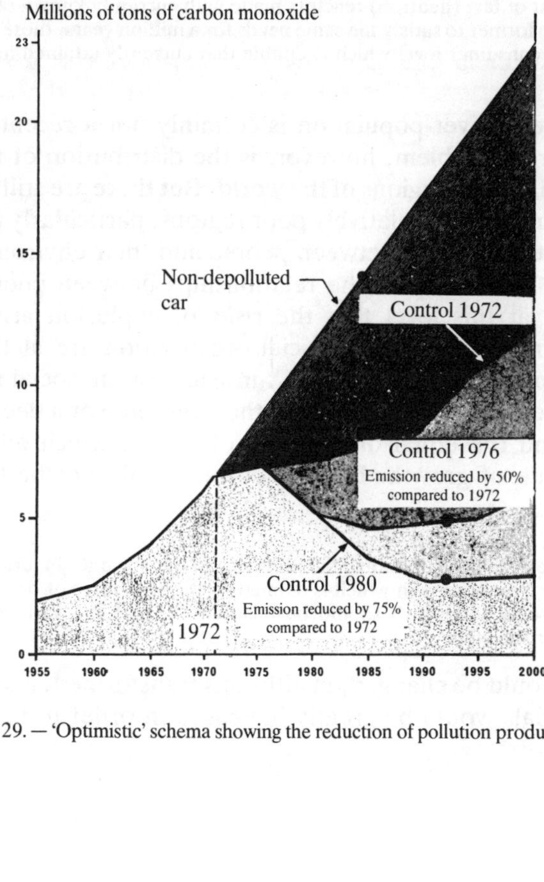

Figure 29. — 'Optimistic' schema showing the reduction of pollution produced by cars.

The finite character of our world is certainly a given fact. But the fateful date remains a hypothesis. The greatest contribution of the M.I.T. report was probably to state that the effect of technical progress has hitherto been underestimated.

Other critiques were less concerned with theory than the actual data. Take for example the response of R. Barre, the then president of the European Community, to a letter sent by his vice-president. His judgement of the M.I.T. report was a harsh one. This particular criticism is aimed against the extrapolations concerning pollution.

It is not inconceivable that if in the future, 5 to 10% of the annual increase of the GNP was devoted to reducing levels of pollution, then not only would the destruction of the environment be checked, but significant improvements would be made.

And here, technical progress and raw materials have an essential role to play.

The development of fast (neutron) reactors made with currently known raw materials would enable the former to satisfy the same needs for a million years, those of 10 billion people having a consumer level which is double that currently attained in the United States.

The spectre of over-population is certainly not a recent preoccupation. The real problem, however, is the distribution of the population in the different regions of the world. But there are still problems for over-populated and relatively poor regions, particularly in parts of Asia. If the relationships between people and their environment can be mastered, then so must the relationships between people themselves. "It is at this level that the risks of explosion arising from political incompetence or bad social organisation are at their most acute." The president of the E.E.C. insisted on the social aspects of growth: the security of employment, the guarantee of a decent standard of life and the equal distribution of wealth which all favour a continuing rate of growth. But expansion should not be forced. In fact,

It would be difficult to pose and solve the problems of our societies in terms of rates of growth. It is the type of growth which is defined by the structures of demand and the conditions of production which should be adapted to meet the various competing objectives of our societies.

Societies would be changed, institutions transformed, and the price of raw materials would be greatly increased in order to increase the financial resources of under-developed countries (on condition that these resources were properly used).

In fact, – and several eminent scholars have taken great delight in emphasising this – the theory of exponential growth which is at the basis of the M.I.T. report, cannot be proved. As P. Uri wrote:

> Can one really project over a hundred years without allowing for unforeseen factors? Other forms of energy might be discovered, more powerful and more abundant, which might enable us to use the poorest minerals, such as clay alumina or titanium from the sea, for example; new plastics could be substituted for rare metals; and it is possible that chemistry might provide us with a new generation of pesticides which would not, unlike DDT – even though it did save the lives of 500 million people – fall under suspicion.

It is clear that over-population could be stopped by raising the living standards of poorer countries. And furthermore, as Barre claimed, by the best possible distribution of resources.

In actual fact, the optimists used the picture of the year 2000 drawn up by a team from the Hudson Institute led by Hermann Kahn in 1968. This team was similarly aware of the dangers of technical overheating and they constructed several pessimistic scenarios, notably in the perspective of world conflict. However, they lingered much longer on the optimistic elements.

The Kahn-Wiener report deals with the last third of the 20th century. It is claimed that by the year 2000, modern technology will have been mastered, and that the growth of the GNP per capita will have increased. One objection that could be made against this report is that certain parameters found in the M.I.T. report are not given any or enough weight in this model. On the subject of demography, for example, Kahn and Wiener estimate that in fifteen years or so: "birth control will have become a universal phenomenon, and will be far more effective than it is today." This is in striking comparison with the pessimistic view of demographic trends to be found among the M.I.T. experts, who included the possibility of "complete birth control," but who then discovered that the pattern of the other curves changed as a result. Naturally, the Hudson Institute rejected the idea of technological unemployment. "In the near future, automation and cybernetics, in contributing to the increase of productivity and the development of the economy, will probably create as many jobs as they have replaced."

These American optimists estimate that technology will be perfected, and they draw up impressive lists of probable inventions. They make a finer analysis of six particular areas of technology where progress will be more marked: nuclear power, strategic warfare, electronics (calculators, data processing and automation), lasers, holography and biological engineering. In each case, they reach the opposite conclusions to those drawn in the M.I.T. report. Growth is not necessarily exponential, and our standard of living must necessarily reach a peak, as it probably is in the United States at the moment. On the other hand, a curve demonstrating rice yields (fig. 30), reveals that the optimists have more faith in agriculture than the researchers at the M.I.T. It is also possible that the rise in prices would enable us to exploit resources which otherwise would have been unprofitable.

The main idea of the Hudson Institute report is that this progress will necessarily bring about the birth of a post-industrial society.

The possibility of economic development, and the path this development is taking, our means of controlling both the interior and exterior environments, and the consequence of this, technological innovation, and finally the application and diffusion of these possibilities is increasing without any apparent limits. The question is, can our culture and institutions adapt themselves to such a fundamental and yet rapid change?

Keynes had already made this point. On a more detailed level, the partisans of the Hudson Institue did not fail to criticise some of the hypotheses used in the M.I.T. report. The treatment of pollution was one of the areas under attack. Despite the lack of knowledge about the effects of pollution, many people believe that it can be curbed. The subject of non-renewable resources was a different problem. Even with constant progress – it is thought that the findings of oil reserves are increasing twice as quickly as the product are being consumed – and even if technology can exploit these possibilities, it is no less true that at some stage there will be shortages in a certain number of essential areas. However, the techniques of recycling can still be called into play, which would give humanity an infinitely longer breathing space before the final catastrophe forecast by the experts at M.I.T. When oil is exhausted, there will be nuclear power, and after that probably solar energy (fig. 31). In other words, the world would be finite within a given technical system, but it might not within another – that is the basis for optimism.

Of course there are some sectors which are saturated, and within which no major invention can take place, but which can still be developed. Two examples of this are the fields of transport and traction.

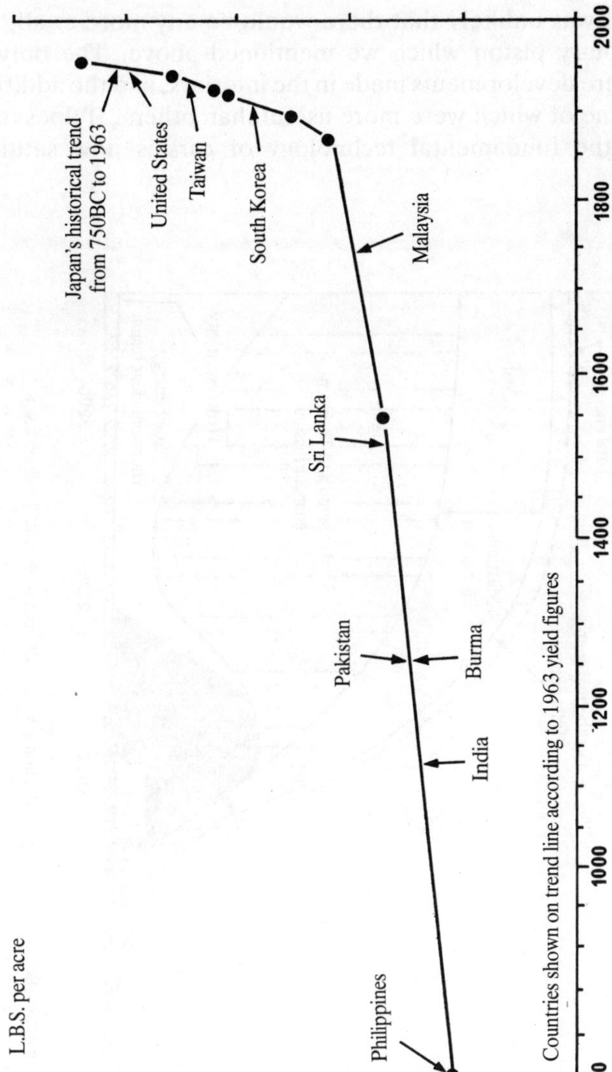

Figure 30. — Current rice yields in selected countries related to Japan's historical trend. (From Kahn and Wiener, *The Year 2 000*, Hudson Inst., 1967)

All specialists are agreed during a car show held recently that there had been no new notable inventions. It was also agreed that furthermore, it was unlikely that there would be any more at all, apart from the rotary piston which we mentioned above. The only new elements were developments made in the interiors, and the addition of gadgets, some of which were more useful than others. It does indeed seem that the fundamental technology of cars is now saturated.

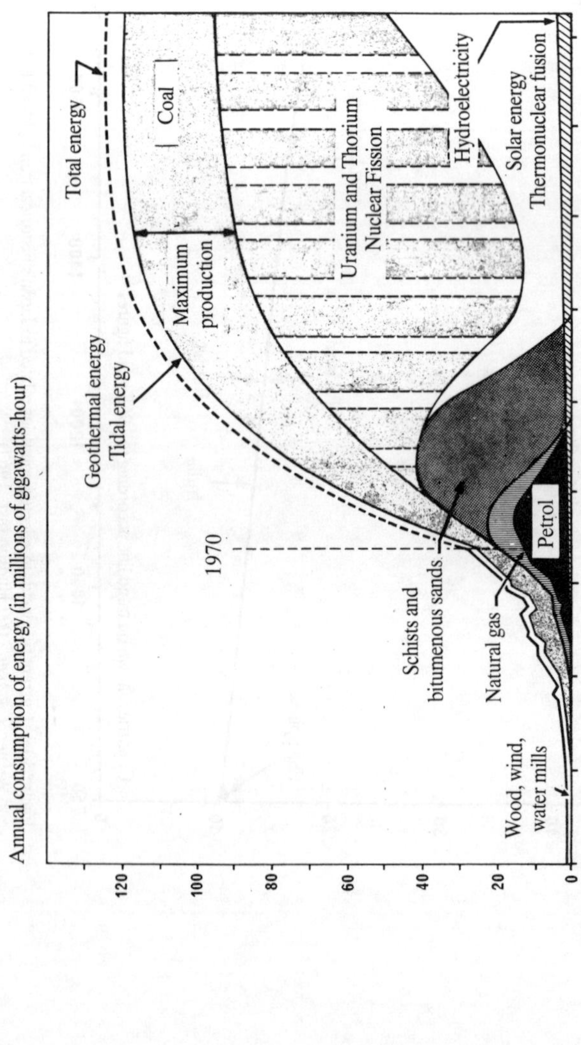

Figure 31. — An optimistic vision of energy production. (A gigawatt-hour represents 1 million k.w.)

Progress is still possible in some areas: bodyworks are now being made of plastic, the designs and accessories can still be improved. These developments can be brought about only by using tributary techniques. And this is all the more so now that component techniques – particularly those concerned with road networks and urban topography – are no longer at the same degree of development.

One might have thought that the same was true of one of the most successful areas of technology in the 19th century, the railways. However, the replacement of steam with electric traction and diesel, and the replacement of the latter with gas turbines introduced new elements. One could have invoked the impossibility of adapting the old superstructures, such as the rails, for instance. But it seems that this field of technology can be more easily adapted than others, and the railway has certainly not had its final word, especially when one considers its importance in everyday life. It is more reliable and safer – at the moment it is the least dangerous of all forms of transport – it has a greater capacity, and the costs are lower. It can even compete with planes in certain respects, especially the latter. Everyone is aware of the problems of getting to airports, while trains can run right into the centre of towns. Furthermore, the trains have been almost completely standardised. For these reasons, and because of the way in which the distribution of activities is evolving, the train of 2000 might still be an important factor in transport. It should be developed for use in certain specific areas: large-scale public transport; links with areas separated by long stretches of land; providing a service for suburbs which in fact are an extension of the urban transport network; and for heavy duty transport. The other side of this development is the construction of a new network which can be used by high-speed trains. This has already begun in some areas, there are now such lines between Florence and Rome, and Paris and Lyons. The number of stations served by express trains will be reduced by half, which means that the road network must be linked with the rail network, as it is with the motorways. The older lines, which are no longer useful in this respect, could accomodate the transport of freight. The train takes two hours to get from Paris to Lyon, and there is one every thirteen minutes. This certainly makes it competitive with the car, and even with the plane. It is fairly easy to calculate the limits of this competition. It is noteworthy, however, that the building and maintenance of airports is very expensive, and their use would soon be limited, the capacity of planes being much less than that of trains.

As far as urban transport is concerned, research is currently directed towards creating a network of automised engines. It is still early days yet, however. Take for example the *Urba* system, or the programmed *Aramis* by Matra transport system, to name but two of such projects. These depend very much on developments in urbanisation and the continuity of older centres. Acceptable solutions can only be found if these two areas of research are linked. It is here that the notion of a technical system is most relevant, and the same is true with regard to the hovertrain and its use in a medium or even small sized network (the latter is particularly important for travelling to airports and satellite towns).

It is possible to forecast the saturation of certain sectors, and we shall consider two very different examples of this. The example of white light is characteristic. Progress was steady in this field, from Edison's lamp to the fluorescent lamps of today. Each of the successive techniques employed is represented by an S-shaped curve, which are superposed on each other. The envelope curve has been simplified here and is represented by a staight line (fig. 32). It is clear that it could not pass the theoretical limit of white light.

The second curve represents the output of power stations (fig. 33). They are at maximum output around the year 2000, and it would be

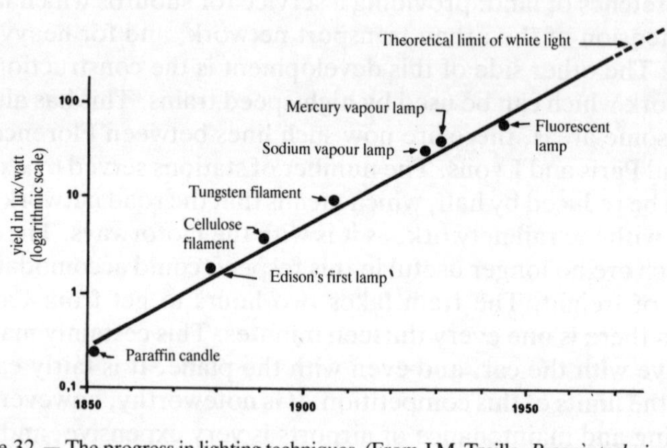

Figure 32. — The progress in lighting techniques. (From J.M. Treille, *Progrès technique et stratégie industrielle*, Paris, 1973)

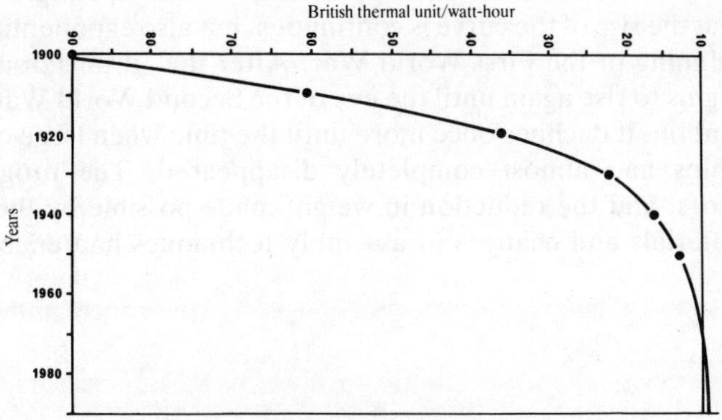

Figure 33. — Yield of thermal electric power plants.

impossible to go beyond this. But here again, it would be possible to substitute old techniques with new ones (fig. 34).

Of course, there is always the solution of finding an alternative for saturated techniques, and this has been done in the techniques used in transport. Consider the following two graphs, which are quite reveal-

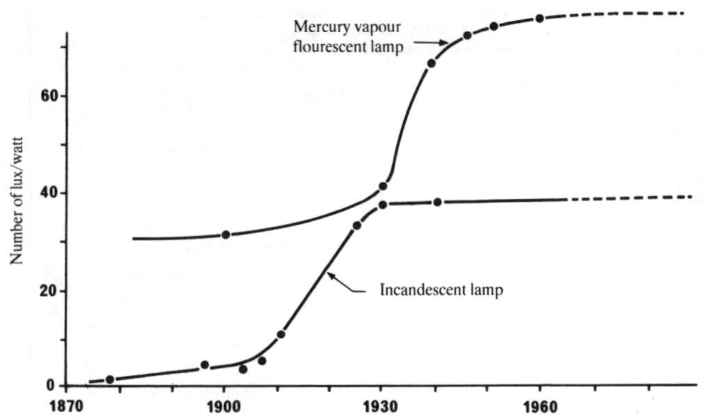

Figure 34. — Curve showing new products and progress made in lighting techniques. Evolution through the years. (From J.M. Treille)

ing. The first concerns the tonnage of large steamships (fig. 35). It is clear that the rise of the curve is continuous, but also exponential, until the beginning of the First World War. After this, it falls drastically, then begins to rise again until the eve of the Second World War. From this point on, it declines once more until the time when large capacity steamships had almost completely disappeared. The progress of generators, and the reduction in weight made possible by the use of new materials and changes in assembly techniques had enabled the

Evolution in the tonnage of steamships

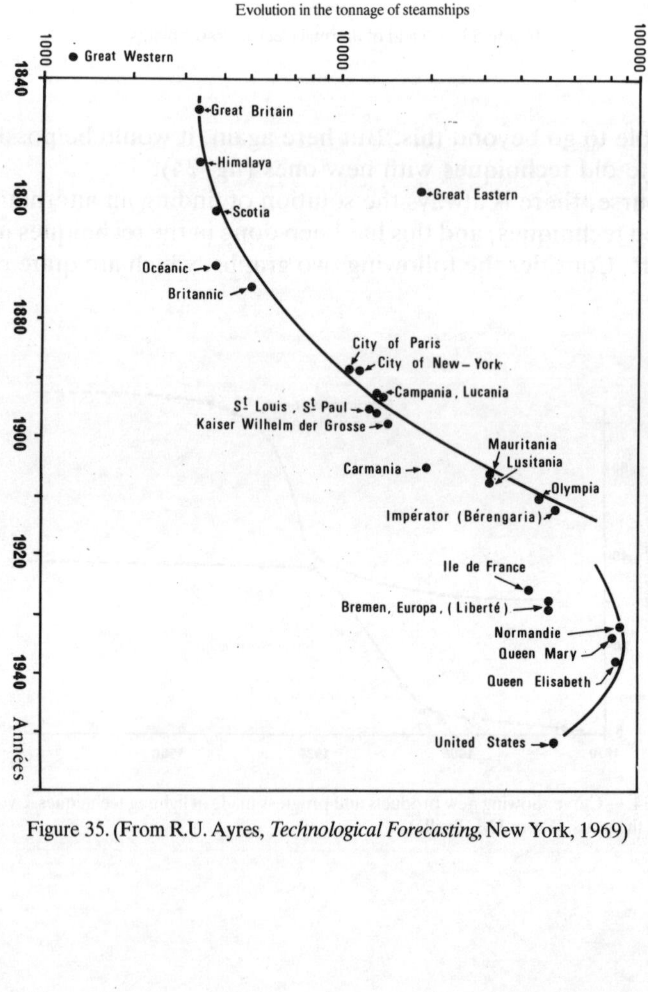

Figure 35. (From R.U. Ayres, *Technological Forecasting*, New York, 1969)

degree of tonnage to increase. But it soon became apparent that technical conditions were not the principle factors: the cost of personnel and of developing steamships since the end of the Second World War was too great to be able to compete with the plane.

The second curve is complementary to the first. This is the famous speed curve, which covers several different types of speed, from transport to missiles (fig. 36). The diagram consists of a series of curves, each one representing a different technique, and which grow flatter as techniques are saturated. The ensemble of curves is repre-

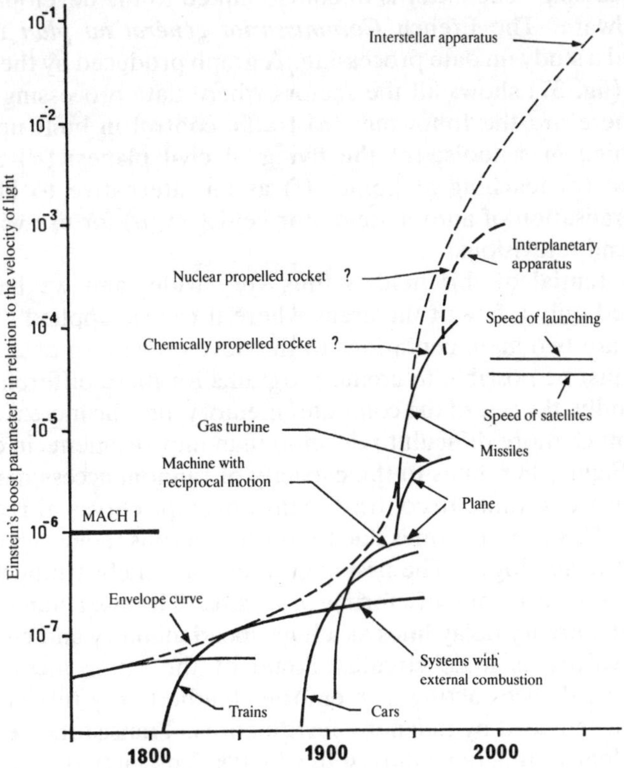

Figure 36. — Comparative evolution of the speed curve representing different techniques evolving towards saturation.

sented by what is called an envelope curve, which indicates the general tendency. Would it be logical to extrapolate from this? Some people have done this, considering that it would be possible to substitute a saturated technique with a new one, since this has been happening now for centuries. This depends of course on the imagination of inventors, and the compatibility of tributary techniques. We shall consider this point further on.

It is not difficult to see how recent techniques could undergo spectacular developments. Take the computer, for example. There are two paths to follow, which have tributary elements. The computer can be considered as an autonomous object, especially in terms of its memory. But one should also consider the long-term development of data processing. The latter is of course linked to the development of the hardware. The French *Commissariat général au plan* recently published a study on data processing. A graph produced by the Delphi method (fig. 37) shows all the sectors where data processing will be used. These are the following: (*a*) traffic control in built up areas; (*b*) teaching in schools; (*c*) the flying of civil planes; (*d*) medical diagnosis; (*e*) teaching at home; (*f*) as an alternative to libraries; (*g*) generalisation of automatic motor vehicles; (*h*) for use with telephones and televisions.

The potential of this field is thus very wide, and we have still considered only a few of the areas where it can be applied.

There are two main conditions to the development of data processing: it must be possible to create programs for these different tasks, and secondly, the size of the computer memory must be increased. The latter is much more difficult to develop than improvements in computations. Figure 38 relates to the capacity of random access memories classed by access time. In contrast to the envelope curves, this curve is flattened. This progression is due to the "numerous radical changes in subjacent technology." The first computer was an electromechanical system, with no means of storing programs. The next models used contained mercury delay lines as a high speed memory and circuits of vacuum valves as logic circuits. Some of the latter incorporated electrostatic devices acting as memories (cathode ray tubes), which were soon replaced by magnetic ferrite cores. Transistors were used for most logical functions during the 1950's. The later stages are not represented in the parameters illustrated here.

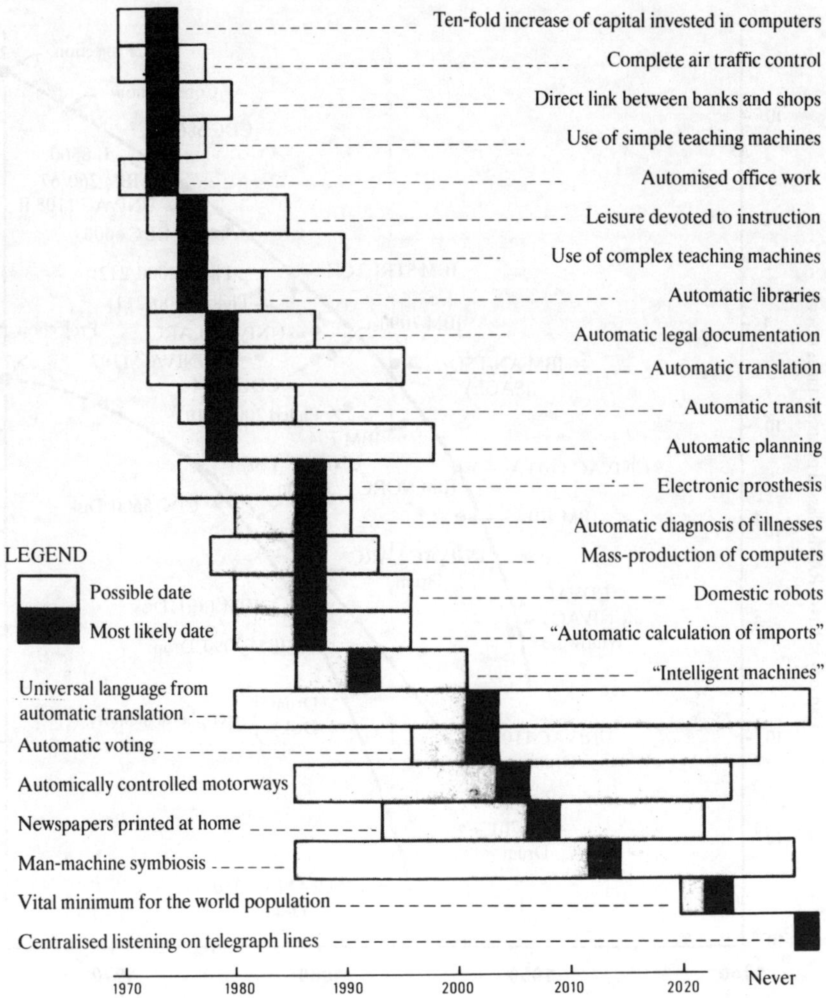

Figure 37. — A few representative dates of the future applications of automisation. (From the Rand Corporation)

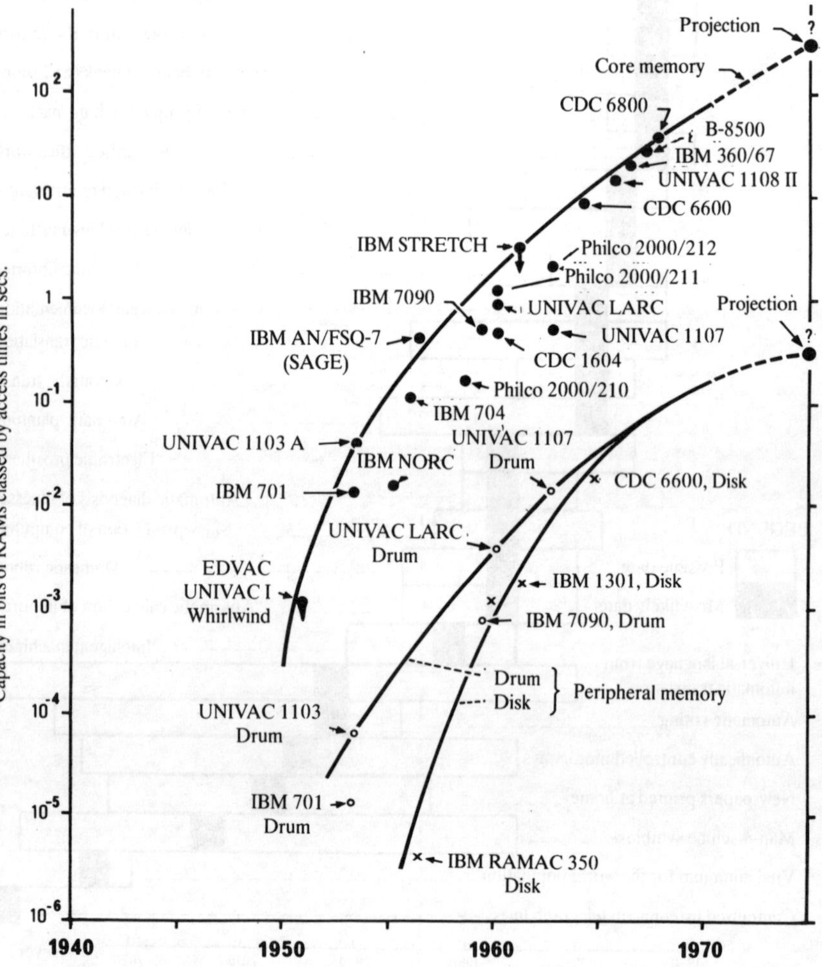

Figure 38. — Capacity of random-access memories classed by access times in microseconds. (From R.U. Ayres)

The growth rate of energy provided by accelerators creates the same type of envelope curve as that which represents the progress made in terms of speed (figs. 39 and 40).

It is not sufficient merely to use envelope curves, however. In a sense, they are only significant within closed systems, which do not allow for exterior variables, whose effects can be fundamental. They serve as a basis for general extrapolation only.

From one industry to another, and from one technology to another, it was felt that technological forecasting on a global level still seemed possible, however, and indeed, this has been practised for longer than is generally thought. The modern era of technological forecasting

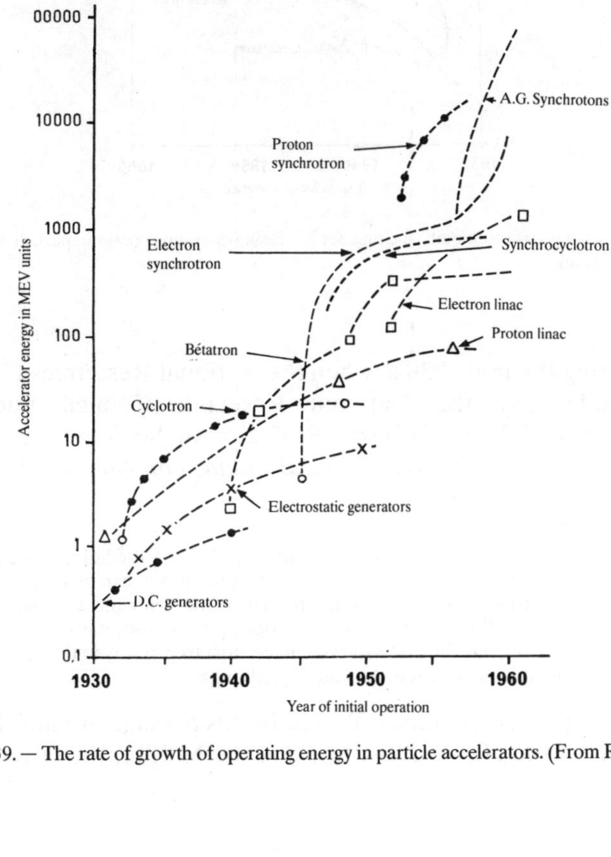

Figure 39. — The rate of growth of operating energy in particle accelerators. (From R.U. Ayres)

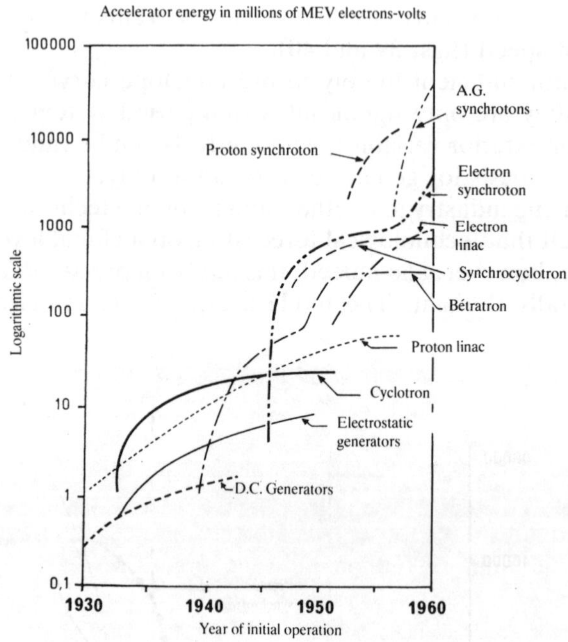

Figure 40. — Change and continuity: the time for increasing operating energy in particle accelerators. (From J.M. Treille)

began during the mid 1930's, when the National Resources Committe was founded by the National Research Council under the chairmanship of W.F. Ogburn, Professor of Sociology at Chicago University. His report, *Technological Trends and National Policy* was published in 1937.

During a period of many changes, it is necessary that those at the helm of the ship of the state should anticipate future events. A study of inventions can reveal future social conditions and a nation's future problems. In fact, of the four most important factors regarding a nation's well-being, namely, invention, population, natural resources and economic organisation, it is the first which changes most frequently in the modern world, and which therefore is most often the cause of change.

The main preoccupations reflected in this passage are also found in the reports of the Hudson Institute and the M.I.T., the latter dealing with technical progress only from certain angles. In fact the Ogburn report was unable to forecast atomic energy, radar, antibiotics and jet propulsion, even though the research for all these fields had begun in the 1920's.

Global planning, on the scale adopted in the U.S.A. does not exist, but it has still been found necessary to organise some form of fairly long-term technological forecasting. This has been done in the army, and in some universities, sometimes actually on behalf of the army. Private industry concerned itself with the leading sectors where techniques were evolving most rapidly: such was the case with the Lockheed Aircraft Corporation, or with the General Electric Tempo laboratory.

International organisations gradually began to follow the same path: the O.E.C.D., the constituent elements of the E.E.C., etc. European countries, who are more in favour of forecasting and economic planning, were fairly slow in adopting the practice of technological forecasting. This was all the more indispensable and restricting because of the limited financial resources which forced choices, apart from the ever-present dilemma of fundamental or applied research.

Since there is a fundamental risk in research, noted the authors of the french 6th Plan, the state organisation of science would be better served by a greater understanding of the extent and the significance of this risk, and by a conscious acceptance of responsibility for it, with the aim of obtaining clearly defined cultural, social, economic and military objectives.

Or, similarly:

The first impulse of a nation which wants to stay ahead in those areas essential to its independence, is to try and understand the relationships between research and the economy, and to determine the criteria by which some scientific works are judged to generate expansion more than others.

The constraints are indicated here. If there is a plan, then there is also technological forecasting. The latter demands that choices be made, and the lack of financial, material and human resources means that the repercussions of these extend right back to pure research.

From this point, there is the *possible* and the *desirable*, which permits the distinction of two complementary types of forecasting. Exploratory forecasting uses information from current science and technology, the main lines of which enable us to forecast fairly long-term possible solutions to any technical or scientific problems which are known to exist. The information used in normative forecasting is based on needs and the tendencies which effect them. This type of forecasting searches for means to supply such needs, and as such represents a quite different approach to the problem. Ideally, these two methods of research should be converged. Today, more than ever, technical innovation is due to precise normative conceptions and to various technological forecasts, a large part of which is normative.

We have already noted on several occasions the use of the envelope curve, or extrapolation, in exploratory forecasting, working from anterior technical data. One author has noted that the growth rate of some performance variables will be indefinitely asymptotic before the year 2000. Such is the case with the speed of vehicles. On the one hand, therefore, the exponential curve is not generalised, and on the other, the method has limits, in that different techniques are being compared.

In the case of a clearly determined technique, we saw that the curves were indefinitely asymptotic: this was the case with electrical power stations. But one could also use intersecting curves. From 1913, S.C. Gilfillan produced the above cited forecast for the size of steamships.

The extrapolation of temporal series (note that the period of extrapolation is no longer than the period over which the phenomena are observed) has given substance to attempts to formulate simple analytical models: this was so for Henry Adams in 1907, and then for the models created by Insenson, Hartman, Holton, and Putnam. The extrapolation of temporal series has also been done on a phenomenological basis: linear growth with saturation, S-curves, exponential curves.

The methods of normative technological forecasting have also been developed. We do not wish to describe, even summarily, from the matrix to the network method, from business management games to the games theory. We have considered relevant graphs, such as the Pattern method, which are currently perhaps the most advanced.

The most spectacular results are probably those produced by the Delphi method, developed by the Rand Corporation, itself born of the forecasting department of American military aviation. This method is one of normative technological forecasting. It enables us to choose social or other objectives of a high level without prejudice, by using scenarios or operational models within social, military, political, economic and of course technical fields. As far as we are concerned, it is an enquiry conducted by different groups, each one concerned with a different problem, and each one composed of technicians and non-technicians. These groups are asked certain questions, and they must try to determine the chronological limits within which the answer can be given. The first study was published in 1965, under the direction of T.J. Gordon and Olaf Helmer, and it dealt with the foreseeable development of science and techniques in the next three decades. In fact, research was limited to a few possible scientific and technical breakthroughs in a given future. In the schema which we have included here (fig. 41), the length of each bar represents the average extent of the difference in the experts' estimations. In each case, a quarter of the experts put the date before the point at which the bar begins, and a quarter put the date after the point where the bar ends. In eight cases, indicated at the bottom of the graph, some experts put "never" as a date, while the dates forwarded by others are quite precise.

The result of this enquiry – whose limits are clear – is all the more interesting in that it is being assessed eight years after it was first published, so that some of its forecasts will have had a chance to become reality. Apart from the somewhat simplistic schema published by a French magazine to attract public attention to a little-known science, it is possible to draw up several annex tables showing these forecasts. The date printed in italics represents the opinion of the majority of the experts. The absence of a third date means that in the opinion of most of those consulted, it was considered impossible that the event should take place in the foreseeable future. Do not misunderstand us however: in effect, this is a straightforward list of possible progress, of progress which is likely within given chronological limits. However simplistic this method might appear, its aim, within a context which is to be considered scientific, is to reveal the desires of certain technicians. In other words, we find ourselves confronted with a sketch of the future development of a technical system which is already evolving, with more difficulty, certainly, under our very eyes.

This is how, a few years ago, the experts of the Rand Corporation envisaged the future in four different fields:

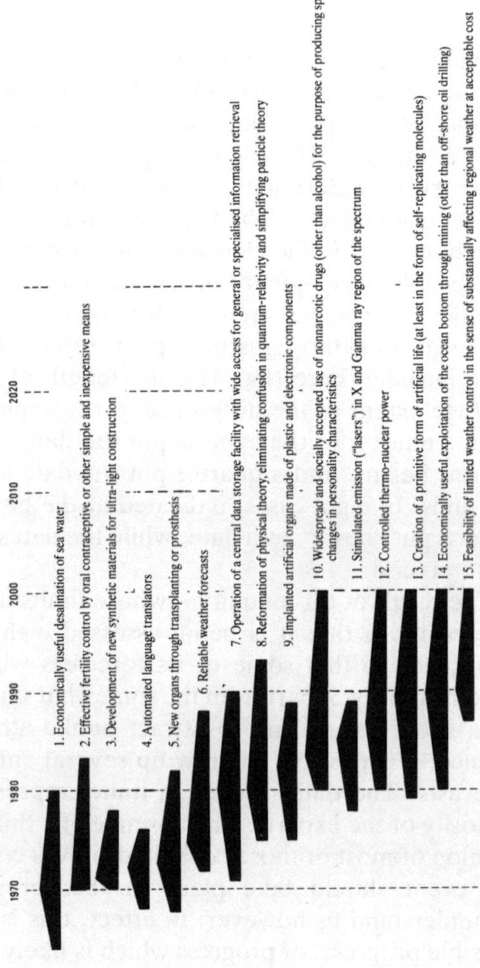
1960
1970
1980
1990
2000
2010
2020
1. Economically useful desalination of sea water
2. Effective fertility control by oral contraceptive or other simple and inexpensive means
3. Development of new synthetic materials for ultra-light construction
4. Automated language translators
5. New organs through transplanting or prosthesis
6. Reliable weather forecasts
7. Operation of a central data storage facility with wide access for general or specialised information retrieval
8. Reformation of physical theory, eliminating confusion in quantum-relativity and simplifying particle theory
9. Implanted artificial organs made of plastic and electronic components
10. Widespread and socially accepted use of nonnarcotic drugs (other than alcohol) for the purpose of producing specific changes in personality characteristics
11. Simulated emission ("lasers") in X and Gamma ray region of the spectrum
12. Controlled thermo-nuclear power
13. Creation of a primitive form of artificial life (at least in the form of self-replicating molecules)
14. Economically useful exploitation of the ocean bottom through mining (other than off-shore oil drilling)
15. Feasibility of limited weather control in the sense of substantially affecting regional weather at acceptable cost

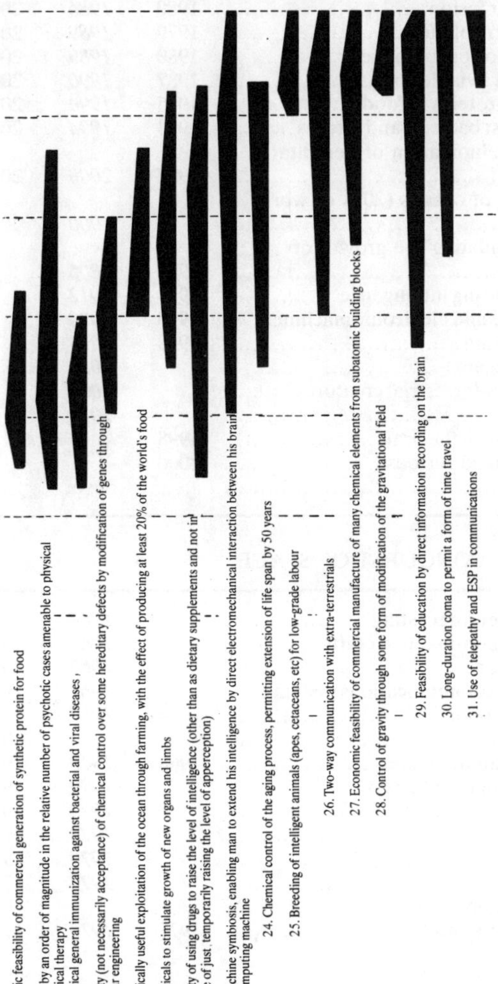

Figure 41. – Evaluation of future technological perspectives. (From Olaf Helmer and T.J. Gordon)

SCIENTIFIC AND TECHNICAL BREAKTHROUGHS

"Economic" desalination of sea	1964	*1970*	1980
Oral contraceptives		*1970*	1980
Automatic translation	1970	*1971*	1978
Accurate weather forecasting	1972	*1975*	1988
Use of central units of electronic control	1971	*1980*	1988
Revision of physics theories, notably, concerning elementary particles	1975	*1980*	1992
Use of artificial prostheses	1975	*1982*	1988
Chemical products for modifying personality	1980	*1983*	2000
X-ray and gamma ray lasers	1978	*1985*	1989
Controlled thermonuclear fusion	1980	*1986*	2000
Creation of a primitive form of life	1979	*1989*	2000
Exploitation of mines at bottom of the sea	1980	*1989*	2000
Control of the weather (at a viable cost)	1987	*1990*	2000
General use of synthetic proteins in food	1985	*1990*	2020
Biochemical agents against bacteria and viruses	1983	*1994*	2000
Chemical agents for the elimination of hereditary defects	1997	*2000*	2010
Economic developement of oceans (20% of world food supply)		*2000*	2018
Chemical agents for stimulating the growth of organs	1985	*2006*	
Chemical agents for increasing intelligence	1984	*2012*	
Symbiosis between people and electronic machines	1990	*2020*	
Chemical control over ageing	1995	*2025*	
Training of animals (cetaceans)		*2020*	
Communication with extra-terrestrial creatures		*2000*	
Control of gravity		*2020*	
Direct input of information in the brain	1998		
Putting to sleep for prolonged periods	2006		
Use of telepathy		never	

CONQUEST OF SPACE

Use of satellites for weather forecasting		*1967*	1970
Inspection of satellites by non-manned crafts		*1967*	1970
Soviet flight round the moon		*1967*	1970
Installation of a world telecommunications system		*1967*	1970
United States flight to the moon		*1967*	1970
Moon landing by a manned craft			1970
Use of lasers in spatial communications	1968	*1970*	1975
Inspection of satellites by manned crafts		*1970*	1974
Ten-crew orbital station		*1970*	1975
Re-useable jet		*1970*	1975
Nuclear-propelled jets		*1970*	1975
Ionically-propelled jets		*1972*	1975
Temporary base on the moon			1975
Manned space-flight to Mars and Venus	1975	*1978*	1979

Experiments of fundamental physics conducted in space	1973	*1981*	1995
Permanent lunar bases	1982		
Control of the weather	1978	*1982*	2002
Landing on Mars	1980	*1985*	1990
Crafts launched out of the solar system	1978	*1987*	1992
Fabrication of material on the moon	1980	*1990*	
Establishment of permanent stations on neighbouring planets		*1990*	2013
Public transport with ballistic rockets	1985	*2000*	
Landing on Jupiter	1993	*2020*	
Space-flight to Pluto	2016	*2023*	
Flight to several generations out of the solar system		*2050*	
Regular traffic to the moon		*2050*	
Anti-gravity system of propulsion		*2050*	
Landing on Venus	2050	*2300*	
Space bombers flotillas orbiting around the earth		*1990*	
Military forces on the moon		*1999*	
Space bombers flotillas orbiting around the sun		*2100*	
Immunisation against radiation		never	

PROGRESS IN AUTOMATION

Ten-fold increase of capital invested in computers	1970	*1973*	1976
Complete air-traffic control	1970	*1974*	1977
Direct links between banks and shops	1972	*1975*	1980
Use of machines in teaching	1971	*1974*	1977
Automation of household work	*1970*	1975	
Education becoming a pastime	1972	*1975*	1985
General use of machines in teaching	*1975*	1990	
Automatic libraries	1971	*1976*	1982
Automatic rapid translation	1973	*1979*	1985
Use of apparatuses for automatic decision-making	1977	*1979*	1995
Electronic prosthesis (radars for the blind)	1976	*1985*	1990
Automatic diagnosis of symptoms of illness	1980	*1985*	1990
Use of domestic robots	1980	*1988*	1995
Definition of a new language from automatic translation	1980	*2000*	
Automatic voting	1996	*2000*	
Automatically controlled motorways	1985	*2002*	
Newspapers printed in the home	1992	*2005*	

THE WEAPONS OF THE FUTURE

Strategic nuclear bombs	1964	*1965*	1967
Non-fatal weapons (gas)		*1968*	1970
Miniaturised apparatuses for weapons control		*1968*	1970

Chemical "disablement"	1965	*1970*	1975
Lasers in radars and communications	1968	*1970*	1975
Biological "disablement"	1968	*1970*	1976
Rockets used for small arms	1966	*1970*	1980
Fatal biological agents	1966	*1970*	1980
Research stations in orbit	1970	*1972*	1974
Advanced propaganda techniques	1970	*1972*	1980
Increased intelligence by the use of computations	1970	*1972*	1985
Possibility of destroying existing submarines	1970	*1975*	1976
Long-range planes	1972	*1975*	1979
Biochemical agents destroying will to resist	1970	*1975*	1980
Use of automatic tactical vectors	1973	*1975*	1980
Anti-missile missiles launched from earth	1975	*1976*	1982
Submarines able to dive to indetectable depths	1975	*1980*	1982
Concentrated energy weapons (lasers)	1973	*1980*	1989
Organisation of world civil defence	1975	*1980*	1990
Use of weather for military objectives	1980	*1990*	2000
Anti-missile missiles launched from aircraft	1984	*1990*	
Anti-aircraft defence by concentrated energy weapons	1989	*1995*	2000
Bombs in orbit	1973	*1992*	
Use of dolphins to detect submarines		*1980*	
Use of hypnosis against enemy forcesic		*2035*	
Mind-reading		*2035*	

These, then, are the tables supplied by the Rand Corporation regarding scientific breakthroughs, the conquest of space, automisation and military art. The reader is bound to be somewhat surprised by the results of the enquiry. Of course, this enquiry was not a systematic one, simply a selection of opinion polls concerning particularly important areas, which no doubt will bring much influence to bear on the world of tomorrow. Some parts may make people smile – such as the control of gravity, or communication with extra-terrestrial beings. Other parts are more worrying, such as the use of mind reading for military purposes. But who has not already thought of these in reality?

What conclusions can we draw from all this0 The Hudson Institute's *Year 2000*, the tables of the Rand Corporation, and the works written about our civilization in the last twenty or thirty years all have common characteristics, even if their methods differ as fundamentally as their results. One of these is the necessity for global forecasting, within which technological forecasting holds an important place, if not the essential one. The primacy of techniques is apparent in all of these texts, to a greater or lesser degree, but it is always clear – even if the

Hudson Institute did consider other sectors in its enquiry and analysis. It is only natural, then, that such a great insistence has been placed on "leading techniques," symbolised in a way by electronic and nuclear techniques which are considered, rightly or wrongly, as the two major poles of technical progress. And it is around this entirely new technical system – that has almost nothing in common with the technical system created in the second half of the 19th century and which continued to develop until the eve of the Second World War – that all the other systems must organise themselves: the economic system, the social system, etc.

The graph in figure 42 naturally has only an indicative value. One can easily see that it has not been mathematically formulated. But it is helpful in that it provides a visual representation of that which has been observed and forecast, apart from possible corrections. The curve of scientific progress incorporates the progress of techniques, which

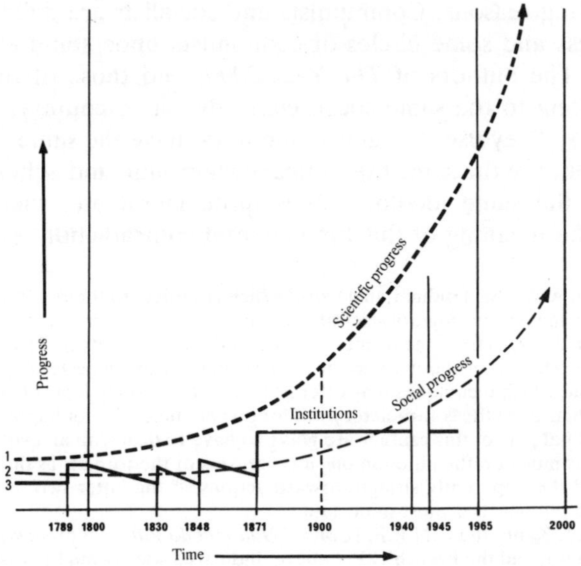

Figure 42. Hypothetical evolution of the curve of social progress and the curve of scientific technical progress.

emerges from applied science. This curve is moderately exponential. The curve representing social progress shows a much more feeble growth, even though the prospective part is on the upturn. The line representing institutions reveals a number of downward trends. It is easily demonstrable that as techniques progress, institutions wear out. At irregular intervals, then, readjustments are necessary. To represent the relationships between institutions and scientific progress between 187 and 1945 by a straight line is no doubt a bit excessive.

It is not impossible that in the following decades, this graph may be able to offer more compatibility between the various elements of which it is composed. This is a work of the imagination.

At the end of these two short and necessarily fragmented analyses of hopes and fears, the reader may well be perplexed – but then he inhabits a perplexing reality. This perplexity was clearly expressed by a technician, M. Giuliani, in *Le Monde* in 1972. Since 1967 when the Hudson Institute published its report, or even from 1965, when the report by the Rand Corporation was published, everything has been said for and against. Differences of opinion have remained, sometimes for surprising reasons. Communists and socialists are divided, while big business and some circles of economists understand each other perfectly. The authors of *The Year 2000*, and those of the M.I.T. report belong to the same social class, the same country, the same universities. They use the same computers, have the same sources of information, use the same theoretical background and adhere, grosso modo, to the same ideology. It is quite legitimate, therefore, to consider the meaning of this fundamental contradiction.

We are living, writes M. Giuliani, in a world which is subject to three different types of logic. There is the rational logic of scientific and technical development; there is the logic of need which diverges from the former because of the irrationality of desire, but which approaches it because of the inability of the imagination to conceive of anything else other than that which techniques may offer it; finally, there is the logic of decisions and choices which inspires the contradictory will for power and desire for happiness. People living in the final part of this century are going to have to deal with an evolution which will depend as much on the autonomous mechanisms of the former, as on the obscure reactions and the apparently straightforward actions of the latter two, and it will be impossible to tell which is most important.

Either, as H. Kahn, the C.N.P.F. (*Conseil National du Patronat Francais*, the french industrial lobby), and the French C.P. believe, industrial society might possibly be able to maintain the objectives of more or less exponential growth. To do this would require an unprecedented technical revolution, a whole host of new inventions, and a complete change in methods and morals. Or, it is possible that this revolution might never take place or be sufficient, in which case, as the M.I.T. and Sisco Mansholt say, the objectives must be changed, which would bring about an entirely different type of revolution. Who knows which one would create the most upheaval. And what face would one or the other give to capitalism or socialism?

In any case, the question must be posed within as large a context as possible, and the authors of *The Year 2000* understood this better than the researchers at the M.I.T. Technological forecasting by itself is not sufficient, it requires an understanding of the sort of world with which it might be compatible, and consequently, the transformations to be effected in a large number of areas. But is it not characteristic of humans to put the cart before the horse: the desire to understand matter is infinitely stronger than the need to understand oneself. Attitudes of mind and established situations are all part of a certain comfort which people are reluctant to abandon, indeed, they often refuse to do so.